Current Topics in Behavioral Neurosciences

Volume 50

Current Topics in Behavioral Neurosciences provides critical and comprehensive discussions of the most significant areas of behavioral neuroscience research, written by leading international authorities. Each volume in the series represents the most informative and contemporary account of its subject available, making it an unrivalled reference source. Each volume will be made available in both print and electronic form.

With the development of new methodologies for brain imaging, genetic and genomic analyses, molecular engineering of mutant animals, novel routes for drug delivery, and sophisticated cross-species behavioral assessments, it is now possible to study behavior relevant to psychiatric and neurological diseases and disorders on the physiological level. The Behavioral Neurosciences series focuses on translational medicine and cutting-edge technologies. Preclinical and clinical trials for the development of new diagnostics and therapeutics as well as prevention efforts are covered whenever possible. Special attention is also drawn on epigenetical aspects, especially in psychiatric disorders.

CTBN series is indexed in PubMed and Scopus.

More information about this series at http://www.springer.com/series/7854

Lucette A. Cysique • Sean B. Rourke
Editors

Neurocognitive Complications of HIV-Infection

Neuropathogenesis to Implications for Clinical Practice

Springer

Editors
Lucette A. Cysique
School of Psychology
The University of New South Wales
Sydney, NSW, Australia

Sean B. Rourke
MAP Centre for Urban Health Solutions
Li Ka Shing Knowledge Institute
St. Michael's Hospital and Unity Health Toronto
Toronto, ON, Canada

ISSN 1866-3370 ISSN 1866-3389 (electronic)
Current Topics in Behavioral Neurosciences
ISBN 978-3-030-80758-0 ISBN 978-3-030-80759-7 (eBook)
https://doi.org/10.1007/978-3-030-80759-7

This Springer imprint is published by the registered company Springer Nature Switzerland AG.
The registered company address is: Gewerbestrasse 11, 6330 Cham, Switzerland

We dedicate this book to our families and friends for their loving support, to our esteemed colleagues for their intellectual engagement and exchanges, and to all people living with HIV, including those who have participated in our research studies. Their contributions and lived experiences have enriched our own scientific inquiries and have been critical to the advancement of HIV awareness and research. A special recognition and tribute to the memories of Professors Kevin Robertson and Ned Sacktor who passed away in 2019 and 2020. They were pioneers in NeuroHIV research and clinical practice and uniquely contributed to the international expansion of this field of research. We are honoured they contributed to this book.

Preface

This *Current Topics in Behavioral Neurosciences* volume presents the latest international research and scientific discoveries in the neurocognitive aspects and complications of HIV.

By bringing together this scope and breadth of evidence base, we hope to further inform our understanding of the neurocognitive complications of HIV infection; advance clinical practice and patient care in HIV to reduce the social and health consequences of HIV; and to enhance the overall wellbeing of all people living with HIV.

What's Unique About This Volume

It represents the most comprehensive update of NeuroHIV research within the last 15 years. It was envisioned to review the leading, cutting-edge research in HIV-associated neurocognitive impairments and consequences with implications for putative neuropathogenesis. This volume was also designed to show the related implications of this work for future research directions.

The book has four parts and a total of 18 chapters from leading NeuroHIV scientists and clinicians from across the world. Together, their expertise covers neurology; neuroscience; clinical HIV science and HIV treatments; immunology and virology; infectious diseases; neuroimaging; public health; research and clinical neuro/psychology; cross-cultural neuropsychology; psychiatry; cardiovascular health; and geriatrics.

Unique to this volume the authors were asked to include a dedicated part at the end of their chapters about the clinical implications and potential translational aspects of their work, in addition to future research directions. In this final part, the authors were encouraged to consider how the latest scientific evidence covered in their chapter may have clinical practice relevance (for both patients and care delivery). They were also asked to reflect on how this evidence may contribute to and improve HIV-associated neurocognitive disorders (HAND) criteria, nosology, and treatment implications.

We hope you find the work collected in here as enlightening as we did and that it enriches your research and clinical practice.

The First Part

The first part covers research on the neuropathogenesis and biomarkers of HIV-related neurocognitive complications.

Thomas A. Angelovich, Melissa J. Churchill, Edwina J. Wright, and Bruce J. Brew
Thomas A. Angelovich and colleagues give an overview of the research into new potential axes of HIV neuropathogenesis with relevance to biomarkers and treatment. This chapter focuses on the neuropathogenesis of HAND in the context of viral suppression and considers the role of neuroinflammation as central. It considers it as driven by a combination of chronic intermittent low-level replication of whole virus or viral components, latent HIV infection, and peripheral inflammation, possibly from a disturbed-gut microbiome or chronic cellular dysfunction in the central nervous system.

Daniela Gomez, Christopher Power, and Esther Fujiwara
Daniela Gomez and colleagues focus on neurocognitive impairment and associated genetic aspects of HIV infection. This chapter reviews primary host genomic changes (immune-related genes, genes implicated in cognitive changes in primary neurodegenerative diseases), epigenetic mechanisms, and genetic interactions with ART implicated in HIV progression or HAND/neurocognitive complications of HIV.

Nicole Fernandes and Lynn Pulliam
Nicole Fernandes and Lynn Pulliam explicate the inflammatory mechanisms and cascades contributing to neurocognitive impairment. This chapter discusses the role of HIV, viral proteins, and virally infected cells on the development of neuroinflammation, as well as the effect of viral proteins on the cells of the central nervous system. The authors comprehensively examine how biological and behavioral factors result in an inflammatory context that triggers the development of neurocognitive impairment in HIV, including the contributions of antiretrovirals and drugs of abuse (methamphetamine, cannabis, and opioids), circulating biomarkers, NF-L, sCD163, sCD14, exosomes, and the microbiome.

Talia M. Nir, Elizabeth Haddad, Paul M. Thompson, and Neda Jahanshad
This first part finishes with a chapter on the neuroimaging advances in the diagnosis and differentiation of HIV, comorbidities, and aging, by Neda Jahanshad and colleagues. This chapter summarizes over two decades of NeuroHIV research evaluating brain volumetric differences and their correlations in adults living

with HIV. The authors highlight brain mapping technologies that go beyond understanding morphometric patterns and delve into brain circuitry, including functional brain mapping and brain microstructure quantification. They discuss the role of neuroimaging in determining aging processes and the effects of interacting comorbidities on brain structure and functions. In doing so, they also highlight how infection may contribute to the risk for late-onset dementias. Finally, the authors describe how new neuroimaging and analytic (artificial intelligence) technologies and large-scale international collaborations may help disentangle the effect of genetic and environmental risk factors on brain aging and disease and could ultimately contribute to better clinical outcomes for people living with HIV.

The Second Part

The second major part opens with questions of incidence, prevalence, and contexts in the research on neurocognitive impairments in low- and high-income countries. It critically includes research on children with HIV and on women with HIV, as well as issues relevant to cross-cultural neuropsychology in HIV research and clinical practice. The part also covers innovative research methods for deep phenotyping (e.g., machine learning) and statistics for the assessment of individual neurocognitive trajectories.

Sarah Benki-Nugent and Michael J. Boivin
The first chapter of this part by Sarah Benki-Nugent and Michael J. Boivin covers the latest regarding the neurocognitive complications in pediatric HIV infections. The authors review the biological mechanisms and complex interaction with environmental factors, associated with cognitive impairment of learning difficulties in HIV-infected children receiving antiretroviral treatment.

The authors also discuss the role of multipronged approaches that encompass both pharmacological and psychosocial approaches, for optimizing neurodevelopmental outcomes in children living with HIV infection. They present evidence for the use of innovative neuropsychological testing methods and neuroinflammatory biomarkers, in addition to brain development neuroprotective factors (BDNFs), to evaluate the brain/behavior integrity of children in response to new treatment options.

Leah H. Rubin and Pauline M. Maki
The next chapter by Leah H. Rubin and Pauline M. Maki in this part is dedicated to neurocognitive complications of HIV infection in women, with insights from the *Women's Interagency HIV Study (WIHS)*, the largest study of the natural and treated history of women living with HIV. The authors provide evidence that women living with HIV are more cognitively vulnerable than men living with HIV and that there are key differences in the pattern of cognitive impairment, which also need more attention.

They discuss the factors that contribute to these differences, including biological factors (e.g., inflammation, hormonal, genetic) as well as common comorbidities (mental health; substance use; vascular and metabolic risk factors; coinfections and liver function; non-antiretroviral medications; and genetic markers). The authors emphasize the importance of considering "sex" as a biological factor in studies of cognitive dysfunction and suggest avenues for future research.

Monica G. Rivera Mindt, Desiree A. Byrd, Emily P. Morris, Kayla Tureson, Vanessa Guzman, Angela C. Summers, Cara Crook, Micah J. Savin, and Maral Aghvinian

The following chapter by Monica G. Rivera Mindt and colleagues in the part gives an in-depth review of cultural neuropsychology considerations in the diagnosis of HAND. The authors review the evidence of how HIV can be viewed through a lens of health disparities, specifically those that can affect culturally and linguistically diverse (CALD) and underrepresented minority populations to a greater degree than non-Hispanic white populations. They delineate how CALD populations can experience worse HIV-related neurological and health outcomes, which can be exacerbated by inadequate neurocognitive measures, poor normative samples, and the complex interplay of sociocultural factors that may affect test interpretation.

The authors provide a careful explication of how the most well-studied CALD groups in NeuroHIV research are of African American/Black and Latinx adults in the USA (including when it comes to the provision of appropriate normative neuropsychological data about their HIV-negative counterparts). The authors also discuss and emphasize that there is a lack of research in CALD populations outside of the USA, despite their disproportionate HIV burden (e.g., First Peoples of Australia and Indigenous and First Nations People of Canada; migrant populations in Europe; and large multi-cultural populations in South America, Caribbean countries, parts of Asia and Africa). They examine and describe a range of sociocultural and health factors, including global and regional (e.g., rural versus urban) considerations, migration, and gender. The authors conclude by providing guidelines, with international relevance, on how to incorporate sociocultural consideration into the assessment and interpretation of neurocognitive data and HAND diagnosis when working with HIV-positive CALD populations.

Robert Paul, Paola Garcia-Egan, Jacob Bolzenius, Rebecca Preston-Campbell, and Julie Mannarino

Robert Paul and colleagues illustrate that while HIV-infected individuals residing in high-income countries are positioned to benefit from having access to antiretroviral therapy – and a lower burden of HIV disease mortality and morbidity, including fewer neurocognitive complications – these advantages are not universal for all residents of high-income countries.

The authors discuss that the growing population of HIV-infected individuals who are now reaching advanced age in high-income countries represents a new dimensional risk for the persistence and development of incident neurocognitive complications among individuals receiving suppressive ART. Within this context, the authors outline the diverse and highly dimensional nature of risk factors for neurocognitive complications of HIV, described in recent studies conducted in high-income countries. They propose how innovative data science methods may help to advance the existing conceptual framework of HAND and catalyze the development and implementation of much-needed neurocognitive interventions to achieve global HIV treatment and eradication efforts.

Alyssa Vecchio, Ned Sacktor, Deanna Saylor, and Kevin Robertson
Alyssa Vecchio and colleagues in their chapter highlight the paucity of information on neurocognitive dysfunction in individuals with HIV in resource-limited regions, despite these areas having the greatest burden of infection. The authors show that HAND is a major cause of morbidity of people living with HIV and is estimated to be the most prevalent form of neurocognitive impairment worldwide in young adults. The authors further illustrate how this finding has drastic implications for the productivity and social engagement of young adults in the development of industry, education, and healthcare – which is particularly relevant in low-income countries. The authors end by providing insights and critical knowledge on how to build an infrastructure for neurocognitive testing in resource-limited setting.

Lucette A. Cysique, Kaitlin B. Casaletto, and Robert K. Heaton
This part concludes with a chapter by Cysique and colleagues on the issues of reliably measuring cognitive change in the era of chronic HIV infection and chronic HAND. After providing a rationale for the importance of longitudinal studies in the understanding of HIV-related neurocognitive complications, Cysique et al. present several statistical frameworks to quantify cognitive change. The authors provide a critical review of naturalistic longitudinal studies and select randomized clinical trials, conducted since the advent of the combined antiretroviral therapy era, that assessed neurocognitive change as a primary outcome in people living with HIV. In doing so they discuss how specific study design and statistical factors impacted the studies' findings.

The authors conclude by emphasizing the need for longitudinal studies to include more diverse sets of people living with HIV from high HIV burden countries, and how a longitudinal study framework would improve the current criteria for HAND diagnosis as well as the clinical management of HAND.

The Third Part

The third part focuses on the comorbidities and complications associated with HAND.

Julian Falutz, Susan Kirkland, and Giovanni Guaraldi
Julian Falutz and colleagues delineate the scope and questions of geriatric syndromes in HIV: aging and increasing comorbidities, and their implications for the neurocognitive complications of HIV infection. After reviewing the epidemiology of the aging HIV epidemic, the authors review evidence for premature and accelerated aging as well as the multifactorial etiology of this process. In doing so, the authors address the question of whether common geriatric syndromes in people living with

HIV contribute to cognitive impairment, and whether common risk factors may provide clues to limiting or delaying cognitive decline.

Victoria M. Kordovski, Savanna M. Tierney, and Steven Paul Woods
The following chapter, by Victoria M. Kordovski and colleagues, provides a comprehensive review of questions of conceptualizing and assessing everyday functioning in the context of HAND.

The authors highlight that although impairments in everyday functioning are a hallmark of HAND diagnoses and can adversely influence quality of life, there are no gold standard measures of this fundamentally important and complex construct. The authors provide a review of the various self-reported, clinician-rated, and performance-based methods by which everyday functioning is measured in the setting of HIV disease. This includes global activities of daily living and the specific domains of medication adherence, financial management, automobile driving, and vocational functioning. The authors conclude on novel methods to potentially improve the HAND diagnostic criteria.

Jose A. Muñoz-Moreno, Lucette A. Cysique, and Sean B. Rourke
Jose Muñoz-Moreno and colleagues provide an up-to-date overview of the research into the neuropsychiatric disorders and cognitive symptoms associated with the neurocognitive complications of HIV infection. The chapter briefly reviews epidemiological data for major depressive disorders, anxiety disorders, and apathy in people living with HIV. Then, the authors review research into the connections between a range of neuropsychiatric disorders, including depressive and anxiety disorders – but also emotional dysregulations (apathy, alexithymia, and emotional processing impairment), which are distinguishable from depression and anxiety.

In doing so, they also include research into the evaluation and interpretation of cognitive symptoms. The authors review research regarding the roles of coping skills, perceived stress, and response to stressful life events in contributing to neurocognitive impairment in people living with HIV. Non-pharmacological interventions are then briefly reviewed. The authors conclude with recommendations on how to best consider neuropsychiatric disorders and cognitive symptoms for the diagnosis of HAND, as well as future research directions.

Antoine Moulignier and Dominique Costagliola
Antoine Moulignier and Dominique Costagliola address the evidence about the questions of metabolic syndrome and cardiovascular diseases, and their impacts on the pathophysiology and phenotype of HAND. To evaluate the brain-aging processes and vascular brain injury, the authors highlight the correspondence observed between the findings about a range of cardiovascular diseases and metabolic syndrome in the general population and virus-suppressed cART-treated people living with HIV. The authors conclude with the inconsistent findings in the relevant literature and discuss how inconsistent definitions of neurocognitive impairment in people living with HIV may have contributed to this effect. The authors also provide evidence to emphasize that, given the growing evidence that cardiovascular diseases

and metabolic syndrome are associated with HAND, it is crucial to improve early detection and ensure appropriate management of these conditions.

Sindhura Kompella, Thabit Al-Khateeb, Ossama Abid Riaz, Sylvester Olubolu Orimaye, Patrick Olumuyiwa Sodeke, Adeola Olubukola Awujoola, Joseph Ikekwere, and Karl Goodkin
The last chapter of this part, by Sindhura Kompella and colleagues, provides a review on HAND and the relative risk factors for dementia risk. The authors review a comprehensive set of factors and processes, including neurodegeneration and type of dementia; non-modifiable factors (aging, genetics, ethnicity, and gender); modifiable factors (cardiovascular health, depressive and substance use disorders, education, and socio-economic status); and specific risk factors for HAND (ART neurotoxicity, HIV effects, neuropathological findings).

The authors also include a part on animal models of dementia with relevance to HAND. They review a range of biomarkers which may help to detect all-type dementia risk in aging people living with HIV. In addition, they further include specific research from fronto-temporal dementia and dementia with Lewy body disease. The authors conclude by presenting some of the most promising interventional research in dementia risk reduction.

The Fourth Part

The fourth and last part focuses on screening, interventions, and the clinical management of HIV-related neurocognitive complications.

Reuben N. Robbins, Travis M. Scott, Hetta Gouse, Thomas D. Marcotte, and Sean B. Rourke
This part opens with a chapter by Reuben N. Robbins and colleagues on screening for HAND and issues of sensitivity and specificity. After reviewing issues of needs versus practice in screening for HAND, the authors provide a comprehensive review of the NeuroHIV research into cognitive screenings. The authors report on the strengths, limitations, and psychometric properties of the currently available screening tests. They then examine issues of the screens' application to low- and middle-income countries. Their future directions part highlights the most innovative screening tools and their potential.

Jessica L. Montoya, Brook Henry, and David J. Moore
The next chapter by Jessica L. Montoya and colleagues of this part summarizes the latest evidence and reviews the behavioral and physical activity interventions for HAND. The authors provide an in-depth review of the research and development work that has gone into behavioral interventions for HAND, including physical activity, diet, sleep, and promotion of ART adherence. The authors conclude on strategies to best implement this research.

David E. Vance, Pariya L. Fazeli, John Cheatwood, Chance Nicholson, Shannon Morrison, and Linda D. Moneyham

David E. Vance and colleagues then provide a comprehensive review on the questions of cognitive training strategies for HIV-related neurocognitive impairments, complete with insights from the cognitive aging literature. The authors give the scientific context in which computerized cognitive training approaches have been successfully used in older adults and provide examples of how these approaches have been, and could be better, translated to adults with HIV. The authors also provide evidence from ongoing clinical trials that suggest reversing HAND may be possible. The authors conclude by providing recommendations for both clinical and research practice.

Shih-Ping Lin, Andrea Calcagno, Scott L. Letendre, and Qing Ma

The last part, and the book volume by Shih-Ping Lin and colleagues, finishes with an up-to-date and detailed review of the clinical treatment options and randomized clinical trials for neurocognitive complications of HIV infection. This review, covers combination antiretroviral therapy, central nervous system penetration effectiveness, and adjuvants.

Our Sincerest Thanks

In closing, we sincerely thank all of the authors for their exceptional contributions to the field of NeuroHIV research and particularly for their contributions to this series. We are also thankful to the Springer editorial team, who have given us immense support throughout the development of the series.

We ardently hope that scientists and clinicians alike will find the scope of this volume to be a clear and compelling summary of the latest scientific evidence; evidence that will support, inform – and even improve – care, support, and practice for HAND.

Editor's/author's note: This series was started before the emergence of the COVID-19 pandemic and so this topic was not covered. However, we recognize that it will be very important to consider and explore the potential impact of COVID-19 for people living with HIV, especially for those who are aging with comorbidities.

Sydney, NSW, Australia — Lucette A. Cysique
Toronto, ON, Canada — Sean B. Rourke

Contents

Part I
Neuropathogenesis and Biomarkers

New Potential Axes of HIV Neuropathogenesis with Relevance to Biomarkers and Treatment

Thomas A. Angelovich, Melissa J. Churchill, Edwina J. Wright, and Bruce J. Brew

Contents

Abstract Human immunodeficiency virus (HIV)-associated neurocognitive disorders (HAND) affect approximately half of people living with HIV despite viral suppression with antiretroviral therapies and represent a major cause of morbidity. HAND affects activities of daily living including driving, using the Internet and,

T. A. Angelovich (✉)
Chronic Infectious and Inflammatory Diseases Research, School of Health and Biomedical Sciences, RMIT University, Melbourne, VIC, Australia

Life Sciences Discipline, Burnet Institute, Melbourne, VIC, Australia
e-mail: thomas.angelovich@rmit.edu.au

Curr Topics Behav Neurosci (2021) 50: 3–40
DOI 10.1007/7854_2019_126
Published Online: 11 February 2020

importantly, maintaining drug adherence. Whilst viral suppression with antiretroviral therapies (ART) has reduced the incidence of severe dementia, mild neurocognitive impairments continue to remain prevalent. The neuropathogenesis of HAND in the context of viral suppression remains ill-defined, but underlying neuroinflammation is likely central and driven by a combination of chronic intermittent low-level replication of whole virus or viral components, latent HIV infection, peripheral inflammation possibly from a disturbed gut microbiome or chronic cellular dysfunction in the central nervous system. HAND is optimally diagnosed by clinical assessment with imaging and neuropsychological testing, which can be difficult to perform in resource-limited settings. Thus, the identification of biomarkers of disease is a key focus of the field. In this chapter, recent advances in the pathogenesis of HAND and biomarkers that may aid its diagnosis and treatment will be discussed.

Keywords Biomarkers · Human immunodeficiency virus · Inflammation · Neurocognitive disorders

1 Introduction

Despite viral suppression with antiretroviral therapy (ART), no cure for HIV exists, and comorbid disease now represents the major challenge for people living with HIV (PLWH). HIV-associated neurocognitive disorders (HAND) affect approximately

M. J. Churchill
Chronic Infectious and Inflammatory Diseases Research, School of Health and Biomedical Sciences, RMIT University, Melbourne, VIC, Australia

Life Sciences Discipline, Burnet Institute, Melbourne, VIC, Australia

Department of Microbiology and Immunology, Monash University, Melbourne, VIC, Australia
e-mail: melissa.churchill@rmit.edu.au

E. J. Wright
Burnet Institute, Melbourne, VIC, Australia

Department of Infectious Diseases, Alfred Hospital and Central Clinical School, Monash University, Melbourne, Victoria, Australia

The Peter Doherty Institute for Infection and Immunity, Melbourne, VIC, Australia
e-mail: edwina.wright@monash.edu

B. J. Brew
Applied Neurosciences Program, Peter Duncan Neurosciences Research Unit, St. Vincent's Centre for Applied Medical Research, Sydney, NSW, Australia

Faculty of Medicine, St. Vincent's Clinical School, University of New South Wales, Sydney, NSW, Australia

Departments of Neurology and Immunology, St. Vincent's Hospital, Sydney, NSW, Australia
e-mail: b.brew@unsw.edu.au

18–69% (variability related to characteristics of study population with differing risk factors) of PLWH despite viral suppression with ART (Heaton et al. 2010; Simioni et al. 2010; Sacktor et al. 2016; dos Santos-Silva et al. 2017; Robertson et al. 2007; Cysique and Brew 2011) and are independent risk factors of mortality (Ellis et al. 1997; Tambussi et al. 2000). HAND in the context of virally suppressive ART mainly impairs cognitive function, although mild motor and behavioural deficits are still common. HAND induces brain injury including total and regional brain volume reduction (Cardenas et al. 2009; Küper et al. 2011) as well as neurodegeneration at least in its broadest sense (Boban et al. 2017). However, the complete neuropathological characterisation of HAND with viral suppression is still evolving (Gelman 2015). Thus, HAND may dramatically affect a person's ability to function and live independently. Typical examples include impacting a person's ability to drive and use the Internet which may result in social withdrawal and relationship breakdowns (Woods et al. 2017). Furthermore, there is strong evidence to suggest that individuals with HAND are more prone to poor medication adherence (Hinkin et al. 2002; Kamal et al. 2017) which may cause antiviral drug resistance and HIV disease progression. Together with cardiovascular disease and non-AIDS-cancers, HAND represent the major burden that continue to affect virally suppressed PLWH.

HAND is subdivided into three categories (by increasing cognitive and functional severity): (1) asymptomatic neurocognitive impairment (ANI), (2) mild neurocognitive disorders (MND) and (3) HIV-associated dementia (HAD) (Antinori et al. 2007). Effective viral suppression with current ART regimens has reduced the incidence of HAD (once a major cause of HIV-associated mortality) from ~20% to 2–4% of PLWH (Heaton et al. 2010). However, the incidence of ANI and MND remains high in virologically suppressed individuals resulting in impairments that can predominantly affect learning and/or memory functions and complex attention (Heaton et al. 2010, 2011; Robertson et al. 2004; Harezlak et al. 2011; Cysique et al. 2004). Furthermore, the presence of ANI has also been linked to a twofold to sixfold increased risk of earlier development and more rapid onset of symptomatic HAND (Grant et al. 2014), highlighting the potential progressive nature of the disease. Thus, HAND is a significant health, social and financial burden affecting PLWH worldwide.

Traditionally, the pathogenesis of HAND, especially HAD, has been strongly associated with HIV viremia (i.e. low nadir $CD4^+$ T cell count; high plasma and/or CSF viral load). However, HAND in the context of virally suppressive ART is likely related to other factors such as low-level viral replication of whole virus or viral components, epigenetic changes and even systemic inflammation driven by gastrointestinal dysfunction, factors which may present greater prognostic/diagnostic value in this era of suppressive ART. Furthermore, HAND does not affect all PLWH, indicating that some individuals are, for as yet undefined reasons, more susceptible to disease possibly implying that host cell factors may also play a role in disease pathogenesis. Here, we assess the potential mechanisms driving the pathogenesis of HAND in chronically HIV-infected virally suppressed individuals and how they may be targeted to improve health outcomes for people living with HIV (Fig. 1).

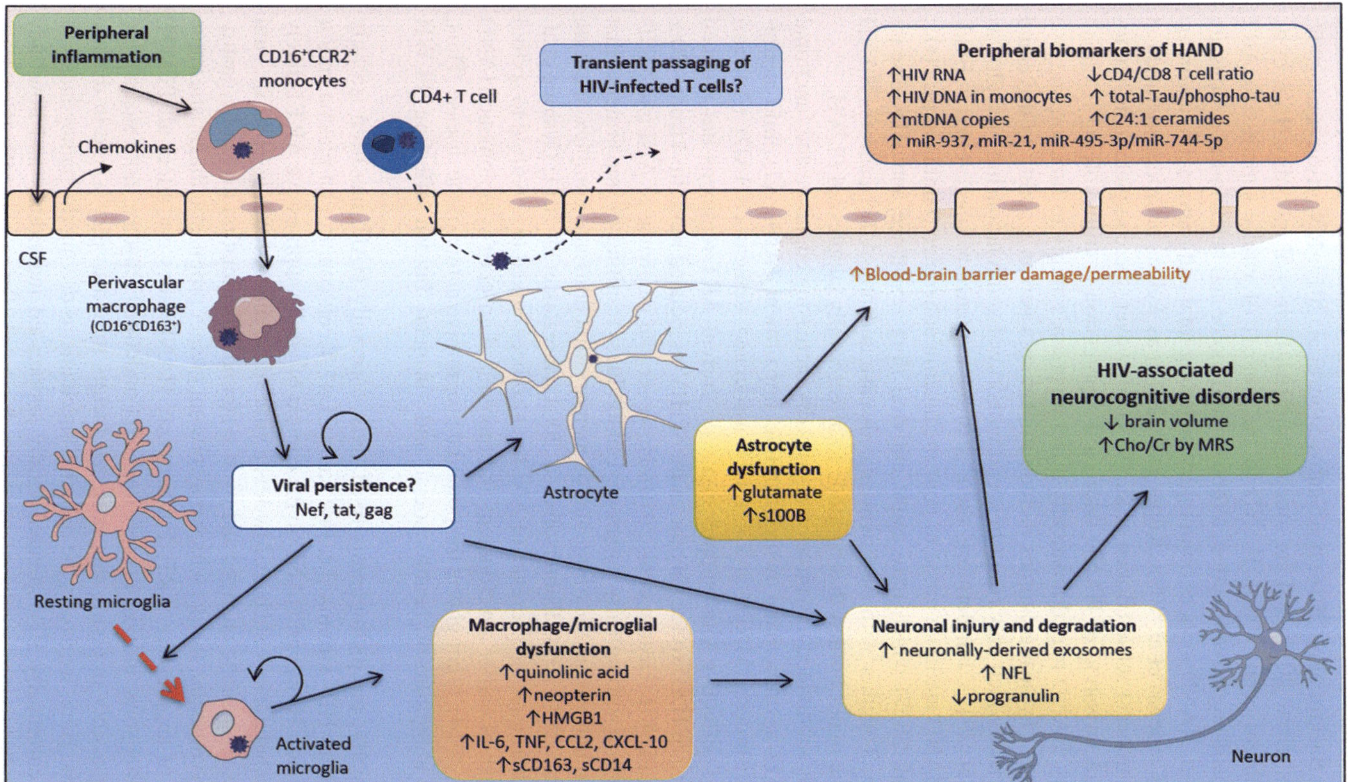

Fig. 1 Insight from biomarkers of HAND into disease pathogenesis in the presence of ART. HIV-infected monocytes (and T cells) transmigrate the blood-brain barrier (BBB) to seed perivascular macrophages, astrocytes and microglia. Viral persistence may drive activation of microglia and astrocytes inducing pro-inflammatory cytokine production, altered glutamate uptake and altered metabolism which drives neuronal injury and degradation contributing to BBB damage, brain atrophy and ultimately HIV-associated neurocognitive disorders. Both CNS-derived and peripheral biomarkers of HAND are shown

2 HAND Prevalence, Diagnosis, Prognosis and Treatment

HAND is diagnosed by clinical neurological assessment with neuropsychological examination followed by radiology such as magnetic resonance imaging of the brain and CSF analysis. Since 2007 the 'Frascati' criteria – a consensus proposed by the National Institute of Mental Health and the National Institute of Neurological Diseases and Stroke (Antinori et al. 2007) – have been used to assess individuals' neurocognitive capabilities, by scoring their (1) motor skills, (2) sensory-perceptual capacity, (3) speed of information processing, (4) memory (learning and recall), (5) abstraction and executive function, (6) attention and working memory, (7) verbal and language skills and (8) whether individuals need assistance with daily living (Antinori et al. 2007), with impairments in two or more cognitive areas indicative of disease. Scores are compared to the mean scores of age- and education-matched neurologically unimpaired individuals, and diagnosis is made as below:

HAD Similar to dementia in HIV-uninfected individuals, HAD causes significant deficits in cognitive function and is the most severe form of HAND (Antinori et al. 2007; Vivithanaporn et al. 2010; Tozzi et al. 2005). As such, affected individuals have neurological impairments reflected of at least two standard deviations below those of neurologically unimpaired individuals in at least two cognitive domains, and they show mark decline in everyday function. Although the incidence of HAD has reduced globally (Heaton et al. 2010), HAD remains a significant burden in regions where access to efficacious ART is limited or non-existent with a 31–38% incidence rate (Wong et al. 2007; Lawler et al. 2010). Notably, HAD diagnosis is an independent predicator of mortality in individuals with advanced HIV/AIDS (Sevigny et al. 2007).

MND MND affects ~11.7% of virally suppressed PLWH receiving ART (Heaton et al. 2010) and interferes with daily functioning mildly to moderately. It requires acquired impairment of at least one standard deviation below that of matched neurologically unimpaired individuals in two cognitive domains (Antinori et al. 2007). Typically, cognitive impairments associated with MND are detected by either self-reporting or observation by close relations. Impairments may present as mild interference in mental acuity; absenteeism; issues regarding relationships and social functioning; and/or poor adherence to ART, for which some individuals may require professional assistance (Hinkin et al. 2002).

ANI ANI is the most recent subcategory of HAND (Antinori et al. 2007) and has been estimated to affect ~32.7% of PLWH despite viral suppression, representing ~70% of HAND cases (Heaton et al. 2010). Similar to the diagnosis of MND, individuals with ANI exhibit acquired impairment of at least one standard deviation below those of neurologically unimpaired individuals in two cognitive domains (Antinori et al. 2007). However, unlike individuals with MND, those with ANI do not exhibit interference with everyday living (Antinori et al. 2007). As such, ANI remains difficult to diagnose, and the clinical implications of ANI remain unclear and controversial (Torti et al. 2011; Nightingale et al. 2014). For example, some suggest that due to the asymptomatic nature of ANI, the 'diagnosis' of ANI may be a result of suboptimal performance in neurocognitive tests due to subjects being 'tired' or 'distracted' (Levine et al. 2017) and impart an unnecessary categorisation that could incur a degree of stigmatism (Chiao et al. 2013).

Conversely, growing evidence suggests that individuals with ANI are prone to disease progression with a twofold to sixfold increased risk of earlier development of symptomatic HAND (Grant et al. 2014), highlighting the importance of early identification affected individuals. These findings are supported by large longitudinal clinical studies describing disease progression including in patients with ANI (Sacktor et al. 2016; Gott et al. 2017; Heaton et al. 2015). Results from the CNS HIV Antiretroviral Therapy Effects Research (CHARTER) cohort found that 22.7% of participants (436 individuals in total) declined whilst 60.8% remained stable (Heaton et al. 2015). Similarly, the Multicentre AIDS Cohort Study (MACS) showed that whilst the majority of individuals (77%) remained neurocognitively stable, 13% deteriorated to a more severe form of HAND and 10% improved (Sacktor et al. 2016). Further, our recent data have shown that ANI has a biological underpinning of focal frontal brain atrophy, thereby validating its importance (Nichols et al. 2019). Despite these findings, the direct pathogenic mechanisms driving disease progression remain unclear.

2.1 Treatment

One of the early issues with HIV antiretroviral therapies was the variable and often poor ability of drugs to cross the blood-brain barrier (BBB) and target virus in the CNS. Treatment with antiretrovirals with the capacity to penetrate the CNS well (including abacavir, lamivudine, maraviroc, nevirapine, darunavir, lopinavir, dolutegravir and possibly other integrase inhibitors) is associated with improved neurocognitive functions (Cysique et al. 2009) and increased suppression of CSF HIV RNA in comparison to treatment with other drugs with poorer CNS penetrance scores (Carvalhal et al. 2016). However, some evidence suggests that standard ART regimens in some patients are sufficient (Lanoy et al. 2011; Ellis et al. 2014), possibly because there is already BBB impairment which allows drug access to the brain. Definitive results from randomised controlled trials have been hampered by under-recruitment, unexpected drug resistance, confounds and underappreciated delayed time to maximum efficacy of pre-trial entry therapies (sometimes up to 6–12 months) (Ellis et al. 2014). Current guidelines recommend the immediate initiation of therapy following the diagnosis of HIV (Panel on Antiretroviral Guidelines for Adults and Adolescents). Nonetheless, recent results from the START neurology sub-study have shown no neurocognitive advantage of immediate vs delayed initiation of therapy in individuals with $CD4^+$ T cell counts >500 cells/μL (Wright et al. 2018). Novel therapeutic strategies, regimens and limitations are further discussed later in this chapter.

3 Risk Factors of HAND

HAND is influenced by many risk factors including current and nadir $CD4^+$ T cell count, duration of HIV disease, age, co-infection, other comorbidities, notable diabetes and substance use (Saylor et al. 2016). Although viral suppression with ART has improved the life expectancy of PLWH to almost that of uninfected individuals (Trickey et al. 2017), chronic HIV-infection presents its own issues as age is an independent risk factor for HAND and chronic HIV infection is hypothesised to induce 'premature ageing' as people living with HIV exhibit higher risk and possibly earlier onset of age-related disease than uninfected individuals. Whether the mechanisms underlying the 'premature ageing' phenomenon reflect true 'accelerated ageing' (i.e. earlier onset and/or faster progression of disease) or 'accentuated ageing' (i.e. heightened severity of disease with same time of onset) is unclear, but work assessing onset/severity of neurocognitive disorders in people living with HIV supports a combination of both scenarios (Sheppard et al. 2017). Older PLWH (>50 years) are 3.26 times more likely to develop HAD than younger PLWH (<40 years) (Valcour et al. 2004) and show more rapid rates of progressive atrophy despite therapy, particularly in subcortical regions, than age-matched individuals (Clifford et al. 2017; Pfefferbaum et al. 2014; Nir et al. 2019). Older PLWH also show deficits in episodic memory and motor function relative to age-matched HIV-uninfected individuals (Goodkin et al. 2017), and functional brain changes in blood flow and brain age appear similar to those observed in ~15–20 year older HIV-uninfected individuals as measured by functional MRI (Ances et al. 2010). Furthermore, recent findings from a longitudinal study assessing 549 people living with HIV for up to 14 years have shown that these individuals exhibit accelerated age-related decline in brain volume, especially in the frontal cortex as measured by MRI (Pfefferbaum et al. 2014, 2018). Importantly, whilst these age-related changes were accentuated in individuals with a history of substance abuse and/or HCV co-infection, they persist even in individuals without these cofounders (Pfefferbaum et al. 2014, 2018).

The discrete mechanisms driving accelerated/accentuated age-related changes are unknown, but persistent inflammation and immune activation are at higher levels than the ageing process alone can explain (Angelovich et al. 2015). Specifically, measures of glial activation, neurotransmitters and ketone bodies in the CSF of PLWH mimic levels observed in older HIV-uninfected individuals and also correlated with age and adverse neurocognitive scores (Cassol et al. 2014). Recently, CSF levels of galectin-9, an immune modulatory protein linked to HIV persistence, was also found to be higher in virally suppressed individuals (Premeaux et al. 2019). However, levels were only indicative of neurocognitive impairment in older individuals, but not younger individuals. Furthermore, structural neuroimaging analyses by MRI suggests that virally suppressed PLWH experience accentuated brain atrophy, indicative of age-related changes, in comparison to HIV-uninfected controls (Cole et al. 2017). Epigenetic changes have also been implicated as the 'brain age' of PLWH with HAND, estimated by the epigenetic clock using DNA methylation analysis, appears 3.5 years older than individuals without HAND (Levine et al. 2016). Circulating HIV DNA in PBMCs also correlates with neurocognitive impairment, namely, in executive function, in older PLWH despite viral suppression (de Oliveira et al. 2015), also suggesting the role of HIV reservoirs in disease pathogenesis.

It is of note that whilst the extended duration of HIV infection is known to influence the pathogenesis of HAND, the median age of HIV infection may also influence neurocognitive disease prognosis. In 2016 over 17% of new HIV notifications in the USA were from individuals older than 50 years of age (Centers for Disease Control and Prevention 2017). This is problematic as older individuals tend to present with more advanced clinical disease as they are not routinely tested for HIV despite exhibiting some similar risk factors as younger persons such as multiple sexual partners (Asher et al. 2016). Furthermore, the dynamics of disease pathogenesis, and the use of biomarkers used to identify disease, may also change with age-related risk factors. Specifically, in a study by Fogel and colleagues, lipid dysfunction was found to be more highly associated with HAND in older PLWH, whilst methamphetamine use was more strongly associated with HAND in younger people (Fogel et al. 2015). Thus, chronological age, duration of infection and age of infection are significant confounders in HAND pathogenesis and may influence the efficacy of prognostic biomarkers of disease.

HAND is often associated with several comorbidities as HIV disease progresses (Guaraldi et al. 2011). As such, cardiovascular disease with risk factors such as hypercholesterolemia and hypertension is associated with neurocognitive impairment in virally suppressed PLWH (Wright et al. 2010). HAND has also been associated with dyslipidaemia in the CSF (Bandaru et al. 2013) as well as glomerular filtration rate, an indicator of both kidney function and CVD (Yuen et al. 2017). Non-HAND neurocognitive diseases such as Alzheimer's disease (AD) may also confound HAND diagnosis. Until recently it was controversial whether PLWH have a propensity to develop AD. Beta-amyloid, characteristic of AD, has been found in brain tissue of PLWH (Green et al. 2005), and phosphorylated tau has been found in the CSF and plasma of PLWH and was associated with HAND (Brew et al. 2005), however, not in all cases (Krut et al. 2017; Clifford et al. 2009). PET scans have provided the first conclusive evidence of beta-amyloid accumulation in the CNS of a HIV-infected individual (Turner et al. 2016), suggesting the possible co-presence of HAND and/or AD. Thus, delineating HAND from AD, and more importantly treating appropriately, is of the utmost importance. Whilst PLWH appear at increased risk of Parkinson's disease (Tisch and Brew 2009, 2010), the issue remains controversial and requires further study (Moulignier et al. 2015). It seems likely that the increased risk only pertains to those who already have HIV brain involvement as the mechanism is probably a 'double hit' to the basal ganglia and dopaminergic pathways both of which are also targets for HIV.

Host gene factors pose a risk to developing HAND with several immune-related genotypes associated with higher risk including CCR5-wt/wt (Singh et al. 2003), 195ApoE ε4, MBL2-O/O (Spector et al. 2010) and CCL2-2578G (Thames et al.

2015), although these genotypes do not encompass all individuals with HAND. Co-infection also contributes to HAND pathology. Cytomegalovirus (CMV) is a common co-infection in people living with HIV, and anti-CMV IgG levels have recently been linked with neurocognitive impairment in PLWH (Brunt et al. 2016; Letendre et al. 2018). Hepatitis C virus co-infection also confers ~twofold increased risk of cognitive impairment than HIV-infection alone (Ciccarelli et al. 2013).

4 HAND Pathophysiology

4.1 Natural HIV Infection of the CNS

HIV probably infects the CNS via a 'Trojan horse'-like mechanism, whereby HIV-infected peripheral blood monocytes, particularly the $CD14^{+}CD16^{+}$ subset (Veenstra et al. 2017; Ellery et al. 2007), and T cells cross the BBB (Fischer-Smith et al. 2001; Honeycutt et al. 2018) (Fig. 1). Current thinking considers that BBB damage is a result of CNS infection though the reverse may be true to an extent – that is, systemic infection may also directly damage the BBB. During acute infection, HIV-1 RNA is detected in human CSF as early as 8 days postinfection (Valcour et al. 2012) and probably continues throughout natural infection in response to chemokines produced by microglia and astrocytes, as seen in individuals diagnosed with HAD (Persidsky et al. 1999). Furthermore, monocyte DNA content in untreated individuals is associated with progressive stages of HAND (Valcour et al. 2013). As such, recruitment of HIV-infected monocytes from sites such as the bone marrow to the CNS is thought to be a major contributor to CNS infection and neurological impairment (Burdo et al. 2010; Strickland et al. 2014).

Following transmigration to the CNS, monocytes mature into perivascular macrophages, which may (1) sustain HIV replication and mediate infection of surrounding cells, such as CD4-expressing macrophage and microglia (Fischer-Smith et al. 2001; Thompson et al. 2011; Burdo et al. 2013a; Kim et al. 2006), and (2) harbour latent HIV reservoirs (Thompson et al. 2011). We and others have shown that up to 19% of astrocytes, which express little to no CD4, are infected with HIV in individuals with HAD (Churchill et al. 2009), can latently harbour HIV and may mediate trans-infection HIV to $CD4^{+}$ T cells (Gray et al. 2014). A recent report also suggests that pericytes forming the BBB may also be a site of HIV infection (Cho et al. 2019). Although HIV-infected $CD4^{+}$ T cells and cell-free virus also enter the CNS, likely due to breakdown of the tight junctions forming the BBB (Strazza et al. 2011), macrophage-driven viral replication is thought to be the primary source of HIV replication in the CNS of patients at least those with severe HAND (Schnell et al. 2009).

Following CNS infection, HIV viral proteins, such as nef, vpr, gp120 and, in particular, extracellular tat (Bagashev and Sawaya 2013), activate surrounding macrophages, microglia and astrocytes to release cytokines and chemokines that drive localised neuroinflammation (Shah et al. 2011a, b; Shah and Kumar 2010; Sami Saribas et al. 2017); the effects of which target multiple sites in the CNS. Firstly, cytokines and chemokines can activate and actively damage the BBB, resulting in enhanced permeability and monocyte recruitment (driven by MCP-1 and CCL2 (Conant et al. 1998)), particularly of $CD14^{+}CD16^{+}$ monocytes (Williams et al. 2013). Secondly, cellular activation can act to enhance HIV replication (osteopontin, TNF and CXCL-10 (Brown et al. 2011; Williams et al. 2009)). Finally, components of the inflammatory milieu such as CXCL-10 (Mehla et al. 2012) and quinolinic acid (Kandanearatchi and Brew 2012) ultimately have neurotoxic effects on neurons via the generation of reactive oxygen species and lipid peroxidation; disruption of calcium homeostasis of neurons (Haughey et al. 1999) leading to neuronal apoptosis (IL-1β, IL-8, TNF (Guha et al. 2012)) and dysfunction of astrocytes ability to buffer glutamate via the N-methyl-D-aspartate receptor; all of these contribute to neuronal dysfunction, apoptosis and neurodegeneration. Neuronal injury occurs early in infection with PLWH with primary, untreated infection displaying heightened measures of neuronal injury such as CSF levels of neurofilament light chain protein (NFL) and neurometabolite dysfunction as determined by proton-magnetic resonance spectroscopy (MRS) (Peluso et al. 2013). Thus, HIV-induced neuroinflammation acts as a positive feedback loop driving HAND and other pathological injuries.

One major complication of uncontrolled HIV infection is HIV-associated encephalitis (HIVE), a neuropathological correlate of HAD. HIVE is strongly associated with reduced brain volume (as measured by MRI) and chronic immune activation and inflammation. Specifically individuals with HIVE express a higher percentage of $CD163^{+}CD14^{+}CD16^{+}$ perivascular macrophages in the brain (Fischer-Smith et al. 2001; Kim et al. 2006); an expansion of 'inflammatory' $CD16^{+}$-expressing monocytes in the periphery (Ancuta et al. 2008); and multinucleated giant cells, comprising aggregated microglia and macrophage cells (Brew et al. 1995; Takahashi et al. 1996). Furthermore, individuals with HIVE have widespread dysregulation of genes involved in synapto-dendritic functioning and integrity, toll-like receptors, interferon responses, mitochondrial genes, synaptic transmission and cell-to-cell signalling (Masliah et al. 2004; Gelman et al. 2012). Recent evidence also suggests that microglia in individuals with HIVE are also dysfunctional, with lower gene expression levels of functional markers (Ginsberg et al. 2018). Thus, HIVE is strongly associated with advanced neurocognitive impairment such as HAD.

Finally, neurotropic HIV variants, geno- and phenotypically distinct from peripheral viruses, exist in PLWH and may drive disease (Gorry et al. 2001, 2002; Haggerty and Stevenson 1991). These variants are detectable at different stages of infection and induce inflammatory responses such as increased neopterin and pleocytosis in CSF (Sturdevant et al. 2015; Hagberg et al. 2010), suggesting that a compartment of neurotropic/neurovirulent viral strains may play a more critical role in HAND than absolute viral load in the CNS.

4.2 HAND Pathophysiology Following Viral Suppression with ART

The persistence of HAND despite viral suppression with ART indicates an underlying pathology in the CNS. However, the exact pathophysiology of disease is unclear mainly due to the inability to perform biopsy analysis of the CNS from individuals who are virally suppressed with ART and limited numbers of autopsy samples from such patients. Therefore, most evidence of ongoing pathology in virally suppressed individuals is based on peripheral indicators of CNS disease or MRI/PET scans. Individuals with HAND have heightened indicators of persistent neuronal injury (as measured by cerebral metabolites) (Young et al. 2014), degradation of cortical grey matter, BBB dysfunction (Chaganti et al. 2019), microglial activation (Rubin et al. 2018; Garvey et al. 2014) and a continued reduction of brain volume despite therapy (Cardenas et al. 2009; Küper et al. 2011). Indicators of axonal damage such as CSF neurofilament light chain are also elevated in virologically suppressed PLWH (Jessen Krut et al. 2014) and are associated with cognitive decline (Sun et al. 2017; Sailasuta et al. 2016). Notably, osteopontin, a pro-inflammatory bone-matrix protein, is also found in high levels in virally suppressed individuals with HAND (Brown et al. 2011; Yu et al. 2017; Vera et al. 2016). These biomarkers of disease, discussed further below, are indicative of chronic neuroinflammation and neurodegeneration despite ART (Figure 1).

Virally suppressed individuals also show peripheral indicators of chronic cellular activation and dysfunction that may indicate/contribute to neuropathology. Individuals with HAND have a higher number of activated macrophages/microglia in the CNS relative to neurologically unimpaired individuals (Tavazzi et al. 2014). Plasma levels of MCP-1 and CXCL-10, produced by activated macrophages and monocytes, are associated with neuronal injury and higher inflammation in virally suppressed individuals with HAND (Letendre et al. 2011). Levels of quinolinic acid, a product of tryptophan metabolism, are also elevated in virally suppressed PLWH and are directly associated with levels of pTau that are predictive of neurocognitive impairment in this population (Anderson et al. 2018). Some evidence suggests that the perivascular macrophage phenotype may be predisposed by the precursor monocyte phenotype as monocyte activation (as measured by CSF sCD14) remains elevated in virologically suppressed individuals with HAND (Eden et al. 2007; Spudich et al. 2011; Yilmaz et al. 2013) and is associated with significantly worse neurological outcomes for those on non-suppressive ART (Kamat et al. 2012). Furthermore, PBMCs from individuals with HAD or MND express higher levels of pro-inflammatory cytokine genes (such as TNF, IL-6, IL-27) and lower levels of associated miRNAs (miR-124-3p, miR-210) than PLWH who were neurologically unimpaired (Venkatachari et al. 2017).

HIV-associated astrocyte dysfunction – CNS cells that normally help to maintain cerebral immunity and homeostasis – has been associated with HAND. Specifically, astrocyte dysfunction results in improper buffering of glutamate, dysregulation of BBB permeability, and release of pro-inflammatory cytokines (such as TNFα).

Moreover, because astrocytes are susceptible to HIV infection, they may serve as a viral reservoir. We have further shown that astrocytes are less responsive to some antiretroviral drugs (Gray et al. 2013), thus potentially contributing to continued viral presence in the CNS.

5 Mechanisms Driving HAND in Virally Suppressed Patients

It is well accepted that HIV viremia is not the sole driver of neurocognitive impairment as incidence of HAND continues to affect between 18 and 69% of people living with HIV despite viral suppression with ART. Therefore, other factors such as persistent viral transcription in CNS reservoirs and/or chronic peripheral inflammation outside of the CNS play pertinent roles in determining the HAND status of virally suppressed individuals. These mechanisms will be discussed below.

5.1 HIV Viral Presence/Replication in the CNS

HIV is considered to persist in latent reservoirs in the CNS such as macrophages and astrocytes (Thompson et al. 2011) and is controversially thought to directly contribute to HAND pathogenesis (Churchill et al. 2015). HIV RNA (Dahl et al. 2014) and anti-HIV antibodies (Burbelo et al. 2018) are detectable in the CSF after 10 years of ART, potentially indicating viral persistence and ongoing replication in the CNS. Notably, increasing ratios of 'HIV RNA in CSF'-to-'HIV RNA in plasma' are associated with worsening neurological outcomes in virologically suppressed PLWH (Anderson et al. 2017; Canestri et al. 2010) and elevated neopterin levels (Dahl et al. 2014), indicating that localised replication in the CNS may contribute to neuroinflammation and disease progression. A recent longitudinal study identified CNS escape in 6% of cases (6/101 individuals) in the absence of peripheral viremia following 3 years of ART (Joseph et al. 2019). Further genotypic analysis suggests that the virus was derived from persistent CNS replication, likely from macrophage/microglia, linked to emtricitabine resistance and low nadir CD4 count (10 copies/mm^3). Interestingly, we have shown that some CNS-derived viruses have impaired transcriptional activity due to polymorphisms in the transcription factor Sp1 (Gray et al. 2016). Thus, whether virus in the CSF truly represents active replication in the CNS of virologically suppressed individuals remains unclear, possibly confounded by poor suppression in the periphery associated with a blip in plasma HIV levels (Eden et al. 2016). HIV resistance to lamivudine and emtricitabine (Mukerji et al. 2017) and also treatment with protease inhibitor + NRTI regimens (Mukerji et al. 2018) are also associated with CSF viral escape. Finally in vivo mouse studies have shown that chronic low levels of neurotoxic tat, a step in the HIV life cycle not affected by current ART, can cause astrocyte activation (Dickens et al. 2017) which may contribute to HAND pathogenesis as tat has been found in both CNS tissue and the CSF of virally suppressed individuals (Johnson et al. 2013). Thus, further studies are required to establish the role of active replication and/or viral persistence in the CNS in HAND associated pathology.

Whilst active replication is an important focus of current research in defining the CNS as a reservoir of HIV, persistence of non-replication competent virus in the CNS is also detrimental to the patient. Studies in the periphery have identified that the bulk (~93%) of the HIV reservoir in T cells consists of non-replication competent viruses harbouring hyper-deletions that may generate short abortive transcripts or viral proteins that are immunogenic (Bruner et al. 2016). As such, generation of highly neurotoxic HIV nef and tat in the CNS by persistent non-productive infection in cells such as astrocytes may drive neuroinflammation, cellular dysfunction and neuronal degradation (as discussed above). Thus, persistent non-replication competent virus in the CNS may play as important a role in HAND pathogenesis as active viral replication and must be considered in appropriate therapeutic strategies.

5.2 *Host Cell and Epigenetic Factors*

Host cell and environmental factors induced by microRNAs (miRNA) – short non-coding repressors of transcription – and DNA methylation have been identified to be associated with HAND pathogenesis (Tatro et al. 2010), suggesting a role for epigenetic changes to components of the immune system in driving HAND. Large genomic studies have found strong associations between dysregulated miRNAs and mRNA in the brains of individuals with HAD in comparison to neurocognitively unimpaired individuals (Zhou et al. 2012). Specifically, miR-21, known to repress transcription factors involved in neuronal functioning, is present in high levels in the CNS of PLWH with HAD (Yelamanchili et al. 2010). Furthermore, both miR-128a (Eletto et al. 2008) and miR-34a (Mukerjee et al. 2011) are upregulated by HIV Tat or Vpr, respectively, resulting in adverse effects on neuronal function in vitro. HIV Tat also influences histone acetylation/deacetylation via chromatin structure remodelling by histone deacetylases in neuronal cells in vitro, resulting in synaptic plasticity and neuronal dysfunction (Saiyed et al. 2011). Epigenetic changes derived by alterations in DNA methylation sites in monocytes are also strongly associated with worse neurocognitive scores (Corley et al. 2016), implying that epigenetic changes in peripheral cells transmigrating to the CNS may also influence HAND.

5.3 *Chronic Peripheral Inflammation*

Chronic immune activation and inflammation is a hallmark of HIV infection despite viral suppression, and measures of inflammation/immune activation are reliable indicators of HIV disease progression and have been linked with HAND pathogenesis (Hazenberg et al. 2003; Somsouk et al. 2015; Kuller et al. 2008; Boulware et al. 2011; Hunt et al. 2011; Pandrea et al. 2012). The causes of persistent immune activation in chronic HIV infection are multifactorial, but the host response to translocated microbial products from the gastrointestinal (GI) tract, resulting from the depletion of gut-associated lymphoid tissue $CD4^+$ T cells during acute infection, is thought to be a major contributor (Brenchley et al. 2004, 2006; Estes et al. 2010). This is exemplified in elegant in vivo studies where experimental damage of the GI tract of African green monkeys (AGMs), a natural host of SIV that does not develop AIDS, induces local and systemic inflammation similar to that of HIV infection in humans (Hao et al. 2015). Specifically, bacterial lipopolysaccharide (LPS) – a cell wall component of gram-negative bacteria and potent immune activator – remains elevated in HIV/SIV infection despite ART and is thought to drive chronic inflammation (Deeks et al. 2004). Progression to AIDS in PLWH may be driven (in part) in response to LPS as sooty mangabeys, natural hosts of SIV that do not develop AIDS, have recently been shown to have a C-terminal frameshift in the LPS receptor (TLR-4) which renders it non-functional (Palesch et al. 2018).

Measures of microbial translocation, especially measures of monocyte activation following exposure to LPS such as sCD14 and sCD163, have been linked with neurocognitive impairment, indicating a central role for monocyte activation in neuropathology (Vassallo et al. 2013). Notably, elevated levels of circulating $CD16^+$ monocytes, $CD14^+$ monocyte HIV DNA content, sCD14 and sCD163 are associated with microbial translocation and correlate with an increased risk of neurocognitive impairment (Valcour et al. 2013). sCD14 levels in CSF are also associated with NFL, indicative of axonal damage, in untreated PLWH (Jespersen et al. 2016). Moreover, plasma LPS levels are higher in HIV-HCV co-infected individuals with HAND, relative to HIV-HCV co-infected neurologically unimpaired individuals (Vassallo et al. 2013). GI tract damage in PLWH is also associated with more pathogenic microbial populations, compared to uninfected individuals (Mutlu et al. 2014). Notably, similar alterations in the gut microbiome have been linked to other CNS disorders, namely, multiple sclerosis (Glenn and Mowry 2016) and Parkinson's disease (Houser and Tansey 2017). Whilst it follows that, within PLWH, changes in the gut microbiome and the resultant inflammatory response may contribute to HAND, studies are required to test this hypothesis.

5.4 ART Neurotoxicity

Whilst current antiretrovirals are less neurotoxic than earlier iterations, ART neurotoxicity continues to be linked to HAND pathogenesis, and many ARVs may have lasting adverse 'legacy' effects that may contribute to disease decades after use (Jonathan et al. 2015). Efavirenz, a commonly prescribed non-nucleoside reverse-transcriptase inhibitor, is still in use despite known neurological side effects such as psychiatric episodes and neurocognitive effects on information processing and executive functioning (Ma et al. 2016). The underlying pathology of efavirenz is most likely due to either the induction of pro-inflammatory cytokines or mitochondrial toxicity (Funes et al. 2015) and endoplasmic reticulum stress (Funes et al. 2014; Bertrand and Toborek 2015). Treatment of neuronal cell lines with 15 different ARVs at physiological concentrations identified modest neuronal toxicity in vitro (Robertson et al. 2012). In vitro raltegravir has also been linked with enhanced production of macrophage-derived IL-8 (Tatro et al. 2014).

Protease inhibitors also have a higher risk of drug-drug interactions by metabolism via CYP450 enzymes and have also been linked to neurotoxicity. Treatment of human astrocytes with the protease inhibitors amprenavir and lopinavir induced dysregulation of glutamate transport in vitro and was associated with worse neurocognitive performance in mouse models (Vivithanaporn et al. 2016). In vitro ritonavir or lopinavir treatment also impaired oligodendrocyte maturation, and similar treatment in mice reduced myelin protein levels in the frontal cortex (Jensen et al. 2015). Importantly Jensen and colleagues further showed that levels of myelin basic protein in the frontal cortex of PLWH were lower than those present from untreated PLWH or HIV-uninfected patients. Finally, recent evidence has found that individuals treated with integrase stand transfer inhibitors (INSTIs) such as dolutegravir are associated with poorer neurocognitive performance and reduced brain volume than those on non-INSTI containing regimens (O'Halloran et al. 2019). As many patients transition to regimens including INSTIs, the long-term neurocognitive effects of these drugs need to be closely monitored. Thus, there remains a clinical demand for highly effective and non-neurotoxic antiretrovirals as well as improved treatment strategies to treat HAND.

6 Biomarkers of CNS Infection and HAND

Identifying biomarkers of cognitive impairment, and more specifically HAND, is a major research focus due to the clinical demand for reliable means to (1) predict HAND before it occurs, (2) diagnose HAND, (3) distinguish the early stages of HAND (ANI and MND) and (4) predict and monitor changes in cognition for individuals with HAND. Biomarkers must be readily accessible, cheap and reliable, and as such most emphasis has been placed on indicators of neuroinflammation or neuronal damage in the blood and CSF. Identifying biomarkers of HAND has proven to be a difficult task as well-established plasma and CSF biomarkers of inflammation (such as sCD14, osteopontin (Brown et al. 2011), TNFα, IFNγ, sCD14, IL-1β, IL-6, S100β, TGF-β levels) or HIV viremia (such as the surrogate marker nadir $CD4^+$ T cell count) that are predictive of HAND in viremic or ART-naïve PLWH (Abassi et al. 2017; Lyons et al. 2011) are not predictive of HAND in virologically suppressed individuals receiving ART (Kamat et al. 2012; Lyons et al. 2011). Biomarkers of HAND may be indicative of neuronal damage, altered metabolism, generally neuroinflammation or even HIV itself. Specific biomarkers of HAND and their related pathological implications are discussed in detail below (Fig. 1).

6.1 Measures of HIV Viremia and Viral Latency

Although viral suppression with ART has reduced the predictive value of many traditional HIV-related biomarkers of HAND such as CSF HIV RNA (Ellis et al. 2002; Brew et al. 1997), some measures of HIV remain indicative of HAND despite therapy with ART. Low $CD4^+$ T cell counts continue to be associated with HAND, albeit mainly in individuals failing therapy or with moderate immune suppression ($CD4^+$ T cell count 200–349 cells/mm^3) (Bhaskaran et al. 2008). $CD4^+$ T cell nadir may offer some prognostic value (Ellis et al. 2011). However, a recent study found that immediate therapy initiation with a CD4 nadir >500 cells/mm^3 had no neurocognitive benefit over delayed treatment initiation (Wright et al. 2018), questioning the predictive value of CD4 nadir at high CD4 T cell counts. Similarly, CSF and/or plasma levels of HIV RNA are also indicative of neurocognitive decline in individuals who exhibit therapy failure or drug resistance (Canestri et al. 2010; Peluso et al. 2012; Bingham et al. 2011).

Additionally, biomarkers of persistent viral replication/presence in the CNS are gaining attention as possible prognostic markers of disease. HIV DNA persists in the CNS despite viral suppression and studies using research tools including PCR and the in situ hybridisation technique DNAscope have offered insight into HIV persistence in the CNS (Churchill et al. 2009; Ko et al. 2019; Lamers et al. 2016; Estes et al. 2017). However, its application as a true diagnostic/prognostic biomarker is limited as these assays generally investigate autopsy brain tissue. Therefore, HIV DNA content in peripheral cells such as T cells and monocytes may be a more practical approach to assessing the CNS reservoir. HIV DNA in peripheral blood monocytes (Valcour et al. 2013; Cysique et al. 2015a) and CSF cells (Oliveira et al. 2017; Shaunak et al. 1990) is associated with risk of HAND and brain atrophy in non-suppressed individuals (Kallianpur et al. 2014) and recent evidence supports that this may also be true in suppressed patients. In a recent study of 69 people living with HIV who were virally suppressed for a median of 8.6 years with ART, Spudich and colleagues found that ~50% of individuals harboured cell-associated HIV DNA in the CSF (Spudich et al. 2019). Furthermore, cell-associated HIV DNA was associated with adverse neurocognitive outcomes, suggesting that persistent HIV in the CSF despite ART may contribute to HAND pathogenesis. BCL11B, an inhibitor of viral transcription, is upregulated in latently infected cells of the CNS (Desplats et al. 2013), and higher CSF levels are associated with lower creatine levels in frontal white matter (Cysique et al. 2019), potentially validating BCL11B as an indicator of latent HIV-related pathology. However, the associations of CSF BCL11B and HAND remain to be defined. CSF escape is also associated with higher neopterin levels (Eden et al. 2016), although as this is a rare phenomenon and neopterin is produced by multiple cells types during cellular activation, this may not be the most robust method of confirming latency. A recent study also found that HIV gp120 protein sequences are predictive of HAD and/or HIVE in silico (Ogishi and Yotsuyanagi 2018), highlighting the possibility of HIV quasi-species as a predictive tool for HAND. However, the methodology required to characterise sequences from isolated immune cells is more challenging than standard biochemical assays which may limit its practical application as a biomarker.

6.2 Neuronal Damage

NFL, a component of myelinated axons, is readily detected in the CSF of neurodegenerative diseases and is indicative of neuronal axon damage. Multiple studies have found that CSF levels of NFL are present at higher levels in untreated individuals with or without neurocognitive impairment and CSF NFL levels in individuals with HAD are significantly higher than those in more mild disease (McGuire et al. 2015; Gisslén et al. 2007, 2015; Peterson et al. 2014). Moreover, NFL is associated with worse CD4 T cell counts and is positively associated with plasma viral load, suggesting that it is indicative of HIV-directed neuropathology. Correlative analysis in individuals with primary infection further shows associations between higher NFL levels in CSF and neuroinflammatory markers (CXCL-10, neopterin) and adverse levels of CNS metabolites (Peluso et al. 2013). NFL has also been shown to be a more sensitive neuronal marker than total and phosphorylated tau and amyloid precursor proteins (Peterson et al. 2014). Furthermore, CSF NFL levels correlate with plasma NFL levels, a significant advantage due to ease of blood draw vs lumbar puncture required for CSF analyses. Therapy reduces NFL levels; however, they remain elevated in comparison to HIV-uninfected controls (Jessen Krut et al. 2014). However, abnormal NFL levels have only been detected in ~16% of virally suppressed subjects, questioning whether this is sensitive enough to detect ANI or MND (Peterson et al. 2014).

Exosomes are gaining attention as biomarkers of neuroinflammation and neurocognitive disorders including AD and HAND (Pulliam et al. 2019). Exosomes are 30–150 nm vesicles containing host cell proteins and DNA that are released following cellular activation into plasma. As plasma exosomes harbour protein and RNA from their host cell such as neurons, they potentially offer a method of interrogating cell-specific pathology from different, difficult-to-access sites such as the brain. Studies in AD have identified the presence of neuronally derived exosomes (NDE) from patients with preclinical AD that predicted AD 10 years prior to onset (Fiandaca et al. 2015). More recently, NDEs have been detected in the plasma of PLWH with cognitive impairment (Sun et al. 2017) and are indicative of neuronal health. Sun and colleagues identified that neurocognitively impaired individuals had fewer NDEs than non-neurocognitively impaired individuals, possibly due to neuronal stress or death. Furthermore, NDEs from these individuals contained higher levels of proteins associated with neuronal damage such as high motility group box 1 (HMGB1), NFL and amyloid β than exosomes from HIV-infected neurocognitively unimpaired individuals (Sun et al. 2017). Thus, exosomes, particularly NDEs, may offer some prognostic benefit over other inflammatory biomarkers that are not truly specific to the CNS.

Mitochondrial DNA (mtDNA) is indicative of the energy levels/demands of a cell. Each cell contains mitochondria with a number of copies of mtDNA; thus mtDNA copy number may be indicative of cellular dysfunction and apoptosis. mtDNA damage is present in approximately 45% of total cells in the frontal cortex of patients with HAND without viral suppression (Zhang et al. 2012). Furthermore, individuals with HAND in this cohort had less mtDNA than those who were not cognitively impaired. Similarly, mtDNA damage can be found in the periphery. Following viral suppression, mtDNA copy number from peripheral blood is associated with worse cognitive performance in CHARTER patients (Hulgan et al. 2018). Furthermore, CSF levels are higher in MND patients in comparison to neurocognitively normal individuals (Mehta et al. 2017); however, no difference was observed in ANI patients – this may just be a reflection of the insensitivity of the CSF as a marker of mild events in a discreet part of the frontal lobe.

6.3 Cellular Activation

Monocytes, and markers of monocyte of activation, are readily measurable indicators of HAND due to their key role in pathogenesis. CCR2 expression on $CD14^{+}CD16^{+}$ monocytes is negatively associated with neurometabolite levels, and by extension pathology, as measured by MRS imaging (Veenstra et al. 2019). CCR2 expression on $CD14^{+}CD16^{+}$ monocytes is also associated with HIV DNA copies per 10^{6} PBMCs (Veenstra et al. 2019). CCR2's ligand, CCL2, is a chemokine that acts to recruit monocytes into the brain. In individuals with the CCL2-2578G, CSF levels of CCL2 are higher than in non-carriers, and higher levels are associated with poor cognitive impairment, albeit mainly in individuals with high plasma viral (Thames et al. 2015). Plasma CCL2 levels are also elevated in individuals with either HAD or minor cognitive motor disorder (precursor to MND) in viremic individuals (Ancuta et al. 2008). However, it is of note that levels are similarly elevated in individuals with non-HIV-related neurocognitive impairments and no difference in ANI patients, and neurocognitively normal individuals were observed in this study.

Soluble indicators of monocyte activation have also been linked to HAND. LPS activates monocytes via TLR-4 as described above, and plasma levels are associated with HAD in individuals with AIDS (Ancuta et al. 2008). LPS levels are known to remain elevated following viral suppression; however, to date no evidence links plasma LPS levels with HAND in the absence of co-infection such as HCV. Nonetheless, following monocyte activation by immunogens such as LPS, monocytes shed CD14, the co-receptor for LPS, which can be measured and has been associated with worse global deficit scores, lower CD4 T cell counts and worse learning, attention and motor T scores in a cohort of generally unsuppressed individuals (Lyons et al. 2011). The reliability and sensitivity of sCD14 to identify HAND in virally suppressed individuals is less clear. Similarly, soluble CD163 – a scavenger receptor – is shed from activated macrophage/microglia and detectable in both plasma and CSF. sCD163 levels are associated with adverse neurocognitive scores (Burdo et al. 2013b; Royal et al. 2016).

Neopterin, a pteridine produced by activated macrophages and monocytes, is commonly found in higher levels in the plasma and CSF of PLWH, despite viral suppression. Neopterin levels have been shown to predict HAND and are associated with HAND severity in therapy-naïve individuals (Brew et al. 1990, 1996); however, there is limited evidence supporting the prognostic value of neopterin in virally suppressed individuals. Plasma levels, but not cellular gene expression in monocytes (Quach et al. 2018), of CCL2, HMGB1, IL-8, CXCL-10 and neopterin are predictive of HAND in virally suppressed individuals (Sun et al. 2017; Eden et al. 2016; Yuan et al. 2013). CSF fractalkine levels are higher in individuals with MND/HAD than ANI (Letendre et al. 2011). CSF levels of HMGB1, produced by activated macrophages and astrocytes, may also be predictive of HAND, even very minor disease, in virally suppressed PLWH (Gougeon et al. 2017).

Whilst many peripheral biomarkers of monocyte/macrophage activation are associated with adverse neurocognitive outcomes (as discussed above), individual biomarkers (such as sCD14, MCP-1, CXCL-10, etc.) are associated with unique neuropathology and inflammation at different locations in the brain (such as frontal white matter, basal ganglia, grey matter). For example, MRS studies assessing levels of neurometabolites in the CNS of virally suppressed individuals found that CSF levels of sCD14 were negatively associated with Cho/Cr ratios in frontal white matter (Anderson et al. 2015), indicative of monocyte-induced neuronal damage. However, MCP-1 levels, indicative of monocyte recruitment, were positively associated with Cho/Cr ratios in the basal ganglia (Anderson et al. 2015), supporting findings from similar studies (Harezlak et al. 2011). Thus, it is important to acknowledge that each biomarker of myeloid activation may be indicative of different stages and/or sites of neurocognitive diseases which must be considered when interpreting results.

S100β is a calcium binding protein mainly expressed on glial cells such as astrocytes and thus is useful as a biomarker of astrocyte activation. S100β levels in the CSF have been associated with cognitive impairment in PLWH and were higher in those with more severe AIDS dementia complex (precursor terminology to HAND) or progressed to more advanced disease more quickly (Pemberton and Brew 2001). S100β levels are also inversely associated with executive function including the rapid generation of verbs testing 'action fluency' (Woods et al. 2010).

Finally, measures of T cell activation are also altered with symptomatic HAND (HAD and MND) exhibiting different T-cell immune activation profiles to those with ANI (Vassallo et al. 2015) (diagnosis of MND was associated with a CD4:CD8 ratio <1), suggesting the pathophysiological mechanism underlying symptomatic- and asymptomatic-HAND is different or the difference could reflect a longer time with symptomatic disease given that ANI leads to MND. T-cell phenotype/function and HIV itself are also indicative of neurocognitive impairment. The percentages of IFN-γ producing $CD8^+$ T-cells in the CSF are directly associated with the severity of neurocognitive decline as measured by global deficit scores in virally supressed

PLWH (Schrier et al. 2015). However, whether these observations are indicative or causative in disease pathogenesis is unknown.

6.4 Metabolic Dysfunction

PLWH with neurocognitive impairment experience dysregulated kynurenine pathways of tryptophan catabolism, skewing to the production of neurotoxic metabolites. As such CSF levels of quinolinic acid/tryptophan ratios are indicative of CNS neurocognitive impairment in SIV-infected macaques, particularly early in infection (Drewes et al. 2015; Heyes et al. 1990). Furthermore, a recent study showed that virally suppressed individuals had higher CSF levels of quinolinic acid and kynurenine/tryptophan ratios than HIV-uninfected controls (Anderson et al. 2018). Similarly, HAND has been associated with lipid dysfunction and cholesterol accumulation in the brain (Mielke et al. 2010). In a large multi-centre study, HIV-infected virally suppressed individuals had altered levels of cholesterol species in the CSF; namely, an increase in the ceramide species C24:1 that was associated was higher likelihood of neurocognitive decline (Bandaru et al. 2013). Furthermore, a shift in lipid species was associated with progressive disease.

6.5 Novel Non-traditional Biomarkers

Progranulin is produced by microglia and neurons and is thought to have neuroprotective effects by modulating neuroinflammation whilst also acting as a neuronal growth factor. Progranulin also has known anti-viral effects that inhibit HIV replication, and its production by microglia is upregulated during active HIV infection (Suh et al. 2014a). However, during viral suppression with ART, lower levels of progranulin in the CSF are present and are associated with worse cognitive impairment and elevated levels of other pro-inflammatory biomarkers (Suh et al. 2014b). This may be because downregulation of the production of neuroprotective progranulin by microglia may leave neurons more susceptible to damage (Suh et al. 2014a). Thus, low progranulin levels in the CSF may offer insight into underlying CNS neuronal damage and HAND pathogenesis.

Total tau (t-tau) and phosphorylated tau (p-tau) levels in the CSF are commonly used as an indicator of Alzheimer's disease; however, levels have also been assessed in the context of HAND. We and others have found that CSF levels of p-tau (as well as total-tau levels) are higher in individuals with AIDS dementia complex and HAND than those who are uninfected (Brew et al. 2005; Cysique et al. 2015b), which is in line with more recent reports where p-tau levels in CSF were inversely associated with scores of prospective memory in PLWH (Anderson et al. 2018). However, others have found a difference in only CSF levels of t-tau, but not p-tau (Krut et al. 2017; Steinbrink et al. 2013) in comparison to the HIV-uninfected population, possibly suggesting that patient viral load and/or ongoing neuronal damage at the time of sampling may influence these associations.

MicroRNAs are non-coding molecules that regulate both viral and host gene expression by binding and repressing RNA and can be detected in PBMCs, plasma and CSF. A small cross-sectional study found that HIVE patients express altered CSF levels of 11 different miRNAs in comparison to subjects without HIVE (Pacifici et al. 2013), with miR-937 being highly upregulated in the CSF. Furthermore, levels of miRNAs in plasma are different between individuals with or with HAND (Kadri et al. 2016; Asahchop et al. 2016) making miRNAs a potentially useful biomarker of disease. The Veterans Aging Cohort Study (VACS) index, consisting of traditional HIV biomarkers of disease and non-traditional biomarkers of comorbidity, is also predictive of neurological decline (Marquine et al. 2016).

6.6 *Limitations of Biomarkers and Current Challenges*

Whilst the aforementioned biomarkers are indicative of cognitive decline in PLWH, no clinically validated biomarker can reliably differentiate the stages of HAND, especially ANI. This may be because many are broad inflammatory components that are not uniquely specific to the brain, or HIV infection, and thus they may be influenced by inflammation from other sites or other diseases/disorders or even age. Furthermore, it is likely that 'waves' of neuroinflammation may take place during disease pathogenesis, confounding the accuracy and specificity of biomarkers with HAND. This is evidenced by diffusion tensor imaging assessing white matter damage where reductions in white matter in chronic infection were associated with markers of inflammation (Wright et al. 2015), whilst changes in white matter in primary infection were associated with breakdown of the BBB, thus, representing a change in neuropathogenesis and neuroinflammation with disease progression. Therefore, different biomarkers of disease may be representative of stages of neuropathogenesis that are not strictly related to Frascati Criteria stages of disease. In summary, whilst much work has been conducted evaluating biomarkers of HAND, a need remains for cheap, accessible, and highly accurate biomarkers that accurately diagnoses, distinguishes and compares distinct forms of HAND. Combinations of biomarkers are more likely to be more predictive or diagnostic of disease.

7 Biomarkers of Neurocognitive Decline as Indicators of ART Success

Due to effective ART regimens suppressing HIV viremia to undetectable levels in CSF/plasma, biomarkers of HIV disease such as CSF HIV RNA offer little insight into underlying pathology and neurocognitive disease in virally suppressed individuals. Therefore, biomarkers of underlying neuropathology or cellular activation such as those described above provide critical information regarding the success of novel treatment strategies that complement traditional indicators of HIV viremia.

Intensification of ART to treat CNS infection has been evaluated with some success. Intensive therapy with maraviroc, a CCR5 inhibitor, for 24–52 weeks in combination with standard ART reduced the proportion of $CD16^+$ monocytes harbouring HIV DNA and was associated with better neuropsychological performance in the majority of patients (Ndhlovu et al. 2014; Gates et al. 2016). Early initiation of ART (i.e. <30 days postinfection) also has elicited some neuroprotective effects in both small cross-sectional (Evering et al. 2016) and short-term longitudinal studies (Hellmuth et al. 2016). Furthermore, early initiation also reduced the diversity of HIV DNA in PBMCs, cells isolated from CSF and levels of pro-inflammatory cytokine (IL-6 and TNF) in the CSF, compared to delayed ART initiatives (>14 months post-exposure) (Oliveira et al. 2017). However, the lack of large long-term longitudinal studies limits our understanding of the efficacy of such treatment regimens.

Targeting neuroinflammation using adjunctive therapy in concert with ART also may have beneficial effects. However, to date all attempts to reduce the risk of HAND by targeting neuroinflammation in HIV infection have been unsuccessful (Meulendyke et al. 2014; Sacktor et al. 2011, 2014, 2018; Schifitto et al. 2007). Treatment with the anti-oxidant minocycline reduced CSF levels of some oxidative stress markers in comparison to placebo (Sacktor et al. 2014), but did not improve neurocognitive impairment over 24 weeks (Sacktor et al. 2011). Similarly treatment with selegiline, a MAO-B inhibitor and anti-oxidant, had no neurocognitive or functional benefit in PLWH with mild to moderate neurological impairment (Schifitto et al. 2007). Although treatment with a combination of paroxetine and fluconazole reduced CSF neurofilament light chain and amyloid precursor protein levels in SIV-infected RMs (Meulendyke et al. 2014), a recent double-blind placebo-controlled study using this regimen showed neurocognitive improvement in some, but not all, domains (Sacktor et al. 2018), highlighting that further validation is required. Interestingly, these drugs target HIV gp120 and tat-related neurotoxicity and did not show any improvement in neuroinflammatory markers in plasma or CNS (Meulendyke et al. 2014), suggesting that peripheral inflammation may be more important in human HAND. However, in other studies, drugs specifically targeting extracellular Tat have been shown to counteract its production of IL-1β and TNF in glial cell lines in vitro (Mediouni et al. 2015), and several novel Tat/TAR (and Tat/cyclin T1/CDk9) small molecule inhibitors have been described to date (Serena et al. 2013). These initiatives are important as current therapies do not stop tat secretion from CNS cells (Sonia et al. 2012).

Therapies directed against metabolic dysfunction in the CNS such as intranasal insulin have also been shown to improve neurobehavioural performance in animal models and are currently guiding clinical studies in humans. Specifically, cats who were infected with feline immunodeficiency virus and treated with 20 IU of intranasal insulin for 6 weeks had reduced markers of glial activation and neuroinflammation and improved neurobehavioural performance than placebo controls (Mamik et al. 2016). Furthermore, intranasal insulin treatment of chimeric EcoHIV-infected mice, who exhibit neurocognitive impairment similar to HAND, was found to transiently reverse hippocampal dendritic injury and improve learning and retention ability in some animals; however, this only occurred in the presence of the drug (Kim et al. 2019). These findings support other studies testing intranasal insulin in neurocognitive disorders such as Alzheimer's disease and have led to Phase I/II clinical trials (NCT03277222, NCT03081117) that are underway testing the effects of intranasal insulin treatment in combination with ART with findings expected in 2020–2021.

Finally, targeting peripheral inflammation is also a possible adjunctive therapy to ART in both HAND and other HIV-associated comorbidities. Targeting monocyte activation using atorvastatin and simvastatin treatment, generally used for treating hyperlipidaemia, reduced the number of circulating $CD16^+$ monocytes (Yadav et al. 2016) and treatment with the dual CCR2 and CCR5 antagonist cenicriviroc; thus migration of $CD14^+CD16^+$ monocytes improved neurocognitive performance following 24 weeks of treatment (D'Antoni et al. 2018). Recent in vitro evidence also suggests that treatment with buprenorphine, an opioid derivate used as a replacement for heroin in addicts, may have anti-inflammatory properties by binding monocytes and reducing monocyte transmigration across a BBB model (Jaureguiberry-Bravo et al. 2018). Whilst this treatment may be beneficial in injecting drug using PLWH, its efficacy and safety needs to be confirmed in vivo in non-heroin addicts. Thus, targeting cellular inflammation and immune activation in the periphery may be a viable treatment strategy with beneficial effects across multiple HIV-associated co-morbidities.

8 Where to from Here?

HAND remains a significant burden affecting PLWH globally; however, due to the relatively specialised nature of neurocognitive testing required to diagnose HAND, especially ANI, people living with HIV and neurocognitive impairment are either being misdiagnosed or missed completely. The limited ability to accurately diagnose mild HAND coupled with a change in the HIV epidemic to a chronic inflammatory disease whereby individuals are advancing to older age with an elevated risk of age-related comorbidities represents a difficult problem. Thus, early and accurate diagnoses of HAND are essential in improving health outcomes.

Significant steps forward in understanding the neuropathology of HAND have provided some clues into possible biomarkers for diagnosis and prognosis. Continuous viral presence in the CNS, chronic peripheral immune activation and inflammation as well as underlying drug toxicity and cellular dysfunction all pose viable mechanisms driving neurological diseases in PLWH. However, the discrete effects of each remain unclear. It is likely that a fine balance exists, whereby chronic systemic inflammation, which persists despite viral suppression, may amplify virally induced disease to contribute to severe diseases outcomes such as HAD. This, however, complicates the clinical specificity of many biomarkers that are in effect broad indicators of inflammation such as sCD14 and sCD163. Thus, an added focus on biomarker research delineating underlying CNS pathology such as neuronally derived exosomes, NFL or even MRI of neuroinflammation may be more useful clinically. Furthermore, combining a number of biomarkers to form a 'risk score' or 'index' using an algorithmic approach may be more predictive than any one biomarker. Whilst significant advances have been made in identifying biomarkers of HAND, the need remains for truly specific, sensitive and reliable biomarkers to reduce the risk of misdiagnosis and delayed treatment and ultimately improve health outcomes in people living with HIV.

References

Abassi M, Morawski BM, Nakigozi G, Nakasujja N, Kong X, Meya DB et al (2017) Cerebrospinal fluid biomarkers and HIV-associated neurocognitive disorders in HIV-infected individuals in Rakai, Uganda. J Neurovirol 23(3):369–375

Ances BM, Vaida F, Yeh MJ, Liang CL, Buxton RB, Letendre S et al (2010) HIV and aging independently affect brain function as measured by functional magnetic resonance imaging. J Infect Dis 201(3):336–340

Ancuta P, Kamat A, Kunstman KJ, Kim E-Y, Autissier P, Wurcel A et al (2008) Microbial translocation is associated with increased monocyte activation and dementia in AIDS patients. PLoS One 3(6):e2516

Anderson AM, Fennema-Notestine C, Umlauf A, Taylor MJ, Clifford DB, Marra CM et al (2015) CSF biomarkers of monocyte activation and chemotaxis correlate with magnetic resonance spectroscopy metabolites during chronic HIV disease. J Neurovirol 21(5):559–567

Anderson AM, Muñoz-Moreno JA, McClernon D, Ellis RJ, Cookson D, Clifford DB et al (2017) Prevalence and correlates of persistent HIV-1 RNA in cerebrospinal fluid during antiretroviral therapy. J Infect Dis 215:105–113

Anderson AM, Croteau D, Ellis RJ, Rosario D, Potter M, Guillemin GJ et al (2018) HIV, prospective memory, and cerebrospinal fluid concentrations of quinolinic acid and phosphorylated Tau. J Neuroimmunol 319:13–18

Angelovich TA, Hearps AC, Maisa A, Martin GE, Lichtfuss GF, Cheng W-J et al (2015) Viremic and virologically suppressed HIV infection increases age-related changes to monocyte activation equivalent to 12 and 4 years of aging, respectively. J Acquir Immune Defic Syndr 69 (1):11–17

Antinori A, Arendt G, Becker J, Brew B, Byrd D, Cherner M et al (2007) Updated research nosology for HIV-associated neurocognitive disorders. Neurology 69(18):1789–1799

Asahchop EL, Akinwumi SM, Branton WG, Fujiwara E, Gill MJ, Power C (2016) Plasma microRNA profiling predicts HIV-associated neurocognitive disorder. AIDS 30(13):2021–2031

Asher I, Guri KM, Elbirt D, Bezalel SR, Maldarelli F, Mor O et al (2016) Characteristics and outcome of patients diagnosed with HIV at older age. Medicine 95(1):e2327

Bagashev A, Sawaya BE (2013) Roles and functions of HIV-1 Tat protein in the CNS: an overview. Virol J 10:358

Bandaru VVR, Mielke MM, Sacktor N, McArthur JC, Grant I, Letendre S et al (2013) A lipid storage–like disorder contributes to cognitive decline in HIV-infected subjects. Neurology 81 (17):1492–1499

Bertrand L, Toborek M (2015) Dysregulation of endoplasmic reticulum stress and autophagic responses by the antiretroviral drug efavirenz. Mol Pharmacol 88(2):304–315

Bhaskaran K, Mussini C, Antinori A, Walker AS, Dorrucci M, Sabin C et al (2008) Changes in the incidence and predictors of human immunodeficiency virus–associated dementia in the era of highly active antiretroviral therapy. Ann Neurol 63(2):213–221

Bingham R, Ahmed N, Rangi P, Johnson M, Tyrer M, Green J (2011) HIV encephalitis despite suppressed viraemia: a case of compartmentalized viral escape. Int J STD AIDS 22 (10):608–609

Boban J, Kozic D, Turkulov V, Ostojic J, Semnic R, Lendak D et al (2017) HIV-associated neurodegeneration and neuroimmunity: multivoxel MR spectroscopy study in drug-naive and treated patients. Eur Radiol 27(10):4218–4236

Boulware DR, Hullsiek KH, Puronen CE, Rupert A, Baker JV, French MA et al (2011) Higher levels of CRP, D-dimer, IL-6, and hyaluronic acid before initiation of antiretroviral therapy (ART) are associated with increased risk of AIDS or death. J Infect Dis 203(11):1637–1646

Brenchley JM, Schacker TW, Ruff LE, Price DA, Taylor JH, Beilman GJ et al (2004) CD4+ T cell depletion during all stages of HIV disease occurs predominantly in the gastrointestinal tract. J Exp Med 200(6):749–759

Brenchley JM, Price DA, Schacker TW, Asher TE, Silvestri G, Rao S et al (2006) Microbial translocation is a cause of systemic immune activation in chronic HIV infection. Nat Med 12 (12):1365–1371

Brew BJ, Bhalla RB, Paul M, Gallardo H, McArthur JC, Schwartz MK et al (1990) Cerebrospinal fluid neopterin in human immunodeficiency virus type 1 infection. Ann Neurol 28(4):556–560

Brew BJ, Rosenblum M, Cronin K, Price RW (1995) AIDS dementia complex and HIV-1 brain infection: clinical-virological correlations. Ann Neurol 38(4):563–570

Brew BJ, Dunbar N, Pemberton L, Kaldor J (1996) Predictive markers of AIDS dementia complex: CD4 cell count and cerebrospinal fluid concentrations of β2-microglobulin and neopterin. J Infect Dis 174(2):294–298

Brew BJ, Pemberton L, Cunningham P, Law MG (1997) Levels of human immunodeficiency virus type 1 RNA in cerebrospinal fluid correlate with AIDS dementia stage. J Infect Dis 175 (4):963–966

Brew BJ, Pemberton L, Blennow K, Wallin A, Hagberg L (2005) CSF amyloid β42 and tau levels correlate with AIDS dementia complex. Neurology 65(9):1490–1492

Brown A, Islam T, Adams R, Nerle S, Kamara M, Eger C et al (2011) Osteopontin enhances HIV replication and is increased in the brain and cerebrospinal fluid of HIV-infected individuals. J Neurovirol 17(4):382–392

Bruner KM, Murray AJ, Pollack RA, Soliman MG, Laskey SB, Capoferri AA et al (2016) Defective proviruses rapidly accumulate during acute HIV-1 infection. Nat Med 22(9):1043–1049

Brunt SJ, Cysique LA, Lee S, Burrows S, Brew BJ, Price P (2016) Short communication: do cytomegalovirus antibody levels associate with age-related syndromes in HIV patients stable on antiretroviral therapy? AIDS Res Hum Retrovir 32(6):567–572

Burbelo PD, Price RW, Hagberg L, Hatano H, Spudich S, Deeks SG et al (2018) Anti-human immunodeficiency virus antibodies in the cerebrospinal fluid: evidence of early treatment impact on central nervous system reservoir? J Infect Dis 217(7):1024–1032

Burdo TH, Soulas C, Orzechowski K, Button J, Krishnan A, Sugimoto C et al (2010) Increased monocyte turnover from bone marrow correlates with severity of SIV encephalitis and CD163 levels in plasma. PLoS Pathog 6(4):e1000842

Burdo TH, Lackner A, Williams KC (2013a) Monocyte/macrophages and their role in HIV neuropathogenesis. Immunol Rev 254(1):102–113

Burdo TH, Weiffenbach A, Woods SP, Letendre S, Ellis RJ, Williams KC (2013b) Elevated sCD163 in plasma but not cerebrospinal fluid is a marker of neurocognitive impairment in HIV infection. AIDS (London, England) 27(9):1387–1395

Canestri A, Lescure F-X, Jaureguiberry S, Moulignier A, Amiel C, Marcelin A et al (2010) Discordance between cerebral spinal fluid and plasma HIV replication in patients with neurological symptoms who are receiving suppressive antiretroviral therapy. Clin Infect Dis 50 (5):773–778

Cardenas VA, Meyerhoff DJ, Studholme C, Kornak J, Rothlind J, Lampiris H et al (2009) Evidence for ongoing brain injury in human immunodeficiency virus-positive patients treated with antiretroviral therapy. J Neurovirol 15(4):324–333
Carvalhal A, Gill MJ, Letendre SL, Rachlis A, Bekele T, Raboud J et al (2016) Central nervous system penetration effectiveness of antiretroviral drugs and neuropsychological impairment in the Ontario HIV Treatment Network Cohort Study. J Neurovirol 22(3):349–357
Cassol E, Misra V, Dutta A, Morgello S, Gabuzda D (2014) Cerebrospinal fluid metabolomics reveals altered waste clearance and accelerated aging in HIV patients with neurocognitive impairment. AIDS 28(11):1579–1591
Centers for Disease Control and Prevention (2017) HIV Surveillance Report, 2016, vol 28. http://www.cdc.gov/hiv/library/reports/hiv-surveillance.html. Accessed 2 May 2018
Chaganti J, Marripundi K, Staub LP, Rae CD, Gates T, Moffat KJ et al (2019) Imaging correlates of the Blood Brain Barrier disruption in HIV associated neurocognitive disorder and therapeutic implications. AIDS 33(12):1843–1852
Chiao S, Rosen HJ, Nicolas K, Wendelken LA, Alcantar O, Rankin KP et al (2013) Deficits in self-awareness impact the diagnosis of asymptomatic neurocognitive impairment in HIV. AIDS Res Hum Retrovir 29(6):949–956
Cho HJ, Bertrand L, Toborek M (2019) Blood–brain barrier pericytes as a target for HIV-1 infection. Brain 142(3):502–511
Churchill MJ, Wesselingh SL, Cowley D, Pardo CA, McArthur JC, Brew BJ et al (2009) Extensive astrocyte infection is prominent in human immunodeficiency virus-associated dementia. Ann Neurol 66(2):253–258
Churchill MJ, Deeks SG, Margolis DM, Siliciano RF, Swanstrom R (2015) HIV reservoirs: what, where and how to target them. Nat Rev Microbiol 14:55
Ciccarelli N, Fabbiani M, Grima P, Falasca K, Tana M, Baldonero E et al (2013) Comparison of cognitive performance in HIV or HCV mono-infected and HIV–HCV co-infected patients. Infection 41(6):1103–1109
Clifford DB, Fagan AM, Holtzman DM, Morris JC, Teshome M, Shah AR et al (2009) CSF biomarkers of Alzheimer disease in HIV-associated neurologic disease. Neurology 73 (23):1982–1987
Clifford KM, Samboju V, Cobigo Y, Milanini B, Marx GA, Hellmuth JM et al (2017) Progressive brain atrophy despite persistent viral suppression in HIV patients older than 60 years. J Acquir Immune Defic Syndr 76(3):289–297
Cole JH, Underwood J, Caan MWA, De Francesco D, van Zoest RA, Leech R et al (2017) Increased brain-predicted aging in treated HIV disease. Neurology 88(14):1349–1357
Conant K, Garzino-Demo A, Nath A, McArthur JC, Halliday W, Power C et al (1998) Induction of monocyte chemoattractant protein-1 in HIV-1 Tat-stimulated astrocytes and elevation in AIDS dementia. Proc Natl Acad Sci U S A 95(6):3117–3121
Corley MJ, Dye C, D'Antoni ML, Byron MM, Yo KL-A, Lum-Jones A et al (2016) Comparative DNA methylation profiling reveals an immunoepigenetic signature of HIV-related cognitive impairment. Sci Rep 6:33310
Cysique LA, Brew BJ (2011) Prevalence of non-confounded HIV-associated neurocognitive impairment in the context of plasma HIV RNA suppression. J Neurovirol 17(2):176–183
Cysique LA, Maruff P, Brew BJ (2004) Prevalence and pattern of neuropsychological impairment in human immunodeficiency virus-infected/acquired immunodeficiency syndrome (HIV/AIDS) patients across pre- and post-highly active antiretroviral therapy eras: a combined study of two cohorts. J Neurovirol 10(6):350–357
Cysique LA, Vaida F, Letendre S, Gibson S, Cherner M, Woods SP et al (2009) Dynamics of cognitive change in impaired HIV-positive patients initiating antiretroviral therapy. Neurology 73(5):342–348
Cysique LA, Hey-Cunningham WJ, Dermody N, Chan P, Brew BJ, Koelsch KK (2015a) Peripheral blood mononuclear cells HIV DNA levels impact intermittently on neurocognition. PLoS One 10(4):e0120488
Cysique LA, Hewitt T, Croitoru-Lamoury J, Taddei K, Martins RN, Chew CSN et al (2015b) APOE ε4 moderates abnormal CSF-abeta-42 levels, while neurocognitive impairment is

associated with abnormal CSF tau levels in HIV+ individuals – a cross-sectional observational study. BMC Neurol 15:51

Cysique LA, Juge L, Lennon MJ, Gates TM, Jones SP, Lovelace MD et al (2019) HIV brain latency as measured by CSF BcL11b relates to disrupted brain cellular energy in virally suppressed HIV infection. AIDS 33(3):433–441

D'Antoni ML, Paul RH, Mitchell BI, Kohorn L, Fischer L, Lefebvre E et al (2018) Improved cognitive performance and reduced monocyte activation in virally suppressed chronic HIV after dual CCR2 and CCR5 antagonism. J Acquir Immune Defic Syndr 79(1):108–116

Dahl V, Peterson J, Fuchs D, Gisslen M, Palmer S, Price RW (2014) Low levels of HIV-1 RNA detected in the cerebrospinal fluid after up to 10 years of suppressive therapy are associated with local immune activation. AIDS 28(15):2251–2258

de Oliveira MF, Murrell B, Pérez-Santiago J, Vargas M, Ellis RJ, Letendre S et al (2015) Circulating HIV DNA correlates with neurocognitive impairment in older HIV-infected adults on suppressive ART. Sci Rep 5:17094

Deeks SG, Kitchen CMR, Liu L, Guo H, Gascon R, Narváez AB et al (2004) Immune activation set point during early HIV infection predicts subsequent CD4+ T-cell changes independent of viral load. Blood 104(4):942–947

Desplats P, Dumaop W, Smith D, Adame A, Everall I, Letendre S et al (2013) Molecular and pathologic insights from latent HIV-1 infection in the human brain. Neurology 80 (15):1415–1423

Dickens AM, Yoo SW, Chin AC, Xu J, Johnson TP, Trout AL et al (2017) Chronic low-level expression of HIV-1 Tat promotes a neurodegenerative phenotype with aging. Sci Rep 7:7748

dos Santos-Silva AF, Alves JMF, Ramos I, del Cármen Piñeiro-Calvo M, Sousa C, do Rosário Serrão M et al (2017) Neurocognitive disorders in patients with HIV infection with virologic suppression for more than 10 years. World J AIDS 7(01):59

Drewes JL, Meulendyke KA, Liao Z, Witwer KW, Gama L, Ubaida-Mohien C et al (2015) Quinolinic acid/tryptophan ratios predict neurological disease in SIV-infected macaques and remain elevated in the brain under cART. J Neurovirol 21(4):449–463

Eden A, Price RW, Spudich S, Fuchs D, Hagberg L, Gisslen M (2007) Immune activation of the central nervous system is still present after >4 years of effective highly active antiretroviral therapy. J Infect Dis 196(12):1779–1783

Eden A, Nilsson S, Hagberg L, Fuchs D, Zetterberg H, Svennerholm B et al (2016) Asymptomatic cerebrospinal fluid HIV-1 viral blips and viral escape during antiretroviral therapy: a longitudinal study. J Infect Dis 214(12):1822–1825

Eletto D, Russo G, Passiatore G, Valle LD, Giordano A, Khalili K et al (2008) Inhibition of SNAP25 expression by HIV-1 Tat Involves the activity of mir-128a. J Cell Physiol 216 (3):764–770

Ellery PJ, Tippett E, Chiu Y-L, Paukovics G, Cameron PU, Solomon A et al (2007) The CD16+ monocyte subset is more permissive to infection and preferentially harbors HIV-1 in vivo. J Immunol 178(10):6581–6589

Ellis RJ, Deutsch R, Heaton RK et al (1997) Neurocognitive impairment is an independent risk factor for death in HIV infection. Arch Neurol 54(4):416–424

Ellis RJ, Moore DJ, Childers ME et al (2002) Progression to neuropsychological impairment in human immunodeficiency virus infection predicted by elevated cerebrospinal fluid levels of human immunodeficiency virus RNA. Arch Neurol 59(6):923–928

Ellis RJ, Badiee J, Vaida F, Letendre S, Heaton RK, Clifford D et al (2011) CD4 nadir is a predictor of HIV neurocognitive impairment in the era of combination antiretroviral therapy. AIDS 25 (14):1747–1751

Ellis RJ, Letendre S, Vaida F, Haubrich R, Heaton RK, Sacktor N et al (2014) Randomized trial of central nervous system-targeted antiretrovirals for HIV-associated neurocognitive disorder. Clin Infect Dis 58(7):1015–1022

Estes JD, Harris LD, Klatt NR, Tabb B, Pittaluga S, Paiardini M et al (2010) Damaged intestinal epithelial integrity linked to microbial translocation in pathogenic simian immunodeficiency virus infections. PLoS Pathog 6(8):e1001052

Estes JD, Kityo C, Ssali F, Swainson L, Makamdop KN, Del Prete GQ et al (2017) Defining total-body AIDS-virus burden with implications for curative strategies. Nat Med 23:1271–1276
Evering TH, Applebaum A, La Mar M, Garmon D, Dorfman D, Markowitz M (2016) Rates of non-confounded HIV associated neurocognitive disorders in men initiating cART during primary infection. AIDS 30(2):203–210
Fiandaca MS, Kapogiannis D, Mapstone M, Boxer A, Eitan E, Schwartz JB et al (2015) Identification of preclinical Alzheimer's disease by a profile of pathogenic proteins in neurally derived blood exosomes: a case-control study. Alzheimers Dement 11(6):600–7.e1
Fischer-Smith T, Croul S, Sverstiuk AE, Capini C, L'Heureux D, Régulier EG et al (2001) CNS invasion by CD14+/CD16+ peripheral blood-derived monocytes in HIV dementia: perivascular accumulation and reservoir of HIV infection. J Neurovirol 7(6):528–541
Fogel GB, Lamers SL, Levine AJ, Valdes-Sueiras M, McGrath MS, Shapshak P et al (2015) Factors related to HIV-associated neurocognitive impairment differ with age. J Neurovirol 21(1):56–65
Funes HA, Apostolova N, Alegre F, Blas-Garcia A, Alvarez A, Marti-Cabrera M et al (2014) Neuronal bioenergetics and acute mitochondrial dysfunction: a clue to understanding the central nervous system side effects of efavirenz. J Infect Dis 210(9):1385–1395
Funes HA, Blas-Garcia A, Esplugues JV, Apostolova N (2015) Efavirenz alters mitochondrial respiratory function in cultured neuron and glial cell lines. J Antimicrob Chemother 70 (8):2249–2254
Garvey LJ, Pavese N, Politis M, Ramlackhansingh A, Brooks DJ, Taylor-Robinson SD et al (2014) Increased microglia activation in neurologically asymptomatic HIV-infected patients receiving effective ART. AIDS 28(1):67–72
Gates TM, Cysique LA, Siefried KJ, Chaganti J, Moffat KJ, Brew BJ (2016) Maraviroc-intensified combined antiretroviral therapy improves cognition in virally suppressed HIV-associated neurocognitive disorder. AIDS 30(4):591–600
Gelman BB (2015) Neuropathology of HAND with suppressive antiretroviral therapy: encephalitis and neurodegeneration reconsidered. Curr HIV/AIDS Rep 12(2):272–279
Gelman BB, Chen T, Lisinicchia JG, Soukup VM, Carmical JR, Starkey JM et al (2012) The national NeuroAIDS Tissue Consortium brain gene array: two types of HIV-associated neurocognitive impairment. PLoS One 7(9):e46178
Ginsberg SD, Alldred MJ, Gunnam SM, Schiroli C, Lee SH, Morgello S et al (2018) Expression profiling suggests microglial impairment in human immunodeficiency virus neuropathogenesis. Ann Neurol 83(2):406–417
Gisslén M, Hagberg L, Brew BJ, Cinque P, Price RW, Rosengren L (2007) Elevated cerebrospinal fluid neurofilament light protein concentrations predict the development of AIDS dementia complex. J Infect Dis 195(12):1774–1778
Gisslén M, Price RW, Andreasson U, Norgren N, Nilsson S, Hagberg L et al (2015) Plasma concentration of the neurofilament light protein (NFL) is a biomarker of CNS injury in HIV infection: a cross-sectional study. EBioMedicine 3:135–140
Glenn JD, Mowry EM (2016) Emerging concepts on the gut microbiome and multiple sclerosis. J Interf Cytokine Res 36(6):347–357
Goodkin K, Miller EN, Cox C, Reynolds S, Becker JT, Martin E et al (2017) Effect of aging on neurocognitive function by stage of HIV infection: evidence from the Multi-Center AIDS Cohort Study. Lancet HIV 4(9):e411–ee22
Gorry PR, Bristol G, Zack JA, Ritola K, Swanstrom R, Birch CJ et al (2001) Macrophage tropism of human immunodeficiency virus type 1 isolates from brain and lymphoid tissues predicts neurotropism independent of coreceptor specificity. J Virol 75(21):10073–10089
Gorry PR, Zhang C, Wu S, Kunstman K, Trachtenberg E, Phair J et al (2002) Persistence of dual-tropic HIV-1 in an individual homozygous for the CCR5 Delta 32 allele. Lancet 359 (9320):1832–1834
Gott C, Gates T, Dermody N, Brew BJ, Cysique LA (2017) Cognitive change trajectories in virally suppressed HIV-infected individuals indicate high prevalence of disease activity. PLoS One 12 (3):e0171887

Gougeon M-L, Poirier-Beaudouin B, Durant J, Lebrun-Frenay C, Saïdi H, Seffer V et al (2017) HMGB1/anti-HMGB1 antibodies define a molecular signature of early stages of HIV-Associated Neurocognitive Isorders (HAND). Heliyon 3:e00245
Grant I, Franklin DR, Deutsch R, Woods SP, Vaida F, Ellis RJ et al (2014) Asymptomatic HIV-associated neurocognitive impairment increases risk for symptomatic decline. Neurology 82(23):2055–2062
Gray LR, Tachedjian G, Ellett AM, Roche MJ, Cheng WJ, Guillemin GJ et al (2013) The NRTIs lamivudine, stavudine and zidovudine have reduced HIV-1 inhibitory activity in astrocytes. PLoS One 8(4):e62196
Gray LR, Turville SG, Hitchen TL, Cheng W-J, Ellett AM, Salimi H et al (2014) HIV-1 entry and trans-infection of astrocytes involves CD81 vesicles. PLoS One 9(2):e90620
Gray LR, Cowley D, Welsh C, Lu HK, Brew BJ, Lewin SR et al (2016) CNS-specific regulatory elements in brain-derived HIV-1 strains affect responses to latency-reversing agents with implications for cure strategies. Mol Psychiatry 21(4):574–584
Green DA, Masliah E, Vinters HV, Beizai P, Moore DJ, Achim CL (2005) Brain deposition of beta-amyloid is a common pathologic feature in HIV positive patients. AIDS 19(4):407–411
Guaraldi G, Orlando G, Zona S, Menozzi M, Carli F, Garlassi E et al (2011) Premature age-related comorbidities among HIV-infected persons compared with the general population. Clin Infect Dis 53(11):1120–1126
Guha D, Nagilla P, Redinger C, Srinivasan A, Schatten GP, Ayyavoo V (2012) Neuronal apoptosis by HIV-1 Vpr: contribution of proinflammatory molecular networks from infected target cells. J Neuroinflammation 9(1):138
Hagberg L, Cinque P, Gisslen M, Brew BJ, Spudich S, Bestetti A et al (2010) Cerebrospinal fluid neopterin: an informative biomarker of central nervous system immune activation in HIV-1 infection. AIDS Res Ther 7:15
Haggerty S, Stevenson M (1991) Predominance of distinct viral genotypes in brain and lymph node compartments of HIV-1-infected individuals. Viral Immunol 4(2):123–131
Hao XP, Lucero CM, Turkbey B, Bernardo ML, Morcock DR, Deleage C et al (2015) Experimental colitis in SIV-uninfected rhesus macaques recapitulates important features of pathogenic SIV infection. Nat Commun 6:8020
Harezlak J, Buchthal S, Taylor M, Schifitto G, Zhong J, Daar ES et al (2011) Persistence of HIV–associated cognitive impairment, inflammation and neuronal injury in era of highly active antiretroviral treatment. AIDS 25(5):625–633
Haughey NJ, Holden CP, Nath A, Geiger JD (1999) Involvement of inositol 1,4,5-trisphosphate-regulated stores of intracellular calcium in calcium dysregulation and neuron cell death caused by HIV-1 protein Tat. J Neurochem 73(4):1363–1374
Hazenberg MD, Otto SA, van Benthem BH, Roos MT, Coutinho RA, Lange JM et al (2003) Persistent immune activation in HIV-1 infection is associated with progression to AIDS. AIDS 17(13):1881–1888
Heaton R, Clifford D, Franklin D, Woods S, Ake C, Vaida F et al (2010) HIV-associated neurocognitive disorders persist in the era of potent antiretroviral therapy CHARTER Study. Neurology 75(23):2087–2096
Heaton RK, Franklin DR, Ellis RJ, McCutchan JA, Letendre SL, LeBlanc S et al (2011) HIV-associated neurocognitive disorders before and during the era of combination antiretroviral therapy: differences in rates, nature, and predictors. J Neurovirol 17(1):3–16
Heaton RK, Franklin DR Jr, Deutsch R, Letendre S, Ellis RJ, Casaletto K et al (2015) Neurocognitive change in the era of HIV combination antiretroviral therapy: the longitudinal CHARTER study. Clin Infect Dis 60(3):473–480
Hellmuth J, Fletcher JLK, Valcour V, Kroon E, Ananworanich J, Intasan J et al (2016) Neurologic signs and symptoms frequently manifest in acute HIV infection. Neurology 87(2):148–154
Heyes MP, Mefford IN, Quearry BJ, Dedhia M, Lackner A (1990) Increased ratio of quinolinic acid to kynurenic acid in cerebrospinal fluid of D retrovirus-infected rhesus macaques: relationship to clinical and viral status. Ann Neurol 27(6):666–675

Hinkin C, Castellon S, Durvasula R, Hardy D, Lam M, Mason K et al (2002) Medication adherence among HIV+ adults effects of cognitive dysfunction and regimen complexity. Neurology 59 (12):1944–1950

Honeycutt JB, Liao B, Nixon CC, Cleary RA, Thayer WO, Birath SL et al (2018) T cells establish and maintain CNS viral infection in HIV-infected humanized mice. J Clin Invest 128 (7):2862–2876

Houser MC, Tansey MG (2017) The gut-brain axis: is intestinal inflammation a silent driver of Parkinson's disease pathogenesis? NPJ Parkinsons Dis 3(1):3

Hulgan T, Kallianpur AR, Guo Y, Barnholtz-Sloan JS, Gittleman H, Brown TT et al (2018) Peripheral blood mitochondrial DNA copy number obtained from genome-wide genotype data is associated with neurocognitive impairment in persons with chronic HIV infection. J Acquir Immune Defic Syndr 80(4):e95–e102

Hunt PW, Landay AL, Sinclair E, Martinson JA, Hatano H, Emu B et al (2011) A low T regulatory cell response may contribute to both viral control and generalized immune activation in HIV controllers. PLoS One 6(1):e15924

Jaureguiberry-Bravo M, Lopez L, Berman JW (2018) Frontline science: buprenorphine decreases CCL2-mediated migration of CD14+CD16+ monocytes. J Leukoc Biol 104(6):1049–1059

Jensen BK, Monnerie H, Mannell MV, Gannon PJ, Espinoza CA, Erickson MA et al (2015) Altered oligodendrocyte maturation and myelin maintenance: the role of antiretrovirals in HIV-associated neurocognitive disorders. J Neuropathol Exp Neurol 74(11):1093–1118

Jespersen S, Pedersen KK, Anesten B, Zetterberg H, Fuchs D, Gisslén M et al (2016) Soluble CD14 in cerebrospinal fluid is associated with markers of inflammation and axonal damage in untreated HIV-infected patients: a retrospective cross-sectional study. BMC Infect Dis 16 (1):176

Jessen Krut J, Mellberg T, Price RW, Hagberg L, Fuchs D, Rosengren L et al (2014) Biomarker evidence of axonal injury in neuroasymptomatic HIV-1 patients. PLoS One 9(2):e88591

Johnson TP, Patel K, Johnson KR, Maric D, Calabresi PA, Hasbun R et al (2013) Induction of IL-17 and nonclassical T-cell activation by HIV-Tat protein. Proc Natl Acad Sci U S A 110 (33):13588–13593

Jonathan U, Kevin RR, Alan W (2015) Could antiretroviral neurotoxicity play a role in the pathogenesis of cognitive impairment in treated HIV disease? AIDS 29(3):253

Joseph SB, Kincer LP, Bowman NM, Evans C, Vinikoor MJ, Lippincott CK et al (2019) HIV-1 RNA detected in the CNS after years of suppressive antiretroviral therapy can originate from a replicating CNS reservoir or clonally expanded cells. Clin Infect Dis 69:1345–1352

Kadri F, LaPlante A, De Luca M, Doyle L, Velasco-Gonzalez C, Patterson JR et al (2016) Defining plasma microRNAs associated with cognitive impairment in HIV-infected patients. J Cell Physiol 231(4):829–836

Kallianpur KJ, Valcour VG, Lerdlum S, Busovaca E, Agsalda M, Sithinamsuwan P et al (2014) HIV DNA in CD14+ reservoirs is associated with regional brain atrophy in patients naive to combination antiretroviral therapy. AIDS (London, England) 28(11):1619–1624

Kamal S, Locatelli I, Wandeler G, Sehhat A, Bugnon O, Metral M et al (2017) The presence of human immunodeficiency virus-associated neurocognitive disorders is associated with a lower adherence to combined antiretroviral treatment. Open Forum Infect Dis 4(2):ofx070

Kamat A, Lyons JL, Misra V, Uno H, Morgello S, Singer EJ et al (2012) Monocyte activation markers in cerebrospinal fluid associated with impaired neurocognitive testing in advanced HIV infection. J Acquir Immune Defic Syndr Hum Retrovirol 60(3):234–243

Kandanearatchi A, Brew BJ (2012) The kynurenine pathway and quinolinic acid: pivotal roles in HIV associated neurocognitive disorders. FEBS J 279(8):1366–1374

Kim WK, Alvarez X, Fisher J, Bronfin B, Westmoreland S, McLaurin J et al (2006) CD163 identifies perivascular macrophages in normal and viral encephalitic brains and potential precursors to perivascular macrophages in blood. Am J Pathol 168(3):822–834

Kim B-H, Kelschenbach J, Borjabad A, Hadas E, He H, Potash MJ et al (2019) Intranasal insulin therapy reverses hippocampal dendritic injury and cognitive impairment in a model of HIV-associated neurocognitive disorders in EcoHIV-infected mice. AIDS 33(6):973–984

Ko A, Kang G, Hattler JB, Galadima HI, Zhang J, Li Q et al (2019) Macrophages but not astrocytes harbor HIV DNA in the brains of HIV-1-infected aviremic individuals on suppressive antiretroviral therapy. J Neuroimmune Pharmacol 14(1):110–119

Krut JJ, Price RW, Zetterberg H, Fuchs D, Hagberg L, Yilmaz A et al (2017) No support for premature central nervous system aging in HIV-1 when measured by cerebrospinal fluid phosphorylated tau (p-tau). Virulence 8(5):599–604

Kuller LH, Tracy R, Belloso W, Wit SD, Drummond F, Lane HC et al (2008) Inflammatory and coagulation biomarkers and mortality in patients with HIV infection. PLoS Med 5(10):e203

Küper M, Rabe K, Esser S, Gizewski ER, Husstedt IW, Maschke M et al (2011) Structural gray and white matter changes in patients with HIV. J Neurol 258(6):1066–1075

Lamers SL, Rose R, Maidji E, Agsalda-Garcia M, Nolan DJ, Fogel GB et al (2016) HIV DNA is frequently present within pathologic tissues evaluated at autopsy from combined antiretroviral therapy-treated patients with undetectable viral loads. J Virol 90:8968–8983

Lanoy E, Guiguet M, Bentata M, Rouveix E, Dhiver C, Poizot-Martin I et al (2011) Survival after neuroAIDS: association with antiretroviral CNS penetration-effectiveness score. Neurology 76 (7):644–651

Lawler K, Mosepele M, Ratcliffe S, Seloilwe E, Steele K, Nthobatsang R et al (2010) Neurocognitive impairment among HIV-positive individuals in Botswana: a pilot study. J Int AIDS Soc 13(1):15

Letendre SL, Zheng JC, Kaul M, Yiannoutsos CT, Ellis RJ, Taylor MJ et al (2011) Chemokines in cerebrospinal fluid correlate with cerebral metabolite patterns in HIV-infected individuals. J Neurovirol 17(1):63–69

Letendre S, Bharti A, Perez-Valero I, Hanson B, Franklin D, Woods SP et al (2018) Higher anti-cytomegalovirus immunoglobulin G concentrations are associated with worse neurocognitive performance during suppressive antiretroviral therapy. Clin Infect Dis 67(5):770–777

Levine AJ, Quach A, Moore DJ, Achim CL, Soontornniyomkij V, Masliah E et al (2016) Accelerated epigenetic aging in brain is associated with pre-mortem HIV-associated neurocognitive disorders. J Neurovirol 22(3):366–375

Levine AJ, Martin E, Sacktor N, Munro C, Becker J, For the multicenter ACS-NWG (2017) Predictors and impact of self-reported suboptimal effort on estimates of prevalence of HIV-associated neurocognitive disorders. J Acquir Immune Defic Syndr 75(2):203–210

Lyons JL, Uno H, Ancuta P, Kamat A, Moore DJ, Singer EJ et al (2011) Plasma sCD14 is a biomarker associated with impaired neurocognitive test performance in attention and learning domains in HIV infection. J Acquir Immune Defic Syndr 57(5):371–379

Ma Q, Vaida F, Wong J, Sanders CA, Kao Y-T, Croteau D et al (2016) Long-term efavirenz use is associated with worse neurocognitive functioning in HIV-infected patients. J Neurovirol 22 (2):170–178

Mamik MK, Asahchop EL, Chan WF, Zhu Y, Branton WG, McKenzie BA et al (2016) Insulin treatment prevents neuroinflammation and neuronal injury with restored neurobehavioral function in models of HIV/AIDS neurodegeneration. J Neurosci 36(41):10683–10695

Marquine MJ, Montoya JL, Umlauf A, Fazeli PL, Gouaux B, Heaton RK et al (2016) The Veterans Aging Cohort Study (VACS) Index and neurocognitive change: a longitudinal study. Clin Infect Dis 63(5):694–702

Masliah E, Roberts ES, Langford D, Everall I, Crews L, Adame A et al (2004) Patterns of gene dysregulation in the frontal cortex of patients with HIV encephalitis. J Neuroimmunol 157 (1):163–175

McGuire JL, Gill AJ, Douglas SD, Kolson DL (2015) Central and peripheral markers of neurodegeneration and monocyte activation in HIV-associated neurocognitive disorders. J Neurovirol 21(4):439–448

Mediouni S, Jablonski J, Paris JJ, Clementz MA, Thenin-Houssier S, McLaughlin JP et al (2015) Didehydro-cortistatin A inhibits HIV-1 Tat mediated neuroinflammation and prevents potentiation of cocaine reward in Tat transgenic mice. Curr HIV Res 13(1):64–79

Mehla R, Bivalkar-Mehla S, Nagarkatti M, Chauhan A (2012) Programming of neurotoxic cofactor CXCL-10 in HIV-1-associated dementia: abrogation of CXCL-10-induced neuro-glial toxicity in vitro by PKC activator. J Neuroinflammation 9:239

Mehta SR, Pérez-Santiago J, Hulgan T, Day TRC, Barnholtz-Sloan J, Gittleman H et al (2017) Cerebrospinal fluid cell-free mitochondrial DNA is associated with HIV replication, iron transport, and mild HIV-associated neurocognitive impairment. J Neuroinflammation 14(1):72

Meulendyke KA, Queen SE, Engle EL, Shirk EN, Liu J, Steiner JP et al (2014) Combination fluconazole/paroxetine treatment is neuroprotective despite ongoing neuroinflammation and viral replication in an SIV model of HIV neurological disease. J Neurovirol 20(6):591–602

Mielke MM, Bandaru VVR, McArthur JC, Chu M, Haughey NJ (2010) Disturbance in cerebral spinal fluid sphingolipid content is associated with memory impairment in subjects infected with the human immunodeficiency virus. J Neurovirol 16(6):445–456

Moulignier A, Gueguen A, Lescure F-X, Ziegler M, Girard P-M, Cardon B et al (2015) Does HIV infection alter Parkinson disease? J Acquir Immune Defic Syndr 70(2):129–136

Mukerjee R, Chang JR, Del Valle L, Bagashev A, Gayed MM, Lyde RB et al (2011) Deregulation of microRNAs by HIV-1 Vpr protein leads to the development of neurocognitive disorders. J Biol Chem 286(40):34976–34985

Mukerji SS, Misra V, Lorenz D, Cervantes-Arslanian AM, Lyons J, Chalkias S et al (2017) Temporal patterns and drug resistance in CSF viral escape among ART-experienced HIV-1 infected adults. J Acquir Immune Defic Syndr 75(2):246–255

Mukerji SS, Misra V, Lorenz DR, Uno H, Morgello S, Franklin D et al (2018) Impact of antiretroviral regimens on cerebrospinal fluid viral escape in a prospective multicohort study of antiretroviral therapy-experienced human immunodeficiency virus-1–infected adults in the United States. Clin Infect Dis 67(8):1182–1190

Mutlu EA, Keshavarzian A, Losurdo J, Swanson G, Siewe B, Forsyth C et al (2014) A compositional look at the human gastrointestinal microbiome and immune activation parameters in HIV infected subjects. PLoS Pathog 10(2):e1003829

Ndhlovu LC, Umaki T, Chew GM, Chow DC, Agsalda M, Kallianpur KJ et al (2014) Treatment intensification with maraviroc (CCR5 antagonist) leads to declines in CD16-expressing monocytes in cART-suppressed chronic HIV-infected subjects and is associated with improvements in neurocognitive test performance: implications for HIV-associated neurocognitive disease (HAND). J Neurovirol 20(6):571–582

Nichols MJ, Gates TM, Soares JR, Moffat KJ, Rae CD, Brew BJ et al (2019) Atrophic brain signatures of mild forms of neurocognitive impairment in virally suppressed HIV infection. AIDS 33(1):55–66

Nightingale S, Winsto A, Letendre S, Michael BD, McArthur JC, Khoo S et al (2014) Controversies in HIV-associated neurocognitive disorders. Lancet Neurol 13(11):1139–1151

Nir TM, Jahanshad N, Ching CRK, Cohen RA, Harezlak J, Schifitto G et al (2019) Progressive brain atrophy in chronically infected and treated HIV+ individuals. J Neurovirol 25(3):342–353

O'Halloran J, Cooley SA, Strain JF, Boerwinkle A, Paul R, Presti RM et al (2019) Altered neuropsychological performance and reduced brain volumetrics in people living with HIV on integrase strand transfer inhibitors. AIDS 33(9):1477

Ogishi M, Yotsuyanagi H (2018) Prediction of HIV-associated neurocognitive disorder (HAND) from three genetic features of envelope gp120 glycoprotein. Retrovirology 15(1):12

Oliveira MF, Chaillon A, Nakazawa M, Vargas M, Letendre SL, Strain MC et al (2017) Early antiretroviral therapy is associated with lower HIV DNA molecular diversity and lower inflammation in cerebrospinal fluid but does not prevent the establishment of compartmentalized HIV DNA populations. PLoS Pathog 13(1):e1006112

Pacifici M, Delbue S, Ferrante P, Jeansonne D, Kadri F, Nelson S et al (2013) Cerebrospinal fluid miRNA profile in HIV-encephalitis. J Cell Physiol 228(5):1070–1075

Palesch D, Bosinger SE, Tharp GK, Vanderford TH, Paiardini M, Chahroudi A et al (2018) Sooty mangabey genome sequence provides insight into AIDS resistance in a natural SIV host. Nature 553:77

Pandrea I, Cornell E, Wilson C, Ribeiro RM, Ma D, Kristoff J et al (2012) Coagulation biomarkers predict disease progression in SIV-infected nonhuman primates. Blood 120(7):1357–1366

Panel on Antiretroviral Guidelines for Adults and Adolescents. Guidelines for the use of Antiretroviral agents in adults and adolescents living with HIV. Department of Health and Human Services. Available at http://www.aidsinfo.nih.gov/ContentFiles/AdultandAdolescentGL.pdf. Accessed 30 July 2019

Peluso MJ, Ferretti F, Peterson J, Lee E, Fuchs D, Boschini A et al (2012) Cerebrospinal fluid HIV escape associated with progressive neurologic dysfunction in patients on antiretroviral therapy with well-controlled plasma viral load. AIDS (London, England) 26(14)

Peluso MJ, Meyerhoff DJ, Price RW, Peterson J, Lee E, Young AC et al (2013) Cerebrospinal fluid and neuroimaging biomarker abnormalities suggest early neurological injury in a subset of individuals during primary HIV infection. J Infect Dis 207(11):1703–1712

Pemberton LA, Brew BJ (2001) Cerebrospinal fluid S-100β and its relationship with AIDS dementia complex. J Clin Virol 22(3):249–253

Persidsky Y, Ghorpade A, Rasmussen J, Limoges J, Liu XJ, Stins M et al (1999) Microglial and astrocyte chemokines regulate monocyte migration through the blood-brain barrier in human immunodeficiency virus-1 encephalitis. Am J Pathol 155(5):1599–1611

Peterson J, Gisslen M, Zetterberg H, Fuchs D, Shacklett BL, Hagberg L et al (2014) Cerebrospinal fluid (CSF) neuronal biomarkers across the spectrum of HIV infection: hierarchy of injury and detection. PLoS One 9(12):e116081

Pfefferbaum A, Rogosa DA, Rosenbloom MJ, Chu W, Sassoon SA, Kemper CA et al (2014) Accelerated aging of selective brain structures in human immunodeficiency virus infection: a controlled, longitudinal magnetic resonance imaging study. Neurobiol Aging 35(7):1755–1768

Pfefferbaum A, Zahr NM, Sassoon SA, Kwon D, Pohl KM, Sullivan EV (2018) Accelerated and premature aging characterizing regional cortical volume loss in human immunodeficiency virus infection: contributions from alcohol, substance use, and hepatitis C coinfection. Biol Psychiatry 3(10):844–859

Premeaux TA, D'Antoni ML, Abdel-Mohsen M, Pillai SK, Kallianpur KJ, Nakamoto BK et al (2019) Elevated cerebrospinal fluid Galectin-9 is associated with central nervous system immune activation and poor cognitive performance in older HIV-infected individuals. J Neurovirol 25(2):150–161

Pulliam L, Sun B, Mustapic M, Chawla S, Kapogiannis D (2019) Plasma neuronal exosomes serve as biomarkers of cognitive impairment in HIV infection and Alzheimer's disease. J Neurovirol 25:702–709

Quach A, Horvath S, Nemanim N, Vatakis D, Witt MD, Miller EN et al (2018) No reliable gene expression biomarkers of current or impending neurocognitive impairment in peripheral blood monocytes of persons living with HIV. J Neurovirol 24(3):350–361

Robertson KR, Robertson WT, Ford S, Watson D, Fiscus S, Harp AG et al (2004) Highly active antiretroviral therapy improves neurocognitive functioning. J Acquir Immune Defic Syndr 36 (1):562–566

Robertson KR, Smurzynski M, Parsons TD, Wu K, Bosch RJ, Wu J et al (2007) The prevalence and incidence of neurocognitive impairment in the HAART era. AIDS 21(14):1915–1921

Robertson K, Liner J, Meeker RB (2012) Antiretroviral neurotoxicity. J Neurovirol 18(5):388–399

Royal W III, Cherner M, Burdo TH, Umlauf A, Letendre SL, Jumare J et al (2016) Associations between cognition, gender and monocyte activation among HIV infected individuals in Nigeria. PLoS One 11(2):e0147182

Rubin LH, Sacktor N, Creighton J, Du Y, Endres CJ, Pomper MG et al (2018) Microglial activation is inversely associated with cognition in individuals living with HIV on effective antiretroviral therapy. AIDS (London, England) 32(12):1661–1667

Sacktor N, Miyahara S, Deng L, Evans S, Schifitto G, Cohen BA et al (2011) Minocycline treatment for HIV-associated cognitive impairment: results from a randomized trial. Neurology 77 (12):1135–1142

Sacktor N, Miyahara S, Evans S, Schifitto G, Cohen B, Haughey N et al (2014) Impact of minocycline on cerebrospinal fluid markers of oxidative stress, neuronal injury, and inflammation in HIV seropositive individuals with cognitive impairment. J Neurovirol 20(6):620–626

Sacktor N, Skolasky RL, Seaberg E, Munro C, Becker JT, Martin E et al (2016) Prevalence of HIV-associated neurocognitive disorders in the Multicenter AIDS Cohort Study. Neurology 86 (4):334–340

Sacktor N, Skolasky RL, Moxley R, Wang S, Mielke MM, Munro C et al (2018) Paroxetine and fluconazole therapy for HIV-associated neurocognitive impairment: results from a double-blind, placebo-controlled trial. J Neurovirol 24(1):16–27

Sailasuta N, Ananworanich J, Lerdlum S, Sithinamsuwan P, Fletcher JLK, Tipsuk S et al (2016) Neuronal-glia markers by magnetic resonance spectroscopy in HIV before and after combination antiretroviral therapy. J Acquir Immune Defic Syndr 71(1):24–30

Saiyed ZM, Gandhi N, Agudelo M, Napuri J, Samikkannu T, Reddy PVB et al (2011) HIV-1 Tat upregulates expression of histone deacetylase-2 (HDAC2) in human neurons: implication for HIV-associated neurocognitive disorder (HAND). Neurochem Int 58(6):656–664

Sami Saribas A, Cicalese S, Ahooyi TM, Khalili K, Amini S, Sariyer IK (2017) HIV-1 Nef is released in extracellular vesicles derived from astrocytes: evidence for Nef-mediated neurotoxicity. Cell Death Dis 8:e2542

Saylor D, Dickens AM, Sacktor N, Haughey N, Slusher B, Pletnikov M et al (2016) HIV-associated neurocognitive disorder – pathogenesis and prospects for treatment. Nat Rev Neurol 12 (4):234–248

Schifitto G, Zhang J, Evans SR, Sacktor N, Simpson D, Millar LL et al (2007) A multicenter trial of selegiline transdermal system for HIV-associated cognitive impairment. Neurology 69 (13):1314–1321

Schnell G, Spudich S, Harrington P, Price RW, Swanstrom R (2009) Compartmentalized human immunodeficiency virus type 1 originates from long-lived cells in some subjects with HIV-1–associated dementia. PLoS Pathog 5(4):e1000395

Schrier RD, Hong S, Crescini M, Ellis R, Pérez-Santiago J, Spina C et al (2015) Cerebrospinal fluid (CSF) CD8+ T-cells that express interferon-gamma contribute to HIV Associated Neurocognitive Disorders (HAND). PLoS One 10(2):e0116526

Serena M, Stefano S, Oriana T (2013) Blocking HIV-1 replication by targeting the Tat-hijacked transcriptional machinery. Curr Pharm Des 19(10):1860–1879

Sevigny JJ, Albert SM, McDermott MP et al (2007) An evaluation of neurocognitive status and markers of immune activation as predictors of time to death in advanced HIV infection. Arch Neurol 64(1):97–102

Shah A, Kumar A (2010) HIV-1 gp120-mediated increases in IL-8 production in astrocytes are mediated through the NF-κB pathway and can be silenced by gp120-specific siRNA. J Neuroinflammation 7(1):96

Shah A, Singh DP, Buch S, Kumar A (2011a) HIV-1 envelope protein gp120 up regulates CCL5 production in astrocytes which can be circumvented by inhibitors of NF-κB pathway. Biochem Biophys Res Commun 414(1):112–117

Shah A, Verma AS, Patel KH, Noel R, Rivera-Amill V, Silverstein PS et al (2011b) HIV-1 gp120 induces expression of IL-6 through a nuclear factor-kappa B-dependent mechanism: suppression by gp120 specific small interfering RNA. PLoS One 6(6):e21261

Shaunak S, Albright RE, Klotman ME, Henry SC, Bartlett JA, Hamilton JD (1990) Amplification of HIV-1 provirus from cerebrospinal fluid and its correlation with neurologic disease. J Infect Dis 161(6):1068–1072

Sheppard DP, Iudicello JE, Morgan EE, Kamat R, Clark LR, Avci G et al (2017) Accelerated and accentuated neurocognitive aging in HIV infection. J Neurovirol 23(3):492–500

Simioni S, Cavassini M, Annoni J-M, Abraham AR, Bourquin I, Schiffer V et al (2010) Cognitive dysfunction in HIV patients despite long-standing suppression of viremia. AIDS 24 (9):1243–1250
Singh KK, Barroga CF, Hughes MD, Chen J, Raskino C, McKinney RE et al (2003) Genetic influence of CCR5, CCR2 and SDF1 variants on human immunodeficiency virus 1 (HIV-1)–related disease progression and neurological impairment, in children with symptomatic HIV-1 infection. J Infect Dis 188(10):1461–1472
Somsouk M, Estes JD, Deleage C, Dunham RM, Albright R, Inadomi JM et al (2015) Gut epithelial barrier and systemic inflammation during chronic HIV infection. AIDS 29(1):43–51
Sonia M, Albert D, Gilbert B, Isabelle R, Catherine D, Herve T-D et al (2012) Antiretroviral therapy does not block the secretion of the human immunodeficiency virus Tat protein. Infect Disord Drug Targets 12(1):81–86
Spector SA, Singh KK, Gupta S, Cystique LA, Jin H, Letendre S et al (2010) 195APOE ε4 and MBL-2 O/O genotypes are associated with neurocognitive impairment in HIV-infected plasma donors from Anhui Province, China. AIDS 24(10):1471–1479
Spudich S, Gisslen M, Hagberg L, Lee E, Liegler T, Brew B et al (2011) Central nervous system immune activation characterizes primary human immunodeficiency virus 1 infection even in participants with minimal cerebrospinal fluid viral burden. J Infect Dis 204(5):753–760
Spudich S, Robertson KR, Bosch RJ, Gandhi RT, Cyktor JC, Mar H et al (2019) Persistent HIV-infected cells in cerebrospinal fluid are associated with poorer neurocognitive performance. J Clin Invest 129(8):3339–3346
Steinbrink F, Evers S, Buerke B, Young P, Arendt G, Koutsilieri E et al (2013) Cognitive impairment in HIV infection is associated with MRI and CSF pattern of neurodegeneration. Eur J Neurol 20(3):420–428
Strazza M, Pirrone V, Wigdahl B, Nonnemacher MR (2011) Breaking down the barrier: the effects of HIV-1 on the blood-brain barrier. Brain Res 1399:96–115
Strickland SL, Rife BD, Lamers SL, Nolan DJ, Veras NMC, Prosperi MCF et al (2014) Spatiotemporal dynamics of simian immunodeficiency virus brain infection in CD8(+) lymphocyte-depleted rhesus macaques with neuroAIDS. J Gen Virol 95(Pt 12):2784–2795
Sturdevant CB, Joseph SB, Schnell G, Price RW, Swanstrom R, Spudich S (2015) Compartmentalized replication of R5 T cell-tropic HIV-1 in the central nervous system early in the course of infection. PLoS Pathog 11(3):e1004720
Suh H-S, Gelman BB, Lee SC (2014a) Potential roles of microglial cell progranulin in HIV-associated CNS pathologies and neurocognitive impairment. J Neuroimmune Pharmacol 9(2):117–132
Suh H-S, Lo Y, Choi N, Letendre S, Lee SC (2014b) Evidence of the innate antiviral and neuroprotective properties of progranulin. PLoS One 9(5):e98184
Sun B, Dalvi P, Abadjian L, Tang N, Pulliam L (2017) Blood neuron-derived exosomes as biomarkers of cognitive impairment in HIV. AIDS 31(14):F9–F17
Takahashi K, Wesselingh SL, Griffin DE, McArthur JC, Johnson RT, Glass JD (1996) Localization of HIV-1 in human brain using polymerase chain reaction/in situ hybridization and immunocytochemistry. Ann Neurol 39(6):705–711
Tambussi G, Gori A, Capiluppi B, Balotta C, Papagno L, Morandini B et al (2000) Neurological symptoms during primary human immunodeficiency virus (HIV) infection correlate with high levels of HIV RNA in cerebrospinal fluid. Clin Infect Dis 30(6):962–965
Tatro ET, Scott ER, Nguyen TB, Salaria S, Banerjee S, Moore DJ et al (2010) Evidence for alteration of gene regulatory networks through microRNAs of the HIV-infected brain: novel analysis of retrospective cases. PLoS One 5(4):e10337
Tatro ET, Soontornniyomkij B, Letendre SL, Achim CL (2014) Cytokine secretion from brain macrophages infected with human immunodeficiency virus in vitro and treated with raltegravir. BMC Infect Dis 14:386
Tavazzi E, Morrison D, Sullivan P, Morgello S, Fischer T (2014) Brain inflammation is a common feature of HIV-infected patients without HIV encephalitis or productive brain infection. Curr HIV Res 12(2):97–110

Thames AD, Briones MS, Magpantay LI, Martinez-Maza O, Singer EJ, Hinkin CH et al (2015) The role of chemokine C-C motif ligand 2 genotype and cerebrospinal fluid chemokine C-C motif ligand 2 in neurocognition among HIV-infected patients. AIDS 29(12):1483–1491

Thompson KA, Cherry CL, Bell JE, McLean CA (2011) Brain cell reservoirs of latent virus in presymptomatic HIV-infected individuals. Am J Pathol 179(4):1623–1629

Tisch S, Brew B (2009) Parkinsonism in HIV-infected patients on highly active antiretroviral therapy. Neurology 73(5):401–403

Tisch S, Brew BJ (2010) HIV, HAART, and Parkinson's disease: co-incidence or pathogenetic link? Mov Disord 25(13):2257–2258

Torti C, Focà E, Cesana BM, Lescure FX (2011) Asymptomatic neurocognitive disorders in patients infected by HIV: fact or fiction? BMC Med 9(1):138

Tozzi V, Balestra P, Serraino D, Bellagamba R, Corpolongo A, Piselli P et al (2005) Neurocognitive impairment and survival in a cohort of HIV-infected patients treated with HAART. AIDS Res Hum Retrovir 21(8):706–713

Trickey A, May MT, Vehreschild J-J, Obel N, Gill MJ, Crane HM et al (2017) Survival of HIV-positive patients starting antiretroviral therapy between 1996 and 2013: a collaborative analysis of cohort studies. Lancet HIV 4(8):e349–ee56

Turner RS, Chadwick M, Horton WA, Simon GL, Jiang X, Esposito G (2016) An individual with human immunodeficiency virus, dementia, and central nervous system amyloid deposition. Alzheimers Dement 4:1–5

Valcour V, Shikuma C, Shiramizu B, Watters M, Poff P, Selnes O et al (2004) Higher frequency of dementia in older HIV-1 individuals: the Hawaii Aging with HIV-1 Cohort. Neurology 63 (5):822–827

Valcour V, Chalermchai T, Sailasuta N, Marovich M, Lerdlum S, Suttichom D et al (2012) Central nervous system viral invasion and inflammation during acute HIV infection. J Infect Dis 206 (2):275–282

Valcour VG, Ananworanich J, Agsalda M, Sailasuta N, Chalermchai T, Schuetz A et al (2013) HIV DNA reservoir increases risk for cognitive disorders in cART-Naïve patients. PLoS One 8(7): e70164

Vassallo M, Dunais B, Durant J, Carsenti-Dellamonica H, Harvey-Langton A, Cottalorda J et al (2013) Relevance of lipopolysaccharide levels in HIV-associated neurocognitive impairment: the Neuradapt study. J Neurovirol 19(4):376–382

Vassallo M, Durant J, Lebrun-Frenay C, Fabre R, Ticchioni M, Andersen S et al (2015) Virologically suppressed patients with asymptomatic and symptomatic HIV-associated neurocognitive disorders do not display the same pattern of immune activation. HIV Med 16(7):431–440

Veenstra M, León-Rivera R, Li M, Gama L, Clements JE, Berman JW (2017) Mechanisms of CNS viral seeding by HIV+ CD14+ CD16+ monocytes: establishment and reseeding of viral reservoirs contributing to HIV-associated neurocognitive disorders. MBio 8(5)

Veenstra M, Byrd DA, Inglese M, Buyukturkoglu K, Williams DW, Fleysher L et al (2019) CCR2 on peripheral blood CD14+CD16+ monocytes correlates with neuronal damage, HIV-associated neurocognitive disorders, and peripheral HIV DNA: reseeding of CNS reservoirs? J Neuroimmune Pharmacol 14:120–133

Venkatachari NJ, Jain S, Walker L, Bivalkar-Mehla S, Chattopadhyay A, Bar-Joseph Z et al (2017) Transcriptome analyses identify key cellular factors associated with HIV-1 associated neuropathogenesis in infected men. AIDS 31(5):623–633

Vera JH, Guo Q, Cole JH, Boasso A, Greathead L, Kelleher P et al (2016) Neuroinflammation in treated HIV-positive individuals: a TSPO PET study. Neurology 86(15):1425–1432

Vivithanaporn P, Heo G, Gamble J, Krentz H, Hoke A, Gill M et al (2010) Neurologic disease burden in treated HIV/AIDS predicts survival a population-based study. Neurology 75 (13):1150–1158

Vivithanaporn P, Asahchop EL, Acharjee S, Baker GB, Power C (2016) HIV protease inhibitors disrupt astrocytic glutamate transporter function and neurobehavioral performance. AIDS (London, England) 30(4):543–552

Williams R, Dhillon NK, Hegde ST, Yao H, Peng F, Callen S et al (2009) Proinflammatory cytokines and HIV-1 synergistically enhance CXCL10 expression in human astrocytes. Glia 57(7):734–743

Williams DW, Calderon TM, Lopez L, Carvallo-Torres L, Gaskill PJ, Eugenin EA et al (2013) Mechanisms of HIV entry into the CNS: increased sensitivity of HIV infected CD14+CD16+ monocytes to CCL2 and key roles of CCR2, JAM-A, and ALCAM in diapedesis. PLoS One 8 (7):e69270

Wong M, Robertson K, Nakasujja N, Skolasky R, Musisi S, Katabira E et al (2007) Frequency of and risk factors for HIV dementia in an HIV clinic in sub-Saharan Africa. Neurology 68 (5):350–355

Woods SP, Iudicello JE, Dawson MS, Weber E, Grant I, Letendre SL et al (2010) HIV-associated deficits in action (verb) generation may reflect astrocytosis. J Clin Exp Neuropsychol 32 (5):522–527

Woods SP, Iudicello JE, Morgan EE, Verduzco M, Smith TV, Cushman C et al (2017) Household everyday functioning in the internet age: online shopping and banking skills are affected in HIV-associated neurocognitive disorders. J Int Neuropsychol Soc 23(7):605–615

Wright EJ, Grund B, Robertson K, Brew BJ, Roediger M, Bain MP et al (2010) Cardiovascular risk factors associated with lower baseline cognitive performance in HIV-positive persons. Neurology 75(10):864–873

Wright PW, Vaida FF, Fernández RJ, Rutlin J, Price RW, Lee E et al (2015) Cerebral white matter integrity during primary HIV infection. AIDS 29(4):433–442

Wright EJ, Grund B, Robertson KR, Cysique L, Brew BJ, Collins GL et al (2018) No neurocognitive advantage for immediate antiretroviral treatment in adults with greater than 500 CD4+ T-cell counts. AIDS 32(8):985–997

Yadav A, Betts MR, Collman RG (2016) Statin modulation of monocyte phenotype and function: implications for HIV-1 associated neurocognitive disorders. J Neurovirol 22(5):584–596

Yelamanchili SV, Chaudhuri AD, Chen LN, Xiong H, Fox HS (2010) MicroRNA-21 dysregulates the expression of MEF2C in neurons in monkey and human SIV/HIV neurological disease. Cell Death Dis 1(9):e77

Yilmaz A, Yiannoutsos CT, Fuchs D, Price RW, Crozier K, Hagberg L et al (2013) Cerebrospinal fluid neopterin decay characteristics after initiation of antiretroviral therapy. J Neuroinflammation 10(1):62

Young AC, Yiannoutsos CT, Hegde M, Lee E, Peterson J, Walter R et al (2014) Cerebral metabolite changes prior to and after antiretroviral therapy in primary HIV infection. Neurology 83 (18):1592–1600

Yu C, Seaton M, Letendre S, Heaton R, Al-Harthi L (2017) Plasma dickkopf-related protein 1, an antagonist of the Wnt pathway, is associated with HIV-associated neurocognitive impairment. AIDS 31(10):1379–1385

Yuan L, Qiao L, Wei F, Yin J, Liu L, Ji Y et al (2013) Cytokines in CSF correlate with HIV-associated neurocognitive disorders in the post-HAART era in China. J Neurovirol 19 (2):144–149

Yuen T, Brouillette M-J, Fellows LK, Ellis RJ, Letendre S, Heaton R et al (2017) Personalized risk index for neurocognitive decline among people with well-controlled HIV infection. J Acquir Immune Defic Syndr 76(1):48–54

Zhang Y, Wang M, Li H, Zhang H, Shi Y, Wei F et al (2012) Accumulation of nuclear and mitochondrial DNA damage in the frontal cortex cells of patients with HIV-associated neurocognitive disorders. Brain Res 1458:1–11

Zhou L, Pupo GM, Gupta P, Liu B, Tran SL, Rahme R et al (2012) A parallel genome-wide mRNA and microRNA profiling of the frontal cortex of HIV patients with and without HIV-associated dementia shows the role of axon guidance and downstream pathways in HIV-mediated neurodegeneration. BMC Genomics 13(1):677

Neurocognitive Impairment and Associated Genetic Aspects in HIV Infection

Daniela Gomez, Christopher Power, and Esther Fujiwara

Contents

Abstract HIV enters the central nervous system (CNS) early after infection. HIV-associated neurocognitive disorders (HAND) remain a serious complication of HIV infection despite available antiretroviral therapy (ART). Neurocognitive deficits observed in HAND are heterogeneous, suggesting a variability in individuals' susceptibility or resiliency to the detrimental CNS effects of HIV infection.

D. Gomez and E. Fujiwara (✉)
Department of Psychiatry, University of Alberta, Edmonton, AB, Canada
e-mail: efujiwara@ualberta.ca

C. Power
Departments of Psychiatry and Medicine, University of Alberta, Edmonton, AB, Canada

Curr Topics Behav Neurosci (2021) 50: 41–76
DOI 10.1007/7854_2018_69
Published Online: 7 December 2018

This chapter reviews primary host genomic changes (immune-related genes, genes implicated in cognitive changes in primary neurodegenerative diseases), epigenetic mechanisms, and genetic interactions with ART implicated in HIV progression or HAND/neurocognitive complications of HIV. Limitations of the current findings include diversity of the HAND phenotype and limited replication of findings across cohorts. Strategies to improve the precision of future (epi)genetic studies of neurocognitive consequences of HIV infection are offered.

Keywords Antiretroviral toxicity · Epigenetics · Genetics · HIV · HIV-associated neurocognitive disorders

1 Introduction

Human immunodeficiency virus type I (HIV) enters the central nervous system (CNS) early after infection (Davis et al. 1992). The mechanisms underlying HIV-related brain damage remain uncertain although direct (cytotoxic) and indirect (immune-mediated) viral and host factors likely contribute (Elbirt et al. 2015). HIV access to the brain occurs through HIV-infected trafficking macrophages and lymphocytes that cross the brain-blood barrier (BBB) and the blood-choroid plexus barrier. Viral replication within CNS macrophages and microglia activates immune cells, leading to the secretion of host immune (pro-inflammatory) and viral (neurotoxic) molecules. In the CNS, inflammation and viral replication can lead to progressive neurodegeneration with accompanying cognitive deficits and behavioral changes, resulting in a spectrum syndrome termed HIV-associated neurocognitive disorders (HAND; Elbirt et al. 2015; Hong and Banks 2015). Despite available antiretroviral therapy (ART), HAND remains a serious complication of HIV infection that affects between 20 and over 60% of patients (Nightingale et al. 2014). HAND is a pressing health concern for both patients and clinicians, as even in its mildest form HAND can adversely affect activities of daily life, financial management, independence, employment, and treatment adherence (Andrade et al. 2013; Doyle et al. 2016). However, the neurocognitive deficits observed in HAND are heterogeneous, suggesting variability in individuals' susceptibility or resiliency to the detrimental CNS effects of HIV infection. Host genetic diversity (i.e., immune-related genes and genes associated with neurocognition), epigenetic changes, and a genetic susceptibility to ART toxicity might contribute to the heterogeneity of HAND. The current chapter focuses on reviewing primary host genomic changes, epigenetic mechanisms, and genetic interactions with ART implicated with onset and progression of HAND/neurocognitive impairment in HIV. Study outcomes are partly limited by current diagnostic ambiguities of the HAND phenotype, which will be briefly discussed. We conclude with strategies to improve the precision of future genetic studies on neurocognitive complications of HIV.

2 HIV-Associated Neurocognitive Disorders

The phenotypic expression of CNS complications associated with HIV infection underwent a dramatic change after the introduction of modern ART. In the pre-ART era, the terms AIDS dementia complex (ADC), HIV encephalopathy, or HIV dementia were used to describe the marked motor deficits and dementia-like signs and symptoms frequently observed in a subset of adult patients with the acquired immunodeficiency syndrome (AIDS; Navia et al. 1986). ADC/HIV encephalopathy/HIV dementia was a severe and progressive syndrome, usually observed in patients with marked immunosuppression. The American Academy of Neurology (AAN) introduced the first consensus criteria in 1991 (Janssen et al. 1991), distinguishing HIV-associated minor cognitive/motor disorder (MCMD) from severe HIV-associated dementia (HAD). The AAN criteria did not specify the number of cognitive domains that should be examined nor the degree of cognitive impairment, introducing variability between clinicians' appraisals of severity.

With the introduction of ART in 1996, the course of HIV infection evolved from a largely fatal diagnosis to a chronic but manageable disorder (D'Aquila et al. 1996; Collier et al. 1996). Milder forms of neurocognitive impairment became more evident as ART was increasingly effective in suppressing viral burden and improving immune status (Heaton et al. 2011). The current consensus criteria ("Frascati" criteria) were developed in 2007 and introduced HAND as an umbrella term to classify the range of symptoms: asymptomatic neurocognitive impairment, mild neurocognitive disorders, and HAD (Antinori et al. 2007). To arrive at a HAND diagnosis, the Frascati criteria require multi-domain neuropsychological testing, an assessment of patients' functional impairment in daily life and the exclusion of comorbidities that confound the interpretation of the neuropsychological results (e.g., head injury, CNS coinfections, developmental delay, psychosis, etc.). Asymptomatic neurocognitive impairment (ANI) is characterized by abnormal cognitive performance in at least two cognitive domains without functional impairment on a patient's day-to-day activities. Mild neurocognitive disorder (MND) requires cognitive deficits in two or more domains and at least mild functional impairment in activities of daily living and/or work. HAD, the most severe form of HAND, requires severe cognitive impairment with marked functional deficits. The prevalence of HAND and the proportions of patients developing any form of HAND vary widely. For example, diagnoses of HAND reported in cohort studies range from 26 to 74% (Bonnet et al. 2013; Dufouil et al. 2015; Wright et al. 2015; Vassallo et al. 2014; Winston et al. 2013; Simioni et al. 2010; Cysique et al. 2014; Garvey et al. 2011; Sacktor et al. 2016). According to Saylor et al. (2016), most HAND patients receiving current ART have the milder forms, while only 2–8% of patients develop HAD. Variable severity in HAND over time, evidenced by patients bidirectionally transitioning from one category to another, can also complicate and undermine the stability of the HAND diagnosis (Antinori et al. 2007).

In summary, pre-ART ADC/HAD-like phenotypes were associated with subcortical motor symptoms and neuropathological changes such as HIV encephalitis including microglial nodules, multinucleated giants cell, and the presence of viral antigens (e.g., p24, gp41) which are uncommon today (Heaton et al. 2011; Saylor et al. 2016). Instead, milder forms of HAND are associated with changes in cortical brain functions including attention, memory, and executive functions (Heaton et al. 2010; Sacktor 2018). Thus, these phenotypic changes in HAND before and after modern ART must be considered when evaluating (epi)genetic biomarkers of HAND.

3 Primary Host Genetic Variations in HAND: Immune-Related Genes

Studies attempting to identify genome-wide associations to HAND or neurocognitive impairment in HIV have been few, and results have been inconclusive (Jia et al. 2017; Levine et al. 2012). Association studies targeting candidate genes related to host immune function and/or to viral entry and replication have shown more promise. Among these are genes related to chemokines and their receptors, cytokines, and mannose-binding lectin genes. Core findings from studies investigating the effects of immune-related genetic polymorphisms on neurocognitive complications of HIV are summarized below (Table 1).

3.1 C-C Chemokine Receptor Type 5 (CCR5)

The C-C chemokine receptor type 5 (CCR5) is the most common HIV co-receptor for viral entry into memory T cells and macrophages. A 32-base pair deletion in the open reading frame of the *CCR5* gene, referred to as *CCR5-Δ-32* allele (rs333), leads to the production of a truncated protein. The deletion spans nucleotides 794–825 corresponding to the second extracellular loop of the receptor, which precludes insertion of the mutant protein into the membrane. It therefore leads to structural changes to the HIV co-receptor and limits viral entry (Berger et al. 1999; Liu et al. 1996). Homozygosity of the *Δ32* variant confers protection against HIV infection (Samson et al. 1996; Zimmerman et al. 1997). The *Δ32* variant has also been associated with decreased disease progression in heterozygote individuals (Gonzalez et al. 1999; Zimmerman et al. 1997; Ioannidis et al. 2001). Early genetic studies suggested a protective role of *Δ32* heterozygosity against HAD. Boven et al. (1999) reported that from brain tissue of 16 deceased patients from the Netherlands, the *CCR5-Δ-32* allele was absent in all patients diagnosed with ADC ($n = 8$) despite a 10–20% natural occurrence of the allele in individuals with northern European ancestry. van Rij et al. (1999) also reported a reduced frequency of the *CCR5-Δ-32* allele in ADC patients, where only 2 out of 49 (4.1%) patients with ADC were heterozygous, while 27 of 186 (14.5%) AIDS patients without dementia

Table 1 Genetic polymorphisms related to HIV progression and HAND

Gene	Variant	Mechanisms	Effect on HIV progression and/or HAND	Evidence	
				Supportive	Negative
Immune-related genes					
CCR5	*CCR5 Δ-32* (rs333)	Truncated CCR5 receptor protein	Homozygosity for the *Δ-32* variant is associated with decreased disease progression. It has also been associated with low risk for AIDS dementia complex and protective against neurocognitive impairment. Nevertheless, some suggest the protective effect may only be found in individuals who developed AIDS prior to 1991	Zimmerman et al. (1997), Boven et al. (1999), Gonzalez et al. (1999), van Rij et al. (1999), Ioannidis et al. (2001), Singh et al. (2003), and Bol et al. (2012)	Singh et al. (2004) and Spector et al. (2010)
CCR2	*CCR2-V64I*	Linkage disequilibrium with *CCR5*. Heterologous receptor desensitization of CCR5 and CXCR4	Associated with slowed progression to AIDS (by 2–4 years) in a cohort of over 3,000 HIV-infected participants. However, *CCR2-V64I* was also associated with rapid progression to cognitive decline in adults	Smith et al. (1997) and Singh et al. (2004)	Singh et al. (2006)
CCL3	rs1130371	Linkage disequilibrium with *CCL3*, *CCL4*, and *CCL18*	TT genotype associated with a twofold increased risk for HAD	Levine et al. (2009)	
	rs1719134	High linkage disequilibrium with rs1130371	Carriers of the A allele demonstrate steeper cognitive decline than infected noncarriers and uninfected carriers	Levine et al. (2014)	
CCL3L1	*CCL3L1* low copy number	Dominant HIV suppressant. Modulation of CCR5 expression	Possession of low copy numbers in conjunction with *CCR5 Δ-32* genotype associated with over a threefold increased risk for HAD	Gonzalez et al. (2005)	
CXCL12	*SDF1-3′-A* (rs1801157)	Increased expression of CXCL12	The *SDF1-3′A/A* genotype associated with delayed progression to AIDS in	Winkler et al. (1998), Singh et al. (2003),	Meyer et al. (1999), van Rij et al. (1999),

(continued)

Table 1 (continued)

Gene	Variant	Mechanisms	Effect on HIV progression and/or HAND	Evidence	
				Supportive	Negative
			European American participants. Conversely, homozygosity for the A allele is also associated with faster disease progression and cognitive decline in African American children	Modi et al. (2006), and Ding et al. (2018)	Ioannidis et al. (2001), Levine et al. (2009), Spector et al. (2010), and Levine et al. (2014)
	rs754618	Potential regulator role of CXCL12 levels	Associated with increased risk for disease progression in African American participants only	Modi et al. (2006) and Ding et al. (2018)	
	rs2297630	Potential regulator role of CXCL12 levels	Associated with increased risk for disease progression in African American participants only	Modi et al. (2006) and Ding et al. (2018)	
CCL2	*CCL2-2578A>G* (rs1024611)	Increased transcription and production of CCL2. Associated with increased CSF CCL2	Associated with rapid disease progression, decreased cognitive performance, and 4.7-fold increased risk for HAD	Gonzalez et al. (2002) and Levine et al. (2014)	Levine et al. (2009), Bol et al. (2012), and Thames et al. (2015)
PREP1	rs2839619	Preferentially binds to *CCL2-2578G* allele; influences CCL2 transcription	Potentially protective against HAND; however, study replication is required	Bol et al. (2012)	Levine et al. (2012)
MBL2	*MBL-2 O/O* genotype	Structural deficit of MBL, thus reduced levels of functional MBL	The "O" alleles have been associated with higher risk for cognitive decline	Spector et al. (2010) and Levine et al. (2014)	
HLA	*HLA-DR*04*	Low CD4+ T-cell response	Associated with larger cognitive impairment at baseline and steeper cognitive decline over a year	Schrier et al. (2012)	
	HLA class I alleles (*B*27, 57, 58, A*03, 33*)	Specifies CD8+ T-cell response	Potentially neuroprotective; associated with higher neurocognitive performance at baseline and lower rates of cognitive decline over a year	Schrier et al. (2012)	

TNFA	*TNFA-308G>A* (rs1800629)	Higher transcriptional activity leading to increased production of TNF-α	The A allele was associated with an increased risk for ADC/HAD	Quasney et al. (2001) and Pemberton et al. (2008)	Diaz-Arrastia et al. (2004), Levine et al. (2009), and Bol et al. (2012)
Genes associated with neurocognitive performance					
APOE	*APOE* ε4 (rs429358; rs7412)	Dysregulated lipid and sterol metabolism. Enhanced HIV cell entry in vitro	Direct or indirect associations between *APOE* ε4 and HAND/neurocognitive impairment. Further moderation by older age in some studies	Corder et al. (1998), Valcour et al. (2004)[a], Burt et al. (2008), Pomara et al. (2008), Spector et al. (2010), Andres et al. (2011), Chang et al. (2011), Chang et al. (2014), Cysique et al. (2015), Panos et al. (2013)[a], Mukerji et al. (2016)[a], and Wendelken et al. (2016)[a]	Sun et al. (2010), Joska et al. (2010), Morgan et al. (2013), and Becker et al. (2015)
COMT	*COMT Val158Met* (rs4680)	40% less metabolically active enzyme	*Val/Val* genotype associated with worse neurocognitive performance	Sundermann et al. (2015)	Levine et al. (2014)
DRD2	rs6277	Altered dopamine D2 receptor expression	Associated with executive dysfunction	Villalba et al. (2015)	Levine et al. (2014)
DRD4	*DRD4 VNTR-7*	Enhanced dopamine response	Associated with executive dysfunction	Villalba et al. (2015)	Levine et al. (2014)
DRD3	rs6280	Increased release of dopamine, however exact role not known	Linked to neurocognitive impairment in HIV-infected patients with concurrent stimulant addiction	Gupta et al. (2011)	
Genes associated with susceptibility to ART CNS toxicity					
CYP	*CYP2B6* G516T	Reduced CYP2B6 function; well-known link to higher risk of	No association with neurocognition in two studies; however, higher levels of metabolite 8-OH-efavirenz have been		Haas et al. (2004) and Sandkovsky et al. (2017)

(continued)

Table 1 (continued)

Gene	Variant	Mechanisms	Effect on HIV progression and/or HAND	Evidence	
				Supportive	Negative
		neuropsychiatric complications with efavirenz	associated with higher neurocognitive scores		
	*CYP3A5*1*	Faster CYP3A5 clearance of protease inhibitors	Risk for neurocognitive deficits unknown		

[a]Studies that found a moderating effect of older age (>50 years)

were heterozygotes. In HIV-infected children, the *CCR5-Δ-32* allele was found to be protective against neurocognitive disorders (Singh et al. 2003). However, later studies have been unable to replicate these findings (Spector et al. 2010; Singh et al. 2004; Bol et al. 2012). In a case-control study of 86 HAD patients and 246 non-HAD AIDS patient controls, Bol et al. (2012) found a low frequency of the *CCR5-Δ-32* allele in HAD patients compared to non-HAD patients but only in individuals diagnosed with AIDS prior to 1991. Given the changing phenotype of HAND, these results may suggest a fading effect of the protective qualities of the *CCR5-Δ-32* allele with effective ART. However, CCR5 is expressed in multiple brain cells, including neurons (Sorce et al. 2011), and is known to play a role in several noninfectious CNS disorders (e.g., Alzheimer disease, multiple sclerosis, stroke). Across these conditions, the *CCR5-Δ-32* allele has mostly been found to be *deleterious*. For example, *CCR5-Δ-32* has been linked to faster disease progression/early death in multiple sclerosis (Gade-Andavolu et al. 2004), increased risk for ischemic heart disease (Dinh et al. 2015), and increased risk for atherosclerotic disease (Zhang et al. 2015). Thus, *CCR5-Δ-32* deletion may aggravate the course of comorbid (rather than HIV-related) degenerative CNS processes in HIV-infected individuals over time, although the precise interplay between potential protective and harmful consequences of carrying the *CCR5-Δ-32* allele in an aging HIV demographic remains to be explored.

3.2 *Macrophage Inflammatory Protein-1 Alpha (MIP-1α or CCL3)*

Macrophage inflammatory protein-1 alpha (MIP-1α) is encoded by *CCL3* gene and is a natural ligand of the CCR1, CCR3, and CCR5 receptors. Binding of MIP-1α to the CCR5 receptor may block HIV entry, suppressing HIV viral replication (Arenzana-Seisdedos and Parmentier 2006). Levine et al. (2009) examined polymorphisms of 7 immune-related genes in 143 individuals with HIV infection, including 23 HAD patients, from the National NeuroAIDS Tissue Consortium. Only the rs1130371 SNP of the *CCL3* gene significantly differed between the groups. The rs1130371 SNP occurs in the coding region of *CCL3*; however, the synonymous C>T substitution does not change the open reading frame (Modi et al. 2006). In Levine et al. (2009), 16% of unimpaired patients were homozygous for the T allele, whereas 31% of HAD patients were homozygous. The presence of rs1130371 increased the risk of HAD twofold. However, comorbid depression was associated with a fivefold risk of HAD, qualifying the impact of rs1130371 alone. In a longitudinal study, Levine et al. (2014) reported that HIV-infected carriers of the A allele of a *CCL3* rs1719134 SNP – an A>G substitution in noncoding intron 1 region of the CCL3 gene (Modi et al. 2006) – had greater memory decline than both HIV-infected noncarriers and HIV-uninfected carriers. A higher *CCL3L1* gene copy number has also been associated with reduced risk for acquiring HIV, whereas low copy number are associated with increased HIV susceptibility (Liu et al. 2010). Gonzalez et al. (2005) stratified their sample into genetic risk groups based on *CCL3L1* copy numbers (high vs. low) and disease altering *CCR5* genotype (*Δ-32* homozygous vs. wild type). Possession of low copy numbers in combination with *CCR5-Δ-32* carried over a threefold increased risk for HAD.

3.3 Stromal Cell-Derived Factor-1 (SDF-1 or CXCL12)

Stromal cell-derived factor-1 (SDF-1 or C-X-C motif chemokine 12, CXCL12) is a chemokine involved in leukocyte activation. CXCL12 is the chief ligand for CXCR4. CXCL12 may bind to CXCR4 competitively inhibiting HIV binding and limiting viral entry (Agace et al. 2000). Winkler et al. (1998) investigated the association between the *CXCL12* rs1801157 SNP (also SDF1-3′A), a G>A substitution at base pair 801 of the 3′-untranslated region of the *CXCL12* gene (Modi et al. 2006), and HIV disease progression in a sample of HIV-infected men from the Multicenter AIDS Cohort Study (MACS). The rs1801157 AA genotype was associated with delayed progression to AIDS. However, multiple subsequent studies failed to find this association or even reported faster disease progression for the AA genotype (van Rij et al. 1999; Ioannidis et al. 2001; Meyer et al. 1999).

A recent meta-analysis of 16 studies by Ding et al. (2018) investigated the relationship between *CXCL12* polymorphisms, infection vulnerability, and disease progression. No significant association between *CXCL12* polymorphism and vulnerability to HIV infection was observed, and disease progression was related to *CXCL12* polymorphism only in studies from the MACS cohort. Ding et al. (2018) attribute specific demographic features of this cohort to the unique findings. Indeed, directly comparing the MACS with other cohorts regarding associations of *CXCL12* polymorphisms, Modi et al. (2005) reported a significant protective effect of rs1801157 against disease progression in patients of European American descent only in the MACS cohort. A detrimental effect of two different *CXCL12* polymorphisms (rs754618 and rs2297630) on disease progression was observed only in African American participants, pooled across several cohorts. Thus, the rs1801157 polymorphism may only exert protective influences on disease progression in select populations.

Results regarding the neurocognitive impact of *CXCL12* polymorphisms are also mixed. Levine et al. (2014) failed to find an association between rs1801157 and neurocognitive impairment in patients with HIV in a longitudinal MACS cohort study. Similarly, Levine et al. (2009) reported that rs1801157 was not a significant predictor of HAD, using data of 117 non-HAD and 26 HAD patients from the National NeuroAIDS Tissue Consortium (see also Spector et al. 2010). However, in a pediatric HIV-infected cohort ($n = 1{,}049$), Singh et al. (2003) reported that homozygosity for the A allele was associated with faster disease progression and cognitive decline in African American children. The contradictory findings between children and adults may be due to a moderating effect of *CXCL12* genotype on age and cognition, but further studies are needed.

3.4 Monocyte Chemoattractant Protein-1 (MCP-1 or CCL2)

Monocyte chemoattractant protein-1 (MCP-1), also known as CCL2, is the most potent chemokine to induce chemotaxis of myeloid cells into the CNS. CCL2 mediates the early neuroinflammatory response after HIV infection (Dhillon et al. 2008). Increased levels of CCL2 are found in the brain and cerebrospinal fluid (CSF) of infected patients with HAND or HIV encephalopathy compared to cognitively intact patients and uninfected controls (Conant et al. 1998; Dhillon et al. 2008). The *CCL2*-2578G allele has been the most extensively studied allele in relation to HAND. The *CCL2*-2578G allele represents an A>G substitution at position -2578 in the promoter region of the *CCL2* gene (Rovin et al. 1999). However, findings on associations between *CCL2*-2578G and HAND have been contradictory. Gonzalez et al. (2002) observed that homozygous individuals for the *CCL2*-2578G allele not only had an accelerated progression to AIDS and death but also a 4.7-fold increase risk for HAD. These associations remained significant after controlling for the year in which HAD developed, suggesting that newer ART regimens did not diminish the susceptibility of carriers to develop HAD. In a longitudinal study, Levine et al. (2014) reported that patients homozygous for the *CCL2-2578G* allele had lower memory performance over time compared to HIV-positive heterozygotes and uninfected controls. However, others failed to find a direct association between the *CCL2-2578G* allele and HAD (Levine et al. 2009) or cognitive dysfunction (Spector et al. 2010). Thames et al. (2015) reported no difference in cognitive status (Global Deficit Score) between *CCL2-2578G* allele carriers and noncarriers. However, *CCL2-2578G* carriers had increased levels of CCL2 in the CSF, and these in turn correlated with CSF neuroinflammatory markers and lower neurocognitive functioning. Thus, *CCL2* polymorphisms may influence intermediate neurocognitive impairment phenotypes in HIV infection. Bol et al. (2012) studied the association between HAD and several candidate genetic polymorphisms (e.g., *APOE*, *CCL3*, *TNFA*), including *CCL2-2578G*. A polymorphism in another candidate gene, *PREP1*, was also investigated. Prep1 (also known as PKNOX1) is a transcriptional factor that preferentially binds to the *CCL2-2578G* allele, thereby affecting transcription of the protein. Among 17 *PREP1* SNPs, rs2839619 was targeted based on its role in cholesterol metabolism and linkage disequilibrium with nearby intronic SNP rs234720, which had been associated with neurocognitive performance (Cirulli et al. 2010). Although none of the studied genes were associated with HAD, including *CCL2-2578G*, Bol et al. (2012) found that the heterozygous *PREP1* rs2839619 AG genotype was observed in 55% of non-HAD patients with AIDS, but only in 24% of HAD patients, suggesting a potentially protective property of the heterozygous genotype. These results await replication.

A point mutation in the minor HIV co-receptor CCR2, whose natural ligand is CCL2, has also been reported. This *CCR2* polymorphism results from a G>A transition in the coding region, resulting in a valine to isoleucine replacement at amino acid position 64 (*CCR2-V64I*). The *CCR2-V64I* allele has been linked to slowed disease progression, delaying AIDS by 2–4 years in a cohort of over 3,000

HIV-infected patients prior to the introduction of ART (Smith et al. 1997). The same allele was found to slow progression of cognitive impairment in 121 patients with HIV (Singh et al. 2004). However, in a follow-up study with a pediatric cohort ($n = 1{,}059$), Singh et al. (2006) failed to find associations between *CCL2-2578G* or *CCR2-V64I* alleles and HIV disease progression or cognitive impairment.

3.5 Mannose-Binding Lectin 2 (MBL2)

Mannose-binding lectin 2 (MBL2) is a protein integral to the innate immune response, the first line of response against infection. Low concentrations of MBL2 are associated with greater susceptibility to infections and accelerated disease progression in HIV (Ji et al. 2005). The *MBL2* gene has multiple polymorphisms on its coding and promoter regions. Three variations in the coding regions, collectively known as O allele (Garred 2008), alter the structural integrity of *MBL2*. In Chinese men with HIV infection ($n = 201$), those possessing the O/O genotype had a higher risk for cognitive decline compared to patients without mutations or with the A/A genotype (Spector et al. 2010). However, this study lacked an HIV-negative control group. To disentangle whether the neurocognitive decline was due to genotype or HIV status alone, Levine et al. (2014) investigated the influence of *MBL2* genotype, among others, on cognitive decline in a large cohort ($n = 952$) of participants with or without HIV. They found a deleterious influence of the O genotype in uninfected controls only, suggesting that Spector et al. (2010) results may not have been due to HIV status alone.

3.6 Human Leukocyte Antigen (HLA) Alleles

Human leukocyte antigen (HLA) genes encode the major histocompatibility complex, cell surface proteins involved in cell recognition and immune activation. In humans, *HLA* genes are highly polymorphic. Multiple studies have linked specific HLA types with HIV susceptibility, immune control, and disease progression (Borghans et al. 2007; Huang et al. 2009; Schrier et al. 1996; Li et al. 2007; Goulder and Watkins 2008; Telenti 2005). An earlier US cohort showed an association between the *HLA-DR*04* gene and T-cell activation, such that individuals with this genotype had low CD4+ T-cell response and low plasma HIV viral loads (Schrier et al. 1989). Schrier et al. (2012) assessed the association between HLA genotypes with neurocognition and virologic factors in a longitudinal study of plasma donors in China. In this cohort, *HLA-DR*04* was associated with neurocognitive impairment at baseline, and more pronounced neurocognitive decline, as well as lower plasma viral loads at the 12-month follow-up. The association between *HLA-DR*04* genotype and low viral loads may be explained by the fact that HIV best replicates in activated CD4+ T cells. Since *HLA-DR*04* is associated with low CD4+ T-cell activation, individuals with this genotype may be better able to suppress viral replication than individuals with the wild type. Viral load was a robust predictor of CNS involvement and neurocognitive decline prior to ART availability. However, these relationships are more complicated in treated cohorts, in particular if ART are only partially effective (Gelman 2015). Schrier et al. (2012) results suggest that neurocognitive impairment in patients with the *HLA-DR*04* genotype may be driven by a process other than viral replication. The authors also reported that *HLA* class I alleles (*B*27*, *57*, *58*, *A*03*, *33*) were associated with CD8+ T-cell control of HIV infection, and these alleles were related to less neurocognitive impairment (baseline, 12 months) and slower neurocognitive decline (12 months).

3.7 *Tumor Necrosis Factor Alpha (TNF-α)*

Tumor necrosis factor alpha (TNF-α) is an inflammatory cytokine released by macrophages and microglia, associated with apoptosis and viral replication and implicated in neuronal injury after HIV infection (Brabers and Nottet 2006; Vaidya et al. 2014). TNF-α levels are elevated in HIV-infected patients (Roux-Lombard et al. 1989; Brabers and Nottet 2006). Activated macrophages in ADC patients show increased expression *TNFA*-encoded mRNA, and increased expression correlated with ADC progression and severity in early studies (Wesselingh et al. 1993, 1997). The rs1800629 SNP in the promoter region of the *TNFA* gene, resulting in the *TNFA-308* allele, has been associated with enhanced susceptibility to viral infections. Quasney et al. (2001) found that possession of even one copy of the *TNFA-308* allele was associated with an increased risk for ADC compared to non-ADC HIV-infected patients and to non-infected controls. A meta-analysis by Pemberton et al. (2008) reported a significantly higher frequency of the *TNFA-308* alleles in HAD patients compared to both non-HAD HIV-infected patients and healthy controls. However, some studies have failed to find an association between this allele and HAD (Diaz-Arrastia et al. 2004), including more recent investigations (Levine et al. 2009; Bol et al. 2012). The relevance of the *TNFA-308* allele to milder forms of HAND also remains to be examined.

4 Primary Host Genetic Variations in HAND: Genes Associated with Neurocognition

Genetic polymorphisms known to be involved in neurocognitive functions in healthy populations, as well as in the context of neurodegenerative conditions such as Alzheimer disease (AD) and mild cognitive impairment (MCI), have also been studied in the context of neurocognitive impairment in HIV. Core findings are summarized below.

4.1 Apolipoprotein E

Apolipoprotein E (ApoE) is a major receptor for low-density lipids (LDL) and LDL receptor-related proteins. Its main function is to clear plasma of triglycerides and cholesterol-rich glycoproteins. ApoE is expressed at high levels in the brain and is synthesized by astrocytes and infiltrating macrophages. The ApoE ε3 is the most frequent allele (65–70%), followed by ApoE ε4 (15–20%). The ApoE ε2 allele is the least frequent (5–10%). Homozygosity for the ApoE ε4 allele is recognized as a strong genetic factor for late-onset sporadic AD (Liu et al. 2013; Kim et al. 2009). Interest in the potential role of ApoE ε4 in the pathogenesis of HAND has resurged in the last decade. ApoE ε4 is known to alter transcription of amyloid precursor protein (APP), a biomarker for AD. Alterations to the cleavage of APP have been reported in HAD patients (Gisslen et al. 2009; Peterson et al. 2014). In an early study by Corder et al. (1998), HIV-infected ApoE ε4 carriers were twice as likely to be diagnosed with ADC compared to a group of HIV-infected noncarriers over a 5-year period. Cutler et al. (2004) found evidence of dysregulated lipid and sterol metabolism in the brains of HAD patients with ApoE ε4 genotypes, suggesting an increased susceptibility to neurologic insults in ApoE ε4 carriers. ApoE ε4 is also associated with faster disease progression and accelerated progression to death and has been shown to enhance viral fusion to HIV co-receptors (CCR5 and CXCR4) in vitro (Burt et al. 2008). However, the link between ApoE ε4 genotype and HAND is unclear. Whereas some studies have found a deleterious effect of ApoE ε4 on neurocognitive functions in HIV patients (Pomara et al. 2008; Spector et al. 2010; Andres et al. 2011; Chang et al. 2011, 2014; Mukerji et al. 2016), others could not confirm these effects (Sun et al. 2010; Becker et al. 2015; Morgan et al. 2013). The null results might have been caused by low proportions of ApoE ε4 carriers (Sun et al. 2010) or inclusion of participants with possible non-HIV-associated neurocognitive impairment (Morgan et al. 2013). However, the ApoE ε4 genotype may indirectly relate to neurocognitive functions in HIV. Cysique et al. (2015) reported in 43 HIV-infected patients that ApoE ε4 (any allelic variant involving ε4; $n = 13$) was associated with lower CSF levels of Aβ1–42, similar to observations of low Aβ1–42 in AD (Paternico et al. 2012; Popp et al. 2010). Regardless of ApoE ε4 status, abnormal Aβ1–42 levels increased CSF p-tau and t-tau levels, and these in turn were associated with neurocognitive impairment (global deficit scores) in this HIV cohort. Patients with AD-like CSF biomarkers profiles (i.e., combinations of low Aβ1–42 and/or increased CNS-levels of t-tau/ p-tau) were also more likely to have a current or past diagnosis of HAD. These results suggest that CSF biomarkers indicative of neurodegenerative processes – some aspects of which were more likely in ApoE ε4 carriers – could also underlie neurocognitive problems in the context of HIV infection.

Furthermore, some studies report a moderating effect of age on the relationship between ApoE ε4 and neurocognitive functioning in HIV, such that older carriers of the ApoE ε4 allele (age $\geq$50 years) are at higher risk of developing neurocognitive impairment/HAND (Panos et al. 2013; Valcour et al. 2004; Mukerji et al. 2016;

Wendelken et al. 2016). However, this finding has not been replicated in other studies (Joska et al. 2010; Chang et al. 2011; Becker et al. 2015). The inclusion of relatively young participants (mean age <50 years; Joska et al. 2010; Chang et al. 2011; Becker et al. 2015) or small numbers of older ApoE ε4 carriers (Chang et al. 2011) may partly have caused the null findings.

4.2 Dopamine-Related Genes

Dopamine is a critical neurotransmitter of the frontal-striatal-thalamic circuitry preferentially affected by HIV (Gaskill et al. 2013; Chang et al. 2008; Kumar et al. 2009, 2011; Gelman et al. 2012). Primary dopaminergic functions in humans include motor control, reward processing, and executive functions (Chinta and Andersen 2005; Oak et al. 2000). The most commonly studied dopaminergic polymorphisms include the catechol-O-methyltransferase (*COMT*) gene, dopamine receptors (e.g., *DRD4*, *DRD2*), and brain-derived neurotropic factor (*BDNF*) gene. Extracellular dopamine has been shown to act on monocytes to increase HIV replication via dopamine receptor 2 (DR2; Gaskill et al. 2009). Furthermore, a SNP in the *DR3* gene (rs6280) has been linked to neurocognitive impairment among HIV-infected individuals with concurrent stimulant addiction (Gupta et al. 2011).

Thus far, only a few studies have investigated the association between dopamine-related genetic polymorphisms and HAND. The COMT enzyme degrades dopamine (Lewis et al. 2001), and a valine (Val) to methionine (Met) substitution results in an enzyme that is 40% less metabolically active (Chen et al. 2004). Therefore, *Met* carriers metabolize dopamine more slowly. Bousman et al. (2010) examined the effects of COMT genotype on cognition (e.g., executive functions) and sexual risk behavior, a reward-based behavior with a presumed dopaminergic basis, in 192 participants ($n = 107$ HIV infected). Although there was no direct effect of COMT genotype on executive functions, *Met* carriers (i.e., *Met/Met* and *Val/Met*) with lower executive functions engaged in riskier sexual practices than *Val/Val* carriers (Bousman et al. 2010). In a small group of HIV-infected females from the Women's Interagency HIV Study Consortium, Sundermann et al. (2015) observed that *Val/Val* carriers, but not *Met/Met* and *Met/Val* genotypes, performed significantly worse in a working memory task than uninfected controls with the same genotypes. Additionally, HIV-infected *Val/Val* carriers showed increased prefrontal and anterior cingulate cortex activation compared to uninfected *Val/Val* carriers during the working memory task. Thus, detrimental effects of HIV infection on working memory functions specifically may be moderated by COMT genotype, potentially placing *Val/Val* HIV-infected women at a higher risk of working memory deficits.

In a study ($n = 257$ HIV-infected patients) exploring relationships between *DRD2*, *DRD4*, and cognition, Villalba et al. (2015) showed a significant association between executive dysfunction and the *DRD2* rs6277 (also known as C957T) polymorphism. The rs6277 SNP, located on exon 7 of the *DRD2* gene, changes mRNA stability and dopamine-induced upregulation of DRD2 expression (Duan et al. 2003). rs6277 is a well-known neuropsychiatric risk factor, especially for schizophrenia (Betcheva et al. 2009). Villalba et al. (2015) also found that a *DRD4* 48 base-pair variable number tandem repeat (*DRD4 VNTR-7*) polymorphism in the coding sequence was a significant predictor of executive dysfunction in their HIV-infected cohort. However, Levine et al. (2014) failed to find an association between multiple genes related to dopaminergic function and HIV-associated cognitive impairment in a large ($n = 952$) longitudinal study. In this study, the *COMT* genotype influenced longitudinal cognitive functioning only in HIV-uninfected controls. Thus, dopamine genetics may be most relevant to specific neurocognitive (dys)functions (e.g., executive functions, working memory, risk taking, reward processing), rather than global cognitive status or HAND, although HIV-related and HIV-unrelated contributions should be considered.

5 Epigenetic Changes Associated with HAND

Results from the reviewed association studies of candidate genes suggest that individual genes are not robustly associated with HAND and/or HIV progression. Intermediary epigenetic processes may modulate gene expression and thereby indirectly influence the genetics of HAND. Epigenetic studies of HAND pathogenesis are relatively new, with most studies focused on microRNAs. Few studies have investigated the roles of histone modification and DNA methylation in the context of HAND (Table 2). Understanding the epigenetic changes associated with HAND might help better elucidate the neuropathological mechanisms underpinning HAND.

5.1 MicroRNA (miRNA)

MicroRNAs (miRNAs) are small noncoding RNA molecules that can regulate both host and viral gene expression by targeting mRNAs and directing them for cleavage (Guo et al. 2010). Studying miRNAs might contribute to the identification of components of HAND pathogenesis. Eletto et al. (2008) conducted the first study to associate changes in miRNA expression and HIV. The authors observed that the HIV Tat protein promoted miR-128a activity in primary cortical neurons, reducing the expression of SNAP25, a presynaptic protein. Since then, multiple studies have reported altered CNS expression of miRNAs of HAND patients. In brains of patients with HIV encephalitis, Noorbakhsh et al. (2010) observed a downregulation of miRNAs associated with effector caspases involved in cell death pathways. This was accompanied by higher levels of caspase-6 transcripts in HIV encephalitis brains compared to control brains, localized to astrocytes, indicating a linkage between differential expression of miRNAs in advanced stages of HIV and dysregulation of cell death pathways. Yelamanchili et al. (2010) found a significant upregulation of miR-21 in the brains of HAD patients. miR-21 reduced the

Table 2 Epigenetic mechanisms associated with HAND

Epigenetic mechanism	Affected processes associated with HAND	References
miRNA	Downregulation of multiple miRNAs associated with effector caspases involved in cell death pathways in HIV encephalitis	Noorbakhsh et al. (2010)
	Upregulation of miR-146a leading to translational suppression of pro-inflammatory cytokines associated with viral entry inhibition in HIV encephalitis	Rom et al. (2010)
	Upregulation of miRNAs (miR-500a-5p, miR-34c-3p, miR-93-3p, and miR-381-3p) that target peroxisomal genes in HAND patients	Xu et al. (2017)
	Upregulation/downregulation of miRNAs broadly associated with synaptic and neuronal functions	Yelamanchili et al. (2010), Kadri et al. (2016), Asahchop et al. (2016), and Wyczechowska et al. (2017)
Histone modification	Upregulation of HDAC2 associated with the inhibition of translation of genes broadly synaptic and neuronal functions	Saiyed et al. (2011)
Telomere length	Shortened telomere length (LTL) in HIV-infected females compared to controls and LTL positively correlated to memory (Malan-Muller et al. 2013). However, results have not been replicated (Giesbrecht et al. 2014)	Malan-Muller et al. (2013) and Giesbrecht et al. (2014)
DNA methylation	Increased DNA methylation in PBMCs and brain tissue in HAND patients compared to controls. Increased methylation associated with lower neurocognitive function in HIV-infected children	Horvath and Levine (2015), Rickabaugh et al. (2015), Levine et al. (2016), Corley et al. (2016), and Horvath et al. (2018)

expression of myocyte enhancer factor 2C (MEFC2), a transcriptional factor associated with neuronal function. Rom et al. (2010) found an upregulation of miR-146a, associated with the release of pro-inflammatory cytokines such as MCP-2, in frontal lobe brain tissue of HIV encephalitis patients compared to controls. MCP-2 inhibits CCR5-mediated HIV entry and viral replication (Gong et al. 1998; Yang et al. 2002). Rom et al. (2010) reported that in vitro transfection of miR-146a into microglial cells inhibited MCP-2 release. Therefore, the upregulation of mir-146a during HIV infection may impede MCP-2 release, thereby facilitating viral entry and replication. A recent study by Xu et al. (2017) further identified four miRNAs upregulated in HAND patients (miR-500a-5p, miR-34c-3p, miR-93-3p, and miR-381-3p). These miRNAs were identified as targeting mRNAs encoding peroxisomal proteins. Peroxisomes are considered necessary for normal brain functioning (Wanders and

Waterham 2006), and peroxisomal genetic diseases (e.g., adrenoleukodystrophy) have severe neurological consequences (Ferrer et al. 2010; Wanders and Waterham 2006). The loss of peroxisomal proteins signaled by changes in miRNAs may be a novel biological pathway to identify the development of HAND.

Plasma miRNA profiles may also serve as biomarkers for neurocognitive impairment in HIV infection. For example, Kadri et al. (2016) identified and verified the following miRNA pairs that best differentiated between cognitively normal and cognitively impaired (Global Deficit Score $\geq$0.5) HIV-infected patients from the Louisiana State University Health Sciences Center cohort (LSUHSC): miR-744-5p/miR-495-3p, let-7b-5p/miR-495-3p, miR-151a-5p/miR-495-3p, and miR-376a-3p/miR-16-532-3p. Although these miRNA pairs were not associated with other clinical characteristics of the patients, they had been previously identified as brain-rich (Landgraf et al. 2007; Im and Kenny 2012) or as involved in normal brain functions (e.g., miR-495 is associated with neurogenesis and synaptic functioning, Mellios et al. 2008; let-7b is implicated in neurodegeneration by Toll-like receptor seven activation, Lehmann et al. 2012). A follow-up study by the same group with newly enrolled LSUHSC patients ($n = 66$) and patients from the CNS HIV Antiretroviral Therapy Effects Research (CHARTER) cohort ($n = 70$; Wyczechowska et al. 2017) identified several miRNA pairs that distinguished cognitively impaired patients from non-impaired patients. Two miRNA pairs confirmed the previous findings (miR-744-5p/miR-495-3p and let-7b-5p/miR-495-3p; Kadri et al. 2016). However, some miRNA-pairs were upregulated or downregulated differentially in each cohort, likely due to demographic differences between the LSUHSC and CHARTER participants. Both ethnicity and sex can influence miRNA expression (Guo et al. 2017). Another recent study by Asahchop et al. (2016) profiling plasma miRNA in a Canadian cohort found three miRNAs (miR-3665, miR-4516, and miR-4707-5p) distinguishing between HAND ($n = 22$) and non-HAND ($n = 25$) patients with an accuracy of 87%. These miRNAs target genes involved in neural development, cell death, inflammation, cell signaling, and cytokine functions.

These studies provide some evidence for the utility of miRNA profiling as a potential diagnostic tool for HAND or cognitive impairment in the context of HIV. Further studies and larger cohorts are required to continue to achieve broader convergence of the miRNA findings.

5.2 *Histone Modification*

Histone acetylation and deacetylation are prominent epigenetic modifications that can impact gene expression. Altered histone acetylation is present in multiple neurological diseases, and inhibitors of histone deacetylases (HDACs) are associated with neurocognitive improvement in experimental settings (Kazantsev and Thompson 2008; Graff and Tsai 2013). Studies have also shown that HDACs might contribute to eradicating latent HIV reservoirs (Zwergel et al. 2016; Sogaard et al. 2015). To our knowledge, only one study has investigated the role of histone modification in the development of HAND. Saiyed et al. (2011) reported that HIV Tat protein upregulated the expression of HDAC2 in neural cells in vitro, which in turn suppressed the expression of CREB and CaMKIIa genes. Both have been involved in synaptic and neuronal functions and long-term potentiation and are implicated in memory (Vecsey et al. 2007; Silva et al. 1992). Expression of these genes was restored when cells were treated with HDAC inhibitors, emphasizing the potential role of HDAC in the neuropathogenesis of HAND.

5.3 *Other Epigenetic Markers of Biological Aging*

Epigenetic mechanisms regulate telomere dynamics, and telomere length is one of the most popular measures of biological aging (Wong et al. 2014). Telomere shortening is predictive of morbidity and mortality in aging (Cawthon et al. 2003), occurs in multiple neurological (Suchy-Dicey et al. 2018; Martin-Ruiz et al. 2006) and psychiatric (Wolkowitz et al. 2017; Stein et al. 2018; Han et al. 2018) conditions, and may also contribute to accelerated brain aging in HIV. However, results so far have been inconclusive. Malan-Muller et al. (2013) studied the association between leukocyte telomere length (LTL), from DNA extracted from whole blood, and cognition in a group of South African women ($n = 128$, aged 18–50). Compared to uninfected controls, HIV-infected women had significantly shorter LTL. There was also a small but significant correlation between LTL and memory performance ($r = 0.26$, $p = 0.02$), and this association was present only in the HIV-infected group. In contrast, Giesbrecht et al. (2014) found no significant differences in LTL between HIV-negative ($n = 45$, aged 31–67) and HIV-positive women ($n = 81$, aged 30–66), nor was cognitive performance associated with LTL in either group.

DNA methylation levels change in aging. The 'epigenetic clock' is a recently developed method for estimating accelerated epigenetic aging based on age-related DNA methylation levels (Horvath 2013). This approach has been only recently begun to be applied to HAND, with promising results. For example, Horvath and Levine (2015) found that HIV-positive participants ($n = 8$, mean age at death $= 45.9$) showed accelerated epigenetic aging (based on DNA methylation) by 7.4 years in brain tissue compared to data from uninfected controls. Similarly, using two datasets from the MACS (aged 20–56), Rickabaugh et al. (2015) found that HIV-infection accelerated age-related DNA methylation in peripheral blood mononuclear cells (PBMC) by over 10 years.

In relation to HAND, Levine et al. (2016) reported that brain samples from HAND individuals within 1 year of death showed accelerated epigenetic aging by 3.5 years compared to HIV-positive non-HAND patients. Corley et al. (2016) also reported increased DNA methylation in PBMC of neurocognitively impaired patients with HIV infection ($n = 11$, aged 51–72) compared to cognitively unimpaired HIV-positive patients ($n = 10$, aged 50–60). In a large South African cohort of adolescents who perinatally acquired HIV ($n = 204$, mean age $= 10.4$), Horvath et al. (2018) found increased DNA methylation levels (i.e., epigenetic age) in PBMC, in patients compared to age-matched healthy adolescents. In both groups, DNA methylation levels correlated negatively with neurocognition (attention, information processing speed, executive functions). Together, these findings suggest that accelerated epigenetic aging might contribute to poor neurocognitive performance in HIV/AIDS and could serve as a potential biomarker for HAND and/or elderly dementia in HIV-infected adults.

6 Genetic Susceptibility to CNS Toxicity of Antiretroviral Drugs

Just as genotyping has been helpful in identifying potential fatal consequences of ART for some patients (i.e., HLA-B*5701 and abacavir hypersensitivity; Martin et al. 2004), pharmacogenetic studies may help identify genetic variants associated with higher risk for potential ART neurotoxicity which may contribute to HAND.

Most studies have focused on efavirenz (EFV), a non-nucleoside reverse transcriptase inhibitor (NNRTI). EFV is a common component of first-line ART regimens known for its neuropsychiatric side effects, such as dizziness, confusion, psychotic symptoms, memory and attention problems, and sleep abnormalities (Kenedi and Goforth 2011). Although highly effective in reducing viral replication, its side effect profile makes EFV-containing regimens less tolerable compared to other alternatives. As such, EFV is not recommended for patients with HAD to avoid exacerbation of the neurocognitive deficits (Ma et al. 2016; Ciccarelli et al. 2011). EFV is primarily metabolized by cytochrome (CYP) P450, isozymes 2B6, and to a lesser extent 3A4. The *CYP2B6 G516T* mutation is associated with decreased CYP2B6 function, and this polymorphism has been associated with higher EFV plasma concentrations, which in turn can lead to higher incidence of CNS side effects (Haas et al. 2004; Rotger et al. 2005; Gounden et al. 2010; Heil et al. 2012). Haas et al. (2004) conducted a randomized controlled trial to study whether polymorphisms in CYP2B6, CYP3A4, CYP3A5, and multidrug resistance protein 1 (P-glycoprotein) MDR1 genes were associated with higher EFV plasma concentration and CNS side effects in a cohort of 154 HIV-positive patients. Only the *CYP2B6 G516T* TT genotype was significantly associated with higher plasma concentration of EFV. Individuals who were homozygous for the *G516T* allele, which was more common in African Americans (20%) than Caucasians (3%), had three times the plasma concentration of EFV compared to wild-type carriers and heterozygous participants. Carriers of the *G516T* allele also reported a significantly higher incidence of neuropsychiatric side effects after week 1 (see also Rotger et al. 2005; Gounden et al. 2010; Heil et al. 2012).

Although there is ample evidence of a relationship between neuropsychiatric side effects and EFV, the relationship between EFV and neuropsychological performance is less clear. Haas et al. (2004) did not observe a relationship between homozygosity for the *CYP2B6 G516T* allele and neurocognitive performance levels in their cohort.

Clifford et al. (2005) reported a small but statistically significant negative correlation between overall neurocognitive performance and EFV plasma concentrations at week 4 and week 12. However, mean neurocognitive performance levels were the same in patients receiving EFV and those who were not. Conversely, EFV use was a significant predictor of HAND (odds ratio = 4.00, $p = 0.008$) in Ciccarelli et al. (2011). Neither of these studies investigated the influence of *CYP2B6 G516T*, preventing a direct comparison to Haas et al.'s (2004) findings. Sandkovsky et al. (2017) evaluated the relationship between neuropsychological performance, EFV concentration, and *CYP2B6* G516T polymorphism in a small cross-sectional cohort of 30 older patients with HIV infection. The authors failed to find a significant correlation between neuropsychological performance and EFV plasma levels, and they were also unable to find a relationship to the *G516T* genotype. However, *better* neurocognitive functioning was associated with higher plasma concentrations of an EFV metabolite, 8-OH-efavirenz, suggesting that fast metabolizers of EFV (without the *G516T* allele) might be able to avoid EFV neurotoxicity and its neurocognitive complications. There were several limitations to this study, including the small sample size, lack of controls, and only three patients who were homozygous for the *G516T* allele such that the outcomes remain preliminary.

Nevirapine (NVP) is another NNRTI metabolized by CYP2B6 and CYP3A4. It is commonly used in resource-limited settings as a first-choice antiretroviral drug due to its high efficiency/CNS penetrance, at relatively lower cost. There are a few studies investigating the influence of genetic variants such as *CYP2B6 G516T* on pharmacokinetic and patient response to NVP. Like EFV, the *CYP2B6 G516T* allele was associated with higher plasma concentrations of NVP in multiple studies (Heil et al. 2012; Rotger et al. 2005; Penzak et al. 2007; Giacomelli et al. 2018) and has been shown to be neurotoxic in vitro (Robertson et al. 2012; Stauch et al. 2017). However, there is a lack of clinical studies.

Unlike NVP and EFV, protease inhibitors (PI) are mainly metabolized by CYP3A4/CYP3A5 enzymes. Anderson et al. (2006) investigated the pharmacokinetic and pharmacodynamic relationships of two NRTIs (lamivudine-triphosphate, zidovudine-triphosphate) and the highly CNS-penetrant PI indinavir with polymorphisms in multiple candidate genes, including *CYP3A5*. Results showed that CYP3A5 expressors, defined as carriers of at least one *CYP3A5 *1* allele (wild type), showed a 44% faster indinavir oral clearance than non-expressors. At higher concentrations (>10 μmol/L) indinavir has been found to reduce synaptic transmission in the acetylcholine neurotransmitter system through inhibition of α7-nicotinic acetylcholine receptors (Ekins et al. 2017). Based on Anderson et al.'s (2006) findings, non-expressors of the *CYP3A5 *1* allele may be at a higher risk for neurocognitive side effects of indinavir. However, this hypothesis remains to be tested.

Faster clearance of another PI, atazanavir, as a function of *CYP3A5* genotype was seen in CYP3A5 expressors in Savic et al. (2012), whereas Castillo-Mancilla et al. (2016) only observed higher atazanavir metabolite ratios in CYP3A5 expressors compared to non-expressors. However, the clinical relevance of CYP3A5 expressing and non-expressing genotypes on potential atazanavir neurotoxicity are unstudied and appear less likely given the lower CNS penetrance of atazanavir.

6.1 Limitations and Outlook

Apparent from this chapter is that for most genetic biomarker findings remain to be replicated. Differences in the type and severity of neurocognitive changes in HIV infection before and after the introduction of the current ART complicate comparability of findings. Even considering only current cohorts, the HAND phenotypes in the reviewed studies also varied, in part due to diagnostic ambiguities of the current consensus criteria. For example, test batteries and definition of neuropsychological domains vary across cohorts (Bonnet et al. 2013; Dufouil et al. 2015; Wright et al. 2010, 2015; Vassallo et al. 2014; Winston et al. 2013; Simioni et al. 2010; Cysique et al. 2014; Garvey et al. 2011; Sacktor et al. 2016). A HAND diagnosis also requires neurocognitive performance to be classified as "intact" or "impaired," but these categorical labels have been ascertained using different scoring techniques, such as clinical ratings (Carey et al. 2004), domain averaging (Gisslen et al. 2011), or global deficit scores (Carey et al. 2004). Furthermore, although alternative diagnostic criteria exist, they have low agreement with the Frascati criteria (Tierney et al. 2017). Assessment of functional impairment in daily life is needed for HAND staging but is challenging. The most commonly used self-report scales are subjective, were originally developed for other neurological disorders, and hence are not HIV-specific (Clifford and Ances 2013). Even when more quantitative measures are used (e.g., the Columbia Medication Management Test; Heaton et al. 2004), both self-reported and performance-based measures are related to patients' educational and sociocultural background and do not predict progression of HAND and/or neurocognitive performance.

These uncertainties in defining, assessing, and delineating HAND limit the ability to detect robust biomarkers, including the genetic and epigenetic markers discussed in this chapter. To avoid the phenotypic variations of HAND, especially the milder forms, revised classification criteria should develop clearer guidelines on staging and interpreting neurocognitive test performance. Perhaps most importantly, clinically accessible methods to distinguish HAND from neurocognitive impairment related to other factors are needed to either exclude or specify neurocognitive phenotypes due to comorbidities, especially in the context of aging and HIV. Likewise, cost-effective and easy-to-perform techniques to screen for HAND have yet to be established. Within such screening, mandatory inclusion of two neuropsychological tests per domain, focusing on the core neuropsychological deficits shared among cohorts, may avoid overdiagnosis of the milder forms of HAND. The use of one instead of multiple scoring methods (e.g., global deficit scores) would further reduce phenotypic variability. Regardless of the scoring method, categorical classification of neurocognitive deficits requires demographically appropriate normative data from large groups of non-infected controls, which are currently not universally available for all HIV cohorts.

Definition and measurement of HAND may also benefit from the integration of empirical statistical methods to classify neurocognitive impairment (Devlin and Giovannetti 2017). Such tools could be used to arrive at neurocognitive profiles reflective of HAND to help validate and potentially adjust the clinical staging. For example, mixture modeling techniques such as latent profile analysis (LPA) and latent class analysis (LCA) are statistically robust tools (Muthen and Muthen 2000) that have been used in neurological conditions other than in HIV infection to differentiate between multiple neurocognitive subtypes (Flensborg Damholdt et al. 2012; Frndak et al. 2016; McGuinness et al. 2015; Kohler et al. 2013). Assignment of patients to subtypes is guided by rigorous statistical tests, model fit statistics, a probability of correct assignment, and can also be used for longitudinal modeling of change in neurocognitive performance (Nylund et al. 2007). Establishing empirical neurocognitive subtypes of HAND would require multiple, large cohorts but may render more robust phenotypic targets for genetic studies. Lastly, the identification of genetic and epigenetic biomarkers for HAND may also benefit from individual and joint assessment of candidate markers across multiple cohorts, using the same HAND phenotype, to ensure replicability and robustness of outcomes. For these types of approaches, shared public and collaborative global data repositories, including neuropsychological, clinical-demographic, genetic, and epigenetic data, will be needed to maximize statistical power.

References

Agace WW, Amara A, Roberts AI, Pablos JL, Thelen S, Uguccioni M, Li XY, Marsal J, Arenzana-Seisdedos F, Delaunay T, Ebert EC, Moser B, Parker CM (2000) Constitutive expression of stromal derived factor-1 by mucosal epithelia and its role in HIV transmission and propagation. Curr Biol 10(6):325–328. https://doi.org/10.1016/S0960-9822(00)00380-8

Anderson PL, Lamba J, Aquilante CL, Schuetz E, Fletcher CV (2006) Pharmacogenetic characteristics of indinavir, zidovudine, and lamivudine therapy in HIV-infected adults: a pilot study. J Acquir Immune Defic Syndr 42(4):441–449. https://doi.org/10.1097/01.qai.0000225013.53568.69

Andrade AS, Deutsch R, Celano S, Duarte NA, Marcotte TD, Umlauf A, Atkinson JH, McCutchan JA, Franklin D, Alexander TJ, McArthur JC, Marra C, Grant I, Collier AC (2013) Relationships among neurocognitive status, medication adherence measured by pharmacy refill records, and virologic suppression in HIV-infected persons. J Acquir Immune Defic Syndr 62(3):282–292. https://doi.org/10.1097/QAI.0b013e31827ed678

Andres MA, Feger U, Nath A, Munsaka S, Jiang CS, Chang L (2011) APOE epsilon 4 allele and CSF APOE on cognition in HIV-infected subjects. J Neuroimmune Pharmacol 6(3):389–398. https://doi.org/10.1007/s11481-010-9254-3

Antinori A, Arendt G, Becker JT, Brew BJ, Byrd DA, Cherner M, Clifford DB, Cinque P, Epstein LG, Goodkin K, Gisslen M, Grant I, Heaton RK, Joseph J, Marder K, Marra CM, McArthur JC, Nunn M, Price RW, Pulliam L, Robertson KR, Sacktor N, Valcour V, Wojna VE (2007) Updated research nosology for HIV-associated neurocognitive disorders. Neurology 69 (18):1789–1799. https://doi.org/10.1212/01.WNL.0000287431.88658.8b

Arenzana-Seisdedos F, Parmentier M (2006) Genetics of resistance to HIV infection: role of co-receptors and co-receptor ligands. Semin Immunol 18(6):387–403. https://doi.org/10.1016/j.smim.2006.07.007

Asahchop EL, Akinwumi SM, Branton WG, Fujiwara E, Gill MJ, Power C (2016) Plasma microRNA profiling predicts HIV-associated neurocognitive disorder. AIDS 30(13):2021–2031. https://doi.org/10.1097/QAD.0000000000001160

Becker JT, Martinson JJ, Penugonda S, Kingsley L, Molsberry S, Reynolds S, Aronow A, Goodkin K, Levine A, Martin E, Miller EN, Munro CA, Ragin A, Sacktor N (2015) No association between Apoε4 alleles, HIV infection, age, neuropsychological outcome, or death. J Neurovirol 21(1):24–31. https://doi.org/10.1007/s13365-014-0290-2

Berger EA, Murphy PM, Farber JM (1999) Chemokine receptors as HIV-1 coreceptors: roles in viral entry, tropism, and disease. Annu Rev Immunol 17:657–700. https://doi.org/10.1146/annurev.immunol.17.1.657

Betcheva ET, Mushiroda T, Takahashi A, Kubo M, Karachanak SK, Zaharieva IT, Vazharova RV, Dimova II, Milanova VK, Tolev T, Kirov G, Owen MJ, O'Donovan MC, Kamatani N, Nakamura Y, Toncheva DI (2009) Case-control association study of 59 candidate genes reveals the DRD2 SNP rs6277 (C957T) as the only susceptibility factor for schizophrenia in the Bulgarian population. J Hum Genet 54(2):98–107. https://doi.org/10.1038/jhg.2008.14

Bol SM, Booiman T, van Manen D, Bunnik EM, van Sighem AI, Sieberer M, Boeser-Nunnink B, de Wolf F, Schuitemaker H, Portegies P, Kootstra NA, van't Wout AB (2012) Single nucleotide polymorphism in gene encoding transcription factor Prep1 is associated with HIV-1-associated dementia. PLoS One 7(2):e30990. https://doi.org/10.1371/journal.pone.0030990

Bonnet F, Amieva H, Marquant F, Bernard C, Bruyand M, Dauchy FA, Mercie P, Greib C, Richert L, Neau D, Catheline G, Dehail P, Dabis F, Morlat P, Dartigues JF, Chene G, ANRS CO3 Aquitaine Study Group (2013) Cognitive disorders in HIV-infected patients: are they HIV-related? AIDS 27(3):391–400. https://doi.org/10.1097/QAD.0b013e32835b1019

Borghans JA, Molgaard A, de Boer RJ, Kesmir C (2007) HLA alleles associated with slow progression to AIDS truly prefer to present HIV-1 p24. PLoS One 2(9):e920. https://doi.org/10.1371/journal.pone.0000920

Bousman CA, Cherner M, Atkinson JH, Heaton RK, Grant I, Everall IP, The HNRC Group (2010) COMT Val158Met polymorphism, executive dysfunction, and sexual risk behavior in the context of HIV infection and methamphetamine dependence. Interdiscip Perspect Infect Dis 2010:678648. https://doi.org/10.1155/2010/678648

Boven LA, van der Bruggen T, van Asbeck BS, Marx JJ, Nottet HS (1999) Potential role of CCR5 polymorphism in the development of AIDS dementia complex. FEMS Immunol Med Microbiol 26(3–4):243–247. https://doi.org/10.1111/j.1574-695X.1999.tb01395.x

Brabers NA, Nottet HS (2006) Role of the pro-inflammatory cytokines TNF-alpha and IL-1beta in HIV-associated dementia. Eur J Clin Investig 36(7):447–458. https://doi.org/10.1111/j.1365-2362.2006.01657.x

Burt TD, Agan BK, Marconi VC, He W, Kulkarni H, Mold JE, Cavrois M, Huang Y, Mahley RW, Dolan MJ, McCune JM, Ahuja SK (2008) Apolipoprotein (apo) E4 enhances HIV-1 cell entry in vitro, and the APOE epsilon4/epsilon4 genotype accelerates HIV disease progression. Proc Natl Acad Sci U S A 105(25):8718–8723. https://doi.org/10.1073/pnas.0803526105

Carey CL, Woods SP, Gonzalez R, Conover E, Marcotte TD, Grant I, Heaton RK, The HNRC Group (2004) Predictive validity of global deficit scores in detecting neuropsychological impairment in HIV infection. J Clin Exp Neuropsychol 26(3):307–319. https://doi.org/10.1080/13803390490510031

Castillo-Mancilla JR, Aquilante CL, Wempe MF, Smeaton LM, Firnhaber C, LaRosa AM, Kumarasamy N, Andrade A, Baheti G, Fletcher CV, Campbell TB, Haas DW, MaWhinney S, Anderson PL (2016) Pharmacogenetics of unboosted atazanavir in HIV-infected individuals in resource-limited settings: a sub-study of the AIDS Clinical Trials Group (ACTG) PEARLS study (NWCS 342). J Antimicrob Chemother 71(6):1609–1618. https://doi.org/10.1093/jac/dkw005

Cawthon RM, Smith KR, O'Brien E, Sivatchenko A, Kerber RA (2003) Association between telomere length in blood and mortality in people aged 60 years or older. Lancet 361 (9355):393–395. https://doi.org/10.1016/S0140-6736(03)12384-7

Chang L, Wang GJ, Volkow ND, Ernst T, Telang F, Logan J, Fowler JS (2008) Decreased brain dopamine transporters are related to cognitive deficits in HIV patients with or without cocaine abuse. NeuroImage 42(2):869–878. https://doi.org/10.1016/j.neuroimage.2008.05.011

Chang L, Andres M, Sadino J, Jiang CS, Nakama H, Miller E, Ernst T (2011) Impact of apolipoprotein E epsilon4 and HIV on cognition and brain atrophy: antagonistic pleiotropy and premature brain aging. NeuroImage 58(4):1017–1027. https://doi.org/10.1016/j.neuroimage.2011.07.010

Chang L, Jiang C, Cunningham E, Buchthal S, Douet V, Andres M, Ernst T (2014) Effects of APOE epsilon4, age, and HIV on glial metabolites and cognitive deficits. Neurology 82 (24):2213–2222. https://doi.org/10.1212/WNL.0000000000000526

Chen J, Lipska BK, Halim N, Ma QD, Matsumoto M, Melhem S, Kolachana BS, Hyde TM, Herman MM, Apud J, Egan MF, Kleinman JE, Weinberger DR (2004) Functional analysis of genetic variation in catechol-O-methyltransferase (COMT): effects on mRNA, protein, and enzyme activity in postmortem human brain. Am J Hum Genet 75(5):807–821. https://doi.org/10.1086/425589

Chinta SJ, Andersen JK (2005) Dopaminergic neurons. Int J Biochem Cell Biol 37(5):942–946. https://doi.org/10.1016/j.biocel.2004.09.009

Ciccarelli N, Fabbiani M, di Giambenedetto S, Fanti I, Baldonero E, Bracciale L, Tamburrini E, Cauda R, de Luca A, Silveri MC (2011) Efavirenz associated with cognitive disorders in otherwise asymptomatic HIV-infected patients. Neurology 76(16):1403–1409. https://doi.org/10.1212/WNL.0b013e31821670fb

Cirulli ET, Kasperaviciute D, Attix DK, Need AC, Ge D, Gibson G, Goldstein DB (2010) Common genetic variation and performance on standardized cognitive tests. Eur J Hum Genet 18 (7):815–820. https://doi.org/10.1038/ejhg.2010.2

Clifford DB, Ances BM (2013) HIV-associated neurocognitive disorder. Lancet Infect Dis 13 (11):976–986. https://doi.org/10.1016/S1473-3099(13)70269-X

Clifford DB, Evans S, Yang Y, Acosta EP, Goodkin K, Tashima K, Simpson D, Dorfman D, Ribaudo H, Gulick RM (2005) Impact of efavirenz on neuropsychological performance and symptoms in HIV-infected individuals. Ann Intern Med 143(10):714–721. https://doi.org/10.7326/0003-4819-143-10-200511150-00008

Collier AC, Coombs RW, Schoenfeld DA, Bassett RL, Timpone J, Baruch A, Jones M, Facey K, Whitacre C, McAuliffe VJ, Friedman HM, Merigan TC, Reichman RC, Hooper C, Corey L (1996) Treatment of human immunodeficiency virus infection with saquinavir, zidovudine, and zalcitabine. AIDS Clinical Trials Group. N Engl J Med 334(16):1011–1017. https://doi.org/10.1056/NEJM199604183341602

Conant K, Garzino-Demo A, Nath A, McArthur JC, Halliday W, Power C, Gallo RC, Major EO (1998) Induction of monocyte chemoattractant protein-1 in HIV-1 Tat-stimulated astrocytes and elevation in AIDS dementia. Proc Natl Acad Sci U S A 95(6):3117–3121. https://doi.org/10.1073/pnas.95.6.3117

Corder EH, Robertson K, Lannfelt L, Bogdanovic N, Eggertsen G, Wilkins J, Hall C (1998) HIV-infected subjects with the E4 allele for APOE have excess dementia and peripheral neuropathy. Nat Med 4(10):1182–1184. https://doi.org/10.1038/2677

Corley MJ, Dye C, D'Antoni ML, Byron MM, Yo KL, Lum-Jones A, Nakamoto B, Valcour V, SahBandar I, Shikuma CM, Ndhlovu LC, Maunakea AK (2016) Comparative DNA methylation profiling reveals an immunoepigenetic signature of HIV-related cognitive impairment. Sci Rep 6:33310. https://doi.org/10.1038/srep33310

Cutler RG, Haughey NJ, Tammara A, McArthur JC, Nath A, Reid R, Vargas DL, Pardo CA, Mattson MP (2004) Dysregulation of sphingolipid and sterol metabolism by ApoE4 in HIV dementia. Neurology 63(4):626–630. https://doi.org/10.1212/01.WNL.0000134662.19883.06

Cysique LA, Heaton RK, Kamminga J, Lane T, Gates TM, Moore DM, Hubner E, Carr A, Brew BJ (2014) HIV-associated neurocognitive disorder in Australia: a case of a high-functioning and optimally treated cohort and implications for international neuroHIV research. J Neurovirol 20 (3):258–268. https://doi.org/10.1007/s13365-014-0242-x

Cysique LA, Hewitt T, Croitoru-Lamoury J, Taddei K, Martins RN, Chew CS, Davies NN, Price P, Brew BJ (2015) APOE epsilon4 moderates abnormal CSF-abeta-42 levels, while neurocognitive impairment is associated with abnormal CSF tau levels in HIV+ individuals – a cross-sectional observational study. BMC Neurol 15:51. https://doi.org/10.1186/s12883-015-0298-0

D'Aquila RT, Hughes MD, Johnson VA, Fischl MA, Sommadossi JP, Liou SH, Timpone J, Myers M, Basgoz N, Niu M, Hirsch MS (1996) Nevirapine, zidovudine, and didanosine compared with zidovudine and didanosine in patients with HIV-1 infection. A randomized, double-blind, placebo-controlled trial. National Institute of Allergy and Infectious Diseases AIDS Clinical Trials Group Protocol 241 Investigators. Ann Intern Med 124(12):1019–1030. https://doi.org/10.7326/0003-4819-124-12-199606150-00001

Davis LE, Hjelle BL, Miller VE, Palmer DL, Llewellyn AL, Merlin TL, Young SA, Mills RG, Wachsman W, Wiley CA (1992) Early viral brain invasion in iatrogenic human immunodeficiency virus infection. Neurology 42(9):1736–1739. https://doi.org/10.1212/WNL.42.9.1736

Devlin KN, Giovannetti T (2017) Heterogeneity of neuropsychological impairment in HIV infection: contributions from mild cognitive impairment. Neuropsychol Rev 27(2):101–123. https://doi.org/10.1007/s11065-017-9348-2

Dhillon NK, Williams R, Callen S, Zien C, Narayan O, Buch S (2008) Roles of MCP-1 in development of HIV-dementia. Front Biosci 13:3913–3918. https://doi.org/10.2741/2979

Diaz-Arrastia R, Gong Y, Kelly CJ, Gelman BB (2004) Host genetic polymorphisms in human immunodeficiency virus-related neurologic disease. J Neurovirol 10(1):67–73. https://doi.org/10.1080/753312755

Ding J, Zhao J, Zhou J, Li X, Wu Y, Ge M, Cen S (2018) Association of gene polymorphism of SDF1(CXCR12) with susceptibility to HIV-1 infection and AIDS disease progression: a meta-analysis. PLoS One 13(2):e0191930. https://doi.org/10.1371/journal.pone.0191930

Dinh KM, Pedersen OB, Petersen MS, Sorensen E, Sorensen CJ, Kaspersen KA, Larsen MH, Moller B, Hjalgrim H, Ullum H, Erikstrup C (2015) The impact of CCR5-Delta32 deletion on C-reactive protein levels and cardiovascular disease: results from the Danish Blood Donor Study. Atherosclerosis 242(1):222–225. https://doi.org/10.1016/j.atherosclerosis.2015.07.031

Doyle KL, Woods SP, Morgan EE, Iudicello JE, Cameron MV, Gilbert PE, Beltran J, HIV Neurobehavioral Research Program Group (2016) Health-related decision-making in HIV disease. J Clin Psychol Med Settings 23(2):135–146. https://doi.org/10.1007/s10880-016-9455-x

Duan J, Wainwright MS, Comeron JM, Saitou N, Sanders AR, Gelernter J, Gejman PV (2003) Synonymous mutations in the human dopamine receptor D2 (DRD2) affect mRNA stability and synthesis of the receptor. Hum Mol Genet 12(3):205–216. https://doi.org/10.1093/hmg/ddg055

Dufouil C, Richert L, Thiebaut R, Bruyand M, Amieva H, Dauchy FA, Dartigues JF, Neau D, Morlat P, Dehail P, Dabis F, Bonnet F, Chene G, ANRS CO3 Aquitaine Study Group (2015) Diabetes and cognitive decline in a French cohort of patients infected with HIV-1. Neurology 85 (12):1065–1073. https://doi.org/10.1212/WNL.0000000000001815

Ekins S, Mathews P, Saito EK, Diaz N, Naylor D, Chung J, McMurtray AM (2017) alpha7-Nicotinic acetylcholine receptor inhibition by indinavir: implications for cognitive dysfunction in treated HIV disease. AIDS 31(8):1083–1089. https://doi.org/10.1097/QAD.0000000000001488

Elbirt D, Mahlab-Guri K, Bezalel-Rosenberg S, Gill H, Attali M, Asher I (2015) HIV-associated neurocognitive disorders (HAND). Isr Med Assoc J 17(1):54–59

Eletto D, Russo G, Passiatore G, del Valle L, Giordano A, Khalili K, Gualco E, Peruzzi F (2008) Inhibition of SNAP25 expression by HIV-1 Tat involves the activity of mir-128a. J Cell Physiol 216(3):764–770. https://doi.org/10.1002/jcp.21452

Ferrer I, Aubourg P, Pujol A (2010) General aspects and neuropathology of X-linked adrenoleukodystrophy. Brain Pathol 20(4):817–830. https://doi.org/10.1111/j.1750-3639.2010.00390.x

Flensborg Damholdt M, Shevlin M, Borghammer P, Larsen L, Ostergaard K (2012) Clinical heterogeneity in Parkinson's disease revisited: a latent profile analysis. Acta Neurol Scand 125(5):311–318. https://doi.org/10.1111/j.1600-0404.2011.01561.x

Frndak SE, Smerbeck AM, Irwin LN, Drake AS, Kordovski VM, Kunker KA, Khan AL, Benedict RH (2016) Latent profile analysis of regression-based norms demonstrates relationship of compounding MS symptom burden and negative work events. Clin Neuropsychol 30 (7):1050–1062. https://doi.org/10.1080/13854046.2016.1200144

Gade-Andavolu R, Comings DE, MacMurray J, Rostamkhani M, Cheng LS, Tourtellotte WW, Cone LA (2004) Association of CCR5 delta32 deletion with early death in multiple sclerosis. Genet Med 6(3):126–131 doi:10.109701.GIM.0000127274.45301.54

Garred P (2008) Mannose-binding lectin genetics: from A to Z. Biochem Soc Trans 36:1461–1466. https://doi.org/10.1042/BST0361461

Garvey L, Surendrakumar V, Winston A (2011) Low rates of neurocognitive impairment are observed in neuro-asymptomatic HIV-infected subjects on effective antiretroviral therapy. HIV Clin Trials 12(6):333–338. https://doi.org/10.1310/hct1206-333

Gaskill PJ, Calderon TM, Luers AJ, Eugenin EA, Javitch JA, Berman JW (2009) Human immunodeficiency virus (HIV) infection of human macrophages is increased by dopamine: a bridge between HIV-associated neurologic disorders and drug abuse. Am J Pathol 175(3):1148–1159. https://doi.org/10.2353/ajpath.2009.081067

Gaskill PJ, Calderon TM, Coley JS, Berman JW (2013) Drug induced increases in CNS dopamine alter monocyte, macrophage and T cell functions: implications for HAND. J Neuroimmune Pharmacol 8(3):621–642. https://doi.org/10.1007/s11481-013-9443-y

Gelman BB (2015) Neuropathology of HAND with suppressive antiretroviral therapy: encephalitis and neurodegeneration reconsidered. Curr HIV/AIDS Rep 12(2):272–279. https://doi.org/10.1007/s11904-015-0266-8

Gelman BB, Lisinicchia JG, Chen T, Johnson KM, Jennings K, Freeman DH Jr, Soukup VM (2012) Prefrontal dopaminergic and enkephalinergic synaptic accommodation in HIV-associated neurocognitive disorders and encephalitis. J Neuroimmune Pharmacol 7(3):686–700. https://doi.org/10.1007/s11481-012-9345-4

Giacomelli A, Rusconi S, Falvella FS, Oreni ML, Cattaneo D, Cozzi V, Renisi G, Monge E, Cheli S, Clementi E, Riva A, Galli M, Ridolfo AL (2018) Clinical and genetic determinants of nevirapine plasma trough concentration. SAGE Open Med 6:2050312118780861. https://doi.org/10.1177/2050312118780861

Giesbrecht CJ, Thornton AE, Hall-Patch C, Maan EJ, Cote HC, Money DM, Murray M, Pick N (2014) Select neurocognitive impairment in HIV-infected women: associations with HIV viral load, hepatitis C virus, and depression, but not leukocyte telomere length. PLoS One 9(3): e89556. https://doi.org/10.1371/journal.pone.0089556

Gisslen M, Krut J, Andreasson U, Blennow K, Cinque P, Brew BJ, Spudich S, Hagberg L, Rosengren L, Price RW, Zetterberg H (2009) Amyloid and tau cerebrospinal fluid biomarkers in HIV infection. BMC Neurol 9:63. https://doi.org/10.1186/1471-2377-9-63

Gisslen M, Price RW, Nilsson S (2011) The definition of HIV-associated neurocognitive disorders: are we overestimating the real prevalence? BMC Infect Dis 11:356. https://doi.org/10.1186/1471-2334-11-356

Gong W, Howard OM, Turpin JA, Grimm MC, Ueda H, Gray PW, Raport CJ, Oppenheim JJ, Wang JM (1998) Monocyte chemotactic protein-2 activates CCR5 and blocks CD4/CCR5-mediated HIV-1 entry/replication. J Biol Chem 273(8):4289–4292. https://doi.org/10.1074/jbc.273.8.4289

Gonzalez E, Bamshad M, Sato N, Mummidi S, Dhanda R, Catano G, Cabrera S, McBride M, Cao XH, Merrill G, O'Connell P, Bowden DW, Freedman BI, Anderson SA, Walter EA, Evans JS, Stephan KT, Clark RA, Tyagi S, Ahuja SS, Dolan MJ, Ahuja SK (1999) Race-specific HIV-1 disease-modifying effects associated with CCR5 haplotypes. Proc Natl Acad Sci U S A 96 (21):12004–12009. https://doi.org/10.1073/pnas.96.21.12004

Gonzalez E, Rovin BH, Sen L, Cooke G, Dhanda R, Mummidi S, Kulkarni H, Bamshad MJ, Telles V, Anderson SA, Walter EA, Stephan KT, Deucher M, Mangano A, Bologna R, Ahuja SS, Dolan MJ, Ahuja SK (2002) HIV-1 infection and AIDS dementia are influenced by a mutant MCP-1 allele linked to increased monocyte infiltration of tissues and MCP-1 levels. Proc Natl Acad Sci U S A 99(21):13795–13800. https://doi.org/10.1073/pnas.202357499

Gonzalez E, Kulkarni H, Bolivar H, Mangano A, Sanchez R, Catano G, Nibbs RJ, Freedman BI, Quinones MP, Bamshad MJ, Murthy KK, Rovin BH, Bradley W, Clark RA, Anderson SA, O'Connell RJ, Agan BK, Ahuja SS, Bologna R, Sen L, Dolan MJ, Ahuja SK (2005) The influence of CCL3L1 gene-containing segmental duplications on HIV-1/AIDS susceptibility. Science 307(5714):1434–1440. https://doi.org/10.1126/science.1101160

Goulder PJ, Watkins DI (2008) Impact of MHC class I diversity on immune control of immunodeficiency virus replication. Nat Rev Immunol 8(8):619–630. https://doi.org/10.1038/nri2357

Gounden V, van Niekerk C, Snyman T, George JA (2010) Presence of the CYP2B6 516G> T polymorphism, increased plasma Efavirenz concentrations and early neuropsychiatric side effects in South African HIV-infected patients. AIDS Res Ther 7:32. https://doi.org/10.1186/1742-6405-7-32

Graff J, Tsai LH (2013) The potential of HDAC inhibitors as cognitive enhancers. Annu Rev Pharmacol Toxicol 53:311–330. https://doi.org/10.1146/annurev-pharmtox-011112-140216

Guo H, Ingolia NT, Weissman JS, Bartel DP (2010) Mammalian microRNAs predominantly act to decrease target mRNA levels. Nature 466(7308):835–840. https://doi.org/10.1038/nature09267

Guo L, Zhang Q, Ma X, Wang J, Liang T (2017) miRNA and mRNA expression analysis reveals potential sex-biased miRNA expression. Sci Rep 7:39812. https://doi.org/10.1038/srep39812

Gupta S, Bousman CA, Chana G, Cherner M, Heaton RK, Deutsch R, Ellis RJ, Grant I, Everall IP (2011) Dopamine receptor D3 genetic polymorphism (rs6280TC) is associated with rates of cognitive impairment in methamphetamine-dependent men with HIV: preliminary findings. J Neurovirol 17(3):239–247. https://doi.org/10.1007/s13365-011-0028-3

Haas DW, Ribaudo HJ, Kim RB, Tierney C, Wilkinson GR, Gulick RM, Clifford DB, Hulgan T, Marzolini C, Acosta EP (2004) Pharmacogenetics of efavirenz and central nervous system side effects: an Adult AIDS Clinical Trials Group study. AIDS 18(18):2391–2400

Han LKM, Aghajani M, Clark SL, Chan RF, Hattab MW, Shabalin AA, Zhao M, Kumar G, Xie LY, Jansen R, Milaneschi Y, Dean B, Aberg KA, van den Oord E, Penninx B (2018) Epigenetic aging in major depressive disorder. Am J Psychiatry 175:774. https://doi.org/10.1176/appi.ajp.2018.17060595

Heaton RK, Marcotte TD, Mindt MR, Sadek J, Moore DJ, Bentley H, McCutchan JA, Reicks C, Grant I, HNRC Group (2004) The impact of HIV-associated neuropsychological impairment on everyday functioning. J Int Neuropsychol Soc 10(3):317–331. https://doi.org/10.1017/S1355617704102130

Heaton RK, Clifford DB, Franklin DR Jr, Woods SP, Ake C, Vaida F, Ellis RJ, Letendre SL, Marcotte TD, Atkinson JH, Rivera-Mindt M, Vigil OR, Taylor MJ, Collier AC, Marra CM, Gelman BB, McArthur JC, Morgello S, Simpson DM, McCutchan JA, Abramson I, Gamst A, Fennema-Notestine C, Jernigan TL, Wong J, Grant I, CHARTER Group (2010) HIV-associated neurocognitive disorders persist in the era of potent antiretroviral therapy: CHARTER Study. Neurology 75(23):2087–2096. https://doi.org/10.1212/WNL.0b013e318200d727

Heaton RK, Franklin DR, Ellis RJ, McCutchan JA, Letendre SL, Leblanc S, Corkran SH, Duarte NA, Clifford DB, Woods SP, Collier AC, Marra CM, Morgello S, Mindt MR, Taylor MJ, Marcotte TD, Atkinson JH, Wolfson T, Gelman BB, McArthur JC, Simpson DM, Abramson I, Gamst A, Fennema-Notestine C, Jernigan TL, Wong J, Grant I, CHARTER Group, HNRC Group (2011) HIV-associated neurocognitive disorders before and during the era of combination antiretroviral therapy: differences in rates, nature, and predictors. J Neurovirol 17(1):3–16. https://doi.org/10.1007/s13365-010-0006-1

Heil SG, van der Ende ME, Schenk PW, van der Heiden I, Lindemans J, Burger D, van Schaik RH (2012) Associations between ABCB1, CYP2A6, CYP2B6, CYP2D6, and CYP3A5 alleles in relation to efavirenz and nevirapine pharmacokinetics in HIV-infected individuals. Ther Drug Monit 34(2):153–159. https://doi.org/10.1097/FTD.0b013e31824868f3

Hong S, Banks WA (2015) Role of the immune system in HIV-associated neuroinflammation and neurocognitive implications. Brain Behav Immun 45:1–12. https://doi.org/10.1016/j.bbi.2014.10.008

Horvath S (2013) DNA methylation age of human tissues and cell types. Genome Biol 14(10): R115. https://doi.org/10.1186/gb-2013-14-10-r115

Horvath S, Levine AJ (2015) HIV-1 infection accelerates age according to the epigenetic clock. J Infect Dis 212(10):1563–1573. https://doi.org/10.1093/infdis/jiv277

Horvath S, Phillips N, Heany SJ, Kobor MS, Lin DT, Myer L, Zar HJ, Stein DJ, Levine AJ, Hoare J (2018) Perinatally acquired HIV infection accelerates epigenetic aging in South African adolescents. AIDS 32:1465. https://doi.org/10.1097/QAD.0000000000001854

Huang X, Ling H, Feng L, Ding X, Zhou Q, Han M, Mao W, Xiong H (2009) Human leukocyte antigen profile in HIV-1 infected individuals and AIDS patients from Chongqing, China. Microbiol Immunol 53(9):512–523. https://doi.org/10.1111/j.1348-0421.2009.00150.x

Im HI, Kenny PJ (2012) MicroRNAs in neuronal function and dysfunction. Trends Neurosci 35 (5):325–334. https://doi.org/10.1016/j.tins.2012.01.004

Ioannidis JP, Rosenberg PS, Goedert JJ, Ashton LJ, Benfield TL, Buchbinder SP, Coutinho RA, Eugen-Olsen J, Gallart T, Katzenstein TL, Kostrikis LG, Kuipers H, Louie LG, Mallal SA, Margolick JB, Martinez OP, Meyer L, Michael NL, Operskalski E, Pantaleo G, Rizzardi GP, Schuitemaker H, Sheppard HW, Stewart GJ, Theodorou ID, Ullum H, Vicenzi E, Vlahov D, Wilkinson D, Workman C, Zagury JF, O'Brien TR, International Meta-Analysis of HIV Host Genetics (2001) Effects of CCR5-Delta32, CCR2-64I, and SDF-1 3′A alleles on HIV-1 disease progression: an international meta-analysis of individual-patient data. Ann Intern Med 135 (9):782–795. https://doi.org/10.7326/0003-4819-135-9-200111060-00008

Janssen RS, Cornblath DR, Epstein LG, Foa RP, McArthur JC, Price RW, Asbury AK, Beckett A, Benson DF, Bridge TP, Leventhal CM, Satz P, Saykin AJ, Sidtis JJ, Tross S (1991) Nomenclature and research case definitions for neurologic manifestations of human immunodeficiency virus-type 1 (HIV-1) infection. Report of a working group of the American Academy of Neurology AIDS Task Force. Neurology 41(6):778–785

Ji X, Gewurz H, Spear GT (2005) Mannose binding lectin (MBL) and HIV. Mol Immunol 42 (2):145–152. https://doi.org/10.1016/j.molimm.2004.06.015

Jia P, Zhao Z, Hulgan T, Bush WS, Samuels DC, Bloss CS, Heaton RK, Ellis RJ, Schork N, Marra CM, Collier AC, Clifford DB, Gelman BB, Sacktor N, Morgello S, Simpson DM, McCutchan JA, Barnholtz-Sloan JS, Franklin DR, Rosario D, Letendre SL, Grant I, Kallianpur AR, CHARTER Group (2017) Genome-wide association study of HIV-associated neurocognitive disorder (HAND): a CHARTER group study. Am J Med Genet B Neuropsychiatr Genet 174 (4):413–426. https://doi.org/10.1002/ajmg.b.32530

Joska JA, Combrinck M, Valcour VG, Hoare J, Leisegang F, Mahne AC, Myer L, Stein DJ (2010) Association between apolipoprotein E4 genotype and human immunodeficiency virus-associated dementia in younger adults starting antiretroviral therapy in South Africa. J Neurovirol 16 (5):377–383. https://doi.org/10.3109/13550284.2010.513365

Kadri F, LaPlante A, de Luca M, Doyle L, Velasco-Gonzalez C, Patterson JR, Molina PE, Nelson S, Zea AH, Parsons CH, Peruzzi F (2016) Defining plasma microRNAs associated with cognitive impairment in HIV-infected patients. J Cell Physiol 231(4):829–836. https://doi.org/10.1002/jcp.25131

Kazantsev AG, Thompson LM (2008) Therapeutic application of histone deacetylase inhibitors for central nervous system disorders. Nat Rev Drug Discov 7(10):854–868. https://doi.org/10.1038/nrd2681

Kenedi CA, Goforth HW (2011) A systematic review of the psychiatric side-effects of efavirenz. AIDS Behav 15(8):1803–1818. https://doi.org/10.1007/s10461-011-9939-5

Kim J, Basak JM, Holtzman DM (2009) The role of apolipoprotein E in Alzheimer's disease. Neuron 63(3):287–303. https://doi.org/10.1016/j.neuron.2009.06.026

Kohler S, Hamel R, Sistermans N, Koene T, Pijnenburg YA, van der Flier WM, Scheltens P, Visser PJ, Aalten P, Verhey FR, Ramakers I (2013) Progression to dementia in memory clinic patients without dementia: a latent profile analysis. Neurology 81(15):1342–1349. https://doi.org/10.1212/WNL.0b013e3182a82536

Kumar AM, Fernandez JB, Singer EJ, Commins D, Waldrop-Valverde D, Ownby RL, Kumar M (2009) Human immunodeficiency virus type 1 in the central nervous system leads to decreased dopamine in different regions of postmortem human brains. J Neurovirol 15(3):257–274. https://doi.org/10.1080/13550280902973952

Kumar AM, Ownby RL, Waldrop-Valverde D, Fernandez B, Kumar M (2011) Human immunodeficiency virus infection in the CNS and decreased dopamine availability: relationship with neuropsychological performance. J Neurovirol 17(1):26–40. https://doi.org/10.1007/s13365-010-0003-4

Landgraf P, Rusu M, Sheridan R, Sewer A, Iovino N, Aravin A, Pfeffer S, Rice A, Kamphorst AO, Landthaler M, Lin C, Socci ND, Hermida L, Fulci V, Chiaretti S, Foa R, Schliwka J, Fuchs U, Novosel A, Muller RU, Schermer B, Bissels U, Inman J, Phan Q, Chien M, Weir DB, Choksi R, de Vita G, Frezzetti D, Trompeter HI, Hornung V, Teng G, Hartmann G, Palkovits M, di Lauro R, Wernet P, Macino G, Rogler CE, Nagle JW, Ju J, Papavasiliou FN, Benzing T, Lichter P, Tam W, Brownstein MJ, Bosio A, Borkhardt A, Russo JJ, Sander C, Zavolan M, Tuschl T (2007) A mammalian microRNA expression atlas based on small RNA library sequencing. Cell 129(7):1401–1414. https://doi.org/10.1016/j.cell.2007.04.040

Lehmann SM, Kruger C, Park B, Derkow K, Rosenberger K, Baumgart J, Trimbuch T, Eom G, Hinz M, Kaul D, Habbel P, Kalin R, Franzoni E, Rybak A, Nguyen D, Veh R, Ninnemann O, Peters O, Nitsch R, Heppner FL, Golenbock D, Schott E, Ploegh HL, Wulczyn FG, Lehnardt S (2012) An unconventional role for miRNA: let-7 activates Toll-like receptor 7 and causes neurodegeneration. Nat Neurosci 15(6):827–835. https://doi.org/10.1038/nn.3113

Levine AJ, Singer EJ, Sinsheimer JS, Hinkin CH, Papp J, Dandekar S, Giovanelli A, Shapshak P (2009) CCL3 genotype and current depression increase risk of HIV-associated dementia. Neurobehav HIV Med 1:1–7. https://doi.org/10.2147/NBHIV.S6820

Levine AJ, Service S, Miller EN, Reynolds SM, Singer EJ, Shapshak P, Martin EM, Sacktor N, Becker JT, Jacobson LP, Thompson P, Freimer N (2012) Genome-wide association study of neurocognitive impairment and dementia in HIV-infected adults. Am J Med Genet B Neuropsychiatr Genet 159B(6):669–683. https://doi.org/10.1002/ajmg.b.32071

Levine AJ, Reynolds S, Cox C, Miller EN, Sinsheimer JS, Becker JT, Martin E, Sacktor N, Neuropsychology Working Group of the Multicenter AIDS Cohort Study (2014) The longitudinal and interactive effects of HIV status, stimulant use, and host genotype upon neurocognitive functioning. J Neurovirol 20(3):243–257. https://doi.org/10.1007/s13365-014-0241-y

Levine AJ, Quach A, Moore DJ, Achim CL, Soontornniyomkij V, Masliah E, Singer EJ, Gelman B, Nemanim N, Horvath S (2016) Accelerated epigenetic aging in brain is associated with pre-mortem HIV-associated neurocognitive disorders. J Neurovirol 22(3):366–375. https://doi.org/10.1007/s13365-015-0406-3

Lewis DA, Melchitzky DS, Sesack SR, Whitehead RE, Auh S, Sampson A (2001) Dopamine transporter immunoreactivity in monkey cerebral cortex: regional, laminar, and ultrastructural localization. J Comp Neurol 432(1):119–136. https://doi.org/10.1002/cne.1092

Li S, Jiao H, Yu X, Strong AJ, Shao Y, Sun Y, Altfeld M, Lu Y (2007) Human leukocyte antigen class I and class II allele frequencies and HIV-1 infection associations in a Chinese cohort. J Acquir Immune Defic Syndr 44(2):121–131. https://doi.org/10.1097/01.qai.0000248355.40877.2a

Liu R, Paxton WA, Choe S, Ceradini D, Martin SR, Horuk R, MacDonald ME, Stuhlmann H, Koup RA, Landau NR (1996) Homozygous defect in HIV-1 coreceptor accounts for resistance of some multiply-exposed individuals to HIV-1 infection. Cell 86(3):367–377. https://doi.org/10.1016/S0092-8674(00)80110-5

Liu S, Yao L, Ding D, Zhu H (2010) CCL3L1 copy number variation and susceptibility to HIV-1 infection: a meta-analysis. PLoS One 5(12):e15778. https://doi.org/10.1371/journal.pone.0015778

Liu CC, Liu CC, Kanekiyo T, Xu H, Bu G (2013) Apolipoprotein E and Alzheimer disease: risk, mechanisms and therapy. Nat Rev Neurol 9(2):106–118. https://doi.org/10.1038/nrneurol.2012.263

Ma Q, Vaida F, Wong J, Sanders CA, Kao YT, Croteau D, Clifford DB, Collier AC, Gelman BB, Marra CM, McArthur JC, Morgello S, Simpson DM, Heaton RK, Grant I, Letendre SL, CHARTER Group (2016) Long-term efavirenz use is associated with worse neurocognitive functioning in HIV-infected patients. J Neurovirol 22(2):170–178. https://doi.org/10.1007/s13365-015-0382-7

Malan-Muller S, Hemmings SM, Spies G, Kidd M, Fennema-Notestine C, Seedat S (2013) Shorter telomere length – a potential susceptibility factor for HIV-associated neurocognitive impairments in South African women. PLoS One 8(3):e58351. https://doi.org/10.1371/journal.pone.0058351

Martin AM, Nolan D, Gaudieri S, Almeida CA, Nolan R, James I, Carvalho F, Phillips E, Christiansen FT, Purcell AW, McCluskey J, Mallal S (2004) Predisposition to abacavir hypersensitivity conferred by HLA-B*5701 and a haplotypic Hsp70-Hom variant. Proc Natl Acad Sci U S A 101(12):4180–4185. https://doi.org/10.1073/pnas.0307067101

Martin-Ruiz C, Dickinson HO, Keys B, Rowan E, Kenny RA, von Zglinicki T (2006) Telomere length predicts poststroke mortality, dementia, and cognitive decline. Ann Neurol 60(2):174–180. https://doi.org/10.1002/ana.20869

McGuinness B, Barrett SL, McIlvenna J, Passmore AP, Shorter GW (2015) Predicting conversion to dementia in a memory clinic: a standard clinical approach compared with an empirically defined clustering method (latent profile analysis) for mild cognitive impairment subtyping. Alzheimers Dement (Amst) 1(4):447–454. https://doi.org/10.1016/j.dadm.2015.10.003

Mellios N, Huang HS, Grigorenko A, Rogaev E, Akbarian S (2008) A set of differentially expressed miRNAs, including miR-30a-5p, act as post-transcriptional inhibitors of BDNF in prefrontal cortex. Hum Mol Genet 17(19):3030–3042. https://doi.org/10.1093/hmg/ddn201

Meyer L, Magierowska M, Hubert JB, Theodorou I, van Rij R, Prins M, de Roda Husman AM, Coutinho R, Schuitemaker H (1999) CC-chemokine receptor variants, SDF-1 polymorphism, and disease progression in 720 HIV-infected patients. SEROCO Cohort. Amsterdam Cohort Studies on AIDS. AIDS 13(5):624–626. https://doi.org/10.1097/00002030-199904010-00015

Modi WS, Scott K, Goedert JJ, Vlahov D, Buchbinder S, Detels R, Donfield S, O'Brien SJ, Winkler C (2005) Haplotype analysis of the SDF-1 (CXCL12) gene in a longitudinal HIV-1/AIDS cohort study. Genes Immun 6(8):691–698. https://doi.org/10.1038/sj.gene.6364258

Modi WS, Lautenberger J, An P, Scott K, Goedert JJ, Kirk GD, Buchbinder S, Phair J, Donfield S, O'Brien SJ, Winkler C (2006) Genetic variation in the CCL18-CCL3-CCL4 chemokine gene cluster influences HIV Type 1 transmission and AIDS disease progression. Am J Hum Genet 79(1):120–128. https://doi.org/10.1086/505331

Morgan EE, Woods SP, Letendre SL, Franklin DR, Bloss C, Goate A, Heaton RK, Collier AC, Marra CM, Gelman BB, McArthur JC, Morgello S, Simpson DM, McCutchan JA, Ellis RJ, Abramson I, Gamst A, Fennema-Notestine C, Smith DM, Grant I, Vaida F, Clifford DB, Group C (2013) Apolipoprotein E4 genotype does not increase risk of HIV-associated neurocognitive disorders. J Neurovirol 19(2):150–156. https://doi.org/10.1007/s13365-013-0152-3

Mukerji SS, Locascio JJ, Misra V, Lorenz DR, Holman A, Dutta A, Penugonda S, Wolinsky SM, Gabuzda D (2016) Lipid profiles and APOE4 allele impact midlife cognitive decline in HIV-infected men on antiretroviral therapy. Clin Infect Dis 63(8):1130–1139. https://doi.org/10.1093/cid/ciw495

Muthen B, Muthen LK (2000) Integrating person-centered and variable-centered analyses: growth mixture modeling with latent trajectory classes. Alcohol Clin Exp Res 24(6):882–891. https://doi.org/10.1111/j.1530-0277.2000.tb02070.x

Navia BA, Jordan BD, Price RW (1986) The AIDS dementia complex: I. Clinical features. Ann Neurol 19(6):517–524. https://doi.org/10.1002/ana.410190602

Nightingale S, Winston A, Letendre S, Michael BD, McArthur JC, Khoo S, Solomon T (2014) Controversies in HIV-associated neurocognitive disorders. Lancet Neurol 13(11):1139–1151. https://doi.org/10.1016/S1474-4422(14)70137-1

Noorbakhsh F, Ramachandran R, Barsby N, Ellestad KK, LeBlanc A, Dickie P, Baker G, Hollenberg MD, Cohen EA, Power C (2010) MicroRNA profiling reveals new aspects of HIV neurodegeneration: caspase-6 regulates astrocyte survival. FASEB J 24(6):1799–1812. https://doi.org/10.1096/fj.09-147819

Nylund KL, Asparouhov T, Muthen B (2007) Decising on the number of classes in latent class analysis and growth mixture modeling. A Monte Carlo simulation study. Struct Equ Model 14:535–569

Oak JN, Oldenhof J, van Tol HH (2000) The dopamine D(4) receptor: one decade of research. Eur J Pharmacol 405(1–3):303–327. https://doi.org/10.1016/S0014-2999(00)00562-8

Panos SE, Hinkin CH, Singer EJ, Thames AD, Patel SM, Sinsheimer JS, del Re AC, Gelman BB, Morgello S, Moore DJ, Levine AJ (2013) Apolipoprotein-E genotype and human immunodeficiency virus-associated neurocognitive disorder: the modulating effects of older age and disease severity. Neurobehav HIV Med 5:11–22. https://doi.org/10.2147/NBHIV.S39573

Paternico D, Galluzzi S, Drago V, Bocchio-Chiavetto L, Zanardini R, Pedrini L, Baronio M, Amicucci G, Frisoni GB (2012) Cerebrospinal fluid markers for Alzheimer's disease in a cognitively healthy cohort of young and old adults. Alzheimers Dement 8(6):520–527. https://doi.org/10.1016/j.jalz.2011.10.003

Pemberton LA, Stone E, Price P, van Bockxmeer F, Brew BJ (2008) The relationship between ApoE, TNFA, IL1a, IL1b and IL12b genes and HIV-1-associated dementia. HIV Med 9(8):677–680. https://doi.org/10.1111/j.1468-1293.2008.00614.x

Penzak SR, Kabuye G, Mugyenyi P, Mbamanya F, Natarajan V, Alfaro RM, Kityo C, Formentini E, Masur H (2007) Cytochrome P450 2B6 (CYP2B6) G516T influences nevirapine plasma concentrations in HIV-infected patients in Uganda. HIV Med 8(2):86–91. https://doi.org/10.1111/j.1468-1293.2007.00432.x

Peterson J, Gisslen M, Zetterberg H, Fuchs D, Shacklett BL, Hagberg L, Yiannoutsos CT, Spudich SS, Price RW (2014) Cerebrospinal fluid (CSF) neuronal biomarkers across the spectrum of HIV infection: hierarchy of injury and detection. PLoS One 9(12):e116081. https://doi.org/10.1371/journal.pone.0116081

Pomara N, Belzer KD, Silva R, Cooper TB, Sidtis JJ (2008) The apolipoprotein E epsilon4 allele and memory performance in HIV-1 seropositive subjects: differences at baseline but not after acute oral lorazepam challenge. Psychopharmacology 201(1):125–135. https://doi.org/10.1007/s00213-008-1253-1

Popp J, Lewczuk P, Frommann I, Kolsch H, Kornhuber J, Maier W, Jessen F (2010) Cerebrospinal fluid markers for Alzheimer's disease over the lifespan: effects of age and the APOEepsilon4 genotype. J Alzheimers Dis 22(2):459–468. https://doi.org/10.3233/JAD-2010-100561

Quasney MW, Zhang Q, Sargent S, Mynatt M, Glass J, McArthur J (2001) Increased frequency of the tumor necrosis factor-alpha-308 A allele in adults with human immunodeficiency virus dementia. Ann Neurol 50(2):157–162

Rickabaugh TM, Baxter RM, Sehl M, Sinsheimer JS, Hultin PM, Hultin LE, Quach A, Martinez-Maza O, Horvath S, Vilain E, Jamieson BD (2015) Acceleration of age-associated methylation patterns in HIV-1-infected adults. PLoS One 10(3):e0119201. https://doi.org/10.1371/journal.pone.0119201

Robertson K, Liner J, Meeker RB (2012) Antiretroviral neurotoxicity. J Neurovirol 18(5):388–399. https://doi.org/10.1007/s13365-012-0120-3

Rom S, Rom I, Passiatore G, Pacifici M, Radhakrishnan S, del Valle L, Pina-Oviedo S, Khalili K, Eletto D, Peruzzi F (2010) CCL8/MCP-2 is a target for mir-146a in HIV-1-infected human microglial cells. FASEB J 24(7):2292–2300. https://doi.org/10.1096/fj.09-143503

Rotger M, Colombo S, Furrer H, Bleiber G, Buclin T, Lee BL, Keiser O, Biollaz J, Decosterd L, Telenti A, Swiss HIV Cohort Study (2005) Influence of CYP2B6 polymorphism on plasma and intracellular concentrations and toxicity of efavirenz and nevirapine in HIV-infected patients. Pharmacogenet Genomics 15(1):1–5. https://doi.org/10.1097/01213011-200501000-00001

Roux-Lombard P, Modoux C, Cruchaud A, Dayer JM (1989) Purified blood monocytes from HIV 1-infected patients produce high levels of TNF alpha and IL-1. Clin Immunol Immunopathol 50(3):374–384

Rovin BH, Lu L, Saxena R (1999) A novel polymorphism in the MCP-1 gene regulatory region that influences MCP-1 expression. Biochem Biophys Res Commun 259(2):344–348. https://doi.org/10.1006/bbrc.1999.0796

Sacktor N (2018) Changing clinical phenotypes of HIV-associated neurocognitive disorders. J Neurovirol 24(2):141–145. https://doi.org/10.1007/s13365-017-0556-6

Sacktor N, Skolasky RL, Seaberg E, Munro C, Becker JT, Martin E, Ragin A, Levine A, Miller E (2016) Prevalence of HIV-associated neurocognitive disorders in the Multicenter AIDS Cohort Study. Neurology 86(4):334–340. https://doi.org/10.1212/WNL.0000000000002277

Saiyed ZM, Gandhi N, Agudelo M, Napuri J, Samikkannu T, Reddy PV, Khatavkar P, Yndart A, Saxena SK, Nair MP (2011) HIV-1 Tat upregulates expression of histone deacetylase-2 (HDAC2) in human neurons: implication for HIV-associated neurocognitive disorder (HAND). Neurochem Int 58(6):656–664. https://doi.org/10.1016/j.neuint.2011.02.004

Samson M, Libert F, Doranz BJ, Rucker J, Liesnard C, Farber CM, Saragosti S, Lapoumeroulie C, Cognaux J, Forceille C, Muyldermans G, Verhofstede C, Burtonboy G, Georges M, Imai T, Rana S, Yi Y, Smyth RJ, Collman RG, Doms RW, Vassart G, Parmentier M (1996) Resistance to HIV-1 infection in caucasian individuals bearing mutant alleles of the CCR-5 chemokine receptor gene. Nature 382(6593):722–725. https://doi.org/10.1038/382722a0

Sandkovsky U, Podany AT, Fletcher CV, Owen A, Felton-Coleman A, Winchester LC, Robertson K, Swindells S (2017) Impact of efavirenz pharmacokinetics and pharmacogenomics on neuropsychological performance in older HIV-infected patients. J Antimicrob Chemother 72(1):200–204. https://doi.org/10.1093/jac/dkw403

Savic RM, Barrail-Tran A, Duval X, Nembot G, Panhard X, Descamps D, Verstuyft C, Vrijens B, Taburet AM, Goujard C, Mentre F, ANRS 134–COPHAR 3 Study Group (2012) Effect of adherence as measured by MEMS, ritonavir boosting, and CYP3A5 genotype on atazanavir pharmacokinetics in treatment-naive HIV-infected patients. Clin Pharmacol Ther 92 (5):575–583. https://doi.org/10.1038/clpt.2012.137

Saylor D, Dickens AM, Sacktor N, Haughey N, Slusher B, Pletnikov M, Mankowski JL, Brown A, Volsky DJ, McArthur JC (2016) HIV-associated neurocognitive disorder – pathogenesis and prospects for treatment. Nat Rev Neurol 12(4):234–248. https://doi.org/10.1038/nrneurol.2016.27

Schrier RD, Gnann JW Jr, Landes R, Lockshin C, Richman D, McCutchan A, Kennedy C, Oldstone MB, Nelson JA (1989) T cell recognition of HIV synthetic peptides in a natural infection. J Immunol 142(4):1166–1176

Schrier RD, Wiley CA, Spina C, McCutchan JA, Grant I (1996) Pathogenic and protective correlates of T cell proliferation in AIDS. HNRC Group. HIV Neurobehavioral Research Center. J Clin Invest 98(3):731–740. https://doi.org/10.1172/JCI118845

Schrier RD, Gupta S, Riggs P, Cysique LA, Letendre S, Jin H, Spector SA, Singh KK, Wolfson T, Wu Z, Hong KX, Yu X, Shi C, Heaton RK, The HNRC Group (2012) The influence of HLA on HIV-associated neurocognitive impairment in Anhui, China. PLoS One 7(5):e32303. https://doi.org/10.1371/journal.pone.0032303

Silva AJ, Stevens CF, Tonegawa S, Wang Y (1992) Deficient hippocampal long-term potentiation in alpha-calcium-calmodulin kinase II mutant mice. Science 257(5067):201–206. https://doi.org/10.1126/science.1378648

Simioni S, Cavassini M, Annoni JM, Rimbault Abraham A, Bourquin I, Schiffer V, Calmy A, Chave JP, Giacobini E, Hirschel B, du Pasquier RA (2010) Cognitive dysfunction in HIV patients despite long-standing suppression of viremia. AIDS 24(9):1243–1250. https://doi.org/10.1097/QAD.0b013e3283354a7b

Singh KK, Barroga CF, Hughes MD, Chen J, Raskino C, McKinney RE, Spector SA (2003) Genetic influence of CCR5, CCR2, and SDF1 variants on human immunodeficiency virus 1 (HIV-1)-related disease progression and neurological impairment, in children with symptomatic HIV-1 infection. J Infect Dis 188(10):1461–1472. https://doi.org/10.1086/379038

Singh KK, Ellis RJ, Marquie-Beck J, Letendre S, Heaton RK, Grant I, Spector SA (2004) CCR2 polymorphisms affect neuropsychological impairment in HIV-1-infected adults. J Neuroimmunol 157(1–2):185–192. https://doi.org/10.1016/j.jneuroim.2004.08.027

Singh KK, Hughes MD, Chen J, Spector SA (2006) Impact of MCP-1-2518-G allele on the HIV-1 disease of children in the United States. AIDS 20(3):475–478. https://doi.org/10.1097/01.aids.0000200540.09856.58

Smith MW, Dean M, Carrington M, Winkler C, Huttley GA, Lomb DA, Goedert JJ, O'Brien TR, Jacobson LP, Kaslow R, Buchbinder S, Vittinghoff E, Vlahov D, Hoots K, Hilgartner MW, Hemophilia Growth and Development Study (HGDS), Multicenter AIDS Cohort Study (MACS), Multicenter Hemophilia Cohort Study (MHCS), San Francisco City Cohort (SFCC), ALIVE Study, O'Brien SJ (1997) Contrasting genetic influence of CCR2 and CCR5 variants on HIV-1 infection and disease progression. Science 277(5328):959–965. https://doi.org/10.1126/science.277.5328.959

Sogaard OS, Graversen ME, Leth S, Olesen R, Brinkmann CR, Nissen SK, Kjaer AS, Schleimann MH, Denton PW, Hey-Cunningham WJ, Koelsch KK, Pantaleo G, Krogsgaard K, Sommerfelt M, Fromentin R, Chomont N, Rasmussen TA, Ostergaard L, Tolstrup M (2015) The depsipeptide romidepsin reverses HIV-1 latency in vivo. PLoS Pathog 11(9):e1005142. https://doi.org/10.1371/journal.ppat.1005142

Sorce S, Myburgh R, Krause KH (2011) The chemokine receptor CCR5 in the central nervous system. Prog Neurobiol 93(2):297–311. https://doi.org/10.1016/j.pneurobio.2010.12.003

Spector SA, Singh KK, Gupta S, Cystique LA, Jin H, Letendre S, Schrier R, Wu Z, Hong KX, Yu X, Shi C, Heaton RK, The HNRC Group (2010) APOE epsilon4 and MBL-2 O/O genotypes are associated with neurocognitive impairment in HIV-infected plasma donors. AIDS 24(10):1471–1479. https://doi.org/10.1097/QAD.0b013e328339e25c

Stauch KL, Emanuel K, Lamberty BG, Morsey B, Fox HS (2017) Central nervous system-penetrating antiretrovirals impair energetic reserve in striatal nerve terminals. J Neurovirol 23(6):795–807. https://doi.org/10.1007/s13365-017-0573-5

Stein JY, Levin Y, Uziel O, Abumock H, Solomon Z (2018) Traumatic stress and cellular senescence: the role of war-captivity and homecoming stressors in later life telomere length. J Affect Disord 238:129–135. https://doi.org/10.1016/j.jad.2018.05.037

Suchy-Dicey AM, Muller CJ, Madhyastha TM, Shibata D, Cole SA, Zhao J, Longstreth WTJ, Buchwald D (2018) Telomere length and magnetic resonance imaging findings of vascular brain injury and central brain atrophy: the Strong Heart Study. Am J Epidemiol 187(6):1231–1239. https://doi.org/10.1093/aje/kwx368

Sun B, Abadjian L, Rempel H, Calosing C, Rothlind J, Pulliam L (2010) Peripheral biomarkers do not correlate with cognitive impairment in highly active antiretroviral therapy-treated subjects with human immunodeficiency virus type 1 infection. J Neurovirol 16(2):115–124. https://doi.org/10.3109/13550280903559789

Sundermann EE, Bishop JR, Rubin LH, Little DM, Meyer VJ, Martin E, Weber K, Cohen M, Maki PM (2015) Genetic predictor of working memory and prefrontal function in women with HIV. J Neurovirol 21(1):81–91. https://doi.org/10.1007/s13365-014-0305-z

Telenti A (2005) Adaptation, co-evolution, and human susceptibility to HIV-1 infection. Infect Genet Evol 5(4):327–334. https://doi.org/10.1016/j.meegid.2004.11.001

Thames AD, Briones MS, Magpantay LI, Martinez-Maza O, Singer EJ, Hinkin CH, Morgello S, Gelman BB, Moore DJ, Heizerling K, Levine AJ (2015) The role of chemokine C-C motif ligand 2 genotype and cerebrospinal fluid chemokine C-C motif ligand 2 in neurocognition among HIV-infected patients. AIDS 29(12):1483–1491. https://doi.org/10.1097/QAD.0000000000000706

Tierney SM, Sheppard DP, Kordovski VM, Faytell MP, Avci G, Woods SP (2017) A comparison of the sensitivity, stability, and reliability of three diagnostic schemes for HIV-associated neurocognitive disorders. J Neurovirol 23(3):404–421. https://doi.org/10.1007/s13365-016-0510-z

Vaidya SA, Korner C, Sirignano MN, Amero M, Bazner S, Rychert J, Allen TM, Rosenberg ES, Bosch RJ, Altfeld M (2014) Tumor necrosis factor alpha is associated with viral control and early disease progression in patients with HIV type 1 infection. J Infect Dis 210(7):1042–1046. https://doi.org/10.1093/infdis/jiu206

Valcour V, Shikuma C, Shiramizu B, Watters M, Poff P, Selnes OA, Grove J, Liu Y, Abdul-Majid KB, Gartner S, Sacktor N (2004) Age, apolipoprotein E4, and the risk of HIV dementia: the Hawaii Aging with HIV Cohort. J Neuroimmunol 157(1–2):197–202. https://doi.org/10.1016/j.jneuroim.2004.08.029

van Rij RP, Portegies P, Hallaby T, Lange JM, Visser J, de Roda Husman AM, van't Wout AB, Schuitemaker H (1999) Reduced prevalence of the CCR5 delta32 heterozygous genotype in human immunodeficiency virus-infected individuals with AIDS dementia complex. J Infect Dis 180(3):854–857. https://doi.org/10.1086/314940

Vassallo M, Durant J, Biscay V, Lebrun-Frenay C, Dunais B, Laffon M, Harvey-Langton A, Cottalorda J, Ticchioni M, Carsenti H, Pradier C, Dellamonica P (2014) Can high central nervous system penetrating antiretroviral regimens protect against the onset of HIV-associated neurocognitive disorders? AIDS 28(4):493–501. https://doi.org/10.1097/QAD.0000000000000096

Vecsey CG, Hawk JD, Lattal KM, Stein JM, Fabian SA, Attner MA, Cabrera SM, McDonough CB, Brindle PK, Abel T, Wood MA (2007) Histone deacetylase inhibitors enhance memory and synaptic plasticity via CREB: CBP-dependent transcriptional activation. J Neurosci 27(23):6128–6140. https://doi.org/10.1523/JNEUROSCI.0296-07.2007

Villalba K, Devieux JG, Rosenberg R, Cadet JL (2015) DRD2 and DRD4 genes related to cognitive deficits in HIV-infected adults who abuse alcohol. Behav Brain Funct 11:25. https://doi.org/10.1186/s12993-015-0072-x

Wanders RJ, Waterham HR (2006) Biochemistry of mammalian peroxisomes revisited. Annu Rev Biochem 75:295–332. https://doi.org/10.1146/annurev.biochem.74.082803.133329

Wendelken LA, Jahanshad N, Rosen HJ, Busovaca E, Allen I, Coppola G, Adams C, Rankin KP, Milanini B, Clifford K, Wojta K, Nir TM, Gutman BA, Thompson PM, Valcour V (2016) ApoE epsilon4 is associated with cognition, brain integrity, and atrophy in HIV over age 60. J Acquir Immune Defic Syndr 73(4):426–432. https://doi.org/10.1097/QAI.0000000000001091

Wesselingh SL, Power C, Glass JD, Tyor WR, McArthur JC, Farber JM, Griffin JW, Griffin DE (1993) Intracerebral cytokine messenger RNA expression in acquired immunodeficiency syndrome dementia. Ann Neurol 33(6):576–582. https://doi.org/10.1002/ana.410330604

Wesselingh SL, Takahashi K, Glass JD, McArthur JC, Griffin JW, Griffin DE (1997) Cellular localization of tumor necrosis factor mRNA in neurological tissue from HIV-infected patients by combined reverse transcriptase/polymerase chain reaction in situ hybridization and immunohistochemistry. J Neuroimmunol 74(1–2):1–8. https://doi.org/10.1016/S0165-5728(96)00160-9

Winkler C, Modi W, Smith MW, Nelson GW, Wu X, Carrington M, Dean M, Honjo T, Tashiro K, Yabe D, Buchbinder S, Vittinghoff E, Goedert JJ, O'Brien TR, Jacobson LP, Detels R, Donfield S, Willoughby A, Gomperts E, Vlahov D, Phair J, ALIVE Study, Hemophilia Growth and Development Study (HGDS), Multicenter AIDS Cohort Study (MACS), Multicenter Hemophilia Cohort Study (MHCS), San Francisco City Cohort (SFCC), O'Brien SJ (1998) Genetic restriction of AIDS pathogenesis by an SDF-1 chemokine gene variant. Science 279(5349):389–393. https://doi.org/10.1126/science.279.5349.389

Winston A, Arenas-Pinto A, Stohr W, Fisher M, Orkin CM, Aderogba K, de Burgh-Thomas A, O'Farrell N, Lacey CJ, Leen C, Dunn D, Paton NI, Pivot Trial Team (2013) Neurocognitive function in HIV infected patients on antiretroviral therapy. PLoS One 8(4):e61949. https://doi.org/10.1371/journal.pone.0061949

Wolkowitz OM, Jeste DV, Martin AS, Lin J, Daly RE, Reuter C, Kraemer H (2017) Leukocyte telomere length: effects of schizophrenia, age, and gender. J Psychiatr Res 85:42–48. https://doi.org/10.1016/j.jpsychires.2016.10.015

Wong JY, de Vivo I, Lin X, Grashow R, Cavallari J, Christiani DC (2014) The association between global DNA methylation and telomere length in a longitudinal study of boilermakers. Genet Epidemiol 38(3):254–264. https://doi.org/10.1002/gepi.21796

Wright EJ, Grund B, Robertson K, Brew BJ, Roediger M, Bain MP, Drummond F, Vjecha MJ, Hoy J, Miller C, Penalva de Oliveira AC, Pumpradit W, Shlay JC, El-Sadr W, Price RW, INSIGH SMART Study Group (2010) Cardiovascular risk factors associated with lower baseline cognitive performance in HIV-positive persons. Neurology 75(10):864–873. https://doi.org/10.1212/WNL.0b013e3181f11bd8

Wright EJ, Grund B, Cysique LA, Robertson KR, Brew BJ, Collins G, Shlay JC, Winston A, Read TR, Price RW, INSIGHT START Study Group (2015) Factors associated with neurocognitive test performance at baseline: a substudy of the INSIGHT Strategic Timing of AntiRetroviral Treatment (START) trial. HIV Med 16(Suppl 1):97–108. https://doi.org/10.1111/hiv.12238

Wyczechowska D, Lin HY, LaPlante A, Jeansonne D, Lassak A, Parsons CH, Molina PE, Peruzzi F (2017) A miRNA signature for cognitive deficits and alcohol use disorder in persons living with HIV/AIDS. Front Mol Neurosci 10:385. https://doi.org/10.3389/fnmol.2017.00385

Xu Z, Asahchop EL, Branton WG, Gelman BB, Power C, Hobman TC (2017) MicroRNAs upregulated during HIV infection target peroxisome biogenesis factors: implications for virus biology, disease mechanisms and neuropathology. PLoS Pathog 13(6):e1006360. https://doi.org/10.1371/journal.ppat.1006360

Yang OO, Garcia-Zepeda EA, Walker BD, Luster AD (2002) Monocyte chemoattractant protein-2 (CC chemokine ligand 8) inhibits replication of human immunodeficiency virus type 1 via CC chemokine receptor 5. J Infect Dis 185(8):1174–1178. https://doi.org/10.1086/339678

Yelamanchili SV, Chaudhuri AD, Chen LN, Xiong H, Fox HS (2010) MicroRNA-21 dysregulates the expression of MEF2C in neurons in monkey and human SIV/HIV neurological disease. Cell Death Dis 1:e77. https://doi.org/10.1038/cddis.2010.56

Zhang Z, Liu J, Wang H, Wu H, Wu X, Dong J, Liao L (2015) Association between chemokine receptor 5 (CCR5) delta32 gene variant and atherosclerosis: a meta-analysis of 13 studies. Int J Clin Exp Med 8(1):658–665

Zimmerman PA, Buckler-White A, Alkhatib G, Spalding T, Kubofcik J, Combadiere C, Weissman D, Cohen O, Rubbert A, Lam G, Vaccarezza M, Kennedy PE, Kumaraswami V, Giorgi JV, Detels R, Hunter J, Chopek M, Berger EA, Fauci AS, Nutman TB, Murphy PM (1997) Inherited resistance to HIV-1 conferred by an inactivating mutation in CC chemokine receptor 5: studies in populations with contrasting clinical phenotypes, defined racial background, and quantified risk. Mol Med 3(1):23–36

Zwergel C, Stazi G, Valente S, Mai A (2016) Histone deacetylase inhibitors: updated studies in various epigenetic-related diseases. J Clin Epigenetics 2(1):7

Inflammatory Mechanisms and Cascades Contributing to Neurocognitive Impairment in HIV/AIDS

Nicole Fernandes and Lynn Pulliam

Contents

Abstract Neurocognitive impairment caused by chronic human immunodeficiency virus (HIV) infection is a growing concern. In this chapter we discuss the inflammatory mechanisms underlying the pathology of asymptomatic and mild

N. Fernandes and L. Pulliam (✉)
Department of Laboratory Medicine, San Francisco VA Health Care System, San Francisco, CA, USA

University of California, San Francisco, San Francisco, CA, USA
e-mail: lynn.pulliam@ucsf.edu

Curr Topics Behav Neurosci (2021) 50: 77–104
DOI 10.1007/7854_2019_100
Published Online: 6 August 2019

neurocognitive impairment in the context of antiretroviral therapy. We discuss the role of HIV, viral proteins, and virally infected cells on the development of neuroinflammation and the effect of viral proteins on the cells of the central nervous system.

We examine how these collective factors result in an inflammatory context that triggers the development of neurocognitive impairment in HIV. We assess the contribution of antiretrovirals and drugs of abuse, including methamphetamine, cannabis, and opioids, to the neurotoxic and neuroinflammatory milieu that leads to the development of neurocognitive impairment in HIV-infected individuals. We also examined circulating biomarkers, NF-L, sCD163, and sCD14, pertinent to identifying changes in the CNS that could indicate real-time changes in patient physiology. Lastly, we discuss future studies, such as exosomes and the microbiome, which could play a role in the HIV-induced neuroinflammation that eventually manifests as cognitive impairment.

Keywords ART · Biomarkers · HIV · Neurocognitive impairment · Neuroinflammation

1 Neurocognitive Impairment in HIV

1.1 What Is Neurocognitive Impairment?

Human immunodeficiency virus (HIV) affects 36.9 million people worldwide (UNAIDS 2018). With the widespread use of antiretroviral therapy (ART), survival rates have dramatically increased, and HIV-infected persons are able to live decades longer (Woods et al. 2009). Of all the individuals currently living with HIV, around 21.7 million (58.8%) have access to ART (UNAIDS 2018). With the increase in long-term survival, the incidence of HIV-associated neurocognitive disorders (HAND) has also increased in HIV-positive individuals. Although the more severe form of HAND, HIV-associated dementia (HAD), has dramatically decreased with the use of antiretrovirals (ARV), studies estimate that between 30 and 50% of all HIV-infected persons present with neurocognitive impairment (Woods et al. 2009; Rizzo et al. 2018; Heaton et al. 2010; Jaureguiberry-Bravo et al. 2016). Because ART are known to have poor penetration across the blood-brain barrier (BBB), the virus is not well controlled in the central nervous system (CNS) (Weber et al. 2013). Thus, in the CNS, the virus contributes to a chronic neuroinflammatory state that eventually results in pathological changes and cognitive impairment. HAND is diagnosed by a battery of neuropsychological tests, which evaluate executive function, episodic memory, speed of information processing, motor skills, attention/working memory, language, and sensoriperception (Rizzo et al. 2018; Weber et al. 2013; McGuire et al. 2015; Soontornniyomkij et al. 2016).

Less severe HAND is often divided into two categories based on neuropsychological testing, i.e., asymptomatic neurocognitive impairment (ANI) and mild neurocognitive disorder (MND) (McGuire et al. 2015). In a study evaluating HAND using a population of HIV-infected participants who did not have comorbid risks of CNS dysfunction, over 30% were diagnosed with ANI, and around 12% and 2% presented with MND and HAD, respectively (Woods et al. 2009; Heaton et al. 2010). It is likely that HIV-infected subjects may develop ANI as a transition point to MND, so it may be important to not only identify affected individuals but also to ascertain the progression of the impairment and potential timepoints for therapeutic intervention. The etiology of HAND is not fully elucidated, but its presentation is attributed to dysfunction, damage, and death of neurons, without productive HIV infection of neurons (Rizzo et al. 2018). Studies focused on the neurobiological mechanisms of neurocognitive impairment have found many pathological changes, such as synaptic loss and dendritic simplification (McGuire et al. 2015). Some of these progressive pathological changes can be detected in the cerebrospinal fluid (CSF) and blood of HIV patients. Diagnosis of ANI and MND with objective biomarkers in "real time" continues to be difficult in HAND and remains as a research aim.

2 Peripheral Inflammation Sets Up Processes That Lead to Neurocognitive Impairment

2.1 Monocyte/Macrophage (M/Mϕ) Activation

Monocytes are phagocytic myeloid cells that are distinguished by the heterogeneous expression of several cell surface molecules, including CD14, CD16, and Toll-like receptor 4 (Jakubzick et al. 2017; Rempel et al. 2010; Zhen et al. 2014). Monocytes are typically classified as $CD14^{dim}CD16^{+}$ (nonclassical monocytes), $CD14^{high}CD16^{+}$ (intermediate monocytes), and $CD14^{high}CD16^{-}$ (classical monocytes). While a facet of monocyte function involves differentiation into tissue-specific macrophages, monocytes are also capable of synthesizing and secreting inflammatory mediators, in a manner similar to macrophages, upon activation (Jakubzick et al. 2017). Meanwhile, HIV and its proteins, such as gp120 and Tat, are also capable of activating monocytes, which subsequently upregulate CD16 and change their phenotype to more "macrophage-like" $CD14^{dim}CD16^{+}$ and $CD14^{high}CD16^{+}$ populations. We will collectively refer to these monocytes as CD14/CD16 M/Mϕ in this review. Ironically, these phenotypes are more susceptible to further HIV infection and produce a variety of inflammatory factors. CD14/CD16 monocytes enter the brain, most likely in response to CCL2 produced by astrocytes (Chompre et al. 2013; Dubovy et al. 2018), and studies show that these cells preferentially migrate across the BBB (Veenstra et al. 2017). Once in the CNS, they can contribute to the events that produce chronic neuroinflammation.

While monocytes are fairly refractile to HIV infection, a small subset of CD14/CD16 monocytes are more susceptible to HIV and preferentially harbor HIV

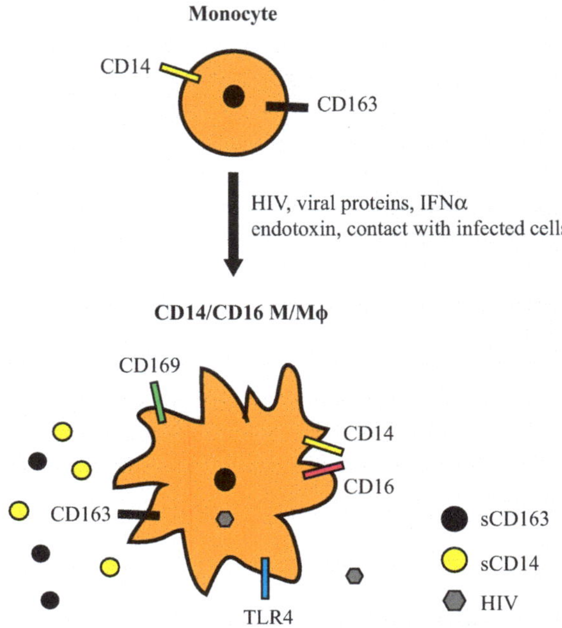

Fig. 1 Monocyte activation in HIV

DNA (Ellery et al. 2007). Monocyte activation can trigger differentiation into macrophages, and, upon CD4 ligation and CCR5 binding to HIV, most monocytes differentiate into macrophages (Zhen et al. 2014). Monocyte activation also results in the release of soluble(s) CD14 and sCD163 and the formation of monocyte-platelet aggregates which triggers proinflammatory cytokine secretion and upregulates adhesion molecules and tissue factors and have additional roles in immune activation (Liang et al. 2015). In addition, individuals with significant decline in memory performance had lower expression of CD163 on CD14/CD16 monocytes that corresponded well with the higher release of soluble CD163 during cognitive impairment (Liang et al. 2015; Fabbiani et al. 2017). The role of inflammation in the development of neurocognitive disorders among HIV-infected people is of renewed importance since previously there was a positive correlation between the degree of immunosuppression and risk for HAND (Heaton et al. 2010). Importantly, persons with controlled viral loads and partial recovery of the immune system with ART continue to have detectable levels of inflammation (Heaton et al. 2010) (Fig. 1).

2.2 *CD14/16 Monocytes Facilitate BBB Crossing*

Because macrophages are long-lived cells, infection continues despite adherence to ART and is a key obstacle to HIV cure (Clayton et al. 2017). Macrophages can be infected through the viral synapse, receptor-mediated endocytosis, or phagocytosis of HIV-infected living, dying, or dead cells (Baxter et al. 2014). The well-known "Trojan Horse" model of CNS infection suggests that HIV-infected M/Mϕ cross the blood-brain barrier and release the virus into the CNS, where the virus subsequently infects microglia (Mediouni et al. 2015). This hypothesis is supported by data showing higher levels of E-selectin and vascular cell adhesion molecule-1 in HIV-infected M/Mϕ (Nottet et al. 1996) and thus a higher affinity for endothelial adhesion molecules. A recent study showed a decreased trend in circulating intermediate monocytes in patients with memory deficits, which suggests that loss of CD14/CD16 M/Mϕ in the periphery may reflect the transmigration of this monocyte subset into the CNS (Fabbiani et al. 2017).

2.3 Chronic Interferon-α (IFN-α) Monocyte Phenotype

An interferon spike would be the first response to any viral infection, including HIV pathogenesis; however, with treatment and undetectable viral loads, the normal inflammatory response to pathogens would be to revert to a quiescent state. However, IFN-α is a double-edged sword such that when chronically activated can have deleterious effects. In HIV infection, the presence of interferon-induced Siglec 1 (CD169) expression on monocytes signals ongoing monocyte activation in HIV-infected persons (Rempel et al. 2008). CD169 is an interferon-stimulated gene (ISG) that is induced in response to IFN-α, is present on the surface of M/Mϕ, and is highly expressed during HIV infection (Patro et al. 2016). Ironically, circulating interferon cannot be measured in treated, chronically infected HIV-infected persons. However, an IFN-α "alarm" signature, defined as monocytes expressing CD169, signals that a subset of HIV-infected persons may progress to cognitive impairment (Pulliam 2014). A sustained type 1 interferon response in HIV infection can cause cognitive impairment (Pulliam et al. 2014). Interruption of antiretrovirals increases viremia, and HIV-infected individuals revert to an untreated increased type I IFN monocyte profile (Rempel et al. 2010). Treatment to decrease IFN-α expression may result in increased viral load; however strategies to lower the chronic IFN-α response to a normal threshold are warranted. Recent studies on the neurotoxicity of IFN-α showed a disruption of neuronal dendritic spines (Koneru et al. 2018). Using an inhibitor of IFN-α in vitro and in a mouse HAND model, both brain pathology and behavioral abnormalities improved, suggesting that this approach may be beneficial. Further studies to dampen the IFN-α response in HIV infection may prove to be an effective adjunctive therapy for cognitive impairment.

2.4 Contribution of Lipopolysaccharide (LPS) to Inflammation Is Diminished

The gastrointestinal (GI) tract is a major site of HIV replication. As a result, intestinal permeability is fivefold greater in HIV-infected individuals, and increased GI inflammation is a common sequela (Brenchley and Douek 2008; Allam et al. 2018; Marchetti et al. 2013). LPS released from the gut during this process enters circulation and triggers strong innate immune responses by activating a variety of cells, including monocytes and macrophages (Marchetti et al. 2013). Although current ART regimens have suppressed HIV replication to undetectable levels, low levels of virus, undetectable by available techniques, may still be present, even with ARV adherence (Marchetti et al. 2013). The replication of residual virus and release of viral proteins cause chronic, systemic, low-grade inflammation, which contributes toward accelerated aging, cardiovascular disease, and cancer in ART and HIV-infected individuals (Allam et al. 2018). Moreover, although the initiation of ART reduces the endotoxin level, residual LPS persists in circulation and continues to propagate peripheral inflammation (Marchetti et al. 2013). The role of LPS was investigated more thoroughly after some reports indicated that LPS levels may be an independent marker of neurocognitive impairment (Marchetti et al. 2013), but other studies have not replicated this data, showing instead that cognitive impairment correlated with monocyte activation and not plasma LPS (Rempel et al. 2013). In addition to endotoxin, microbial translocation is also common in HIV-infected patients and is known to contribute to rapid liver disease progression in individuals who are coinfected with hepatitis C (HCV) (Marchetti et al. 2013). Indeed, LPS levels in HIV infection were most likely a result of HCV coinfection and/or drug abuse (Ancuta et al. 2008). However, the new generation of HCV therapies has been incredibly successful at treating HCV in HIV/HCV-coinfected population and has significantly lowered the chronic inflammatory environment in the periphery (Lopez-Cortes et al. 2018).

2.5 Blood-Brain Barrier (BBB)

HIV is thought to enter the CNS by several routes: through infected cells and through infection/endocytosis of endothelial cells across the BBB. Monocytes are thought to be the major contributor to viral neuroinvasion when infected CD14/CD16 monocytes cross the BBB in response to CCL2 and bring the virus into the CNS (Veenstra et al. 2017). In fact, HIV-infected CD14/CD16 monocytes expressing CXCR5, CXCR1, and CD11b and CD169 can cross the BBB with a higher efficiency than uninfected monocytes (Veenstra et al. 2017; Fabbiani et al. 2017; Persidsky et al. 1999; Pulliam et al. 1997). In addition, during the acute phase of viral infection, the massive peripheral immune response could induce an inflammatory environment that could affect the integrity of the BBB and contribute to CNS infection. However, with the widespread use of ARV for the treatment of HIV infection, peripheral viral loads are fairly well suppressed (Fabbiani et al. 2017). Due to treatment efficacy, peripheral inflammation may not be the major factor affecting BBB integrity or contributing to neuroinflammation currently. Nevertheless, HIV replication persists in tissue sites, such as the CNS, despite ART, and low-level viral replication continues to drive immune activation and inflammation by the production of HIV viral proteins (McGuire et al. 2015). The poor penetrability of ARV across the BBB likely contributes to the detectable levels of HIV RNA and DNA in the CNS, despite long-term adherence to ART, and also explains instances of HIV escape detected in the CSF of HIV-infected patients (Peluso et al. 2012; Hellmuth et al. 2015). Moreover, ART may not be effective at controlling BBB disruption, further contributing to an exchange of inflammatory mediators between the periphery and the CNS (Nair et al. 2015).

3 Neuroinflammation: Transition of Inflammation from the Periphery to the CNS and the Consequences on Neurodegeneration

The CNS becomes infected with HIV quickly after initial exposure to the virus, with reports estimating that the virus can cross the BBB as soon as 4 days after peripheral infection (Veenstra et al. 2017), likely through CD14/CD16 M/Mϕ. Microglia, in particular, are highly susceptible to HIV infection, as they express CD4 receptors and chemokine co-receptors, including CCR3, CCR5, and CXCR4, through which the virus infects cells (Ginsberg et al. 2018; Nedellec et al. 2009; He et al. 1997; Cenker et al. 2017). Microglia are also the only CNS-specific cells capable of supporting productive infection (Ginsberg et al. 2018; Cenker et al. 2017; Chen et al. 2017). Microglial infection results in the release of neurotoxic viral proteins and cytokines, which together contribute to an inflammatory environment in the CNS. While neuroinflammation is normally a neuroprotective tactic, chronic microglial activation, which occurs as a consequence of unsuppressed CNS viral infection, establishes a chronic neuroinflammatory environment that eventually causes neurotoxicity. Microglia-derived neuroinflammation can further exacerbate HIV entry and replication in microglia, through the IL-4- and IL-10-dependent upregulation of CD4 and CCR5 expression, respectively, and propagate neuroinflammation viral infection throughout the CNS. In fact, markers of neuroinflammation persist and are detectable in autopsy brain tissues of HIV-infected individuals despite ARV and in the absence of productive virus (Ginsberg et al. 2018). Importantly, HAD and HAND both have strong correlations with microglia activation, presumably due to the combined deleterious effects of viral proteins and microglia-derived cytokines on neurons (Watkins and Treisman 2015). HIV also causes mitochondrial dysfunction and subsequent secretion of neurotoxic mediators in a variety of cell types. Microglia-derived neuroinflammation is sustained by HIV infection at multiple levels (Potula et al. 2010; Var et al. 2016; Gelman et al. 2012). Moreover, microglia are thought to serve as sites of latent infection and function as long-term viral reservoirs, which may be continuously reestablished by infected monocytes (Fabbiani et al. 2017; Cenker et al. 2017). This is one way that HIV persists despite ARV and underscores the challenges to controlling viral replication in the CNS. Over time, chronic microglia-derived neuroinflammation and neurotoxicity collectively result in the cognitive, motor, and behavioral deficits seen in at-risk patients (Chen et al. 2017; Gill et al. 2014).

Microglial infection leads to released virions and viral products that cause astrocyte activation. HIV can infect astrocytes, and examination of postmortem brain slices has shown that up to 19% of astrocytes contain HIV DNA, indicating that astrocytes could also serve as viral reservoirs in addition to microglia (Chompre et al. 2013; Veenstra et al. 2017; Churchill et al. 2009; Eugenin et al. 2011). However, unlike microglia, astrocytic infection does not result in productive replication (Chompre et al. 2013; Eugenin et al. 2007, 2011). Restricted infection in astrocytes, also described as nonproductive infection, is a product of inefficient HIV replication due to blocks at various levels in the viral lifecycle (Kovalevich and Langford 2012; Liu et al. 2004). In spite of this restricted infection, astrocytes still produce and release viral proteins that negatively impact cell viability and homeostasis in the CNS (Chompre et al. 2013; Dubovy et al. 2018; Churchill et al. 2009; Eugenin et al. 2011; Kovalevich and Langford 2012; Liu et al. 2004). Moreover, astrocytic infection can still result in major damage to the CNS, as the small number of HIV-infected astrocytes trigger alterations in gap junction proteins, spread toxic signals to uninfected astrocytes, and alter the integrity of the BBB (Eugenin et al. 2011; Eugenin et al. 2007). As normal astrocytic function is regulated by neurons, HIV-induced changes in neuronal signaling, caused by excitotoxicity, energy failure, ischemia, and neurodegeneration, alter astrocytic function independently of HIV infection or astrocytic activation by microglia and viral proteins (Eugenin et al. 2011). Together these processes produce aberrant glial-vascular signaling and cause apoptosis of BBB endothelial cells, which leads to increased vascular permeability and allows the influx of proinflammatory cells and mediators from the periphery into the CNS (Eugenin et al. 2011). In addition, as viral load is less controlled in the CNS, these mechanisms are ongoing, even in the context of ART, and could be another mechanism through which neuroinflammation is chronically sustained in HIV-infected people (Eugenin et al. 2011).

4 HIV Proteins as Activators of Neuroinflammation

In autopsy samples of HIV-infected subjects, neuronal apoptosis is less prevalent in milder neurological diseases. Although there are conflicting reports as to whether or not HIV directly infects neurons, the virus is able to cause significant neuronal toxicity through induction of inflammatory mediators from microglia, astrocytes, and peripheral immune cells and by release of soluble viral proteins (Eugenin et al. 2007; Kovalevich and Langford 2012). In addition, HIV proteins have a toxic effect on neurons and glia that is independent of the infectious and lytic properties of the whole virus. Although HIV causes significant damage by inducing the release of neurotoxic and neuroinflammatory factors, viral proteins, such as tat, gp120, Nef, and Vpr, also have well-documented cytotoxic effects and can remain elevated despite ART and viral suppression (Mediouni et al. 2015; Kanmogne et al. 2007; Kanmogne 2005; Zhong et al. 2010).

Tat (trans-activator of transcription) is found in the CNS of HIV-infected individuals, even in those with controlled viral levels (Kesby et al. 2017; Mediouni et al. 2012). Tat can enter cells passively or by receptor-mediated endocytosis; is involved in a variety of neuroinflammatory processes, such as induction of oxidative stress; and is associated with irregular neurogenesis and glial loss that is well described in the pathologies of HAND and HAD (Mediouni et al. 2015; Eugenin et al. 2011; Kesby et al. 2017). Tat is known to potentiate glutamate-induced excitotoxicity and is readily detected in brains of persons with HIV-associated dementia and encephalitis (Mediouni et al. 2015; Eugenin et al. 2007; Kesby et al. 2017). Another mechanism through which tat affects HIV pathogenesis is by increasing the function of P-glycoprotein (P-gp), a BBB protein involved in the efflux of small molecules out of the brain. Studies have theorized that low levels of ARV in the brain, and subsequent consequences of CNS HIV infection, could be due to this tat-induced increase in P-gp function (Mediouni et al. 2015; Zhong et al. 2010). In addition, tat is known to decrease the tight junction proteins that preserve the integrity of the BBB by promoting the expression of inflammatory molecules, such as TNF-α, IL-1β, and IL-6, oxidative stress, and expression of the matrix metalloproteinase 9 and may facilitate the translocation of peripheral inflammation into the CNS (Mediouni et al. 2015). In this manner, tat enhances HIV infection and neuroinflammation in the CNS.

Gp120 has both direct and indirect effects on neurotoxicity and with Tat are major HIV proteins causing neuronal disruption. Secretion of gp120 has direct effect on BBB by disrupting the tight junction proteins ZO-1, ZO-2, and occludin in endothelial cells (Kanmogne 2005). Further studies identified gp120 involvement in enhanced monocyte migration across the BBB through the protein kinase pathway (Kanmogne et al. 2007). Direct neurotoxicity is seen when gp120 is exposed to neural cells and macrophages and induces TNF-α and IL-6 (Yeung et al. 1995). In addition, gp120 induces ROS, TNF-α, and MCP-1 in microglia that stimulates neuronal apoptosis (Guo et al. 2013). Gp120 exposure to microglia also induces IL-1β in vitro and in vivo with downstream neuronal loss and neurological impairment (Walsh et al. 2014).

Although astrocytes are unable to produce full virions, HIV proteins, such as *Nef*, are still produced. Indeed, HIV Nef has been found in postmortem brains of HIV-infected patients with dementia and has been implicated as a cause of cognitive impairment (Chompre et al. 2013; Sami Saribas et al. 2017). Although the mechanisms of neurotoxicity remain to be fully identified, Nef is implicated in IP-10-related neurodegeneration (Sami Saribas et al. 2017). Nef compromises astrocytic autophagy pathways and causing neuronal death by inducing oxidative stress (Sami Saribas et al. 2017). In addition, few studies have thoroughly investigated Nef in patients, but in vitro studies suggest that HIV-infected astrocytes and macrophages likely transmit Nef to neurons through extracellular vesicles (Sami Saribas et al. 2017).

Another HIV protein, viral protein R (*Vpr*), is found at extremely high levels in the CSF of people with AIDS and causes neuronal death through axonal disruption or synaptodendritic injury (Kitayama et al. 2008). Vpr is known to contribute to mitochondrial dysfunction, dysregulate ATP synthesis, suppress axonal growth, and affect neuronal differentiation, all of which negatively impact adult neurogenesis and may lead to some of the negative symptoms associated with HAND (Kitayama et al. 2008). This viral protein may also be important to the pediatric HIV population, as developmental delays in children coincide with the presence of a large population of undifferentiated neural progenitor cells, signifying that immature neurons exposed to Vpr may lose the ability to mature (Kitayama et al. 2008). Therefore, it is possible that HIV Vpr-induced neuronal toxicity and complications with hippocampal neurogenesis contribute to memory deficits observed in the manifestation of HAND (Kitayama et al. 2008).

5 Antiretrovirals Contribute to Neuroinflammation

Since their introduction, ART has dramatically altered the prognosis of HIV. Common first-line ART regimens include efavirenz (EFV) with tenofovir disoproxil fumarate/emtricitabine (TDF/FTC); EFV with lamivudine/zidovudine; maraviroc (MVC), the only available CCR5 inhibitor, with TDF/FTC; maraviroc and raltegravir (MVC + RAL) with TDF/FTC; or the new four-drug fixed dose of elvitegravir, tenofovir alafenamide, FTC, and cobicistat (Serrano-Villar et al. 2016; Treisman and Soudry 2016; Gatch et al. 2013; Dalwadi et al. 2016). With treatment, viral load is diminished, HIV-infected individuals are living longer, and their immune systems are less affected. As a result, they are less susceptible to opportunistic infections and progress to AIDS with a lower frequency. As HIV-infected individuals transition into the aging population, a new array of comorbidities including cardiovascular disease, insulin resistance and type II diabetes, osteoporosis, neurocognitive deficits, and cancer occur in part as a result of their HIV status and their history of substance abuse or lifestyle choices and in part due to the normal aging process (Hileman and Funderburg 2017). All of these comorbidities are also associated with inflammation and may elevate the peripheral inflammation caused by low levels of viral replication.

5.1 ART Drives Inflammation

As immune activation and systemic inflammation levels remain slightly elevated despite ART and rarely recover to levels prior to HIV infection, an onus of aging and comorbidities falls on the above-normal, chronic inflammation that persists despite ART and suppression of viral replication. Due to the continued presence of chronic low-grade peripheral inflammation, recent studies have examined the contribution of ART to inflammation (Treisman and Soudry 2016). After treatment initiation, heightened monocyte activation as well as decreased chemokine receptors, CCR2 and CX3CR1, tended to normalize, with levels approaching those measured in HIV-uninfected controls. Thus, the impact of HIV on peripheral inflammation was greatly reduced. Based on the regimens prescribed, various components of ART could independently contribute a low-grade chronic inflammation and neurocognitive impairment seen in HIV-infected people in addition to the virus or viral proteins (Treisman and Soudry 2016; Kahouadji et al. 2013). Raltegravir, in particular, has been associated with an increase in microbial translocation and with HIV persistence in the gut (Serrano-Villar et al. 2016). Thus, the low levels of inflammation may be caused by residual replication and by the ARV themselves.

5.2 ART and Neuroinflammation

As reviewed by Treisman and Soudry, several reports in the past 5 years have noted considerable overlap between the HAND symptoms and toxicity of ART, and it is likely that both factors synergistically contribute to neurocognitive impairment (Treisman and Soudry 2016). Antiretrovirals have typically demonstrated poor permeability into the brain parenchyma, and the benefit of bioavailability in the CNS remains a hotly contested issue, as ART can both inhibit viral replication and cause direct neurotoxicity (Treisman and Soudry 2016; Marra et al. 2009). However, since cognitive impairment has been linked to viral replication in the brain, it has become increasingly important to establish the relationship between HIV replication, ART penetration into the CNS, and the neuroinflammatory processes that arise as a consequence (Treisman and Soudry 2016; Kahouadji et al. 2013). An early study examining the ability of antiretrovirals to cross the BBB reported that despite the low concentrations in the CNS, nelfinavir was able to significantly reduce CSF viral loads (Treisman and Soudry 2016; Aweeka et al. 1999). In contrast, Letendre et al. concluded that it was the greater CNS penetration of ART that resulted in CSF viral suppression and improved neuropsychological scores in HIV-infected individuals with HAND (Letendre et al. 2004). The authors also found that viral load reduction was more pronounced if treatment regimens contained a higher ratio ARV that could cross the BBB and that viral suppression in the CNS, particularly in ART-naïve individuals, improved global deficit scores in impaired HIV-infected individuals (Weber et al. 2013; Marra et al. 2009; Letendre et al. 2004). Yet other studies noted the opposite phenomenon and saw an inverse correlation between CNS penetrability of ART and cognition (Treisman and Soudry 2016; Kahouadji et al. 2013), with evidence pointing to ART-induced neurotoxicity as the cause for the worsened neurocognition (Treisman and Soudry 2016; Marra et al. 2009). Moreover, CSF levels of antiretrovirals may not accurately reflect CNS penetration of the drugs (Tozzi et al. 2009). With many of the newer ARV having higher BBB penetrability, their effects on CNS viral suppression and neuroinflammation will be clear in future studies that monitor HIV-infected individuals over time.

6 Drugs of Abuse Contribute to Peripheral and Neuroinflammation

HIV acquisition in a sizeable portion of the infected population can be attributed to the sharing of contaminated needles in intravenous drug injection and risky sexual behaviors produced by consumption of alcohol and illicit drugs. Many substances of abuse, including methamphetamine (Meth), opiates, cocaine, and alcohol, enhance HIV replication and suppress the immune system (Soontornniyomkij et al. 2016); therefore, individuals with substance abuse disorders are at an increased risk of contracting HIV. HIV infection and substance abuse are independently associated significant neuropathology, but they share similar phenotypes of behavioral and cognitive anomalies (Heaton et al. 2010; Soontornniyomkij et al. 2016; Gelman et al. 2012). Thus, the prevailing hypothesis suggests that concurrent substance abuse and HIV infection could synergistically suppress immune responses, increase viral replication, and alter synaptic transmission in HIV-positive substance abusers (Gelman et al. 2012). In addition, recent studies suggest the presence of sustained neuroinflammation in addiction has prompted the use of anti-inflammatory drugs for treatment of both substance abuse and HAND (Xu et al. 2017). HIV-induced pathological changes could be exacerbated by substance abuse and accelerate the development and progression of HAND (Var et al. 2016; Gelman et al. 2012; Byrd et al. 2011). Moreover, illicit substance use could contribute to HAND through multiple routes as polydrug use is common in people with substance use disorders.

6.1 *Methamphetamine*

Methamphetamine (Meth) is a highly abused psychostimulant that has long been implicated in the pathogenesis of cognitive impairment by altering dopaminergic signaling (Soontornniyomkij et al. 2016; Xu et al. 2017; Volkow et al. 2001). Moreover, reports indicate that between 40 and 60% of HIV-infected individuals abuse Meth, emphasizing the need to examine the contribution of substance abuse to viral infection (Kesby et al. 2017). As altered dopaminergic neurotransmission is also a consequence of HIV infection, populations that abuse Meth may present with worse and accelerated HAND, compared to HIV-infected individuals without substance abuse issues (Soontornniyomkij et al. 2016; Gelman et al. 2012). Meth-induced neurotoxicity involves loss of dopaminergic neurons, high microglia-derived neuroinflammation, and behavioral consequences (Volkow et al. 2001). Moreover, although the striatum is particularly susceptible to Meth toxicity, neuronal injury is noted in many brain regions, and microgliosis is a feature of HIV encephalitis patients that also abuse Meth (Soontornniyomkij et al. 2016; Xu et al. 2017). Microglial activation contributes to the release of proinflammatory mediators and chemokines, such as IL-1α, IL-6, CCL2, and TNF-α, which are known to be neurotoxic with chronic exposure. Meth facilitates HIV-induced damage to the CNS by decreasing patient adherence to ARV, by disrupting the BBB, and by increasing viral replication (Potula et al. 2010; Xu et al. 2017). In addition, viral proteins, such as tat and gp120, can also elicit microglia-derived inflammatory responses and eventually lead to altered cognitive function in individuals that use Meth (Xu et al. 2017).

6.2 *Opioids*

Opioid abuse is a major health concern in the United States, and recent estimates have suggested that treatment for prescription opioid abuse increased 900% (Jaureguiberry-Bravo et al. 2016). Chronic opioid abuse, like Meth, is associated with neurocognitive deficits, triggers the production of proinflammatory cytokines such as IL-1β and TNF-α, and disrupts glutamatergic homeostasis (Cahill and Taylor 2017). The three main classes of opioid receptors, mu, kappa, and delta receptors, are present on monocytes and macrophages, and exposure to exogenous and endogenous opioids may increase monocyte adherence and trafficking into the CNS during HIV infection (Jaureguiberry-Bravo et al. 2016). Furthermore, opioid exposure increases macrophage expression of CXCR4 and CCR5, which then increases the susceptibility of these cells to HIV infection (Jaureguiberry-Bravo et al. 2016). Exogenous opioids, such as morphine, are also able to downregulate antiviral molecules, including IFN-α, in macrophages and anti-HIV miRNAs, which may facilitate HIV infection and inflammation (Jaureguiberry-Bravo et al. 2016). Thus, opioid use can increase the efficiency of and subsequent damage associated with HIV infection.

6.3 *Cannabis*

Estimates suggest that almost 50% of the HIV-infected population use cannabis, and though there is conflicting literature on the effect of cannabinoids on neurocognitive impairment in HIV-infected people, studies using animal models of neuroinflammation-induced cognitive damage support the neuroprotective effects of cannabinoids (Thames et al. 2016). One study determined that HIV-infected individuals that used cannabis had lower levels of CD14/CD16 monocytes and IP-10, an interferon-stimulated cytokine, compared to HIV-infected individuals who did not use cannabis (Rizzo et al. 2018). Therefore, cannabis use decreases immune cell recruitment and chemotaxis to sites of infection or injury, which could lower CNS and peripheral infection and further suppress viral replication by altering the activation of HIV co-receptors CXCR4 and CCR5 and by decreasing the number of M/Mϕ entering the CNS (Cabral and Griffin-Thomas 2009). Elevated IP-10 has been detected in the CSF of cognitively impaired patients and is known to induce neuronal apoptosis in vitro. Moreover, IP-10 can increase HIV replication in macrophages, so it is possible that cannabis use decreases IP-10 induced viral replication in infected microglia and peripheral macrophages. In addition, this effect could be attributed to the known immune suppressive, anti-inflammatory effects of the psychoactive component of cannabis, Δ9-Tetrahydrocannabinol, through CB_2 receptor-expressing immune cells (Rizzo et al. 2018; Cabral and Griffin-Thomas 2009). When taken together, these effects may explain the correlation between cannabis use and lower plasma viral load in injection drug users who were initially HIV negative and seroconverted during a year-long study (Cabral and Griffin-Thomas 2009; Milloy et al. 2015). However, cannabis-induced decreases in viral load was not seen in cannabis users in another larger study of infected people, suggesting that cannabis use may have a greater effect in suppressing HIV during acute infection (Okafor et al. 2017). Various components of cannabis may also be responsible for the conflicting effects seen in the studies. For example, despite its anti-inflammatory properties, Δ9-Tetrahydrocannabinol can impair memory and decrease ARV adherence, and cannabidiol, associated with reduced anxiety and antipsychotic effects, could have positive behavior outcomes (Cabral and Griffin-Thomas 2009; Okafor et al. 2017). Future studies evaluating the effect of cannabis on the HIV-infected population will be able to clarify any benefits associated with cannabis use, as decriminalization and legalization of medical and recreational marijuana increase in the United States (Milloy et al. 2015).

7 Fluid Biomarkers of Neurocognitive Impairment Are Products of Inflammation

Recently, fluid biomarkers have emerged as a means to identify and track disease progression and are a valuable tool to clinicians as they are minimally invasive procedures. As inflammation has been identified as a major contributor to neurocognitive impairment, hallmarks of peripheral inflammation and neuroinflammation could be indicators of damage to the CNS in HIV. We discuss a few of these markers below.

7.1 Neurofilament Light Chain (NF-L)

Neurofilaments are structural proteins in neurons that are released into the CSF and blood following neuronal damage and axonal disruption (McGuire et al. 2015; Gisslen et al. 2016; Sun et al. 2010; Jessen Krut et al. 2014). Of the three neurofilament core chains, the neurofilament light chain protein (NF-L) is elevated in HIV-infected individuals with ANI, MND, and HAD and was identified as one of the earliest biomarkers of neurocognitive impairment (McGuire et al. 2015; Gisslen et al. 2016). Elevated NF-L in the CSF is considered to be a sensitive surrogate marker of neuronal damage, since it is elevated with aging and also associated with pathological white matter changes in several neurodegenerative diseases, including

Alzheimer's disease, subcortical vascular dementia, amyotrophic lateral sclerosis, and multiple sclerosis (McGuire et al. 2015; Gisslen et al. 2016; Abdulle et al. 2007). Although nonspecific for HIV, the positive correlation between plasma viral load and NF-L levels in the CSF further emphasizes a link between neuronal injury and systemic HIV infection and suggests that NF-L could be used as a blood biomarker of neurocognitive impairment in HIV (McGuire et al. 2015; Jessen Krut et al. 2014; Peluso et al. 2013). Because NF-L levels are elevated in both acute and chronic HIV infections and are detected in HIV-infected individuals even before a diagnosis of HAND, neuronal injury likely occurs early in HIV infection and could be useful in diagnosis of subclinical neurocognitive impairment (Gisslen et al. 2016; Peluso et al. 2013). NF-L levels decrease with ART initiation and increase upon ART interruption, and HAD patients have improved neurocognition when CSF NF-L is decreased (Jessen Krut et al. 2014; Abdulle et al. 2007; Peluso et al. 2013). Taken together, these pieces of evidence suggest that NF-L levels are a valid marker for approximating neurocognitive impairment in HIV-infected individuals (McGuire et al. 2015; Mellgren et al. 2007).

7.2 *Soluble CD14 (sCD14)*

While CSF is closer to the CNS pathology, is often used to monitor neurological changes, and is a relatively safe procedure, its sampling is more invasive than a simple blood draw and may not always be feasible (Chan et al. 2018; Kamat et al. 2012; Knudsen et al. 2016). Moreover, CSF may not be the most informational barometer for monitoring the pathological changes that occur during HIV infection in the CNS (Chan et al. 2018); thus, biomarkers detectable in plasma have gained favor as a less invasive method to monitor ANI and MND in HIV-infected patients.

Many studies report an elevated sCD14 in treated and untreated HIV-infected subjects (McGuire et al. 2015; Rempel et al. 2010; Patro et al. 2016; Pulliam et al. 1997; Sun et al. 2010; Castley et al. 2016), which is not surprising since monocytic activation and the increase in CD14/16 monocytes are also established as a facet of immune activation in HIV. Increased sCD14 levels are associated with decreased cognitive performance, particularly in women and men who have suppressed plasma HIV RNA (Ancuta et al. 2008; Imp et al. 2017; Lyons et al. 2011) and with a higher risk of death (Sandler et al. 2011). In contrast, although studies confirmed elevated sCD14 in HIV patients, some reports found little association between sCD14 and neuropsychological impairment, potentially due to differential M/Mϕ activation (Sun et al. 2010; Burdo et al. 2013). Imp et al. also demonstrated that markers of inflammation that were not linked to monocyte activation were less likely to impact cognition in HIV-infected individuals (Imp et al. 2017). They also noted that sCD163 correlated better with cognitive impairment than sCD14 and suggested this phenomenon is likely due to monocyte infection and activation rather than gut microbial translocation in their cohort (Imp et al. 2017). Collectively, these results suggest that although sCD14 is elevated in HIV-infected individuals, it may not be the best plasma biomarker for detecting HAND in ANI.

7.3 Soluble CD163 (sCD163)

CD163 is a receptor expressed predominantly on M/Mϕ, and although the breadth of its function remains to be defined, sCD163 may have dual proinflammatory and anti-inflammatory roles (Knudsen et al. 2016; Fabriek et al. 2009). For example, sCD163 is believed to be important in resolving inflammation but is also associated with disease progression in viral hepatitis and increased mortality after sepsis and tuberculosis (Knudsen et al. 2016). CD163 activation triggers the production of many inflammatory mediators, such as nitric oxide, TNF-α, and interleukins (IL)-1, 6, and 10, confirming its role in the immune response (Fabriek et al. 2009). M/Mϕ activation in response to proinflammatory stimuli also results in the release of sCD163 into circulation (Knudsen et al. 2016).

Although studies have hinted that plasma sCD163 levels could be correlated with HIV-related morbidity, the definitive relationship between sCD163 to disease progression and outcome remains to be determined (Knudsen et al. 2016). What is known is that CD163 expression is increased in HIV-infected subjects (Knudsen et al. 2016; Castley et al. 2016; Imp et al. 2017; Beltran et al. 2014). In addition, CD14/16 expressing M/Mϕ have higher CD163 expression, show greater efficiency in crossing the BBB and entering the brain parenchyma, accumulate in perivascular brain regions in HIV-infected individuals, and can harbor productive infection, indicating that CD163 could indeed contribute to and/or signal neurocognitive impairment (McGuire et al. 2015; Imp et al. 2017; Burdo et al. 2013; Fischer-Smith et al. 2008). Furthermore, sCD163 remains elevated in chronically infected HIV persons, despite ART, possibly due to the lingering, low-level peripheral inflammation that persists even after HIV suppression (Knudsen et al. 2016; Burdo et al. 2013; Beltran et al. 2014). In addition to mortality and HIV replication, sCD163 in plasma but not CSF is also correlated with impairments in executive function and learning, hallmarks of cognitive impairment in the ART-treated HIV-infected individuals with HAND (Burdo et al. 2013). CD14/16 M/Mϕ are implicated as a significant source of sCD163 in plasma. High plasma CD163 is associated with microglial activation and synaptodendritic damage in HIV cognitive impairment (Mediouni et al. 2015; Burdo et al. 2013; Bryant et al. 2017). Only subjects with undetectable levels of HIV RNA showed significantly reduced plasma sCD163 levels, supporting the use of sCD163 as a biomarker for decreased inflammation in HIV infection (Castley et al. 2016). As sCD163 blocks proliferation of activated T cells (Etzerodt et al. 2014; Hogger and Sorg 2001), identifying and treating HIV infection early may better preserve cognition in HIV-infected individuals.

Coinfection with HCV, persistent HIV replication, and the inclusion of a protease inhibitor to treatment could be other mechanisms that sustain inflammation in HIV infection, as they may attenuate ART-dependent decreases of plasma sCD163 levels in treated individuals (Knudsen et al. 2016; Beltran et al. 2014). Moreover, due to the high comorbidity of substance abuse disorders and HIV infection, increased immune activation was hypothesized to be partly a result of substance abuse. However, Knudsen et al. observed the opposite phenomenon; they noted that elevated sCD163 levels and mortality had significantly stronger associations in nonsmokers and non-injection drug users than in smokers and injection drug users (Knudsen et al. 2016). Therefore, sCD163 levels due to substance abuse may not significantly correlate to mortality in HIV (Knudsen et al. 2016; Hunt 2016).

In HIV-infected persons, viral replication was also highly correlated with elevated sCD163 levels in women compared to men (Knudsen et al. 2016; Castley et al. 2016; Imp et al. 2017; Hunt 2016). This phenomenon indicates that there could be sex-based differences in immune activation in HIV infection, particularly since the correlation between sCD163 levels and mortality was also stronger in women than in men (Hunt 2016). Newer studies looking at other markers of neuroinflammation and neurocognitive deficits will likely continue to find sex-based differences (Fig. 2).

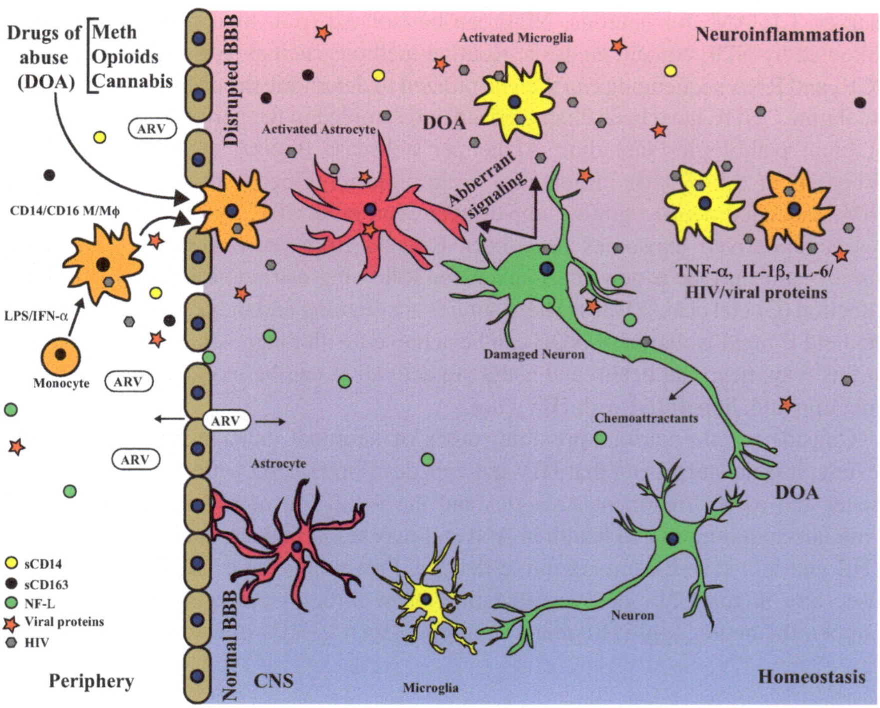

Fig. 2 Changes in CNS homeostasis due to neuroinflammation

7.4 Exosomes as Fluid Biomarkers of Neuroinflammation

We have discussed peripheral inflammation and neuroinflammation and their potential biomarkers in relation to cognitive impairment. To date, classic monocyte markers of inflammation may be appropriate to gauge effective ART as well as sCD163 and sCD14. However, simultaneously assessing neuroinflammation and neuronal health in connection with cognitive impairment is more difficult.

Extracellular vesicles are shed from all cells during homeostasis and pathological conditions. They can be separated by size, and studies typically define exosomes as vesicles around 100 nm in size. The use of plasma or serum to isolate exosomes would be ideal, as sampling blood is less invasive than CSF. Like all cells, M/Mϕ also secrete exosomes normally, and M/Mϕ-derived exosomes can enter resting monocytes and activate them (Tang et al. 2016). M/Mϕ exosomes preferentially contain DNA and abundant miRNA from the parent cell, they can migrate across the BBB, and be taken up by other cells, including cells in the CNS (Gupta and Pulliam 2014; Alvarez-Erviti et al. 2011). In vitro, M/Mϕ exosomes are readily taken up by neurons and astrocytes (Pulliam and Gupta 2015). Just as cells take up exosomes, they also release them, and in 2015 neuron-derived exosomes (NDE) were isolated and characterized from the plasma (Goetzl et al. 2015). By using the cell surface marker, L1CAM, for neurons, NDE can be isolated from total plasma exosomes, followed by NDE content analysis. Routine methods such as Western blot, ELISA, PCR, and RNA sequencing can then be utilized to determine the state of the neuron in "real time." NDE have been characterized and sampled in the periphery for a number of brain pathologies including Alzheimer's disease (Goetzl et al. 2016, 2018a; Winston et al. 2016), traumatic brain injury (Goetzl et al. 2019), and HIV-associated neurocognitive impairment (Sun et al. 2017; Pulliam et al. 2019). Astrocyte-derived exosomes have been isolated from persons with Alzheimer's disease and may be a new mechanism for following astrocytic activation in HIV infection (Goetzl et al. 2018b). These studies are ongoing and may reflect the future of the field if rapid isolation of NDE can be achieved with diagnostic signature targets. In this way, neuronal health and astrocyte activation can be monitored during HIV infection and, hopefully, with HIV cure.

Considering the various possible routes of neuronal damage discussed in this review, it is not surprising that HIV patients develop HAND with chronic infection. Better and earlier treatment strategies and the validation of biomarkers will allow clinicians to monitor CNS health in ANI and may result in decreased viral load in the CNS and less severe neurocognitive deficits. Preventing the establishment of viral reservoirs in the CNS or by eradicating them through emerging therapies may dampen the neurocognitive damages currently seen in HIV-infected individuals.

8 Clinical Implications, Translational Aspects, and Future Directions

For those treated and virally suppressed individuals with HIV infection who continue to have peripheral activation, going forward, we need to find an indicator for this peripheral activation, and in the context of cognitive impairment, does the indicator correlate with comorbidities and/or cognitive impairment? In a recent report, protein cargo from NDE differentiated ANI from MND in HIV-infected men and women in a differential manner, suggesting gender differences in the mechanisms of HAND (Sun et al. 2019). If one is in the M/Mϕ camp that believes these activated cells are an initiator of cognitive impairment, then decreasing this activation would be paramount. While an undetectable viral load in most treated HIV-infected individuals also correlates with a diminished peripheral activation, a subset continues to have circulating activated monocytes. Silencing this expression with targeted antibodies might be one approach to diminishing peripheral and thereby cerebral monocyte activation. Microvesicles bind to and can transport functional miRNAs to recipient cells. The field is still looking at the propensity of like to like transfer of exosomes; that is, are activated monocyte exosomes more likely to be transferred to other monocytes or to endothelial cells? One approach might be engineering monocyte exosomes to silence miRNAs associated with monocyte peripheral inflammation (Ismail et al. 2013). Monocytes activated with LPS and IFNα in vitro to mimic activated monocytes in HIV infection release exosomes that can be readily transferred to endothelial cells (Dalvi et al. 2017). The exosomes contain abundant miRNAs that can regulate genes at the posttranscriptional level. The functional expression of miR-155 and miR-222 coincided with an increase in CCL2, VCAM, and ICAM, adhesion molecules that modulate vascular function. Manipulation of monocyte miRNAs to overexpress or silence a particular miRNA associated with inflammation might decrease HIV comorbidities such as metabolic and cardiovascular diseases before the M/Mϕ cross the endothelium.

Another translational aspect for the future of cognitive impairment is the gut-brain axis connection. Increasing data shows that biochemical manipulation of the gut may have implications in the CNS.

Studies on infants show that a vaginal birth has colonization similar to the mother's signature with an abundance in *Lactobacillus* and *Prevotell*a spp., while those born with cesarean section have more skin flora enriched with *Staphylococci* and *Propionibacterium* spp. (Dominguez-Bello et al. 2010). The gut microbiome expands with age and is greatly influenced by diet, disease, environmental factors, drugs, and an inflammatory response. In a recent publication, the interferon response was altered with a twice daily multistrain probiotic supplementation for 6 months (Pinacchio et al. 2018). This regimen was able to change the IFN subtype response for the better. Might a probiotic lower the peripheral IFN phenotype in HIV? In HIV infection, the gut microbiome is altered with some bacteria enriched and other bacteria reduced [Review (Scagnolari and Antonelli 2018)] with an overall conclusion that HIV was associated with reduced bacterial diversity (Dillon et al. 2016; Ribeiro et al. 2017). Since the gut is known to be an HIV reservoir, attention is now focused on reducing the size of this reservoir for cure, and this may be possible by altering the gut microbiome in HIV infection (Koay et al. 2018).

There is now an explosive field looking at how the gut microbiome influences neurodegenerative diseases [Review (Giau et al. 2018)]. Certain bacterial species produce functional extracellular amyloid that forms Aβ fibrils in vitro (Pistollato et al. 2016). In addition, LPS produced from gut bacteria also creates a proinflammatory environment and is associated with breakdown of the BBB (Martin et al. 2018; Michel and Prat 2016). Some of the best evidence for the gut-brain connection comes from behavioral and mental illness studies using probiotics [Review (Gareau 2016)]. Data from irritable bowel syndrome showed that manipulation of the gut flora with probiotics (certain *Bifidobacterium* spp.) improved cognitive function (Pinto-Sanchez et al. 2017; Savignac et al. 2014). Another study using a multistrain probiotic for 6 months reported an improvement in some neurocognitive functions in HIV-infected persons (Ceccarelli et al. 2017). If we could determine the right gut balance to ward off inflammation, this may translate to improved cognition.

The challenge will be to diagnose cognitive impairment early to be able to stop the advancement with these new approaches and, in the case of HIV, stop the further impairment until virus eradication is established.

References

Abdulle S, Mellgren A, Brew BJ, Cinque P, Hagberg L, Price RW et al (2007) CSF neurofilament protein (NFL) – a marker of active HIV-related neurodegeneration. J Neurol 254(8):1026–1032

Allam O, Samarani S, Mehraj V, Jenabian MA, Tremblay C, Routy JP et al (2018) HIV induces production of IL-18 from intestinal epithelial cells that increases intestinal permeability and microbial translocation. PLoS One 13(3):e0194185

Alvarez-Erviti L, Seow Y, Yin H, Betts C, Lakhal S, Wood MJ (2011) Delivery of siRNA to the mouse brain by systemic injection of targeted exosomes. Nat Biotechnol 29(4):341–345

Ancuta P, Kamat A, Kunstman KJ, Kim EY, Autissier P, Wurcel A et al (2008) Microbial translocation is associated with increased monocyte activation and dementia in AIDS patients. PLoS One 3(6):e2516

Aweeka F, Jayewardene A, Staprans S, Bellibas SE, Kearney B, Lizak P et al (1999) Failure to detect nelfinavir in the cerebrospinal fluid of HIV-1-infected patients with and without AIDS dementia complex. J Acquir Immune Defic Syndr Hum Retrovirol 20(1):39–43

Baxter AE, Russell RA, Duncan CJ, Moore MD, Willberg CB, Pablos JL et al (2014) Macrophage infection via selective capture of HIV-1-infected CD4+ T cells. Cell Host Microbe 16 (6):711–721

Beltran LM, Munoz Hernandez R, de Pablo Bernal RS, Garcia Morillo JS, Egido J, Noval ML et al (2014) Reduced sTWEAK and increased sCD163 levels in HIV-infected patients: modulation by antiretroviral treatment, HIV replication and HCV co-infection. PLoS One 9(3):e90541

Brenchley JM, Douek DC (2008) HIV infection and the gastrointestinal immune system. Mucosal Immunol 1(1):23–30

Bryant AK, Moore DJ, Burdo TH, Lakritz JR, Gouaux B, Soontornniyomkij V et al (2017) Plasma soluble CD163 is associated with postmortem brain pathology in human immunodeficiency virus infection. AIDS 31(7):973–979

Burdo TH, Weiffenbach A, Woods SP, Letendre S, Ellis RJ, Williams KC (2013) Elevated sCD163 in plasma but not cerebrospinal fluid is a marker of neurocognitive impairment in HIV infection. AIDS 27(9):1387–1395

Byrd DA, Fellows RP, Morgello S, Franklin D, Heaton RK, Deutsch R et al (2011) Neurocognitive impact of substance use in HIV infection. J Acquir Immune Defic Syndr 58(2):154–162

Cabral GA, Griffin-Thomas L (2009) Emerging role of the cannabinoid receptor CB2 in immune regulation: therapeutic prospects for neuroinflammation. Expert Rev Mol Med 11:e3

Cahill CM, Taylor AM (2017) Neuroinflammation-a co-occurring phenomenon linking chronic pain and opioid dependence. Curr Opin Behav Sci 13:171–177

Castley A, Williams L, James I, Guelfi G, Berry C, Nolan D (2016) Plasma CXCL10, sCD163 and sCD14 levels have distinct associations with antiretroviral treatment and cardiovascular disease risk factors. PLoS One 11(6):e0158169

Ceccarelli G, Fratino M, Selvaggi C, Giustini N, Serafino S, Schietroma I et al (2017) A pilot study on the effects of probiotic supplementation on neuropsychological performance and microRNA-29a-c levels in antiretroviral-treated HIV-1-infected patients. Brain Behav 7(8):e00756

Cenker JJ, Stultz RD, McDonald D (2017) Brain microglial cells are highly susceptible to HIV-1 infection and spread. AIDS Res Hum Retroviruses 33(11):1155–1165

Chan P, Hellmuth J, Colby D, Kroon E, Sacdalan C, Fletcher J et al (2018) Safety of lumbar puncture procedure in an international research setting during acute HIV infection. J Virus Erad 4(1):16–20

Chen NC, Partridge AT, Sell C, Torres C, Martin-Garcia J (2017) Fate of microglia during HIV-1 infection: from activation to senescence? Glia 65(3):431–446

Chompre G, Cruz E, Maldonado L, Rivera-Amill V, Porter JT, Noel RJ Jr (2013) Astrocytic expression of HIV-1 Nef impairs spatial and recognition memory. Neurobiol Dis 49:128–136

Churchill MJ, Wesselingh SL, Cowley D, Pardo CA, McArthur JC, Brew BJ et al (2009) Extensive astrocyte infection is prominent in human immunodeficiency virus-associated dementia. Ann Neurol 66(2):253–258

Clayton KL, Garcia JV, Clements JE, Walker BD (2017) HIV infection of macrophages: implications for pathogenesis and cure. Pathog Immun 2(2):179–192

Dalvi P, Sun B, Tang N, Pulliam L (2017) Immune activated monocyte exosomes alter microRNAs in brain endothelial cells and initiate an inflammatory response through the TLR4/MyD88 pathway. Sci Rep 7(1):9954

Dalwadi DA, Kim S, Amdani SM, Chen Z, Huang RQ, Schetz JA (2016) Molecular mechanisms of serotonergic action of the HIV-1 antiretroviral efavirenz. Pharmacol Res 110:10–24

Dillon SM, Frank DN, Wilson CC (2016) The gut microbiome and HIV-1 pathogenesis: a two-way street. AIDS 30(18):2737–2751

Dominguez-Bello MG, Costello EK, Contreras M, Magris M, Hidalgo G, Fierer N et al (2010) Delivery mode shapes the acquisition and structure of the initial microbiota across multiple body habitats in newborns. Proc Natl Acad Sci U S A 107(26):11971–11975

Dubovy P, Klusakova I, Hradilova-Svizenska I, Joukal M, Boadas-Vaello P (2018) Activation of astrocytes and microglial cells and CCL2/CCR2 upregulation in the dorsolateral and ventrolateral nuclei of periaqueductal gray and rostral ventromedial medulla following different types of sciatic nerve injury. Front Cell Neurosci 12:40

Ellery PJ, Tippett E, Chiu YL, Paukovics G, Cameron PU, Solomon A et al (2007) The CD16+ monocyte subset is more permissive to infection and preferentially harbors HIV-1 in vivo. J Immunol 178(10):6581–6589

Etzerodt A, Rasmussen MR, Svendsen P, Chalaris A, Schwarz J, Galea I et al (2014) Structural basis for inflammation-driven shedding of CD163 ectodomain and tumor necrosis factor-alpha in macrophages. J Biol Chem 289(2):778–788

Eugenin EA, King JE, Nath A, Calderon TM, Zukin RS, Bennett MV et al (2007) HIV-tat induces formation of an LRP-PSD-95- NMDAR-nNOS complex that promotes apoptosis in neurons and astrocytes. Proc Natl Acad Sci U S A 104(9):3438–3443

Eugenin EA, Clements JE, Zink MC, Berman JW (2011) Human immunodeficiency virus infection of human astrocytes disrupts blood-brain barrier integrity by a gap junction-dependent mechanism. J Neurosci 31(26):9456–9465

Fabbiani M, Muscatello A, Perseghin P, Bani M, Incontri A, Squillace N et al (2017) Brief report: peripheral monocyte/macrophage phenotypes associated with the evolution of cognitive performance in HIV-infected patients. J Acquir Immune Defic Syndr 76(2):219–224

Fabriek BO, van Bruggen R, Deng DM, Ligtenberg AJ, Nazmi K, Schornagel K et al (2009) The macrophage scavenger receptor CD163 functions as an innate immune sensor for bacteria. Blood 113(4):887–892

Fischer-Smith T, Bell C, Croul S, Lewis M, Rappaport J (2008) Monocyte/macrophage trafficking in acquired immunodeficiency syndrome encephalitis: lessons from human and nonhuman primate studies. J Neurovirol 14(4):318–326

Gareau MG (2016) Cognitive function and the microbiome. Int Rev Neurobiol 131:227–246

Gatch MB, Kozlenkov A, Huang RQ, Yang W, Nguyen JD, Gonzalez-Maeso J et al (2013) The HIV antiretroviral drug efavirenz has LSD-like properties. Neuropsychopharmacology 38 (12):2373–2384

Gelman BB, Lisinicchia JG, Chen T, Johnson KM, Jennings K, Freeman DH Jr et al (2012) Prefrontal dopaminergic and enkephalinergic synaptic accommodation in HIV-associated neurocognitive disorders and encephalitis. J Neuroimmune Pharmacol 7(3):686–700

Giau VV, Wu SY, Jamerlan A, An SSA, Kim SY, Hulme J (2018) Gut microbiota and their neuroinflammatory implications in Alzheimer's disease. Nutrients 10(11). pii: E1765

Gill AJ, Kovacsics CE, Cross SA, Vance PJ, Kolson LL, Jordan-Sciutto KL et al (2014) Heme oxygenase-1 deficiency accompanies neuropathogenesis of HIV-associated neurocognitive disorders. J Clin Invest 124(10):4459–4472

Ginsberg SD, Alldred MJ, Gunnam SM, Schiroli C, Lee SH, Morgello S et al (2018) Expression profiling suggests microglial impairment in human immunodeficiency virus neuropathogenesis. Ann Neurol 83(2):406–417

Gisslen M, Price RW, Andreasson U, Norgren N, Nilsson S, Hagberg L et al (2016) Plasma concentration of the neurofilament light protein (NFL) is a biomarker of CNS injury in HIV infection: a cross-sectional study. EBioMedicine 3:135–140

Goetzl EJ, Boxer A, Schwartz JB, Abner EL, Petersen RC, Miller BL et al (2015) Altered lysosomal proteins in neural-derived plasma exosomes in preclinical Alzheimer disease. Neurology 85 (1):40–47

Goetzl EJ, Goetzl L, Karliner JS, Tang N, Pulliam L (2016) Human plasma platelet-derived exosomes: effects of aspirin. FASEB J 30(5):2058–2063

Goetzl EJ, Abner EL, Jicha GA, Kapogiannis D, Schwartz JB (2018a) Declining levels of functionally specialized synaptic proteins in plasma neuronal exosomes with progression of Alzheimer's disease. FASEB J 32(2):888–893

Goetzl EJ, Schwartz JB, Abner EL, Jicha GA, Kapogiannis D (2018b) High complement levels in astrocyte-derived exosomes of Alzheimer disease. Ann Neurol 83(3):544–552

Goetzl EJ, Elahi FM, Mustapic M, Kapogiannis D, Pryhoda M, Gilmore A et al (2019) Altered levels of plasma neuron-derived exosomes and their cargo proteins characterize acute and chronic mild traumatic brain injury. FASEB J 33(4):5082–5088. https://doi.org/10.1096/fj.201802319R

Guo L, Xing Y, Pan R, Jiang M, Gong Z, Lin L et al (2013) Curcumin protects microglia and primary rat cortical neurons against HIV-1 gp120-mediated inflammation and apoptosis. PLoS One 8(8):e70565

Gupta A, Pulliam L (2014) Exosomes as mediators of neuroinflammation. J Neuroinflammation 11:68

He J, Chen Y, Farzan M, Choe H, Ohagen A, Gartner S et al (1997) CCR3 and CCR5 are co-receptors for HIV-1 infection of microglia. Nature 385(6617):645–649

Heaton RK, Clifford DB, Franklin DR Jr, Woods SP, Ake C, Vaida F et al (2010) HIV-associated neurocognitive disorders persist in the era of potent antiretroviral therapy: CHARTER Study. Neurology 75(23):2087–2096

Hellmuth J, Valcour V, Spudich S (2015) CNS reservoirs for HIV: implications for eradication. J Virus Erad 1(2):67–71

Hileman CO, Funderburg NT (2017) Inflammation, immune activation, and antiretroviral therapy in HIV. Curr HIV/AIDS Rep 14(3):93–100

Hogger P, Sorg C (2001) Soluble CD163 inhibits phorbol ester-induced lymphocyte proliferation. Biochem Biophys Res Commun 288(4):841–843

Hunt PW (2016) Soluble CD163 and clinical outcomes in treated HIV infection: insights into mechanisms. J Infect Dis 214(8):1132–1133

Imp BM, Rubin LH, Tien PC, Plankey MW, Golub ET, French AL et al (2017) Monocyte activation is associated with worse cognitive performance in HIV-infected women with virologic suppression. J Infect Dis 215(1):114–121

Ismail N, Wang Y, Dakhlallah D, Moldovan L, Agarwal K, Batte K et al (2013) Macrophage microvesicles induce macrophage differentiation and miR-223 transfer. Blood 121(6):984–995

Jakubzick CV, Randolph GJ, Henson PM (2017) Monocyte differentiation and antigen-presenting functions. Nat Rev Immunol 17(6):349–362

Jaureguiberry-Bravo M, Wilson R, Carvallo L, Berman JW (2016) Opioids and opioid maintenance therapies: their impact on monocyte-mediated HIV neuropathogenesis. Curr HIV Res 14 (5):417–430

Jessen Krut J, Mellberg T, Price RW, Hagberg L, Fuchs D, Rosengren L et al (2014) Biomarker evidence of axonal injury in neuroasymptomatic HIV-1 patients. PLoS One 9(2):e88591

Kahouadji Y, Dumurgier J, Sellier P, Lapalus P, Delcey V, Bergmann J et al (2013) Cognitive function after several years of antiretroviral therapy with stable central nervous system penetration score. HIV Med 14(5):311–315

Kamat A, Lyons JL, Misra V, Uno H, Morgello S, Singer EJ et al (2012) Monocyte activation markers in cerebrospinal fluid associated with impaired neurocognitive testing in advanced HIV infection. J Acquir Immune Defic Syndr 60(3):234–243

Kanmogne GD (2005) Noninfectious pulmonary complications of HIV/AIDS. Curr Opin Pulm Med 11(3):208–212

Kanmogne GD, Schall K, Leibhart J, Knipe B, Gendelman HE, Persidsky Y (2007) HIV-1 gp120 compromises blood-brain barrier integrity and enhances monocyte migration across blood-brain barrier: implication for viral neuropathogenesis. J Cereb Blood Flow Metab 27(1):123–134

Kesby JP, Najera JA, Romoli B, Fang Y, Basova L, Birmingham A et al (2017) HIV-1 TAT protein enhances sensitization to methamphetamine by affecting dopaminergic function. Brain Behav Immun 65:210–221

Kitayama H, Miura Y, Ando Y, Hoshino S, Ishizaka Y, Koyanagi Y (2008) Human immunodeficiency virus type 1 Vpr inhibits axonal outgrowth through induction of mitochondrial dysfunction. J Virol 82(5):2528–2542

Knudsen TB, Ertner G, Petersen J, Moller HJ, Moestrup SK, Eugen-Olsen J et al (2016) Plasma soluble CD163 level independently predicts all-cause mortality in HIV-1-infected individuals. J Infect Dis 214(8):1198–1204

Koay WLA, Siems LV, Persaud D (2018) The microbiome and HIV persistence: implications for viral remission and cure. Curr Opin HIV AIDS 13(1):61–68

Koneru R, Bimonte-Nelson H, Ciavatta V, Haile W, Elmore K, Ward J et al (2018) Reversing interferon-alpha neurotoxicity in a HIV-associated neurocognitive disorder mouse model. AIDS 32(11):1403–1411

Kovalevich J, Langford D (2012) Neuronal toxicity in HIV CNS disease. Future Virol 7 (7):687–698

Letendre SL, McCutchan JA, Childers ME, Woods SP, Lazzaretto D, Heaton RK et al (2004) Enhancing antiretroviral therapy for human immunodeficiency virus cognitive disorders. Ann Neurol 56(3):416–423
Liang H, Duan Z, Li D, Li D, Wang Z, Ren L et al (2015) Higher levels of circulating monocyte-platelet aggregates are correlated with viremia and increased sCD163 levels in HIV-1 infection. Cell Mol Immunol 12(4):435–443
Liu Y, Liu H, Kim BO, Gattone VH, Li J, Nath A et al (2004) CD4-independent infection of astrocytes by human immunodeficiency virus type 1: requirement for the human mannose receptor. J Virol 78(8):4120–4133
Lopez-Cortes LF, Trujillo-Rodriguez M, Baez-Palomo A, Benmarzouk-Hidalgo OJ, Dominguez-Molina B, Milanes-Guisado Y et al (2018) Eradication of hepatitis C virus (HCV) reduces immune activation, microbial translocation, and the HIV DNA level in HIV/HCV-coinfected patients. J Infect Dis 218(4):624–632
Lyons JL, Uno H, Ancuta P, Kamat A, Moore DJ, Singer EJ et al (2011) Plasma sCD14 is a biomarker associated with impaired neurocognitive test performance in attention and learning domains in HIV infection. J Acquir Immune Defic Syndr 57(5):371–379
Marchetti G, Tincati C, Silvestri G (2013) Microbial translocation in the pathogenesis of HIV infection and AIDS. Clin Microbiol Rev 26(1):2–18
Marra CM, Zhao Y, Clifford DB, Letendre S, Evans S, Henry K et al (2009) Impact of combination antiretroviral therapy on cerebrospinal fluid HIV RNA and neurocognitive performance. AIDS 23(11):1359–1366
Martin CR, Osadchiy V, Kalani A, Mayer EA (2018) The brain-gut-microbiome axis. Cell Mol Gastroenterol Hepatol 6(2):133–148
McGuire JL, Gill AJ, Douglas SD, Kolson DL, CNS HIV Anti-Retroviral Therapy Effects Research (CHARTER) Group (2015) Central and peripheral markers of neurodegeneration and monocyte activation in HIV-associated neurocognitive disorders. J Neurovirol 21(4):439–448
Mediouni S, Darque A, Baillat G, Ravaux I, Dhiver C, Tissot-Dupont H et al (2012) Antiretroviral therapy does not block the secretion of the human immunodeficiency virus tat protein. Infect Disord Drug Targets 12(1):81–86
Mediouni S, Marcondes MC, Miller C, McLaughlin JP, Valente ST (2015) The cross-talk of HIV-1 Tat and methamphetamine in HIV-associated neurocognitive disorders. Front Microbiol 6:1164
Mellgren A, Price RW, Hagberg L, Rosengren L, Brew BJ, Gisslen M (2007) Antiretroviral treatment reduces increased CSF neurofilament protein (NFL) in HIV-1 infection. Neurology 69(15):1536–1541
Michel L, Prat A (2016) One more role for the gut: microbiota and blood brain barrier. Ann Transl Med 4(1):15
Milloy MJ, Marshall B, Kerr T, Richardson L, Hogg R, Guillemi S et al (2015) High-intensity cannabis use associated with lower plasma human immunodeficiency virus-1 RNA viral load among recently infected people who use injection drugs. Drug Alcohol Rev 34(2):135–140
Nair M, Maria JM, Agudelo M, Yndart A, Vargas-Rivera ME (2015) Platelets contribute to BBB disruption induced by HIV and alcohol. J Alcohol Drug Depend 3(1):182
Nedellec R, Coetzer M, Shimizu N, Hoshino H, Polonis VR, Morris L et al (2009) Virus entry via the alternative coreceptors CCR3 and FPRL1 differs by human immunodeficiency virus type 1 subtype. J Virol 83(17):8353–8363
Nottet HS, Persidsky Y, Sasseville VG, Nukuna AN, Bock P, Zhai QH et al (1996) Mechanisms for the transendothelial migration of HIV-1-infected monocytes into brain. J Immunol 156 (3):1284–1295
Okafor CN, Zhou Z, Burrell LE 2nd, Kelso NE, Whitehead NE, Harman JS et al (2017) Marijuana use and viral suppression in persons receiving medical care for HIV-infection. Am J Drug Alcohol Abuse 43(1):103–110

Patro SC, Azzoni L, Joseph J, Fair MG, Sierra-Madero JG, Rassool MS et al (2016) Antiretroviral therapy in HIV-1-infected individuals with CD4 count below 100 cells/mm3 results in differential recovery of monocyte activation. J Leukoc Biol 100(1):223–231

Peluso MJ, Ferretti F, Peterson J, Lee E, Fuchs D, Boschini A et al (2012) Cerebrospinal fluid HIV escape associated with progressive neurologic dysfunction in patients on antiretroviral therapy with well controlled plasma viral load. AIDS 26(14):1765–1774

Peluso MJ, Meyerhoff DJ, Price RW, Peterson J, Lee E, Young AC et al (2013) Cerebrospinal fluid and neuroimaging biomarker abnormalities suggest early neurological injury in a subset of individuals during primary HIV infection. J Infect Dis 207(11):1703–1712

Persidsky Y, Ghorpade A, Rasmussen J, Limoges J, Liu XJ, Stins M et al (1999) Microglial and astrocyte chemokines regulate monocyte migration through the blood-brain barrier in human immunodeficiency virus-1 encephalitis. Am J Pathol 155(5):1599–1611

Pinacchio C, Scheri GC, Statzu M, Santinelli L, Ceccarelli G, Innocenti GP et al (2018) Type I/II interferon in HIV-1-infected patients: expression in gut mucosa and in peripheral blood mononuclear cells and its modification upon probiotic supplementation. J Immunol Res 2018:1738676

Pinto-Sanchez MI, Hall GB, Ghajar K, Nardelli A, Bolino C, Lau JT et al (2017) Probiotic bifidobacterium longum NCC3001 reduces depression scores and alters brain activity: a pilot study in patients with irritable bowel syndrome. Gastroenterology 153(2):448–59.e8

Pistollato F, Sumalla Cano S, Elio I, Masias Vergara M, Giampieri F, Battino M (2016) Role of gut microbiota and nutrients in amyloid formation and pathogenesis of Alzheimer disease. Nutr Rev 74(10):624–634

Potula R, Hawkins BJ, Cenna JM, Fan S, Dykstra H, Ramirez SH et al (2010) Methamphetamine causes mitochondrial oxidative damage in human T lymphocytes leading to functional impairment. J Immunol 185(5):2867–2876

Pulliam L (2014) Cognitive consequences of a sustained monocyte type 1 IFN response in HIV-1 infection. Curr HIV Res 12(2):77–84

Pulliam L, Gupta A (2015) Modulation of cellular function through immune-activated exosomes. DNA Cell Biol 34(7):459–463

Pulliam L, Gascon R, Stubblebine M, McGuire D, McGrath MS (1997) Unique monocyte subset in patients with AIDS dementia. Lancet 349(9053):692–695

Pulliam L, Calosing C, Sun B, Grunfeld C, Rempel H (2014) Monocyte activation from interferon-alpha in HIV infection increases acetylated LDL uptake and ROS production. J Interferon Cytokine Res 34(10):822–828

Pulliam L, Sun B, Mustapic M, Chawla S, Kapogiannis D (2019) Plasma neuronal exosomes serve as biomarkers of cognitive impairment in HIV infection and Alzheimer's disease. J Neurovirol. https://doi.org/10.1007/s13365-018-0695-4

Rempel H, Calosing C, Sun B, Pulliam L (2008) Sialoadhesin expressed on IFN-induced monocytes binds HIV-1 and enhances infectivity. PLoS One 3(4):e1967

Rempel H, Sun B, Calosing C, Pillai SK, Pulliam L (2010) Interferon-alpha drives monocyte gene expression in chronic unsuppressed HIV-1 infection. AIDS 24(10):1415–1423

Rempel H, Sun B, Calosing C, Abadjian L, Monto A, Pulliam L (2013) Monocyte activation in HIV/HCV coinfection correlates with cognitive impairment. PLoS One 8(2):e55776

Ribeiro A, Heimesaat MM, Bereswill S (2017) Changes of the intestinal microbiome-host homeostasis in HIV-infected individuals – a focus on the bacterial gut microbiome. Eur J Microbiol Immunol (Bp) 7(3):158–167

Rizzo MD, Crawford RB, Henriquez JE, Aldhamen YA, Gulick P, Amalfitano A et al (2018) HIV-infected cannabis users have lower circulating CD16+ monocytes and IFN-gamma-inducible protein 10 levels compared with nonusing HIV patients. AIDS 32(4):419–429

Sami Saribas A, Cicalese S, Ahooyi TM, Khalili K, Amini S, Sariyer IK (2017) HIV-1 Nef is released in extracellular vesicles derived from astrocytes: evidence for Nef-mediated neurotoxicity. Cell Death Dis 8(1):e2542

Sandler NG, Wand H, Roque A, Law M, Nason MC, Nixon DE et al (2011) Plasma levels of soluble CD14 independently predict mortality in HIV infection. J Infect Dis 203(6):780–790
Savignac HM, Kiely B, Dinan TG, Cryan JF (2014) Bifidobacteria exert strain-specific effects on stress-related behavior and physiology in BALB/c mice. Neurogastroenterol Motil 26 (11):1615–1627
Scagnolari C, Antonelli G (2018) Type I interferon and HIV: subtle balance between antiviral activity, immunopathogenesis and the microbiome. Cytokine Growth Factor Rev 40:19–31
Serrano-Villar S, Sainz T, Ma ZM, Utay NS, Chun TW, Mann S et al (2016) Effects of combined CCR5/integrase inhibitors-based regimen on mucosal immunity in HIV-infected patients naive to antiretroviral therapy: a pilot randomized trial. PLoS Pathog 12(1):e1005381
Soontornniyomkij V, Kesby JP, Morgan EE, Bischoff-Grethe A, Minassian A, Brown GG et al (2016) Effects of HIV and methamphetamine on brain and behavior: evidence from human studies and animal models. J Neuroimmune Pharmacol 11(3):495–510
Sun B, Abadjian L, Rempel H, Calosing C, Rothlind J, Pulliam L (2010) Peripheral biomarkers do not correlate with cognitive impairment in highly active antiretroviral therapy-treated subjects with human immunodeficiency virus type 1 infection. J Neurovirol 16(2):115–124
Sun B, Dalvi P, Abadjian L, Tang N, Pulliam L (2017) Blood neuron-derived exosomes as biomarkers of cognitive impairment in HIV. AIDS 31(14):F9–F17
Sun B, Fernandes N, Pulliam L (2019) Profile of neuronal exosomes in HIV cognitive impairment exposes gender differences. AIDS. https://doi.org/10.1097/QAD.0000000000002272. [Epub ahead of print]
Tang N, Sun B, Gupta A, Rempel H, Pulliam L (2016) Monocyte exosomes induce adhesion molecules and cytokines via activation of NF-kappaB in endothelial cells. FASEB J 30 (9):3097–3106
Thames AD, Mahmood Z, Burggren AC, Karimian A, Kuhn TP (2016) Combined effects of HIV and marijuana use on neurocognitive functioning and immune status. AIDS Care 28(5):628–632
Tozzi V, Balestra P, Salvatori MF, Vlassi C, Liuzzi G, Giancola ML et al (2009) Changes in cognition during antiretroviral therapy: comparison of 2 different ranking systems to measure antiretroviral drug efficacy on HIV-associated neurocognitive disorders. J Acquir Immune Defic Syndr 52(1):56–63
Treisman GJ, Soudry O (2016) Neuropsychiatric effects of HIV antiviral medications. Drug Saf 39 (10):945–957
UNAIDS (2018) Global HIV & AIDS statistics – 2018 fact sheet
Var SR, Day TR, Vitomirov A, Smith DM, Soontornniyomkij V, Moore DJ et al (2016) Mitochondrial injury and cognitive function in HIV infection and methamphetamine use. AIDS 30 (6):839–848
Veenstra M, Leon-Rivera R, Li M, Gama L, Clements JE, Berman JW (2017) Mechanisms of CNS viral seeding by HIV(+) CD14(+) CD16(+) monocytes: establishment and reseeding of viral reservoirs contributing to HIV-associated neurocognitive disorders. MBio 8(5). pii: e01280-17
Volkow ND, Chang L, Wang GJ, Fowler JS, Franceschi D, Sedler M et al (2001) Loss of dopamine transporters in methamphetamine abusers recovers with protracted abstinence. J Neurosci 21 (23):9414–9418
Walsh JG, Reinke SN, Mamik MK, McKenzie BA, Maingat F, Branton WG et al (2014) Rapid inflammasome activation in microglia contributes to brain disease in HIV/AIDS. Retrovirology 11:35
Watkins CC, Treisman GJ (2015) Cognitive impairment in patients with AIDS – prevalence and severity. HIV AIDS (Auckl) 7:35–47
Weber E, Blackstone K, Woods SP (2013) Cognitive neurorehabilitation of HIV-associated neurocognitive disorders: a qualitative review and call to action. Neuropsychol Rev 23(1):81–98
Winston CN, Goetzl EJ, Akers JC, Carter BS, Rockenstein EM, Galasko D et al (2016) Prediction of conversion from mild cognitive impairment to dementia with neuronally derived blood exosome protein profile. Alzheimers Dement (Amst) 3:63–72
Woods SP, Moore DJ, Weber E, Grant I (2009) Cognitive neuropsychology of HIV-associated neurocognitive disorders. Neuropsychol Rev 19(2):152–168

Xu E, Liu J, Liu H, Wang X, Xiong H (2017) Role of microglia in methamphetamine-induced neurotoxicity. Int J Physiol Pathophysiol Pharmacol 9(3):84–100

Yeung MC, Pulliam L, Lau AS (1995) The HIV envelope protein gp120 is toxic to human brain-cell cultures through the induction of interleukin-6 and tumor necrosis factor-alpha. AIDS 9 (2):137–143

Zhen A, Krutzik SR, Levin BR, Kasparian S, Zack JA, Kitchen SG (2014) CD4 ligation on human blood monocytes triggers macrophage differentiation and enhances HIV infection. J Virol 88 (17):9934–9946

Zhong Y, Hennig B, Toborek M (2010) Intact lipid rafts regulate HIV-1 Tat protein-induced activation of the Rho signaling and upregulation of P-glycoprotein in brain endothelial cells. J Cereb Blood Flow Metab 30(3):522–533

Neuroimaging Advances in Diagnosis and Differentiation of HIV, Comorbidities, and Aging in the cART Era

Talia M. Nir, Elizabeth Haddad, Paul M. Thompson, and Neda Jahanshad

Contents

Abstract In the "cART era" of more widely available and accessible treatment, aging and HIV-related comorbidities, including symptoms of brain dysfunction, remain common among HIV-infected individuals on suppressive treatment. A better understanding of the neurobiological consequences of HIV infection is essential for

T. M. Nir, E. Haddad, P. M. Thompson, and N. Jahanshad (✉)
Imaging Genetics Center, Mark and Mary Stevens Neuroimaging and Informatics Institute, Keck School of Medicine of USC, University of Southern California, Marina del Rey, CA, USA
e-mail: neda.jahanshad@usc.edu

Curr Topics Behav Neurosci (2021) 50: 105–144
https://doi.org/10.1007/7854_2021_221
Published Online: 30 March 2021

developing thorough treatment guidelines and for optimizing long-term neuropsychological outcomes and overall brain health. In this chapter, we first summarize magnetic resonance imaging (MRI) methods used in over two decades of neuroHIV research. These methods evaluate brain volumetric differences and circuitry disruptions in adults living with HIV, and help map clinical correlations with brain function and tissue microstructure. We then introduce and discuss aging and associated neurological complications in people living with HIV, and processes by which infection may contribute to the risk for late-onset dementias. We describe how new technologies and large-scale international collaborations are helping to disentangle the effect of genetic and environmental risk factors on brain aging and neurodegenerative diseases. We provide insights into how these advances, which are now at the forefront of Alzheimer's disease research, may advance the field of neuroHIV. We conclude with a summary of how we see the field of neuroHIV research advancing in the decades to come and highlight potential clinical implications.

Keywords Aging · Brain MRI · Cognitive impairment · Diffusion MRI · Neurodegeneration · Neuroimaging · Population studies

1 Neurological Consequences of HIV Infection (NeuroHIV)

An estimated 36.9 million people are currently living with HIV worldwide (UNAIDS 2018). Approximately 60% of infected individuals have access to combination antiretroviral therapy (cART), which has reduced transmission rates and increased survival rates. While cART may improve the overall quality of life for those infected, HIV continues to be a major global health issue. Neurological and cognitive impairments are of particular concern to HIV+ individuals on and off treatment, for which the long-term effects are not well understood.

After transmission, HIV rapidly replicates and crosses the blood brain barrier (BBB), either through endothelial cells as free virions or, more likely, through infected CD4+ T-lymphocytes or infected macrophage-monocytes (i.e., the "Trojan horse" hypothesis). The virus then settles and replicates in perivascular macrophages and parenchymal microglial cells that express both the CD4 receptor and the CCR5 or CXCR4 chemokine co-receptor (Peluso et al. 1985; Gonzalez-Scarano and Martin-Garcia 2005; Spudich and Gonzalez-Scarano 2012). Although evidence suggests that neurons are not directly infected by HIV, once HIV is in the central nervous system (CNS), neuronal injury does occur via indirect mechanisms. Inflammatory cytokines, chemokines, and other neurotoxins from the immune system's inflammatory cascade response, along with glutamate excitotoxicity resulting from glial cell death (Gonzalez-Scarano and Martin-Garcia 2005), and viral proteins such as gp120, Tat, and Vpr, can all contribute to injury and functional impairment.

Neuronal dysfunction is often accompanied by neurocognitive deficits (McArthur et al. 2010). One in two adults infected with HIV will experience a range of neuropsychological symptoms, which can include a conjunction of motor, behavioral, and cognitive facets (Navia et al. 1986a, b, Heaton et al. 2010). Common motor symptoms include slowness and loss of balance, behavioral symptoms include apathy and mood disturbances, and cognitive deficits often include mental slowing and deficits in attention or memory. The prevalence of HIV-associated dementia has greatly diminished with increased cART access (Ances and Ellis 2007), but many infected adults continue to experience a broad set of neuropsychological symptoms, often termed HIV-associated neurocognitive disorders or HAND (Navia et al. 1986a, b; Sacktor et al. 2002; Antinori et al. 2007; Tozzi et al. 2007; Brew et al. 2009; Heaton et al. 2010, 2011; Robertson et al. 2012). The incidence of neurocognitive impairments in people living with HIV may, paradoxically, be increasing in the cART era; a spectrum of abnormalities has been observed for nearly every cognitive domain (Cysique et al. 2004, 2006; McArthur 2004; Ances and Ellis 2007; Woods et al. 2009; Sacktor et al. 2016). These neurocognitive deficits persist even with treatment adherence and may reflect distinct underlying HIV neuropathologies in the setting of long-term survival. They may be attributable to factors such as (1) prolonged cART exposure, which may be neurotoxic (Robertson et al. 2012); (2) a reservoir of ongoing low-grade viral replication in the CNS due to poor cART penetrance (Cysique et al. 2004; Ellis et al. 2007; Anthony and Bell 2008; Hult et al. 2008); (3) accelerated cerebrovascular disease from chronic immune activation and inflammation, coupled with other comorbidities including hypercholesterolemia, diabetes, renal and hepatic dysfunction (Valcour et al. 2005; Becker et al. 2009); and (4) neurodegenerative processes that can occur with aging (Brew et al. 2009). These factors, along with the duration of HIV infection and chronic immune activation, the lag in cART initiation, and the type and duration of treatment, may modulate the impact of the disease on the brain (Cohen et al. 2010; Carvalhal et al. 2016).

In the current "cART era" of more widely available and accessible treatment, aging and HIV-related comorbidities, including symptoms of brain dysfunction, remain common among individuals on suppressive treatment. A better understanding of the neurobiological consequences of HIV infection is essential for developing thorough treatment guidelines for acute care and for optimizing long-term neuropsychological outcomes and overall brain health. It is important to identify and disentangle the effects of comorbid conditions that could potentially be masked by or misattributed to the infection. Monitoring brain aging processes in people living with and without HIV may be able to provide clinicians information on when and how to intervene.

Ongoing works aimed at understanding the neurological complications of HIV capitalize on new and emerging brain imaging acquisition and processing techniques. These advances are often designed to improve sensitivity and specificity to track subtle variations in brain structure and function. Neuroimaging advances, coupled with initiatives that bring together researchers and data from all over the world, will vastly increase the scope and power of neuroHIV studies to discover reliable neurological consequences of HIV infection in the aging populations, along with age-related comorbidities and related therapeutic targets (Thompson and Jahanshad 2015).

In the remaining sections of this chapter, we will summarize over two decades of NeuroHIV research evaluating brain volumetric differences and correlations in adults living with HIV. Next, we highlight brain mapping technologies that go beyond evaluating morphometric patterns and delve into brain circuitry, including functional brain mapping and quantification of brain microstructural properties. We then introduce and discuss aging and associated neurological complications in people living with HIV and how infection contributes to the risk for late-onset dementias. We describe how new technologies and large-scale international collaborations may help disentangle the effect of genetic and environmental risk factors on brain aging and disease. We conclude with a summary of how we see the field of NeuroHIV research advancing in the decades to come and highlight some of the associated clinical implications.

2 Mapping NeuroHIV and Neurological Disease with MR Based Imaging

Non-invasive, in-vivo neuroimaging has played a key role in delineating the spectrum of CNS impairments in HIV+ individuals. Charting the course of disease trajectory by assessing clinical and biological complications associated with infection, or its therapies, is important for effective treatment planning. Reliable imaging biomarkers of HIV-associated brain changes can help objectively measure and quantify the degree of disease neuropathology and ultimately predict neurological decline. Many brain related disorders involve abnormalities in biological processes and measurements before the onset of clinical symptoms. For example, in the case of Alzheimer's disease, magnetic resonance imaging (MRI) markers have been shown to be abnormally altered years prior to the cognitive manifestation of the disease (Jack et al. 2013, 2018). The use of reliable biomarkers in a safe and non-invasive manner may allow for unbiased assessments during clinical practice. While neuroimaging is still an expensive tool that may not be feasible for every clinical visit, it has been shown to be a powerful tool for assessing risk and mapping disease progression.

Over the last 25 years, multimodal brain imaging has helped uncover consistent deficits in brain morphology, wiring, and function in HIV+ individuals, which were initially only reported after autopsy. Early neuropathological and immunohistochemical studies of HIV encephalitis showed the presence of multinucleated giant cells and microglial nodules, as well as viral antigens that had a predilection for the brain's white matter (WM) and subcortical structures (Neuen-Jacob et al. 1993; Brew et al. 1995; Berger and Nath 1997; Morgello 2018). Many of the earliest neuroimaging studies of the HIV-infected brain were performed using computed tomography (CT), which exposes individuals to ionizing radiation and may therefore not be practical for routine study in stable individuals. More recently, the neuroimaging field has benefited from the use of multimodal MRI to study brain structure and function. MRI studies of HIV infection have largely been consistent with the earlier studies, suggesting HIV prominently affects the basal ganglia and WM (Tucker et al. 2004; Cohen et al. 2010; Jernigan et al. 2011; Ances et al. 2012; Heaps et al. 2015).

Quantitative MRI processing methods have given way to larger population studies and allowed for a more extensive examination of specific brain structures and circuits affected, providing new insights into how brain deficits vary across individuals in the cART era. While studies have shown that the brain appears generally more intact in those with restored immune function, growing evidence across MRI modalities suggests that chronic HIV infection can promote continued brain deficits despite effective viral suppression with cART (Cardenas et al. 2009; Cohen et al. 2010; Becker et al. 2011; Harezlak et al. 2011, 2014; Tate et al. 2011; Ances et al. 2012; Hua et al. 2013). In the following subsections, we discuss findings from structural, functional, and diffusion MRI studies of people living with HIV.

2.1 *Mapping Anatomical Brain Morphometry*

T1-weighted brain MRIs are generally considered the most standard type of anatomical image; image analysis methods are commonly used to segment specific brain structures and tissue compartments to estimate global and regional brain volumes or thicknesses and surface areas. When comparing brain regions of interest (ROIs) between HIV-infected individuals and seronegative controls, studies have frequently revealed that, on average, HIV+ populations tend to have significantly lower regional brain volumes – cortical gray matter, white matter, and structures of the basal ganglia are most commonly noted to be smaller while the fluid filled lateral ventricles are reportedly enlarged (Tate et al. 2009; Ances and Hammoud 2014; Masters and Ances 2014; Rahimian and He 2017; Chang and Shukla 2018).

ROI-based studies require the investigator to select regions *a priori*, while 3D brain-wide studies at the voxel-wise level may also be performed to potentially reveal novel insights into the extent of neurological deficits. Some factors may selectively target distributed brain regions or systems that are readily visualized by brain-wide statistical mapping, but may be overlooked if only a number of candidate regions of interest are assessed. One statistical mapping approach – tensor-based morphometry (or TBM) – has been validated as a powerful and unbiased technique to map disease-related regional brain volume differences, not only in HIV (Thompson et al. 2005; Chiang et al. 2007; Wang et al. 2010; Hua et al. 2013), but also in large multi-site studies such as the Alzheimer's Disease Neuroimaging Initiative (Hua et al. 2016). Through non-linear registration of T1-weighted scans, voxel-wise volumetric variations can be identified by examining the gradients of the resulting deformation fields. TBM provides a full 3D Jacobian "expansion/

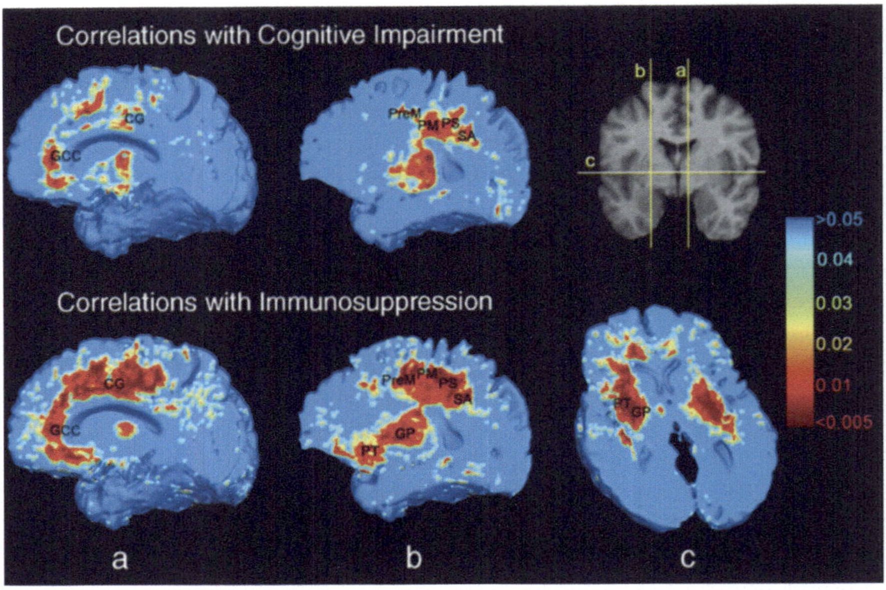

Fig. 1 Brain volume deficits, cognitive impairment, and immune system suppression. Significant regional brain volume deficits are associated with greater cognitive impairment (*upper row*) – e.g., in the middle cingulate, genu of the corpus callosum, medial and basal frontal lobes, and primary and association sensorimotor areas, tested here by Spearman's (non-parametric) rank correlation. Correlations of brain atrophy with CD4+ T-lymphocyte depletion are more extensive in the above regions, as well as in the putamen and globus pallidus (*lower row*). This analysis was performed in a group of HIV/AIDS patients, with a method called "tensor-based morphometry," which maps regional volume differences in a group of individuals by nonlinearly warping their brain MRI scans to an average brain template [Reproduced, with permission, from Chiang et al. 2007]

contraction factor" map of effects, which reflects local voxel-wise differences throughout the brain, without requiring prior anatomical hypotheses. This may be especially helpful in studying factors for which the extent of their impact on different brain regions is not well known.

Cross-sectional TBM studies have linked the extent of structural differences to the most severe immunological impairments, indicated by the lowest, or nadir, CD4+ T-cell count recorded in the individuals' medical history; these brain deficits have also been associated with the degree of cognitive impairment in HIV+ individuals (Fig. 1; Chiang et al. 2007). In recent work that followed a cohort of acutely HIV-infected and immediately treated adults in Thailand over 2 years, Kallianpur et al. (2020) used longitudinal TBM alongside the more standard ROI approach, to map brain changes from the first 2 years of infection and treatment (Fig. 2). They noted that individuals who reached viral suppression within 4 weeks had less atrophy in the basal ganglia structures. They also found that a greater frequency and density of P-selectin glycoprotein ligand-1 (PSGL-1)-expressing inflammatory

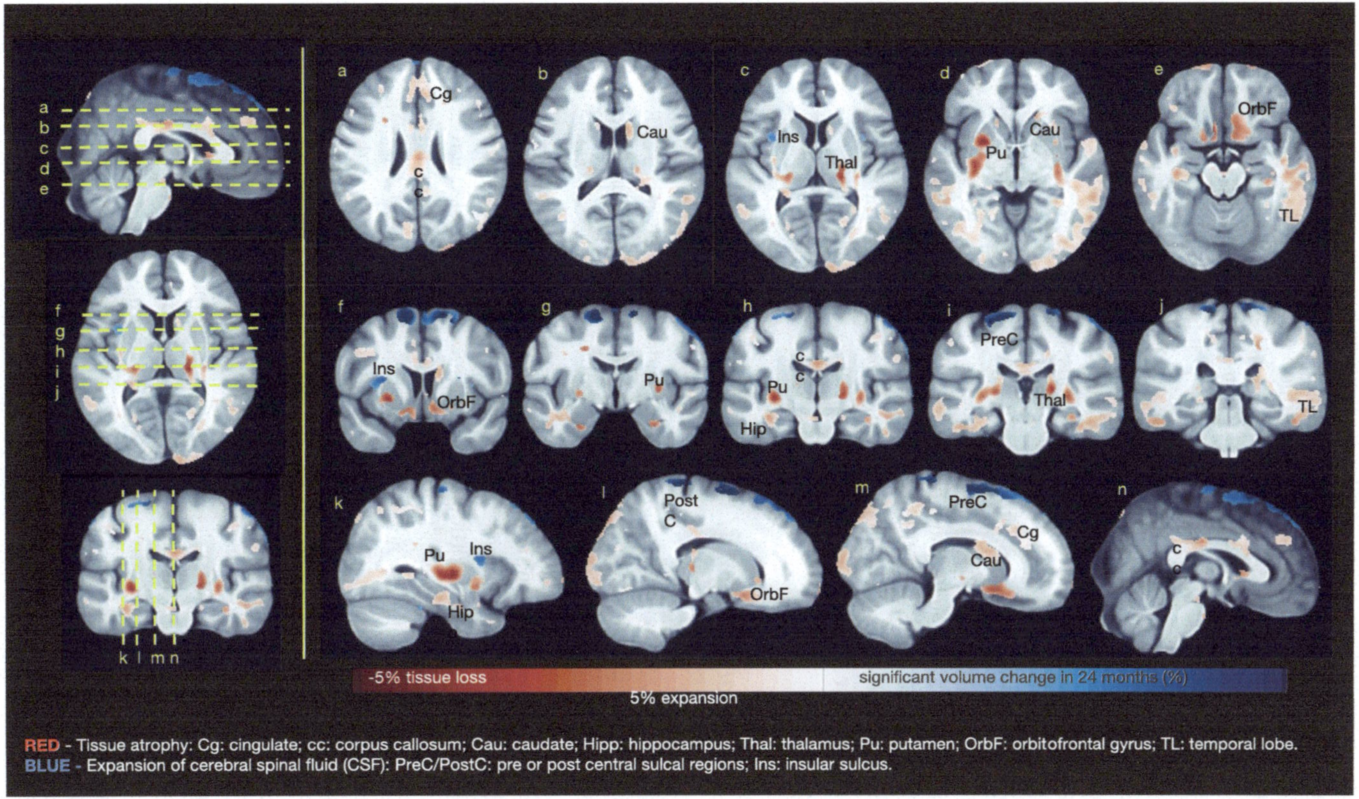

Fig. 2 Brain volume changes over 2 years after acute infection and treatment. 24 months after acute infection with immediate treatment, significant brain tissue atrophy was detected bilaterally in many key brain structures including the putamen (up to 5.2% atrophy in 24 months), the caudate (up to 2.5%), and the thalamus, where atrophy of approximately 4% was detected over the 24 months period. These regions were also identified using more standard region of interest approaches, however, TBM shows effects brain-wide, which, here, also highlighted tissue loss in the orbitofrontal cortex and portions of the occipital lobe, while also suggesting expansion of the CSF around the motor cortex and insula [Reproduced, with permission, from Kallianpur et al. 2020]

monocytes derived from plasma were associated with greater atrophy of the basal ganglia structures including the caudate and the putamen.

In one study, Nir et al. (2019a) used longitudinal TBM to map the annual rate of change of regional brain volumes (mean time interval 1.0 ± 0.5 years), in 155 chronically infected and treated HIV+ participants (mean age: 48 years; 85% male) from the HIV Neuroimaging Consortium (HIVNC). The authors found significant brain tissue loss across HIV+ participants, including those neuro-asymptomatic with undetectable viral loads, largely localized to subcortical regions (Fig. 3). Measures of disease severity (nadir CD4+, current CD4+, and plasma viral load), age, and neurocognitive decline were associated with greater atrophy. This data suggests that chronically HIV-infected and treated individuals may undergo progressive brain tissue loss despite stable and effective cART, which may contribute to neurocognitive decline. These findings are in line with other longitudinal morphometric studies showing atrophy rates in HIV+ individuals on cART are higher than those in a group of age-matched healthy controls (Cardenas et al. 2009; Clifford et al. 2017). The common patterns of brain atrophy in people living with chronic HIV are important to establish, as deviations from such patterns may imply added risk for severe brain disease.

2.2 *Functional Brain MRI Studies of HIV*

Many HIV neuroimaging studies have also included more advanced MRI modalities, such as magnetic resonance spectroscopy (MRS), which can produce maps and measures of cellular metabolites in the living brain, and blood-oxygenation level dependent functional MRI (BOLD fMRI).

In their recent review of functional MRI studies of HIV, Hakkers et al. (2017) noted that many investigations reported differences in activation between HIV-positive and -negative participants when performing tasks across multiple domains, including attention, working memory, and especially executive function. Overall, HIV-positive patients showed hyperactivations in task-related brain regions despite comparable performances to controls; task performance was degraded only for the most complex tasks.

There is also some enthusiasm in the HIV research field for using resting-state functional MRI (rs-fMRI), an approach to study brain synchrony that has matured in recent years to identify and measure a range of fundamental functional networks in the brain. As resting-state fMRI does not require participants to perform a cognitive task, the data is relatively easy to acquire and may offer sensitivity to acute brain changes that occur after HIV infection, as well as treatment response. In one study, Samboju et al. (2018) assessed 49 individuals with acute HIV infection (AHI) and 23 HIV-uninfected Thai participants with both diffusion tensor imaging (DTI) and rs-fMRI, to determine whether white matter microstructure and resting-state functional connectivity (rsFC) are disrupted in AHI. Seed-based voxel-wise rsFC analyses were completed for the default mode (DMN), fronto-parietal, and salience and

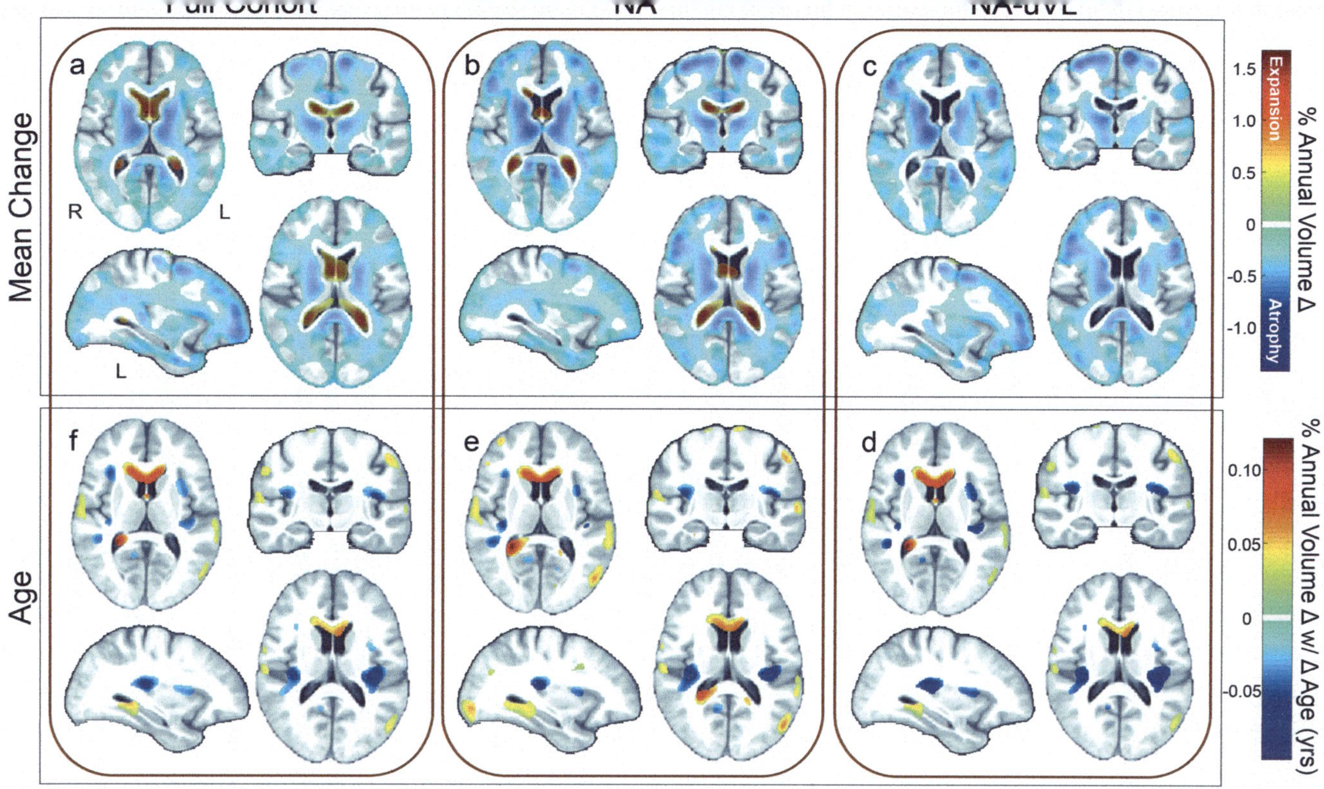

Fig. 3 Progressive brain atrophy in HIV, mapped with longitudinal MRI. Annual volumetric change (%) maps averaged across (**a**) the HIV Neuroimaging Consortium cohort (mean age: 48.0 ± 8.9 years; $N = 155$) revealed significant ventricular expansion (*red*) and tissue atrophy (*blue*) throughout the brain (regions with no change are colored *gray/white*; one-sample *T*-test; P(corrected) ≤ 0.05). Maps of the (**b**) subset of neuro-asymptomatic individuals (NA; ADC Stage = 0; $N = 92$) and (**c**) those who are NA with undetectable viral load (NA-uVL; plasma HIV RNA < 400 copies/mL; $N = 76$) showed a similar distribution and magnitude of change as the full cohort. (**d–f**) In the full cohort and the NA subsets, older age at baseline was significantly associated with greater annual ventricular and sulcal expansion and tissue atrophy (P(corrected) ≤ 0.05). However, the association between brain atrophy and age was relatively localized and did not account for the pervasive brain atrophy detected across participants over time (Reproduced, with permission, from Nir et al. 2019b)

6 subcortical networks. The authors found marginally more intact white matter microstructure among participants who had a very brief exposure to ART compared to those who were imaged just prior to starting ART. The AHI group had reduced rsFC between the left parahippocampal cortex (PHC) of the DMN and left middle frontal gyrus compared to controls. Within AHI, ART status was unrelated to rsFC. However, higher CD4 cell count was associated with higher rsFC for the right lateral parietal and PHC seeds in the DMN. These alterations in large-scale network connectivity, in particular lower rsFC with inferior parietal cortex, may be a marker of impaired cognitive performance in chronic HIV.

2.3 Magnetic Resonance Spectroscopy Studies of HIV

Magnetic resonance spectroscopy (MRS) has been used to detect disturbances in brain metabolites in HIV-infected individuals (Descamps et al. 2008). In their recent review of the clinical utility of MRS, Bartnik-Olson et al. (2020) note that in contrast to MRI, the MR spectrum can be used to measure a range of key metabolites, as the MR signal intensity (amplitude) received from protons in a given metabolite is a function of their frequency (i.e., position on the chemical shift axis). Each functional group of a metabolite resonates at a frequency dependent on its chemical structure. HIV studies have used MRS to detect differences in markers of inflammation and neuronal loss and dysfunction throughout the brain in HIV-infected patients; such differences can be reflected in levels of N-acetyl aspartate (NAA), choline (Cho), creatine (CR), myoinositol (MI), and glutamate and glutamine (GLX). In one such study, Bladowska et al. (2013) analyzed NAA/Cr, Cho/Cr, mI/Cr ratios in HIV-infected individuals and controls, and correlated these MRS measurements with immunologic data, in a cohort of 65 asymptomatic patients. They reported a significant decrease in NAA/Cr ratios in the posterior cingulate gyrus, anterior cingulate gyrus, and parietal white matter regions in HIV-positive cART treated individuals compared to uninfected controls. They also found lower CD4 count associated with lower posterior cingulate gyrus NAA/Cr ratios and higher frontal white matter Cho/Cr ratios. In a subsequent study of 110 neuro-asymptomatic subjects (32 HIV+ subjects on cART, 28 HIV-infected therapy-naïve subjects, and 50 healthy controls), Boban et al. (2017) found significant decreases in the NAA/Cr ratio in HIV-infected subjects in several brain regions, an increase in mI/Cr levels in the anterior cingulate gyrus, but no significant differences in Cho/Cr ratios. They were unable to detect a moderating effect of cART on these abnormalities. Similarly, in a longitudinal study, 226 treated HIV-infected individuals, including 138 neurocognitively asymptomatic individuals, showed significant annual decreases in frontal white matter NAA and Cho, mid-frontal cortical NAA, Cr, Cho, and Glx, and basal ganglia Glx (Gongvatana et al. 2013). In another cross-sectional study of the same treated HIV+ cohort, higher Glx and lower NAA in these regions were associated with lower white matter, putamen and thalamus volumes, and ventricular and CSF expansion (Hua et al. 2013). Together, these findings suggest that HIV-related brain injury may persist even in treated HIV-infected individuals.

2.4 Brain Mapping with Diffusion-Weighted MRI

While anatomical MRI-based measures of HIV pathology are widely reported, diffusion MRI (dMRI) may be more sensitive to subtle white matter (WM) microstructural changes. dMRI is a variant of MRI that measures the hindered or restricted diffusion of water molecules in brain tissue. Water diffusion in the brain is hindered by barriers such as hydrophobic myelin sheaths which promote highly anisotropic water diffusion along axons. By characterizing the diffusion process at the voxel level, it is therefore possible to make tentative inferences about the underlying WM microstructure (Descoteaux and Poupon 2012).

Since the development of dMRI, along with improvements in acquisition protocols – including increases in angular, spatial, and spectral resolution, multi-echo sequences, and acquisitions with multiple diffusion times – multiple mathematical models have been developed to describe diffusion in the brain. One of the first, and still most popular, methods developed to summarize diffusion properties in a specific voxel is the single diffusion tensor model (DTI) (Basser et al. 1994). This model assumes purely Gaussian diffusion and is limited as it can only model a single fiber population at every voxel. It cannot resolve complex features of WM microarchitecture, such as dispersing, crossing, or "kissing" fibers. At the current resolution of typical dMRI acquisitions, at least two-thirds of WM voxels contain multiple fiber crossings (Behrens et al. 2007; Descoteaux 2008; Jeurissen et al. 2013). That said, DTI-derived fractional anisotropy (FA) is still the scalar measure most widely used to characterize WM architecture in HIV and in most clinical populations (O'Connor et al. 2017). Furthermore, although FA may be sensitive, it is somewhat non-specific as it is influenced by many microstructural factors such as axonal diameter, packing density, membrane permeability, myelination, and intra-voxel orientation coherence (Descoteaux 2008).

DTI studies of HIV+ patients have found lower FA and higher mean diffusivity (MD) in the corpus callosum and frontal WM among other regions, suggesting compromised WM tissue and cortical connections (Chang et al. 2008; Jahanshad et al. 2012; Wright et al. 2012; O'Connor et al. 2017; Oh et al. 2018). In one recent study (Cysique et al. 2017), DTI was used to compare WM microstructure in 40 HIV– and 82 HIV+ men with comparable demographics and lifestyle factors. Within the HIV+ sample, the authors found that a higher CNS penetrating antiretroviral treatment, higher current CD4+ T cell count, and immune recovery from the nadir CD4+ T cell count were associated with increased FA and decreased MD; HIV duration, symptomatic, and asymptomatic cognitive impairment were associated with decreased FA and increased MD. Overall, however, the authors concluded that WM DTI measures are near normal except for patients with neurocognitive impairment and longer HIV disease duration.

Many dMRI models have been proposed over the last two decades to overcome limitations of DTI. Diffusion-propagator based methods including high angular resolution diffusion imaging (HARDI) – q-ball imaging (Tuch 2004) and spherical deconvolution (SD; Tournier et al. 2004) – diffusion spectrum imaging (DSI)

(Wedeen et al. 2005), and multishell diffusion kurtosis imaging (DKI; Jensen et al. 2005), and mean apparent propagator MRI (MAP-MRI; Özarslan et al. 2013) have helped resolve multiple dominant fiber directions within voxels. However, few published studies of HIV take advantage of these newer models which may result in more robust measurements of disease effects and give a richer understanding of the underlying HIV-related WM microstructural changes and pathology.

Often as part of larger data collection paradigms, time constraints are often placed on imaging protocols to reduce patient discomfort, as well as patient attrition or motion, and ensure adequate sample sizes. This may prevent reliable reconstruction of many of the aforementioned models, which can require extremely dense or multishell dMRI acquisitions. However, multi-tensor models, such as the tensor distribution function (TDF), which models crossing fibers as a probabilistic ensemble of Gaussian tensors (Leow et al. 2009), may still be feasible and derived measures more sensitive to disease-related microstructural differences than DTI – in studies of both HIV and Alzheimer's disease. To emphasize the benefit of such advanced, "beyond DTI" methods, it is important to note that WM microstructural differences in people with HIV infection compared to uninfected controls have been frequently reported in the corpus callosum. The corpus callosum is the large interhemispheric WM bundle that mainly contains highly coherent WM fiber organization along a dominant direction, between the left and right hemispheres. However, it is unlikely that HIV infection affects only this tract. Building on work showing more widespread differences in Alzheimer's disease related WM microstructure using FA derived from TDF compared to DTI (Nir et al. 2017; Zavaliangos-Petropulu et al. 2019), a TDF study of HIV infection was conducted. Nir et al. (2019b) tested the hypothesis of widespread WM alterations in HIV and noted that WM pathways known to have fiber crossings, particularly the corona radiata, do indeed show significant differences in HIV+ individuals compared to HIV- individuals. This effect was only detected with FA estimated with the TDF model and not the DTI model, while both models did identify significant differences in the corpus callosum. As imaging methods advance, it becomes possible to detect more specific in-vivo pathological changes related to infection. In subsequent sections, we elaborate on state-of-the-art methods that have been applied to understand neurodegeneration outside the context of HIV infection, paving new avenues for neuroHIV researchers.

The abovementioned neuroimaging methods have begun to identify HIV-related disruptions in brain structure, microstructure, and function in cases of both acute infection and chronic disease. Even with viral suppression, people living and aging with chronic HIV often present with additional comorbidities. Understanding and disentangling the neurological consequences of these complications are active areas of study, particularly in the context of aging and late-onset neurological diseases.

3 HIV-Related Comorbidities and Aggregated Risk for Brain Dysfunction in Aging: The Role of Neuroimaging

3.1 Aging with HIV Infection

An estimated 50% of HIV-infected individuals in the USA are over age 50, and this is also the fastest growing age group (CDC 2017). Combination ART has improved survival in HIV-infected adults to near-normal longevity (Rodger et al. 2013; Costagliola 2014), yet despite viral suppression, people with chronic HIV infection have a higher risk of multiple health conditions linked to advancing age – including geriatric syndromes and frailty, cardiovascular disease (hypertension and stroke), diabetes, cancers, liver, renal diseases and chronic neurological complications (Brew et al. 2009; Wing 2016). This suggests that common age and HIV-related pathological processes, including immune dysregulation, immunosenescence and inflammation, may accelerate aspects of the aging process (Pathai et al. 2014).

HIV-associated neurocognitive impairments and those associated with age related disorders share many similarities (Brew et al. 2009). For example, late-onset neurodegenerative disorders and HIV often include executive function deficits and memory dysfunction as well as brain abnormalities in fronto-striatal and fronto-temporal networks and hippocampal tissue (Cherner et al. 2004; Tucker et al. 2004; Raz and Rodrigue 2006; Brew et al. 2009; Tate et al. 2009; Woods et al. 2009; Schouten et al. 2011; Cysique and Brew 2014; Pfefferbaum et al. 2014; Kamkwalala and Newhouse 2017). Active areas of study involve investigating whether HIV facilitates age-related neurodegeneration and cognitive impairments, in other words causes accelerated brain aging, or whether it is age-related brain decline that puts the brain at greater risk for HIV-related damage (Brew et al. 2009; Holt et al. 2012; Pathai et al. 2014). Either way, there is evidence to suggest a complex interaction and that chronic infection and increasing age together exacerbate brain injury and result in deficits in cognitive function (Goodkin et al. 2001; Wendelken and Valcour 2012; Cysique and Brew 2014; Cohen et al. 2015). Older HIV-infected individuals are twice as likely as younger infected adults to have some form of cognitive impairment, even with the same duration of infection (Valcour et al. 2004a). Furthermore, older age at the time of seroconversion increases the risk for cognitive impairment (Valcour et al. 2004b; Bhaskaran et al. 2008). Many studies report independent detrimental effects of age and HIV on the brain (Gongvatana et al. 2011; Valcour et al. 2011; Ances et al. 2012; Becker et al. 2012; Nir et al. 2014; Cohen et al. 2015), while some have successfully detected the additional effect of an age-by-HIV interaction (Harezlak et al. 2011; Scott et al. 2011; Cysique et al. 2013; Seider et al. 2016; Kuhn et al. 2017).

As age is the greatest risk factor for dementia, it is especially important to study brain health in HIV+ individuals over the age of 60 to determine whether the independent or interactive effects of age and infection contribute to the risk for dementia. Healthy aging studies have shown that age-associated brain atrophy rates are non-linear and may accelerate in older individuals (DeCarli et al. 2005). To date, few studies have been conducted in elder HIV+ individuals, with much of the literature focusing on middle-aged or younger cohorts. Clifford et al. (2017) compare the brain tissue atrophy rates in treated HIV+ individuals over age 60 with and without HAND to those of age-matched controls and found higher atrophy rates in infected individuals; atrophy rates, however, did not differ by cognitive status. There is a need to replicate findings and expand on such studies to include clinically and demographically heterogeneous infected individuals. Comorbidities and additional risk factors may exacerbate the rate of brain tissue loss and identify those at heightened risk for severe degenerative disorders or dementias such as AD.

3.2 Risk Factors and Interactions Contributing to Complex Brain Disorders

Several models of disease onset in psychiatry – such as the so-called diathesis-stress model – suggest an early trauma or adverse event increases the brain's vulnerability to later brain injury from other independent causes or via other mechanisms. The same concept may apply to age and HIV-associated cognitive decline as HIV-related pathology and neurodegenerative processes may interact.

Aging HIV+ individuals face increased risk for cerebrovascular injury and disrupted blood flow in the brain. Cerebrovascular disease (CVD) can contribute to cognitive dysfunction and dementias, including vascular cognitive impairment (VCI), vascular dementia, and Alzheimer's disease (AD). As persistent brain injury is seen in treated individuals living with HIV, vascular disruptions may be a key mechanism for developing HAND in the cART era, with VCI being an underlying risk factor (Cysique and Brew 2019). A recent meta-analysis of 2,139 HIV+ individuals across 11 studies found vascular risk factors, including type 2 diabetes, hyperlipidemia, and smoking, and subclinical cardiovascular disease were consistently associated with neurocognitive impairment in HIV-infected individuals (McIntosh et al. 2020). The specific symptoms and domains of the cognitive deficits partially overlap across late-onset disorders and dementias; complicated by mixed pathologies and co-occurrence of disorders, differential diagnoses would be an almost impossible task were it not for specific biological indicators of disease. While autopsy is still needed to confirm most age-related neurodegenerative disorders, factors such as genetic predispositions, cardiovascular risk factors, and plasma or CSF markers are essential for guiding clinical diagnoses and devising a treatment or maintenance plan. In conjunction with other biomarkers, advanced in-vivo imaging methods offer additional mechanistic insights into disease complexities and can suggest the degree of pathological complications and co-occurrences as the disease, or diseases, progresses. Gross brain injury from CVD, such as strokes, can result in large morphological deviations that are visible on T1-weighted structural scans along with commonly acquired T2-weighted images. However, advanced microstructural and functional neuroimaging methods may capture the underlying pathophysiology of vascular disease. We refer the reader to Zwanenburg and van Osch (2017) for a more exhaustive list and description of these modalities in highlighting cerebrovascular disruptions. In upcoming sections, we discuss many of these methods and focus on how measures that can be derived from advanced neuroimaging modalities, those less commonly seen in clinical practice, have helped reveal brain injury associated with cognitive impairment and AD.

In line with the VCI interaction hypothesis for HIV, several lines of work relate viral infection to a greater risk of AD pathogenesis. Complementary work further examines how risk factors for AD – such as the APOE4 genotype – might complicate the effects of aging with HIV infection. In the following sections, we review and give examples of each of these factors – how viral infection might facilitate AD pathogenesis and how HIV-related brain decline may be exacerbated in carriers of the APOE4 genotype.

3.3 *Effects of Viral Infection on the Pathogenesis of Neurodegenerative Disorders*

HIV infection may promote the appearance and progression of age-related neurodegenerative diseases, such as Alzheimer's disease (AD). Levels of phosphorylated-tau and beta-amyloid (Aβ), classical biomarkers of AD, may be elevated in HIV+ individuals compared to controls, particularly in older HIV+ adults with neurocognitive impairment (Brew et al. 2005; Green et al. 2005; Anthony et al. 2006; Clifford et al. 2009; Cohen et al. 2015). Some studies suggest that the HIV inflammatory cascade may lead to an overproduction of beta-amyloid precursor protein (APP), as well as factors that degrade APP into neurotoxic Aβ (Forloni et al. 1992; Stanley et al. 1994; Adle-Biassette et al. 1999; Nebuloni et al. 2001; Liao et al. 2004). The HIV Tat protein may further inhibit effective amyloid-beta degradation (Rempel and Pulliam 2005).

HIV is not the only viral infection thought to play a role in neurodegeneration and risk for late-onset dementias. In fact, members of the *Herpesviridae* family – which are far more common than HIV in human populations – have been linked at varying levels to the pathogenesis of AD. Hypotheses first emerging in 1982 posited a role of HHV-1 (Human herpesvirus 1) reactivation of latent infection in the development of AD (Ball 1982; Gannicliffe et al. 1986). When infected with HHV-1, the virus infects nerve endings and translocates to sensory or autonomic ganglia where it establishes latency and the viral genome remains in an episomal state (Whitley et al. 2011). If the immune system is compromised, as it is in HIV infection and to some extent as a result of normal aging, HHV-1 is able to propagate freely to infect other nerve cells, produce viral proteins, and activate inflammatory processes. This cascade of events is thought to promote the formation of Aβ plaques and the accumulation of tau, suggesting a role for HHV-1 reactivation in the pathogenesis of AD

(Sochocka et al. 2017). More recently, work using human-induced pluripotent stem cell (hiPSC) technology reported a HHV-1 induced tissue model of sporadic AD (Cairns et al. 2020). Whereas high levels of HHV-1 infection resulted in cell death, low-level viral inoculations led to the development of large, multicellular, dense Aβ + fibrillar plaque-formations (PLFs), upregulation of *PSEN1* and *PSEN2*, reactive gliosis, and neuroinflammation. AD genetic risk scores have also been shown to interact with HHV-1 antibodies in affecting AD risk, suggesting a role for the host's genetic background in HHV-1-associated AD (Lopatko Lindman et al. 2019). HHV-2, another member of the *alphaherpesvirinae* family alongside HHV-1 and HHV-3, is another neurotropic virus that establishes lifelong latent infections. Kristen et al. (2015) demonstrated that HHV-2 infection can lead to the prominent accumulation of hyperphosphorylated tau, Aβ40, and Aβ42 in human neuroblastoma cells. Bubak et al. (2020) also reported a possible effect on AD pathogenesis of HHV-3 – also known as the *varicella zoster* virus, which causes chicken pox in children and shingles in adults. When HHV-3 was used experimentally to infect quiescent primary human spinal cord astrocytes, the infection also produced intracellular amyloid. This suggests that HHV-3 infection may increase toxic amyloid burden and play a role in amyloid-associated disease progression.

To complicate matters further, co-infection with HHV-5, also known as human cytomegalovirus (CMV) is common in people living with HIV. Interestingly, several mechanisms have been put forth on how HHV-5 might promote HIV persistence. Christensen-Quick et al. (2017) describe how latent HIV-infected cells can increase through HHV-5 associated inflammation, altered trafficking, inhibitory signaling, proliferation, or inhibition of the apoptosis of HIV-infected cells. HHV-5 can even directly transactivate latent HIV by inducing ongoing HIV RNA expression. HHV-5 and other members of the *betaherpesvirinae* family – including HHV-6 and HHV-7 (*roseola viruses*) – also have a postulated role in AD. Barnes et al. (2015) found that HHV-5 seropositivity was associated with an increased risk of AD and a faster rate of decline in global cognition in a diverse population adjusted for age, sex, education duration, and race while also observing that HHV-5 seropositivity was higher in Black populations compared to White, possibly partially accounting for racial differences seen in AD burden. Bu et al. (2015) reported a similar finding where HHV-5 seropositivity was associated with AD risk, even after adjusting for other risk factors and comorbidities.

3.4 Effects of Alzheimer's Disease Genetic Risk in the Context of HIV

Genetic susceptibility to neurodegenerative diseases may also play a significant role in the effects of HIV on the brain. The apolipoprotein E4 (APOE4) polymorphism is the greatest known genetic risk factor for late-onset AD (Raber et al. 2004; Lambert et al. 2013) and the strongest genetic risk factor for any complex neuropsychiatric disease. The genotype, which impacts amyloid metabolism, has been repeatedly associated with lower brain volumes and faster brain tissue atrophy rates, in individuals with and without cognitive impairment. Within the central nervous system, *APOE* genotype is associated with the onset and amount of cerebral amyloid angiopathy (CAA; Biffi and Greenberg 2011; Nelson et al. 2013). CAA is caused by amyloid β build up on blood vessels in the brain and it is often associated with hemorrhagic lesions, ischemic lesions, encephalopathy, and dementia. The cerebrovascular effects of APOE4 include accelerated breakdown of the blood brain barrier (BBB), which has been shown to be more severe in APOE4 carriers with cognitive impairment, independent of amyloid and tau (Montagne et al. 2020). In HIV+ individuals, the APOE4 genotype has been associated with faster disease progression (Burt et al. 2008), increased neurotoxicity associated with the HIV Tat protein (Turchan-Cholewo et al. 2006), higher brain beta-amyloid deposits (Green et al. 2005; Soontornniyomkij et al. 2012), greater brain atrophy (Wendelken et al. 2016), and increased risk of HAND (Corder et al. 1998; Valcour et al. 2004b; Spector et al. 2010; Andres et al. 2011; Chang et al. 2011). However, some studies have found no effect of APOE4 risk on HAND (Morgan et al. 2013). The genotype may itself have an age-related effect on degeneration, as more recent large-scale studies in younger and middle-aged adults have found little evidence for APOE4's role in explaining a significant degree of population variance in brain morphometry. Lyall et al. (2020) examined brain MRI and genetic data from over 8,000 individuals scanned as part of a large public biobank database, the UK Biobank (Miller et al. 2016), and found a statistically significant association between APOE4 genotype and greater WM hyperintensity volumes – a marker of poor cerebrovascular health. However, they found no evidence for associations with left or right hippocampal, total gray matter (GM) or WM volumes, or WM DTI FA and MD. Interestingly, despite being such an important risk factor, APOE4 genotypes may not play a significant role in preclinical cognitive function for much of the adult lifespan (Jorm et al. 2007). Even so, in people living with HIV, amyloid pathways – implicated in aging and AD – may also be perturbed by HIV-related inflammation, BBB disruption, and neurotoxic proteins and may accelerate pathogenesis (Milanini and Valcour 2017).

The increased life expectancy of HIV+ individuals in the cART era compared to the pre-cART era is a direct result of successful viral suppression and immune restoration in the majority of treated individuals. Yet, the aging process is now exacerbating the neurological complications of HIV infection, and cognitive issues are becoming one of the primary, unaddressed, morbidities associated with the infection. Neuroimaging studies are beginning to shed light on age-related brain alterations in the context of chronic HIV infection. A first step in disentangling HIV-related neurological deficits from age-related neurodegenerative disorders and late-life dementias is to map the typical course of brain aging throughout the lifespan. Normative data from healthy populations provides a valuable reference to understand how genetic and environmental risk factors, including HIV infection and related comorbidities, contribute to altered or accelerated brain decline.

4 Advanced Imaging Methods to Identify Biomarkers of Aging, Alzheimer's Disease and Related Dementias

Relative to HIV, a larger body of literature has shown how neuroimaging can play a key role in the assessment and differential diagnosis of people with AD and related disorders. Many ongoing initiatives are advancing imaging acquisition methods and creating and testing new analysis methods to differentiate normal age-related brain changesquo;s from immune and other neuropathological processes involved in Alzheimer's disease, Parkinson's disease, and vascular causes of dementia.

4.1 A Biomarker Approach to Diagnosing Alzheimer's Disease

Recently, Jack et al. (2018) proposed a novel biological framework for understanding Alzheimer's disease and other dementias, by using biomarkers in the blood or cerebrospinal fluid along with neuroimaging. Progress in identifying treatment effects on AD had been stalled by a lack of objective biological markers of the disease – clinical trials had typically enrolled patients based on clinical and cognitive criteria, and inevitably these patients often had unknown combinations of vascular pathology, varying degrees of abnormal accumulation of proteins such as amyloid and tau in the brain, and regional volume loss indicative of cellular atrophy and overt neuronal loss. In the new National Institute on Aging and Alzheimer's Association (NIA-AA) Research Framework, AD is now defined using objective measures of underlying pathological processes in living persons using biomarkers. Biomarkers are grouped into those assessing β amyloid deposition ("A"), pathologic tau ("T"), and neurodegeneration ("N"). This so-called AT(N) classification system groups different biomarkers (imaging and biofluids) by the pathologic process that each measures. Amyloid-sensitive PET ligands and tau-sensitive PET tracers document characteristic trajectories for the accumulation and spread of amyloid and tau in vivo (Braskie et al. 2010; Protas et al. 2012). Operationally, the "positivity" of A, T, or N biomarkers is defined using standard cut-offs, with some efforts to reconcile differences among various radiotracers using a norming approach called the centiloid system (Klunk et al. 2015; Rowe et al. 2017). This method converts the uptake value of any tracer into the same standardized units.

Biomarkers of typical AD have historically been tracked using PET scans or CSF. There is existing evidence that changes in CSF may precede those detectable with PET (Palmqvist et al. 2016). Palmqvist et al. (2015) noted that Aβ42/total tau (t-tau) and Aβ42/hyperphosphorylated tau (p-tau), derived from CSF, were able to identify individuals with cognitive impairment who converted to AD over a 3-year period (area under the curve [AUC] 0.93–0.94). PET measures provided similar predictive power, but the resulting brain maps could additionally localize regional disruptions (AUC 0.92–0.93; anterior cingulate, posterior cingulate/precuneus, and global neocortical uptake). Understanding the patterns of effects and the trajectory of decline for different brain regions and pathways will likely be essential for differential diagnoses and identifying subtypes of patients, who may respond differently to various treatments. The most recent work in AD diagnostics has pursued the development of plasma assays that can evaluate the levels of A, T, and N pathology from a minimally invasive blood draw (Zetterberg and Burnham 2019), yet there is still a need to identify biomarkers from non-invasive imaging methods to map pathology and disease trajectory in the brain.

In AD, abnormal beta-amyloid (Aβ) protein deposits and neurofibrillary tangles (NFTs) spread in a characteristic pattern in the brain, typically targeting the entorhinal and temporal cortices at early stages and spreading as the disease progresses. These changes, originally noted by Braak and Braak (1995) in post-mortem data, can also be inferred to some extent using brain MRI, and the sequence of cortical atrophy typically mirrors the underlying trajectory of pathology (Thompson et al. 2003). As Jack et al. (2013) note, the spread of amyloid plaques in the brain can occur while individuals are still cognitively normal, often along with *reductions* in plasma levels of Aβ42; after a lag period, which varies across patients, neuronal dysfunction and neurodegeneration become dominant pathological processes, evident from progressive atrophy on brain MRI. Atrophy on MRI has been found to better correlate with Braak NFT stage than Aβ (Jack et al. 2013). A recent longitudinal study of amyloid positive individuals with mild cognitive impairment (MCI) or mild dementia found that baseline tau PET, but not amyloid PET, predicted the rate of cortical atrophy and that the spatial distribution of baseline tau and future atrophy was correlated (La Joie et al. 2020).

Only recently have a small number of neuroimaging studies examined brain amyloid and tau accumulation in HIV+ individuals using PET. Mohamed et al. (2020) used [18F] AV-45 (florbetapir) PET to determine if beta-amyloid deposition was increased in older HIV+ individuals compared to HIV- individuals. The proportion of HIV+ participants in their 50s with elevated global amyloid uptake (SUVR > 1.40) was significantly higher than the proportion in HIV- participants (67% versus 25%, $p = 0.04$), and selected regional SUVR values were also higher. However, these group differences were not seen in participants in their 60s; the authors noted that the conclusions require replication, given the small sample size of the study. In a subsequent study, Howdle et al. (2020) used a different amyloid tracer – 11C-labeled Pittsburgh compound B – in conjunction with PET, to study 10 men with virally suppressed HAND, aged 46–68 years, and a matched group with no cognitive deficits; the two groups had similar amyloid deposition, which was lower than that in matched MCI and AD groups. Although studies with larger samples are clearly needed, the current evidence suggests that brain amyloid burden does not differ substantially, at the group level, in people with virally suppressed HAND, relative to cognitively normal older controls. Such distinctions may be vital for differentiating disease and identifying individuals with HAND who may also be at risk for other age-related neurodegenerative disorders.

Brown et al. (2014) also pointed to the importance of studying tau pathology in people living with HIV. They examined the seven human studies over the preceding 14 years measuring p-tau and/or total tau (t-tau) in HIV-infected patients, either in cerebrospinal fluid (CSF) or via *post-mortem* brain immunohistochemistry. In one study they reviewed (Steinbrink et al. 2013), HAND severity as measured by the Memorial Sloan-Kettering scale, HIV dementia scale and Mosaic test correlated significantly with the total t-tau level in CSF but not p-tau levels. The authors suggested that t-tau might be a non-specific marker of ongoing subcortical CNS damage in the periventricular WM and basal ganglia. Recently available tau-sensitive PET tracers may be valuable tools to provide additional evidence for or against premature pathology in people living with HIV. These markers are of growing interest given the role of tau as a future treatment target across multiple neurodegenerative conditions.

4.2 Uncovering the Integrity of the Blood Brain Barrier with Dynamic Contrast-Enhanced MRI and Susceptibility Imaging

In Alzheimer's disease, the "two-hit" vascular hypothesis suggests that blood vessel damage occurs first, impairing the blood brain barrier (BBB) which can lead to the accumulation of neurotoxins and reduced brain perfusion; each of these, in turn, promotes neuronal injury and increases in cerebral amyloid load (Zlokovic 2011). Dynamic contrast-enhanced MRI (DCE-MRI) is an in-vivo method used to study BBB breakdown. The extravasation of contrast agents in areas where the BBB is disrupted enables the localization of affected tissues by increasing longitudinal relaxation thereby increasing T1 signal intensity (Heye et al. 2014). In this way, DCE-MRI has been able to measure subtle changes in BBB permeability in normal aging, MCI, and AD (Sweeney et al. 2018). Recently, Montagne et al. (2020) distinguished APOE4 carriers by assessing BBB breakdown in the hippocampus and the medial temporal lobe independent of tau and amyloid accumulation. Interestingly, DCE-MRI has also revealed impaired BBB in HAND patients in the basal ganglia and anterior frontal WM, compared to controls (Chaganti et al. 2019). Another by-product of BBB disruption is the leakage of blood into the brain parenchyma, also known as microbleeds. Microbleeds can be visualized with hemosiderin-sensitive sequences that take advantage of their paramagnetic nature and susceptibility effects. Methods to image these include susceptibility weighted imaging (SWI) or T2* MRI. These microbleeds are often observed in patients with AD, MCI, or various types of cerebral small vessel diseases (CSVD), as well as those at genetic risk for AD. Using T2* derived microbleeds to classify the presence of CSVD, Moulignier et al. (2018) found HIV to be an independent risk factor for CSVD, and CSVD prevalence was twice as high in people living with HIV versus matched seronegative controls. DCE-MRI and SWI will help researchers better understand how viral infections disrupt the integrity of the BBB.

4.3 Components of Brain Tissue Microstructure Assessed In Vivo with Multishell Diffusion MRI

To better understand the histopathological processes underlying age-related disorders, there is a need for improved neuroimaging methods that examine brain tissue microstructure in more detail. Previously, we discussed how dMRI, mainly DTI, has expanded our understanding of the neurological consequences of HIV infection, and how advanced modeling techniques, such as multi-tensor modeling with TDF, may shed more light on the extent of WM affected. In addition to capturing multiple, often crossing fiber bundles in WM voxels, the relatively low spatial resolution of a single voxel in dMRI typically captures partial volumes from different tissue components. This can include, for example, intracellular, extracellular, vascular, CSF, or myelin compartments. Biophysical or multi-compartment modeling approaches go beyond the standard tensor fitting and attempt to fit different diffusion models to distinct tissue types. Quantifying various compartment contributions to the measured diffusion signal aims to provide greater specificity, if not sensitivity, to underlying pathology. For example, the perivascular spaces (PVS), known to expand with neurodegenerative and inflammatory disorders (Wardlaw et al. 2020), have been modeled in a two-compartment, bi-tensor model, with one tensor capturing the tissue compartment and the other, the PVS, modeled as an anisotropic tensor with an axis aligned to the tissue principal direction with a higher diffusivity (Sepehrband et al. 2019).

Multishell diffusion MRI acquisitions (i.e., with multiple *b*-values) are needed for biophysical modeling approaches, as low *b*-value acquisitions are more sensitive to non-restricted diffusion, compared to high *b*-values that are more sensitive to restricted diffusion. Early models included the ball and stick model (Behrens et al. 2003), composite hindered and restricted model of diffusion (CHARMED; Assaf and Basser 2005), and neurite orientation dispersion and density imaging (NODDI; Zhang et al. 2012). NODDI is currently the most widely used "multi-compartment" biophysical model, estimating brain microarchitecture assuming three compartments: the intracellular restricted water diffusion in neurites (ICVF; modeled as cylinders or sticks), the surrounding extracellular hindered diffusion (ECVF; modeled as an anisotropic Gaussian "zeppelin") and CSF free water (ISOVF; a large isotropic sphere); the neurite orientation dispersion index (ODI) is also extracted. While NODDI provides greater insight into the underlying tissue microstructure, the validity of biological interpretations of data derived from clinical scanner acquisitions has been challenged (Novikov et al. 2018; Jelescu et al. 2020). As with results from all in-vivo estimations, findings should be interpreted with caution. Replication studies as well as histopathological follow-ups would be needed to truly understand the biological underpinnings of observed in-vivo associations.

Since the original formulation in 2012, tissue-specific modifications to the NODDI model have been proposed. The original NODDI model was (1) designed with fixed diffusivity values appropriate for the analysis of WM tissue and (2) assumed one tissue response across the whole brain. As the spatial resolution of diffusion MRI has improved over the years, the field can now attempt to fit multi-compartment models in the gray matter, with modified assumptions and diffusivity parameters (Fukutomi et al. 2018; Guerrero et al. 2019). Furthermore, as different brain tissue types – WM, GM, and CSF – have different T2 relaxation times, a multi-tissue NODDI model was recently proposed to incorporate tissue-specific responses (Jeurissen et al. 2014) and improve microstructural volume fraction estimates throughout the brain (Frigo et al. 2020).

Clinical datasets from aging populations are now being collected with multishell dMRI, allowing researchers more specific insights into the microstructural effects of disease. A recent review highlighting the use of advanced dMRI for studying Alzheimer's disease and Parkinson's disease describes several works evaluating multishell DKI and NODDI measures in older individuals with or at risk for dementia (Kamagata et al. 2020). These lines of work have identified pathological changes beyond those seen with standard DTI and volumetric methods. Decreases in gray matter NODDI, ICVF, and ODI have been reported in AD patients compared to controls, even after adjusting for differences in commonly used thickness estimates of the cortex, suggesting NODDI provides information beyond volumetric measurements; moreover, in patients, ICVF was a more sensitive marker for cognitive impairment than cortical thickness (Parker et al. 2018). In another study, cortical ODI outperformed cortical thickness and WM DTI FA in the prediction of chronological age (Nazeri et al. 2015). In another study evaluating both DTI and NODDI in the gray matter of a mouse model of human tauopathy (rTg4510), only NODDI neurite density index (i.e., ICVF) was correlated with histologic measurements of hyperphosphorylated tau levels (Colgan et al. 2016). Using the more recent multi-tissue multi-compartment approach, with gray matter specific parameters, Nir et al. (2021) found associations between regional cortical dMRI measures and in vivo PET-derived measures of amyloid burden; in those cognitively impaired, in addition to lower ICVF and higher ISOVF (suggesting neuronal loss), lower ECVF was associated with greater extracellular amyloid.

In this rapidly developing field, advanced dMRI acquisition protocols can be used to estimate additional and more specific compartments in novel biophysical models. In the gray matter, extra-neurite diffusion may combine water from the extracellular space with water in cell bodies or soma. Soma and neurite density imaging (SANDI) makes use of very high *b*-values (e.g., 10,000 s/mm^2) and short diffusion times (<20 ms) to divide the intracellular compartment into intra-neurite and intra-soma compartments (Palombo et al. 2020). In another recent model, Garcia-Hernandez et al. (2020) introduced a paradigm to characterize brain inflammation by modeling time-dependent morphological changes in activated microglia and astrocytes in rats. In this model, astrocytes consist of large round cells, modeled as large spheres, and microglia are composed of small cell somas and thin cellular processes, modeled as small spheres and sticks with dispersion. Using dMRI, they were able to detect increases in cell body size of activated microglia coupled with retraction of cellular processes/reduced dispersion, as well as increases in astrocyte cell body size.

Continued improvements in the speed and availability of these advanced diffusion imaging protocols promise to yield better insights into the role of neuronal loss, demyelination, and inflammation in neurodegeneration and neuroinflammatory conditions and may be particularly well suited for establishing an in-vivo understanding of the neuropathological processes underlying aging with HIV.

4.4 Neuroimaging and AI Methods to Assess Age-Related Decline and for Differential Diagnosis

The growing use of standardized biomarkers that index separable biological processes has also led to improved understanding of the mixed brain pathologies that account for most dementia cases in community-dwelling older persons. As Schneider et al. (2009) note in their assessment of the Rush Memory and Aging Project, a longitudinal community-based clinical-pathologic cohort study, most community-dwelling elderly people have brain pathology, and those with dementia often have multiple brain pathologies including vascular dementia, Parkinson's disease, Lewy body dementia, limbic-predominant age-related TDP-43 encephalopathy (LATE), cerebrovascular pathologies, or other tauopathies such as frontotemporal lobar degeneration and corticobasal degeneration.

The diagnosis and subtyping of age-related diseases have been assisted by rapid developments in machine learning methods and artificial intelligence (AI). Machine learning – and more recently deep learning – can be used to identify and extract features from multiple brain imaging modalities, along with genetic and clinical data, to make inferences about a person's health, prognosis, or likely treatment response. All these applications of AI are rapidly developing.

AI methods now outperform many classical methods for extracting features from brain MRI – in a recent study of stroke patients, Zavaliangos-Petropulu et al. (2020) compared the accuracy of segmentations generated by a convolutional neural network, called "Hippodeep," relative to two well-accepted hippocampal segmentation methods, in T1-weighted MRIs of stroke patients; the deep learning method performed better than the standard methods in terms of producing high-quality segmentations across multiple datasets.

Another such example is the use of machine learning algorithms to estimate a person's age from their brain scan, without any need for expert user input. As Cole and Franke (2017) note in their review of these methods, various diseases and disorders, including HIV, schizophrenia, and diabetes, have been shown to make the brain appear older. The "brain age gap" – the discrepancy between a person's real age and that predicted from their imaging data by a machine learning algorithm – can be used to identify possible protective or deleterious factors for brain health as people age.

In one study, Cole et al. (2018) performed a multicenter study of 134 virologically suppressed people living with HIV and 78 neurological healthy matched controls. At baseline, the HIV+ individuals had poorer global cognitive performance, lower gray matter volume, higher white matter hyperintensity load, abnormal white matter microstructure, and a greater brain age gap. Even so, the authors found no longitudinal evidence that middle-aged individuals living with HIV, when receiving successful treatment, were at increased risk of accelerated aging-related brain changes or cognitive decline over 2 years.

Recent advances in deep learning, such as convolutional neural networks (CNNs), have been shown to predict a person's brain age or diagnosis from raw MRI scans, without the need for human intervention (Lam et al. 2020). Another future application of AI to brain aging research is to identify loci or patterns in the genome that are associated with disease resilience or risk, with particular subtypes, or with treatment response. Recent efforts are pooling neuroimaging, clinical and next generation sequencing data to better predict patient outcomes and treatment response, as well as identify new drug targets, by analyzing data from large deeply phenotyped populations of older adults. The extension of these AI methods to HIV+ cohorts shows promise in identifying subtle and distributed effects of the virus on the brain, as well as sets of predictors in the genome and environment that might influence these patterns.

5 Heterogeneity Across HIV Studies and Discrepancies in Findings

Across HIV neuroimaging studies, inconsistencies in the effect sizes, regional distribution, and even direction of volumetric and microstructural brain associations across studies limit the generalizability of the conclusions (O'Connor et al. 2017; O'Connor et al. 2018). Sources of heterogeneity in findings from single cohort studies include methodological variability as well as differences in study participants, including age, sex, and environmental, socioeconomic or lifestyle attributes of the cohorts, and other differences in study inclusion and exclusion criteria. While this is true for studies across multiple diagnostic conditions, studies of HIV are further complicated by differences in viral load status, comorbidities and co-infections, drug use, age at infection, mode of transmission, duration of infection, treatment regimen and timing, and degree of neurocognitive impairment, among other factors, which can all differ drastically across studies, and all influence the perceived magnitude of effect that HIV infection has on the brain. For example, the mode of transmission, whether via intravenous drug use or perinatal infection, may play an important role in the profile of the disease. The pattern of brain deficits in perinatally infected HIV+ children may differ from the pattern in those who acquired

HIV in adulthood, for instance, due to HIV and potentially cART exposure during the time of critical brain development and reorganization (Tardieu et al. 2000; Hoare et al. 2014). On the other end of the lifespan, infected individuals surviving into old age are faced with numerous comorbidities as previously discussed (Brew et al. 2009).

While the majority of biomedical research, including brain imaging, has been performed in populations of individuals living in high-income countries near research-based medical institutions, HIV disproportionately affects individuals in low- and middle-income countries. Even outside of access to cART and treatment compliance (Falagas et al. 2008), factors related to socioeconomic status play a role in HIV prognosis (Hogg et al. 1994; Perry 1998). People are often exposed to several cumulative risk factors, including comorbid illness, poor nutrition, adverse living conditions, and educational disadvantages (WHO 2003; Hackman et al. 2010). Childhood health and social factors, often associated with poverty, can impact brain function and psychological health later in life (Duncan et al. 2010; Walker et al. 2011; Mani et al. 2013; Blair and Raver 2016), and therefore may affect HIV prognosis.

It is not then surprising that, for example, both hypertrophy (Castelo et al. 2007) and hypotrophy (Jernigan et al. 2005) have been reported in the basal ganglia of HIV+ individuals or that there is disagreement whether brain changes are always linked to the degree of immunosuppression (e.g., nadir or current CD4+ count) or specific cognitive domain deficits (Chiang et al. 2007; Gongvatana et al. 2009; Cohen et al. 2010; Becker et al. 2011; Jernigan et al. 2011; Ances et al. 2012). Ultimately, the inconsistent findings in many single cohort neuroimaging studies may be explained by variations in study design including differences in the inclusion and exclusion criteria for enrolling participants, variable MRI acquisition protocols and analysis techniques, limited power to detect differences in small samples, or incomplete consideration of the many confounds associated with the heterogeneous HIV-infected population.

To address variations in methods and boost statistical power, the HIV Working Group was established within the Enhancing Neuro Imaging Genetics through Meta-Analysis (ENIGMA) consortium to harmonize data analysis from neuroimaging studies around the world. By pooling datasets from independent studies of diverse HIV cohorts worldwide, well-powered ENIGMA-HIV studies may determine sources of brain differences that are otherwise difficult to disentangle and assess whether these factors are specific to one cohort or likely to generalize to HIV patients globally. It is important to not only identify biomarkers that are consistently related to disease burden and functional impairment, but also distinguish disease modulators and brain effects that may differ. Understanding common neuropathogenic pathways of HIV infection across international populations could ultimately help lead to improved therapeutic targets and surrogate markers to evaluate treatment effects in clinical trials.

In one of the largest coordinated brain imaging studies of HIV+ individuals worldwide, the ENIGMA HIV Working Group recently pooled data from 1,203 HIV+ individuals across Africa, Asia, Australia, Europe, and North America, to find associations between subcortical brain volumes and two plasma markers routinely collected to monitor immune function and treatment response (Nir et al. 2020). Brain volume associations with both plasma CD4+ counts and detectable viral loads largely implicated the limbic system, extending beyond the classically implicated regions of the basal ganglia. Lower CD4+ counts were associated with smaller hippocampal and thalamic volumes, in addition to larger ventricles, while a detectable viral load was also associated with smaller hippocampal volumes. Sensitivity analyses stratifying the data by cART-status revealed limbic associations were driven by individuals on cART. However, in the subset of individuals who were not on cART, CD4+ associations with basal ganglia volumes, specifically putamen volumes, were significant (Fig. 4). These findings may represent a generalizable brain signature of HIV infection in the cART era and suggest that these regions remain an important target of cART era HIV research, especially given their heightened vulnerability to age-associated atrophy and neurodegeneration.

6 Future Directions in HIV and Aging Research and Clinical Implications

The neuroHIV community is set to gain an enormous set of computational and neuroimaging tools, already vetted in the AD community. As people living with HIV are aging and at increased risk for more complex pathology and cognitive impairment, adopting more advanced neuroimaging techniques – such as multishell diffusion imaging and dynamic contrast-enhanced MRI – may help identify complications unique to chronic infection and those compounded by aging. NeuroHIV researchers can now make use of recent advances in statistical and AI methods that model variation, determine subtypes, and predict patient-specific trajectories in individuals living with HIV. Well-powered studies are now possible by bridging efforts across international studies, but efforts are still needed to expand on the limited number of longitudinal studies of individuals aging with chronic infection. As the neuroHIV field works towards understanding the complex relationship between disease modifying risk factors and cognitive impairment in people living with HIV, identifying pathways of brain disruption may suggest targets for therapeutic intervention. Neuroimaging can be used alongside other biomarkers to both objectively monitor progressive neurodegeneration and evaluate treatment efficacy and response. Although more research is urgently needed, the clinical utility of neuroimaging in assessing cognitive decline and dementia risk in people aging with chronic HIV infection is very promising.

Acknowledgments The authors were funded, in part, by NIH grants: R01 AG059874, U54 EB020403, T32 AG058507, and U01 AG068057. The authors also received partial grant support from Biogen, Inc., for research unrelated to this work.

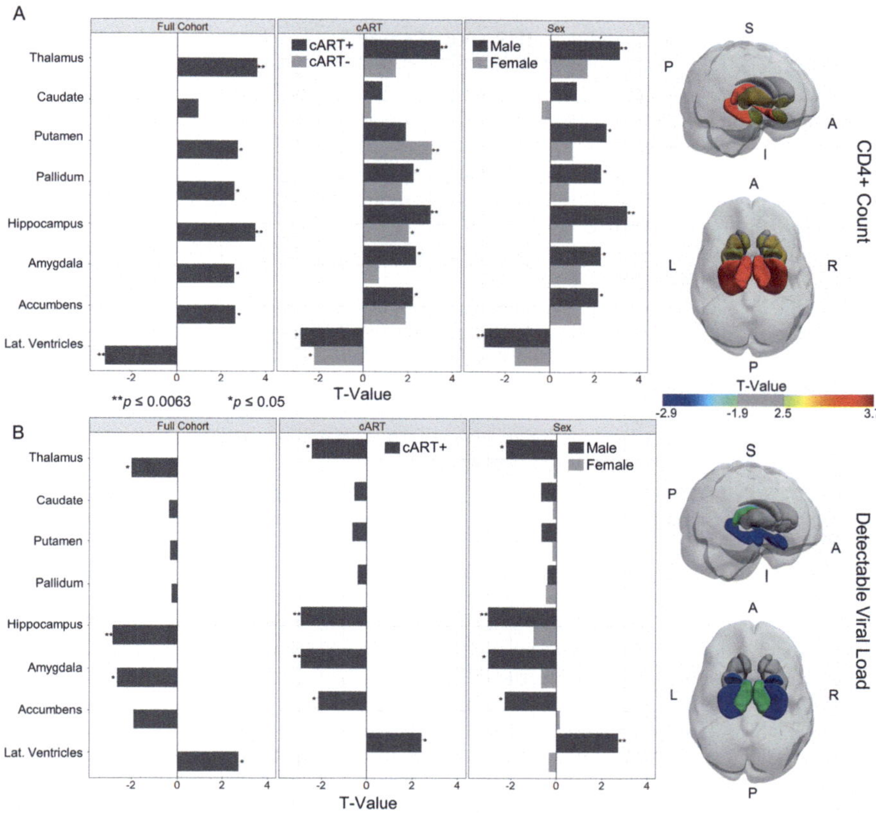

Fig. 4 *T*-values for associations between subcortical volumes and (**a**) CD4+ count, or (**b**) detectable viral load in a cohort of people living with HIV. Bar graphs show *T*-values for associations across all HIV+ participants, and separately in the subset of cART+ participants, cART- participants, males, and females. Structures for which association results indicated an uncorrected significance level of $p \leq 0.05$ are marked with a *, while regions that survive Bonferroni correction across multiple comparisons ($p < 0.05/8$) are indicated with **. Viral load in those not on treatment (cART-) at the time of scanning was not assessed due to the limited number of individuals in this subgroup with undetectable viral load ($n = 8$). In the brain plots, *T*-values are shown for subcortical structures for which association results indicated an uncorrected significance level of $p \leq 0.05$ across all HIV+ participants; those with $p > 0.05$ are shown in *gray* (Reproduced from Nir et al. 2020)

References

Adle-Biassette H, Chretien F, Wingertsmann L, Hery C, Ereau T, Scaravilli F et al (1999) Neuronal apoptosis does not correlate with dementia in HIV infection but is related to microglial activation and axonal damage. Neuropathol Appl Neurobiol 25(2):123–133

Ances BM, Ellis RJ (2007) Dementia and neurocognitive disorders due to HIV-1 infection. Semin Neurol 27(1):86–92. https://doi.org/10.1055/s-2006-956759

Ances BM, Hammoud DA (2014) Neuroimaging of HIV-associated neurocognitive disorders (HAND). Curr Opin HIV AIDS 9(6):545–551. https://doi.org/10.1097/COH.0000000000000112

Ances BM, Ortega M, Vaida F, Heaps J, Paul R (2012) Independent effects of HIV, aging, and HAART on brain volumetric measures. J Acquir Immune Defic Syndr 59(5):469–477. https://doi.org/10.1097/QAI.0b013e318249db17

Andres MA, Feger U, Nath A, Munsaka S, Jiang CS, Chang L (2011) APOE ε 4 allele and CSF APOE on cognition in HIV-infected subjects. J Neuroimmune Pharmacol 6(3):389–398. https://doi.org/10.1007/s11481-010-9254-3

Anthony IC, Bell JE (2008) The neuropathology of HIV/AIDS. Int Rev Psychiatry 20(1):15–24. https://doi.org/10.1080/09540260701862037

Anthony IC, Ramage SN, Carnie FW, Simmonds P, Bell JE (2006) Accelerated Tau deposition in the brains of individuals infected with human immunodeficiency virus-1 before and after the advent of highly active anti-retroviral therapy. Acta Neuropathol 111(6):529–538. https://doi.org/10.1007/s00401-006-0037-0

Antinori A, Arendt G, Becker JT, Brew BJ, Byrd DA, Cherner M et al (2007) Updated research nosology for HIV-associated neurocognitive disorders. Neurology 69(18):1789–1799. https://doi.org/10.1212/01.WNL.0000287431.88658.8b

Assaf Y, Basser PJ (2005) Composite hindered and restricted model of diffusion (CHARMED) MR imaging of the human brain. NeuroImage 27(1):48–58. https://doi.org/10.1016/j.neuroimage.2005.03.042

Ball MJ (1982) Limbic predilection in Alzheimer dementia: is reactivated herpesvirus involved? Can J Neurol Sci 9(3):303–306

Barnes LL, Capuano AW, Aiello AE, Turner AD, Yolken RH, Torrey EF et al (2015) Cytomegalovirus infection and risk of Alzheimer disease in older black and white individuals. J Infect Dis 211(2):230–237. https://doi.org/10.1093/infdis/jiu437

Bartnik-Olson BL, Alger JR, Babikian T, Harris AD, Holshouser B, Kirov II et al (2020) The clinical utility of proton magnetic resonance spectroscopy in traumatic brain injury: recommendations from the ENIGMA MRS working group. Brain Imag Behav. https://doi.org/10.1007/s11682-020-00330-6

Basser PJ, Mattiello J, LeBihan D (1994) MR diffusion tensor spectroscopy and imaging. Biophys J 66(1):259–267. https://doi.org/10.1016/S0006-3495(94)80775-1

Becker JT, Kingsley L, Mullen J, Cohen B, Martin E, Miller EN et al (2009) Vascular risk factors, HIV serostatus, and cognitive dysfunction in gay and bisexual men. Neurology 73(16):1292–1299. https://doi.org/10.1212/WNL.0b013e3181bd10e7

Becker JT, Sanders J, Madsen SK, Ragin A, Kingsley L, Maruca V et al (2011) Subcortical brain atrophy persists even in HAART-regulated HIV disease. Brain Imaging Behav 5(2):77–85. https://doi.org/10.1007/s11682-011-9113-8

Becker JT, Maruca V, Kingsley LA, Sanders JM, Alger JR, Barker PB et al (2012) Factors affecting brain structure in men with HIV disease in the post-HAART era. Neuroradiology 54(2):113–121. https://doi.org/10.1007/s00234-011-0854-2

Behrens TE, Woolrich MW, Jenkinson M, Johansen-Berg H, Nunes RG, Clare S et al (2003) Characterization and propagation of uncertainty in diffusion-weighted MR imaging. Magn Reson Med 50(5):1077–1088. https://doi.org/10.1002/mrm.10609

Behrens TE, Berg HJ, Jbabdi S, Rushworth MF, Woolrich MW (2007) Probabilistic diffusion tractography with multiple fibre orientations: what can we gain? NeuroImage 34(1):144–155. https://doi.org/10.1016/j.neuroimage.2006.09.018

Berger JR, Nath A (1997) HIV dementia and the basal ganglia. Intervirology 40(2-3):122–131. https://doi.org/10.1159/000150539

Bhaskaran K, Mussini C, Antinori A, Walker AS, Dorrucci M, Sabin C et al (2008) Changes in the incidence and predictors of human immunodeficiency virus-associated dementia in the era of highly active antiretroviral therapy. Ann Neurol 63(2):213–221. https://doi.org/10.1002/ana.21225

Biffi A, Greenberg SM (2011) Cerebral amyloid angiopathy: a systematic review. J Clin Neurol 7 (1):1–9. https://doi.org/10.3988/jcn.2011.7.1.1

Bladowska J, Zimny A, Kołtowska A, Szewczyk P, Knysz B, Gąsiorowski J et al (2013) Evaluation of metabolic changes within the normal appearing gray and white matters in neurologically asymptomatic HIV-1-positive and HCV-positive patients: magnetic resonance spectroscopy and immunologic correlation. Eur J Radiol 82(4):686–692. https://doi.org/10.1016/j.ejrad.2012.11.029

Blair C, Raver CC (2016) Poverty, stress, and brain development: new directions for prevention and intervention. Acad Pediatr 16(3 Suppl):S30–SS6. https://doi.org/10.1016/j.acap.2016.01.010

Boban J, Kozic D, Turkulov V, Ostojic J, Semnic R, Lendak D et al (2017) HIV-associated neurodegeneration and neuroimmunity: multivoxel MR spectroscopy study in drug-naive and treated patients. Eur Radiol 27(10):4218–4236. https://doi.org/10.1007/s00330-017-4772-5

Braak H, Braak E (1995) Staging of Alzheimer's disease-related neurofibrillary changes. Neurobiol Aging 16(3):271–278; discussion 8-84. https://doi.org/10.1016/0197-4580(95)00021-6

Braskie MN, Klunder AD, Hayashi KM, Protas H, Kepe V, Miller KJ et al (2010) Plaque and tangle imaging and cognition in normal aging and Alzheimer's disease. Neurobiol Aging 31 (10):1669–1678. https://doi.org/10.1016/j.neurobiolaging.2008.09.012

Brew BJ, Rosenblum M, Cronin K, Price RW (1995) AIDS dementia complex and HIV-1 brain infection: clinical-virological correlations. Ann Neurol 38(4):563–570. https://doi.org/10.1002/ana.410380404

Brew BJ, Pemberton L, Blennow K, Wallin A, Hagberg L (2005) CSF amyloid beta42 and tau levels correlate with AIDS dementia complex. Neurology 65(9):1490–1492. https://doi.org/10.1212/01.wnl.0000183293.95787.b7

Brew BJ, Crowe SM, Landay A, Cysique LA, Guillemin G (2009) Neurodegeneration and ageing in the HAART era. J Neuroimmune Pharmacol 4(2):163–174. https://doi.org/10.1007/s11481-008-9143-1

Brown LAM, Scarola J, Smith AJ, Sanberg PR, Tan J, Giunta B (2014) The role of tau protein in HIV-associated neurocognitive disorders. Mol Neurodegener 9:40. https://doi.org/10.1186/1750-1326-9-40

Bu XL, Yao XQ, Jiao SS, Zeng F, Liu YH, Xiang Y et al (2015) A study on the association between infectious burden and Alzheimer's disease. Eur J Neurol 22(12):1519–1525. https://doi.org/10.1111/ene.12477

Bubak AN, Como CN, Coughlan CM, Johnson NR, Hassell JE, Mescher T et al (2020) Varicella-zoster virus infection of primary human spinal astrocytes produces intracellular amylin, amyloid-beta, and an amyloidogenic extracellular environment. J Infect Dis 221 (7):1088–1097. https://doi.org/10.1093/infdis/jiz560

Burt TD, Agan BK, Marconi VC, He W, Kulkarni H, Mold JE et al (2008) Apolipoprotein (apo) E4 enhances HIV-1 cell entry in vitro, and the APOE epsilon4/epsilon4 genotype accelerates HIV disease progression. Proc Natl Acad Sci U S A 105(25):8718–8723. https://doi.org/10.1073/pnas.0803526105

Cairns DM, Rouleau N, Parker RN, Walsh KG, Gehrke L, Kaplan DL (2020) A 3D human brain-like tissue model of herpes-induced Alzheimer's disease. Sci Adv 6(19):eaay8828. https://doi.org/10.1126/sciadv.aay8828

Cardenas VA, Meyerhoff DJ, Studholme C, Kornak J, Rothlind J, Lampiris H et al (2009) Evidence for ongoing brain injury in human immunodeficiency virus-positive patients treated with antiretroviral therapy. J Neurovirol 15(4):324–333. https://doi.org/10.1080/13550280902973960

Carvalhal A, Gill MJ, Letendre SL, Rachlis A, Bekele T, Raboud J et al (2016) Central nervous system penetration effectiveness of antiretroviral drugs and neuropsychological impairment in the Ontario HIV treatment network cohort study. J Neurovirol 22(3):349–357. https://doi.org/10.1007/s13365-015-0404-5

Castelo JM, Courtney MG, Melrose RJ, Stern CE (2007) Putamen hypertrophy in nondemented patients with human immunodeficiency virus infection and cognitive compromise. Arch Neurol 64(9):1275–1280. https://doi.org/10.1001/archneur.64.9.1275

CDC (2017) HIV surveillance report, 2016. http://www.cdc.gov/hiv/library/reports/hiv-surveillance.html

Chaganti J, Marripudi K, Staub LP, Rae CD, Gates TM, Moffat KJ et al (2019) Imaging correlates of the blood-brain barrier disruption in HIV-associated neurocognitive disorder and therapeutic implications. AIDS 33(12):1843–1852. https://doi.org/10.1097/QAD.0000000000002300

Chang L, Shukla DK (2018) Imaging studies of the HIV-infected brain. Handb Clin Neurol 152:229–264. https://doi.org/10.1016/b978-0-444-63849-6.00018-9

Chang L, Wong V, Nakama H, Watters M, Ramones D, Miller EN et al (2008) Greater than age-related changes in brain diffusion of HIV patients after 1 year. J Neuroimmune Pharmacol 3 (4):265–274. https://doi.org/10.1007/s11481-008-9120-8

Chang L, Andres M, Sadino J, Jiang CS, Nakama H, Miller E et al (2011) Impact of apolipoprotein E epsilon4 and HIV on cognition and brain atrophy: antagonistic pleiotropy and premature brain aging. NeuroImage 58(4):1017–1027. https://doi.org/10.1016/j.neuroimage.2011.07.010

Cherner M, Ellis RJ, Lazzaretto D, Young C, Mindt MR, Atkinson JH et al (2004) Effects of HIV-1 infection and aging on neurobehavioral functioning: preliminary findings. AIDS 18(Suppl 1): S27–S34

Chiang MC, Dutton RA, Hayashi KM, Lopez OL, Aizenstein HJ, Toga AW et al (2007) 3D pattern of brain atrophy in HIV/AIDS visualized using tensor-based morphometry. NeuroImage 34 (1):44–60. https://doi.org/10.1016/j.neuroimage.2006.08.030

Christensen-Quick A, Vanpouille C, Lisco A, Gianella S (2017) Cytomegalovirus and HIV persistence: pouring gas on the fire. AIDS Res Hum Retrovir 33(S1):S23–S30. https://doi.org/10.1089/aid.2017.0145

Clifford DB, Fagan AM, Holtzman DM, Morris JC, Teshome M, Shah AR et al (2009) CSF biomarkers of Alzheimer disease in HIV-associated neurologic disease. Neurology 73 (23):1982–1987. https://doi.org/10.1212/WNL.0b013e3181c5b445

Clifford KM, Samboju V, Cobigo Y, Milanini B, Marx GA, Hellmuth JM et al (2017) Progressive brain atrophy despite persistent viral suppression in HIV patients older than 60 years. J Acquired Immune Deficiency Syndrom 76(3)

Cohen RA, Harezlak J, Schifitto G, Hana G, Clark U, Gongvatana A et al (2010) Effects of nadir CD4 count and duration of human immunodeficiency virus infection on brain volumes in the highly active antiretroviral therapy era. J Neurovirol 16(1):25–32. https://doi.org/10.3109/13550280903552420

Cohen RA, Seider TR, Navia B (2015) HIV effects on age-associated neurocognitive dysfunction: premature cognitive aging or neurodegenerative disease? Alzheimers Res Ther 7(1):37. https://doi.org/10.1186/s13195-015-0123-4

Cole JH, Franke K (2017) Predicting age using neuroimaging: innovative brain ageing biomarkers. Trends Neurosci 40(12):681–690. https://doi.org/10.1016/j.tins.2017.10.001

Cole JH, Caan MWA, Underwood J, De Francesco D, van Zoest RA, Wit FWNM et al (2018) No evidence for accelerated aging-related brain pathology in treated human immunodeficiency virus: longitudinal neuroimaging results from the comorbidity in relation to AIDS (COBRA) project. Clin Infect Dis 66(12):1899–1909. https://doi.org/10.1093/cid/cix1124

Colgan N, Siow B, O'Callaghan JM, Harrison IF, Wells JA, Holmes HE et al (2016) Application of neurite orientation dispersion and density imaging (NODDI) to a tau pathology model of Alzheimer's disease. NeuroImage 125:739–744. https://doi.org/10.1016/j.neuroimage.2015.10.043

Corder EH, Robertson K, Lannfelt L, Bogdanovic N, Eggertsen G, Wilkins J et al (1998) HIV-infected subjects with the E4 allele for APOE have excess dementia and peripheral neuropathy. Nat Med 4(10):1182–1184. https://doi.org/10.1038/2677

Costagliola D (2014) Demographics of HIV and aging. Curr Opin HIV AIDS 9(4):294–301. https://doi.org/10.1097/coh.0000000000000076

Cysique LA, Brew BJ (2014) The effects of HIV and aging on brain functions: proposing a research framework and update on last 3 years' findings. Curr Opin HIV AIDS 9(4):355–364. https://doi.org/10.1097/COH.0000000000000078

Cysique LA, Brew BJ (2019) Vascular cognitive impairment and HIV-associated neurocognitive disorder: a new paradigm. J Neurovirol 25(5):710–721. https://doi.org/10.1007/s13365-018-0706-5

Cysique LA, Maruff P, Brew BJ (2004) Prevalence and pattern of neuropsychological impairment in human immunodeficiency virus-infected/acquired immunodeficiency syndrome (HIV/AIDS) patients across pre- and post-highly active antiretroviral therapy eras: a combined study of two cohorts. J Neurovirol 10(6):350–357. https://doi.org/10.1080/13550280490521078

Cysique LA, Maruff P, Brew BJ (2006) The neuropsychological profile of symptomatic AIDS and ADC patients in the pre-HAART era: a meta-analysis. J Int Neuropsychol Soc 12(3):368–382

Cysique LA, Moffat K, Moore DM, Lane TA, Davies NWS, Carr A et al (2013) HIV, vascular and aging injuries in the brain of clinically stable HIV-infected adults: A (1)H MRS study. PLoS One 8(4):e61738. https://doi.org/10.1371/journal.pone.0061738

Cysique LA, Soares JR, Geng G, Scarpetta M, Moffat K, Green M et al (2017) White matter measures are near normal in controlled HIV infection except in those with cognitive impairment and longer HIV duration. J Neurovirol 23(4):539–547. https://doi.org/10.1007/s13365-017-0524-1

DeCarli C, Massaro J, Harvey D, Hald J, Tullberg M, Au R et al (2005) Measures of brain morphology and infarction in the Framingham heart study: establishing what is normal. Neurobiol Aging 26(4):491–510. https://doi.org/10.1016/j.neurobiolaging.2004.05.004

Descamps M, Hyare H, Stebbing J, Winston A (2008) Magnetic resonance imaging and spectroscopy of the brain in HIV disease. J HIV Ther 13(3):55–58

Descoteaux M (2008) High angular resolution diffusion MRI: from local estimation to segmentation and tractography. Université Nice Sophia Antipolis

Descoteaux M, Poupon C (2012) Diffusion-weighted MRI. Comprehensive Biomed Phys 2:3

Duncan GJ, Ziol-Guest KM, Kalil A (2010) Early-childhood poverty and adult attainment, behavior, and health. Child Dev 81(1):306–325. https://doi.org/10.1111/j.1467-8624.2009.01396.x

Ellis R, Langford D, Masliah E (2007) HIV and antiretroviral therapy in the brain: neuronal injury and repair. Nat Rev Neurosci 8:33–44. https://doi.org/10.1038/nrn2040

Falagas ME, Zarkadoulia EA, Pliatsika PA, Panos G (2008) Socioeconomic status (SES) as a determinant of adherence to treatment in HIV infected patients: a systematic review of the literature. Retrovirology 5:13. https://doi.org/10.1186/1742-4690-5-13

Forloni G, Demicheli F, Giorgi S, Bendotti C, Angeretti N (1992) Expression of amyloid precursor protein mRNAs in endothelial, neuronal and glial cells: modulation by interleukin-1. Brain Res Mol Brain Res 16(1-2):128–134

Frigo M, Fick RHJ, Zucchelli M, Deslauriers-Gauthier S, Deriche R (2020) Multi tissue modelling of diffusion MRI signal reveals volume fraction bias. In: 2020 IEEE 17th international symposium on biomedical imaging (ISBI); 3–7 Apr 2020

Fukutomi H, Glasser MF, Zhang H, Autio JA, Coalson TS, Okada T et al (2018) Neurite imaging reveals microstructural variations in human cerebral cortical gray matter. NeuroImage 182:488–499. https://doi.org/10.1016/j.neuroimage.2018.02.017

Gannicliffe A, Sutton RN, Itzhaki RF (1986) Viruses, brain and immunosuppression. Psychol Med 16(2):247–249. https://doi.org/10.1017/s0033291700009053

Garcia-Hernandez R, Carpena AT, Drakesmith M, Koller K, Jones DK, Canals S, et al. (2020) Imaging microglia and astrocytes non-invasively using Diffusion MRI. bioRxiv. 2020.02.07.938910. https://doi.org/10.1101/2020.02.07.938910

Gongvatana A, Schweinsburg BC, Taylor MJ, Theilmann RJ, Letendre SL, Alhassoon OM et al (2009) White matter tract injury and cognitive impairment in human immunodeficiency virus-infected individuals. J Neurovirol 15(2):187–195. https://doi.org/10.1080/13550280902769756

Gongvatana A, Cohen RA, Correia S, Devlin KN, Miles J, Kang H et al (2011) Clinical contributors to cerebral white matter integrity in HIV-infected individuals. J Neurovirol 17(5):477–486. https://doi.org/10.1007/s13365-011-0055-0

Gongvatana A, Harezlak J, Buchthal S, Daar E, Schifitto G, Campbell T et al (2013) Progressive cerebral injury in the setting of chronic HIV infection and antiretroviral therapy. J Neurovirol 19 (3):209–218. https://doi.org/10.1007/s13365-013-0162-1

Gonzalez-Scarano F, Martin-Garcia J (2005) The neuropathogenesis of AIDS. Nat Rev Immunol 5 (1):69–81. https://doi.org/10.1038/nri1527

Goodkin K, Wilkie FL, Concha M, Hinkin CH, Symes S, Baldewicz TT et al (2001) Aging and neuro-AIDS conditions and the changing spectrum of HIV-1-associated morbidity and mortality. J Clin Epidemiol 54(Suppl 1):S35–S43

Green DA, Masliah E, Vinters HV, Beizai P, Moore DJ, Achim CL (2005) Brain deposition of beta-amyloid is a common pathologic feature in HIV positive patients. AIDS 19(4):407–411

Guerrero JM, Adluru N, Bendlin BB, Goldsmith HH, Schaefer SM, Davidson RJ et al (2019) Optimizing the intrinsic parallel diffusivity in NODDI: an extensive empirical evaluation. PloS One 14(9):e0217118. https://doi.org/10.1371/journal.pone.0217118

Hackman DA, Farah MJ, Meaney MJ (2010) Socioeconomic status and the brain: mechanistic insights from human and animal research. Nat Rev Neurosci 11(9):651–659. https://doi.org/10.1038/nrn2897

Hakkers CS, Arends JE, Barth RE, Du Plessis S, Hoepelman AI, Vink M (2017) Review of functional MRI in HIV: effects of aging and medication. J Neurovirol 23(1):20–32. https://doi.org/10.1007/s13365-016-0483-y

Harezlak J, Buchthal S, Taylor M, Schifitto G, Zhong J, Daar E et al (2011) Persistence of HIV-associated cognitive impairment, inflammation, and neuronal injury in era of highly active antiretroviral treatment. AIDS 25(5):625–633. https://doi.org/10.1097/QAD.0b013e3283427da7

Harezlak J, Cohen R, Gongvatana A, Taylor M, Buchthal S, Schifitto G et al (2014) Predictors of CNS injury as measured by proton magnetic resonance spectroscopy in the setting of chronic HIV infection and CART. J Neurovirol 20(3):294–303. https://doi.org/10.1007/s13365-014-0246-6

Heaps JM, Sithinamsuwan P, Paul R, Lerdlum S, Pothisri M, Clifford D et al (2015) Association between brain volumes and HAND in cART-naïve HIV+ individuals from Thailand. J Neurovirol 21(2):105–112. https://doi.org/10.1007/s13365-014-0309-8

Heaton RK, Clifford DB, Franklin DR Jr, Woods SP, Ake C, Vaida F et al (2010) HIV-associated neurocognitive disorders persist in the era of potent antiretroviral therapy: charter study. Neurology 75(23):2087–2096. https://doi.org/10.1212/WNL.0b013e318200d727

Heaton RK, Franklin DR, Ellis RJ, McCutchan JA, Letendre SL, Leblanc S et al (2011) HIV-associated neurocognitive disorders before and during the era of combination antiretroviral therapy: differences in rates, nature, and predictors. J Neurovirol 17(1):3–16. https://doi.org/10.1007/s13365-010-0006-1

Heye AK, Culling RD, Hernandez MDV, Thrippleton MJ, Wardlaw JM (2014) Assessment of blood-brain barrier disruption using dynamic contrast-enhanced MRI. A systematic review. Neuroimage Clin 6:262–274. https://doi.org/10.1016/j.nicl.2014.09.002

Hoare J, Ransford GL, Phillips N, Amos T, Donald K, Stein DJ (2014) Systematic review of neuroimaging studies in vertically transmitted HIV positive children and adolescents. Metab Brain Dis 29(2):221–229. https://doi.org/10.1007/s11011-013-9456-5

Hogg R, Strathdee S, Craib K, O'Shaughnessy M, JSG M, Schechter M (1994) Lower socioeconomic status and shorter survival following HIV infection. Lancet 344:1120–1124

Holt JL, Kraft-Terry SD, Chang L (2012) Neuroimaging studies of the aging HIV-1-infected brain. J Neurovirol 18(4):291–302. https://doi.org/10.1007/s13365-012-0114-1

Howdle GC, Quidé Y, Kassem MS, Johnson K, Rae CD, Brew BJ et al (2020) Brain amyloid in virally suppressed HIV-associated neurocognitive disorder. Neurology Neuroimmunol Neuroinflam 7(4):e739. https://doi.org/10.1212/NXI.0000000000000739

Hua X, Boyle CP, Harezlak J, Tate DF, Yiannoutsos CT, Cohen R et al (2013) Disrupted cerebral metabolite levels and lower nadir CD4 + counts are linked to brain volume deficits in 210 HIV-infected patients on stable treatment. Neuroimage Clin 3:132–142. https://doi.org/10.1016/j.nicl.2013.07.009

Hua X, Ching CR, Mezher A, Gutman BA, Hibar DP, Bhatt P et al (2016) MRI-based brain atrophy rates in ADNI phase 2: acceleration and enrichment considerations for clinical trials. Neurobiol Aging 37:26–37. https://doi.org/10.1016/j.neurobiolaging.2015.09.018

Hult B, Chana G, Masliah E, Everall I (2008) Neurobiology of HIV. Int Rev Psychiatry 20(1):3–13. https://doi.org/10.1080/09540260701862086

Jack CR Jr, Knopman DS, Jagust WJ, Shaw LM, Aisen PS, Weiner MW et al (2010) Hypothetical model of dynamic biomarkers of the Alzheimer's pathological cascade. Lancet Neurol 9 (1):119–128

Jack CR Jr, Knopman DS, Jagust WJ, Petersen RC, Weiner MW, Aisen PS et al (2013) Tracking pathophysiological processes in Alzheimer's disease: an updated hypothetical model of dynamic biomarkers. Lancet Neurol 12(2):207–216. https://doi.org/10.1016/s1474-4422(12)70291-0

Jack CR Jr, Bennett DA, Blennow K, Carrillo MC, Dunn B, Haeberlein SB et al (2018) NIA-AA research framework: toward a biological definition of Alzheimer's disease. Alzheimers Dement 14(4):535–562. https://doi.org/10.1016/j.jalz.2018.02.018

Jahanshad N, Valcour VG, Nir TM, Kohannim O, Busovaca E, Nicolas K et al (2012) Disrupted brain networks in the aging HIV+ population. Brain Connect 2(6):335–344. https://doi.org/10.1089/brain.2012.0105-Rev

Jelescu IO, Palombo M, Bagnato F, Schilling KG (2020) Challenges for biophysical modeling of microstructure. J Neurosci Methods 344:108861. https://doi.org/10.1016/j.jneumeth.2020.108861

Jensen JH, Helpern JA, Ramani A, Lu H, Kaczynski K (2005) Diffusional kurtosis imaging: the quantification of non-gaussian water diffusion by means of magnetic resonance imaging. Magn Reson Med 53(6):1432–1440. https://doi.org/10.1002/mrm.20508

Jernigan TL, Gamst AC, Archibald SL, Fennema-Notestine C, Mindt MR, Marcotte TD et al (2005) Effects of methamphetamine dependence and HIV infection on cerebral morphology. Am J Psychiatry 162(8):1461–1472. https://doi.org/10.1176/appi.ajp.162.8.1461

Jernigan TL, Archibald SL, Fennema-Notestine C, Taylor MJ, Theilmann RJ, Julaton MD et al (2011) Clinical factors related to brain structure in HIV: the CHARTER study. J Neurovirol 17 (3):248–257. https://doi.org/10.1007/s13365-011-0032-7

Jeurissen B, Leemans A, Tournier JD, Jones DK, Sijbers J (2013) Investigating the prevalence of complex fiber configurations in white matter tissue with diffusion magnetic resonance imaging. Hum Brain Mapp 34(11):2747–2766. https://doi.org/10.1002/hbm.22099

Jeurissen B, Tournier JD, Dhollander T, Connelly A, Sijbers J (2014) Multi-tissue constrained spherical deconvolution for improved analysis of multi-shell diffusion MRI data. NeuroImage 103:411–426. https://doi.org/10.1016/j.neuroimage.2014.07.061

Jorm AF, Mather KA, Butterworth P, Anstey KJ, Christensen H, Easteal S (2007) APOE genotype and cognitive functioning in a large age-stratified population sample. Neuropsychology 21 (1):1–8. https://doi.org/10.1037/0894-4105.21.1.1

Kallianpur KJ, Jahanshad N, Sailasuta N, Benjapornpong K, Chan P, Pothisri M et al (2020) Regional brain volumetric changes despite 2 years of treatment initiated during acute HIV infection. AIDS 34(3):415–426. https://doi.org/10.1097/qad.0000000000002436

Kamagata K, Andica C, Hatano T, Ogawa T, Takeshige-Amano H, Ogaki K et al (2020) Advanced diffusion magnetic resonance imaging in patients with Alzheimer's and Parkinson's diseases. Neural Regen Res 15(9):1590–1600. https://doi.org/10.4103/1673-5374.276326

Kamkwalala A, Newhouse P (2017) Mechanisms of cognitive aging in the HIV-positive adult. Curr Behav Neurosci Rep 4(3):188–197. https://doi.org/10.1007/s40473-017-0122-9

Klunk WE, Koeppe RA, Price JC, Benzinger TL, Devous MD Sr, Jagust WJ et al (2015) The Centiloid Project: standardizing quantitative amyloid plaque estimation by PET. Alzheimers Dement 11(1):1–15.e1-4. https://doi.org/10.1016/j.jalz.2014.07.003

Kristen H, Santana S, Sastre I, Recuero M, Bullido MJ, Aldudo J (2015) Herpes simplex virus type 2 infection induces AD-like neurodegeneration markers in human neuroblastoma cells. Neurobiol Aging 36(10):2737–2747. https://doi.org/10.1016/j.neurobiolaging.2015.06.014

Kuhn T, Schonfeld D, Sayegh P, Arentoft A, Jones JD, Hinkin CH et al (2017) The effects of HIV and aging on subcortical shape alterations: a 3D morphometric study. Hum Brain Mapp 38 (2):1025–1037. https://doi.org/10.1002/hbm.23436

La Joie R, Visani AV, Baker SL, Brown JA, Bourakova V, Cha J et al (2020) Prospective longitudinal atrophy in Alzheimer's disease correlates with the intensity and topography of baseline tau-PET. Sci Trans Med 12(524):eaau5732. https://doi.org/10.1126/scitranslmed.aau5732

Lam PK, Santhalingam V, Suresh P, Baboota R, Zhu AH, Thomopoulos SI, et al (2020) Accurate brain age prediction using recurrent slice-based networks. bioRxiv. 2020.08.04.235069. https://doi.org/10.1101/2020.08.04.235069

Lambert JC, Ibrahim-Verbaas CA, Harold D, Naj AC, Sims R, Bellenguez C et al (2013) Meta-analysis of 74,046 individuals identifies 11 new susceptibility loci for Alzheimer's disease. Nat Genet 45(12):1452–1458. https://doi.org/10.1038/ng.2802

Leow AD, Zhu S, Zhan L, McMahon K, de Zubicaray GI, Meredith M et al (2009) The tensor distribution function. Magn Reson Med 61(1):205–214. https://doi.org/10.1002/mrm.21852

Liao YF, Wang BJ, Cheng HT, Kuo LH, Wolfe MS (2004) Tumor necrosis factor-alpha, interleukin-1beta, and interferon-gamma stimulate gamma-secretase-mediated cleavage of amyloid precursor protein through a JNK-dependent MAPK pathway. J Biol Chem 279 (47):49523–49532. https://doi.org/10.1074/jbc.M402034200

Lopatko Lindman K, Weidung B, Olsson J, Josefsson M, Kok E, Johansson A et al (2019) A genetic signature including apolipoprotein Eepsilon4 potentiates the risk of herpes simplex-associated Alzheimer's disease. Alzheimers Dement 5(1):697–704. https://doi.org/10.1016/j.trci.2019.09.014

Lyall DM, Cox SR, Lyall LM, Celis-Morales C, Cullen B, Mackay DF et al (2020) Association between APOE e4 and white matter hyperintensity volume, but not total brain volume or white matter integrity. Brain Imaging Behav 14(5):1468–1476. https://doi.org/10.1007/s11682-019-00069-9

Mani A, Mullainathan S, Shafir E, Zhao J (2013) Poverty impedes cognitive function. Science 341 (6149):976–980. https://doi.org/10.1126/science.1238041

Masters MC, Ances BM (2014) Role of neuroimaging in HIV-associated neurocognitive disorders. Semin Neurol 34(1):89–102. https://doi.org/10.1055/s-0034-1372346

McArthur JC (2004) HIV dementia: an evolving disease. J Neuroimmunol 157(1-2):3–10. https://doi.org/10.1016/j.jneuroim.2004.08.042

McArthur JC, Steiner J, Sacktor N, Nath A (2010) Human immunodeficiency virus-associated neurocognitive disorders: mind the gap. Ann Neurol 67(6):699–714. https://doi.org/10.1002/ana.22053

McIntosh EC, Tureson K, Rotblatt LJ, Singer EJ, Thames AD (2020) HIV, vascular risk factors, and cognition in the combination antiretroviral therapy era: a systematic review and meta-analysis. J Int Neuropsychol Soc:1–17. https://doi.org/10.1017/S1355617720001022

Milanini B, Valcour V (2017) Differentiating HIV-associated neurocognitive disorders from Alzheimer's disease: an emerging issue in geriatric NeuroHIV. Curr HIV/AIDS Rep 14 (4):123–132. https://doi.org/10.1007/s11904-017-0361-0

Miller KL, Alfaro-Almagro F, Bangerter NK, Thomas DL, Yacoub E, Xu J et al (2016) Multimodal population brain imaging in the UK biobank prospective epidemiological study. Nat Neurosci 19(11):1523–1536. https://doi.org/10.1038/nn.4393

Mohamed M, Skolasky RL, Zhou Y, Ye W, Brasic JR, Brown A et al (2020) Beta-amyloid (Aβ) uptake by PET imaging in older HIV+ and HIV- individuals. J Neurovirol 26(3):382–390. https://doi.org/10.1007/s13365-020-00836-1

Montagne A, Nation DA, Sagare AP, Barisano G, Sweeney MD, Chakhoyan A et al (2020) APOE4 leads to blood–brain barrier dysfunction predicting cognitive decline. Nature 581(7806):71–76. https://doi.org/10.1038/s41586-020-2247-3

Morgan EE, Woods SP, Letendre SL, Franklin DR, Bloss C, Goate A et al (2013) Apolipoprotein E4 genotype does not increase risk of HIV-associated neurocognitive disorders. J Neurovirol 19 (2):150–156. https://doi.org/10.1007/s13365-013-0152-3

Morgello S (2018) Chapter 2 – HIV neuropathology. In: Brew BJ (ed) Handbook of clinical neurology, vol 152. Elsevier, Amsterdam, pp 3–19

Moulignier A, Savatovsky J, Assoumou L, Lescure FX, Lamirel C, Godin O et al (2018) Silent cerebral small-vessel disease is twice as prevalent in middle-aged individuals with well-controlled, combination antiretroviral therapy-treated human immunodeficiency virus (HIV) than in HIV-uninfected individuals. Clin Infect Dis 66(11):1762–1769. https://doi.org/10.1093/cid/cix1075

Navia BA, Cho ES, Petito CK, Price RW (1986a) The AIDS dementia complex: II. Neuropathology. Ann Neurol 19(6):525–535. https://doi.org/10.1002/ana.410190603

Navia BA, Jordan BD, Price RW (1986b) The AIDS dementia complex: I. clinical features. Ann Neurol 19(6):517–524. https://doi.org/10.1002/ana.410190602

Nazeri A, Chakravarty MM, Rotenberg DJ, Rajji TK, Rathi Y, Michailovich OV et al (2015) Functional consequences of neurite orientation dispersion and density in humans across the adult lifespan. J Neurosci 35(4):1753–1762. https://doi.org/10.1523/jneurosci.3979-14.2015

Nebuloni M, Pellegrinelli A, Ferri A, Bonetto S, Boldorini R, Vago L et al (2001) Beta amyloid precursor protein and patterns of HIV p24 immunohistochemistry in different brain areas of AIDS patients. AIDS 15(5):571–575

Nelson PT, Pious NM, Jicha GA, Wilcock DM, Fardo DW, Estus S et al (2013) APOE-ε2 and APOE-ε4 correlate with increased amyloid accumulation in cerebral vasculature. J Neuropathol Exp Neurol 72(7):708–715. https://doi.org/10.1097/NEN.0b013e31829a25b9

Neuen-Jacob E, Arendt G, Wendtland B, Jacob B, Schneeweis M, Wechsler W (1993) Frequency and topographical distribution of CD68-positive macrophages and HIV-1 core proteins in HIV-associated brain lesions. Clin Neuropathol 12(6):315–324

Nir TM, Jahanshad N, Villalon-Reina JE, Toga AW, Jack CR, Weiner MW et al (2013) Effectiveness of regional DTI measures in distinguishing Alzheimer's disease, MCI, and normal aging. Neuroimage Clin 3:180–195. https://doi.org/10.1016/j.nicl.2013.07.006

Nir TM, Jahanshad N, Busovaca E, Wendelken L, Nicolas K, Thompson PM et al (2014) Mapping white matter integrity in elderly people with HIV. Hum Brain Mapp 35(3):975–992. https://doi.org/10.1002/hbm.22228

Nir TM, Jahanshad N, Villalon-Reina JE, Isaev D, Zavaliangos-Petropulu A, Zhan L et al (2017) Fractional anisotropy derived from the diffusion tensor distribution function boosts power to detect Alzheimer's disease deficits. Magn Reson Med 78(6):2322–2333. https://doi.org/10.1002/mrm.26623

Nir TM, Jahanshad N, CRK C, Cohen RA, Harezlak J, Schifitto G et al (2019a) Progressive brain atrophy in chronically infected and treated HIV+ individuals. J Neurovirol 25(3):342–353. https://doi.org/10.1007/s13365-019-00723-4

Nir TM, Lam HY, Ananworanich J, Boban J, Brew BJ, Cysique L et al (2019b) Effects of diffusion MRI model and harmonization on the consistency of findings in an international multi-cohort HIV neuroimaging study. Computational diffusion MRI. Springer, Cham

Nir TM, Fouche JP, Ananworanich J, Ances B, Boban J, Brew B, Changanti J et al (2020) Association of immunosuppression and viral load with subcortical brain volume in an international sample of people living with HIV. JAMA Netw Open. (In Press)

Nir TM, Villalon-Reina JE, Zhu A, Thomopolous SI, Maiti P, Salminen LE, et al. (2021) Multi-compartment diffusion measures capture microstructural associations with amyloid and tau PET measures in the cortex. In: 15th international conference on Alzheimer's and Parkinson's diseases. Virtual conference

Novikov DS, Kiselev VG, Jespersen SN (2018) On modeling. Magnet Resonance Med 79 (6):3172–3193. https://doi.org/10.1002/mrm.27101

O'Connor EE, Jaillard A, Renard F, Zeffiro TA (2017) Reliability of white matter microstructural changes in HIV infection: meta-analysis and confirmation. Am J Neuroradiol 38(8):1510–1519. https://doi.org/10.3174/ajnr.A5229

O'Connor EE, Zeffiro TA, Zeffiro TA (2018) Brain structural changes following HIV infection: meta-analysis. AJNR Am J Neuroradiol 39(1):54–62. https://doi.org/10.3174/ajnr.A5432

Oh SW, Shin NY, Choi JY, Lee SK, Bang MR (2018) Altered white matter integrity in human immunodeficiency virus-associated neurocognitive disorder: a tract-based spatial statistics study. Korean J Radiol 19(3):431–442. https://doi.org/10.3348/kjr.2018.19.3.431

Özarslan E, Koay CG, Shepherd TM, Komlosh ME, İrfanoğlu MO, Pierpaoli C et al (2013) Mean apparent propagator (MAP) MRI: a novel diffusion imaging method for mapping tissue microstructure. NeuroImage 78:16–32. https://doi.org/10.1016/j.neuroimage.2013.04.016

Palmqvist S, Zetterberg H, Mattsson N, Johansson P, Minthon L, Blennow K et al (2015) Detailed comparison of amyloid PET and CSF biomarkers for identifying early Alzheimer disease. Neurology 85(14):1240–1249. https://doi.org/10.1212/wnl.0000000000001991

Palmqvist S, Mattsson N, Hansson O (2016) Alzheimer's disease neuroimaging I. cerebrospinal fluid analysis detects cerebral amyloid-β accumulation earlier than positron emission tomography. Brain 139(Pt 4):1226–1236. https://doi.org/10.1093/brain/aww015

Palombo M, Ianus A, Guerreri M, Nunes D, Alexander DC, Shemesh N et al (2020) SANDI: a compartment-based model for non-invasive apparent soma and neurite imaging by diffusion MRI. NeuroImage 215:116835. https://doi.org/10.1016/j.neuroimage.2020.116835

Parker TD, Slattery CF, Zhang J, Nicholas JM, Paterson RW, Foulkes AJM et al (2018) Cortical microstructure in young onset Alzheimer's disease using neurite orientation dispersion and density imaging. Hum Brain Mapp 39(7):3005–3017. https://doi.org/10.1002/hbm.24056

Pathai S, Bajillan H, Landay AL, High KP (2014) Is HIV a model of accelerated or accentuated aging? J Gerontol Ser A Biol Sci Med Sci 69(7):833–842. https://doi.org/10.1093/gerona/glt168

Peluso R, Haase A, Stowring L, Edwards M, Ventura P (1985) A Trojan horse mechanism for the spread of visna virus in monocytes. Virology 147(1):231–236

Perry MJ (1998) Gender, race and economic perspectives on the social epidemiology of HIV infection: implications for prevention. J Prim Prev 19(2):97–104. https://doi.org/10.1023/A:1022688827012

Pfefferbaum A, Rogosa DA, Rosenbloom MJ, Chu W, Sassoon SA, Kemper CA et al (2014) Accelerated aging of selective brain structures in human immunodeficiency virus infection: a controlled, longitudinal magnetic resonance imaging study. Neurobiol Aging 35(7):1755–1768. https://doi.org/10.1016/j.neurobiolaging.2014.01.008

Protas HD, Kepe V, Hayashi KM, Klunder AD, Braskie MN, Ercoli L et al (2012) Prediction of cognitive decline based on hemispheric cortical surface maps of FDDNP PET. NeuroImage 61 (4):749–760. https://doi.org/10.1016/j.neuroimage.2012.02.056

Raber J, Huang Y, Ashford JW (2004) ApoE genotype accounts for the vast majority of AD risk and AD pathology. Neurobiol Aging 25(5):641–650. https://doi.org/10.1016/j.neurobiolaging.2003.12.023

Rahimian P, He JJ (2017) HIV/neuroAIDS biomarkers. Prog Neurobiol 157:117–132. https://doi.org/10.1016/j.pneurobio.2016.04.003

Raz N, Rodrigue KM (2006) Differential aging of the brain: patterns, cognitive correlates and modifiers. Neurosci Biobehav Rev 30(6):730–748. https://doi.org/10.1016/j.neubiorev.2006.07.001

Rempel HC, Pulliam L (2005) HIV-1 tat inhibits neprilysin and elevates amyloid beta. AIDS 19 (2):127–135

Robertson K, Liner J, Meeker RB (2012) Antiretroviral neurotoxicity. J Neurovirol 18(5):388–399. https://doi.org/10.1007/s13365-012-0120-3

Rodger AJ, Lodwick R, Schechter M, Deeks S, Amin J, Gilson R et al (2013) Mortality in well controlled HIV in the continuous antiretroviral therapy arms of the SMART and ESPRIT trials

compared with the general population. AIDS 27(6):973–979. https://doi.org/10.1097/QAD.0b013e32835cae9c
Rowe CC, Dore V, Jones G, Baxendale D, Mulligan RS, Bullich S et al (2017) (18)F-Florbetaben PET beta-amyloid binding expressed in Centiloids. Eur J Nucl Med Mol Imaging 44 (12):2053–2059. https://doi.org/10.1007/s00259-017-3749-6
Sacktor N, McDermott MP, Marder K, Schifitto G, Selnes OA, McArthur JC et al (2002) HIV-associated cognitive impairment before and after the advent of combination therapy. J Neurovirol 8(2):136–142. https://doi.org/10.1080/13550280290049615
Sacktor N, Skolasky RL, Seaberg E, Munro C, Becker JT, Martin E et al (2016) Prevalence of HIV-associated neurocognitive disorders in the multicenter AIDS cohort study. Neurology 86 (4):334–340. https://doi.org/10.1212/WNL.0000000000002277
Samboju V, Philippi CL, Chan P, Cobigo Y, Fletcher JLK, Robb M et al (2018) Structural and functional brain imaging in acute HIV. Neuroimage Clin 20:327–335. https://doi.org/10.1016/j.nicl.2018.07.024
Schneider JA, Aggarwal NT, Barnes L, Boyle P, Bennett DA (2009) The neuropathology of older persons with and without dementia from community versus clinic cohorts. J Alzheimers Dis 18 (3):691–701. https://doi.org/10.3233/JAD-2009-1227
Schouten J, Cinque P, Gisslen M, Reiss P, Portegies P (2011) HIV-1 infection and cognitive impairment in the cART era: a review. AIDS 25(5):561–575. https://doi.org/10.1097/QAD.0b013e3283437f9a
Scott JC, Woods SP, Carey CL, Weber E, Bondi MW, Grant I (2011) Neurocognitive consequences of HIV infection in older adults: an evaluation of the "cortical" hypothesis. AIDS Behav 15 (6):1187–1196. https://doi.org/10.1007/s10461-010-9815-8
Seider TR, Gongvatana A, Woods AJ, Chen H, Porges EC, Cummings T et al (2016) Age exacerbates HIV-associated white matter abnormalities. J Neurovirol 22(2):201–212. https://doi.org/10.1007/s13365-015-0386-3
Sepehrband F, Cabeen RP, Choupan J, Barisano G, Law M, Toga AW (2019) Perivascular space fluid contributes to diffusion tensor imaging changes in white matter. NeuroImage 197:243–254. https://doi.org/10.1016/j.neuroimage.2019.04.070
Sochocka M, Zwolinska K, Leszek J (2017) The infectious etiology of Alzheimer's disease. Curr Neuropharmacol 15(7):996–1009. https://doi.org/10.2174/1570159X15666170313122937
Soontornniyomkij V, Moore DJ, Gouaux B, Soontornniyomkij B, Tatro ET, Umlauf A et al (2012) Cerebral beta-amyloid deposition predicts HIV-associated neurocognitive disorders in APOE epsilon4 carriers. AIDS 26(18):2327–2335. https://doi.org/10.1097/QAD.0b013e32835a117c
Spector SA, Singh KK, Gupta S, Cystique LA, Jin H, Letendre S et al (2010) APOE epsilon4 and MBL-2 O/O genotypes are associated with neurocognitive impairment in HIV-infected plasma donors. AIDS 24(10):1471–1479. https://doi.org/10.1097/QAD.0b013e328339e25c
Spudich S, Gonzalez-Scarano F (2012) HIV-1-related central nervous system disease: current issues in pathogenesis, diagnosis, and treatment. Cold Spring Harb Perspect Med 2(6):a007120. https://doi.org/10.1101/cshperspect.a007120
Stanley LC, Mrak RE, Woody RC, Perrot LJ, Zhang S, Marshak DR et al (1994) Glial cytokines as neuropathogenic factors in HIV infection: pathogenic similarities to Alzheimer's disease. J Neuropathol Exp Neurol 53(3):231–238
Steinbrink F, Evers S, Buerke B, Young P, Arendt G, Koutsilieri E et al (2013) Cognitive impairment in HIV infection is associated with MRI and CSF pattern of neurodegeneration. Eur J Neurol 20(3):420–428. https://doi.org/10.1111/ene.12006
Sweeney MD, Kisler K, Montagne A, Toga AW, Zlokovic BV (2018) The role of brain vasculature in neurodegenerative disorders. Nat Neurosci 21(10):1318–1331. https://doi.org/10.1038/s41593-018-0234-x
Tardieu M, Le Chenadec J, Persoz A, Meyer L, Blanche S, Mayaux MJ (2000) HIV-1-related encephalopathy in infants compared with children and adults. French pediatric HIV infection study and the SEROCO group. Neurology 54(5):1089–1095

Tate DF, Conley JJ, Meier DS, Navia BA, Cohen R, Guttmann CRG (2009) Neuroimaging among HIV-infected patients: current knowledge and future directions. In: Paul HR, Sacktor CN, Valcour V, Tashima TK (eds) HIV and the brain: new challenges in the modern era. Humana Press, Totowa, pp 75–107

Tate DF, Sampat M, Harezlak J, Fiecas M, Hogan J, Dewey J et al (2011) Regional areas and widths of the midsagittal corpus callosum among HIV-infected patients on stable antiretroviral therapies. J Neurovirol 17(4):368–379. https://doi.org/10.1007/s13365-011-0033-6

Thompson PM, Jahanshad N (2015) Novel neuroimaging methods to understand how HIV affects the brain. Curr HIV/AIDS Rep 12(2):289–298. https://doi.org/10.1007/s11904-015-0268-6

Thompson PM, Hayashi KM, de Zubicaray G, Janke AL, Rose SE, Semple J et al (2003) Dynamics of gray matter loss in Alzheimer's disease. J Neurosci 23(3):994–1005

Thompson PM, Dutton RA, Hayashi KM, Toga AW, Lopez OL, Aizenstein HJ et al (2005) Thinning of the cerebral cortex visualized in HIV/AIDS reflects CD4+ T lymphocyte decline. Proc Natl Acad Sci U S A 102(43):15647–15652. https://doi.org/10.1073/pnas.0502548102

Tournier JD, Calamante F, Gadian DG, Connelly A (2004) Direct estimation of the fiber orientation density function from diffusion-weighted MRI data using spherical deconvolution. NeuroImage 23(3):1176–1185. https://doi.org/10.1016/j.neuroimage.2004.07.037

Tozzi V, Balestra P, Bellagamba R, Corpolongo A, Salvatori MF, Visco-Comandini U et al (2007) Persistence of neuropsychologic deficits despite long-term highly active antiretroviral therapy in patients with HIV-related neurocognitive impairment: prevalence and risk factors. J Acquir Immune Defic Syndr 45(2):174–182. https://doi.org/10.1097/QAI.0b013e318042e1ee

Tuch DS (2004) Q-ball imaging. Magn Reson Med 52(6):1358–1372. https://doi.org/10.1002/mrm.20279

Tucker KA, Robertson KR, Lin W, Smith JK, An H, Chen Y et al (2004) Neuroimaging in human immunodeficiency virus infection. J Neuroimmunol 157(1-2):153–162. https://doi.org/10.1016/j.jneuroim.2004.08.036

Turchan-Cholewo J, Liu Y, Gartner S, Reid R, Jie C, Peng X et al (2006) Increased vulnerability of ApoE4 neurons to HIV proteins and opiates: protection by diosgenin and L-deprenyl. Neurobiol Dis 23(1):109–119. https://doi.org/10.1016/j.nbd.2006.02.005

UNAIDS (2018) Global HIV & AIDS statistics – 2018 fact sheet 2018. http://www.unaids.org/en/resources/fact-sheet

Valcour V, Shikuma C, Shiramizu B, Watters M, Poff P, Selnes O et al (2004a) Higher frequency of dementia in older HIV-1 individuals: the Hawaii aging with HIV-1 cohort. Neurology 63 (5):822–827

Valcour V, Shikuma C, Shiramizu B, Watters M, Poff P, Selnes OA et al (2004b) Age, apolipoprotein E4, and the risk of HIV dementia: the Hawaii aging with HIV cohort. J Neuroimmunol 157(1-2):197–202. https://doi.org/10.1016/j.jneuroim.2004.08.029

Valcour VG, Shikuma CM, Shiramizu BT, Williams AE, Watters MR, Poff PW et al (2005) Diabetes, insulin resistance, and dementia among HIV-1-infected patients. J Acquir Immune Defic Syndr 38(1):31–36

Valcour V, Paul R, Neuhaus J, Shikuma C (2011) The effects of age and HIV on neuropsychological performance. J Int Neuropsychol Soc JINS 17(1):190–195. https://doi.org/10.1017/S1355617710001438

Walker SP, Wachs TD, Grantham-McGregor S, Black MM, Nelson CA, Huffman SL et al (2011) Inequality in early childhood: risk and protective factors for early child development. Lancet 378(9799):1325–1338. https://doi.org/10.1016/s0140-6736(11)60555-2

Wang Y, Zhang J, Gutman B, Chan TF, Becker JT, Aizenstein HJ et al (2010) Multivariate tensor-based morphometry on surfaces: application to mapping ventricular abnormalities in HIV/AIDS. NeuroImage 49(3):2141–2157. https://doi.org/10.1016/j.neuroimage.2009.10.086

Wardlaw JM, Benveniste H, Nedergaard M, Zlokovic BV, Mestre H, Lee H et al (2020) Perivascular spaces in the brain: anatomy, physiology and pathology. Nat Rev Neurol 16 (3):137–153. https://doi.org/10.1038/s41582-020-0312-z

Wedeen VJ, Hagmann P, Tseng WY, Reese TG, Weisskoff RM (2005) Mapping complex tissue architecture with diffusion spectrum magnetic resonance imaging. Magn Reson Med 54 (6):1377–1386. https://doi.org/10.1002/mrm.20642

Wendelken LA, Valcour V (2012) Impact of HIV and aging on neuropsychological function. J Neurovirol 18(4):256–263. https://doi.org/10.1007/s13365-012-0094-1

Wendelken LA, Jahanshad N, Rosen HJ, Busovaca E, Allen I, Coppola G et al (2016) ApoE epsilon4 is associated with cognition, brain integrity, and atrophy in HIV over age 60. J Acquir Immune Defic Syndr 73(4):426–432. https://doi.org/10.1097/QAI.0000000000001091

Whitley R, Kimberlin DW, Prober CG (2011) Pathogenesis and disease. In: Arvin A, Campadelli-Fiume G, Mocarski E, Moore PS, Roizman B, Whitley R et al (eds) Human herpesviruses: biology, therapy, and immunoprophylaxis. Cambridge University Press, Cambridge

WHO (2003) Diet, nutrition and the prevention of chronic diseases. World Health Organization technical report series. 916:i-viii, 1-149, backcover

Wing EJ (2016) HIV and aging. Int J Infectious Dis 53:61–68. https://doi.org/10.1016/j.ijid.2016.10.004

Woods SP, Moore DJ, Weber E, Grant I (2009) Cognitive neuropsychology of HIV-associated neurocognitive disorders. Neuropsychol Rev 19(2):152–168. https://doi.org/10.1007/s11065-009-9102-5

Wright PW, Heaps JM, Shimony JS, Thomas JB, Ances BM (2012) The effects of HIV and combination antiretroviral therapy on white matter integrity. AIDS 26(12):1501–1508. https://doi.org/10.1097/QAD.0b013e3283550bec

Zavaliangos-Petropulu A, Nir TM, Thomopoulos SI, Reid RI, Bernstein MA, Borowski B et al (2019) Diffusion MRI indices and their relation to cognitive impairment in brain aging: the updated multi-protocol approach in ADNI3. Front Neuroinform 13:2. https://doi.org/10.3389/fninf.2019.00002

Zavaliangos-Petropulu A, Tubi MA, Haddad E, Zhu A, Braskie MN, Jahanshad N et al (2020) Testing a convolutional neural network-based hippocampal segmentation method in a stroke population. Human Brain Map. https://doi.org/10.1002/hbm.25210

Zetterberg H, Burnham SC (2019) Blood-based molecular biomarkers for Alzheimer's disease. Mol Brain 12(1):26. https://doi.org/10.1186/s13041-019-0448-1

Zhang H, Schneider T, Wheeler-Kingshott CA, Alexander DC (2012) NODDI: practical in vivo neurite orientation dispersion and density imaging of the human brain. NeuroImage 61 (4):1000–1016. https://doi.org/10.1016/j.neuroimage.2012.03.072

Zlokovic BV (2011) Neurovascular pathways to neurodegeneration in Alzheimer's disease and other disorders. Nat Rev Neurosci 12(12):723–738. https://doi.org/10.1038/nrn3114

Zwanenburg JJM, van Osch MJP (2017) Targeting cerebral small vessel disease with MRI. Stroke 48(11):3175–3182. https://doi.org/10.1161/strokeaha.117.016996

Part II
Neurocognitive Impairments in Low-, Middle- and High-Income Countries – Incidence/Prevalence and Contexts

Neurocognitive Complications of Pediatric HIV Infections

Sarah Benki-Nugent and Michael J. Boivin

Contents

S. Benki-Nugent (✉)
Department of Global Health, University of Washington, Seattle, WA, USA
e-mail: benki@uw.edu

M. J. Boivin
Departments of Psychiatry and Neurology and Ophthalmology, College of Osteopathic Medicine, Michigan State University, East Lansing, MI, USA

Department of Psychiatry, School of Medicine, University of Michigan, Ann Arbor, MI, USA
e-mail: boivin@msu.edu

Curr Topics Behav Neurosci (2021) 50: 147–174
DOI 10.1007/7854_2019_102
Published Online: 15 September 2019

Abstract The prevalence of cognitive impairment or learning difficulties in HIV-infected children is high despite access to antiretroviral treatment (ART). Several biological mechanisms, including latent HIV reservoir persistence in the brain, local inflammation within the central nervous system (CNS), disruption of neuronal function and integrity, and ongoing infiltration of activated HIV target cells to the CNS during brain development, may each dampen efficacy of ART. Development of therapeutics to target these mechanisms will be crucial, and potential candidates may include chemokine receptor antagonists. Separately, psychosocial approaches to support caregiving may leverage neuroplasticity and support brain development during critical developmental windows in spite of perinatal HIV infection. Multipronged approaches that encompass both approaches are crucial for optimizing neurodevelopmental outcomes in these extraordinarily vulnerable children. Dynamic neuropsychological assessments, such as what can be obtained with the use of computerized cognitive games intervention, may prove more sensitive to the brain/behavior benefits of ART. Dynamic neuropsychological outcomes, neuroinflammatory biomarkers, and brain development neuroprotective factors (BDNFs) may each be used to evaluate brain/behavior integrity of children in response to new treatment options.

Keywords Antiretroviral · Caregiving · Cognitive rehabilitation · Neurodevelopment · Neurocognitive · Monocyte · Perinatal HIV

1 Neurocognitive and Neurologic Manifestations of Perinatal HIV Infection and Advances in Behavioral Interventions

Perinatal HIV results in a spectrum of mild to severe cognitive impairments and neurologic abnormalities. The most severe form of HIV-related neurological disease is HIV encephalopathy, which may be progressive or static and involves loss of or failure to attain developmental milestones, cognitive impairment, microcephaly, and neurologic abnormalities (Belman 1992; Wiley et al. 1990). With the advent of combination antiretroviral therapy (ART), the rate of HIV encephalopathy substantially declined (Patel et al. 2009; Chiriboga et al. 2005); however a constellation of language, neurocognitive and motor impairments, and behavioral and psychiatric problems have remained common in perinatally HIV-infected children (Nozyce et al. 1994a; Cohen et al. 2015; Malee et al. 2011), even among subsets who are long-term treated and virologically and immunologically stable (Jeremy et al. 2005). In some select cohorts, untreated and surviving children and adolescents do not have any apparent CNS impairment (Gay et al. 1995; Drotar et al. 1997; Dollfus et al. 2010;

Bagenda et al. 2006). However, in a detailed study of long-term slow progressors, defined by survival with no or limited ART for the first 8–12 years of life on the basis of their ability to remain asymptomatic and maintain a robust CD4+ T cell count, the proportion with low scores on neuropsychological testing and white matter micro-structural damage, was substantial (Hoare et al. 2012a).

1.1 Extent and Nature of Cognitive Impairments in Perinatally HIV-Infected Children

In carefully followed birth cohorts of perinatally HIV-infected children, the rates of cognitive and motor neurodevelopmental delays are high and range from 26 to 36% and 30 to 36%, respectively (Gay et al. 1995; Drotar et al. 1997; Chase et al. 2000). Likewise in cross-sectional studies, the prevalence of cognitive impairment or learning difficulties in HIV-infected children is quite high (16–42%) despite access to ART (Cohen et al. 2015; Jeremy et al. 2005; Wood et al. 2009). In a South African study that applied the HIV-associated neurocognitive disorders (HAND) criteria (Phillips et al. 2016), often used to characterize neurocognitive impairment in HIV-infected adults (Antinori et al. 2007), the prevalence of neurocognitive disorders was 44% (Hoare et al. 2016). Interpretation of these and other (reviewed by Laughton et al. 2013; van Wyhe et al. 2017) observational cohort studies is limited because of differences in socioeconomic environments between HIV-infected children and controls, differences in caregiver status among children, differences in ART regimens, and duration of treatment. In low- and middle-income countries in particular, HIV-infected children often have higher malnutrition or poor weight gain, predictive of significant motor, language, and other developmental delay (Benki-Nugent et al. 2017; Puthanakit et al. 2010; Ferguson and Jelsma 2009; Smith et al. 2006).

In addition, appropriate comparison groups are challenging to define. HIV-infected children face a myriad of psychosocial challenges including poverty, chronic parental illness, stigma, social isolation, and orphanhood (Donald et al. 2014), each of which could substantially impact their neurodevelopmental trajectory and which limit comparisons to HIV-unexposed uninfected cohorts. In turn, comparisons to HIV-exposed uninfected cohorts are limited because perinatal HIV and antiretroviral exposure may also cause risk of neurodevelopmental impairment (Tran et al. 2016; Le Doare et al. 2012). For studies that focus on HIV-infected children in sub-Saharan Africa, where most HIV-infected children reside, culturally relevant neurocognitive assessments with local norms for comparison have generally not been available. Nevertheless, existing estimates of neurocognitive impairment are sobering and suggest that perinatally HIV-infected children may face a multitude of challenges in school achievement and attainment of economic independence and high quality of life.

1.2 Preliminary Findings Using Diffusion Tensor Imaging (DTI) to Evaluate Neuronal Injury in Perinatal HIV

A recent systematic review of neuroimaging studies in perinatally infected HIV-infected children found that the relationship between cognitive impairment and CNS damage in HIV as seen by neuroimaging remains incompletely understood due to the paucity of neuroimaging studies in pediatric HIV (Hoare et al. 2014). Dr. Hoare's team has conducted detailed DTI studies which have found evidence of white matter damage in both slow progressors and ART-treated children (Hoare et al. 2011, 2012b, 2015). "Slow progressors" (SP) are typically defined as children or adolescents who were vertically infected with HIV but who survived and remained asymptomatic for several years with no or minimal ART. Hoare and colleagues compared asymptomatic HIV-positive children (8–12 years) with matched controls on a neuropsychological battery as well as DTI in a masked region of interest analysis (ROI) focusing on the corpus callosum, internal capsule, and superior longitudinal fasciculus. The SP group performed significantly worse than controls on the Wechsler Abbreviated Scale of Intelligence Verbal and Performance IQ scales and on standardized tests of visual-spatial processing, visual memory, and executive functioning. "Slow progressors" had lower fractional anisotropy (FA), higher mean diffusivity (MD), and radial diffusivity (RD) in the corpus callosum ($p = <0.05$) and increased MD in the superior longitudinal fasciculus, compared to controls. A correlation was found between poor performance on a test of executive function and a test of attention with corpus callosum FA and a test of executive function with lowered FA in the superior longitudinal fasciculus. These findings suggest that demyelination as reflected by the increase in RD may be a prominent disease process in pediatric HIV infection.

Associations were explored between a number of clinical variables and DTI measures in 50 ART-treated children aged 6–15 years. FA, MD, RD, and axial diffusion (AD) were derived from 48 cerebral white matter regions. Significant effects of clinical variables were found with white matter integrity in a number of brain regions (Hoare et al. 2015). Decreased FA, a measure of neuronal damage, was associated with being on second-line ART, low hemoglobin, and younger age in the fornix, cerebellar peduncles, fronto-occipital fasciculus, and cingulum. Children with increased MD, a measure of neuronal damage, in the corpus callosum, fornix, fronto-occipital fasciculus, and superior longitudinal fasciculus, were younger and had reduced albumin and hemoglobin and increased total protein and viral load. Decreased AD, a measure of axonal damage, was associated with increased viral load and total protein, decreased albumin and hemoglobin, younger age, poorer fronto-striatal cognition, and being on second-line ART in the corpus callosum, superior longitudinal fasciculus, cerebellar peduncle, and sagittal stratum. Increased RD, a measure of myelin loss, was associated with younger age, low current CD4 count, low albumin and hemoglobin, and higher viral load and total protein in the superior fronto-occipital fasciculus, the corpus callosum, and the cerebellar peduncle. The current findings underline the possible association of first-line treatment failure on white matter brain dysfunction in pediatric neuroAIDS and the importance of future interventions to improve adherence (Hoare et al. 2014), perhaps including cognitive training.

1.3 Dynamic Versus Static Neuropsychological Assessment of Neurocognitive Function

Sternberg and Grigorenko argue that dynamic assessment is more sensitive to brain/behavior integrity than traditional static assessment (Sternberg and Grigorenko 2002). *Dynamic assessment* uses a cognitive ability test to measure active learning ability *across numerous learning sessions*. *Static assessment* uses the same test to measure neuropsychological function in a single testing session. In dynamic assessment, children learn skills needed for a given type of test item during testing through teaching and feedback on performance. Dynamic assessment then notes improvements in cognitive performance in response to multiple learning sessions. By evaluating brain/behavior learning as part of testing, dynamic assessment evaluates a higher level of positive biocultural plasticity and brain integrity. Dynamic assessment does this by evaluating the child's ability to adapt and improve from feedback and learning (Sternberg 2004; Sternberg et al. 2002). In a cross-cultural assessment in Tanzania, conventional tests for working memory and analogous reasoning (e.g., Raven's Progressive Matrices) did not adequately assess the full range of cognitive skills children could demonstrate – whereas dynamic assessment did (Sternberg et al. 2002).

1.4 Brain Training and Neuroimaging Evidence Support the Need for Dynamic Assessment

Duncan E. Astle and colleagues published the first conclusive evidence that computerized cognitive rehabilitation training (CCRT) in children strengthens cortical connective pathways for working memory, as well as the attentional processes targeted by CCRT (Astle et al. 2015). Using magnetoencephalography (MEG), they confirmed that the right hemispheric frontoparietal seed network is considerably strengthened in resting state after 24 sessions of CCRT for visual-spatial working memory. Placebo (active) controls did not show strengthening. The degree to which these neural connections were strengthened significantly correlated with improved CCRT working memory performance from beginning until completion of 24 sessions of CCRT training. When dynamic assessment detects improved neuropsychological performance over multiple CCRT sessions, we propose we are seeing a reflection of the integrity of neurocognitive function. On this basis we believe dynamic measures can provide more valid and sensitive assessments in children affected by HIV. We propose that CCRT can be used as a dynamic assessment tool, readily available and easily implemented in resource-constrained settings by community health workers and clinical officers and nurses on tablets and smartphones for "real-time" cognitive assessment monitoring (or surveillance) for HIV disease progression and treatment.

In school-age Ugandan children with HIV, computerized cognitive rehabilitation training (CCRT) significantly enhanced *global cognitive ability and executive functioning* as measured by the Kaufman Assessment Battery for Children (second ed.) (KABC-II) and the maze learning measure of the CogState computerized program of cognitive ability (see www.cogstate.com) (Boivin et al. 2010a, 2017). In a separate test, Boivin and colleagues evaluated the neuropsychological benefits of Brain Powered Games (BPG), a set of games developed for a tablet-based mobile network platform and which has an African village motif (Novak et al. 2017). In a preliminary study with Ugandan school-age HIV-infected children in an impoverished rural setting, this novel game package elicited significant improvements on measures of *attention and processing speed* from the Test of Variables of Attention (TOVA; see www.tovatest.com) and on cognitive ability processing speed for learning on the CogState battery (Giordani et al. 2015). One study found substantial deficits in attention, memory, and visual spatial processing speed in perinatally HIV-infected children (Hoare et al. 2012b). These are the very domains of neurocognitive development in African children that have been most effectively improved through CCRT (Bangirana et al. 2013).

1.5 CCRT Positive Neuroplasticity Intervention Model

According to Mahncke et al., CCRT is effective because it is designed to redress the four principal causes of negative plasticity: reduced schedules of activity, noisy processing, weakened neuromodulatory control, and negative learning (Mahncke et al. 2006a, b). To evaluate CCRT treatment benefit for each of these negative neuroplasticity factors, CCRT intervention could be used to evaluate the corresponding neuropsychological benefits and MRI and DTI correlates corresponding to these with our South African study cohort. To illustrate, Dr. Hoare and her team documented a correlation between poor performance on executive function and attention test, with corpus callosum fractional anisotropy (FA) (Hoare et al. 2015). They also observed that poorer executive function was related to lower FA in the superior longitudinal fasciculus. These findings suggest that demyelination as reflected by the increase in RD may be a prominent disease process in pediatric HIV infection (Hoare et al. 2014, 2012b). We anticipate that integrity of white matter in these regions will specifically mediate neuropsychological benefits as they pertain to each of the four principal neuroplasticity domains described in Mahncke and colleagues (Ackermann et al. 2016; Jankiewicz et al. 2017). In other disease contexts, brain development neuroprotective factors (BDNFs) may be a marker for resilience in neurodegenerative disease and improved response to behavioral interventions to enhance neurocognitive function.

BDNF might also be sensitive to positive neuroplasticity in neurocognitive rehabilitation intervention with computer cognitive games for children with HIV (Angelucci et al. 2015; Kuster et al. 2017).

1.6 Threats to Quality of Caregiving in African HIV-Affected Children

Enhanced access to antiretroviral (ARV) medications for children in the developing world, coupled with advances in the treatment of HIV and AIDS with new ARV drugs, has changed the prognosis for infected children from a uniformly deadly disease early in childhood to one in which survival well into adolescence is not uncommon (Armstrong et al. 2003). Programs such as the UN Global Fund and the US President's Emergency Plan for AIDS Relief (PEPFAR) have dramatically enhanced access to ARV treatment for HIV-infected children in participating African countries such as Uganda (O'Hare et al. 2005). As a result, pediatric HIV illness is increasingly becoming a subacute, chronic disease (Bass 2005; Cohen 2005). HIV-infected African children are now able to survive longer, but they remain at significant risk developmentally because of the progressive encephalopathy of this disease at the neuronal level (Okamoto et al. 2007). Therefore, it is important to consider strategies for enhancing their cognitive and psychosocial development in the face of HIV disease encephalopathy and neurocognitive impairment. This is especially true given that the CNS effects of the disease are compounded by limited medical services for comorbid and opportunistic infections, compromised parental caregiving in the home, and a lack of social/educational support outside the home (Bose 1997).

Over 90% of pediatric HIV infections and AIDS deaths occur in Africa (Foster and Williamson 2000), and more than 11 million children have lost at least one parent to AIDS (UNAIDS 2006). In Uganda, about one million children are orphans with one or both parents dead (UNICEF definition), and a new child is orphaned every 14 s (Ronald and Sande 2005). When considering how to best address the global public health burden of the developmental effects of HIV on children, the African context is clearly paramount. In rural areas of sub-Sahara Africa, the family must rely upon labor-intensive subsistence agriculture to provide for the nutritional needs of the family. Because of this, maternal HIV disease and illness can severely disrupt not only the nurturing capacity of the mother for her children but also food security for the entire family. Chronic nutritional hardship can severely undermine early childhood development (Foster and Williamson 2000; Boivin et al. 1995). Because of this, the AIDS epidemic can also have devastating consequences for non-infected children of HIV-infected parents.

As evidence of the emotional impact of HIV on the Ugandan child, a study of Ugandan AIDS orphans in a rural district found that compared to non-orphans, AIDS orphans had greater risk for higher levels of anxiety (odds ratios (OR) = 6.4), depression (OR = 6.6), and anger (OR = 5.1). These children also had significantly higher scores on individual items in the Beck Youth Depression Inventory, indicative of clinical depressive disorder. These include vegetative symptoms, feelings of hopelessness, and suicidal ideation (Atwine et al. 2005). Maternal HIV infection can lead to stigmatization, abandonment, isolation, and uncertainty about future care for the African child. Furthermore, these sources of emotional trauma can diminish immuno-resilience through psychoneuroimmunological processes. These processes can accelerate the progression of CNS infection and further disrupt the health and neuropsychological development of the HIV-infected child (Balbin et al. 1999). These findings support the need for caregiving interventions with HIV-infected children that extend beyond provision for basic medical and nutritional care only. These findings support the urgent need for programs that can minister to the emotional and psychosocial needs of HIV-affected children.

1.7 Quality of Home Environment, Caregiving, and the Developmental Effects of HIV

A 2-year follow-up study of HIV-infected school-age children concluded that there tended to be a decrease in positive social self-concept over time. However, parental support exerted a stress-buffering effect for depression in both infected and non-infected children of HIV-infected mothers, and this benefit extended to improvements in both psychosocial adjustment and cognitive ability (Forehand et al. 2002; Kotchick et al. 1997). The importance of parenting quality for infants born to HIV-infected mothers was also documented in a longitudinal attempt to describe the development of infants of mothers with HIV. Higher mental ability, motor, and adaptive behavior scores were associated with more consistent caregiving, especially if typified by positive attention and less negative control. Better language ability was associated with more positive attention of the child by caregivers (Holditch et al. 2001). We have replicated these findings with Ugandan preschool-age children born to mothers with HIV in an impoverished rural setting (Bass et al. 2016).

In a 4-year longitudinal study of child psychosocial adjustment and parenting, non-infected children ages 6–11 with HIV-infected mothers were compared to children whose mothers were not infected. Although the patterns of psychosocial adjustment across assessments over time were similar for the two groups, children of HIV-infected mothers were more depressed. Also, parenting style was related to the strength of mother-child relationship and subsequent psychosocial adjustment of the child irrespective of the maternal infection group (Forehand et al. 2002). Thus, even in the absence of HIV infection for the child born to an infected mother, compromised caregiving from HIV can significantly diminish the cognitive and behavioral development of the child over the long term (Mellins et al. 1994).

Others have also concluded that because both parenting quality and consistency of the primary caregiver influenced developmental outcomes, interventions with the mothers of these infants need to focus both on improving the quality of parenting and reducing the frequency of changes in the primary caregiver (Forehand et al. 2002). This is especially true in African countries like Uganda where financial and medical support for maternal HIV is usually not readily accessible. In such settings, the health and nurturing capability of the mother are much more tenuous in buffering the HIV child from the developmental effects of the disease (Nozyce et al. 1994a, b).

1.8 Caregiver Training Intervention to Enhance Early Childhood Development Affected by HIV

Evidence is provided by Tardieu that the cognitive and psychosocial need of HIV school-age children can be further exacerbated when coupled with immunological deficiencies in early childhood. While 22 out of 33 HIV-positive French schoolchildren had normal school achievement and normal IQ, almost half had speech and/or language delay or articulation disorders (Tardieu 1998; Tardieu and Janabi 1994; Tardieu et al. 2000). Furthermore, over half of the children had visual-spatial and time orientation test deficits, and 29% of the children and 42% of the parents had emotional and psychosocial disturbances of intermediate to high severity. For the study group as a whole, cognitive test performance was predictive of school performance, and CD4+ lymphocyte levels in early childhood were predictive of both.

More recent findings from the PACTG pediatric AIDS clinical trials group with 489 HIV-infected children aged 4 months to 17 years also observed that their overall poorer cognitive and neuropsychological functioning was associated with higher viral loads (Jeremy et al. 2005). However, Coscia found that for 43 HIV-infected children between the ages of 2.5 and 12 years, quality of home environment mediated the relationship between socioeconomic status (SES) and overall cognitive ability performance (Coscia et al. 2001). In fact, SES and quality of home environment were much more significantly related to IQ than degree of illness and health status for the children. Quality of home environment was defined by the organization of the environment, play materials, parental involvement, variety of stimulation, and parental attitudes toward the provision of a cognitively stimulating environment. Furthermore, the impact of quality of home environment on IQ was stronger for children who were more ill as determined by health status and CD4 counts (Coscia et al. 2001). This was especially the case for the younger children. To us these findings suggest that clinical stability of young HIV-infected children might best be considered as a factor that moderates the impact of quality of home environment on cognitive and psychosocial development for HIV-infected children.

In rural areas of sub-Sahara Africa, the family must rely upon labor-intensive subsistence agriculture to provide for the nutritional needs of the family. Because of this, maternal HIV disease and illness can severely disrupt not only the nurturing capacity of the mother for her children but also food security for the entire family (Caruso 2006; Fabiani et al. 2006; Orach and De Brouwere 2006). Chronic nutritional hardship can severely undermine early childhood development (Foster and Williamson 2000; Boivin et al. 1995). Because of this, the AIDS epidemic can also have devastating consequences for non-infected children of HIV-infected parents. The Mediational Intervention for Sensitizing Caregivers (MISC) model developed by Professor Pnina Klein has been used to enhance the development of Ugandan children affected by HIV (Boivin et al. 2017, 2013a, b; Bass et al. 2017). Through their efforts, Boivin and colleagues documented that MISC parent/caregiver training improves cognitive and psychosocial development in HIV-infected and HIV-exposed uninfected children in low-resource rural districts in central and in eastern Uganda.

1.9 A Brain/Behavior Model for the Benefits of MISC in Early Childhood Development

One to five years of age is a critical developmental period for children, during which time they develop the dynamic capacity to benefit from new learning experiences. There is a general consensus from developmental research that adult-child interactions are of central importance in this process (Bonnier 2008). Farah and colleagues observed a relationship between parental nurturance and memory development. This relationship was consistent with the animal literature on maternal buffering of stress hormone effects on hippocampal development (Farah et al. 2008). Rao et al. observed that parental nurturance at age 4 predicts the volume of the left hippocampus in adolescence, with warmer and more loving nurturance associated with smaller hippocampal volume. Also, the association between parental nurturance and hippocampal volume disappears at 8 years of age. They concluded that this supports the existence of a sensitive developmental period for brain maturation, especially before 4 years of age (Rao et al. 2009). The caregiver provides for secure emotional attachments in a nurturing environment, creating learning experiences that allow a child's neurocognitive ability to blossom (Feuerstein 1980; Vygotsky 1978). Effective mediational behaviors by caregivers were found to be significantly related to children's social-emotional stability and the willingness to explore and learn about the world around them (Feurerstein 1979, 1980). While the role of effective caregiving in fostering optimal neurocognitive development during sensitive periods in early childhood has been studied in a wide range of cultural settings and across various populations of children with special needs, evaluation of interventions to improve caregiving has not been done in low-resource non-Western settings. Nor have they been done with HIV African children. Caregiver training can help the caregiver interact with their child in a way that promotes development and growth even in the face of adversity.

1.10 Theoretical Foundation of MISC Training

The MISC approach has a clear and well-developed theoretical foundation. Unlike models based on simple direct learning through stimulating the senses with an enriched environment (Grantham-McGregor et al. 2007, 1983, 1987, 1980; Klein 2001; Klein and Rye 2004), MISC is a mediational approach based on Feuerstein's theory of cognitive modifiability (Feuerstein 1979, 1980). The fundamental premise of this approach is that mediated learning best occurs interactively, when the caregiver interprets the environment for the child. To do so, the caregiver must be sensitive to the child's cognitive and emotional needs, interests, and capacities. As such, MISC has a strong emphasis on the importance of the social/interactive/emotional domains as integrally linked to intellectual and cognitive development.

MISC learning is accomplished by training caregivers in mediational processes as *focusing* (gaining the child's attention and directing them to the learning experience in an engaging manner); *exciting* (communicating emotional excitement, appreciation, and affection with the learning experience); *expanding* (making the child aware of how that learning experience transcends the present situation and can include past and future needs and issues, therefore extending beyond the immediate need of the moment); *encouraging* (emotional support of the child to foster a sense of security and competence); and *regulating* (helping direct and shape the child's behavior in constructive ways with a goal toward self-regulation).

Most of the MISC training of caregivers is devoted to helping parents become aware and develop practical strategies for focusing, exciting, expanding, encouraging, and regulating the child as learning opportunities arise in the course of natural everyday caregiver/child interactions (Klein 1985, 1996, 2001; Klein and Rye 2004). It begins by trying to understand and highlight the caregivers' objectives for child-rearing and their goals for the ideal child and ideal parent. It asks parents/caregivers what outcomes they hope to achieve. This process raises parental awareness regarding their own attitudes about child-rearing, perception of the child, perception of themselves as caregivers, awareness of the child's emotional and cognitive needs, and awareness of the impact of parental/caregiver interactive behavior. Because of the facilitative nature of the program, it does not rely on outside resources or materials and can be implemented with most children in a variety of contexts where caregiver/child interactions naturally take place.

The Ugandan ethnic groups within which we evaluated MISC caregiver training for HIV-affected households traditionally highly value children and emphasize the importance of the effective and loving nurture of children for the future betterment of families and communities (Minde 1975). The cultural emphasis on the nurturing of children is a good fit for the MISC, which is a method for sensitizing mothers to the positive aspects of their current child-rearing interactions. Similarly to what Klein found in Ethiopia, our findings with HIV-affected caregivers and their children indicated that the principles of MISC are simple and can be easily understood by caregivers and associated with their own child-rearing goals (Klein and Rye 2004). As a result, caregivers noted how MISC principles can be readily translated into actions within the cultural and contextual constraints of everyday living in each of their families (Klein 2001; Klein and Rye 2004).

2 Cellular and Molecular Mechanisms of Inflammation and Neuronal Injury in Perinatal HIV and Implications for Therapeutics

During infancy and early childhood, dramatic changes in brain volume and structure occur, including neuronal proliferation, migration, myelination, synaptogenesis, apoptosis, and pruning (Tau and Peterson 2010). Although the precise cellular and molecular mechanisms of HIV neuropathogenesis are not known, data from both HIV-infected children and adults suggest that many of these crucial developmental processes may be compromised in an HIV-infected developing child. HIV infection can often result in more rapid disease progression (Richardson et al. 2003) and higher rates of neurologic disease in children than in adults (Mintz 1994). Developmental host factors including differences in immune maturity and function availability of HIV target cells may in part explain these differences.

2.1 *Viral Entry to the CNS*

Perinatal HIV infection may occur in utero, during delivery or postpartum, through exposure to breast milk. The precise timing of infiltration of HIV to the fetal or neonatal brain is not known; however, HIV can traverse the blood-brain barrier within days of infection in adults (Valcour et al. 2012). Likewise, in rhesus macaque neonates inoculated with SIV within 24 h of birth, HIV DNA was detectable in brain parenchyma within 3–8 days of infection (Westmoreland et al. 1999). In a histologic and molecular study of fetal brain tissue provided by HIV seropositive women, HIV DNA was detected by PCR, but immunohistochemistry failed to detect gp41 (Kure et al. 1991). In this study of fetal tissue, there were no pathologies typical of perinatal HIV, such as inflammatory infiltrates or calcification (Kure et al. 1991). The earliest events following viral entry across the blood-brain barrier and whether there are entry mechanisms unique to perinatal HIV acquisition are not known.

Peripheral blood T lymphocytes and monocytes, the main target cells for HIV, are both posited to be potential cellular vehicles for seeding the brain (Spudich and Gonzalez-Scarano 2012). However, several lines of evidence support the importance of monocytes and monocyte-derived cells as the key drivers of HIV replication, persistence, and neuronal injury in the CNS. Lymphocytes are less abundant than monocyte-derived cells in the brain, and HIV has generally been detected in macrophages and other cells of the macrophage lineage in postmortem brain tissue in adults (Wiley et al. 1986; Thompson et al. 2011). Brain viral isolates replicate well in cultured macrophages (Spudich and Gonzalez-Scarano 2012). HIV-infected monocyte-derived macrophages produce neurotoxic factors in vitro (Giulian et al. 1990; Pulliam et al. 1991). In both adult and pediatric histopathology studies, HIV proteins indicative of HIV replication are detected primarily in microglia and macrophages and to a lesser extent in astrocytes (Wiley et al. 1990; Kure et al. 1991; Spudich and Gonzalez-Scarano 2012; Vallat et al. 1998; Blumberg et al. 1994; Vazeux et al. 1992).

2.2 Mediators of HIV Replication, Persistence, and Neuronal Injury in the CNS

Early postmortem studies of pediatric HIV reported profound brain tissue damage and atrophy, with typical lesions including inflammatory infiltrates consisting of glia, microglia, lymphocytes, and mononuclear cells, calcification of tissue and of blood vessels, multinuclear cells, and both focal areas with poor myelination and diffuse myelin pallor (Kure et al. 1991; Blumberg et al. 1994). In a postmortem comparison of adults and children who had died of AIDS, CNS disease was more common in infants and children than in adults and was less often associated with a CNS opportunistic infection (Kure et al. 1991). Paradoxically, HIV protein was detected less often in infants and children than adults with comparable encephalopathy (Kure et al. 1991) and was not detectable in some infants despite of signs of CNS disease (Kure et al. 1991; Vazeux et al. 1992). In macaque neonates, the level of HIV DNA positive cells was lower than in adults with comparable inoculum and time since infection (Westmoreland et al. 1999). Mechanisms underlying these apparent differences remain unclear.

Importantly, early during infection, inflammatory markers are detectable in CSF commensurate with detection of HIV (Valcour et al. 2012) and in the brain (Sailasuta et al. 2012; Lentz et al. 2009). Neurons are generally not considered to support viral replication. Rather, low-level virus replication is believed to generate and amplify a neurotoxic inflammatory cascade involving small molecules, cytokines, and chemokines which may trigger neuronal damage, dysfunction, and cell death (reviewed by Spudich and Gonzalez-Scarano 2012). Apoptotic neurons associated with NF kappa B positive monocytes/macrophages have been observed in situ in the cerebral cortex and basal ganglia in postmortem tissue specimens collected from HIV-infected children with encephalopathy (Gelbard et al. 1995; Dollard et al. 1995). In addition, HIV proteins, including gp120 and Tat, are neurotoxic in vitro and in vivo and may exert neurodevelopmental damage even in the absence of inflammatory mediators (Moran et al. 2014). Recently, intra-hippocampal injection of Tat delayed neurodevelopmental milestones in rat neonate pups (Moran et al. 2014). Notwithstanding general agreement that neurons are not permissive for HIV infection, it is noteworthy that HIV DNA has been isolated from neural progenitor cells, which are crucial for neurogenesis, in pediatric brain tissue (Schwartz et al. 2007), suggesting these cells may be permissive for HIV infection. In addition, neuronal damage in neonates may occur rapidly: SIV-infected neonate macaques had substantially reduced hippocampal pyramidal neurons within 6–18 weeks of infection (Curtis et al. 2014). Early histopathological studies of postmortem brain tissue provided by perinatally HIV-infected children reported presence of HIV DNA and proteins in astrocytes (Vallat et al. 1998; Blumberg et al. 1994; Tornatore et al. 1991; Saito et al. 1994). The question of whether astrocytes support viral replication or play an important role in HIV neuropathogenesis has been controversial but has recently received more attention (Zayyad and Spudich 2015).

2.3 *Brain as a Latent Reservoir in Perinatal HIV*

HIV may be detected in the brains of individuals with undetectable virus in blood and CSF (Desplats et al. 2013), and case reports demonstrate neurologic deterioration and detection of HIV RNA in CSF despite long-term virologic suppression, referred to as "CSF escape" (Canestri et al. 2010; Peluso et al. 2012). Likewise in children, worsening of cognitive function (Tamula et al. 2003) and appearance of encephalopathy (Innes et al. 2017) can occur despite virologic suppression, and in four children with encephalopathy, HIV RNA remained undetectable in CSF (Innes et al. 2017).

Careful analysis of HIV genetic sequences in HIV-infected individuals has been used to characterize compartmentalization or sequestered virus replication in specific anatomic sites, tropism for specific viral entry constraints, and other selective pressures. The appearance of a compartmentalized population of viral variants in CSF suggests local HIV replication in CNS and has been associated with HIV-associated dementia (HAD) (Ritola et al. 2005; Harrington et al. 2009). HIV sequences for Env, Nef, and the LTR promotor that are specific for the brain compartment have been observed (Gray et al. 2016; Churchill and Nath 2013). Integrated HIV provirus is detectable in perivascular macrophages, microglia, and astrocytes, each of which is long-lived cell population with half-lives of 3 months, years to lifelong, and months to years, respectively (Gray et al. 2016). However, both macrophage-tropic and T-cell tropic Env sequences have been observed in CSF in adults (Schnell et al. 2011), suggesting the existence of a CNS T-cell compartment as well (Zayyad and Spudich 2015). In a detailed study of CNS viral variants in HIV-infected children in Malawi, very young infants (age 0–6 months) often had equilibration of viral variants in blood and CSF, whereas compartmentalized virus was more often observed in older infants and appeared at 13.5 months on average (Sturdevant et al. 2012).

2.4 Virologic Factors Can Potentially Impact the Pathogenesis of the CNS

When comparing subtype A to subtype D in perinatally infected school-age Ugandan children who were treatment-naïve, children infected with subtype A did more poorly neuropsychologically (Boivin et al. 2010b). However, there were not significant neuropsychologically differences by subtype for school-age Ugandan children who were on ARV treatment from early childhood, with good viral suppression (Bangirana et al. 2017). The same was true for neurodevelopment in preschool-age Ugandan children perinatally infected with HIV (Ruisenor-Escudero et al. 2018). In a study of children with HIV from Myanmar, subtypes B and C were associated with reduced brain volumetrics (Ortega et al. 2013; Paul et al. 2018).

2.5 Biomarkers for Immune Activation and Neuronal Injury in the CNS

Soluble markers for immune and inflammatory response to HIV and neuronal injury mediated by HIV can be quantified in CSF, offering substantial utility for understanding consequences of HIV reservoir, replication, and immune response in adults (Price et al. 2013); however studies in children have been scant and are challenging due to ethical issues involved in collection of CSF. In one study of 23 ART-experienced children participating in a clinical trial, MCP-1 and matrix metalloproteinase were most often detected; other markers evaluated included RANTES, MIP-1-alpha, and MIP-1-beta (McCoig et al. 2002). In adults, the macrophage activation marker neopterin, which is a small molecule produced only by activated macrophages (Hagberg et al. 2010), is a key marker for immune activation as a result of HIV and HIV-associated dementia, disease progression, and HIV in CSF; other key markers include monocyte chemokine protein 1, CXCL10, and beta-2-microglobulin (reviewed by Price et al. 2013).

Studies of monocyte cell populations and soluble markers of monocyte activation in the blood compartment could provide some insight on the relation between circulating monocyte/macrophage cell populations and potential trafficking of HIV to the brain and subsequent neuronal injury. One study examined monocytes in relation to CNS disease in young HIV-infected children and found that children with HIV encephalopathy had higher peripheral monocyte concentration and percentages than those without CNS disease (Sanchez-Ramon et al. 2003). In contrast, in older Thai and Cambodian HIV-infected children newly starting ART and participating in the PREDICT study, children with better neurocognitive outcomes had higher frequency of certain monocyte subsets (Ananworanich et al. 2015). This unexpected finding was hypothesized to reflect protective mechanisms in untreated children (Ananworanich et al. 2015).

The CD14 and CD163 molecules are both monocyte markers and are shed in response to activation of the innate immune system (Weaver et al. 2007; Moller 2012). In a postmortem study of individuals who also had provided antemortem blood and CSF, HIV-infected individuals with higher plasma soluble (s)CD163 had lower markers of synapto-dendritic damage and higher markers of microglial activation (Bryant et al. 2017). HIV-infected adults with higher plasma sCD14 had evidence of cerebral atrophy (Ryan et al. 2001). Both sCD14 and sCD163 have been associated with poor neurocognitive outcomes in HIV-infected adults (Lyons et al. 2011; Kamat et al. 2012; Royal et al. 2016; Burdo et al. 2013a; Imp et al. 2016). Soluble CD163 may have unique utility in ART-treated individuals. Multiple studies have found high sCD163 was associated with poor cognitive outcomes despite of virologic suppression (Burdo et al. 2013a, b; Imp et al. 2016; Cassol et al. 2013). Benki-Nugent and colleagues found that in Kenyan HIV-infected infants, high levels of sCD163 prior to ART were associated with earlier age at attainment of developmental milestones (Benki-Nugent et al. 2019), a result that mirrored evidence of protective immune function in the PREDICT cohort (Ananworanich et al. 2015). In the Kenyan study, infants who maintained high sCD163 after starting ART went on to have worse neurodevelopmental outcomes, including later age at attainment of developmental milestones and worse cognitive outcomes at school age (Benki-Nugent et al. 2019). These seemingly contradictory findings suggest that high sCD163 may be a marker for immune function that is initially protective but that may also be detrimental if prolonged activation occurs.

2.6 Developmental Differences in Monocyte Turnover and Function in the Neonatal Window and Implications for Perinatal HIV

Intriguingly, multiple studies suggest that there are age-related differences in monocyte turnover and function and these differences may help explain more rapid disease progression in perinatally HIV-infected infants and adults (Merino et al. 2017). In healthy neonates, monocytes had significantly higher cytokine and toll-like receptor expression in response to lipopolysaccharide than both older infants and adults (Yerkovich et al. 2007). Monocyte/macrophages isolated from cord blood had greater susceptibility to HIV infection than did monocyte macrophages isolated from adults (Sperduto et al. 1993). In this study, differences in the capacity for cord monocytes/macrophages to proliferate may have explained higher susceptibility to viral infection. In a comparison of monocyte BrdU incorporation in uninfected neonate, infant, and adult macaques, monocyte turnover was substantially higher in neonates and infants than in adults (Sugimoto et al. 2017). In infant animals, monocyte turnover further increased following SIV infection and remained significantly higher than in newly infected adult animals (Sugimoto et al. 2017). SIV-infected infant macaques with higher monocyte turnover and immigration of

CD163+ monocytes to tissues had more rapid disease progression than infants with a less rapid disease course (Sugimoto et al. 2017). The authors posited that differences in turnover during the neonatal window were due to relative immaturity of the neonate immune system, which, in turn, compromised cell susceptibility to infection and led to increased demand for emigration of newly differentiated monocytes (Sugimoto et al. 2017).

3 Conclusion: Multipronged Approaches for Optimizing Neurodevelopmental Outcomes in Perinatal HIV

Following the scale-up of combination ART supported by PEPFAR, survival of perinatally HIV-infected children in sub-Saharan Africa has dramatically improved. While initial focus was rightly placed on improved survival, there is increasing need to focus on the quality of life for African children living with HIV. Cognitive, psychiatric, and behavioral (neuropsychological) disorders are emerging as a major concern in ART-treated perinatally infected children as they progress into adolescence (Laughton et al. 2013; Mellins et al. 2013; Mellins and Malee 2013). Boivin and colleagues recently confirmed such findings in African children with HIV at six study sites in four different countries using the same validated neuropsychological and behavioral assessment protocols (Boivin et al. 2018; Chernoff et al. 2018). Such problems can seriously undermine academic and social achievement and therefore require urgent attention.

3.1 Mixed Evidence for Benefit of ART Alone

In spite of combination ART, mild to moderate neurocognitive and fine motor deficits are evident in many studies of HIV-infected children, particularly in those with a prior diagnosis of encephalopathy (Jeremy et al. 2005; Smith et al. 2006, 2012; Boivin et al. 2018; Blanchette et al. 2002; Puthanakit et al. 2013). However, some observational studies have suggested limited benefit of ART. In a retrospective analyses of US children, children treated with ART regimen containing a protease inhibitor had modest improvement in cognition mean scores (Jeremy et al. 2005), and children who achieved virologic suppression by 5 years of age had better IQ scores than those with later age at suppression (Crowell et al. 2015). In this latter study, CNS penetration effectiveness (CPE) ranking did not impact IQ score outcome (Crowell et al. 2015). In Kenyan infants, Benki-Nugent and colleagues found that HIV-infected children receiving protease-inhibitor-based versus non-nucleoside reverse transcriptase inhibitor-based ART had earlier attainment of developmental milestones (Benki-Nugent et al. 2015).

Only a few studies have prospectively followed HIV-infected children from the time of ART initiation and measured neurodevelopmental outcomes. These longitudinal cohort studies suggest a mixed picture for benefit from ART, with some evidence for significant improvements but most data suggesting only modest or limited benefit in certain domains, or that benefits may be limited to prevention of further cognitive decline. In the largest study, Thai and Cambodian HIV-infected children ($N = 139$) who initiated ART at a median of 9 years and without a history of an AIDS-defining illness had no improvement in neurocognitive function after 3 years of follow-up on ART (Puthanakit et al. 2013). In the Democratic Republic of the Congo, children who were HIV diagnosed at median age of 44 months had significant 1-year improvement in motor and cognitive function after starting ART (Van Rie et al. 2009). However, South African children who started ART at a mean of 60 months experienced no improvement in neurodevelopmental outcomes (Eley et al. 2006). In Kenya, hospitalized children who initiated ART at a median of 20 months had gains in fine and gross motor skills but no improvement in either language or social functioning (Gomez et al. 2018). In these latter studies that were focused on early childhood and infancy, children had severe immunosuppression at the time of starting ART (Eley et al. 2006; Gomez et al. 2018), suggesting *some* benefits, in spite of symptomatic disease. Importantly, children treated from a very early age (<12 weeks of age) and without advanced disease at the time of initiation of ART had lower visual perception scores but otherwise generally similar neurodevelopmental scores versus HIV-uninfected children at 5 years post start of ART (Laughton et al. 2018).

3.2 New Treatment Strategies

Given rapid HIV entry and commensurate inflammation following HIV acquisition and rapid brain development during infancy, ensuring access to early and effective ART during infancy is likely the most critical step optimizing neurodevelopmental trajectories in HIV-infected children. In addition, new treatment options may also confer particular benefit. Dolutegravir (DTG), an integrase inhibitor approved for children over 6 years of age, has good CNS penetration, exceeding that of ritonavir-boosted lopinavir (Letendre et al. 2008, 2014) and a higher threshold for resistance than raltegravir. Neuropsychiatric symptoms may occur with treatment with integrase inhibitors, including DTG, possibly relating to the high penetrance into the CNS (Kheloufia and Boucherie 2017). Maraviroc (MVC), a CCR-5 inhibitor, prevents HIV attachment to CCR-5 expressing macrophages and CD4+ T-cells. MVC has a CPE of "3," equivalent to EFV and LPV/r. MVC achieves total CSF concentrations 3× >0.57 ng/ml, the level required to inhibit viral replication (Yilmaz et al. 2009). As noted above, randomized and one-arm studies in HIV-infected adults suggest that MVC may regulate the monocyte compartment in blood and may elicit neurocognitive benefits. Outside of HIV, recent work in mice suggests that CCR-5 inhibition also increases brain plasticity and learning independent of effects on HIV and inflammation (Zhou et al. 2016). Intracranial injections with MVC in mice reduced CCL3-mediated CCR5 activation which otherwise diminished memory performance on maze tasks (Marciniak et al. 2015).

3.3 Concluding Remarks

We propose that the use of dynamic neuropsychological assessments, such as what can be obtained with the use of computerized cognitive games intervention, will prove more sensitive to the brain/behavior benefits of new ART treatment strategies. This is especially true during critical periods of brain development (e.g., the first 1,000 days during gestation and in infancy). Dynamic CCRT-based neuropsychological outcomes can better evaluate the value-added benefit of new ARV treatment options in children, such as MVC and DTG. These new ARV treatment options and dynamic neuropsychological outcomes can be correlated with new and more sensitive neuroinflammatory biomarkers such as sCD163, sCD14, and neopterin. Such potentially more sensitive biomarkers can be used along with such standard measures in pediatric HIV research such as viral load, CD4, CD8, CD4 activation, and CD8 activation measures in serum and CSF. Finally, BDNFs might also be sensitive to positive neuroplasticity in neurocognitive rehabilitation intervention with computer cognitive games for children with HIV (Angelucci et al. 2015; Kuster et al. 2017).

References

Ackermann C, Andronikou S, Saleh MG, Laughton B, Alhamud AA, van der Kouwe A et al (2016) Early antiretroviral therapy in HIV-infected children is associated with diffuse white matter structural abnormality and corpus callosum sparing. AJNR Am J Neuroradiol 37 (12):2363–2369

Ananworanich J, Kerr SJ, Jaimulwong T, Vibol U, Hansudewechakul R, Kosalaraksa P et al (2015) Soluble CD163 and monocyte populations in response to antiretroviral therapy and in relationship with neuropsychological testing among HIV-infected children. J Virus Erad 1(3):196–202

Angelucci F, Peppe A, Carlesimo GA, Serafini F, Zabberoni S, Barban F et al (2015) A pilot study on the effect of cognitive training on BDNF serum levels in individuals with Parkinson's disease. Front Hum Neurosci 9:130

Antinori A, Arendt G, Becker JT, Brew BJ, Byrd DA, Cherner M et al (2007) Updated research nosology for HIV-associated neurocognitive disorders. Neurology 69(18):1789–1799

Armstrong F, Willen E, Surgen K (2003) HIV/AIDS in children and adolescents. In: Roberts M (ed) Handbook of pediatric psychology, 3rd edn. Guilford Press, New York, pp 359–374

Astle DE, Barnes JJ, Baker K, Colclough GL, Woolrich MW (2015) Cognitive training enhances intrinsic brain connectivity in childhood. J Neurosci 35(16):6277–6283

Atwine B, Cantor-Graae E, Bajunirwe F (2005) Psychological distress among AIDS orphans in rural Uganda. Soc Sci Med 61(3):555–564

Bagenda D, Nassali A, Kalyesubula I, Sherman B, Drotar D, Boivin MJ et al (2006) Health, neurologic, and cognitive status of HIV-infected, long-surviving, and antiretroviral-naive Ugandan children. Pediatrics 117(3):729–740

Balbin EG, Ironson GH, Solomon GF (1999) Stress and coping: the psychoneuroimmunology of HIV/AIDS. Baillieres Best Pract Res Clin Endocrinol Metab 13(4):615–633

Bangirana P, Boivin MJ, Giordani B (2013) Computerized cognitive rehabilitation therapy (CCRT) for African children: evidence for neuropsychological benefit and future directions. In: Boivin MJ, Giordani B (eds) Neuropsychology of children in Africa: perspectives on risk and resilience. Springer, New York, pp 277–298

Bangirana P, Ruel TD, Boivin MJ, Pillai SK, Giron LB, Sikorskii A et al (2017) Absence of neurocognitive disadvantage associated with paediatric HIV subtype A infection in children on antiretroviral therapy. J Int AIDS Soc 20(2):e25015

Bass E (2005) The two sides of PEPFAR in Uganda. Lancet 365(9477):2077–2078

Bass JK, Nakasujja N, Familiar-Lopez I, Sikorskii A, Murray SM, Opoka R et al (2016) Association of caregiver quality of care with neurocognitive outcomes in HIV-affected children aged 2-5 years in Uganda. AIDS Care 28(Suppl 1):76–83

Bass JK, Opoka R, Familiar I, Nakasujja N, Sikorskii A, Awadu J et al (2017) Randomized controlled trial of caregiver training for HIV-infected child neurodevelopment and caregiver well being. AIDS 31(13):1877–1883

Belman AL (1992) Acquired immunodeficiency syndrome and the child's central nervous system. Pediatr Clin North Am 39(4):691–714

Benki-Nugent S, Eshelman C, Wamalwa D, Langat A, Tapia K, Okinyi HM et al (2015) Correlates of age at attainment of developmental milestones in HIV-infected infants receiving early antiretroviral therapy. Pediatr Infect Dis J 34(1):55–61

Benki-Nugent S, Wamalwa D, Langat A, Tapia K, Adhiambo J, Chebet D et al (2017) Comparison of developmental milestone attainment in early treated HIV-infected infants versus HIV-unexposed infants: a prospective cohort study. BMC Pediatr 17(1):24

Benki-Nugent SF, Martopullo IS, Laboso T, Tamasha N, Wamalwa DC, Tapia K et al (2019) High plasma soluble CD163 during infancy is a marker for neurocognitive outcomes in early treated HIV-infected children. J Acquir Immune Defic Syndr 81(1):102–109

Blanchette N, Smith ML, King S, Fernandes-Penney A, Read S (2002) Cognitive development in school-age children with vertically transmitted HIV infection. Dev Neuropsychol 21 (3):223–241

Blumberg BM, Gelbard HA, Epstein LG (1994) HIV-1 infection of the developing nervous system: central role of astrocytes in pathogenesis. Virus Res 32(2):253–267

Boivin MJ, Green SD, Davies AG, Giordani B, Mokili JK, Cutting WA (1995) A preliminary evaluation of the cognitive and motor effects of pediatric HIV infection in Zairian children. Health Psychol 14(1):13–21

Boivin MJ, Busman RA, Parikh SM, Bangirana P, Page CF, Opoka RO et al (2010a) A pilot study of the neuropsychological benefits of computerized cognitive rehabilitation in Ugandan children with HIV. Neuropsychology 24(5):667–673

Boivin MJ, Ruel TD, Boal HE, Bangirana P, Cao H, Eller LA et al (2010b) HIV-subtype A is associated with poorer neuropsychological performance compared with subtype D in antiretroviral therapy-naive Ugandan children. AIDS 24(8):1163–1170

Boivin MJ, Bangirana P, Nakasujja N, Page CF, Shohet C, Givon D et al (2013a) A year-long caregiver training program to improve neurocognition in preschool Ugandan HIV-exposed children. J Dev Behav Pediatr 34(2):269–278

Boivin MJ, Bangirana P, Nakasujja N, Page CF, Shohet C, Givon D et al (2013b) A year-long caregiver training program improves cognition in preschool Ugandan children with human immunodeficiency virus. J Pediatr 163:1409–1416

Boivin MJ, Nakasujja N, Familiar-Lopez I, Murray SM, Sikorskii A, Awadu J et al (2017) Effect of caregiver training on the neurodevelopment of HIV-exposed uninfected children and caregiver mental health: a ugandan cluster-randomized controlled trial. J Dev Behav Pediatr 38 (9):753–764

Boivin MJ, Barlow-Mosha L, Chernoff MC, Laughton B, Zimmer B, Joyce C et al (2018) Neuropsychological performance in African children with HIV enrolled in a multisite antiretroviral clinical trial. AIDS 32(2):189–204

Bonnier C (2008) Evaluation of early stimulation programs for enhancing brain development. Acta Paediatr 97(7):853–858

Bose S (1997) An examination of adaptive functioning in HIV infected children: exploring the relationships with HIV disease, neurocognitive functioning, and psychosocial characteristics. ProQuest Information and Learning, Jul 1997 AAM9719709 1997; Dissertation Abstracts International: Section B: The Sciences and Engineering (58):409

Bryant AK, Moore DJ, Burdo TH, Lakritz JR, Gouaux B, Soontornniyomkij V et al (2017) Plasma soluble CD163 is associated with postmortem brain pathology in human immunodeficiency virus infection. AIDS 31(7):973–979

Burdo TH, Weiffenbach A, Woods SP, Letendre S, Ellis RJ, Williams KC (2013a) Elevated sCD163 in plasma but not cerebrospinal fluid is a marker of neurocognitive impairment in HIV infection. AIDS 27(9):1387–1395

Burdo TH, Lackner A, Williams KC (2013b) Monocyte/macrophages and their role in HIV neuropathogenesis. Immunol Rev 254(1):102–113

Canestri A, Lescure FX, Jaureguiberry S, Moulignier A, Amiel C, Marcelin AG et al (2010) Discordance between cerebral spinal fluid and plasma HIV replication in patients with neurological symptoms who are receiving suppressive antiretroviral therapy. Clin Infect Dis 50 (5):773–778

Caruso N (2006) Refuge from the Lord's Resistance Army in Uganda: a report from a Medecins Sans Frontieres team leader. Emerg Med Australas 18(3):295–298

Cassol E, Misra V, Morgello S, Gabuzda D (2013) Applications and limitations of inflammatory biomarkers for studies on neurocognitive impairment in HIV infection. J Neuroimmune Pharmacol 8(5):1087–1097

Chase C, Ware J, Hittelman J, Blasini I, Smith R, Llorente A et al (2000) Early cognitive and motor development among infants born to women infected with human immunodeficiency virus. Women and Infants Transmission Study Group. Pediatrics 106(2):E25

Chernoff MC, Laughton B, Ratswana M, Familiar I, Fairlie L, Vhembo T et al (2018) Validity of neuropsychological testing in young African children affected by HIV. J Pediatr Infect Dis 13 (3):185–201

Chiriboga CA, Fleishman S, Champion S, Gaye-Robinson L, Abrams EJ (2005) Incidence and prevalence of HIV encephalopathy in children with HIV infection receiving highly active antiretroviral therapy (HAART). J Pediatr 146(3):402–407

Churchill M, Nath A (2013) Where does HIV hide? A focus on the central nervous system. Curr Opin HIV AIDS 8(3):165–169

Cohen J (2005) ABC in Uganda: success or subterfuge. HIV AIDS Policy Law Rev 10(2):23–24

Cohen S, Ter Stege JA, Geurtsen GJ, Scherpbier HJ, Kuijpers TW, Reiss P et al (2015) Poorer cognitive performance in perinatally HIV-infected children versus healthy socioeconomically matched controls. Clin Infect Dis 60(7):1111–1119

Coscia JM, Christensen BK, Henry RR, Wallston K, Radcliffe J, Rutstein R (2001) Effects of home environment, socioeconomic status, and health status on cognitive functioning in children with HIV-1 infection. J Pediatr Psychol 26(6):321–329

Crowell CS, Huo Y, Tassiopoulos K, Malee KM, Yogev R, Hazra R et al (2015) Early viral suppression improves neurocognitive outcomes in HIV-infected children. AIDS 29(3):295–304

Curtis K, Rollins M, Carryl H, Bradshaw K, Van Rompay KK, Abel K et al (2014) Reduction of pyramidal and immature hippocampal neurons in pediatric simian immunodeficiency virus infection. Neuroreport 25(13):973–978

Desplats P, Dumaop W, Smith D, Adame A, Everall I, Letendre S et al (2013) Molecular and pathologic insights from latent HIV-1 infection in the human brain. Neurology 80 (15):1415–1423

Dollard SC, James HJ, Sharer LR, Epstein LG, Gelbard HA, Dewhurst S (1995) Activation of nuclear factor kappa B in brains from children with HIV-1 encephalitis. Neuropathol Appl Neurobiol 21(6):518–528

Dollfus C, Le Chenadec J, Faye A, Blanche S, Briand N, Rouzioux C et al (2010) Long-term outcomes in adolescents perinatally infected with HIV-1 and followed up since birth in the French perinatal cohort (EPF/ANRS CO10). Clin Infect Dis 51(2):214–224

Donald KA, Hoare J, Eley B, Wilmshurst JM (2014) Neurologic complications of pediatric human immunodeficiency virus: implications for clinical practice and management challenges in the African setting. Semin Pediatr Neurol 21(1):3–11

Drotar D, Olness K, Wiznitzer M, Guay L, Marum L, Svilar G et al (1997) Neurodevelopmental outcomes of Ugandan infants with human immunodeficiency virus type 1 infection. Pediatrics 100(1):E5

Eley B, Davies MA, Apolles P, Cowburn C, Buys H, Zampoli M et al (2006) Antiretroviral treatment for children. S Afr Med J 96(9 Pt 2):988–993

Fabiani M, Nattabi B, Opio AA, Musinguzi J, Biryahwaho B, Ayella EO et al (2006) A high prevalence of HIV-1 infection among pregnant women living in a rural district of North Uganda severely affected by civil strife. Trans R Soc Trop Med Hyg 100(6):586–593

Farah MJ, Betancourt L, Shera DM, Savage JH, Giannetta JM, Brodsky NL et al (2008) Environmental stimulation, parental nurturance and cognitive development in humans. Dev Sci 11 (5):793–801

Ferguson G, Jelsma J (2009) The prevalence of motor delay among HIV infected children living in Cape Town, South Africa. Int J Rehabil Res 32(2):108–114

Feuerstein R (1979) The dynamic assessment of retarded performers. University Park Press, New York

Feuerstein R (1980) Instrumental enrichment: redevelopment of cognitive functions of retarded performers. University Park Press, New York

Feurerstein R (1979) Then dynamic assessment of retarded performers. University Park Press, Baltimore

Feurerstein R (1980) Instrumental enrichment: redevelopment of cognitive functions of retarded performers. University Park Press, New York

Forehand RJD, Kotchick B, Armistad L, Morse E, Morse P, Stock M (2002) Noninfected children of HIV-infected mothers: a 4-year longitudinal study of child psychosocial adjustment and parenting. Behav Ther 33:579–600

Foster G, Williamson J (2000) A review of current literature on the impact of HIV/AIDS on children in sub-Saharan Africa. AIDS 14(Suppl 3):S275–S284

Gay CL, Armstrong FD, Cohen D, Lai S, Hardy MD, Swales TP et al (1995) The effects of HIV on cognitive and motor development in children born to HIV-seropositive women with no reported drug use: birth to 24 months. Pediatrics 96(6):1078–1082

Gelbard HA, James HJ, Sharer LR, Perry SW, Saito Y, Kazee AM et al (1995) Apoptotic neurons in brains from paediatric patients with HIV-1 encephalitis and progressive encephalopathy. Neuropathol Appl Neurobiol 21(3):208–217

Giordani B, Novak B, Sikorskii A, Bangirana P, Nakasujja N, Winn BW et al (2015) Designing and evaluating Brain Powered Games for cognitive training and rehabilitation in at-risk African children. Global Mental Health 2(e6):1–14

Giulian D, Vaca K, Noonan CA (1990) Secretion of neurotoxins by mononuclear phagocytes infected with HIV-1. Science 250(4987):1593–1596

Gomez LA, Crowell CS, Njuguna I, Cranmer LM, Wamalwa D, Chebet D et al (2018) Improved neurodevelopment following antiretroviral therapy in HIV-infected children. Pediatr Infect Dis J 37(9):916–922

Grantham-McGregor S, Stewart ME, Schofield WN (1980) Effect of long-term psychosocial stimulation on mental development of severely malnourished children. Lancet 2(8198):785–789
Grantham-McGregor S, Schofield W, Harris L (1983) Effect of psychosocial stimulation on mental development of severely malnourished children: an interim report. Pediatrics 72(2):239–243
Grantham-McGregor S, Schofield W, Powell C (1987) Development of severely malnourished children who received psychosocial stimulation: six-year follow-up. Pediatrics 79(2):247–254
Grantham-McGregor S, Cheung YB, Cueto S, Glewwe P, Richter L, Strupp B et al (2007) Developmental potential in the first 5 years for children in developing countries. Lancet 369 (9555):60–70
Gray LR, Brew BJ, Churchill MJ (2016) Strategies to target HIV-1 in the central nervous system. Curr Opin HIV AIDS 11(4):371–375
Hagberg L, Cinque P, Gisslen M, Brew BJ, Spudich S, Bestetti A et al (2010) Cerebrospinal fluid neopterin: an informative biomarker of central nervous system immune activation in HIV-1 infection. AIDS Res Ther 7:15
Harrington PR, Schnell G, Letendre SL, Ritola K, Robertson K, Hall C et al (2009) Cross-sectional characterization of HIV-1 env compartmentalization in cerebrospinal fluid over the full disease course. AIDS 23(8):907–915
Hoare J, Fouche JP, Spottiswoode B, Sorsdahl K, Combrinck M, Stein DJ et al (2011) White-Matter damage in Clade C HIV-positive subjects: a diffusion tensor imaging study. J Neuropsychiatry Clin Neurosci 23(3):308–315
Hoare J, Fouche JP, Spottiswoode B, Donald K, Philipps N, Bezuidenhout H et al (2012a) A diffusion tensor imaging and neurocognitive study of HIV-positive children who are HAART-naive "slow progressors". J Neurovirol 18(3):205–212
Hoare J, Westgarth-Taylor J, Fouche JP, Spottiswoode B, Paul R, Thomas K et al (2012b) A diffusion tensor imaging and neuropsychological study of prospective memory impairment in South African HIV positive individuals. Metab Brain Dis 27(3):289–297
Hoare J, Ransford GL, Phillips N, Amos T, Donald K, Stein DJ (2014) Systematic review of neuroimaging studies in vertically transmitted HIV positive children and adolescents. Metab Brain Dis 29(2):221–229
Hoare J, Fouche JP, Phillips N, Joska JA, Donald KA, Thomas K et al (2015) Clinical associations of white matter damage in cART-treated HIV-positive children in South Africa. J Neurovirol 21 (2):120–128
Hoare J, Phillips N, Joska JA, Paul R, Donald KA, Stein DJ et al (2016) Applying the HIV-associated neurocognitive disorder diagnostic criteria to HIV-infected youth. Neurology 87(1):86–93
Holditch D, Miles MS, Burchinal M, O'Donnell K (2001) Parental caregiving and developmental outcomes of infants and mothers with HIV. Nurs Res 50:5–14
Imp BM, Rubin LH, Tien PC, Plankey MW, Golub ET, French AL et al (2016) Monocyte activation is associated with worse cognitive performance in virologically suppressed HIV-infected women. J Infect Dis 215(1):114
Innes S, van Toorn R, Otwombe K, Dobbels E, van Zyl G, Cotton MF et al (2017) Late-onset hiv encephalopathy in children with long-standing virologic suppression followed by slow spontaneous recovery despite no change in antiretroviral therapy: 4 case reports. Pediatr Infect Dis J 36 (11):e264–e267
Jankiewicz M, Holmes MJ, Taylor PA, Cotton MF, Laughton B, van der Kouwe AJW et al (2017) White matter abnormalities in children with HIV infection and exposure. Front Neuroanat 11:88
Jeremy RJ, Kim S, Nozyce M, Nachman S, McIntosh K, Pelton SI et al (2005) Neuropsychological functioning and viral load in stable antiretroviral therapy-experienced HIV-infected children. Pediatrics 115(2):380–387
Kamat A, Lyons JL, Misra V, Uno H, Morgello S, Singer EJ et al (2012) Monocyte activation markers in cerebrospinal fluid associated with impaired neurocognitive testing in advanced HIV infection. J Acquir Immune Defic Syndr 60(3):234–243

Kheloufia F, Boucherie Q (2017) Neuropsychiatric events and dolutegravir in HIV patients: a worldwide issue involving a class effect. AIDS 31:1775–1800
Klein P (1985) More intelligent and sensitive child. Bar-Ilan University, Ramat-Gan
Klein P (1996) Early intervention: cross-cultural experiences with a mediational approach. Garland Press, New York
Klein P (2001) Seeds of hope: twelve years of early intervention in Africa. Unipub Forlag, Oslo
Klein PS, Rye H (2004) Interaction-oriented early intervention in Ethiopia: the MISC approach. Infants Young Child 17(4):340–354
Kotchick BA, Summers P, Forehand R, Steele RG (1997) The role of parental and extrafamilial social support in the psychosocial adjustment of children with a chronically ill father. Behav Modif 21(4):409–432
Kure K, Llena JF, Lyman WD, Soeiro R, Weidenheim KM, Hirano A et al (1991) Human immunodeficiency virus-1 infection of the nervous system: an autopsy study of 268 adult, pediatric, and fetal brains. Hum Pathol 22(7):700–710
Kuster OC, Laptinskaya D, Fissler P, Schnack C, Zugel M, Nold V et al (2017) Novel blood-based biomarkers of cognition, stress, and physical or cognitive training in older adults at risk of dementia: preliminary evidence for a role of BDNF, irisin, and the kynurenine pathway. J Alzheimers Dis 59(3):1097–1111
Laughton B, Cornell M, Boivin M, Van Rie A (2013) Neurodevelopment in perinatally HIV-infected children: a concern for adolescence. J Int AIDS Soc 16:18603
Laughton B, Cornell M, Kidd M, Springer PE, Dobbels EFM, Rensburg AJV et al (2018) Five year neurodevelopment outcomes of perinatally HIV-infected children on early limited or deferred continuous antiretroviral therapy. J Int AIDS Soc 21(5):e25106
Le Doare K, Bland R, Newell ML (2012) Neurodevelopment in children born to HIV-infected mothers by infection and treatment status. Pediatrics 130(5):e1326–e1344
Lentz MR, Kim WK, Lee V, Bazner S, Halpern EF, Venna N et al (2009) Changes in MRS neuronal markers and T cell phenotypes observed during early HIV infection. Neurology 72 (17):1465–1472
Letendre S, Marquie-Beck J, Capparelli E, Best B, Clifford D, Collier AC et al (2008) Validation of the CNS penetration-effectiveness rank for quantifying antiretroviral penetration into the central nervous system. Arch Neurol 65(1):65–70
Letendre SL, Mills AM, Tashima KT, Thomas DA, Min SS, Chen S et al (2014) ING116070: a study of the pharmacokinetics and antiviral activity of dolutegravir in cerebrospinal fluid in HIV-1-infected, antiretroviral therapy-naive subjects. Clin Infect Dis 59(7):1032–1037
Lyons JL, Uno H, Ancuta P, Kamat A, Moore DJ, Singer EJ et al (2011) Plasma sCD14 is a biomarker associated with impaired neurocognitive test performance in attention and learning domains in HIV infection. J Acquir Immune Defic Syndr 57(5):371–379
Mahncke HW, Bronstone A, Merzenich MM (2006a) Brain plasticity and functional losses in the aged: scientific bases for a novel intervention. In: Moller AR (ed) Progress in brain research. Elsevier, Amsterdam, pp 81–109
Mahncke HW, Connor BB, Appelman J, Ahsanuddin ON, Hardy JL, Wood RA et al (2006b) Memory enhancement in healthy older adults using a brain plasticity-based training program: a randomized, controlled study. PNAS 103(33):12523–12528
Malee KM, Tassiopoulos K, Huo Y, Siberry G, Williams PL, Hazra R et al (2011) Mental health functioning among children and adolescents with perinatal HIV infection and perinatal HIV exposure. AIDS Care 23(12):1533–1544
Marciniak E, Faivre E, Dutar P, Alves Pires C, Demeyer D, Caillierez R et al (2015) The Chemokine MIP-1alpha/CCL3 impairs mouse hippocampal synaptic transmission, plasticity and memory. Sci Rep 5:15862
McCoig C, Castrejon MM, Castano E, De Suman O, Baez C, Redondo W et al (2002) Effect of combination antiretroviral therapy on cerebrospinal fluid HIV RNA, HIV resistance, and clinical manifestations of encephalopathy. J Pediatr 141(1):36–44

Mellins CA, Malee KM (2013) Understanding the mental health of youth living with perinatal HIV infection: lessons learned and current challenges. J Int AIDS Soc 16:18593

Mellins C, Levenson R, Zawadzki R, Kairam R, Weston M (1994) Effects of pediatric HIV infection and prenatal drug exposure on mental and psychomotor development. J Pediatr Psychol 19:617–627

Mellins CA, Elkington KS, Leu CS, Santamaria EK, Dolezal C, Wiznia A et al (2013) Prevalence and change in psychiatric disorders among perinatally HIV-infected and HIV-exposed youth. AIDS Care 24(8):953–962

Merino KM, Allers C, Didier ES, Kuroda MJ (2017) Role of monocyte/macrophages during HIV/SIV infection in adult and pediatric acquired immune deficiency syndrome. Front Immunol 8:1693

Minde KK (1975) Psychological problems in Ugandan school children: a controlled evaluation. J Child Psychol Psychiatry 16(1):49–59

Mintz M (1994) Clinical comparison of adult and pediatric NeuroAIDS. Adv Neuroimmunol 4 (3):207–221

Moller HJ (2012) Soluble CD163. Scand J Clin Lab Invest 72(1):1–13

Moran LM, Fitting S, Booze RM, Webb KM, Mactutus CF (2014) Neonatal intrahippocampal HIV-1 protein Tat(1-86) injection: neurobehavioral alterations in the absence of increased inflammatory cytokine activation. Int J Dev Neurosci 38:195–203

Novak B, Giordani B, Boivin M, Winn B (2017) Potential uses of computer-based cognitive rehabilitation programs. In: Abubakar A, van de Vijver FJR (eds) Handbook of applied developmental science in sub-Saharan Africa. Springer, New York, pp 281–290

Nozyce M, Hittelman J, Muenz L, Durako SJ, Fischer ML, Willoughby A (1994a) Effect of perinatally acquired human immunodeficiency virus infection on neurodevelopment in children during the first two years of life. Pediatrics 94(6 Pt 1):883–891

Nozyce M, Hoberman M, Arpadi S, Wiznia A, Lambert G, Dobroszycki J et al (1994b) A 12-month study of the effects of oral zidovudine on neurodevelopmental functioning in a cohort of vertically HIV-infected inner-city children. AIDS 8(5):635–639

O'Hare BA, Venables J, Southall D (2005) Child health in Africa: 2005 a year of hope? Arch Dis Child 90(8):776–781

Okamoto S-I, Kang Y-J, Brechtel CW, Siviglia E, Russo R, Clemente A et al (2007) HIV/gp120 decreases adult neural progenitor cell proliferation via checkpoint kinase-mediated cell-cycle withdrawal and G1 arrest. Cell Stem Cell 1(2):230–236

Orach CG, De Brouwere V (2006) Integrating refugee and host health services in West Nile districts, Uganda. Health Policy Plan 21(1):53–64

Ortega M, Heaps JM, Joska J, Vaida F, Seedat S, Stein DJ et al (2013) HIV clades B and C are associated with reduced brain volumetrics. J Neurovirol 19(5):479–487

Patel K, Ming X, Williams PL, Robertson KR, Oleske JM, Seage GR 3rd (2009) Impact of HAART and CNS-penetrating antiretroviral regimens on HIV encephalopathy among perinatally infected children and adolescents. AIDS 23(14):1893–1901

Paul R, Apornpong T, Prasitsuebsai W, Puthanakit T, Saphonn V, Aurpibul L et al (2018) Cognition, emotional health, and immunological markers in children with long-term nonprogressive HIV. J Acquir Immune Defic Syndr 77(4):417–426

Peluso MJ, Ferretti F, Peterson J, Lee E, Fuchs D, Boschini A et al (2012) Cerebrospinal fluid HIV escape associated with progressive neurologic dysfunction in patients on antiretroviral therapy with well controlled plasma viral load. AIDS 26(14):1765–1774

Phillips N, Amos T, Kuo C, Hoare J, Ipser J, Thomas KG et al (2016) HIV-associated cognitive impairment in perinatally infected children: a meta-analysis. Pediatrics 138(5):e20160893

Price RW, Peterson J, Fuchs D, Angel TE, Zetterberg H, Hagberg L et al (2013) Approach to cerebrospinal fluid (CSF) biomarker discovery and evaluation in HIV infection. J Neuroimmune Pharmacol 8(5):1147–1158

Pulliam L, Herndier BG, Tang NM, McGrath MS (1991) Human immunodeficiency virus-infected macrophages produce soluble factors that cause histological and neurochemical alterations in cultured human brains. J Clin Invest 87(2):503–512

Puthanakit T, Aurpibul L, Louthrenoo O, Tapanya P, Nadsasarn R, Insee-ard S et al (2010) Poor cognitive functioning of school-aged children in Thailand with perinatally acquired HIV infection taking antiretroviral therapy. AIDS Patient Care STDs 24(3):141–146

Puthanakit T, Ananworanich J, Vonthanak S, Kosalaraksa P, Hansudewechakul R, van der Lugt J et al (2013) Cognitive function and neurodevelopmental outcomes in HIV-infected children older than 1 year of age randomized to early versus deferred antiretroviral therapy: the PREDICT neurodevelopmental study. Pediatr Infect Dis J 32(5):501–508

Rao H, Betancourt L, Giannetta JM, Brodsky NL, Korczykowski M, Avants BB et al (2009) Early parental care is important for hippocampal maturation: evidence from brain morphology in humans. NeuroImage 49(1):1144–1150

Richardson BA, Mbori-Ngacha D, Lavreys L, John-Stewart GC, Nduati R, Panteleeff DD et al (2003) Comparison of human immunodeficiency virus type 1 viral loads in Kenyan women, men, and infants during primary and early infection. J Virol 77(12):7120–7123

Ritola K, Robertson K, Fiscus SA, Hall C, Swanstrom R (2005) Increased human immunodeficiency virus type 1 (HIV-1) env compartmentalization in the presence of HIV-1-associated dementia. J Virol 79(16):10830–10834

Ronald AR, Sande MA (2005) HIV/AIDS care in Africa today. Clin Infect Dis 40(7):1045–1048

Royal W 3rd, Cherner M, Burdo TH, Umlauf A, Letendre SL, Jumare J et al (2016) Associations between cognition, gender and monocyte activation among HIV infected individuals in Nigeria. PLoS One 11(2):e0147182

Ruisenor-Escudero H, Sikorskii A, Familiar-Lopez I, Persaud D, Ziemniak C, Nakasujja N et al (2018) Neurodevelopmental outcomes in pre-school children living with HIV-1 subtypes A and D in Uganda. Pediatr Infect Dis J 37(12):e298–e303

Ryan LA, Zheng J, Brester M, Bohac D, Hahn F, Anderson J et al (2001) Plasma levels of soluble CD14 and tumor necrosis factor-alpha type II receptor correlate with cognitive dysfunction during human immunodeficiency virus type 1 infection. J Infect Dis 184(6):699–706

Sailasuta N, Ross W, Ananworanich J, Chalermchai T, DeGruttola V, Lerdlum S et al (2012) Change in brain magnetic resonance spectroscopy after treatment during acute HIV infection. PLoS One 7(11):e49272

Saito Y, Sharer LR, Epstein LG, Michaels J, Mintz M, Louder M et al (1994) Overexpression of nef as a marker for restricted HIV-1 infection of astrocytes in postmortem pediatric central nervous tissues. Neurology 44(3 Pt 1):474–481

Sanchez-Ramon S, Bellon JM, Resino S, Canto-Nogues C, Gurbindo D, Ramos JT et al (2003) Low blood CD8+ T-lymphocytes and high circulating monocytes are predictors of HIV-1-associated progressive encephalopathy in children. Pediatrics 111(2):E168–E175

Schnell G, Joseph S, Spudich S, Price RW, Swanstrom R (2011) HIV-1 replication in the central nervous system occurs in two distinct cell types. PLoS Pathog 7(10):e1002286

Schwartz L, Civitello L, Dunn-Pirio A, Ryschkewitsch S, Berry E, Cavert W et al (2007) Evidence of human immunodeficiency virus type 1 infection of nestin-positive neural progenitors in archival pediatric brain tissue. J Neurovirol 13(3):274–283

Smith R, Malee K, Leighty R, Brouwers P, Mellins C, Hittelman J et al (2006) Effects of perinatal HIV infection and associated risk factors on cognitive development among young children. Pediatrics 117(3):851–862

Smith R, Chernoff M, Williams PL, Malee KM, Sirois PA, Kammerer B et al (2012) Impact of HIV severity on cognitive and adaptive functioning during childhood and adolescence. Pediatr Infect Dis J 31(6):592–598

Sperduto AR, Bryson YJ, Chen IS (1993) Increased susceptibility of neonatal monocyte/macrophages to HIV-1 infection. AIDS Res Hum Retrovir 9(12):1277–1285

Spudich S, Gonzalez-Scarano F (2012) HIV-1-related central nervous system disease: current issues in pathogenesis, diagnosis, and treatment. Cold Spring Harb Perspect Med 2(6):a007120

Sternberg RJ (2004) Culture and intelligence. Am Psychol 59(5):325–338
Sternberg RJ, Grigorenko EL (2002) Dynamic testing: the nature and measurement of learning potential. Cambridge University Press, New York
Sternberg RJ, Grigorenko EL, Ngorosho D, Tantufuye E, Mbise A, Nokes CJ (2002) Assessing intellectual potential in rural Tanzanian school children. Intelligence 30(2):141–162
Sturdevant CB, Dow A, Jabara CB, Joseph SB, Schnell G, Takamune N et al (2012) Central nervous system compartmentalization of HIV-1 subtype C variants early and late in infection in young children. PLoS Pathog 8(12):e1003094
Sugimoto C, Merino KM, Hasegawa A, Wang X, Alvarez XA, Wakao H et al (2017) Critical role for monocytes/macrophages in rapid progression to AIDS in pediatric simian immunodeficiency virus-infected rhesus macaques. J Virol 91(17):e00379
Tamula MA, Wolters PL, Walsek C, Zeichner S, Civitello L (2003) Cognitive decline with immunologic and virologic stability in four children with human immunodeficiency virus disease. Pediatrics 112(3 Pt 1):679–684
Tardieu M (1998) HIV-1 and the developing central nervous system. Dev Med Child Neurol 40 (12):843–846
Tardieu M, Janabi N (1994) HIV-1 and the developing human nervous system: in vivo and in vitro aspects. Dev Neurosci 16(3–4):137–144
Tardieu M, Le Chenadec J, Persoz A, Meyer L, Blanche S, Mayaux MJ (2000) HIV-1-related encephalopathy in infants compared with children and adults. French Pediatric HIV Infection Study and the SEROCO Group. Neurology 54(5):1089–1095
Tau GZ, Peterson BS (2010) Normal development of brain circuits. Neuropsychopharmacology 35 (1):147–168
Thompson KA, Cherry CL, Bell JE, McLean CA (2011) Brain cell reservoirs of latent virus in presymptomatic HIV-infected individuals. Am J Pathol 179(4):1623–1629
Tornatore C, Nath A, Amemiya K, Major EO (1991) Persistent human immunodeficiency virus type 1 infection in human fetal glial cells reactivated by T-cell factor(s) or by the cytokines tumor necrosis factor alpha and interleukin-1 beta. J Virol 65(11):6094–6100
Tran LT, Roos A, Fouche JP, Koen N, Woods RP, Zar HJ et al (2016) White matter microstructural integrity and neurobehavioral outcome of HIV-exposed uninfected neonates. Medicine (Baltimore) 95(4):e2577
UNAIDS (2006) Report on the global AIDS epidemic. In: Joint United Nations Programmes on HIV/AIDS. UNAIDS, Geneva
Valcour V, Chalermchai T, Sailasuta N, Marovich M, Lerdlum S, Suttichom D et al (2012) Central nervous system viral invasion and inflammation during acute HIV infection. J Infect Dis 206 (2):275–282
Vallat AV, De Girolami U, He J, Mhashilkar A, Marasco W, Shi B et al (1998) Localization of HIV-1 co-receptors CCR5 and CXCR4 in the brain of children with AIDS. Am J Pathol 152 (1):167–178
Van Rie A, Dow A, Mupuala A, Stewart P (2009) Neurodevelopmental trajectory of HIV-infected children accessing care in Kinshasa, Democratic Republic of Congo. J Acquir Immune Defic Syndr 52(5):636–642
van Wyhe KS, van de Water T, Boivin MJ, Cotton MF, Thomas KG (2017) Cross-cultural assessment of HIV-associated cognitive impairment using the Kaufman assessment battery for children: a systematic review. J Int AIDS Soc 20(1):1–11
Vazeux R, Lacroix-Ciaudo C, Blanche S, Cumont MC, Henin D, Gray F et al (1992) Low levels of human immunodeficiency virus replication in the brain tissue of children with severe acquired immunodeficiency syndrome encephalopathy. Am J Pathol 140(1):137–144
Vygotsky LS (1978) Mind in society: the development of higher psychological processes. Harvard University Press, Cambridge
Weaver LK, Pioli PA, Wardwell K, Vogel SN, Guyre PM (2007) Up-regulation of human monocyte CD163 upon activation of cell-surface Toll-like receptors. J Leukoc Biol 81 (3):663–671

Westmoreland SV, Williams KC, Simon MA, Bahn ME, Rullkoetter AE, Elliott MW et al (1999) Neuropathogenesis of simian immunodeficiency virus in neonatal rhesus macaques. Am J Pathol 155(4):1217–1228

Wiley CA, Schrier RD, Nelson JA, Lampert PW, Oldstone MB (1986) Cellular localization of human immunodeficiency virus infection within the brains of acquired immune deficiency syndrome patients. Proc Natl Acad Sci U S A 83(18):7089–7093

Wiley CA, Belman AL, Dickson DW, Rubinstein A, Nelson JA (1990) Human immunodeficiency virus within the brains of children with AIDS. Clin Neuropathol 9(1):1–6

Wood SM, Shah SS, Steenhoff AP, Rutstein RM (2009) The impact of AIDS diagnoses on long-term neurocognitive and psychiatric outcomes of surviving adolescents with perinatally acquired HIV. AIDS 23(14):1859–1865

Yerkovich ST, Wikstrom ME, Suriyaarachchi D, Prescott SL, Upham JW, Holt PG (2007) Postnatal development of monocyte cytokine responses to bacterial lipopolysaccharide. Pediatr Res 62(5):547–552

Yilmaz A, Watson V, Else L, Gisslen M (2009) Cerebrospinal fluid maraviroc concentrations in HIV-1 infected patients. AIDS 23(18):2537–2540

Zayyad Z, Spudich S (2015) Neuropathogenesis of HIV: from initial neuroinvasion to HIV-associated neurocognitive disorder (HAND). Curr HIV/AIDS Rep 12(1):16–24

Zhou M, Greenhill S, Huang S, Silva TK, Sano Y, Wu S et al (2016) CCR5 is a suppressor for cortical plasticity and hippocampal learning and memory. eLife 5:e20985

Neurocognitive Complications of HIV Infection in Women: Insights from the WIHS Cohort

Leah H. Rubin and Pauline M. Maki

Contents

Abstract Although sex differences in brain function and brain disorders are well documented, very few studies have had adequate number of women to address sex-related factors contributing to HIV-associated brain dysfunction. Compared to men living with HIV (MLWH), women living with HIV (WLWH) may be at greater risk for cognitive dysfunction and decline due to biological factors (e.g., hormonal, immunologic) and issues common in underserved communities including poverty, low literacy levels, mental health and substance abuse, barriers to health-care services, and environmental exposures. To address this issue, we review relevant cross-sectional and longitudinal findings from the Women's Interagency HIV Study

L. H. Rubin (✉)
Department of Neurology and Psychiatry, Johns Hopkins School of Medicine, Baltimore, MD, USA

Department of Epidemiology, Johns Hopkins School of Public Health, Baltimore, MD, USA
e-mail: lrubin@jhu.edu

P. M. Maki
Department of Psychiatry, University of Illinois at Chicago, Chicago, IL, USA

Department of Psychology, University of Illinois at Chicago, Chicago, IL, USA

Curr Topics Behav Neurosci (2021) 50: 175–192
DOI 10.1007/7854_2019_101
Published Online: 9 August 2019

(WIHS), the largest study of the natural and treated history of WLWH, as well as other studies focusing on cognitive complications of HIV in women. We provide evidence that WLWH are more cognitively vulnerable than MLWH and that there are differences in the pattern of cognitive impairment. We next discuss factors that contribute to these differences, including biological factors (e.g., inflammation, hormonal, genetic) as well as common comorbidities (mental health, substance use, vascular and metabolic risk factors, coinfections and liver function, non-antiretroviral medications, and genetic markers). These findings demonstrate the importance of considering sex as a biological factor in studies of cognitive dysfunction and suggest avenues for future research.

Keywords Cognition · HIV · Neurocognitive function · neuroHIV · Women

In recent years, the field of neuroscience has emphasized the critical importance of biological sex in understanding healthy and abnormal brain function and recognizes that contrary to popular sentiment, sex differences in brain function and brain disorders are considerable (Cahill 2006). Our understanding of the influence of biological sex and gender-related factors in HIV-associated central nervous system (CNS) dysfunction is limited because few studies have focused on women living with HIV (WLWH) (Durvasula et al. 2001; Wojna et al. 2006; Cohen et al. 2001; Mason et al. 1998; Stern et al. 1998; Richardson et al. 2002, 2005; Maki et al. 2009). Yet, this subgroup comprises approximately 25% of HIV cases (Centers for Disease Control and Prevention 2017) in the USA and half of global cases (UN Women 2018).

CNS dysfunction including cognitive complications of HIV remains high, with estimates that 30 to 50% of individuals living with HIV will exhibit some form of neurocognitive impairment (NCI) during their lifetime (Grant 2008) even if they remain virally suppressed (VS) (Cysique et al. 2014). Notably, these estimates are based on studies either entirely or predominantly comprised of men. Few early studies had adequate numbers of women to sufficiently address questions about cognitive complications of HIV, but rates of NCI among women was estimated to be 42% (Richardson et al. 2002). WLWH may be at greater risk for cognitive NCI than men living with HIV (MLWH) due to poverty, low literacy levels, low educational attainment, substance abuse, poor mental health, barriers to health-care services, and adverse environmental exposures (Maki and Martin-Thormeyer 2009). Similarly, as in healthy women, WLWH are likely to show a different pattern of cognitive test performance than their male counterparts because of biological factors such as sex steroid hormones (e.g., estrogen, testosterone).

The Women's Interagency HIV Study (WIHS), the largest multisite longitudinal study of the natural and treated history of WLWH and at-risk HIV-uninfected (HIV−) women, would have been the ideal cohort in which to understand cognitive complications of HIV in women earlier in the epidemic (mid-to-late 1990s and on). WIHS is a cohort that is representative of AIDS and HIV cases reported among women in the US (Barkan et al. 1998). Until 2009, the cognitive outcomes in the WIHS were limited to an abbreviated longitudinal cognitive battery. Initial neurocognitive studies conducted in the WIHS were limited by sample size (Richardson et al. 2002, 2005; Maki et al. 2009) or the number of neuropsychological tests (Rubin et al. 2014; Meyer et al. 2013). Until 2009, the cognitive outcomes in the WIHS were limited to an abbreviated longitudinal cognitive battery. The abbreviated battery only included measures of processing speed (Symbol Digit Modalities Test (SDMT)), attention (Trail Making Test (TMT) Part A), and executive function (TMT Part B). In the 2005 pilot study, the two measures administered were the Hopkins Verbal Learning Test (HVLT-R), a measure of learning and memory, and the Stroop Color-Word Test, a measure of attention/concentration (Word Reading Trial 1, Color Naming Trial 2) and executive function (Color-Word trial 3). In 2009, the WIHS Neurocognitive Working Group (NCWG) implemented the largest comprehensive evaluation of cognitive function in women living with HIV and controls to date and began to pursue a set of research aims directed at the uniqueness of an all-female HIV-infected cohort. The battery was explicitly designed to assess seven cognitive domains – executive function (TMT Part B, Stroop Color-Word trial 3), processing speed (SDMT, Stroop Color Naming Trial), attention and working memory (Letter Number Sequencing (LNS) control and experimental conditions), learning (HVLT-R total learning), memory (HVLT-R delayed free recall), language (Controlled Oral Word Association Test (COWAT) and category fluency), and motor function (Grooved Pegboard (GPEG)). In September 2011, the efforts of the NCWG were realized when the first cross-sectional dataset in approximately 1,400 WIHS participants became available and provided an unparalleled opportunity to begin to understand how HIV influences cognition in women. Findings from this all-female cohort have yielded novel and important findings regarding cognitive complications in WLWH. The purpose of this chapter will be to discuss these findings as well as highlight other studies focusing on the cognitive complications of WLWH outside of the WIHS.

1 Are WLWH More Cognitively Vulnerable Than MLWH?

Converging evidence suggests that WLWH may be more vulnerable to NCI than MLWH. Although not all studies demonstrate differences between males and females (Burlacu et al. 2018; Behrman-Lay et al. 2016), studies do find either that females show greater NCI than males overall or that females and males differ in the pattern of NCI (Royal et al. 2016; Robertson et al. 1996, 2014; Heaton et al. 2015; Failde-Garrido et al. 2008; Keutmann et al. 2016; Hestad et al. 2012; Maki et al. 2018). For example, in the two longest-running multisite, longitudinal studies of HIV progression in the United States (US), the WIHS and the Multicenter AIDS Cohort Study (MACS), performance was consistently worse among WLWH versus MLWH even after adjusting for HIV-related clinical characteristics (e.g., current and nadir CD4 count, viral load, ARV use). These differences were evident in measures of executive function (TMT Part B), attention (TMT Part A, psychomotor speed (SDMT), and motor function (GPEG) in a sample of 1,420 individuals (Maki et al. 2018). A recent large cross-sectional study of 1,361 people living with HIV (PLWH;

204 WLWH) and 702 HIV− (214 women) community-based individuals in the HIV Neurobehavioral Research Program (HNRP) demonstrated that the association between HIV seropositivity and a higher likelihood of NCI was stronger in women than men (Sundermann et al. 2018). A greater female vulnerability to NCI was also seen in among substance-dependent PLWH (Keutmann et al. 2016; Fogel et al. 2017; Martin et al. 2016).

2 Pattern of Cognitive Impairment Among Women Living with HIV

Cross-sectional results from the comprehensive cognitive test battery were first reported in a sample of 1,521 (1,019 WLWH) WIHS women (Maki et al. 2015). In this sample, 80% were ≥40 years of age, 21% self-identified as Latina or Hispanic and more than 60% as African American, and 45% were living below the federally defined poverty level. Unique to women, the greatest cognitive vulnerabilities in WLWH were on measures of verbal learning, memory, and attention (Fig. 1). Moreover, WLWH demonstrated impaired semantic clustering, a key component of verbal learning and memory reliant on the prefrontal cortex (PFC) (Becker and Lim 2003; Baker et al. 2001). The effect sizes for HIV serostatus were small (<0.20 standard deviations) and were smaller than effects of educational quality, chronological age, poverty, and depressive symptoms. These data suggested that memory,

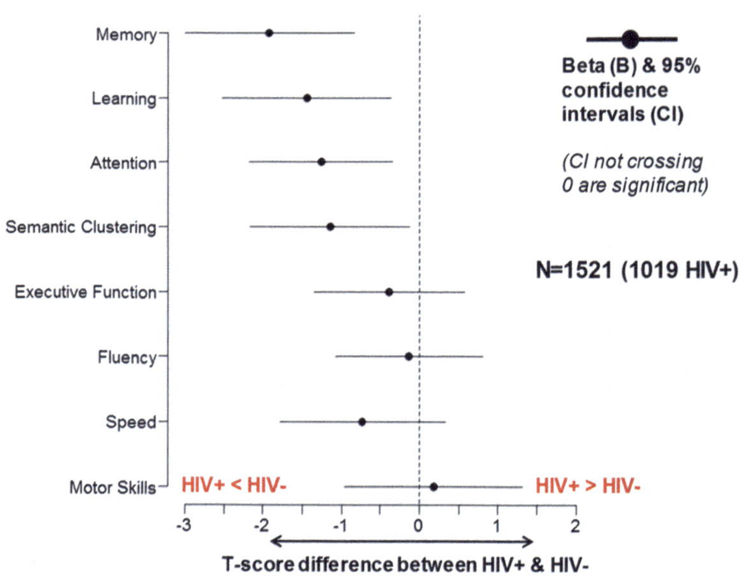

Fig. 1 Among HIV-infected women, verbal learning and memory and simple attention are more sensitive to HIV than standard measures of executive function

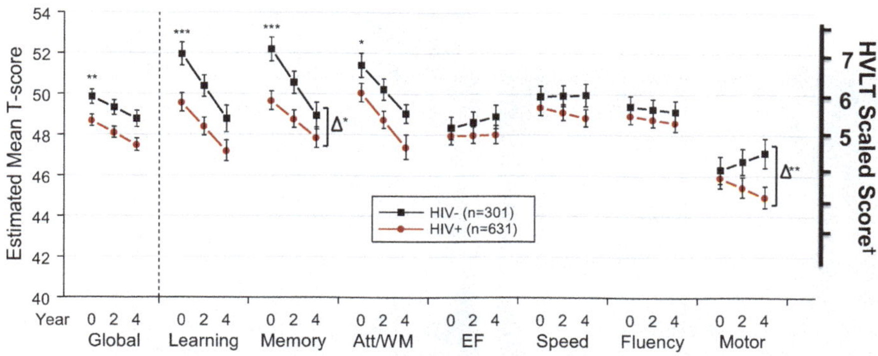

Fig. 2 Learning, memory, and attention remain a long-standing issue among WLWH. Att/WM, attention/working memory; EF, executive function; Δ, group difference in slopes at $p < 0.05$; ***$p < 0.001$; **$p < 0.01$; $p < 0.05$; †(Norman et al. 2011); HVLT, Hopkins Verbal Learning Test

learning, and attention may be the cognitive domains most susceptible among WLWH. Interestingly, these findings contrast with studies in male-dominant HIV cohorts (MACS, CHARTER) where deficits are most prominent in executive function, complex attention, and learning (Heaton et al. 2011; Cysique et al. 2004).

In 2017, the first longitudinal analyses in the WIHS were published in WLWH to HIV− women at three time points over 4 years (Rubin et al. 2017c). WLWH showed persistent vulnerabilities in verbal learning, memory, attention/working memory, and executive function over time, while motor declined over time (Fig. 2). A functional magnetic resonance imaging (fMRI) study in WIHS implicated adverse alterations in the hippocampus (HI) and PFC with learning and memory deficits on the HVLT (Fig. 3). The data shown in Figs. 1 and 2 include a mixed group of VS and unsuppressed individuals. Our WIHS work (Rubin et al. 2017c) demonstrates that NCI persists in WLWH with consistent viral suppression on continuous ARV therapy.

3 Inflammatory Contributors to NCI Among WLWH

Similar to other studies (Burdo et al. 2013; Royal et al. 2016), evidence from WIHS indicates that cognitive vulnerabilities in VS WLWH are associated with soluble markers of monocyte-driven inflammatory markers (e.g., sCD163, sCD14) (Imp et al. 2017), as well as with more general markers of systemic low-grade inflammation (Rubin et al. 2017a), including interleukin (IL)-6, C-reactive protein (CRP), soluble tumor necrosis factor receptor (TNFR) I, matrix metalloproteinase (MMP)-9, but most significantly variability in CRP which had much stronger associations across cognitive domains in WLWH compared to HIV− women. S100 calcium-binding protein A9 (S100A9) has also been found to be decreased in Hispanic WLWH demonstrating NCI (Colon et al. 2016). Such findings show the importance

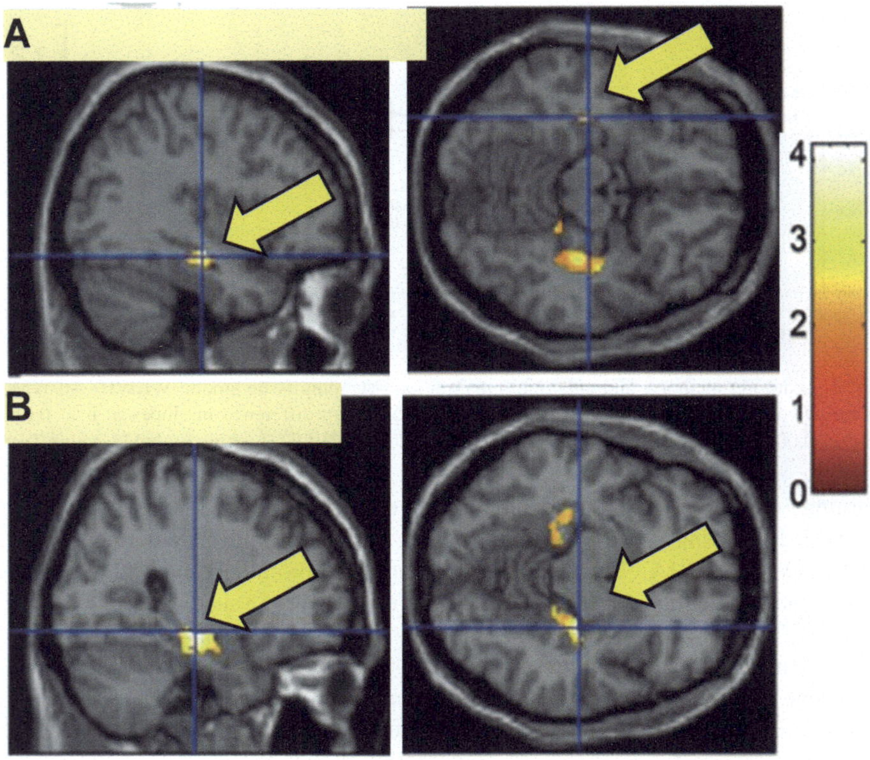

Fig. 3 (**a**) During encoding, WLWH showed decreased HI activity vs. HIV− women. Decreased HI activity was associated with *lower* HVLT performance across groups (r's > 0.54). (**b**) During recognition, HIV+ women showed increased HI activity vs. HIV− women. Increased HI activity was associated with *lower* HVLT performance across groups (r's < −0.62)

of peripheral markers of inflammation as a key mechanism contributing to NCI in WLWH.

4 Common Comorbidities Among WLWH and Cognitive Correlates

Other factors that may contribute to the female-specific cognitive vulnerability in HIV and/or exacerbate the pattern of domain-specific NCI include mental health, substance use vascular and metabolic factors, female-specific risk factors (menopause), coinfections and liver function, non-ARV medications, genetic, and HIV-related factors.

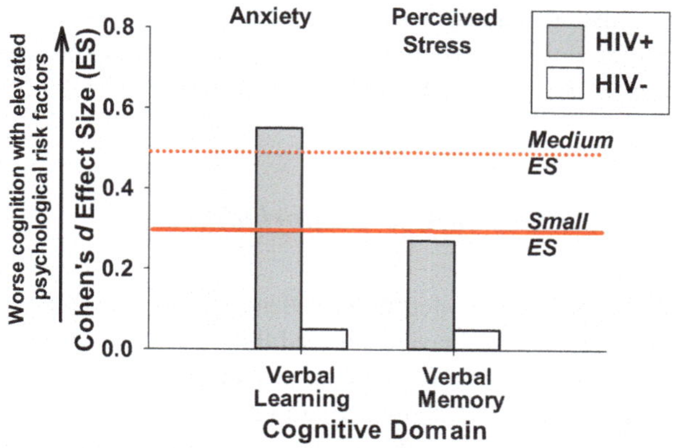

Fig. 4 Anxiety and stress are associated with greater deficits in verbal learning and memory in WLWH versus HIV− women

4.1 Mental Health Comorbidities

Mental health risk factors and disorders are strongly associated with cognitive function in WLWH. In a series of cross-sectional epidemiological studies in the WIHS, we examined associations between post-traumatic stress, anxiety, perceived stress, depression, and cognitive function. Post-traumatic stress, anxiety, perceived stress, and depression were each associated with deficits in verbal learning and memory as well as attention and semantic clustering (Rubin et al. 2014, 2015a, 2016a; Maki et al. 2015). Notably, high levels of perceived stress and elevated anxiety were associated with decreased verbal learning and memory only among WLWH (Fig. 4) (Rubin et al. 2014, 2015a). In a longitudinal investigation of mental health factors and NCI in the WIHS, higher perceived stress and PTSD were associated with a greater cognitive decline in verbal fluency performance compared to those with lower stress and PTSD only among WLWH (Rubin et al. 2017b). Consistent with the cross-sectional analyses, regardless of time, perceived stress and PTSD were negatively associated with verbal learning and memory only among WLWH. Irrespective of time or HIV serostatus, depression, perceived stress, and PTSD were associated with lower processing speed, executive function, and global neuropsychological function. A longitudinal study in South Africa found less improvement over time in executive function (e.g., Wisconsin Card Sorting Test, Stroop) and fluency (COWAT, category fluency) among WLWH exposed to trauma compared to women not exposed to trauma (regardless of serostatus) (Spies et al. 2017).

HIV and mental health factors individually and in combination impair cognition in WLWH through effects on PFC. Among WLWH, those with higher levels of stress showed more PFC atrophy (including inferior frontal, middle frontal, and superior frontal gyri) and deactivation in the medial PFC and posterior cingulate cortex compared to those with lower stress (Rubin et al. 2015b, 2016b). Importantly, PFC atrophy and deactivation were associated with less semantic clustering; atrophy was also associated with worse verbal memory (Rubin et al. 2015b, 2016c). These studies point to the PFC as a region of particular vulnerability among WLWH particularly to mental health factors.

4.2 Substance Use Comorbidities

In the WIHS, the lifetime prevalence of any substance use disorder is 58%. Relatively few studies have examined the association between substance use and cognition among WLWH. Opioids, stimulants, and alcohol exacerbate the neurotoxic effects of HIV including acceleration of disease progression and increases in the risk of NCI (Nath et al. 2001; Carey et al. 2006). In the WIHS, WLWH who reported using crack, cocaine, and/or heroin in the past 6 months demonstrated lower performance on learning and memory compared to WLWH women who reported never using these substances (Meyer et al. 2013). There were no differences in memory performance by illicit substance use among the HIV− women. No interactions between illicit drug use and HIV serostatus were found on measures of attention or behavioral inhibition. Cocaine-associated deficits in verbal learning and memory among WLWH may in part be driven by alterations in PFC regions. In an fMRI study in WLWH, both current and former use of cocaine was associated with decreased activation in medial PFC during the encoding phase of a verbal memory task (Meyer et al. 2014). During the recognition phase of the verbal memory task, WLWH nonusers showed greater activation than current and former cocaine users in prefrontal regions (left dorsal medial PFC, right dorsal lateral PFC, anterior PFC). Other work in substance-dependent individuals shows greater impairment in risky decision-making and visual memory in WLWH compared with MLWH and HIV− individuals (Martin et al. 2016; Keutmann et al. 2016).

4.3 Vascular and Metabolic Risk Factors

The increasing effectiveness of ARVs has led to the aging of the HIV population and, with it, the increasing presence of age-related diseases including cardiovascular and cerebrovascular comorbidities. These comorbidities often result in the dysregulation of multiple systems including negative consequences for brain structure and function. In an early cross-sectional WIHS study, carotid lesion and carotid intima-media thickness (CIMT) were associated with lower performance on the Stroop interference trial but not with performance on the SDMT. Another WIHS study found an association between ultrasound-based measures of carotid artery stiffness and cognitive decline in TMT Parts A and B and SDMT (Huck et al. 2018). Specifically, greater baseline carotid stiffness was associated with greater decline on all outcomes over a median of 8.5 years, but these associations did not differ by HIV serostatus. In another early WIHS study, body mass index (BMI), waist circumference (WC), and waist-to-hip ratio (WHR) were examined in relation to cognition and showed some stronger relationships with cognition in WLWH compared to HIV− women (Gustafson et al. 2013). On TMT A and B and SDMT, being underweight (BMI <18.5 kg/m^2) was associated with worse performance compared to having a normal BMI in WLWH but not in HIV− women. For WLHV but not HIV− women, being obese (body mass index 30 kg/m^2 or higher) was associated with better performance on TMT Part A compared to having a normal BMI. In WLWH only, higher WC was associated with better performance on the Stroop Color-Word Trial and SDMT. Higher leptin levels were associated with lower performance on TMT Part A in both WLWH and HIV− women (Gustafson et al. 2015). Three gut hormones (ghrelin, amylin, gastric inhibitory peptide) were examined to better understand the mechanisms underlying the obesity-cognition association (McFarlane et al. 2017). In general, lower gut hormone levels were associated with lower cognitive performance. Gastric inhibitory peptide and ghrelin were associated with cognitive performance among WLWH, whereas only ghrelin was associated with cognitive performance among HIV− women.

A number of WIHS studies have also examined the link between insulin resistance (IR) and cognition. In the initial cross-sectional study, insulin resistance as measured using the homeostasis model assessment (HOMA) was associated in the overall sample with Stroop Color Naming and not TMT Parts A and B, SDMT, or other Stroop trials (Valcour et al. 2012). There was also an interaction between HOMA-IR and HIV serostatus on the Stroop Color-Word Trial such that among those with average HOMA-IR, WLWH demonstrated worse performance than HIV− women. A follow-up study of HOMA in relation to performance on the comprehensive cognitive test battery (Valcour et al. 2015) revealed an association between increasing HOMA and worse performance on the LNS control condition (attention), HVLT-R recognition, and phonemic fluency across serostatus groups. Interactions were also noted on three measures of attention – LNS control condition, Stroop Word Reading, and Color Naming trials – with worse performance in WLWH versus HIV in women as HOMA values increased. In two non-WIHS studies conducted in Hispanic WLWH, levels of soluble and cell-associated insulin receptor levels, IR substrate-1 (IRS-1) levels, and IRS-1 tyrosine phosphorylation were assayed in plasma and CSF in association with NCI (Gerena et al. 2012). IR secretion was higher in WLWH than HIV− women, and higher IR secretion was associated with increasing NCI severity. Further, higher binding of free insulin to the soluble insulin receptor was also associated with NCI (Gerena et al. 2015).

4.4 *Female-Specific Reproductive Risk Factors*

Despite the wealth of literature on normative changes in women's health across the menopausal transition, very little is known about the natural history of menopause in WLWH (Bull et al. 2018), particularly with regard to cognitive changes. In a cross-sectional WIHS study, the menopause stage was not associated with cognitive functioning on HVLT-R and Stroop (Rubin et al. 2014). However, across serostatus groups, depressive symptoms were associated with worse learning, memory, attention, and executive function, and anxiety symptoms were associated with worse learning and memory. Vasomotor symptoms were also associated with worse attention. Notably, there was an interaction between HIV serostatus and anxiety symptoms on verbal learning such that elevated anxiety was associated with worse verbal learning in WLWH only. Thus, menopause symptoms are associated with cognitive performance in WLWH.

4.5 Coinfections and Liver Function

Hepatitis C (HCV) infection is relatively common in WLWH and the WIHS cohort. There are inconsistent findings regarding the association between HCV infection and cognition. In an early WIHS study ($n = 200$), WLWH who were HCV positive demonstrated greater odds of NCI compared to women not infected with either (Richardson et al. 2005). In a later WIHS study, HCV was not associated with cognition in 1,338 women (Crystal et al. 2012). However, in a smaller non-WIHS sample of WLWH, hepatitis C virus coinfection was associated with lower motor function, processing speed, attention, working memory, and planning (Giesbrecht et al. 2014). Given the significant overlap between HCV and liver fibrosis, a follow-up study examined the association between liver function and cognition (Valcour et al. 2016). Liver fibrosis (APRI) was associated with worse performance in learning, executive function, memory, psychomotor speed, fluency, and fine motor skills in the overall sample. The severity of fibrosis measured via Fibroscan was associated with worse performance in attention, executive function, and fluency. These associations held after controlling for HCV and HIV status, and the associations were not moderated by these factors.

4.6 Non-ARV Medications

The use of non-ARV medications with adverse cognitive effects (NC-AE medications) is more common among WLWH compared to HIV− women (Radtke et al. 2018). Non-ARV medications are associated with worse cognition in the WIHS overall, but HIV serostatus did not moderate these associations (Rubin et al. 2018b). However, for women taking anticholinergic-acting medications, HIV serostatus differences were most pronounced (WLWH < HIV−) in global, learning, fluency, and motor function. For women taking anxiolytics/anticonvulsants or opioids, HIV serostatus differences were also more pronounced (WLWH < HIV−) in learning and processing speed, respectively.

4.7 Genetic Markers

To date, our understanding of the genetics of NCI is based primarily on male-dominant studies. To date, only two studies have been conducted in WLWH that have looked at genetic markers. One study in the WIHS examined the association between catechol-o-methyltransferase (COMT) Val158Met genotype and working memory as well as PFC function in WLWH (Sundermann et al. 2015). The COMT Val158Met (rs4680) single nucleotide polymorphism (SNP) influences executive function and PFC function through its effect on dopamine metabolism. Both HIV and the Val allele of the Val158Met SNP were associated with compromised executive function and inefficient PFC function. Among Val/Val but not Met allele carriers, WLWH performed worse than HIV− women on a measure of working memory. Val/Val carriers also showed greater PFC activations during performance of an n-back working memory task compared to HIV− Val/Val carriers. However, HIV− Met allele carriers demonstrated greater PFC activation versus WLWH Met allele carriers. Together the findings suggested that suboptimal dopamine levels associated with the Val/Val COMT genotype leads to working memory deficits and inefficient PFC function in WLWH. A study in Hispanic women examined apolipoprotein E (ApoE) allele status, HIV serostatus, and CSF APoE protein levels in relation to spatial learning and memory (measure of HI function) (Morales et al. 2012). The ApoE gene produces a protein responsible for the metabolism and transport of lipoproteins and cholesterol. The presence of the e4 allele is a well-known risk factor in Alzheimer's disease and has been linked to NCI in some (Valcour et al. 2004; Wendelken et al. 2016) but not all (Morgan et al. 2013) studies in older PLWH ($\geq$50 years of age). In the small sample of 20 WLWH and 16 controls approximately 40 years of age, the e4 allele was associated with cognitive performance on a standard neuropsychological test battery among HIV− but not WLWH (Morales et al. 2012). However, the e4 allele was associated with an experimental measure of spatial learning and memory (Memory Island task) in WLWH but not HIV− women.

5 Summary

Overall, findings from WIHS and other studies suggest that WLWH demonstrate persistent NCI despite ARV and that WLWH may be more cognitively vulnerable compared to MLWH. Two brain regions that appear to be particularly susceptible to HIV infection and female sex and may contribute to the prominent deficits in verbal learning and memory observed in WLWH are PFC-HI regions. Additional work is needed to understand brain regions that are susceptible to other domain-specific impairment that persists among WLWH including attention/working memory and executive function. Similar to MLWH, inflammation is a key contributor to NCI among WLWH. There are many factors (e.g., mental health, substance use) that contribute or exacerbate cognitive complications among WLWH.

Assessing sex differences in the context of HIV may help to elucidate novel therapeutics for CNS dysfunction in PLWH. We cannot assume that WLWH and MLWH will respond similarly to the same treatment or that the mechanisms underlying cognitive problems are the same for each sex (Rubin et al. 2017d, 2018a). Such results underscore the need to at least stratify cognitive analyses by sex to determine whether the patterns are the same between WLWH and MLWH or whether they qualitatively differ between the sexes (sex-dependent) or are present in one sex and not the other (sex-specific).

6 Future Directions

From the review above, we propose three key directions for future research. Arguably, the most critical research gap is a sufficiently powered cohort study of neuropsychological test performance in men and women living with HIV and matched controls. We were able to identify parallel neuropsychological measures between the WIHS and the MACS and had sufficient power to examine sex differences in HIV-seropositive participants versus HIV-seronegative controls (Maki et al. 2018). However, critical measures, including verbal memory, differed between the two cohorts so it is unknown whether WLWH show vulnerabilities in those measures as well. With the merging of the WIHS and the MACS into the MACS-WIHS Combined Cohort Study (MWCCS), we are now poised to directly compare a sufficiently large sample of men and women in a longitudinal prospective cohort study. With an estimated sample size 4,400 former MACS/WIHS recruits and 1,600–1,700 new recruits, the MWCCS will be unique in providing sufficient statistical power to determine whether such factors as depression, substance abuse, and trauma, as well as vascular, metabolic, and menopause-related risk factors, contribute to sex differences in cognitive performance. Similarly, the MWCCS will be able to draw on GWAS data to investigate genetic contributions to sex differences in cognitive function.

The second key direction is a longitudinal multimodal neuroimaging study comparing WLWH and MLWH in relation to seronegative controls. We are undertaking such a study in the Baltimore and Washington DC sites of the MACS-WIHS Combined Cohort Study (MWCCS). Building on our cross-sectional neuroimaging studies, we are using both task-based and resting state fMRI in HIV+ virally suppressed (HIV + VS) men and women and HIV-uninfected individuals to identify the neural circuitry contributing to deficits in two targeted domains – declarative memory and cognitive control. We are also using [11C]DPA-713 (DPA) PET to assess HIV-related alterations in chronic neuroinflammation and NCI. We are also assessing structural MRI and diffusion-weighted MRI. Imaging assessments will be conducting annually for 3 years. The longitudinal design allows an assessment of the reproducibility of key findings over time and the sensitivity of these neuroimaging measures to changes in cognitive performance. Such work will inform our understanding of the mechanisms linked to neurological comorbidity and to provide novel, more sensitive neuroimaging biomarkers to guide testing of new cognitive therapies for HIV+ individuals.

While it is critical to extend sex differences research in neuroAIDS in the USA, a third and critical question is whether any differences observed in American cohort studies are generalizable to international cohorts, including those in sub-Saharan Africa. Gender differences in factors such as mental health have been posited to contribute to differential cognitive impairment in WLWH compared to MLWH in sub-Saharan Africa, but sample sizes are limited to fewer than 210 people total (Royal et al. 2016; Hestad et al. 2012).

References

Baker JT, Sanders AL, Maccotta L, Buckner RL (2001) Neural correlates of verbal memory encoding during semantic and structural processing tasks. Neuroreport 12:1251–1256

Barkan SE, Melnick SL, Preston-Martin S, Weber K, Kalish LA, Miotti P, Young M, Greenblatt R, Sacks H, Feldman J (1998) The Women's Interagency HIV Study. WIHS Collaborative Study Group. Epidemiology 9:117–125

Becker S, Lim J (2003) A computational model of prefrontal control in free recall: strategic memory use in the California Verbal Learning Task. J Cogn Neurosci 15:821–832

Behrman-Lay AM, Paul RH, Heaps-Woodruff J, Baker LM, Usher C, Ances BM (2016) Human immunodeficiency virus has similar effects on brain volumetrics and cognition in males and females. J Neurovirol 22:93–103

Bull L, Tittle V, Rashid T, Nwokolo N (2018) HIV and the menopause: a review. Post Reprod Health 24:19–25

Burdo TH, Weiffenbach A, Woods SP, Letendre S, Ellis RJ, Williams KC (2013) Elevated sCD163 in plasma but not cerebrospinal fluid is a marker of neurocognitive impairment in HIV infection. AIDS 27:1387–1395

Burlacu R, Umlauf A, Luca A, Gianella S, Radoi R, Ruta SM, Marcotte TD, Ene L, Achim CL (2018) Sex-based differences in neurocognitive functioning in HIV-infected young adults. AIDS 32:217–225

Cahill L (2006) Why sex matters for neuroscience. Nat Rev Neurosci 7:477–484

Carey CL, Woods SP, Rippeth JD, Gonzalez R, Heaton RK, Grant I (2006) Additive deleterious effects of methamphetamine dependence and immunosuppression on neuropsychological functioning in HIV infection. AIDS Behav 10:185–190

Centers for Disease Control and Prevention (2017) Monitoring selected national HIV prevention and care objectives by using HIV surveillance data – United States and 6 dependent areas, 2015. HIV Surveill Suppl Rep 22(2)

Cohen RA, Boland R, Paul R, Tashima KT, Schoenbaum EE, Celentano DD, Schuman P, Smith DK, Carpenter CC (2001) Neurocognitive performance enhanced by highly active antiretroviral therapy in HIV-infected women. AIDS 15:341–345

Colon K, Perez-Laspiur J, Quiles R, Rodriguez Y, Wojna V, Shaffer SA, Leszyk J, Skolasky RL Jr, Melendez LM (2016) Macrophage secretome from women with HIV-associated neurocognitive disorders. Proteomics Clin Appl 10:136–143

Crystal H, Kleyman I, Anastos K, Lazar J, Cohen M, Liu C, Pearce L, Golub E, Valcour V, Ho A, Strickler H, Peters M, Kovacs A, Holman S, Kreek MJ, Manly J (2012) Effects of hepatitis C and HIV on cognition in women: data from The Women's Interagency HIV Study. J Acquir Immune Defic Syndr 59:149–154

Cysique LA, Maruff P, Brew BJ (2004) Prevalence and pattern of neuropsychological impairment in human immunodeficiency virus-infected/acquired immunodeficiency syndrome (HIV/AIDS) patients across pre- and post-highly active antiretroviral therapy eras: a combined study of two cohorts. J Neurovirol 10:350–357

Cysique LA, Heaton RK, Kamminga J, Lane T, Gates TM, Moore DM, Hubner E, Carr A, Brew BJ (2014) HIV-associated neurocognitive disorder in Australia: a case of a high-functioning and optimally treated cohort and implications for international neuro HIV research. J Neurovirol 20:258–268

Durvasula RS, Miller EN, Myers HF, Wyatt GE (2001) Predictors of neuropsychological performance in HIV positive women. J Clin Exp Neuropsychol 23:149–163

Failde-Garrido JM, Alvarez MR, Simon-Lopez MA (2008) Neuropsychological impairment and gender differences in HIV-1 infection. Psychiatry Clin Neurosci 62:494–502

Fogel J, Rubin LH, Maki P, Keutmann MK, Gonzalez R, Vassileva J, Martin EM (2017) Effects of sex and HIV serostatus on spatial navigational learning and memory among cocaine users. J Neurovirol 23:855–863

Gerena Y, Skolasky RL, Velez JM, Toro-Nieves D, Mayo R, Nath A, Wojna V (2012) Soluble and cell-associated insulin receptor dysfunction correlates with severity of HAND in HIV-infected women. PLoS One 7:e37358

Gerena Y, Menendez-Delmestre R, Skolasky RL, Hechavarria RM, Perez S, Hilera C, Gonzalez C, Nath A, Wojna V (2015) Soluble insulin receptor as a source of insulin resistance and cognitive impairment in HIV-seropositive women. J Neurovirol 21:113–119

Giesbrecht CJ, Thornton AE, Hall-Patch C, Maan EJ, Cote HC, Money DM, Murray M, Pick N (2014) Select neurocognitive impairment in HIV-infected women: associations with HIV viral load, hepatitis C virus, and depression, but not leukocyte telomere length. PLoS One 9:e89556

Grant I (2008) Neurocognitive disturbances in HIV. Int Rev Psychiatry 20:33–47

Gustafson DR, Mielke MM, Tien PC, Valcour V, Cohen M, Anastos K, Liu C, Pearce L, Golub ET, Minkoff H, Crystal HA (2013) Anthropometric measures and cognition in middle-aged HIV-infected and uninfected women. The Women's Interagency HIV Study. J Neurovirol 19:574–585

Gustafson DR, Mielke MM, Keating SA, Holman S, Minkoff H, Crystal HA (2015) Leptin, adiponectin and cognition in middle-aged HIV-infected and uninfected women. The Brooklyn Women's Interagency HIV Study. J Gerontol Geriatr Res 4(5). pii: 240

Heaton RK, Franklin DR, Ellis RJ, Mccutchan JA, Letendre SL, Leblanc S, Corkran SH, Duarte NA, Clifford DB, Woods SP, Collier AC, Marra CM, Morgello S, Mindt MR, Taylor MJ, Marcotte TD, Atkinson JH, Wolfson T, Gelman BB, Mcarthur JC, Simpson DM, Abramson I, Gamst A, Fennema-Notestine C, Jernigan TL, Wong J, Grant I (2011) HIV-associated neurocognitive disorders before and during the era of combination antiretroviral therapy: differences in rates, nature, and predictors. J Neurovirol 17:3–16

Heaton RK, Franklin DR Jr, Deutsch R, Letendre S, Ellis RJ, Casaletto K, Marquine MJ, Woods SP, Vaida F, Atkinson JH, Marcotte TD, Mccutchan JA, Collier AC, Marra CM, Clifford DB, Gelman BB, Sacktor N, Morgello S, Simpson DM, Abramson I, Gamst AC, Fennema-Notestine C, Smith DM, Grant I, CHARTER Group (2015) Neurocognitive change in the era of HIV combination antiretroviral therapy: the longitudinal CHARTER study. Clin Infect Dis 60:473–480

Hestad KA, Menon JA, Silalukey-Ngoma M, Franklin DR Jr, Imasiku ML, Kalima K, Heaton RK (2012) Sex differences in neuropsychological performance as an effect of human immunodeficiency virus infection: a pilot study in Zambia, Africa. J Nerv Ment Dis 200:336–342

Huck DM, Hanna DB, Rubin LH, Maki P, Valcour V, Springer G, Xue X, Lazar J, Hodis HN, Anastos K, Kaplan RC, Kizer JR (2018) Carotid artery stiffness and cognitive decline among women with or at risk for HIV infection. J Acquir Immune Defic Syndr 78:338–347

Imp BM, Rubin LH, Tien PC, Plankey MW, Golub ET, French AL, Valcour VG (2017) Monocyte activation is associated with worse cognitive performance in HIV-infected women with virologic suppression. J Infect Dis 215:114–121

Keutmann MK, Gonzalez R, Maki PM, Rubin LH, Vassileva J, Martin EM (2016) Sex differences in HIV effects on visual memory among substance-dependent individuals. J Clin Exp Neuropsychol:1–13

Maki PM, Martin-Thormeyer E (2009) HIV, cognition and women. Neuropsychol Rev 19:204–214

Maki PM, Cohen MH, Weber K, Little DM, Fornelli D, Rubin LH, Perschler P, Gould F, Martin E (2009) Impairments in memory and hippocampal function in HIV-positive vs HIV-negative women: a preliminary study. Neurology 72:1661–1668

Maki PM, Rubin LH, Valcour V, Martin E, Crystal H, Young M, Weber KM, Manly J, Richardson J, Alden C, Anastos K (2015) Cognitive function in women with HIV: findings from the Women's Interagency HIV Study. Neurology 84:231–240

Maki PM, Rubin LH, Springer G, Seaberg EC, Sacktor N, Miller EN, Valcour V, Young MA, Becker JT, Martin EM, Neuropsychology Working Groups of the Women�s Interagency HIV Study and the Multicenter AIDS Cohort Study (2018) Differences in cognitive function between women and men with HIV. J Acquir Immune Defic Syndr 79:101–107

Martin E, Gonzalez R, Vassileva J, Maki PM, Bechara A, Brand M (2016) Sex and HIV serostatus differences in decision making under risk among substance-dependent individuals. J Clin Exp Neuropsychol 38:404–415

Mason KI, Campbell A, Hawkins P, Madhere S, Johnson K, Takushi-Chinen R (1998) Neuropsychological functioning in HIV-positive African-American women with a history of drug use. J Natl Med Assoc 90:665–674

Mcfarlane SI, Mielke MM, Uglialoro A, Keating SM, Holman S, Minkoff H, Crystal HA, Gustafson DR (2017) Ghrelin, amylin, gastric inhibitory peptide and cognition in middle-aged HIV-infected and uninfected women: The Women's Interagency HIV Study. J Neurol Neurophysiol 8(1). pii: 413

Meyer VJ, Rubin LH, Martin E, Weber KM, Cohen MH, Golub ET, Valcour V, Young MA, Crystal H, Anastos K, Aouizerat BE, Milam J, Maki PM (2013) HIV and recent illicit drug use interact to affect verbal memory in women. J Acquir Immune Defic Syndr 63:67–76

Meyer VJ, Little DM, Fitzgerald DA, Sundermann EE, Rubin LH, Martin EM, Weber KM, Cohen MH, Maki PM (2014) Crack cocaine use impairs anterior cingulate and prefrontal cortex function in women with HIV infection. J Neurovirol 20:352–361

Morales D, Acevedo SF, Skolasky RL, Hechavarria R, Santiago S, De La Torre T, Maldonado E, Wojna V (2012) Translational spatial task and its relationship to HIV-associated neurocognitive disorders and apolipoprotein E in HIV-seropositive women. J Neurovirol 18:488–502

Morgan EE, Woods SP, Letendre SL, Franklin DR, Bloss C, Goate A, Heaton RK, Collier AC, Marra CM, Gelman BB, Mcarthur JC, Morgello S, Simpson DM, Mccutchan JA, Ellis RJ, Abramson I, Gamst A, Fennema-Notestine C, Smith DM, Grant I, Vaida F, Clifford DB, CNS HIV Antiretroviral Therapy Effects Research (CHARTER) Group (2013) Apolipoprotein E4 genotype does not increase risk of HIV-associated neurocognitive disorders. J Neurovirol 19:150–156

Nath A, Maragos WF, Avison MJ, Schmitt FA, Berger JR (2001) Acceleration of HIV dementia with methamphetamine and cocaine. J Neurovirol 7:66–71

Norman MA, Moore DJ, Taylor M, Franklin D Jr, Cysique L, Ake C, Lazarretto D, Vaida F, Heaton RK (2011) Demographically corrected norms for African Americans and Caucasians on the Hopkins verbal learning test-revised, brief visuospatial memory test-revised, stroop color and word test, and wisconsin card sorting test 64-card version. J Clin Exp Neuropsychol 33:793–804

Radtke KK, Bacchetti P, Anastos K, Merenstein D, Crystal H, Karim R, Weber KM, Edmonds A, Sheth AN, Fischl MA, Vance D, Greenblatt RM, Rubin LH (2018) Use of nonantiretroviral medications that may impact neurocognition: patterns and predictors in a large, long-term HIV cohort study. J Acquir Immune Defic Syndr 78:202–208

Richardson JL, Martin EM, Jimenez N, Danley K, Cohen M, Carson VL, Sinclair B, Racenstein JM, Reed RA, Levine AM (2002) Neuropsychological functioning in a cohort of HIV infected women: importance of antiretroviral therapy. J Int Neuropsychol Soc 8:781–793
Richardson JL, Nowicki M, Danley K, Martin EM, Cohen MH, Gonzalez R, Vassileva J, Levine AM (2005) Neuropsychological functioning in a cohort of HIV− and hepatitis C virus-infected women. AIDS 19:1659–1667
Robertson K, Fiscus S, Wilkins J, van der Horst C, Hall C (1996) Viral load and neuropsychological functioning in HIV seropositive individuals: a preliminary descriptive study. J Neuro AIDS 1:7–15
Robertson K, Bayon C, Molina JM, Mcnamara P, Resch C, Munoz-Moreno JA, Kulasegaram R, Schewe K, Burgos-Ramirez A, De Alvaro C, Cabrero E, Guion M, Norton M, Van Wyk J (2014) Screening for neurocognitive impairment, depression, and anxiety in HIV-infected patients in Western Europe and Canada. AIDS Care 26:1555–1561
Royal W 3rd, Cherner M, Burdo TH, Umlauf A, Letendre SL, Jumare J, Abimiku A, Alabi P, Alkali N, Bwala S, Okwuasaba K, Eyzaguirre LM, Akolo C, Guo M, Williams KC, Blattner WA (2016) Associations between cognition, gender and monocyte activation among HIV infected individuals in Nigeria. PLoS One 11:e0147182
Rubin LH, Sundermann EE, Cook JA, Martin EM, Golub ET, Weber KM, Cohen MH, Crystal H, Cederbaum JA, Anastos K, Young M, Greenblatt RM, Maki PM (2014) Investigation of menopausal stage and symptoms on cognition in human immunodeficiency virus-infected women. Menopause 21(9):997–1006
Rubin LH, Cook JA, Weber KM, Cohen MH, Martin E, Valcour V, Milam J, Anastos K, Young MA, Alden C, Gustafson DR, Maki PM (2015a) The association of perceived stress and verbal memory is greater in HIV-infected versus HIV-uninfected women. J Neurovirol 21:422–432
Rubin LH, Meyer VJ, Conant R, Sundermann EE, Wu M, Weber KM, Cohen MH, Little DM, Maki PM (2015b) Prefrontal cortical volume loss is associated with stress-related deficits in verbal learning and memory in HIV-infected women. Neurobiol Dis 92(Pt B):166–174
Rubin LH, Pyra M, Cook JA, Weber KM, Cohen MH, Martin E, Valcour V, Milam J, Anastos K, Young MA, Alden C, Gustafson DR, Maki PM (2016a) Post-traumatic stress is associated with verbal learning, memory, and psychomotor speed in HIV-infected and HIV-uninfected women. J Neurovirol 22:159–169
Rubin LH, Wu M, Sundermann EE, Meyer VJ, Smith R, Weber KM, Cohen MH, Little DM, Maki PM (2016b) Elevated stress is associated with prefrontal cortex dysfunction during a verbal memory task in women with HIV. J Neurovirol 22(6):840–851
Rubin LH, Wu M, Sundermann EE, Meyer VJ, Smith R, Weber KM, Cohen MH, Little DM, Maki PM (2016c) Elevated stress is associated with prefrontal cortex dysfunction during a verbal memory task in women with HIV. J Neurovirol 22(6):840–851
Rubin LH, Benning L, Keating SM, Norris PJ, Burke-Miller J, Savarese A, Kumanan KN, Awadalla S, Springer G, Anastos K, Young M, Milam J, Valcour VG, Weber KM, Maki PM (2017a) Variability in C-reactive protein is associated with cognitive impairment in women living with and without HIV: a longitudinal study. J Neurovirol 24(1):41–51
Rubin LH, Cook JA, Springer G, Weber KM, Cohen MH, Martin EM, Valcour VG, Benning L, Alden C, Milam J, Anastos K, Young MA, Gustafson DR, Sundermann EE, Maki PM (2017b) Perceived and post-traumatic stress are associated with decreased learning, memory, and fluency in HIV-infected women. AIDS 31:2393–1401
Rubin LH, Maki PM, Springer G, Benning L, Anastos K, Gustafson D, Villacres MC, Jiang X, Adimora AA, Waldrop-Valverde D, Vance DE, Bolivar H, Alden C, Martin EM, Valcour VG, Women's Interagency HIV Study (2017c) Cognitive trajectories over 4 years among HIV-infected women with optimal viral suppression. Neurology 89:1594–1603
Rubin LH, Phan KL, Keating SM, Weber KM, Maki PM (2017d) Brief report: low-dose hydrocortisone has acute enhancing effects on verbal learning in HIV-infected men. J Acquir Immune Defic Syndr 75:e65–e70

Rubin LH, Phan KL, Keating SM, Maki PM (2018a) A single low-dose of hydrocortisone enhances cognitive functioning in HIV-infected women. AIDS 32:1983–1993

Rubin LH, Radtke KK, Eum S, Tamraz B, Kumanan KN, Springer G, Maki PM, Anastos K, Merenstein D, Karim R, Weber KM, Gustafson D, Greenblatt RM, Bishop JR (2018b) Cognitive burden of common non-antiretroviral medications in HIV-infected women. J Acquir Immune Defic Syndr 79(1):83–91

Spies G, Fennema-Notestine C, Cherner M, Seedat S (2017) Changes in cognitive function in women with HIV infection and early life stress. AIDS Care 29:14–23

Stern RA, Arruda JE, Somerville JA, Cohen RA, Boland RJ, Stein MD, Martin EM (1998) Neurobehavioral functioning in asymptomatic HIV-1 infected women. J Int Neuropsychol Soc 4:172–178

Sundermann EE, Bishop JR, Rubin LH, Little DM, Meyer VJ, Martin E, Weber K, Cohen M, Maki PM (2015) Genetic predictor of working memory and prefrontal function in women with HIV. J Neurovirol 21:81–91

Sundermann EE, Heaton RK, Pasipanodya E, Moore RC, Paolillo EW, Rubin LH, Ellis R, Moore DJ, HNRP Group (2018) Sex differences in HIV-associated cognitive impairment. AIDS 32:2719–2726

UN Women (2018) Facts and figures: HIV and AIDS [Online]. http://www.unwomen.org/en/what-we-do/hiv-and-aids/facts-and-figures#notes. Accessed 15 Aug 2018

Valcour V, Shikuma C, Shiramizu B, Watters M, Poff P, Selnes OA, Grove J, Liu Y, Abdul-Majid KB, Gartner S, Sacktor N (2004) Age, apolipoprotein E4, and the risk of HIV dementia: the Hawaii Aging with HIV Cohort. J Neuroimmunol 157:197–202

Valcour V, Maki P, Bacchetti P, Anastos K, Crystal H, Young M, Mack WJ, Cohen M, Golub ET, Tien PC (2012) Insulin resistance and cognition among HIV-infected and HIV-uninfected adult women: the Women's Interagency HIV Study. AIDS Res Hum Retroviruses 28:447–453

Valcour V, Rubin LH, Tien P, Anastos K, Young M, Mack W, Cohen M, Golub ET, Crystal H, Maki PM (2015) Human immunodeficiency virus (HIV) modulates the associations between insulin resistance and cognition in the current combination antiretroviral therapy (cART) era: a study of the Women's Interagency HIV Study (WIHS). J Neurovirol 21:415–421

Valcour VG, Rubin LH, Obasi MU, Maki PM, Peters MG, Levin S, Crystal HA, Young MA, Mack WJ, Cohen MH, Pierce CB, Adimora AA, Tien PC (2016) Liver fibrosis linked to cognitive performance in HIV and hepatitis C. J Acquir Immune Defic Syndr 72(3):266–273

Wendelken LA, Jahanshad N, Rosen HJ, Busovaca E, Allen I, Coppola G, Adams C, Rankin KP, Milanini B, Clifford K, Wojta K, Nir TM, Gutman BA, Thompson PM, Valcour V (2016) ApoE epsilon4 is associated with cognition, brain integrity, and atrophy in HIV over age 60. J Acquir Immune Defic Syndr 73:426–432

Wojna V, Skolasky RL, Hechavarria R, Mayo R, Selnes O, Mcarthur JC, Melendez LM, Maldonado E, Zorrilla CD, Garcia H, Kraiselburd E, Nath A (2006) Prevalence of human immunodeficiency virus-associated cognitive impairment in a group of Hispanic women at risk for neurological impairment. J Neurovirol 12:356–364

Cultural Neuropsychology Considerations in the Diagnosis of HIV-Associated Neurocognitive Disorders

Monica G. Rivera Mindt, Desiree A. Byrd, Emily P. Morris, Kayla Tureson, Vanessa Guzman, Angela C. Summers, Cara Crook, Micah J. Savin, and Maral Aghvinian

Contents

M. G. Rivera Mindt (✉), E. P. Morris, K. Tureson, V. Guzman, A. C. Summers, C. Crook, M. J. Savin, and M. Aghvinian
Department of Psychology, Fordham University, New York, NY, USA

Department of Neurology, Icahn School of Medicine at Mount Sinai, New York, NY, USA
e-mail: riveramindt@fordham.edu; emorris24@fordham.edu; ktureson@fordham.edu; vguzman4@fordham.edu; asummers4@fordham.edu; ccrook@fordham.edu; msavin@fordham.edu; maghvinian@fordham.edu

D. A. Byrd
Department of Neurology, Icahn School of Medicine at Mount Sinai, New York, NY, USA

Department of Psychology, Queens College, Queens, NY, USA
e-mail: desiree.byrd@mssm.edu

Curr Topics Behav Neurosci (2021) 50: 193–224
DOI 10.1007/7854_2019_121
Published Online: 11 March 2020

Abstract Human Immunodeficiency Virus Type-I (HIV) is a health disparities issue that affects culturally and linguistically diverse (CALD) and underrepresented minority populations to a greater degree than non-Hispanic white populations. Neurologically speaking, CALD populations experience worse HIV-related health outcomes, which are exacerbated by inadequate neurocognitive measures, poor normative samples, and the complex interplay of sociocultural factors that may affect test interpretation. Although cross-cultural neuropsychologists are working diligently to correct this gap in the literature, currently, studies examining neurocognitive outcomes among CALD populations are sparse. The most well-studied CALD groups are of African American/Black and Latinx adults in the US, and the chapter therefore focuses on these studies. There is more limited work among other populations in the US, such as Asians, Native Hawaiians, Pacific Islanders, and American Indians/Alaskan Natives, and even fewer studies for many CALD populations outside of the US. For example, HIV neuropsychology data is rare or nonexistent in the First Peoples of Australia and Indigenous People of Canada. It is often not adequately reported in Europe for the migrant populations within those countries or other world regions that have historically large multicultural populations (e.g., South America, Caribbean countries, Asia, and Africa). Therefore, this chapter reviews HIV-related health disparities faced by CALD populations with focus on North American research where it has been specifically studied, with particular attention given to disparities in HIV-Associated Neurocognitive Disorders (HAND). International data was also included for research with focus on First Peoples of Australia and Indigenous People of Canada. The chapter also examines other sociocultural and health factors, including global and regional (e.g., rural versus urban) considerations, migration, and gender. Further, guidelines for incorporating sociocultural consideration into assessment and interpretation of neurocognitive data and HAND diagnosis when working with HIV-positive CALD populations that would be relevant internationally are provided.

Keywords Culture · Health disparities · HIV · HIV-associated neurocognitive disorder · Racial/ethnic

1 Introduction

Human Immunodeficiency Virus Type-I (HIV) disproportionately affects culturally and linguistically diverse (CALD) populations (e.g., African American and Latinx), who are frequently medically underserved and lack representation in research. Further, HIV-Associated Neurocognitive Disorder (HAND) continues to affect a large portion of HIV-positive adults, despite significant advances in antiretroviral treatments. Given the disproportionate nature in which CALD populations are affected by HIV, it is essential to consider sociocultural influences of neurocognitive dysfunction related to HIV. Accurate neurocognitive assessment and interpretation among CALD populations present a particular challenge in the context of HIV infection because of the influence of culture on neurocognitive function. Despite these challenges, it is essential to strive for culturally responsive evaluations and interventions in CALD HIV-positive populations to best serve those most disproportionately affected by HIV.

This chapter aims to critically analyze the health disparities in epidemiology and neurocognitive assessment relevant to HAND according to the Frascati criteria (described below).[1] Importantly, this chapter will include information and literature regarding health disparities, epidemiology, and sociocultural factors central to neurocognitive assessment and diagnosis of HAND. Throughout this chapter, we will review the literature relevant to diagnosing HAND in CALD populations. This chapter will further present a critique of the literature and propose future directions for research regarding neurocognitive assessment and HAND in CALD populations.

The majority of literature covered in this chapter has been conducted among CALD populations in the US – primarily among African American/Black or Latinx adults – but international research is also discussed to the extent possible. After reading the chapter, readers should be more familiar with the integration of sociocultural factors in CALD populations that influence HAND diagnosis.

2 Health Disparities in HIV

Despite significant advancements in HIV treatment and care, HIV-related health disparities remain prevalent, especially among underrepresented minority populations. HIV-related health disparities are associated with disproportionate rates of infection as well as higher rates of morbidity, mortality, and adverse opportunistic infections (Cargill and Stone 2005; Chu and Selwyn 2008; Earnshaw et al. 2015). As a result, HIV-related health disparities often lead to worse health outcomes which may increase risk for HAND, particularly among those from CALD populations due to greater disease burden and severity (Chu and Selwyn 2008; McArthur et al. 2010). Research in cultural neuropsychology emphasizes the importance of HIV-related health disparities in the treatment and evaluation of HIV-positive adults. Therefore, we will provide an overview of salient health disparities relevant to HAND as well as current findings and recommendations for the diagnosis of HAND in CALD groups, including African American/Black, Latinx, Native Hawaiian/Other Pacific Islanders, and American Indian/Alaskan Native populations.

[1]Note: All subsequent references to HAND refer to the neurocognitive diagnosis using the Frascati criteria.

2.1 African American/Black and Latinx Populations

African American/Black and Latinx populations are disproportionately affected by HIV in the United States (US). They comprise approximately 30% of the total US population, but account for 70% of those currently living with HIV in the US (Centers for Disease Control and Prevention 2017). Increased risk of infection, as well as higher rates of transmission, are observed among African American/Black and Latinx populations, compared with their non-Hispanic white counterparts (Gant et al. 2017; Laurencin et al. 2018). Regarding treatment and care, African American/Black and Latinx HIV-positive adults report experiencing more barriers accessing treatment and poorer quality healthcare (Cargill and Stone 2005; Laurencin et al. 2018; Oramasionwu et al. 2009). They also demonstrate lower rates of optimal medication adherence and viral suppression (Cargill and Stone 2005; McArthur et al. 2010; Centers for Disease Control and Prevention 2017; Laurencin et al. 2018; Oramasionwu et al. 2009; Johnson et al. 2003; Rivera Mindt et al. under review). As a result, African American/Black and Latinx HIV-positive adults often exhibit greater risk and rates of both HIV-related comorbid conditions and non-HIV-related disorders, many of which have also been associated with increased risk of neurocognitive impairment and/or HAND (Barbaro and Klatt 2002; Goodkin et al. 2017; Goulet et al. 2007; Office of Minority Health 2017). Taken together, the health disparities and significantly worse health outcomes among African American/Black and Latinx HIV-positive adults have significant implications for public health and policy.

2.2 Asian/Native Hawaiian/Other Pacific Islanders

Rates of HIV among Asian populations in the US are low. In 2016, Asians represented 6% of the US population, but only 2% of new HIV diagnoses (Centers for Disease Control and Prevention 2017), and incidence has remained stable since 2010 (Centers for Disease Control and Prevention 2018b). Nevertheless, there are important caveats regarding HIV epidemiology in this population. Incidence among this diverse group is variable. From 2011 to 2015, rates of gay and bisexual HIV-positive Asian men increased by 35% (Centers for Disease Control and Prevention 2018b). Rates are also rising in Native Hawaiians and Other Pacific Islanders, who represent only 0.2% of the population (Centers for Disease Control and Prevention 2018c). From 2011 to 2015, HIV diagnoses among Native Hawaiian/Other Pacific Islanders increased by 51% overall and 50% among gay and bisexual men (Centers for Disease Control and Prevention 2018c). Moreover, HIV-positive Asian American and Native Hawaiian/Other Pacific Islanders have limited healthcare access and utilization. In 2014, only 57% of HIV-positive Asian American adults received HIV-related medical care, while 51% reached viral suppression, and just 46% were retained in care (Centers for Disease Control and Prevention 2018b, c). These factors not only bias prevalence and incidence estimates, but also worsen prognosis, including neurocognitive outcomes. In sum, while overall HIV prevalence and incidence may be low in the Asian American population, these numbers fail to capture subgroup health disparities and suboptimal healthcare.

2.3 *American Indians/Alaskan Natives*

Epidemiologic studies among American Indians/Alaskan Natives likely underestimate HIV rates, highlighting the importance of community-based participatory research (Walker et al. 2015; Bertolli et al. 2004). Regardless, HIV remains a significant public health issue for American Indians/Alaskan Natives (Centers for Disease Control and Prevention 2018a). While prevalence rates estimate that American Indians/Alaskan Natives constitute 1% of all HIV-positive adults in the US (Centers for Disease Control and Prevention 2017), incidence rates from 2010 to 2016 rose among American Indians/Alaskan Natives from 46 to 81% among Two-Spirit[2] American Indians/Alaskan Natives (Centers for Disease Control 2018a). American Indians/Alaskan Natives may be at increased risk for HIV given elevated rates of other STDs, condomless sex, and intravenous drug use in these communities (Eng and Butler 1997; Cohen et al. 1997; McClelland et al. 2001; Nelson et al. 2011), as well as an increased likelihood to be younger, less educated, unemployed, and of lower socioeconomic status (Johnson et al. 1989; Vernon 2007).

Stark clinical disparities in HIV are reported among American Indians/Alaskan Natives and demonstrate the deleterious implications of barriers to healthcare access and utilization. Among HIV-positive American Indians/Alaskan Natives in 2014, only 60% received medical care for HIV. Of those that received such care, only 48% reached viral suppression and 43% were later retained in care (Centers for Disease Control 2018a). Furthermore, American Indians/Alaskan Natives had the poorest HIV survival rates of any US racial/ethnic group from 2010 to 2015 (Centers for Disease Control and Prevention 2015). Marginalization of Two-Spirit American Indians/Alaskan Natives may partially contribute to existing clinical disparities in HIV. Thirty-four percent of Two-Spirit American Indians/Alaskan Natives report being denied medical care, and 65% report avoiding seeking care due to fear of discrimination (Fitzgerald et al. 2012). Current screening, treatment, and follow-up practice in American Indians/Alaskan Natives are suboptimal, and clearly demonstrate the need to integrate clinical care of Two-Spirit American Indians. Given the aforementioned disparity in HIV clinical outcomes and their negative effects upon neurocognitive sequelae, the paucity of research surrounding the prevalence of HIV-associated neurocognitive disorders among American Indians/Alaskan Natives

---

[2]Two-Spirit AI/AN American Indian/Alaskan Native identity herein is utilized as an umbrella term referring to a third, unique gender and LGBT Native-specific identity. The Two-Spirit identity has been formally recognized by the National Congress of American Indians (Resolution #MSP-15-047, 2015).

is remarkable (Centers for Disease Control 2018a; Mitsch et al. 2017). To our knowledge, no studies to date have specifically examined HAND among HIV-positive American Indians/Alaskan Natives populations, presenting a significant gap in the literature. For an outstanding overview of neurocognitive assessment and intervention considerations in this population, please see Verney et al. (2016).

2.4 First Peoples of Australia

The available epidemiological data on HIV in First Peoples of Australia (Aboriginals and Torres Strait Islanders) suggest pronounced health disparities in Australia, similar to those seen among American Indians/Alaskan Natives of the United States.

There were an estimated 582 people living with HIV in Australia at the end of 2017 who were reported to be Aboriginal and Torres Strait Islander at the time of HIV diagnosis (McGregor et al. 2018). The proportion of undiagnosed HIV infection is higher in Aboriginal and Torres Strait Islander people (14%) than in the Australian-born non-Indigenous population (10%). HIV prevalence among Aboriginal and Torres Strait Islander people was estimated to be 0.11% in 2017 versus 0.07% among Australian-born non-indigenous people in 2017. HIV incidence rates rose by 33% among First Peoples of Australia from 2012 to 2016. While this concerns a small number of people, this contrasts with a 22% decline in HIV incidence rates among non-Indigenous Australians (Australian Federation of AIDS Organisations 2018). It should be noted that trends in the proportion of HIV notifications classified as newly acquired need to be interpreted with caution as they could reflect increases in regular testing, rather than actual changes. Furthermore, given that trends in HIV notification rates are based upon small numbers of remote and focally located Indigenous Australians, it remains likely that these statistics are representative of local prevalence and underestimate national HIV prevalence of Indigenous Australians. Rates of HIV transmission greatly differ by gender and route of infection between First Peoples of Australia and non-Indigenous Australians (Wright et al. 2005), such that 53% of newly diagnosed cases of HIV were female among First Peoples of Australia and 41% of these cases were attributed to heterosexual contact or intravenous drug use (Australian Federation of AIDS Organisations 2018; Giele 2004). This starkly differs from newly diagnosed cases of non-Indigenous Australians, of which 92% were male and only 21% were attributed to heterosexual contact or intravenous drug use (Australian Federation of AIDS Organisations 2018; Giele 2004). Also, even if HIV is relatively rare in indigenous Australians, at the individual level, HIV infection would add a significant health burden to those infected given disproportionately worse morbidity and mortality outcomes among First Peoples of Australia compared to non-Indigenous Australians. For example, this is reflected in Hepatitis C infection rates among First Peoples of Australia. In 2017, age-standardized rates of hepatitis C notification were four times as high among the Aboriginal and Torres Strait Islander population (168.1 per 100,000) as in the non-Indigenous population (38.4 per 100,000). Rates of hepatitis C diagnosis among Aboriginal and Torres Strait Islander people have increased by 15%, from 146.4 per 100,000 in 2013 to 168.1 per 100,000 in 2017, but there has been a 7% decline in the last year, from 180.4 per 100,000 in 2016

(Welfare AIoHa 2019). With regard to mortality, life expectancy between First Peoples of Australia is 7.8 years shorter than non-Indigenous Australians (Welfare AIoHa 2019).

Disparate HIV clinical outcomes exist between First Peoples of Australia and their non-Indigenous counterparts. For example, HIV-positive First Peoples of Australia possess a higher likelihood to have increased barriers to healthcare services, an advanced disease stage at diagnosis, progression to AIDS after diagnosis, and one or more comorbid conditions (Wright et al. 2005). Furthermore, one study suggests a high likelihood that HIV-positive First Peoples of Australia forego treatment with combination antiretroviral therapy (cART; 45% were untreated) due to fear of disclosure and/or discrimination, poverty, heavy alcohol use, and/or a need to maintain collectivistic social norms (Newman et al. 2007). Given these disparities, culturally tailoring treatments in a way that will engage HIV-positive First Peoples of Australia is of utmost importance. To the best of our knowledge, there has been no specific study on HAND among First Peoples of Australia. However, given the lack of cART uptake among HIV-positive First Peoples of Australia and their overall worse HIV clinical outcomes, HAND may affect this population to a greater degree than non-Indigenous HIV-positive Australians. Furthermore, gender disparities in HAND may be evident given increased intravenous drug use among female First Peoples of Australia.

2.5 Indigenous People of Canada

Indigenous People of Canada have been subjected to systematic discrimination and historical oppression including genocide, forced migration from sacred lands, residential schooling, and pronounced disparities in HIV (Jaworsky et al. 2016). These adverse events are thought to partially explain striking rates of HIV among Indigenous People of Canada. Indigenous People of Canada comprised nearly 80% of newly diagnosed cases in Saskatchewan and were twice as high as the national HIV rate in Canada (Ministry of Health Population Health Branch 2016). HIV transmission rates among Indigenous People of Canada also associate strongly with physical and sexual abuse in residential schools (Jackson and Reimer 2008). After considering sociocultural factors, HIV disproportionately affects Indigenous Peoples of Canada such that higher prevalence of HIV is observed among Indigenous illicit drug users, homeless youth, and female sex workers compared to their non-Indigenous peers (Duncan et al. 2011). Robust disparities are also observed in HIV clinical outcomes. For example, those of Aboriginal ancestry in British Columbia were found to be twice as likely to be virally unsuppressed from 1997 to 2010. Further, the Canadian Observational Cohort (CANOC) study found that HIV-positive Indigenous People of Canada experienced shorter life expectancy compared to their non-Indigenous counterparts (Patterson et al. 2015). Collectively, this literature suggests that the HIV epidemic continues to grow among Indigenous Peoples of Canada, and highlights the need for culturally responsive prevention and care interventions that leverage Elders and other key leaders in community-based participatory research (Flicker et al. 2015).

3 HAND in Culturally Diverse Populations

3.1 HAND Diagnosis

With advances in cART, the severity of HAND decreased over the past few decades, but prevalence estimates of HAND remain high, ranging from 25 to 78% of HIV-positive adults (Cysique and Brew 2011; Heaton et al. 2010; Marquine et al. 2018; Rivera Mindt et al. 2008a; Sacktor et al. 2016; Wojna and Nath 2006). Currently, the gold standard for HAND diagnosis is the Frascati Criteria, which requires impairment in a minimum of two domains (out of at least five domains) that is not better explained by another condition (Antinori et al. 2007; Tierney et al. 2017). However, a new international working group has been tasked with improving the methods used to diagnose HAND in CALD populations. This is critical because although several studies include diverse populations of HIV-positive adults, many of these studies often report only basic demographic information (e.g., age, years of education, gender, and race/ethnicity; Heaton et al. 2010; Molsberry et al. 2018). These studies do not account for other sociocultural factors (see details below) that are well described in the literature as affecting performance on neurocognitive assessments. This is concerning because of the increasing rates of HIV-positive adults from CALD populations both in the US and abroad. Thus, the need for improved culturally sensitive HAND diagnostic criteria that take into consideration relevant sociocultural factors are critical because CALD populations may be at higher risk for HAND than non-Hispanic white HIV-positive adults (Marquine et al. 2018; Rivera Mindt et al. 2014).

HAND in HIV-Positive African American/Black Adults The pervasiveness of neurocognitive impairment and HAND in African American/Black HIV-positive individuals is at least partially dependent on the use of appropriate normative data. When applying demographically corrected norms based on African American/Black samples, Antinori et al. (2007) found that the prevalence of neurocognitive impairment for African American/Black HIV-positive adults was 44%. In contrast, the use of non-Hispanic white normative data for African American/Black people resulted in a striking 71% prevalence of neurocognitive impairment. These findings suggest that there may be significant variability in HAND prevalence estimates for CALD populations depending on the quality of the normative data used, such as those seen in African American/Black adults. Therefore, results must be interpreted with caution. Though research findings on rates of HAND in this population have been somewhat equivocal, findings generally suggest that HIV-positive African American/Black adults exhibit similar rates of HAND compared to their non-Hispanic white counterparts when demographically appropriate normative data are applied (Heaton et al. 2010, 2011; Robertson et al. 2007). These findings suggest that being African American/Black is not associated with an increased risk of HAND or neurocognitive impairment. Nonetheless, higher rates of comorbid conditions have been found among HIV-positive African American/Black adults (Marquine et al. 2016; Rawlings and Masters 2008). Therefore, it is possible that this group may be at heightened risk for neurocognitive impairment, especially as they age. Based on these findings, future research examining the impact of comorbid conditions on the aging process and the risk of HAND development may be particularly necessary for HIV-positive African American/Black adults since they may exhibit unique vulnerabilities to the adverse effects that comorbid conditions have on neurocognitive functioning (e.g., metabolic syndrome and renal disease).

HAND in HIV-Positive Latinx Adults Minimal research regarding neurocognitive function in Latinx HIV-positive adults has been conducted, despite the health disparities that exist in the prevalence of HIV (Centers for Disease Control and Prevention 2016). Recent studies indicate that Latinx HIV-positive adults living in the US experience increased rates of HAND in comparison to non-Hispanic white and African American/Black HIV-positive adults (Marquine et al. 2018; Rivera Mindt et al. 2014). Between 54 and 78% of Latinx HIV-positive adults experience HAND, with the highest prevalence found in Puerto Rican HIV-positive adults (Marquine et al. 2018; Rivera Mindt et al. 2008a; Wojna and Nath 2006). Further, the longitudinal CNS HIV Antiretroviral Therapy Effects Research (CHARTER) Study found that Latinx ethnicity was the only race or ethnicity that predicted neurocognitive decline over time (Heaton et al. 2015). Importantly older Latinx HIV-positive adults appear to be at greater risk for HAND than both younger Latinx HIV-positive adults and non-Hispanic white HIV-positive adults (Rivera Mindt et al. 2014). The limited but meaningful research conducted among Latinx HIV-positive adults demonstrates the increased risk for and prevalence of HAND in these populations and requires further investigation to better serve these populations disproportionately affected by HIV.

HAND in HIV-Positive Asian/Native Hawaiian/Other Pacific Islanders HAND research in US Asian, Native Hawaiian, and Other Pacific Islanders populations is sparse. To date, only one large-scale study, the Hawaiian Aging with HIV-1 Cohort Study, included Asians/Other Pacific Islanders in a representative manner (Valcour et al. 2004a). Unfortunately, although 32% of the sample was Asian/Other Pacific Islanders, Native Hawaiian representation in the sample was not reported. Further, the use of mainland US-based, non-Hispanic white normative data from the Multicenter AIDS Cohort Study (MACS; Selnes et al. 1991) suggests that caution is warranted regarding the generalizability of the neurocognitive implications for this study (Valcour et al. 2004a). Later studies using the same cohort (Shimizu et al. 2011; Valcour et al. 2004b, 2006) have similar limitations with respect to the application to CALD populations. This is especially problematic because, unlike Valcour et al.'s (2004a) study, both Valcour et al.'s (2006) study and Shimizu et al. (2011) found a significant association between minority status and poorer neurocognitive outcomes.

Per global reports, HIV affects 8.3 million people in Asia-Pacific countries, all of whom are at risk for HAND (Zhou et al. 2005). Over the last 15 years, research in these regions highlighted the fact that Asian/Other Pacific Islanders populations are significantly burdened by HAND; see Ian et al. (2016) for a review. Rates of neurocognitive impairment and/or HAND vary widely among Asia-Pacific regions. Estimates range from 12% in the Asia-Pacific NeuroAIDS Consortium (APNAC; Wright et al. 2008), 26% in a South Korean cohort, 29% memory impairment in HIV-positive adults from Hong Kong (Au et al. 2008), 48% in a Chinese cohort of HIV-positive adults (Wang et al. 2013), to 61% in a South Asian (Indian) sample (Gupta et al. 2007). Wang et al. (2013) also identified several HAND risk factors, including older age, female gender, lower education, longer duration of cART, and taking Efavirenz for 2 years or more.

Among these HIV-positive Asian/Other Pacific Islanders/Native Hawaiian populations healthcare barriers (e.g., limited cognitive screening and treatment) often exacerbate neurocognitive and psychological impairment. Moreover, host and viral trait variation within and across these populations confound HAND estimates and prognosis. Unlike in the US and Europe, where HIV viral clade B predominates, clades A, C, and D dominate in Asia, with clade D demonstrating greater neurovirulence (Gopukumar et al. 2008; Sacktor et al. 2009). In addition, Asian populations are particularly susceptible to co-occurring infections. Neurotropic infections (e.g., toxoplasmosis, cryptococcus, and malaria) can result in residual neurocognitive impairment that may be difficult to differentiate from HAND (Sacktor et al. 2001). Hepatitis C is also common in HIV-positive Asian populations, which may negatively impact disease progression, cART responsiveness, and neurocognition (Letendre et al. 2010; Zhou et al. 2007). While Gupta et al. (2007) found no difference in neurocognitive dysfunction among HIV-positive adults with and without Hepatitis C, Heaton et al. (2008) observed higher rates of neurocognitive impairments in those with comorbid HIV and Hepatitis C compared to those HIV, but not Hepatitis C.

HAND in "Non-White" HIV-Positive Adults Other studies investigating disparities in HAND between racial and ethnic groups have taken a broader approach by grouping all CALD participants together in a "non-white" group to conduct comparisons with a non-Hispanic white reference group. Cysique et al. (2011) conducted one such study to create normative data to predict neurocognitive change in both non-Hispanic white and CALD HIV-positive adults from the CHARTER study and the greater San Diego area. Their large sample consisted of a diverse group of both HIV-positive and HIV-negative adults ($N = 296$; 42% African American). Results demonstrated that membership in the "non-white" or CALD group was significantly predictive of neurocognitive change across six assessment periods (Cysique et al. 2011). However, a more recent CHARTER study contradicted this, finding that "non-white" ethnicity was not associated with neurocognitive impairment in HIV-positive adults with neurocognitive decline over a 3-year period (Brouillette et al. 2016). Therefore, grouping CALD participants together in one "non-white" group may mask subgroup variations and limit the external validity of the results, further highlighting the importance of using appropriate normative data for each CALD population.

3.2 *Risk for Misdiagnosis of HAND in CALD Groups*

The Frascati criteria are well validated for the diagnosis of HAND (Blackstone et al. 2012; Weber et al. 2013), yet as Woods et al. (2004) highlighted, as case complexity increases, the diagnostic validity of the Frascati criteria decreases. In such complex cases (e.g., multiple comorbid conditions, substance use, and low literacy), the term Neuropsychological Impairment-Other is used to indicate that the underlying etiology of the observed neurocognitive impairment is difficult to discern (Woods et al. 2004). The increased risk of such comorbid conditions among CALD populations, coupled with the limited utility of many neurocognitive tools and normative samples, suggests that the term Neuropsychological Impairment-Other may be more appropriate and/or prevalent in these groups compared to their non-Hispanic white counterparts. Therefore, special considerations for the evaluation of HAND in CALD HIV-positive adults are presented below.

Comorbidities According to the Frascati criteria, comorbid conditions that affect neurocognitive function should be considered, and no diagnosis of HAND should be given if the impairment is better explained by another condition (Antinori et al. 2007). However, clinical judgment is necessary to make this diagnostic determination, which can greatly reduce interrater reliability, which is highlighted by a National NeuroAIDS Tissue Consortium (NNTC) study. The NNTC employs a standardized protocol for diagnosing HAND among a diverse group of HIV-positive adults. The authors found that, among HIV-positive participants with few or no confounding comorbidities, there was a high degree of interrater reliability in HAND diagnosis ($\kappa = 0.83$), but interrater reliability dropped ($\kappa = 0.23, p < 0.001$) among those with more comorbidities. Additionally, there was an inverse relationship between the interrater reliability and number of participant comorbidities (Woods et al. 2004). The increased prevalence of conditions that are associated with neurocognitive impairment among CALD populations (e.g., Hepatitis-C, stroke, diabetes, poorly managed heart disease; Brickman et al. 2008; Foley et al. 2010; Office of Minority Health 2016) make the results of the NNTC study particularly salient for CALD populations. Accounting for comorbid conditions when interpreting observed patterns in a CALD patient's (or participant's) neurocognitive profile will likely help to improve diagnostic formulations and differential diagnosis.

Construct Validity Because tests do not always measure the same construct across cultures, careful test selection is critical for the evaluation of HIV-positive CALD populations, particularly in international assessments (Helms 1992; International Test Commission (ITC) 2013). Numerous factors can systematically confound performance on neurocognitive measures across cultures, including gender roles, test-wiseness and test taking attitude, and the concepts of time and speed (Ardila 2005; Manly 2008; Rivera Mindt et al. 2010). Therefore, to reduce the probability of misdiagnosing HAND, ensuring that test selection is based on measures with established construct validity for the population or person being evaluated is essential to the assessment and diagnostic process.

Normative Data Currently, the lack of established normative data for the assessment of CALD individuals presents an ethical issue of neurocognitive evaluation and test interpretation (Rivera Mindt et al. 2010). While the field of neuropsychology increasingly recognizes the importance of age, education, gender, and race/ethnicity normative corrections to better understand a person's performance relative to his/her peers (Rivera Mindt et al. 2008a; Manly et al. 2002, 2004), many of these demographically corrected normative samples do not adequately represent CALD populations (Manly et al. 2002, 2004; Dotson et al. 2009). Fortunately, cross-cultural neuropsychologists have aimed to correct this and there is a growing body of published normative datasets for CALD populations, which are reviewed in greater detail in Rivera Mindt et al. (2019).

Given the increasing number of HIV-positive CALD people and the health disparities observed in these populations, the issue of appropriate demographically corrected normative samples is particularly salient when assessing HIV-positive adults for HAND (Manly and Echemendia 2007; Ponton et al. 1996). Therefore, selection of measures with appropriate normative samples may increase the ability to diagnose HAND in CALD populations in the US. Outside of the US, selecting tests that have been previously normed on HIV-negative locals will also increase the accuracy of HAND diagnosis, as has been done in several studies (e.g., Ferrett et al. 2014; Ruffieux et al. 2010). Despite best efforts, ideal tests and normative datasets may not always be available. Therefore, those individuals who work with CALD HIV-positive populations should be informed of current literature in cross-cultural neurocognitive assessment, as clinical judgment is an essential component of understanding both clinical and research findings, particularly when appropriate normative data is limited. Further, any and all limitations regarding test selection, normative corrections, and data interpretations should be discussed throughout an assessment report when assessing CALD HIV-positive patients and/or throughout a manuscript when assessing CALD HIV-positive research participants (Rivera Mindt et al. 2010).

4 Sociocultural Assessment Considerations

Sociocultural considerations are a crucial facet of culturally responsive assessment, particularly as populations continue to grow more culturally and linguistically diverse. Despite a strong body of literature demonstrating sociocultural influences on neurocognition, sociocultural factors are seldom included in neurocognitive assessments. While a myriad of sociocultural factors may affect neurocognition, some notable and well-researched sociocultural considerations include quality of education, acculturation, socioeconomic status, discrimination and social adversity, and language. Qualitative information should also inform an assessment with a CALD patient, particularly for those who are less acculturated to majority culture or unfamiliar with the structure of Western standardized tests, wherein the use of traditional standardized tests may be inappropriate.

4.1 *Quality of Education*

Current demographic corrections for education are based on the years of education a person receives. However, this incorrectly assumes that similar educational experiences are received across schools, regions, and racial/ethnic populations. In contrast, quality of education may serve as a more meaningful metric when assessing educational experience, particularly among CALD populations. Quality of education is often operationalized as performance on single word reading tests, such as the Wide Range Achievement Test – Third Edition (WRAT-3; Wilkinson 1993) Reading Subtest or the Wechsler Test of Adult Reading (WTAR; Wechsler 2001). The application of single word reading tests as a proxy for quality of education is validated in prior research (Manly et al. 2002), and highlights the vital role of quality of education in neurocognitive performance of HIV-positive adults.

Quality of education has consistently been shown to mitigate differences in neurocognitive performance between African American/Black and Latinx adults, compared to their non-Hispanic white counterparts in HIV-positive and HIV-negative populations (Rivera Mindt et al. 2014; Manly et al. 2002; Ryan et al. 2005; Sisco et al. 2015). Given that half of African American/Black and Latinx HIV-positive adults may have reading levels below what is expected based on their years of educational attainment, and these underrepresented minorities also have fewer years of education compared to their non-Hispanic white peers, the application of reading level as a proxy for quality of education can significantly reduce rates of impairment among African American/Black and Latinx HIV-positive adults (Ryan et al. 2005). These findings underscore the utility of word reading measures as proxies for quality of education among HIV-positive adults, especially individuals with lower educational attainment, to better assess neurocognitive performance. Furthermore, the National Institute of Health (NIH) Toolbox can serve as a resource for addressing this issue by providing a number of tests for use with patients who are illiterate, such as the NIH Toolbox Picture Vocabulary Test (Weintraub et al. 2013).

4.2 *Acculturation*

Acculturation, the process through which exposure to another culture facilitates behavioral and psychological change, is another important sociocultural consideration (Zea et al. 2003). Most acculturation measures are unidimensional, only assessing acculturation to the majority culture. However, some measures (e.g., the Abbreviated Multidimensional Acculturation Scale) take into account acculturation to both the majority and minority cultures. Numerous acculturation measures have been developed for specific racial/ethnic populations. The African American Acculturation Scale is unidimensional scale for use with Black/African American populations (Landrine and Klonoff 1995). Acculturation measures for Latinx groups include the Bidimensional Acculturation Scale for Hispanics, Short Acculturation

Scale for Hispanics, Acculturation Rating Scale for Mexican Americans, and the Abbreviated Multidimensional Acculturation Scale (Zea et al. 2003; Cuellar et al. 1995; Marin and Gamba 1996; Marin et al. 1987). For US Asian immigrant populations, the Suinn-Lew Self-Identity Acculturation Scale is widely used, including in prior research with HIV-positive Asian-American populations (Chen et al. 2014; Suinn et al. 1987). While normative data to date does not incorporate acculturation, prior research suggests that lower acculturation to the majority culture is associated with worse neurocognitive performance across a number of cognitive domains (Arentoft et al. 2012; Manly et al. 1998). Therefore, acculturation should still be taken into account when working with CALD patients, especially if a participant obtains low scores across a number of tests.

4.3 Socioeconomic Status

Socioeconomic status refers to one's access to economic resources and their social standing in relation to others. The rates of socioeconomic disadvantage are especially high among HIV-positive adults, resulting in higher rates of morbidity and mortality (Cunningham et al. 2005; Ibrahim et al. 2008; Rabkin et al. 2004). These incongruities also significantly contribute to health disparities among CALD populations (Oakes and Rossi 2003; Shavers 2007). Though a gold standard measure of socioeconomic status does not presently exist, one of the most widely used measures is the Hollingshead Index of Social Prestige (Cirino et al. 2002; Hollingshead 1975). In diverse HIV-positive adults, positive associations were found between socioeconomic status and neurocognitive domains of processing speed, verbal fluency, learning, memory, attention, and executive functioning (Arentoft et al. 2015). These findings also indicated that adult socioeconomic status was a significant predictor of HAND and that neurocognitively normal participants had higher overall socioeconomic status (Arentoft et al. 2015). The limited research literature on socioeconomic status at the individual level suggests that socioeconomic status may significantly impact neurocognitive performance among HIV-positive adults (Arentoft et al. 2015).

Socioeconomic status at the community and neighborhood level is generally estimated using median income within each zip code. Among HIV-positive adults, utilizing such broad proxy measures of socioeconomic status to assess the potential relationship between neurocognitive performance has rendered conflicting findings. While Latinx HIV-positive adults typically have lower median socioeconomic status compared to non-Hispanic white groups, using median income by zip code yielded no significant associations between socioeconomic status and domain-specific or global neurocognitive functioning (Rivera Mindt et al. 2008a). Nonetheless, the association between lower socioeconomic status and worse health-related outcomes is well documented among HIV-positive adults (e.g., delayed access to care; Krieger et al. 2005). Lower socioeconomic status at the neighborhood level has been associated with delayed access to HIV care globally, highlighting the pervasiveness of these disparities among HIV-positive adults (Joy et al. 2008). Additional research is necessary to improve the specificity of community-level estimates of socioeconomic status in an effort to better understand the relationship with neurocognitive functioning among HIV-positive adults.

4.4 Discrimination and Social Adversity

CALD populations also experience high rates of discrimination and social adversity, which are associated with adverse health outcomes. Prior research has shown that greater perceived racial/ethnic discrimination is associated with poorer neurocognitive test performance, which may be due to elevated stress resulting from perceived threat (Thames et al. 2013). Social adversity is also linked to poorer learning and memory performance and mental health conditions such as depression (Shonkoff et al. 2012; Williamson et al. 2017). Social adversity has also been linked to structural neural changes among a sample of African American HIV-positive adults, indicating reduced hippocampal and amygdala volume relative to HIV-negative controls (Thames et al. 2018). However, the impact of social adversity as it relates to HAND in HIV-positive Latinx populations has not yet been examined.

4.5 Language

The number of bilinguals and multilinguals in the US continues to grow as CALD populations comprise a greater proportion of the US population. Twenty percent of the US population was bilingual in 2010, though this rate is now likely higher (US Census Bureau 2013). Unfortunately, few measures and guidelines exist for the assessment of a bilingual or multilingual patient. Whenever possible, multilingual clinicians and psychometrists with appropriate cultural and linguistic competency should perform evaluations with bilingual and multilingual patients. The use of interpreters should be avoided unless no preferred alternatives are available, as interpreters can disrupt the integrity of the assessment (Rivera Mindt et al. 2008b). Judd et al. (2009) provide a comprehensive review of ethical and professional guidelines for working with bilingual clients and using interpreters when a bilingual or multilingual clinician is not available.

When working with a bilingual or multilingual patient, a comprehensive review of proficiency in each language should be performed to determine whether the assessment should be conducted in one language versus another, or perhaps both languages if the patient is sufficiently balanced between the two with regard to fluency. Prior research has found language dominance to be associated with global and domain-specific neurocognitive performance in a sample of HIV+ Latinx bilinguals (Miranda et al. 2016). Language dominance assessment is crucial, as CALD patients are already at an elevated risk of being misdiagnosed as cognitively impaired when years of education are used as a metric, as poor test performance with a bilingual may reflect a linguistic artifact (i.e., less fluency in the language in which they were tested). Please refer to Rivera Mindt et al. (2008b) for more detailed information for working with bilingual patients.

A number of studies have found that bilingual participants are disadvantaged on measures of verbal fluency relative to monolinguals, as monolinguals have a hyperproficiency in one language (Bialystok and Luk 2012; Gollan et al. 2002, 2008). However, in other neurocognitive domains (e.g., executive functioning), other studies have found bilinguals to perform better relative to their monolingual counterparts (e.g., Bialystok et al. 2004). Further, bilingualism was believed to be neuroprotective to age-related cognitive decline, though recent research in the US has not shown support for this effect (Yeung et al. 2014; Zahodne et al. 2014).

4.6 Qualitative Approach

A standard neurocognitive evaluation based solely on quantitative data may not be appropriate or sufficient for the evaluation of CALD HIV-positive adults due to a number of sociocultural and linguistic factors (Rivera Mindt et al. 2010). When standard measures are not preferred or are unavailable in the patient's primary language, numerous qualitative considerations should be factored into the evaluation. Employing a process approach to the standard neurocognitive evaluation may clarify a participant's testing approach, test-taking strategies, and response to testing limits (Milberg et al. 2009). Additionally, an adjunctive process approach may aid neuropsychologists in determining whether participants are able to complete the assigned task, despite reaching the standard time limit for the measure.

For additional information, an exhaustive review of professional considerations when working with Latinx populations is available from Judd et al. (2009). Moreover, Rivera Mindt et al. (2010) discuss a number of professional considerations relevant to the development of cultural competency, diversification of neuropsychology, and the importance of appropriate referral sources that are better equipped to work with the population of interest.

5 Regional to Global Considerations and Special Populations in HAND

5.1 Urban vs. Rural Disparities in HIV

US-Based Research Despite the fact that the vast majority of HIV-positive adults in the US live in urban centers, the disparities in HIV prevalence in CALD populations living in rural areas are similar to those in urban areas (Centers for

Disease Control and Prevention 2016). According to the CDC, most CALD populations living in rural areas have higher prevalence of HIV in comparison to non-Hispanic whites (Centers for Disease Control and Prevention 2016). The disparity in HIV prevalence between non-Hispanic white and CALD populations in rural areas may be due to systemic inequality experienced by rural CALD populations in access to health care, HIV prevention education, and social support (Dreisbach 2009).

Rural health disparities in the US extend beyond HIV health disparities and include increased rates of smoking and alcohol consumption (Dreisbach 2009; Hartley 2004; Meit et al. 2018). The stigma and discrimination that CALD HIV-positive adults experience in urban areas may be intensified in rural areas, further accentuating health disparities in HIV. Critically, neurocognitive evaluations and other medical services important for HIV care are lacking in rural areas (O'Bryant et al. 2011). Moreover, on average, HIV-positive adults in rural areas made fewer doctors' visits per year than their urban and suburban counterparts, and African American/Black HIV-positive adults in rural areas were less likely to receive cART (Wilson et al. 2011). To the best of our knowledge, there is no research examining disparities in HAND diagnosis among CALD HIV-positive adults in rural areas, with the majority of research on neurocognitive health in rural HIV-positive adults discussing disproportionately high rates of depression (O'Bryant et al. 2011; Sheth et al. 2009; Vyavaharkar et al. 2011).

International Research Internationally, most research examining neurocognitive function in rural areas shows results similar to those conducted in Western urban areas. In a sample of rural Chinese HIV-positive adults, Heaton and colleagues (Heaton et al. 1995, 2008) found rates of neurocognitive impairment comparable to those observed in urban and rural American HIV-positive adults. Unfortunately, there was no urban Chinese HIV-positive comparison group. A study conducted among rural Zambian HIV-positive adults found rates of neurocognitive impairment similar to those observed in the rural Chinese sample (Heaton et al. 2008; Birbeck et al. 2011). In a different sample of HIV-seronegative Chinese adults, urban dwellers evidenced better neurocognitive performance in fluency, memory, and processing speed than rural dwellers (Gupta et al. 2011). The limited research conducted examining rural and urban differences between HIV-positive adults internationally points to a need for more research in this important area.

Despite the importance of international HAND research, several limitations must be considered, including the fact that there are few studies examining both urban and rural HIV-positive adults from the same country or culture, limiting the generalizability of HAND in international populations. Further, most neurocognitive measures are developed, standardized, and normed according to Western neuropsychological standards; thus, they may not sufficiently translate to international settings. In the context of international HIV research conducted in rural areas, researchers and clinicians must also consider that different clades (i.e., strains) of HIV exist in different areas and that particular clades are more neurotropic than others (Heaton et al. 2008). For this reason, international HAND research findings cannot always be applied to US HIV-positive populations or vice versa. Another consideration in applying international HAND research to US HIV-positive populations is that culture affects neurocognitive functioning (Rivera Mindt et al. 2014; Manly et al. 1998, 2004), and culture cannot be parsed out of HAND research. Regardless, international research is helpful to contextualize neurocognitive differences in rural and urban HIV-positive adults. Additional international research should include measurements of access to health care and various sociocultural factors (e.g., stigma, discrimination, and education) that are related to neurocognitive function in HIV.

5.2 Gender Inequalities and HAND

Inequalities in the US HIV prevalence remains greater among US men than women (Centers for Disease Control and Prevention 2017), but reports of HAND prevalence by gender are variable. In fact, few studies investigated gender differences in neurocognition among HIV-positive adults in the US, but those that have are often marked by null results (Heaton et al. 2011; Tozzi et al. 2007). Despite this, there is a strong, widespread consensus that HIV-positive women face greater psychiatric and psychosocial comorbidities than men (Basso and Bornstein 2000; Farinpour et al. 2003; Lichtenstein et al. 2002; Maki and Martin-Thormeyer 2009; Ramjee and Daniels 2013). This stems, in part, from the fact that women in the US/developed countries experience many of the same inequalities as in developing countries. This is critical in the context of HIV and HAND, as psychosocial and cultural stressors can contribute to disease exposure and related neurocognitive impairment.

Consistent with findings from developing countries (Spies et al. 2016, 2017), HIV-positive women in the US experience disparately high rates of post-traumatic stress disorder, which is known to compromise neurocognition, namely learning, memory, and motor functioning (Machtinger et al. 2012; Rubin et al. 2016; Vance et al. 2016). Indeed, gender inequality and related psychosocial variables in developed and developing countries contribute to HAND burden. Of note, African American/Black women are particularly impacted by HIV and HAND. In 2016, African Americans/Black women made up 61% of the new HIV diagnoses among all US women (Centers for Disease Control and Prevention 2018d), and this population shows greater neurocognitive impairment and decline relative to their HIV-negative counterparts (Cysique and Becker 2015, 2017).

Inequalities Internationally Internationally, women are disparately affected by HIV, yet they have been underrepresented in the literature (Maki and Martin-Thormeyer 2009; Ramjee and Daniels 2013; Joint United Nations Programme on HIV/AIDS (UNAIDS) 2017). Challenges typically associated with developing countries that increase HIV risk (e.g., poverty and income inequality) disproportionately affect women globally. Relatedly, women experience high rates of domestic and gender-based violence that increase HIV risk (Ramjee and Daniels 2013). Regardless of a gender disparity in disease prevalence, societal and cultural pressures on women, both in the US and in the developing countries, may exacerbate the adverse consequences of HIV, including HAND (Ramjee and Daniels 2013).

Extant literature in developing countries is also limited. One noteworthy study in Zambia found a large effect of HIV on neurocognition among HIV-positive women, but not HIV-positive men (Hestad et al. 2012). However, the authors attributed this difference to the sociocultural conditions (e.g., gender inequality, sexual violence, stigma, and discrimination associated with being an HIV-positive woman) and healthcare barriers (e.g., transportation and traditional gender role responsibilities). Such a strong interaction between gender and HIV signifies the importance of studying sociocultural factors in developing countries, as they may exacerbate the neurocognitive and psychosocial consequences of HIV, particularly among women. Indeed, in sub-Saharan Africa, gender-based violence is particularly pervasive and represents one of the leading risk factors of HIV infection among women (Andersson et al. 2008). In addition to risk of physical exposure to HIV, the experience of gender-based violence is associated with post-traumatic stress disorder, adverse childhood events, depression, anxiety, and more, all of which may contribute to the neurocognitive effects of HIV (Spies et al. 2017). While gender differences in HAND prevalence in sub-Saharan Africa are unclear (Hestad et al. 2012; Joska et al. 2010), it is undeniable that HIV-positive women endure considerable psychosocial, structural, and sociocultural stressors that have neurological and neurocognitive consequences (Ramjee and Daniels 2013; Spies et al. 2016, 2017). It is notable that most of the research on gender inequality and HAND in developing countries has been conducted in African countries; therefore, future research should expand the focus to other countries.

5.3 *Migration and HAND*

Environmental considerations, including lack of access to education and medical care, make immigrants, migrants, and refugees vulnerable to HIV infection (International Organization for Migration (IOM) The UN Migration Agency (TUMA) 2017). Minimal research has been conducted on HAND and neurocognitive function in HIV among HIV-positive immigrants, migrants, or refugees. One recent study found that migrants experience worse HIV-related health outcomes than non-migrants, as well as increased stigma and poorly tailored care (Ross et al. 2018). However, this study did not examine neurocognitive function. Research on neurocognitive function in HIV-negative migrant populations identified ambiguous findings regarding the "healthy immigrant" phenomenon (Gonzalez et al. 2009; Haan et al. 2011; Hill et al. 2012a, b; Nguyen et al. 2002).

Researching these vulnerable populations is challenging but vital to better understand HAND in this population, particularly given the current global migration crisis, the known neurocognitive effects of HIV, and the inconclusive findings regarding the impact of immigration on neurocognitive function. One particular challenge in conducting additional research may be due to the difficulty in conducting evaluations in languages other than English and the availability of appropriate norms for these specific populations. Neuropsychologists must take care to provide culturally responsive care to immigrants, migrants, and refugees by using bilingual neuropsychologists and the best available norms when possible. Further guidance regarding conducting ethical and culturally responsive neurocognitive evaluations among CALD and immigrant, migrant, and refugee populations whose first language is not English is available in Sect. 4 of this chapter.

6 Clinical Implications, Translational Aspects, and Future Directions

This chapter outlined health disparities relevant to HIV-positive CALD populations in the US (i.e., African American/Black, Latinx, Asian/Other Pacific Islanders, and American Indian/Alaskan Native populations in the US) and internationally (i.e., First Peoples of Australia and Indigenous People of Canada) as well as among women, immigrants, and migrants. Sociocultural and psychosocial factors impacting neurocognitive function as well as psychometric considerations in the evaluation of HAND among CALD populations were discussed. Many of these CALD populations, particularly African American/Black populations, appear to be at greater risk of comorbid conditions that increase the risk of neurocognitive dysfunction. Research in the US indicates that HIV-positive Latinx adults of Caribbean descent are more likely to be diagnosed with HAND than their counterparts who are non-Hispanic white, non-Hispanic African American/Black, and Latinx of Mexican descent. Among Asian and Other Pacific Islanders, research on HAND is limited and variable, both in the US and globally. However, international studies suggest relatively low rates of HAND among HIV-positive adults in China and South Korea, but relatively high rates of HAND in India. With respect to Native Hawaiian and American Indian/Alaska Native populations, HAND research is sparse.

Overall, this chapter highlights the profound gap in the literature investigating HAND in CALD populations, as well as the difficulty in conducting studies with these populations that are methodologically appropriate. These limitations include a lack of adequate information regarding demographic characteristics of the sample (e.g., country of origin and acculturation), underreporting of HAND/neurocognitive function by race/ethnicity, inadequate normative corrections, omission of sociocultural assessments, and small sample sizes. Despite these limitations, the prevalence of HAND health disparities in certain CALD populations is clear, especially the increased risk of comorbid conditions that may confound and/or exacerbate HAND.

Although this chapter did not review HIV-positive pediatric populations, the authors wish to recognize that HAND is a prominent issue for children in non-Westernized countries, and sub-Saharan African countries, in particular.

Because this chapter could not cover this important topic in the HAND literature, the authors refer readers to Bovin, Giordani, and the American Academy of Pediatric Neuropsychology (Boivin and Giordani 2013).

Clinical Implications and Translational Aspects To address these current gaps in the literature, we summarized sociocultural considerations that may help contextualize neurocognitive function among CALD populations to better determine if observed test performance is better explained by HIV-related impairment, sociocultural and/or linguistic factors, comorbid conditions, or a combination of the three. Current literature indicates that quality of education, acculturation, socioeconomic status, discrimination, social adversity, and language (e.g., multilingualism) are among the most salient factors affecting neurocognitive function. Further, both quantitative and qualitative assessment of these characteristics may aid in the diagnostic process of evaluating HAND in HIV-positive CALD populations. Therefore, the following guidelines are recommended for both clinical- and research-based neurocognitive evaluations with HIV-positive CALD examinees (i.e., clinical patients or research participants).

- Prior to the evaluation, ascertaining as much demographic and medical information as possible is essential so that the examiner (i.e., a clinician or researcher) can better prepare to see the examinee. If the examiner is unfamiliar with the cultural background of the client, the examiner should determine whether he or she is the most appropriate person to conduct the evaluation, consult with colleagues who might be more knowledgeable about the culture, language, and salient sociocultural/historical events. Together, this information can better guide the evaluation.
- Prior to the evaluation, research and prepare the most appropriate measures that address the referral or research question and are validated for use in populations similar to the examinee. This includes finding the best tests and normative data for a particular client based on the cultural, demographic, and medical backgrounds.
- A thorough discussion of a person's qualitative and quantitative educational background is essential to any clinical or research evaluation. To the extent possible, ascertain educational quality, consider the geographical context of the person's schooling (e.g., rural, urban, and suburban; historical sociocultural environment of their schooling), the academic strengths and/or weaknesses experienced during school, and overall performance as a student.
- An objective measure of premorbid function (e.g., a word reading test) should be administered during the evaluation to ascertain the examinee's quality of education.
- Assessing a person's linguistic background is crucial to the evaluation. When possible, assessing multilingual participants in multiple languages can help contextualize their familiarity with each language. Understanding in which language a patient/participant received formal education, the language they currently speak most frequently, and their comfort in each language are essential topics to be covered when an examinee is multilingual. Further, directly asking a patient/participant which language they would prefer to use during the evaluation is important. If the examiner does not speak that language, a referral is likely most

appropriate. For more information on considerations and suggestions for conducting neurocognitive evaluations with bilingual examinees, please refer to Rivera Mindt et al. (2008b).

- Whenever possible, formal measures of sociocultural background (e.g., acculturation and SES) should be included in the assessment of participants from diverse cultural backgrounds, particularly those who are living in a country different than the one where they were born. For resources and guidance on available measures, please refer to Rivera Mindt et al. (2010, 2019).
- Providing accurate diagnoses for examinees becomes more difficult as case complexity increases (e.g., medical comorbidities, low literacy, multilingual patients, and high substance use). Therefore, consulting with colleagues with expertise in evaluating these populations can provide valuable and fresh perspectives on these case considerations.
  - Use caution not to over-pathologize an examinee when cultural or linguistic differences may limit the interpretability of particular measures and/or the evaluation.
  - Similarly, use caution so a diagnosis is not missed because a low pattern of scores was attributed solely to cultural and linguistic differences of the examinee.
  - Synthesizing neurocognitive data with other data sources (e.g., demographic, sociocultural, medical, psychiatric/substance use, and functional) is imperative to identifying a converging pattern of evidence to support the appropriate HAND diagnosis.

- For clinical cases, reports should address the neuropsychological impairments/weaknesses observed during testing in a culturally-tailored manner. The biopsychosociocultural factors that may impact neurocognitive function (e.g., taking prescribed medications, ensuring that they are receiving adequate treatments, social support groups to better cope with chronic illness, stress, other needs, referral to a social worker) should also be considered. However, it is important to keep in mind that recommendations are only as effective as they are realistic. Therefore, providing names and contact information for specific services that might be able to best address the current needs of the patient is critical.
- In research settings, anticipating and listening to the needs of your participants may help establish and/or improve rapport. For instance, providing lunch/snacks may help participants focus on the evaluation if they do not have regular access to food, while providing a list of resources that might be useful to participants (e.g., HIV support groups) may assist participants in accessing much needed care.
- In the patient report and/or research publication, all limitations of the evaluation from a sociocultural perspective (e.g., language, poor fit of norms with individual or sample, SES, and acculturation) should be clearly documented. The way in which these limitations were addressed during the evaluation, interpretation, and diagnostic formulation should be stated.

Future Directions Future research is urgently needed to address the methodological issues highlighted by this review, as well as assurance of demographically representative samples to reflect the current global HIV epidemic in the twenty-first century. This chapter serves as a call to action for all HAND researchers to include demographically representative samples and important demographic and sociocultural characterization of their sample (e.g., race/ethnicity, language fluency, country of origin, education quality, acculturation, SES, and social adversity), and for editors to request this information when reviewing manuscripts. Further, HAND/neurocognitive characteristics should be reported by race/ethnicity and the use of standardized, well-validated, and best available measures and normative datasets should always be used when evaluating CALD populations. Given the importance of sociocultural factors on neurocognitive function, the inclusion of objective sociocultural measures is also important in future studies. For those studies that do include CALD populations, oversampling is recommended to provide adequate power to better characterize HAND in these populations. Similarly, within-group differences research for each CALD population is necessary, rather than studies comparing these groups to HIV-positive non-Hispanic white adults. Finally, the authors wish to remind and emphasize to readers that each CALD population (e.g., Latinx and African American/Black) is, in fact, a *diverse* group of individuals who should not be thought of as one group. They are, by definition, from different cultural and linguistic backgrounds, and therefore HAND characterization within these various populations may be as diverse as the term. By understanding this when assessing patients or research participants, clinicians and researchers alike may better contextualize neurocognitive findings and inform treatment in a culturally responsive manner for these underserved and overburdened CALD populations.

References

Andersson N, Cockcroft A, Shea B (2008) Gender-based violence and HIV: relevance for HIV prevention in hyperendemic countries of southern Africa. AIDS 22(Suppl 4):S73–S86

Antinori A, Arendt G, Becker JT, Brew BJ, Byrd DA, Cherner M et al (2007) Updated research nosology for HIV-associated neurocognitive disorders. Neurology 69(18):1789–1799

Ardila A (2005) Cultural values underlying psychometric cognitive testing. Neuropsychol Rev 15 (4):185–195

Arentoft A, Byrd D, Robbins RN, Monzones J, Miranda C, Rosario A et al (2012) Multidimensional effects of acculturation on English-language neuropsychological test performance among HIV+ Caribbean Latinas/os. J Clin Exp Neuropsychol 34(8):814–825

Arentoft A, Byrd D, Monzones J, Coulehan K, Fuentes A, Rosario A et al (2015) Socioeconomic status and neuropsychological functioning: associations in an ethnically diverse HIV+ cohort. Clin Neuropsychol 29(2):232–254

Au A, Cheng C, Chan I, Leung P, Li P, Heaton RK (2008) Subjective memory complaints, mood, and memory deficits among HIV/AIDS patients in Hong Kong. J Clin Exp Neuropsychol 30 (3):338–348

Australian Federation of AIDS Organisations (2018) HIV in Australia 2018. https://www.afao.org.au/wp-content/uploads/2017/12/HIV-in-Australia-2018.pdf

Barbaro G, Klatt EC (2002) HIV infection and the cardiovascular system. AIDS Rev 4(2):93–103

Basso MR, Bornstein RA (2000) Estimated premorbid intelligence mediates neurobehavioral change in individuals infected with HIV across 12 months. J Clin Exp Neuropsychol 22 (2):208–218

Bertolli J, McNaghten A, Campsmith M, Lee L, Leman R, Bryan R et al (2004) Surveillance systems monitoring HIV/AIDS and HIV risk behaviors among American Indians and Alaska natives. AIDS Educ Prev 16(3):218–237

Bialystok E, Luk G (2012) Receptive vocabulary differences in monolingual and bilingual adults. Biling Lang Congn 15(2):397–401

Bialystok E, Craik FI, Klein R, Viswanathan M (2004) Bilingualism, aging, and cognitive control: evidence from the Simon task. Psychol Aging 19(2):290–303

Birbeck GL, Kvalsund MP, Byers PA, Bradbury R, Mang'ombe C, Organek N et al (2011) Neuropsychiatric and socioeconomic status impact antiretroviral adherence and mortality in rural Zambia. Am J Trop Med Hyg 85(4):782–789

Blackstone K, Moore DJ, Heaton RK, Franklin DR Jr, Woods SP, Clifford DB et al (2012) Diagnosing symptomatic HIV-associated neurocognitive disorders: self-report versus performance-based assessment of everyday functioning. J Int Neuropsychol Soc 18(1):79–88

Boivin M, Giordani B (2013) Neuropsychology of children in Africa: perspectives on risk and resilience, Specialty topics in pediatric neuropsychology. Springer, New York

Brickman AM, Schupf N, Manly JJ, Luchsinger JA, Andrews H, Tang MX et al (2008) Brain morphology in older African Americans, Caribbean Hispanics, and whites from northern Manhattan. Arch Neurol 65(8):1053–1061

Brouillette MJ, Yuen T, Fellows LK, Cysique LA, Heaton RK, Mayo NE (2016) Identifying neurocognitive decline at 36 months among HIV-positive participants in the CHARTER cohort using group-based trajectory analysis. PLoS One 11(5):e0155766

Cargill VA, Stone VE (2005) HIV/AIDS: a minority health issue. Med Clin North Am 89 (4):895–912

Centers for Disease Control and Prevention (2015) Monitoring selected national HIV prevention and care objectives by using HIV surveillance data – United States and 6 dependent areas. Centers for Disease Control and Prevention

Centers for Disease Control and Prevention (2016) HIV surveillance in urban and nonurban areas. CDC – Slide Sets – Resource Library HIV/AIDS, Centers for Disease Control and Prevention. https://www.cdc.gov/hiv/pdf/library/slidesets/cdc-hiv-urban-nonurban-2017.pdf

Centers for Disease Control and Prevention (2017) HIV surveillance report, 2016

Centers for Disease Control and Prevention (2018a) American Indians and Alaska natives. CDC. https://www.cdc.gov/hiv/group/racialethnic/aian/index.html

Centers for Disease Control and Prevention (2018b) HIV among Asians. Centers for Disease Control. https://www.cdc.gov/hiv/group/racialethnic/asians/index.html

Centers for Disease Control and Prevention (2018c) Native Hawaiians and other Pacific Islanders. https://www.cdc.gov/hiv/group/racialethnic/nhopi/index.html

Centers for Disease Control and Prevention (2018d) Women. https://www.cdc.gov/hiv/group/gender/women/index.html

Chen WT, Guthrie B, Shiu CS, Yang JP, Weng Z, Wang L et al (2014) Acculturation and perceived stress in HIV+ immigrants: depression symptomatology in Asian and Pacific Islanders. AIDS Care 26(12):1581–1585

Chu C, Selwyn PA (2008) Current health disparities in HIV/AIDS. AIDS Read 18(3):144–146, 52–58, C3

Cirino PT, Chin CE, Sevcik RA, Wolf M, Lovett M, Morris RD (2002) Measuring socioeconomic status: reliability and preliminary validity for different approaches. Assessment 9(2):145–155

Cohen MS, Hoffman IF, Royce RA, Kazembe P, Dyer JR, Daly CC et al (1997) Reduction of concentration of HIV-1 in semen after treatment of urethritis: implications for prevention of sexual transmission of HIV-1. AIDSCAP Malawi Research Group. Lancet 349 (9069):1868–1873

Cuellar I, Arnold B, Maldonado R (1995) Acculturation Rating Scale for Mexican Americans-II: a revision of the original ARSMA Scale. Hisp J Behav Sci 17(3):275–304

Cunningham WE, Hays RD, Duan N, Andersen R, Nakazono TT, Bozzette SA et al (2005) The effect of socioeconomic status on the survival of people receiving care for HIV infection in the United States. J Health Care Poor Underserved 16(4):655–676
Cysique LA, Becker JT (2015) Lessons to be learned from the largest study of cognition in American women with HIV disease. Neurology 84(3):220–221
Cysique LA, Becker JT (2017) HIV-related cognitive decline despite viral suppression and complex confounds in American women. Neurology 89(15):1540–1541
Cysique LA, Brew BJ (2011) Prevalence of non-confounded HIV-associated neurocognitive impairment in the context of plasma HIV RNA suppression. J Neurovirol 17(2):176–183
Cysique LA, Franklin D Jr, Abramson I, Ellis RJ, Letendre S, Collier A et al (2011) Normative data and validation of a regression based summary score for assessing meaningful neuropsychological change. J Clin Exp Neuropsychol 33(5):505–522
Dotson VM, Kitner-Triolo MH, Evans MK, Zonderman AB (2009) Effects of race and socioeconomic status on the relative influence of education and literacy on cognitive functioning. J Int Neuropsychol Soc 15(4):580–589
Dreisbach S (2009) Fact Sheet – HIV/AIDS in rural America: challenges and promising strategies. Rural Center for AIDS/STD Prevention at Indiana University, University of Colorado, and University of Kentucky. https://rcap.indiana.edu/doc/factsheets/Fact_Sheet_23.pdf
Duncan KC, Reading C, Borwein AM, Murray MC, Palmer A, Michelow W et al (2011) HIV incidence and prevalence among aboriginal peoples in Canada. AIDS Behav 15(1):214–227
Earnshaw VA, Smith LR, Cunningham CO, Copenhaver MM (2015) Intersectionality of internalized HIV stigma and internalized substance use stigma: implications for depressive symptoms. J Health Psychol 20(8):1083–1089
Eng TR, Butler WT (1997) The hidden epidemic: confronting sexually transmitted diseases. National Academy Press, Washington
Farinpour R, Miller EN, Satz P, Selnes OA, Cohen BA, Becker JT et al (2003) Psychosocial risk factors of HIV morbidity and mortality: findings from the Multicenter AIDS Cohort Study (MACS). J Clin Exp Neuropsychol 25(5):654–670
Ferrett HL, Thomas KG, Tapert SF, Carey PD, Conradie S, Cuzen NL et al (2014) The cross-cultural utility of foreign- and locally-derived normative data for three WHO-endorsed neuropsychological tests for South African adolescents. Metab Brain Dis 29(2):395–408
Fitzgerald E, Grant J, Harrison-Quintana J (2012) Injustice at every turn: a look at American Indian and Alaskan native respondents in the national transgender discrimination survey. National Center for Transgender Equality, Washington
Flicker S, O'Campo P, Monchalin R, Thistle J, Worthington C, Masching R et al (2015) Research done in "a good way": the importance of indigenous elder involvement in HIV community-based research. Am J Public Health 105(6):1149–1154
Foley J, Ettenhofer M, Wright MJ, Siddiqi I, Choi M, Thames AD et al (2010) Neurocognitive functioning in HIV-1 infection: effects of cerebrovascular risk factors and age. Clin Neuropsychol 24(2):265–285
Gant Z, Dailey A, Hu X, Johnson AS (2017) HIV care outcomes among Hispanics or Latinos with diagnosed HIV infection – United States, 2015. MMWR Morb Mortal Wkly Rep 66 (40):1065–1072
Giele C (2004) The epidemiology of HIV infection in aboriginal and non-aboriginal Western Australians, 1983–2002. University of Western Australia, Perth
Gollan TH, Montoya RI, Werner GA (2002) Semantic and letter fluency in Spanish-English bilinguals. Neuropsychology 16(4):562–576
Gollan TH, Montoya RI, Cera C, Sandoval TC (2008) More use almost always a means a smaller frequency effect: aging, bilingualism, and the weaker links hypothesis. J Mem Lang 58 (3):787–814
Gonzalez HM, Ceballos M, Tarraf W, West BT, Bowen ME, Vega WA (2009) The health of older Mexican Americans in the long run. Am J Public Health 99(10):1879–1885
Goodkin K, Miller EN, Cox C, Reynolds S, Becker JT, Martin E et al (2017) Effect of ageing on neurocognitive function by stage of HIV infection: evidence from the Multicenter AIDS Cohort Study. Lancet HIV 4(9):e411–ee22

Gopukumar K, Rao SL, Satishchandra P, Dasgupta J, Ellis RJ, Subbakrishna DK et al (2008) Cognitive changes in asymptomatic drug-naive human immunodeficiency virus type 1 clade C infection. J Neurovirol 14(6):480–485

Goulet JL, Fultz SL, Rimland D, Butt A, Gibert C, Rodriguez-Barradas M et al (2007) Aging and infectious diseases: do patterns of comorbidity vary by HIV status, age, and HIV severity? Clin Infect Dis 45(12):1593–1601

Gupta JD, Satishchandra P, Gopukumar K, Wilkie F, Waldrop-Valverde D, Ellis R et al (2007) Neuropsychological deficits in human immunodeficiency virus type 1 clade C-seropositive adults from South India. J Neurovirol 13(3):195–202

Gupta S, Vaida F, Riggs K, Jin H, Grant I, Cysique L et al (2011) Neuropsychological performance in mainland China: the effect of urban/rural residence and self-reported daily academic skill use. J Int Neuropsychol Soc 17(1):163–173

Haan MN, Zeki Al-Hazzouri A, Aiello AE (2011) Life-span socioeconomic trajectory, nativity, and cognitive aging in Mexican Americans: the Sacramento Area Latino Study on Aging. J Gerontol B Psychol Sci Soc Sci 66(Suppl 1):i102–i110

Hartley D (2004) Rural health disparities, population health, and rural culture. Am J Public Health 94(10):1675–1678

Heaton RK, Grant I, Butters N, White DA, Kirson D, Atkinson JH et al (1995) The HNRC 500 – neuropsychology of HIV infection at different disease stages. HIV Neurobehavioral Research Center. J Int Neuropsychol Soc 1(3):231–251

Heaton RK, Cysique LA, Jin H, Shi C, Yu X, Letendre S et al (2008) Neurobehavioral effects of human immunodeficiency virus infection among former plasma donors in rural China. J Neurovirol 14(6):536–549

Heaton RK, Clifford DB, Franklin DR Jr, Woods SP, Ake C, Vaida F et al (2010) HIV-associated neurocognitive disorders persist in the era of potent antiretroviral therapy: CHARTER study. Neurology 75(23):2087–2096

Heaton RK, Franklin DR, Ellis RJ, McCutchan JA, Letendre SL, Leblanc S et al (2011) HIV-associated neurocognitive disorders before and during the era of combination antiretroviral therapy: differences in rates, nature, and predictors. J Neurovirol 17(1):3–16

Heaton RK, Franklin DR Jr, Deutsch R, Letendre S, Ellis RJ, Casaletto K et al (2015) Neurocognitive change in the era of HIV combination antiretroviral therapy: the longitudinal CHARTER study. Clin Infect Dis 60(3):473–480

Helms JE (1992) Why is there no study of cultural equivalence in standardized cognitive ability testing? Am Psychol 47:1083–1101

Hestad KA, Menon JA, Silalukey-Ngoma M, Franklin DR Jr, Imasiku ML, Kalima K et al (2012) Sex differences in neuropsychological performance as an effect of human immunodeficiency virus infection: a pilot study in Zambia, Africa. J Nerv Ment Dis 200(4):336–342

Hill TD, Angel JL, Balistreri KS, Herrera AP (2012a) Immigrant status and cognitive functioning in late-life: an examination of gender variations in the healthy immigrant effect. Soc Sci Med 75 (12):2076–2084

Hill TD, Angel JL, Balistreri KS (2012b) Does the "healthy immigrant effect" extend to cognitive aging? In: Angel JL, Torres-Gil F, Markides K (eds) Aging, health, and longevity in the Mexican-origin population. Springer, Boston

Hollingshead AB (1975) Four factor index of social status. Yale University, New Haven

Ian E, Gwen CL, Soo CT, Melissa C, Chun-Kai H, Eosu K et al (2016) The burden of HIV-associated neurocognitive disorder (HAND) in the Asia-Pacific region and recommendations for screening. Asian J Psychiatr 22:182–189

Ibrahim F, Anderson J, Bukutu C, Elford J (2008) Social and economic hardship among people living with HIV in London. HIV Med 9(8):616–624

International Organization for Migration (IOM) TUMA (2017) HIV/AIDS and population mobility fact sheet. International Organization for Migration (IOM), The UN Migration Agency. https://www.iom.int/sites/default/files/our_work/DMM/Migration-Health/Factsheet_HIV_24.11.2017.pdf

International Test Commission (ITC) (2013) ITC guidelines on test use. Contract No.: ITC-G-TU-20131008

Jackson R, Reimer G (2008) Canadian aboriginal people living with HIV/AIDS. Care, treatment and support issues. Canadian Aboriginal AIDS Network, Ottawa

Jaworsky D, Benoit A, Raboud J, O'Brien-Teengs D, Blitz S, Rourke SB et al (2016) Comparison of coping strategies and supports between aboriginal and non-aboriginal people living with HIV in Ontario. AIDS Care 28(1):63–69

Johnson DL, Paisano EL, Levin MJ (1989) We, the first Americans. Bureau of the Census, Public Information Office, Population Division

Johnson MO, Catz SL, Remien RH, Rotheram-Borus MJ, Morin SF, Charlebois E et al (2003) Theory-guided, empirically supported avenues for intervention on HIV medication non-adherence: findings from the Healthy Living Project. AIDS Patient Care STDs 17(12):645–656

Joint United Nations Programme on HIV/AIDS (UNAIDS) (2017) Ending AIDS: progress towards the 90-90-90 targets. Joint United Nations Programme on HIV/AIDS (UNAIDS), Geneva

Joska JA, Fincham DS, Stein DJ, Paul RH, Seedat S (2010) Clinical correlates of HIV-associated neurocognitive disorders in South Africa. AIDS Behav 14(2):371–378

Joy R, Druyts EF, Brandson EK, Lima VD, Rustad CA, Zhang W et al (2008) Impact of neighborhood-level socioeconomic status on HIV disease progression in a universal health care setting. J Acquir Immune Defic Syndr 47(4):500–505

Judd T, Capetillo D, Carrion-Baralt J, Marmol LM, Miguel-Montes LS, Navarrete MG et al (2009) Professional considerations for improving the neuropsychological evaluation of Hispanics: a National Academy of Neuropsychology education paper. Arch Clin Neuropsychol 24 (2):127–135

Krieger N, Chen JT, Waterman PD, Rehkopf DH, Subramanian SV (2005) Painting a truer picture of US socioeconomic and racial/ethnic health inequalities: the Public Health Disparities Geocoding Project. Am J Public Health 95(2):312–323

Landrine H, Klonoff EA (1995) The African American Acculturation Scale II: cross-validation and short form. J Black Psychol 21:124–152

Laurencin CT, Murdock CJ, Laurencin L, Christensen DM (2018) HIV/AIDS and the African-American community 2018: a decade call to action. J Racial Ethn Health Disparities 5 (3):449–458

Letendre SL, Ellis RJ, Ances BM, McCutchan JA (2010) Neurologic complications of HIV disease and their treatment. Top HIV Med 18(2):45–55

Lichtenstein B, Laska MK, Clair JM (2002) Chronic sorrow in the HIV-positive patient: issues of race, gender, and social support. AIDS Patient Care STDs 16(1):27–38

Machtinger EL, Wilson TC, Haberer JE, Weiss DS (2012) Psychological trauma and PTSD in HIV-positive women: a meta-analysis. AIDS Behav 16(8):2091–2100

Maki PM, Martin-Thormeyer E (2009) HIV, cognition and women. Neuropsychol Rev 19 (2):204–214

Manly JJ (2008) Critical issues in cultural neuropsychology: profit from diversity. Neuropsychol Rev 18(3):179–183

Manly JJ, Echemendia RJ (2007) Race-specific norms: using the model of hypertension to understand issues of race, culture, and education in neuropsychology. Arch Clin Neuropsychol 22(3):319–325

Manly JJ, Miller SW, Heaton RK, Byrd D, Reilly J, Velasquez RJ et al (1998) The effect of African-American acculturation on neuropsychological test performance in normal and HIV-positive individuals. The HIV Neurobehavioral Research Center (HNRC) Group. J Int Neuropsychol Soc 4(3):291–302

Manly JJ, Jacobs DM, Touradji P, Small SA, Stern Y (2002) Reading level attenuates differences in neuropsychological test performance between African American and White elders. J Int Neuropsychol Soc 8(3):341–348

Manly JJ, Byrd DA, Touradji P, Stern Y (2004) Acculturation, reading level, and neuropsychological test performance among African American elders. Appl Neuropsychol 11(1):37–46

Marin G, Gamba RJ (1996) A new measurement of acculturation for Hispanics: the bidimensional acculturation scale for Hispanics (BAS). Hisp J Behav Sci 18:297–316

Marin G, Sabogal F, VanOss Marin B, Otero-Sabogal R, Perez-Stable EJ (1987) Development of a short acculturation scale for Hispanics. Hisp J Behav Sci 9:183–205
Marquine MJ, Sakamoto M, Dufour C, Rooney A, Fazeli P, Umlauf A et al (2016) The impact of ethnicity/race on the association between the Veterans Aging Cohort Study (VACS) Index and neurocognitive function among HIV-infected persons. J Neurovirol 22(4):442–454
Marquine MJ, Heaton A, Johnson N, Rivera-Mindt M, Cherner M, Bloss C et al (2018) Differences in neurocognitive impairment among HIV-infected Latinos in the United States. J Int Neuropsychol Soc 24(2):163–175
McArthur JC, Steiner J, Sacktor N, Nath A (2010) Human immunodeficiency virus-associated neurocognitive disorders: mind the gap. Ann Neurol 67(6):699–714
McClelland RS, Wang CC, Mandaliya K, Overbaugh J, Reiner MT, Panteleeff DD et al (2001) Treatment of cervicitis is associated with decreased cervical shedding of HIV-1. AIDS 15 (1):105–110
McGregor S, King J, McManus H, Gray R, Guy R (2018) HIV, viral hepatitis and sexually transmissible infections in Australia: annual surveillance report 2018. Kirby Institute, Sydney
Meit M, Knudson A, Gilbert T, Tzy-Chyi Yu A, Tanenbaum E, Ormson E et al (2018) Rural Health Reform Policy Research Center: The 2014 update of the rural-urban chartbook
Milberg WP, Hebben N, Kaplan E (2009) The Boston Process approach to neuropsychological assessment. In: Grant I, Adams KM (eds) Neuropsychological assessment of neuropsychiatric disorders, 3rd edn. Oxford University Press, New York, pp 42–65
Ministry of Health Population Health Branch (2016) HIV prevention and control report for 2016
Miranda C, Arce Renteria M, Fuentes A, Coulehan K, Arentoft A, Byrd D et al (2016) [Formula: see text]The relative utility of three English language dominance measures in predicting the neuropsychological performance of HIV+ bilingual Latino/a adults. Clin Neuropsychol 30 (2):185–200
Mitsch A, Surendera Babu A, Seneca D, Whiteside YO, Warne D (2017) HIV care and treatment of American Indians/Alaska natives with diagnosed HIV infection – 27 states and the District of Columbia, 2012. Int J STD AIDS 28(10):953–961
Molsberry SA, Cheng Y, Kingsley L, Jacobson L, Levine AJ, Martin E et al (2018) Neuropsychological phenotypes among men with and without HIV disease in the multicenter AIDS cohort study. AIDS 32(12):1679–1688
Nelson KM, Simoni JM, Pearson CR, Walters KL (2011) 'I've had unsafe sex so many times why bother being safe now?': the role of cognitions in sexual risk among American Indian/Alaska native men who have sex with men. Ann Behav Med 42(3):370–380
Newman CE, Bonar M, Greville HS, Thompson SC, Bessarab D, Kippax SC (2007) Barriers and incentives to HIV treatment uptake among aboriginal people in Western Australia. AIDS 21 (Suppl 1):S13–S17
Nguyen HT, Black SA, Ray LA, Espino DV, Markides KS (2002) Predictors of decline in MMSE scores among older Mexican Americans. J Gerontol A Biol Sci Med Sci 57(3):M181–M185
Oakes JM, Rossi PH (2003) The measurement of SES in health research: current practice and steps toward a new approach. Soc Sci Med 56(4):769–784
O'Bryant SE, Hall JR, Cukrowicz KC, Edwards M, Johnson LA, Lefforge D et al (2011) The differential impact of depressive symptom clusters on cognition in a rural multi-ethnic cohort: a Project FRONTIER study. Int J Geriatr Psychiatry 26(2):199–205
Office of Minority Health (2016) Minority population profiles. U.S. Department of Health and Human Services. https://minorityhealth.hhs.gov/omh/browse.aspx?lvl=3&lvlid=64
Office of Minority Health (2017) Profile: Hispanic/Latino Americans. https://minorityhealth.hhs.gov/omh/browse.aspx?lvl=3&lvlid=64
Oramasionwu CU, Hunter JM, Skinner J, Ryan L, Lawson KA, Brown CM et al (2009) Black race as a predictor of poor health outcomes among a national cohort of HIV/AIDS patients admitted to US hospitals: a cohort study. BMC Infect Dis 9:127
Patterson S, Cescon A, Samji H, Chan K, Zhang W, Raboud J et al (2015) Life expectancy of HIV-positive individuals on combination antiretroviral therapy in Canada. BMC Infect Dis 15:274

Ponton MO, Satz P, Herrera L, Ortiz F, Urrutia CP, Young R et al (1996) Normative data stratified by age and education for the Neuropsychological Screening Battery for Hispanics (NeSBHIS): initial report. J Int Neuropsychol Soc 2(2):96–104

Rabkin JG, McElhiney M, Ferrando SJ, Van Gorp W, Lin SH (2004) Predictors of employment of men with HIV/AIDS: a longitudinal study. Psychosom Med 66(1):72–78

Ramjee G, Daniels B (2013) Women and HIV in Sub-Saharan Africa. AIDS Res Ther 10(1):30

Rawlings M, Masters H (2008) Comorbidities and challenges affecting African Americans with HIV infection. J Natl Med Assoc 100(12):1477–1481

Rivera Mindt M, Byrd D, Ryan EL, Robbins R, Monzones J, Arentoft A et al (2008a) Characterization and sociocultural predictors of neuropsychological test performance in HIV+ Hispanic individuals. Cultur Divers Ethnic Minor Psychol 14(4):315–325

Rivera Mindt M, Arentoft A, Kubo Germano K, D'Aquila E, Scheiner D, Pizzirusso M et al (2008b) Neuropsychological, cognitive, and theoretical considerations for evaluation of bilingual individuals. Neuropsychol Rev 18(3):255–268

Rivera Mindt M, Byrd D, Saez P, Manly J (2010) Increasing culturally competent neuropsychological services for ethnic minority populations: a call to action. Clin Neuropsychol 24 (3):429–453

Rivera Mindt M, Miranda C, Arentoft A, Byrd D, Monzones J, Fuentes A et al (2014) Aging and HIV/AIDS: neurocognitive implications for older HIV-positive Latina/o adults. Behav Med 40 (3):116–123

Rivera Mindt M, Arentoft A, Coulehan K, Summers AC, Tureson K, Aghvinian M et al (2019) Neuropsychological evaluation of culturally/linguistically diverse older adults. In: Ravdin LD, Katzen HL (eds) Handbook on the neuropsychology of aging and dementia. Springer, New York, pp 25–48

Rivera Mindt M, Arentoft A, Tureson K, Summers AC, Morris EP, Guzman V et al (under review) Disparities in objectively measured antiretroviral adherence and differential adherence predictors in Latinx and non-Hispanic white persons living with HIV

Robertson KR, Smurzynski M, Parsons TD, Wu K, Bosch RJ, Wu J et al (2007) The prevalence and incidence of neurocognitive impairment in the HAART era. AIDS 21(14):1915–1921

Ross J, Cunningham CO, Hanna DB (2018) HIV outcomes among migrants from low-income and middle-income countries living in high-income countries: a review of recent evidence. Curr Opin Infect Dis 31(1):25–32

Rubin LH, Pyra M, Cook JA, Weber KM, Cohen MH, Martin E et al (2016) Post-traumatic stress is associated with verbal learning, memory, and psychomotor speed in HIV-infected and HIV-uninfected women. J Neurovirol 22(2):159–169

Ruffieux N, Njamnshi AK, Mayer E, Sztajzel R, Eta SC, Doh RF et al (2010) Neuropsychology in Cameroon: first normative data for cognitive tests among school-aged children. Child Neuropsychol 16(1):1–19

Ryan EL, Baird R, Mindt MR, Byrd D, Monzones J, Bank SM et al (2005) Neuropsychological impairment in racial/ethnic minorities with HIV infection and low literacy levels: effects of education and reading level in participant characterization. J Int Neuropsychol Soc 11 (7):889–898

Sacktor N, Lyles RH, Skolasky R, Kleeberger C, Selnes OA, Miller EN et al (2001) HIV-associated neurologic disease incidence changes: Multicenter AIDS Cohort Study, 1990-1998. Neurology 56(2):257–260

Sacktor N, Nakasujja N, Skolasky RL, Rezapour M, Robertson K, Musisi S et al (2009) HIV subtype D is associated with dementia, compared with subtype A, in immunosuppressed individuals at risk of cognitive impairment in Kampala, Uganda. Clin Infect Dis 49(5):780–786

Sacktor N, Skolasky RL, Seaberg E, Munro C, Becker JT, Martin E et al (2016) Prevalence of HIV-associated neurocognitive disorders in the Multicenter AIDS Cohort Study. Neurology 86 (4):334–340

Selnes OA, Jacobson L, Machado AM, Becker JT, Wesch J, Miller EN et al (1991) Normative data for a brief neuropsychological screening battery. Multicenter AIDS Cohort Study. Percept Mot Skills 73(2):539–550

Shavers VL (2007) Measurement of socioeconomic status in health disparities research. J Natl Med Assoc 99(9):1013–1023
Sheth SH, Jensen PT, Lahey T (2009) Living in rural New England amplifies the risk of depression in patients with HIV. BMC Infect Dis 9:25
Shimizu SM, Chow DC, Valcour V, Masaki K, Nakamoto B, Kallianpur KJ et al (2011) The impact of depressive symptoms on neuropsychological performance tests in HIV-infected individuals: a study of the Hawaii Aging with HIV Cohort. World J AIDS 1(4):139–145
Shonkoff JP, Garner AS, Committee on Psychosocial Aspects of C, Family H, Committee on Early Childhood A, Dependent C et al (2012) The lifelong effects of early childhood adversity and toxic stress. Pediatrics 129(1):e232–e246
Sisco S, Gross AL, Shih RA, Sachs BC, Glymour MM, Bangen KJ et al (2015) The role of early-life educational quality and literacy in explaining racial disparities in cognition in late life. J Gerontol B Psychol Sci Soc Sci 70(4):557–567
Spies G, Ahmed-Leitao F, Fennema-Notestine C, Cherner M, Seedat S (2016) Effects of HIV and childhood trauma on brain morphometry and neurocognitive function. J Neurovirol 22 (2):149–158
Spies G, Fennema-Notestine C, Cherner M, Seedat S (2017) Changes in cognitive function in women with HIV infection and early life stress. AIDS Care 29(1):14–23
Suinn RM, Rickard-Figueroa K, Lew S, Vigil P (1987) The Suinn-Lew Asian self-identity acculturation scale: an initial report. Educ Psychol Meas 4:401–407
Thames AD, Hinkin CH, Byrd DA, Bilder RM, Duff KJ, Mindt MR et al (2013) Effects of stereotype threat, perceived discrimination, and examiner race on neuropsychological performance: simple as black and white? J Int Neuropsychol Soc 19(5):583–593
Thames AD, Kuhn TP, Mahmood Z, Bilder RM, Williamson TJ, Singer EJ et al (2018) Effects of social adversity and HIV on subcortical shape and neurocognitive function. Brain Imaging Behav 12(1):96–108
Tierney SM, Sheppard DP, Kordovski VM, Faytell MP, Avci G, Woods SP (2017) A comparison of the sensitivity, stability, and reliability of three diagnostic schemes for HIV-associated neurocognitive disorders. J Neurovirol 23(3):404–421
Tozzi V, Balestra P, Bellagamba R, Corpolongo A, Salvatori MF, Visco-Comandini U et al (2007) Persistence of neuropsychologic deficits despite long-term highly active antiretroviral therapy in patients with HIV-related neurocognitive impairment: prevalence and risk factors. J Acquir Immune Defic Syndr 45(2):174–182
US Census Bureau (2013) Language use of the United States. In: 2011 American Community Survey Reports. https://www.census.gov/prod/2013pubs/acs-22.pdf
Valcour V, Shikuma C, Shiramizu B, Watters M, Poff P, Selnes O et al (2004a) Higher frequency of dementia in older HIV-1 individuals: the Hawaii Aging with HIV-1 Cohort. Neurology 63 (5):822–827
Valcour V, Shikuma C, Shiramizu B, Watters M, Poff P, Selnes OA et al (2004b) Age, apolipoprotein E4, and the risk of HIV dementia: the Hawaii Aging with HIV Cohort. J Neuroimmunol 157(1–2):197–202
Valcour VG, Sacktor NC, Paul RH, Watters MR, Selnes OA, Shiramizu BT et al (2006) Insulin resistance is associated with cognition among HIV-1-infected patients: the Hawaii Aging with HIV Cohort. J Acquir Immune Defic Syndr 43(4):405–410
Vance DE, Rubin LH, Valcour V, Waldrop-Valverde D, Maki PM (2016) Aging and neurocognitive functioning in HIV-infected women: a review of the literature involving the women's interagency HIV study. Curr HIV/AIDS Rep 13(6):399–411
Verney SP, Bennet J, Hamilton JM (2016) Cultural considerations in the neuropsychological assessment of American Indians/Alaska natives. In: Ferraro RF (ed) Minority and cross-cultural aspects of neuropsychological assessment. Taylor & Francis, New York, pp 115–158
Vernon IS (2007) American Indian women, HIV/AIDS, and health disparity. Subst Use Misuse 42 (4):741–752

Vyavaharkar M, Moneyham L, Corwin S, Tavakoli A, Saunders R, Annang L (2011) HIV-disclosure, social support, and depression among HIV-infected African American women living in the rural southeastern United States. AIDS Educ Prev 23(1):78–90

Walker F, Llata E, Doshani M, Taylor M, Bertolli J, Weinstick H et al (2015) HIV, chlamydia, gonorrhea, and primary and secondary syphilis among American Indians and Alaska natives within Indian Health Service areas in the United States, 2007-2010. J Community Health 40 (3):484–492

Wang Z, Zheng Y, Liu L, Shen Y, Zhang R, Wang J et al (2013) High prevalence of HIV-associated neurocognitive disorder in HIV-infected patients with a baseline CD4 count </= 350 cells/muL in Shanghai, China. Biosci Trends 7(6):284–289

Weber E, Blackstone K, Woods SP (2013) Cognitive neurorehabilitation of HIV-associated neurocognitive disorders: a qualitative review and call to action. Neuropsychol Rev 23(1):81–98

Wechsler D (2001) Wechsler test of adult reading. The Psychological Corporation, San Antonio

Weintraub S, Dikmen SS, Heaton RK, Tulsky DS, Zelazo PD, Bauer PJ et al (2013) Cognition assessment using the NIH Toolbox. Neurology 80(11 Suppl 3):S54–S64

Welfare AIoHa (2019) Deaths in Australia. Australian Institute of Health and Welfare, Canberra. Updated 17 July 2019

Wilkinson GS (1993) The wide range achievement test: administration manual. Wide Range Inc., Wilmington

Williamson TJ, Mahmood Z, Kuhn TP, Thames AD (2017) Differential relationships between social adversity and depressive symptoms by HIV status and racial/ethnic identity. Health Psychol 36(2):133–142

Wilson LE, Korthuis T, Fleishman JA, Conviser R, Lawrence PB, Moore RD et al (2011) HIV-related medical service use by rural/urban residents: a multistate perspective. AIDS Care 23(8):971–979

Wojna V, Nath A (2006) Challenges to the diagnosis and management of HIV dementia. AIDS Read 16(11):615–616, 21–24, 26, 29–32

Woods SP, Rippeth JD, Frol AB, Levy JK, Ryan E, Soukup VM et al (2004) Interrater reliability of clinical ratings and neurocognitive diagnoses in HIV. J Clin Exp Neuropsychol 26(6):759–778

Wright MR, Giele CM, Dance PR, Thompson SC (2005) Fulfilling prophecy? Sexually transmitted infections and HIV in indigenous people in Western Australia. Med J Aust 183(3):124–128

Wright E, Brew B, Arayawichanont A, Robertson K, Samintharapanya K, Kongsaengdao S et al (2008) Neurologic disorders are prevalent in HIV-positive outpatients in the Asia-Pacific region. Neurology 71(1):50–56

Yeung CM, St John PD, Menec V, Tyas SL (2014) Is bilingualism associated with a lower risk of dementia in community-living older adults? Cross-sectional and prospective analyses. Alzheimer Dis Assoc Disord 28(4):326–332

Zahodne LB, Schofield PW, Farrell MT, Stern Y, Manly JJ (2014) Bilingualism does not alter cognitive decline or dementia risk among Spanish-speaking immigrants. Neuropsychology 28 (2):238–246

Zea MC, Asner-Self KK, Birman D, Buki LP (2003) The abbreviated multidimensional acculturation scale: empirical validation with two Latino/Latina samples. Cultur Divers Ethnic Minor Psychol 9(2):107–126

Zhou J, Kumarasamy N, Ditangco R, Kamarulzaman A, Lee CK, Li PC et al (2005) The TREAT Asia HIV Observational Database: baseline and retrospective data. J Acquir Immune Defic Syndr 38(2):174–179

Zhou J, Dore GJ, Zhang F, Lim PL, Chen YM, Database TAHO (2007) Hepatitis B and C virus coinfection in the TREAT Asia HIV Observational Database. J Gastroenterol Hepatol 22 (9):1510–1518

Neurocognitive Complications of HIV Infection in Low-Income Countries

Alyssa Vecchio, Ned Sacktor, Deanna Saylor, and Kevin Robertson

Contents

Abstract There is a paucity of information on neurocognitive dysfunction in individuals with HIV in resource-limited regions, despite the fact that these areas have the greatest burden of infection. HIV-associated neurocognitive disorder (HAND) remains a common complication of HIV despite the use of effective antiretroviral therapy (ART). HAND is a major cause of morbidity of HIV+ individuals and is estimated to be the most prevalent form of neurocognitive impairment worldwide in young adults. This finding has drastic implications for the productivity and social engagement of young adults in the development of industry, education, and healthcare, which is particularly relevant in low-income countries. Building an infrastructure to examine the neurological and neuropsychological characteristics of

A. Vecchio (✉)
Vita-Salute San Raffaele University, Milan, Italy
e-mail: alyssacv@neurology.unc.edu

N. Sacktor and D. Saylor
Johns Hopkins University, Baltimore, MD, USA

K. Robertson
AIDS Neurological Center, Neurology, UNC, Chapel Hill, NC, USA

Curr Topics Behav Neurosci (2021) 50: 225–244
DOI 10.1007/7854_2019_92
Published Online: 12 July 2019

HIV+ individuals in resource-limited settings (RLS) can advance the understanding of the unique contributing factors of HIV-1 clades in these regions of high prevalence, improve neurological monitoring, explore the CNS HIV reservoir, and provide key information on prevention/interventions to help manage/improve these neurological and neuropsychological complications.

Keywords ART · HIV · HIV-associated neurocognitive disorder · Low income countries · Resource limited settings

1 Introduction: Global Epidemiology of HIV and HIV-Associated Neurological Disorders

Globally, 36.7 million people were living with HIV at the end of 2016. The epicenter of the epidemic is in sub-Saharan Africa, where almost 70% of the world's HIV+ population resides, and nearly 1 in every 25 adults (4.2%) is living with HIV. Southeast Asia has the second highest regional concentration of people living with HIV (WHO 2016). In addition to the burden of disease, these regions also have the highest concentration of low-income (GNI per capita $1,025 or less) and low-middle-income (GNI per capita between $1,026 and $4,035) countries (World Bank 2017). Furthermore, over half of the global burden of dementia occurs in low- and middle-income countries, with statistics indicating it will only increase with the growing population (Ferri et al. 2005).

HAND is the term used to describe the spectrum of neurocognitive dysfunction caused by HIV infection, which includes asymptomatic neurocognitive impairment (ANI), mild neurocognitive disorder (MND), and HIV-associated dementia (HAD). The graded classification, according to Frascati criteria, is based on abnormal performance on neuropsychological testing and the presence or absence of a patient's impairment in daily functions related to cognitive impairment (Antinori et al. 2007). The changes in memory, attention, motor skills, behavior, and other complications of cognitive impairment are direct results of HIV infection and subsequent impact on the CNS.

In 2015, the WHO recommended that all people living with HIV should receive ART, regardless of their CD4 count based on the results of the START study demonstrating the benefits of early ART initiation (INSIGHT 2015). However, despite the use of effective ART, HAND remains a common complication of HIV and a major cause of morbidity where 15–55% of HIV+ individuals from the United States are affected with HAND (Heaton et al. 2010). With proper adherence to ARTs, there has been a significant decrease in the incidence of the most severe manifestation of HAND (HIV-associated dementia), but in most studies, this has been counterbalanced by an increase in the prevalence of more mild stages of HAND (Heaton et al. 2011). There is a similar range in the prevalence in RLS, with studies showing 15–69% of individuals with HIV on ART are diagnosed with HAND (Table 1). However, there is limited longitudinal, standardized data collected on

Table 1 Prevalence of HAND and HIV-1 subtype in resource-limited settings

Country	Prevalence of HAND, pre-ART	Prevalence of HAND, on ART	Citation	Compared with normalized data	Country income status (World Bank 2017)	ART coverage %	Dominant regional HIV-1 subtype (Liner et al. 2007)
Thailand	26% (acute HIV)	15%	Chan et al. (2018)	Yes	Lower middle	69	A, E
India	56%		Yepthomi et al. (2006)	Yes	Lower middle	49	C
Indonesia	51% (HAD, 19%)		Estiasari and Lastri (2015)	Yes	Lower middle	13	CRF01_AE, CRF's and other recombinants
Malaysia		22.8% (HAD, 0%)	Mukherjee et al. (2018)	Yes	Upper middle	37	CRF01_AE, CRF's and other recombinants
Cameroon	25.69%	17%	Njamnshi et al. (2009)	Yes	Lower middle	37	A, G, L, CRF's and other recombinants
Nigeria		21.5% (HAD, 2.9%)	Yusuf et al. (2017)	No	Lower middle	30	A, G, L, CRF's and other recombinants
Uganda	59% (HAD, 15%)	52% (HAD, 5%)		Yes	Low	67	A, D
Ethiopia		33.30%	Belete et al. (2017)	No	Low	59	C
Zambia		34.60%	Kabuba et al. (2016)	No	Lower middle	65	C
Tanzania		19.30%	Sanmartí et al. (2018)	Yes	Low	62	A, C
Kenya		69% (HAD, 0%)	Awori et al. (2018)	No	Lower middle	64	A
Malawi	HAD 15.56%	HAD 13.43% (for 6 months)		No	Low	66	C
Botswana		38.33%	Lawler et al. (2010)	No	Upper Middle	83	C
South Africa	76.5 (HAD, 25.3%)		Joska et al. (2011)	No	Upper Middle	56	C
Brazil		36.20%	Troncoso and Conterno (2015)	No	Upper Middle	60	B, C

Adapted from Habib et al. (2013), World Bank Income Level (2017), Liner et al. (2007), and specific studies listed with citation the prevalence and severity of HAND in RLS, which are the regions most affected by HIV. Furthermore, RLS still have unequal access to ART, which leads to more advanced cases of HIV and thus increased risk of severe manifestations of HAND.

Another factor to consider is that the prevalence of HIV type 1 (HIV-1) clades varies by region, and their unique neurovirulence factors may have functional implication on onset and progression of HAND. HIV-1 is characterized by genetic diversity and can be divided into three classes, the most common of which is Group M (major), which has nine major clades (A–D, F–H, J, and K) (Liner et al. 2007). Studies in sub-Saharan Africa, where clades A, C, and D predominate, have suggested that clade differences lead to disparate frequencies of HAND. A study in Uganda, where the prevalence of HAD was as high as 31% among antiretroviral-naïve HIV+ individuals (Wong et al. 2007), showed that antiretroviral-naïve HIV+ individuals with subtype D were more likely to develop HAD and at a faster progression than those with subtype A (Sacktor et al. 2009). However, a subsequent study among HIV+ individuals with subtypes D and A with less immunosuppression failed to show a difference in the risk of HAD (Sacktor et al. 2014). Most recently, a longitudinal study of neurocognitive impairment in HIV+ individuals in rural Uganda with both moderate and severe immunosuppression found that those with subtype D infection had more severe neurocognitive impairment than those with subtype A infection regardless of level of immunosuppression (Sacktor et al. 2019). These studies are unique from those in North America, where subtype B is predominant.

2 Neurocognitive Complications of HIV-1 in the ART Era: Prevalence and Persistence in Resource-Limited Settings

The neurological complications of HIV have evolved with the introduction of ART and subsequent prolonged life expectancy. There has been a shift from neurological conditions related to low CD4 cell counts with opportunistic infections to virally suppressed patients with prolonged inflammation and neuronal damage. This change followed the introduction of ART, which halts viral replication, decreasing viral load in plasma and CSF and restoring the systemic immune function. The use of ART has beneficial effects on improving and even preventing the most severe forms of HAND, but the mild-to-moderate stages (such as MND and asymptomatic cases (ANI)) remain prevalent and clinically relevant. The impact on the daily lives of individuals with HIV can be economically and socially devastating in low-income countries, as discussed below.

There have been country-specific studies to estimate the burden of HAND on and off ART (Table 1). Several prospective studies from sub-Saharan Africa using the International HIV Dementia Scale were included in a systematic review and a meta-analysis to demonstrate the significant burden of neurocognitive impairment on and off ART in high HIV-prevalent areas (Habib et al. 2013). While there is a range in the prevalence across the countries, the overall results of the studies indicate an improvement of neurocognitive performance following ART. The high burden of HAND also highlights the need for standardized neurological testing in these regions. Several factors could account for the variability among these studies, including the presence of preexisting clinical infrastructure to assess HAND and the quality of routine HIV care available to patients. Although there has been an increase in standardized HAND research in RLS, limited existing research infrastructure and research funding results in reduced quantity and reach of longitudinal HAND studies compared to high-income regions (Robertson et al. 2009; Kalula and Petros 2011). Studies in North America and Europe represent populations with distinct cultures, HIV-1 subtypes, age of highest risk, gender distribution, access to ART, and education status. These differences make it difficult to apply the significance of their results to the African, Asian, and South American settings.

While the increasing availability and initiation at earlier HIV disease stages of ART in sub-Saharan Africa may result in less prevalent and less severe HAND in the future, it is important to note that guidelines to initiate ART at less severe stages of immunocompromise in RLS are relatively recent. As such, most HIV+ individuals in RLS – including those with currently virally suppressed HIV infection and high CD4 counts – have a history of severe immunocompromise at some point in their infection. Existing evidence suggests both current and prior immunosuppression increase HAND risk. For example, a study of an HIV+ cohort in Uganda measured the impact of low CD4 count on HAD. It found that every 100 cells/μL decrement in CD4 cell count was associated with a 60% increase in the odds of having HAD (Wong et al. 2007). Furthermore, many prior studies have shown that a history of severe immunocompromised (e.g., low CD4 cell nadir) is also associated with an increased risk of HAND and more severe HAND stages (Saylor et al. 2016). Taken together, this evidence reiterates the importance of preventing severe immunocompromise, namely, by initiating individuals with HIV on ART at earlier stages in infection to decrease the risk of HAD.

Widespread availability and earlier initiation of ART in RLS had led to a shifting demographic with an increasing number of older HIV+ adults due to the increased life expectancy of HIV+ individuals on ART. Counterintuitively, this may actually lead to increases in HAND prevalence as older age is itself a risk factor for HAND. For example, a Ugandan study found each additional 10 years of age conferred a greater than twofold risk of HAD, and a South African cohort also found an association between older age and increased risk of HAND (Joska et al. 2012). These findings are potentially confounded by the increased prevalence of cerebrovascular events with older age (Heaton et al. 2012). They require further investigation as to whether prolonged HIV inflammation plays a synergistic role in atherosclerosis and whether the increasing rates of neurocognitive impairment are due to HAND alone or in combination with vascular cognitive impairment.

The first multinational neurological clinical trial to study HAND exclusively in RLS, the International Neurological Study (AIDS Clinical Trial Group (ACTG) A5199), found substantial neuropsychological and neurological improvement following the initiation of first-line ART in previously ART-naïve individuals with

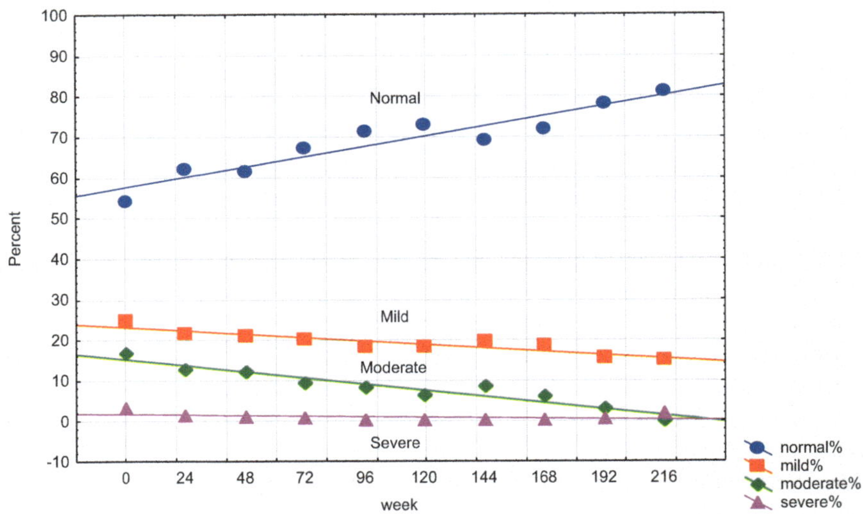

Fig. 1 Neurocognitive impairment over time following art initiation (the International Neurological Study (ACTG A5199))

HIV. Nearly a third of the ART naïve participants had abnormal neurological exams at the beginning of the study (Fig. 1), with significant country variation from 8% in Peru to 74% in Thailand. With the initiation of ART, the estimated odds of impairment were reduced by over 10% for every 24 weeks on ART (Robertson et al. 2018). Neuropsychological test battery improvement, except for semantic verbal fluency, was sustained over 3 years of follow-up while on ARTs, with no differences between treatment regimens detected (Robertson et al. 2012). This further indicates the importance of sustained effective ART in lowering the risk of HAND.

Although ART has decreased the incidence of HAD and moderated the symptoms of MND and HAD in most HIV+ individuals, neurocognitive impairment persists despite long-term administration of ART, as shown in Fig. 1 from the International Neurological Study. Current ART regimens have not been shown to fully reverse milder forms, even though they achieve virological suppression (Robertson et al. 2012; Heaton et al. 2011; Tozzi et al. 2007).

3 HAND in Low-Income Countries: Socioeconomic and Public Health Factors

While HIV/AIDS is a global epidemic with an objective pathology, the illness varies across regions with influences by the social context of HIV, socioeconomic status, capacity of health systems, and gender inequalities. These factors have a negative impact on the course of the infection, contributing to immune dysregulation and subsequent increased risk of neuronal injury. The social inequality and political instability in low-income countries foster factors that influence HIV care and comorbidities that may potentiate neurological inflammation including coinfections, vitamin deficiency, low rate of educational attainment, and psychiatric illness.

3.1 Low Linkage to Care and Poor Health Status

According to the Global Health Workforce Alliance, 1.5 million additional healthcare workers are needed in sub-Saharan Africa to meet basic healthcare needs. Chronic management of HIV exaggerates the deficiency of skilled healthcare workers, clinics, and testing facilities. Access to the limited health centers may be impeded by relatively expensive transport, dangerous roads or access routes, and other obstacles associated with a lack of infrastructure. Even if the patients manage to arrive at the health center, there is the risk that they are not stocked with ART or the most up-to-date ART regimens based on delivery and/or funding. Thus, HIV+ individuals in RLS may present later in the disease course with more advanced neurocognitive impairment and/or interruptions in therapy due to inconsistent follow-up visits.

The interplay of impoverished conditions and the management of HIV can be exemplified by the epidemic in Lesotho, which has nearly a quarter of the general population infected with HIV (Table 2). The mountainous country has limited road infrastructure, making travel difficult, centralizing advancements in education and medicine, and restricting access to healthcare facilities. Most of the population live in rural communities with high unemployment rates, low education attainment, and cultural practices (e.g., lack of medical circumcisions) that perpetuate the spread of HIV. The geographical restrictions, poverty, and stigma provide challenges to increasing HIV testing and expanding updated treatment coverage. The prevalence

Table 2 Lesotho HIV and country statistics (adapted from WHO and UNAIDS)

Indicator	Most recent data from 2012–2016
Adults (older than 15 years old) newly infected with HIV	19,000 [17,000–22,000]
ART coverage among people with HIV infection eligible for ART according to 2010 guidelines (%)	54 [52–57]
Deaths due to HIV/AIDS (per 100,000 population)	755
Gross national income per capita (PPP int. $)	2,990
Prevalence of HIV among adults 15–49 (%)	25.0 [22.7–26.5]
Incidence of tuberculosis (per 100,000 population per year)	852 [551–1,220]
Population living in urban areas (%)	23.74
Hospital beds (per 10,000 population)	13
Psychiatrists working in mental health sector (per 100,000)	0.1

and severity of HAND are relatively unknown in Lesotho due to the lack of research in this area.

The burden of disease is steep in many remote rural areas where in addition to deficient medical care, there can be a lack of running water, electricity, and other essentials for healthy sanitation. Furthermore, many of these RLS are in tropical regions where the climate can harbor several endemic microorganisms. Impoverished urban areas can mimic these sanitation problems with overcrowded living spaces, inadequate plumbing, and lack of clean water. Nutrition in both cases may be insufficient or lack vitamins and protein necessary for proper neurological development and function. A recent multinational study found that neurocognitive impairment among HIV-positive individuals was more prevalent in both overweight/obese and underweight than normal weight individuals in three RLS (Jumare et al. 2018).

Immunocompromised HIV+ patients are susceptible to the gamut of infectious diseases nurtured in these environments. The coinfections can exacerbate the immune dysregulation and ultimately influence the rate of neurological dysfunction. For example, most of the estimated number of TB incident cases in 2016 occurred in the WHO Southeast Asia region (45%). In the WHO African region, where the burden of HIV-associated TB is highest, 82% of TB patients had a documented HIV-positive test result (World Health Organization Global Tuberculosis Report 2017). TB, in HIV+ individuals, has been shown to be associated with more severe cognitive impairment (Robertson et al. 2018; Hestad et al. 2019). Although ART has been able to decrease the incidence of CNS coinfections, long-term HIV+ individuals with a history of immunosuppression may also have a history of brain injury from meningitis or other neurological infection and thus are more susceptible to further neuronal damage.

Early ART therapy has been shown to have promising effects on improving function and even preventing neurocognitive dysfunction. A Peru-based study enrolled HIV+ individuals identified within a month of acute HIV infection who did not yet meet the current national ART treatment initiation guidelines in Peru. Participants were randomized to immediate ART or initiation of ART after 6 months and then monitored for neuropsychologic outcomes. The immediately treated group showed improved neurocognitive performance at 48 weeks as compared to the deferred treatment group (Robertson et al. 2017). For underserved populations, a 6-month deferment of testing and subsequent treatment can be commonplace and may lead to poorer outcomes. The long-term benefits of early treatment were also illustrated in a study of acutely infected (within 30 days of diagnosis) HIV+ adults in Thailand who showed improved neurocognitive performance in the context of ART-induced viral suppression sustained over the 6-year course of treatment. The results exceeded the estimated practice effect and were compared with a group of healthy HIV-uninfected Thai individuals (Chan et al. 2018). Again, this highlights the need for early and reliable sustained treatment.

3.2 *Cognitive Reserve and Education Inequality*

Cognitive reserve can be operationalized as a higher IQ, greater level of educational attainment, highest level of occupation, or a combination of these factors that results in a higher residual cognitive capability. The concept of cognitive reserve is thought to represent a greater capacity to overcome neurodegeneration or at least the expression of such CNS insults (Stern 2002). Pathologically, the threshold of HIV neuronal injury at which an individual with a prominent level of cognitive reserve would develop HAND, particularly a syndromic case, appears to be higher relative to an individual with lower cognitive reserve (Morgan et al. 2012). A recent study in Zambia found that a higher education had a protective effect in young adults infected with HIV-1 clade C against neurocognitive impairments, specifically in the domains of executive functions, learning, and speed of information processing (Kabuba et al. 2018).

Level of education plays a prominent role in cognitive functioning (De Ronchi et al. 2002). Even as primary school enrollments have increased in most low-income countries, levels of educational attainment remain low and highly unequal relative to the developed world. There are areas in the African continent that have less than 50% literacy among children, compared to the nearly 100% youth literacy in South America and Europe (UNESCO 2016). Southern Asia is home to almost half (49%) of the world's adult illiterate population (UNESCO 2016) (Fig. 2). This vast disparity in education in the African and Asian countries, that bare a disproportionally high HIV burden, may contribute to the susceptibility for acquiring HIV, accessing regular healthcare and perhaps accelerated neurocognitive decline.

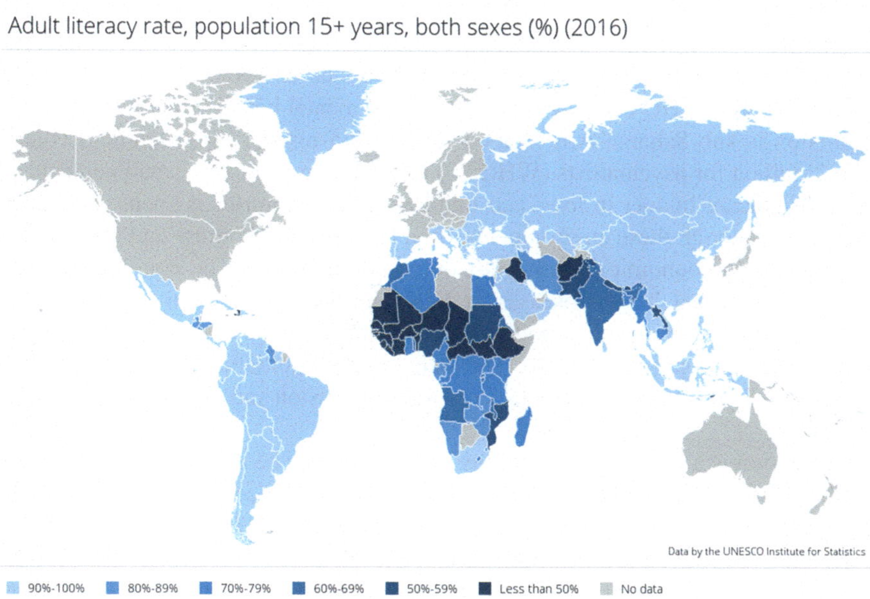

Fig. 2 Global adult literacy rate by country (UNESCO Institute for Statistics 2016)

A recent study from a large East African cohort found a 38% prevalence of cognitive impairment among virally suppressed HIV+ participants. They underwent a 30-min cognitive testing battery and six domain neuropsychological testing with results compared to demographically similar seronegative individuals at the same sites. The inability to read and higher initial WHO stage were strongly associated with increased risk of cognitive impairment (Milanini et al. 2018). A recent study in Uganda comparing cognitive performance in 400 HIV+ and 400 demographically matched HIV-uninfected individuals found cognitive impairment – and especially severe cognitive impairment – was more common among HIV+ than HIV-uninfected participants. However, baseline cognitive impairment was highly prevalent in both groups, occurring in 59% of HIV+ participants and 44% of HIV-uninfected participants (Sacktor et al. 2019). While HAND is a direct result of HIV infection of the CNS, understanding other causes of cognitive impairment and multivariate pathogenic factors can potentially highlight HIV+ individuals at particular risk of more rapid and/or more severe impairment. Furthermore, this highlights another important difference between research conducted in Western countries and RLS, since the participants have vastly different access to education and thus will have distinct levels of cognitive reserve.

3.3 Mental Health and HIV-Related Stigma in Vulnerable Populations

People living with HIV have a higher prevalence of depression and anxiety than non-HIV-infected individuals (Brandt 2009; Mayston et al. 2012). This increased burden of common mental health conditions in areas of high HIV prevalence, namely, low-income countries, often goes untreated (Marwick and Kaaya 2010; Chibanda et al. 2014). The WHO estimated that the ratio of mental health professionals to the population in sub-Saharan Africa stood at 1 per 2.5 million for psychologists and 1 per 2 million for psychiatrists (WHO 2011). Furthermore, a very small percentage of the healthcare budget in low- and middle-income countries is spent on mental health, even though these environments often have persistent stress and trauma due to political and economic instability (Cournos et al. 2014). The consequences of unaddressed mental health diseases in HIV+ individuals can start with delayed diagnosis and continue with suboptimal treatment (Mayston et al. 2016; Parcesepe et al. 2018a, b; Bigna et al. 2018). This delay in ART initiation and then potential poor ART adherence can lead to a lack of viral suppression and overall increased risk for HAND.

Depression is a common comorbidity of HIV and can be difficult to differentiate whether it makes an additional contribution to HAND since it is a confounder of neuropsychological testing performance (Antinori et al. 2007; Tedaldi et al. 2015). Thus, many HAND studies will have depression as an exclusion factor (Robertson et al. 2012; Tedaldi et al. 2015). Nonetheless, depressive symptoms have been studied for the potential association with cognitive impairment among HIV+ individuals (Heaton et al. 2011; Grant et al. 2014). There has been research indicating a particular association between the chronic and unstable forms of major depressive disorder with HAND (Cysique et al. 2016). Depression on its own, without HAND, can impact quality of life, productivity, and medical compliance. For example, depression was found to be associated with a near doubling of HIV viral load in an East African cohort with a point prevalence of depression up to 25% (Meffert et al. 2018). This suggests that HIV itself, depression, and, possibly, HAND would benefit from mental health screening and treatment to improve comprehensive care. As a comorbidity with HAND, depression could be considered in diagnostic strategies aimed at identifying HIV+ individuals with cognitive impairment.

Women are generally at a higher risk of depression and anxiety, and they account for the majority of HIV cases in Africa (UNAIDS 2013). They are twice as likely to contract HIV than their male counterparts and typically seroconvert at a younger age, meaning they can have chronic HIV infection for the majority of their lives. Sexual and gender-based violence contribute to the transmission of HIV, with gender disparities – both cultural and social – still evident in many areas (Teitelman et al. 2016). A study of a Ugandan HIV+ female cohort estimated a prevalence of depressive symptoms to be 47% and associated with a low CD4 count (less than 50) (Kaharuza et al. 2006). With higher rates of HIV infection and mental health diseases, women are at risk for severe manifestations of HAND, compounded by the lack of social support, particularly in patriarchal settings.

In addition to psychological stress, HIV-related stigma and low social support can delay HIV testing and treatment (Parcesepe et al. 2018a, b). Fear of exclusion from community and workplace due to the cultural construct of HIV stigma may be heightened in low-income settings where jobs are limited, and families are central to social support. Key populations including men who have sex with men (MSM), vulnerable youth, transgender people, and sex workers are at particular risk for these gaps in HIV care and mental health services, social isolation, and even political persecution under certain governments. For example, sex workers contributed to between 7 and 11% of new infections in Uganda, Swaziland, and Zambia and up to a third of new infections in West Africa (Gouws et al. 2012). Delayed counseling and testing among these vulnerable groups and a lack of social support could result in more advanced cases of HIV, lack of ART adherence, and again an associated increase in the risk of advanced cases of HAND and other complications related to untreated HIV infection.

4 Consequences of HAND: Productivity, Quality of Life, and Morbidity

As the leading cause of neurocognitive dysfunction in young adults in sub-Saharan Africa, HIV has negative implications on patients' daily activity, including their ability to work, socialize, and overall quality of life. HAND also impacts an HIV+

individuals' daily functioning which can be seen even with early neurocognitive impairment (Tozzi et al. 2007; Doyle et al. 2013).

The majority of HIV-infected individuals in RLS are in their prime working and reproductive age (15–45 years old), when it would be necessary to financially support themselves and potentially a family. Cognitively impaired HIV+ patients with a low socioeconomic status were found to be more likely to be unemployed and fail social planning tasks (Benedict et al. 2000). Being unable to support themselves and their household can lead to further health vulnerability, including depression, decreased access to medical care, and inadequate nutrition.

In RLS, this loss of productivity extends beyond the individual level and debilitates the economic development in countries where poverty is already rampant. In Ethiopia, HIV+ farmers spend between 11.6 and 16.4 h per week farming compared with 33.6 h weekly for healthy farmers (Food and Agriculture Organization 2001). Furthermore, it is expensive to manage millions of HIV+ individuals while preventing the spread of HIV, particularly when there is a lack of infrastructure to meet the increased disease burden and to provide lifesaving but lifelong medication. The national debt and dependence on foreign aid has increased exponentially for the countries most affected by the AIDS epidemic, perpetuating the entanglement of HIV and poverty in areas of limited resources and competing needs. Although there has been tremendous progress in managing HIV, there is still a need for sustainable, affordable interventions to prevent the spread of HIV, retain individuals in care, and improve health outcomes.

Impairment in memory, executive function, and psychomotor functioning can promote significant difficulties with adherence to medication regimens, independence in daily activities, and general health management (e.g., safe sex practices) (Heaton et al. 2004). This impairment can impede independence of HIV+ individuals and require assistance from a caretaker, subtracting another salary from the low-income household. The caretaker's role can have essential medical importance from aiding in the diagnosis of HAND (Kisakye et al. 2018) to ART adherence, which can be essential in RLS where there is limited access to clinics. Thus, the clinical manifestations of HAND do not only plague the patient's quality of life but can also contribute to the long-term stress of a caretaker responsible for a cognitively impaired HIV+ individual (Small et al. 2017).

The most severe consequences conferred by HAND are disabling dementia and increased risk of mortality. HAD has been found to be an independent predictor of death (Lescure et al. 2011; Sevigny et al. 2007), typically seen in advanced stages of HIV, but ANI has also been implicated in earlier mortality (Heaton et al. 2011). Despite greater access to treatment, the HIV epidemic has significantly burdened the most heavily impacted societies, so the morbidity and mortality associated with HAND require urgent attention in RLS.

5 Advancing NeuroAIDS: Building an Infrastructure for Neurocognitive Testing in RLS

Poverty undermines advances in HIV research and treatment, particularly in specialized sectors such as neurology, psychiatry, and psychology that require trained investigators and clinicians, specific tools, and local interest. There is insufficient data on the neurocognitive effects of HIV in RLS since the neuropsychological studies carried out thus far are marked by inconsistent methods, test batteries, and rating systems for levels of cognitive impairment (Robertson et al. 2009).

The neuropsychological (NP) battery and neurological exams are sensitive tools for diagnosing HAND. Screening tools, such as the International HIV Dementia Scale (IHDS), play an essential role for directing limited resources for the more severe cases. The original HIV Dementia Scale (Power et al. 1995) was modified to simplify the administration to patients by eliminating the antisaccades subtest and replacing the timed written alphabet and cube copy time subtests with tests of motor speed (finger tapping) and psychomotor speed (an alternating hand sequence test) (Sacktor et al. 2005). The IHDS was evaluated in both American and Ugandan clinics to determine if it can easily be performed across cultures. However, there have been additional studies in RLS that have raised concerns regarding the psychometric properties across culture and low sensitivity for milder manifestations of HANDS (Joska et al. 2011). A recent study in Brazil found that a higher cutoff point for impairment improved the marginal sensitivity but still compiled data to suggest that the IHDS has limited utility as a screening measure when compared to other commonly used three-test screening batteries (de Almeida et al. 2017). Although IHDS has limitations, it is a publicly available tool that has shifted the focus to simplicity and standardization to allow for easier integration into practice and consistent data among diverse settings.

A major limitation in analyzing the clinical data from studies of HAND in RLS is the lack of local, culturally relevant normative cognitive data. Normative datasets consist of large cohorts stratified by age, gender, and education level to account for region-specific characteristics of language, culture, healthcare barriers, endemic infectious diseases, and genetic variability of both virus and host. The International Neurocognitive Normative Study (ACTG A5271) established normative data to provide a valid interpretation of the results from the study mentioned above on ART naive HIV+ individuals enrolled in the International Neurological Study (ACTG 5199). The substantial variations on the neurocognitive tests between countries indicated the need for country-based normative data for an appropriate comparison with HIV+ cohorts to create a sensitive screening and diagnosis of HAND in specific RLS. It also became apparent that age, education, and, to a lesser extent, gender are important variables in the variance associated with neurocognitive test differences and thus necessary to control for (Robertson et al. 2016). A Thai-based study evaluated this need for local norms by comparing normative data obtained locally in Thailand to Western norms. The Thai and US groups performed significantly differently on all neuropsychological measures except for verbal fluency (Heaps et al. 2012).

Heaton et al. (2008) have refined Western assessment methods to ensure they are suitable in international settings, such as a large study in rural China that developed demographically corrected neuropsychological test norms based upon HIV-negative individuals (Heaton et al. 2008). A similar battery by Heaton has also estimated the prevalence of HAND with demographically matched controls in diverse settings, including Cameroon (Kanmogne et al. 2010), Nigeria (Akolo et al. 2014), and Zambia (Kabuba et al. 2016). A recent study in Southern India (Kamat et al. 2017) confirmed that this neuropsychological battery, when translated into Tamil, was still understood by participants and identified a similar prevalence of HIV neurocognitive deficits as an earlier study in Central India where the battery was available in Marathi (Ghate et al. 2015). Modifying neuropsychological tests to the local language requires more than a translation, since the battery must retain cultural relevance with specific terms understood by the local population. These studies have expanded the versatility of a sensitive battery in regions with prevalent HIV-1 clade C infections and advanced knowledge through the collaboration with international researchers.

At the premise of the gaps in conducting neuropsychological research in RLS is the lack of trained personnel. The WHO Global Burden of Disease 2010 analysis estimates that, together, neurological and psychiatric disorders account for more than 13% of global disease burden, with much of this burden borne by the developing world (Mohammadi 2011). Yet, there are deficient educational resources for physicians and social awareness of these conditions. The International Neurological Study laid infrastructure for future studies by training the site staff to administer the neuropsychological tests and neurological assessments, as have the studies by Heaton et al. These instruments, including the user-friendly IHDS, were previously absent in many of these settings, and now there are translated exams with trained administrators to conduct the assessments for both research and clinical purposes. The ability to conduct neurological and neuropsychological research in RLS will facilitate an expansion of the NeuroAIDS field.

6 Clinical Implications, Translational Aspects, and Future Directions

By mid-2017, 20.9 million people were receiving ART globally (WHO 2017). While this number illustrates a formidable global health success, many are still without treatment. As the population rapidly expands in African and Asian countries, there will be a new, larger generation of adolescent and young adults, who are at the highest risk for HIV. The combination of these factors could lead to an inevitable rebound in the epidemic unless prevention, treatment, and research efforts to combat the infection are enhanced. Furthermore, the burden of dementia is increasing in low-middle-income countries without a proportional increase in the availability of treatment for those with severe cognitive and psychiatric problems

(Ferri et al. 2005). This is in direct contrast to Western Europe and the United States, where a 22–40% decrease in prevalence has been observed and has been attributed to improved education, lifestyle, and living conditions (Wu et al. 2016; Langa et al. 2017).

For those without proper access to testing and treatment of HIV, the symptoms of HAND may be more advanced and could even be a presenting symptom of HIV in those previously undiagnosed. Clinical knowledge and awareness of HAND is thus essential in areas of high HIV prevalence, particularly in low-income countries that have barriers to testing and treatment. Signs of cognitive impairment in young adults should be noted by healthcare professionals as additional incentive for the patient to be tested for HIV infection. As discussed earlier, signs and symptoms of HAND can interfere with an individual's daily activities and financial stability. Thus, physicians need to be informed on the diagnostic process for HAND, along with the resources for their patients, to ensure that ART is started as early as possible to prevent progression of the neurocognitive impairment.

Besides managing HIV infection with ART, there are no other specific treatments for HAND. Innovative studies could aim to identify ART with beneficial effects on the CNS, as well as novel adjuvant therapy. A better understanding of the HIV-1 clades and their neurovirulence factors can contribute to both a more effective treatment as well as the quest for a cure. The immunopathology of these strains has not been fully examined because of the research constraints in these low-income regions, and this may be a missed opportunity for insight into the virus that continues to evade efforts for developing a cure.

Furthermore, the fact that residual mild neurocognitive impairment is unaddressed by effective ART is still a clinical and research concern (Marra et al. 2009). A recent study from the United States study showed continued neurocognitive impairment among a cohort with long-term suppressive ART (median of 8.5 years). The CSF samples collected from this cohort showed that nearly half of the cells had detectable HIV DNA. There was an association between the poor neurocognitive performance and the isolated CSF cells with detectable HIV DNA, suggesting a functional neurocognitive consequence to persistent HIV in the brain (Spudich et al. 2018). Additional research needs to be performed to understand HIV reservoirs in the brain and its clinical manifestation as persistent neurocognitive impairment.

7 Conclusion

Since the advent of ART, HIV+ individuals have a longer life expectancy and less risk of central nervous system opportunistic infections. However, in order for these individuals to not only live a longer but have a productive and improved quality daily life, more attention needs to be focused on preventing and treating milder HIV neurocognitive complications. This is particularly relevant in low-income countries that continue to have barriers to ART availability, psychosocial stressors, high rates of coinfections, low educational attainment, nutritional deficiency, and possibly more neurovirulent HIV-1 clades.

By improving access to equal quality healthcare and education in RLS, the neurocognitive functioning of HIV+ individuals can be recognized and treated earlier. This would minimize the risks associated with a low CD4 count and higher viral loads. However, it will involve both social and medical advocacy to secure the infrastructure necessary to improve the healthcare and educational sectors.

Though there have been strides in improving research and clinical care for HAND in these regions, more steps must be taken to fully understand the dynamics of the virus and disease course in the areas of highest HIV prevalence. Neuropsychological examinations can allow clinicians and researchers to create a better understanding of the causes of neurocognitive impairment and whether it is directly attributable to HIV, comorbid factors, and/or immune factors associated with the HIV disease course. Establishing the causes and severity of the impairment further dictates the treatment and impacts the patient's daily life. Thus, there are meaningful outcomes to implementing standardized, normalized neuropsychological exams in RLS, as several studies from this region have shown, and this offers additional incentive to make them widely available with training opportunities for clinical staff.

Overall, there is a steep burden of disease in low-income countries with insufficient means to meet the medical demand in a manner that offers optimal patient care. The lack of infrastructure, healthcare workers, education, sanitation, and social policies has facilitated the spread of HIV in RLS and perpetuates the complications of the disease, including HAND. It is necessary to continue scale up of ART and build an infrastructure for sustainable healthcare and research to manage the common complication of neurocognitive impairment in HIV+ individuals. Otherwise, HAND will continue to lead to a loss in productivity and more HIV-associated deaths in areas that have already suffered the brunt of the HIV epidemic.

References

Akolo C, Royal W, Cherner M et al (2014) Neurocognitive impairment associated with predominantly early stage HIV infection in Abuja, Nigeria. J Neurovirol 20(4):380–387

Antinori A et al (2007) Updated research nosology for HIV associated neurocognitive disorders. Neurology 69(18):1789–1799

Awori V, Mativo P, Yonga G, Shah R (2018) The association between asymptomatic and mild neurocognitive impairment and adherence to antiretroviral therapy among people living with human immunodeficiency virus. S Afr J HIV Med 19(1):674

Belete T, Medfu G, Yemiyamrew E (2017) Prevalence of HIV associated neurocognitive deficit among HIV positive people in Ethiopia: a cross sectional study at Ayder Referral Hospital. Ethiop J Health Sci 27(1):67–76

Benedict RH, Mezhir JJ, Walsh K, Hewitt RG (2000) Impact of human immunodeficiency virus type-1-associated cognitive dysfunction on activities of daily living and quality of life. Arch Clin Neuropsychol 15(6):535–544

Bigna JJ, Um LN, Asangbeh SL, Sibetcheu AT, Kazé AD, Nansseu JR (2018) Prevalence and incidence of major depressive disorders among people living with HIV residing in Africa: a systematic review and meta-analysis protocol. Syst Rev 7(1):6

Brandt R (2009) The mental health of people living with HIV/AIDS in Africa: a systematic review. Afr J AIDS Res 8:123–133

Chan P, Tokac U, Hellmuth J, Kroon E, Colby D, Sacdalan C, Fletcher J, Tipsuk S, Pinyakorn S, Robb ML, Ananworanich J, Valcour V, Spudich S, Paul R (2018) Longitudinal cognitive outcomes after treatment in acute HIV infection. In: Conf. retroviruses opportunistic infect. Abstract 2018

Chibanda D, Benjamin L, Weiss HA, Abas M (2014) Mental, neurological, and substance use disorders in people living with HIV/AIDS in low- and middle-income countries. J Acquir Immune Defic Syndr 67(Suppl 1):S54–S67

Cournos F, McKinnon K, Pinho B, Waineberg ML (2014) Special populations and public health aspects. In: Joska JA, Stein DJ, Grant I (eds) HIV/AIDS and psychiatry. Wiley, West Essex, pp 211–234

Cysique LA, Dermody N, Carr A, Brew BJ, Teesson M (2016) The role of depression chronicity and recurrence on neurocognitive dysfunctions in HIV-infected adults. J Neurovirol 22(1):56–65

de Almeida SM, Kamat R, Cherner M et al (2017) Improving detection of HIV-associated cognitive impairment: comparison of the international HIV dementia scale and a brief screening battery. J Acquir Immune Defic Syndr 74(3):332–338

De Ronchi D, Faranca I, Berardi D et al (2002) Risk factors for cognitive impairment in HIV-1-infected persons with different risk behaviors. Arch Neurol 59:812–818

Doyle KL, Morgan EE, Morris S et al (2013) Real-world impact of neurocognitive deficits in acute and early HIV infection. J Neurovirol 19(6):565–573

Estiasari R, Lastri N (2015) Cognitive impairment among Indonesia HIV naïve patients. Neurol Asia 20(2):155–160

Ferri CP, Prince M, Brayne C, Brodaty H, Fratiglioni L, Ganguli M, Hall K, Hasegawa K, Hendrie H, Huang Y et al (2005) Global prevalence of dementia: a delphi consensus study. Lancet 366:2112–2117

Food and Agriculture Organization of the United Nations (2001) Rural women carry family burdens. Focus, AIDS – a threat to rural Africa. FAO, Rome

Ghate M, Mehendale S, Meyer R et al (2015) The effects of antiretroviral treatment initiation on cognition in HIV-infected individuals with advanced disease in Pune, India. J Neurovirol 21(4):391–398

Gouws E, Cuchi P, on behalf of the International Collaboration on Estimating HIV Incidence by Modes of Transmission (2012) Focusing the HIV response through estimating the major modes of HIV transmission: a multi-country analysis. Sex Transm Infect 88(Suppl_2):i76–i85. https://doi.org/10.1136/sextrans-2012-050719

Grant I, Franklin DJ, Deutsch R et al (2014) Asymptomatic HIV-associated neurocognitive impairment increases risk for symptomatic decline. Neurology 10:2055–2062

Habib AG, Yakasai AM, Owolabi LF, Ibrahim A, Habib ZG, Gudaji M, Karaye KM, Ibrahim DA, Nashabaru I (2013) Neurocognitive impairment in HIV-1-infected adults in Sub-Saharan Africa: a systematic review and meta-analysis. Int J Infect Dis 17(10):e820–e831

Heaps J, Valcour V, Chalermchai T, Paul R, Rattanamanee S et al (2012) Development of normative neuropsychological performance in Thailand for the assessment of HIV-associated neurocognitive disorders. J Clin Exp Neuropsychol 35(1):1–8

Heaton RK, Marcotte TD, Mindt MR, Sadek J, Moore DJ, Bentley H, McCutchan JA, Reicks C, Grant I (2004) The impact of HIV-associated neuropsychological impairment on everyday functioning. J Int Neuropsychol Soc 10(3):317–331

Heaton RK, Cysique LA, Jin H et al (2008) Neurobehavioral effects of human immunodeficiency virus infection among former plasma donors in rural China. J Neurovirol 14(6):536–549

Heaton R et al (2010) HIV-associated neurocognitive disorders (HAND) persist in the era of potent antiretroviral therapy: the CHARTER study. Neurology 75:2087–2096
Heaton RK, Franklin DR, Ellis RJ et al (2011) HIV-associated neurocognitive disorders before and during the era of combination antiretroviral disorders before and during the era of combination arntiretroviral therapy; differences in rates, nature, and predictors. J Neurovirol 17(1):3–16
Heaton R et al (2012) Aging amplifies HIV neurocognitive impairment: the effects may be related to vascular and metabolic factors. J Neurovirol 17(1):3–16
Hestad KA, Chinyama J, Anitha MJ, Ngoma MS et al (2019) Cognitive impairment in zambians with HIV infection and pulmonary tuberculosis. JAIDS 80(1):110–117
INSIGHT START Study Group, Lundgren JD, Babiker AG, Gordin F, Emery S, Grund B et al (2015) Initiation of antiretroviral therapy in early asymptomatic HIV infection. N Engl J Med 373:795–807
Joska JA, Westgarth-Taylor J, Hoare J et al (2011) Validity of the international HIV dementia scale in South Africa. AIDS Patient Care STDs 25:95–101
Joska JA et al (2012) Neuropsychological outcomes in adults commencing highly active antiretroviral treatment in South Africa: a prospective study. BMC Infect Dis 12(1):39
Jumare J, El-Kamary SS, Magder L et al (2018) Body mass index and cognitive function among HIV-1 infected individuals in China, India and Nigeria. J Acquir Immune Defic Syndr 80(2): e30–e35
Kabuba N, Menon JA, Franklin DR Jr, Heaton RK, Hestad KA (2016) HIV- and AIDS-associated neurocognitive functioning in Zambia – a perspective based on differences between the genders. Neuropsychiatr Dis Treat 12:2021–2028
Kabuba N, Menon JA, Franklin DR Jr, Lydersen S, Heaton RK, Hestad KA (2018) Effect of age and level of education on neurocognitive impairment in HIV positive Zambian adults. Neuropsychology 32(5):519–528
Kaharuza FM, Bunnell R, Moss S et al (2006) Depression and CD4 cell count among persons with HIV infection in Uganda. AIDS Behav 10(Suppl 1):105
Kalula S, Petros G (2011) Responses to dementia in less developed countries with a focus on South Africa. Glob Aging 7:31–40
Kamat R, McCutchan A, Kumarasamy N et al (2017) Neurocognitive functioning among HIV-positive adults in southern India. J Neurovirol 23:750
Kanmogne GD, Kuate CT, Cysique LA et al (2010) HIV-associated neurocognitive disorders in sub-Saharan Africa: a pilot study in Cameroon. BMC Neurol 10:60
Kisakye A, Saylor D, Sacktor N et al (2018) Caregiver versus self-reported activities of daily living among HIV-positive persons in Rakai, Uganda. AIDS Care 15:1–4. https://doi.org/10.1080/09540121.2018.1557591
Langa KM, Larson EB, Crimmins EM et al (2017) A comparison of the prevalence of dementia in the United States in 2000 and 2012. JAMA Intern Med 177(1):51–58
Lawler K et al (2010) Neurocognitive impairment among HIV-positive individuals in Botswana: a pilot study. J Int AIDS Soc 13:15
Lescure FX, Omland LH, Engsig FN, Roed C, Gerstoft J, Pialoux G, Kronborg G, Larsen CS, Obel N (2011) Incidence and impact on mortality of severe neurocognitive disorders in persons with and without HIV infection: a Danish nationwide cohort study. Clin Infect Dis 52(2):235–243
Liner KJ 2nd, Hall CD, Robertson KR (2007) Impact of human immunodeficiency virus (HIV) subtypes on HIV-associated neurological disease. J Neurovirol 13(4):291–304
Marra CM, Zhao Y, Clifford DB, Letendre S, Evans S, Henry K et al (2009) Impact of combination antiretroviral therapy on cerebrospinal fluid HIV RNA and neurocognitive performance. AIDS 23(11):1359–1366
Marwick KF, Kaaya SF (2010) Prevalence of depression and anxiety disorders in HIV-positive outpatients in rural Tanzania. AIDS Care 22(4):415–419
Mayston R, Kinyanda E, Chishinga N, Prince M, Patel V (2012) Mental disorder and the outcome of HIV/AIDS in low-income and middle-income countries: a systematic review. AIDS 26: S117–S135

Mayston R, Lazarus A, Patel V, Abas M, Korgaonkar P, Paranjape R et al (2016) Pathways to HIV testing and care in Goa, India: exploring psychosocial barriers and facilitators using mixed methods. BMC Public Health 16(1):765

Meffert SM, Neylan TC, McCulloch CE, Maswai JJ, Owuoth J, Kiweewa F, Maganga L, Valcour V, Ake J (2018) Depression is independently associated with near doubling of HIV viral load. In: Conf. retroviruses opportunistic infect. Abstract 2018

Milanini B, Allen IE, Bahemana E, Adamu Y, Kiweewa F, Maswai JJ, Owuoth J, Polyak C, Julie Ake J, Valcour V (2018) Predictors of HIV-related cognitive impairment in East Africa. In: Conf. retroviruses opportunistic infect. Abstract 2018

Mohammadi D (2011) Neurology in resource-poor countries: fighting for funding. Lancet Neurol 10:953–954

Morgan EE, Woods SP, Smith C et al (2012) AIDS Behav 16:2279

Mukherjee T, Sakthivel R, Fong HY et al (2018) Utility of using the Montreal cognitive assessment (MoCA) as a screening tool for HIV-associated neurocognitive disorders (HAND) in multi-ethnic Malaysia. AIDS Behav 22(10):3226–3233

Njamnshi AK, Bissek AC, Ongolo-Zogo P, Tabah EN, Lekoubou AZ, Yepnjio FN et al (2009) Risk factors for HIV-associated neurocognitive disorders (HAND) in Sub-Saharan Africa: the case of Yaoundé, Cameroon. J Neurol Sci 285:149–153

Parcesepe AM, Tymejczyk O, Remien RH, Gadisa T, Kulkarni S, Hoffman S, Melaku Z, Elul B, Nash D (2018a) Psychological distress, health and treatment-related factors among individuals initiating ART in Oromia, Ethiopia. AIDS Care 30(3):338–342

Parcesepe A, Tymejczyk O, Remien R, Gadisa T, Kulkarni S, Hoffman S, Melaku Z, Elul B, Nash D (2018b) HIV-related stigma, social support, and psychological distress among individuals initiating ART in Ethiopia. AIDS Behav 22(12):3815–3825

Power C, Selnes OA, Grim JA, McArthur JC (1995) HIV Dementia Scale: a rapid screening test. J Acquir Immune Defic Syndr Hum Retrovirol 8:273–278

Robertson K, Liner J, Heaton R (2009) Neuropsychological assessment of HIV-infected populations in international settings. Neuropsychol Rev 19:232–249

Robertson K, Jiang H, Kumwenda J, Supparatpinyo K, Evans S, Campbell TB, Group, A. C. T et al (2012) Improved neuropsychological and neurological functioning across three antiretroviral regimens in diverse resource-limited settings: AIDS Clinical Trials Group study a5199, the International Neurological Study. Clin Infect Dis 55(6):868–876

Robertson K, Jiang H, Evans S et al (2016) International neurocognitive normative study: neurocognitive comparison data in diverse resource limited settings: AIDS Clinical Trials Group A5271. J Neurovirol 22(4):472–478

Robertson K, Lama J, Pilcher C, Rios J, Brandes P, Ruiz E et al (2017) Can we afford to wait? ART and the CNS. In: Conf. retroviruses opportunistic infect. Abstract 2017

Robertson K, Jiang H, Kumwenda J, Supparatpinyo K, Evans S, Campbell TB et al (2018) HIV-1 and TB co-infection in multinational resource limited settings: increased neurological dysfunction. Clin Infect Dis. https://doi.org/10.1093/cid/ciy718

Sacktor N, Wong M, Nakasujja N, Skolasky RL, Selnes OA, Musisi S, Robertson K, McArthur JC, Ronald A, Katabira E (2005) The International HIV Dementia Scale: a new rapid screening test for HIV dementia. AIDS 19(13):1367–1374

Sacktor N, Nakasujja N, Skolasky RL et al (2009) HIV subtype D is associated with dementia, compared with subtype A, in immunosuppressed individuals at risk of cognitive impairment in Kampala, Uganda. Clin Infect Dis 49(5):780–786

Sacktor N, Nakasujja N, Redd AD, Manucci J, Laeyendecker O, Wendel SK et al (2014) HIV subtype is not associated with dementia among individuals with moderate and advanced immunosuppression in Kampala, Uganda. Metab Brain Dis 29:261–268

Sacktor N, Saylor D, Nakigozi G, Nakasujja N, Robertson KR, Grabwoski K, Kisakye A, Batte J, Mayanja R, Anok A, Gray RH, Wawer MJ (2019) Effect of HIV subtype and antiretroviral therapy on HIV-associated neurocognitive disorder (HAND) in Rakai, Uganda. JAIDS. in press

Sanmartí M, Meyer AC, Jaen A, Robertson K, Tan N, Samson L, Ndaki R, Battegay M, Tanner M, Dalmau D, Letang E (2018) Prevalence of HIV neurocognitive disorders in a rural Tanzanian cohort. In: Conf. retroviruses opportunistic infect. Abstract 2018

Saylor D, Dickens AM, Sacktor N et al (2016) HIV-associated neurocognitive disorder – pathogenesis and prospects for treatment. Nat Rev Neurol 12(4):234–248

Sevigny JJ, Albert S, McDermott M et al (2007) An evaluation of neurocognitive status and markers of immune activation as predictors of time to death in advanced HIV infection. Arch Neurol 64(1):97–102

Small J, Aldwin C, Kowal P, Chatterji S (2017) Aging and HIV-related caregiving in sub-Saharan Africa: a social ecological approach. Gerontologist. https://doi.org/10.1093/geront/gnx159

Spudich S, Gandhi R, Cyktor J, Lalama C, Bosch R, Macatangay B, Rinaldo C, Robertson K, Collier A, Godfrey C, Hogg E, Eron J, McMahon D, Mellors J (2018) HIV-1 persists in CSF cells in half of individuals on long term ART. In: 25th Conference on retroviruses and opportunistic infections (CROI), Boston, Mar 2018

Stern Y (2002) What is cognitive reserve? Theory and research application of the reserve concept. J Int Neuropsychol Soc 8:448–460

Tedaldi EM, Minniti NL, Fischer T (2015) HIV-associated neurocognitive disorders: the relationship of HIV infection with physical and social comorbidities. Biomed Res Int 2015:641913

Teitelman AM, Jemmott JB, Bellamy SL et al (2016) Partner violence, power, and gender differences in South African adolescents' HIV/sexually transmitted infections risk behaviors. Health Psychol 35:751–760

Tozzi V, Balestra P, Bellagamba R, Corpolongo A, Salvatori MF, Visco-Comandini U, Vlassi C, Giulianelli M, Galgani S, Antinori A, Narciso P (2007) Persistence of neuropsychologic deficits despite long-term highly active antiretroviral therapy in patients with HIV-related neurocognitive impairment: prevalence and risk factors. J Acquir Immune Defic Syndr 45(2):174–182

Troncoso FT, Conterno LO (2015) Prevalence of neurocognitive disorders and depression in a Brazilian HIV population. Rev Soc Bras Med Trop 48:390–398

UNAIDS (2013) Report on the global AIDS epidemic 2013. UNAIDS, Geneva

UNESCO Institute for Statistics (UIS) (2016) Youth literacy

Wong M, Robertson K, Nakasujja N et al (2007) Frequency of and risk factors for HIV dementia in an HIV clinic in sub-Saharan Africa. Neurology 68:350–355

World Bank (2017) The world bank atlas. World Bank, Washington, DC

World Health Organization (2011) Mental health atlas. World Health Organization, Geneva

World Health Organization (2016) Estimated number of people living with HIV, by WHO region, static map

World Health Organization (2017) Global tuberculosis report

Wu YT, Fratiglioni L, Matthews FE, Lobo A, Breteler MM, Skoog I, Brayne C (2016) Dementia in western Europe: epidemiological evidence and implications for policy making. Lancet Neurol 15:116–124

Yepthomi T, Paul R, Vallabhaneni S, Kumarasamy N, Tate D, Solomon S, Flanigan T (2006) Neurocognitive consequences of HIV in southern India: a preliminary study of clade C virus. J Int Neuropsychol Soc 12:424–430

Yusuf AJ et al (2017) Prevalence of HIV-associated neurocognitive disorder (HAND) among patients attending a tertiary health facility in Northern Nigeria. J Int Assoc Provid AIDS Care 16(1):48–55

Deep Phenotyping of HIV Neurocognitive Complications Among Individuals Residing in High-Income Countries

Robert Paul, Paola Garcia-Egan, Jacob Bolzenius, and Julie Mannarino

Contents

Abstract People living with HIV (PLWH) residing in high-income countries (HICs) are, in theory, well positioned to benefit from clinical care strategies that predict optimal neurocognitive and neuropsychiatric outcomes. However, there is substantial inter-individual variability in access to clinical care, prevalence of co-occurring risk factors, and comorbid health conditions that represent barriers to achieving the full potential of antiretroviral therapy (ART). Complex interactions between these variables translate into heterogeneity in HIV clinical phenotypes, including abnormalities in brain structure and function. The growing population of PLWH in HICs who are now reaching advanced age introduces additional causal pathways of neurocognitive variability among PLWH receiving ART. These patterns foreshadow trends expected to develop globally in response to increased access to ART. This chapter reviews the combination of highly dimensional risk factors for neurocognitive complications among PLWH residing in HICs. We begin with a brief description of the neuropathological, neuroimaging, and neurocognitive signatures of HIV, followed by a summary of controversies regarding the clinical presentation of HIV-associated neurocognitive disorders (HAND), including putative synergies

R. Paul (✉), P. Garcia-Egan, J. Bolzenius, and J. Mannarino
Missouri Institute of Mental Health, St. Louis, MO, USA
e-mail: paulro@umsl.edu

Curr Topics Behav Neurosci (2021) 50: 245–270
https://doi.org/10.1007/7854_2020_185
Published Online: 14 January 2021

between HIV disease dynamics and advanced age. Finally, we introduce innovative research strategies that have potential to advance the existing conceptual framework of HAND and, ideally, catalyze the development and of clinical interventions needed to achieve HIV treatment and eradication efforts.

Keywords Data science · High-income countries · HIV · HIV-associated neurocognitive disorders (HAND) · Neuroimaging · Neuropsychology

1 Introduction

In 2019, 81 countries were designated by the World Bank as high-income countries (HICs), a designation that is based on per capita income of approximately $12K/year. At the macro level, individuals residing in HICs are positioned to benefit from HIV-related health programs that can reduce the lag time between symptom onset and confirmation of HIV serostatus, access to antiretroviral therapy (ART), and access to ongoing clinical care. However, a large body of evidence indicates that a large percentage of PLWH who reside in HICs exhibit persistent disease complications (e.g., neurocognitive symptoms) despite sustained use of suppressive ART. Understanding these outcomes is facilitated by the recognition that many of the advantages of residing in HICs do not translate equally to all individuals. Rather, there is substantial variability in social, cultural, economic, and related ethnographic factors that moderate the risk of neurocognitive symptoms (Paul 2019). For example, black individuals residing in financially impoverished regions of the USA experience a longer delay between symptom onset and HIV diagnosis, more severe disease status at the time of ART onset, and a lower rate of viral suppression after treatment when compared to white PLWH (Avert 2018). An additional complicating factor is the disproportionate frequency of HIV among individuals who have recently migrated to HICs from lower income regions of the world. For example, studies conducted in Canada, United Kingdom, and Germany reveal that up to 40% of new HIV diagnoses involve individuals who had recently migrated from lower income countries (Alvarez-del Arco et al. 2013; Avert 2017; Krentz and Gill 2009; Prosser et al. 2012; Ross et al. 2018). Clearly, one cannot assume that PLWH who reside in HICs share the same profile of risk and protective mechanisms for neurocognition.

Biological diversity is a cardinal feature of HIV. Studies conducted in HICs report significant inter-individual variability HIV disease dynamics before individuals initiate ART. For example, nadir CD4 T-cell count, current CD4 T-cell count, viremic set point, detectability of HIV in the cerebrospinal fluid (CSF), and plasma, as well as immune activation and inflammatory markers differ substantially across individuals (Gawron et al. 2019; Guha et al. 2016a; Imp et al. 2017; McLaren et al. 2015; Sanford et al. 2018a; Strain et al. 2017). After treatment onset, PLWH continue to exhibit variable degrees of immune discordance (i.e., suboptimal CD4 T-cell reconstitution despite undetectable viral load), viral blips, and treatment failure (Pernas et al. 2016; Robbins et al. 2007; Zoufaly et al. 2011). Differences in host factors including early life adversity, mental distress during adulthood, and substance use are hypothesized to contribute to HIV clinical heterogeneity and risk strata for neurocognitive symptoms (Clark et al. 2018; Hellmuth et al. 2017; Ladak et al. 2019) that comprise HIV-associated neurocognitive disorders (HAND) (Antinori et al. 2007).

This chapter focuses on the multidimensional nature of risk factors for HAND among PLWH residing in HICs. Our work targets studies conducted in the USA, United Kingdom, Canada, Germany, and Australia. We begin with a review of the neuropathological, neuroimaging, and neurocognitive phenotypes of HIV, with comparisons drawn between key studies conducted before vs. after the introduction of ART. We then summarize current controversies regarding the clinical presentation of HAND among individuals residing in HICs, including potential synergies between mechanisms of HIV and advanced age. Lastly, we provide an introduction to data science methods that have potential to inform the conceptual model of HAND and, ideally, offer a new path toward achieving precision health initiatives that are essential to the successful implementation of global HIV treatment initiatives.

2 Variability in Viral-Host-Brain Dynamics Begins During Early Infection

An oft cited observation is that HIV enters the CSF soon after viral infection. A study conducted among individuals with less than 4 weeks of infection revealed detectable levels of HIV RNA in central nervous system (CNS) among adults in Thailand with acute HIV (Valcour et al. 2012). Evidence of HIV penetration was observed as early as Fiebig stage I, which corresponds to approximately 1 week postinfection. Furthermore, most cases exhibited CNS involvement by Fiebig stage III (approximately 3 weeks postinfection), which coincides with the time when viral reservoirs are established. However, HIV RNA was not detected in the CSF in 17% of the sample regardless of Fiebig stage or the degree of neuroinflammation measured by neopterin. These results point toward viral factors associated with brain dysfunction operative prior to the onset of ART. Whether this variability predicts individual risk for the development of neurocognitive impairment represents an important area of future investigation.

3 Neuropathological Signatures of HIV

Neuropathological studies conducted during the early period of the HIV epidemic in HICs described astrocytic reaction and infiltration of multinucleated cells (Navia et al. 1986; Price et al. 1988). Studies also reported high levels of parenchymal macrophages and perivascular lymphocytes in subcortical brain regions, with heavy involvement of brain regions near the third ventricle (e.g., basal ganglia) (Grant et al. 1987; Navia et al. 1986; Price et al. 1988; Price and Brew 1988; Pumarola-Sune et al. 1987). The predilection for frontal-subcortical brain circuits remains the prevailing conceptual model of HIV-associated brain dysfunction (Paul 2019). This phenotype holds in the current era of ART, despite the high rate of cerebrovascular disease among older PLWH (Masia et al. 2019; Moulignier et al. 2018; Vinikoor et al. 2013). While concern has been raised regarding an increased risk of other age-related brain disorders such as Alzheimer's disease (AD) among older PLWH, empirical studies have not demonstrated a clear link between HIV serostatus and AD neuropathology (Ances et al. 2010, 2012a; Cooley et al. 2019; Milanini et al. 2019; Soontornniyomkij et al. 2012; Soontornniyomkij et al. 2019).

4 Neuroimaging Abnormalities of HIV

In the pre-ART era, neuroimaging investigations revealed high signal intensity, ventricular and sulcal dilation (i.e., cortical atrophy), and reduced basal ganglia volumes among PLWH with immunosuppression (Grant et al. 1987; Kramer and Sanger 1990; Navia et al. 1986; R. Paul et al. 2002). Reduced white matter volume and increased burden of white matter hyperintensities (WMH) were also commonly reported (McArthur et al. 1990; Olsen et al. 1988). Results from studies conducted in the post-ART era describe the same general neuroimaging signature of HIV (Ances et al. 2012b; Heaps-Woodruff et al. 2018; Heaps et al. 2012; Ortega et al. 2013; P. W. Wright et al. 2015). Recent studies also reveal alterations in broad cortical and subcortical brain networks (Baker et al. 2017; Guha et al. 2016b; Ortega et al. 2015; Thomas et al. 2013) that correspond to more severe neurocognitive symptoms (du Plessis et al. 2017). Finally, volumetric reductions have been observed in subcortical and cortical brain regions, with modest associations noted across immunological and cognitive indices (Guha et al. 2016b; Sanford et al. 2017; Wade et al. 2015).

Interestingly, Sanford et al. (2018a) reported subcortical brain atrophy among PLWH with less than 1 year of disease duration, with no further atrophy observed over the first 6 months of ART. By contrast, studies conducted in Thailand (a middle-income country) report no evidence of neuroimaging abnormalities in the first 30 days of infection (Samboju et al. 2018), but progressive basal ganglia atrophy over a 2-year period (Kallianpur et al. 2020). These latter findings are concerning, particularly since study participants had initiated suppressive ART in the first weeks of infection. The latter study did not include HIV uninfected individuals as a reference group, but the young age of the sample (~26 years old) suggests that the neuroimaging changes over 2 years are more severe than what would be expected from age alone.

5 The Neurocognitive Phenotype of HIV

Case–control studies conducted in HICs describe a neurocognitive profile of HIV that aligns with neuropathological abnormalities in subcortical brain regions. Investigations revealed that PLWH are prone to experience psychomotor slowing, reduced information processing speed, executive dysfunction, and low verbal and visual learning efficiencies (Kore et al. 2015; Navia 1997; Paul et al. 2002; Price and Brew 1988; Tate et al. 2011; Valcour et al. 2011). In the post-ART era, the neurocognitive phenotype of HIV remains "subcortical" in nature. In alignment with studies focused on biomarkers, there is no clear evidence of a shift in the neurocognitive pattern toward a "cortical" phenotype typical of AD (Paul 2019). Further, most studies report relatively stable neurocognitive symptoms in the context of sustained ART (Alford and Vera 2018; Ananworanich et al. 2016; Liner et al. 2008). Evidence of progressive neurocognitive decline (Grant et al. 2014) is unlikely in younger individuals who are virally suppressed and free of etiologies that complicate the diagnosis of HAND (e.g., psychiatric disease or substance use). The increased prevalence of cerebrovascular disease described above may precipitate progressive worsening of neurocognitive symptoms, but the profile remains subcortical in nature.

6 HIV Disease Correlates of Brain Injury

The degree of immune suppression before ART (indexed by nadir CD4 T-cell count) is associated with neuropathological, neuroimaging, and neurocognitive abnormalities in PLWH (Cohen et al. 2010; Ellis et al. 2011; Joska et al. 2010; Wright et al. 2018). After ART, current CD4 T-cell count is often discordant with the severity of brain injury. This discordance results from the rapid increase in plasma CD4 T-cell count that follows ART (Erb et al. 2000), without a concomitant resolution of neuronal damage (Gates and Cysique 2016; Sanford et al. 2018b). Other HIV disease correlates of brain injury include plasma markers of immune activation (e.g., soluble CD163), monocyte immunophenotypes (e.g., CCL2 receptors), neopterin, neurofilament light chain, and CD4/CD8 T cell ratio <1.0 (Burdo et al. 2011a, b; Du Pasquier et al. 2013; Gisslen et al. 2016; Letendre 2011; McGuire et al. 2015; Spudich 2016; Yilmaz et al. 2017). Nevertheless, a single biomarker or combination of biomarkers of HAND remains elusive.

7 Age-Related Mechanisms of HAND

The average age of HIV-infected individuals in HICs is now over 50 years of age (Centers for Disease Control and Prevention 2019; Clifford et al. 2017). The increase in life expectancy has raised concern about possible synergies between HIV and aging that have potential to exacerbate symptoms of HAND (Clifford et al. 2017; Cohen et al. 2015; Fazeli et al. 2014; Levine et al. 2016; Pfefferbaum et al. 2018; Saloner et al. 2019; Sheppard et al. 2017). Mechanisms of cardiovascular and cerebrovascular diseases, diabetes, and insulin resistance have gained recent attention as potential contributing factors to cerebrovascular disease and associated neurocognitive difficulties (Gallant et al. 2017; Onen et al. 2010; Pelchen-Matthews et al. 2018; Schouten et al. 2014). The severity of cardiovascular disease is associated with immune activation, systemic inflammation, and CD4+ T-cell depletion in older individuals receiving ART (Duprez et al. 2012; Lang et al. 2012; Lichtenstein et al. 2010). Other studies report higher absolute CD8+ T-cell count and lower CD4/CD8 ratio associated with coronary plaque burden and increased risk of myocardial infarction (Lang et al. 2012; Lo et al. 2010). Evidence of cerebrovascular burden on neuroimaging (Chu et al. 2018; Cysique et al. 2013; Jeon et al. 2017; Su et al. 2016) aligns with results from neuropathological studies conducted in the post-ART era, demonstrating increased prevalence of cerebrovascular disease in PLWH receiving ART.

One of the most interesting narratives in the current literature relates to the concept of premature and/or accelerated aging. Premature aging refers to an earlier onset of biological mechanisms that underlie degenerative age conditions (cardiovascular disease, etc.). Accelerated aging, by contrast, refers to a faster progression of related health conditions that begin at normative ages (Margolick and Ferrucci 2015). To date, studies have not demonstrated a clear example of either premature or accelerated aging in HIV-infected individuals. Cole et al. (2017) described lower brain age in HIV-infected individuals residing in HICs based on discrepancies in brain volumes relative to age-adjusted norms. While often cited as evidence of accelerated brain aging, the study reported no age by HIV serostatus interaction. Further, all brain regions were aggregated into a single imaging metric, which did not allow for examination of age-specific patterns. Pfefferbaum et al. (2018) described accelerated brain aging in 68 adults with chronic HIV. Using a mixture of cross-sectional and longitudinal structural neuroimaging data, more severe atrophy was observed in the thalamus and frontal and parietal regions in older individuals, with more severe changes observed among individuals with concurrent substance use disorders. However, longitudinal scans were unavailable for approximately half of the sample, and for the cases with more than one scan, the time interval between scans differed markedly (from 1 month to 8 years). Other studies describe reduced telomere lengths in older HIV-infected individuals receiving ART (Leeansyah et al. 2013; Liu et al. 2015; Quiros-Roldan et al. 2020) and alterations in epigenetic clocks (Horvath and Levine 2015) that implicate mechanisms of premature aging. However, these "aging" biomarkers have also been reported in

HIV-infected children (Cote et al. 2012; Gianesin et al. 2016; Shiau et al. 2018), suggesting that these biomarkers represent biological perturbations across a broader spectrum of disease mechanisms than age per se.

Increased rates of frailty among PLWH have been described as evidence of interactions between advanced age and HIV pathogenicity, but again, the findings have generated debate and controversy. Frailty is defined as a state of significant health vulnerability (Fried et al. 2001). Phenotypically, frailty reflects a constellation of clinical symptoms including reduced ambulation/motor speed, physical weakness, unintended weight loss, exhaustion, and/or reduced activity levels. Correlates of frailty in HIV include higher viral load and lower CD4 count, elevated plasma markers of immune activation, female sex, hepatitis C, depression, diabetes, and chronic obstructive pulmonary disease (Brothers et al. 2014; Desquilbet et al. 2007; Guaraldi et al. 2011; Morgello et al. 2019; Paul et al. 2018). Neuroimaging correlates of frailty are less well understood. However, recent work conducted by our group (Paul et al. 2020) utilized ensemble machine learning to identify a data-driven classifier of frailty in older PLWH. The algorithm was comprised of lower CD4 T-cell count, psychomotor performance, and neuroimaging indices of visual and motor brain systems (see Fig. 1). These results provide proof of concept that data-driven models can identify underlying mechanisms of complex clinical phenotypes among older HIV-infected individuals. Further, the results raise hope that interventions aimed at bolstering visuomotor skills and reducing mild to moderate depressive symptoms could improve health resilience in older individuals with chronic HIV.

8 Application of the Frascati Criteria for HAND

The Frascati criteria (Antinori et al. 2007) were developed to provide a general diagnostic framework for the classification of HAND. Central components of the Frascati criteria include a structured assessment of two or more neurocognitive domains with at least two tests in each domain, comparison of performances to normative references, assessment of real-world impact of neurocognitive difficulties, and assurance that the neurocognitive symptoms reflect mechanisms of HIV rather than comorbid conditions (e.g., depression, substance use). Unfortunately, fidelity to these recommendations varies widely across studies conducted in HICs. Further, standardized methods address the diverse range of socioeconomic, ethnic, and educational factors in the population in these regions (Gisslen et al. 2011; Winston et al. 2013). Hence, the accuracy and utility of the Frascati guidelines depend on the representativeness of the normative data for any given population. Overcoming this hurdle will require methods that allow for more flexibility in neurocognitive phenotyping.

Current methods to ascertain neurocognitive impairment are inflexibly dependent on the details of the assessment process. Demographic (e.g., age, sex, quality of education) and cultural (e.g., language, migrant status) variables, test selection, motivation and effort level of the examinee, fidelity to test administration instruction,

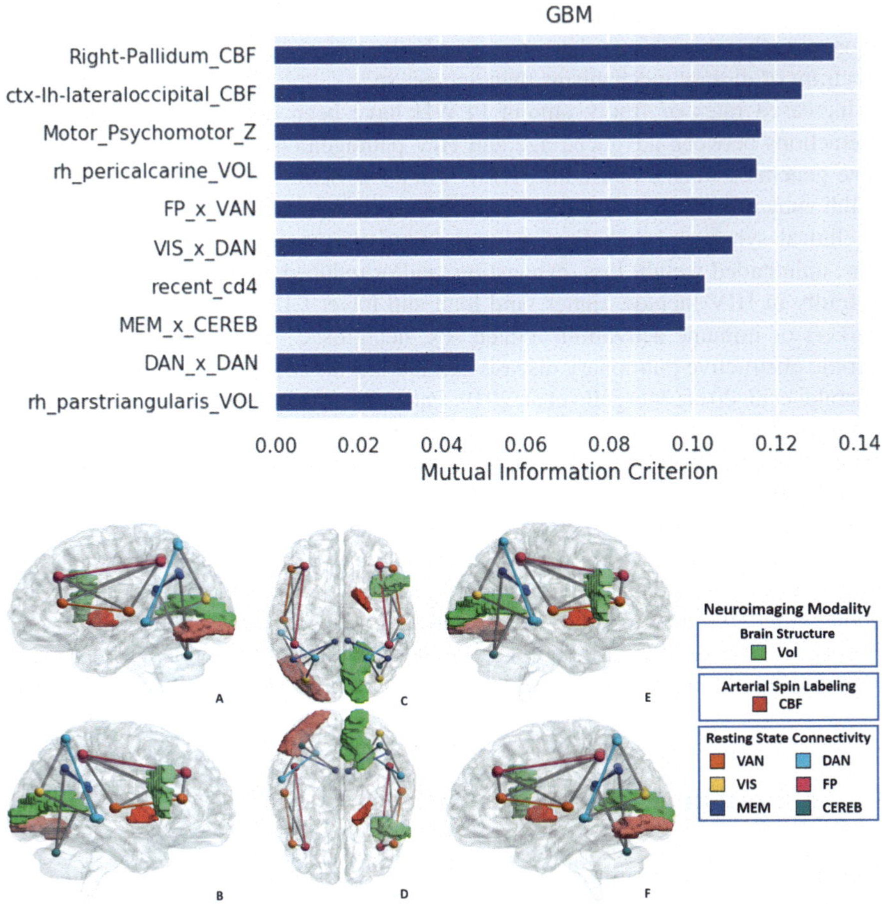

Fig. 1 Predictors of frailty among persons living with HIV individuals depicted by feature importance determined by gradient boosted multivariate regression (top) and multimodal neuroimaging signatures (bottom). Alterations were observed using brain volumetrics, cerebral blood flow (CBF), and resting state functional connectivity. Multiple views include superior (**a**), inferior (**b**), right (**c**), and left (**d**) hemispheres, and rostral (**e**) and caudal (**f**) perspectives

operational definition of impairment, etc. all represent potential sources of errors. These sources of error are equally important in studies of PLWH in HICs as they are to lower income regions of the world (Meyer et al. 2013; Saloner and Cysique 2017). Internal and external validities are also a concern when relying on screening measures (e.g., Mini Mental Status Examination) that lack sensitivity to milder forms of neurocognitive impairment (Brouillette et al. 2015; Elbirt et al. 2015; Janssen et al. 2015; Kim et al. 2016). Reviews suggest that screening tools such as CogState are reasonably sensitive to moderate levels of neurocognitive impairment

(Cysique et al. 2006, 2010; Zipursky et al. 2013). However, these screening tools are less sensitive to mild symptoms. Even with more detailed assessment methods, many systems lack appropriate norms for minority groups and individuals with low education levels (Jacobs et al. 1997; Manly et al. 2004; Nielsen et al. 2018; Perez-Arce and Puente 1996). Research conducted more than 15 years ago revealed that racial differences moderate test performance (Manly et al. 2004), yet we still lack a universal approach to mitigate these challenges. The impact of using inappropriate norms on prevalence rates can be seen in studies reported from the Pharmacokinetic and Clinical Observations in People Over Fifty (POPPY) cohort in the UK (Garvey et al. 2011), which describe exceptionally low rates of HAND based on results obtained from a computerized screening battery with unknown sensitivity to less severe neurocognitive impairment and unknown normative representation for minority groups. As reviewed by Saloner and Cysique (2017), studies that have utilized more comprehensive testing methods and culturally appropriate normative samples report persistent neurocognitive problems in approximately half of PLWH.

Alcohol and illicit drug use, both prevalent in HIV, contribute to the challenges with the current approach to HAND (Anthony et al. 2005; Attonito et al. 2014; Byrd et al. 2011; Devlin et al. 2012; Ferris et al. 2008; Meade et al. 2015; Nath 2010; Schuster and Gonzalez 2012; Tedaldi et al. 2015). Deleterious effects of heavy alcohol use and other substances on structural and functional brain metrics in PLWH have been extensively documented (Carey et al. 2006; Cohen et al. 2019; Gonzalez et al. 2011; Loftis et al. 2011; Meade et al. 2011; Thames et al. 2017). Despite the high prevalence of lifetime and recent polysubstance use among PLWH in HICs, there is no agreed upon method to determine how much use/misuse of a drug, or combination of drugs, meets threshold to disqualify a diagnosis of HAND. Many studies rely on the *Diagnostic and Statistical Manual of Mental Disorders* (*DSM-5*) (American Psychiatric Association 2013) to identify a clinical "use disorder," but this method emphasizes social burden as an indicator (e.g., arrests, loss of employment) vs. brain integrity. A universal approach for considering the potential contribution of recent and remote substance use is needed.

HAND determination also requires evidence of disruption to activities of daily living (ADLs) due to neurocognitive problems. Commonly, studies utilize self-report measures of daily functioning, most of which were developed for other neurological conditions (e.g., Alzheimer's disease). These questionnaires are low cost and low burden and have reasonable face validity (Blackstone et al. 2012; Shirazi et al. 2017). However, the results are susceptible to social desirability, recall bias, and the mental health of the respondent as well as degree of neurocognitive impairment (Blackstone et al. 2012). Further, it is difficult to determine the degree to which a reported decline in ADLs is due to neurocognitive problems rather than concurrent depression, apathy, fatigue, physical illness, and/or other etiological factors (Clifford and Ances 2013; Doyle et al. 2013; Obermeit et al. 2017; Vance et al. 2011). Performance-based measures mitigate some of these concerns, but the external validity of performance-based methods remains unclear. Additionally, a tautological trap exists when performance-based measures of ADLs rely on the same neurocognitive operations assessed in the neuropsychological testing. Unfortunately, variance in real-world functional correlates of either self-report or performance-based ADL measures is compounded by culturally insensitive items and inventories that comprise the majority of current measures (Kordovski et al. 2019).

An international task force was established in 2017 to overcome the challenges described above related to the Frascati criteria (Saloner and Cysique 2017). This task force is charged with ensuring that recommendations for testing consider the quality of the normative data for each task, strength of the psychometric properties (including acceptable floor and ceiling effects), ease of use (portability, brevity), cost, copyright protection, availability of alternate forms for longitudinal studies, and clear evidence of sensitivity to disease neuropathogenesis. The neuroHIV field has lagged behind other disciplines in the adoption of a common, or minimum, neurocognitive assessment protocol. The task force is encouraged to consider exemplars from other fields such as the protocols described by the Alzheimer's Disease Neuroimaging Initiative (ADNI) (Petersen et al. 2010), the Canadian Stroke Best Practice Recommendations for Mood, Cognition and Fatigue following Stroke (Hachinski et al. 2006), and the Minimal Assessment of Cognitive Function in Multiple Sclerosis (MACFIMS) (Benedict et al. 2006). Further, integration of measures of cognitive neuroscience is recommended to align with the ongoing NIH-sponsored Research Domain Constructs or "RDoCS" initiative aimed at dismantling diagnostic categories into dimensional constructs.

The task force will also need to consider the additive value of functional assessment and neuroimaging as qualifiers or nested requirements for a diagnosis of HAND. Intuitively, self-reported measures of decline in ADLs using procedures for dementia resulting from AD might create too much noise in the diagnostic algorithm vs. relying on cognitive performance using a standardized battery. This is an empirical question that should be addressed prior to any effort to revamp the diagnostic algorithm. Furthermore, the creation of normative data with accurate representation of the diverse HIV population should be prioritized. Additional research is needed to determine if neuroimaging signatures of HIV should be a prerequisite in the diagnostic algorithm.

9 Research Opportunities Using Data Science Methods

Traditional data analytic methods are not well suited to discover novel mechanisms of complex clinical phenotypes (see Miller et al. 2016). Inferential methods are restricted by statistical assumptions (e.g., normality), redundancy, and data manipulations that are necessary to address problems that arise from either too little (e.g., imputation) or too much (e.g., data aggregation) information. Traditional analytic methods also require investigators to identify predictors a priori before the underlying structure of the target clinical phenotype has been defined. This is a treacherous task that involves guesswork and/or recycling of significant variables reported in prior studies, a process that is unlikely to yield transformative outcomes. The task is

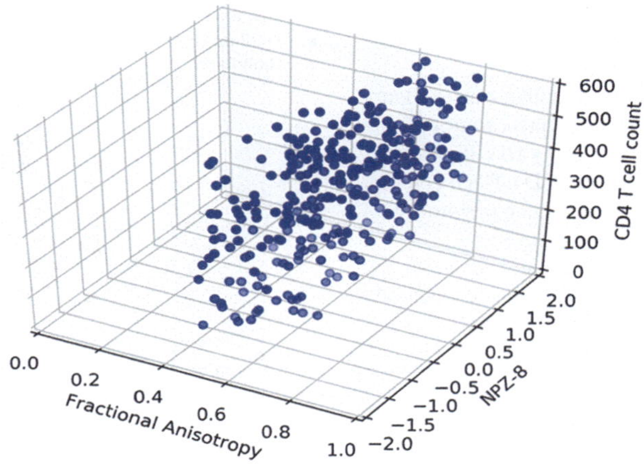

Fig. 2 Theoretical 3D representation of a single HIV neurocognitive phenotype

even more challenging when predictive models are built from longitudinal data and/or clinical phenotypes that morph over time in response to treatment, environmental factors, etc. These limitations have sparked recent interest in the application of analytic approaches that utilize a "top-down" approach that is most consistent with a systems biology framework. Whereas "bottom-up" frameworks require guesswork to build clinical phenotypes, top-down models rely on large amounts of dimensional data to identify patterns of mathematical dependencies without need for a priori assumptions about the most salient features or reliance on categorical labels (e.g., "impairment"). Data-driven top-down models also allow for examination of interactive mechanisms and phenotypes that change over time. This is critical for HIV given the clinical variability that has been reported from one person to the next and from one time point to the next.

Data-driven methods are not constrained by the number of data inputs or the dimensional characteristics of predictor variables, allowing for integration of large amounts of information that cuts across omnibus diagnostic categories to delineate subtypes that are easy to miss using traditional analytic strategies alone. Figures 2 and 3 depict the gain in model specificity, and insights into disease mechanisms, that can be achieved through integrative data-driven methods.

Research on neurocognitive symptoms of HIV frequently employ statistical methods to "control" variables described as "nuisance factors" (age, sex, etc.) that require methodological (e.g., recruitment that excludes subgroups) and/or statistical (e.g., covariate analyses) control. However, to the extent that variables represent key determinants, exclusion of these "nuisance" variables lowers the biological and clinical relevance of the final model. This is problematic for studies focused on HAND because we do not yet understand the underlying risk factors that explain individual risk for neurocognitive symptoms (Paul 2019). Traditional analytic strategies are not well designed to overcome this challenge. This is particularly true for studies that involve multivariate outcomes, in which inference testing is prone to producing spurious results that do not replicate in follow-up studies (Burnham and Anderson 2002; Miller et al. 2016). It is time to adopt a new approach.

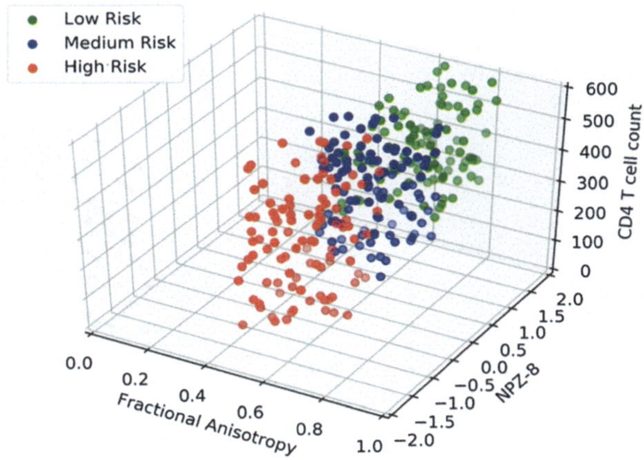

Fig. 3 Theoretical 3D representation of subgroups within the HIV neurocognitive cluster. Here the data represent groups clustered by HIV disease severity (CD4 T-cell count), neurocognitive performance (NPZ-8), and a neuroimaging marker of white matter microstructural integrity (fractional anisotropy)

An exemplar method that our group is currently utilizing to explore deep phenotypes of HAND is an analytic tool referred to as Correlation Explanation (CorEx) (Ver Steeg and Galstyan 2014). CorEX is an unsupervised data-driven clustering algorithm that identifies hierarchical representations of highly dimensional data using information theoretic principles. CorEx maximizes multivariate mutual information (defined as total correlation) by conditioning dependence among observed samples onto constructed latent factors. The hierarchical reconstruction of information interdependence is represented by variable layers, with lower layers referencing local relationships between variables and higher layers representing global or network interactions. This approach is unique among cluster-based approaches because all of the relational information is utilized, rather than a subset of nonoverlapping variables. As a result, results from CorEX more closely approximate the multidimensional approach of systems biology. We have applied CorEx to identify latent phenotypes of individuals with acute HIV, which revealed two highly differentiated subgroups, including a phenotype dominated by worse HIV/inflammatory disease features (immune response, viral load, Fiebig stage; 22%) in addition to emergent mental health features, and a phenotype dominated by mental health features (anxiety, depression; 25%%; Fig. 4).

Interestingly, when comparing disease outcomes 2 years after individuals initiated ART within weeks of infection, the mental health subgroup was less likely than the HIV disease severity subgroup to achieve a favorable clinical phenotype (i.e., no AIDS-defining illness or grade 4 Adverse Events (AEs), CD4 T cells >500 cells/mm^3, CD4/CD8 T-cell ratio >1.0, viral load <20 copies after 6 months of treatment; 48% vs. 33%). Even more intriguing, preliminary work from our group using ensemble machine learning, a form of machine learning revealed that CD4/CD8 T-cell ratio below 1.0 was the strongest classifier of treatment phenotype. We then identified five distinct clusters of CD4/CD8 T-cell ratio trajectories modeled from baseline (pre-ART) to week 96 post ART. The subgroups were comprised of individuals with CD4/CD8 T-cell ratio > 1 at each timepoint (Stable High: 32%); participants with CD4/CD8 T-cell ratio < 1 at baseline with progressive stabilization

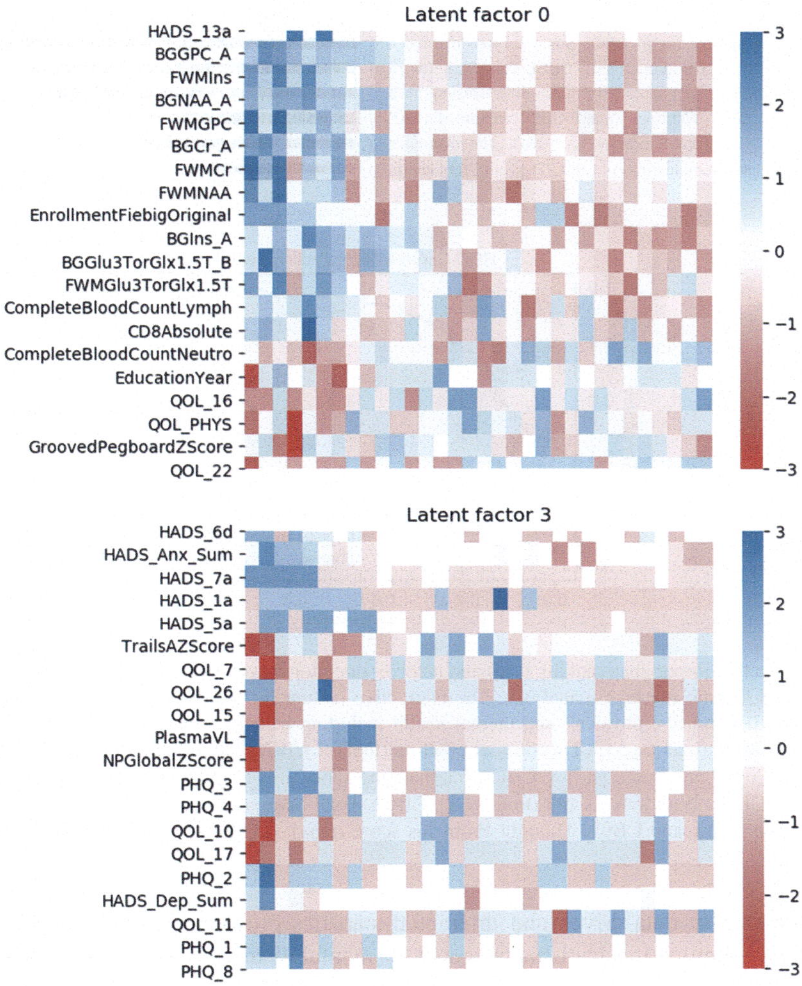

Fig. 4 CorEx heatmaps depicting latent factors among acutely infected individuals at baseline. Participants are classified in a binary fashion (depicted on a scale of blue to red) according to mutual information explained per latent factor

by week 24 (Improving: 18%); those with inconsistent ratios (increases and decreases) over the observation period (Inconsistent:11%); and individuals with a CD4/CD8 T-cell ratio < 1 at each timepoint (Stable Low: 39%). Baseline comparisons between clusters revealed lower levels of mental distress and lower plasma viral load among individuals in the High and Improving subgroups, supporting the CorEx results that focused on baseline values. Results also revealed elevated markers of mental distress, Fiebig stage, and plasma and CSF stress-related cytokines (e.g., IL-6, IL-1β, TNF-α) and markers of HIV-induced immune dysfunction (e.g., IL-5, IL-21, MCP-1). While preliminary, these findings derived from a

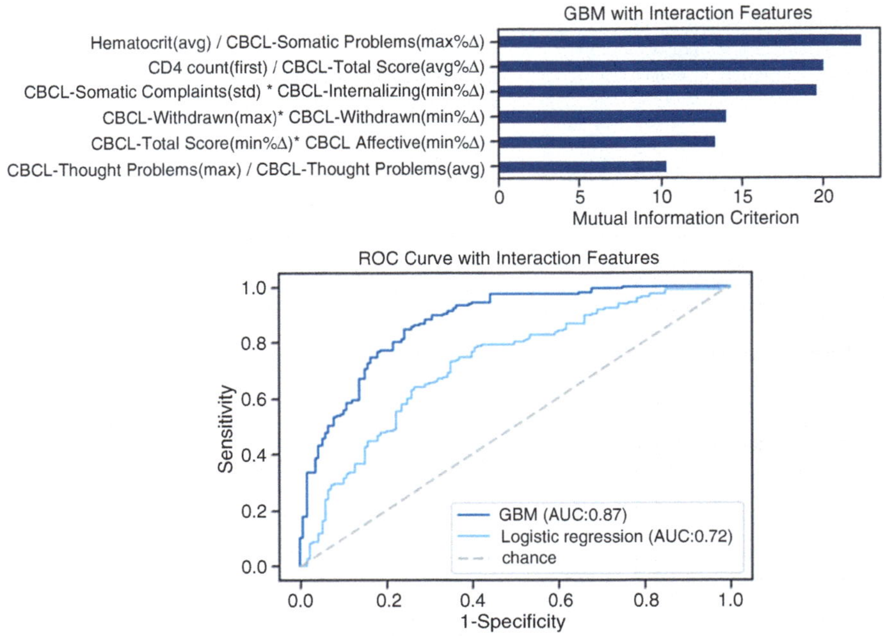

Fig. 5 Feature importance ranking (top) and receiver operator curves (bottom) for the longitudinal GBM and logistic regression analyses. The interactive feature list was comprised of interactions between HIV indices, physical, and mental health including hematocrit (avg) × CBCL Somatic Problems score (max % Δ), CD4 count at baseline (first) × CBCL Total Score (avg % Δ), CBCL Somatic Complaints score (std) × CBCL Internalizing score (min % Δ), CBCL Withdrawn score (max) × CBCL Withdrawn score (min % Δ), CBCL Total Score (min % Δ) × CBCL Affective score (min % Δ), and CBCL Thought Problems score (max) × CBCL Thought Problems score (avg)

combination of data-driven and inferential statistics provide compelling evidence that neuroimmune mechanisms triggered by mental distress are capable of altering HIV disease dynamics and long-term response to ART.

Other studies have demonstrated explanatory and predictive gains using machine learning methods. For example, Ogishi and Yotsuyanagi (2018) utilized machine learning to identify three novel viral proteins that distinguish PLWH with vs. without dementia. More recently, Tu et al. (2019) utilized machine learning to demonstrate the contribution of mental distress on HIV-related neurocognitive complications after ART. Finally, recent studies from our team using ensemble machine learning identified HIV disease mechanisms and mental health difficulties that collectively predicted neurocognitive development in children and adolescents with perinatal HIV (Paul et al. 2019) (Fig. 5). Baseline gradient boosted models classified neurocognitive trajectories with 71% accuracy (AUC 79%), while longitudinal factors improved accuracy to 77% (AUC 87%). Interestingly, the baseline feature list was comprised of mainly HIV disease (viral load) and immune markers (CD4 and CD8 T-cell counts), while the longitudinal model highlighted contributions of physical and mental health to the baseline model. Further, longitudinal models with two-way interactions showed more detailed interactions between physical and mental health and HIV indices (accuracy 80%, AUC 90%). Additional preliminary evidence from our group utilizing advanced clustering techniques also emphasizes the significance of substance use at the time of HIV infection in adults on long-term outcomes, including impacts on brain network connectivity (e.g., default mode network) and mental health phenotypes (e.g., elevated depression and anxiety scale scores).

10 Clinical Implications

It is important to recognize that data science methods, such as machine learning, are vulnerable to overfitting and model interpretation error. Additionally, risks related to human subject's research (e.g., participant privacy, safety, and data fidelity) exist and can even be amplified in research using data-driven models, including the potential for algorithms to exacerbate discriminations that perpetuate health disparities. Additional work is needed to develop tangible guidelines for algorithm training and validation and minimum performance metrics before mathematical algorithms are applied to the clinical setting. Best practices require that multidisciplinary teams work collaboratively across the entire continuum of clinical science, beginning with defining the purpose, available/needed data sources, optimal feature selection strategy, data science model, and potential challenges/alternate approaches. These practices are needed to prevent the development of algorithms that are devoid of clinical relevance. Guidelines have been established (e.g., Transparent Reporting of a Multivariable Prediction Model for Individual Prognosis or Diagnosis; TRIPOD) (Collins et al. 2015; Moons et al. 2015) for scientific transparency, reproducibility, and reliability, which are essential to foster clinical relevance. Additionally, it is important that future studies examine predictive models that differ in complexity of input features (e.g., removal of MRI metrics for some models). This recommendation applies to studies focused on HICs where patient care does not occur solely in the context of research-intensive academic medical centers. Finally, investigators are encouraged to disseminate the processing pipelines, feature selection protocols, extraction methods, final source codes, etc. on Github (and/or similar publicly available sites) to facilitate transparency and reproducibility, as well as post pre-prints on BioRvix or similar sites to shorten the time lag between scientific innovation and clinical implementation. Combined with harmonized study protocols, the recommendations above have potential to transform current clinical care models by identifying high-risk cases, so that tailored interventions can be delivered to improve neurocognitive outcomes and overall health among individuals residing in HICs.

References

Alford K, Vera JH (2018) Cognitive Impairment in people living with HIV in the ART era: a review. Br Med Bull 127(1):55–68. https://doi.org/10.1093/bmb/ldy019

Alvarez-del Arco D, Monge S, Azcoaga A, Rio I, Hernando V, Gonzalez C et al (2013) HIV testing and counselling for migrant populations living in high-income countries: a systematic review. Eur J Public Health 23(6):1039–1045. https://doi.org/10.1093/eurpub/cks130

American Psychiatric Association (2013) Diagnostic and statistical manual of mental disorders, 5th edn. American Psychiatric Association Publishing, Washington

Ananworanich J, Melvin D, Amador JT, Childs T, Medin G, Boscolo V et al (2016) Neurocognition and quality of life after reinitiating antiretroviral therapy in children randomized to planned treatment interruption. AIDS 30(7):1075–1081. https://doi.org/10.1097/QAD.0000000000001011

Ances BM, Christensen JJ, Teshome M, Taylor J, Xiong C, Aldea P et al (2010) Cognitively unimpaired HIV-positive subjects do not have increased 11C-PiB: a case-control study. Neurology 75(2):111–115. https://doi.org/10.1212/WNL.0b013e3181e7b66e

Ances BM, Benzinger TL, Christensen JJ, Thomas J, Venkat R, Teshome M et al (2012a) 11C-PiB imaging of human immunodeficiency virus-associated neurocognitive disorder. Arch Neurol 69 (1):72–77. https://doi.org/10.1001/archneurol.2011.761

Ances BM, Ortega M, Vaida F, Heaps J, Paul R (2012b) Independent effects of HIV, aging, and HAART on brain volumetric measures. J Acquir Immune Defic Syndr 59(5):469–477. https://doi.org/10.1097/QAI.0b013e318249db17

Anthony IC, Ramage SN, Carnie FW, Simmonds P, Bell JE (2005) Does drug abuse alter microglial phenotype and cell turnover in the context of advancing HIV infection? Neuropathol Appl Neurobiol 31(3):325–338. https://doi.org/10.1111/j.1365-2990.2005.00648.x

Antinori A, Arendt G, Becker JT, Brew BJ, Byrd DA, Cherner M et al (2007) Updated research nosology for HIV-associated neurocognitive disorders. Neurology 69(18):1789–1799. https://doi.org/10.1212/01.WNL.0000287431.88658.8b

Attonito JM, Dévieux JG, Lerner BDG, Rosenberg R (2014) Exploring substance use and HIV treatment factors associated with neurocognitive impairment among people living with HIV/AIDS. Front Public Health 2:105. https://doi.org/10.3389/fpubh.2014.00105

Avert (2017) HIV and AIDS in the United Kingdom (UK). Retrieved from https://www.avert.org/professionals/hiv-around-world/western-central-europe-north-america/uk

Avert (2018) HIV and AIDS in the United States of America. Retrieved from https://www.avert.org/professionals/hiv-around-world/western-central-europe-north-america/usa

Baker LM, Cooley SA, Cabeen RP, Laidlaw DH, Joska JA, Hoare J et al (2017) Topological organization of whole-brain white matter in HIV infection. Brain Connect 7(2):115–122. https://doi.org/10.1089/brain.2016.0457

Benedict RH, Cookfair D, Gavett R, Gunther M, Munschauer F, Garg N, Weinstock-Guttman B (2006) Validity of the minimal assessment of cognitive function in multiple sclerosis (MACFIMS). J Int Neuropsychol Soc 12(4):549–558. https://doi.org/10.1017/s1355617706060723

Blackstone K, Moore DJ, Heaton RK, Franklin DR Jr, Woods SP, Clifford DB et al (2012) Diagnosing symptomatic HIV-associated neurocognitive disorders: self-report versus performance-based assessment of everyday functioning. J Int Neuropsychol Soc 18(1):79–88. https://doi.org/10.1017/s135561771100141x

Brothers TD, Kirkland S, Guaraldi G, Falutz J, Theou O, Johnston BL, Rockwood K (2014) Frailty in people aging with human immunodeficiency virus (HIV) infection. J Infect Dis 210 (8):1170–1179. https://doi.org/10.1093/infdis/jiu258

Brouillette M-J, Mayo N, Fellows LK, Lebedeva E, Higgins J, Overton ET et al (2015) A better screening tool for HIV-associated neurocognitive disorders: is it what clinicians need? AIDS 29 (8):895–902. https://doi.org/10.1097/QAD.0000000000000152

Burdo TH, Lentz MR, Autissier P, Krishnan A, Halpern E, Letendre S et al (2011a) Soluble CD163 made by monocyte/macrophages is a novel marker of HIV activity in early and chronic infection prior to and after anti-retroviral therapy. J Infect Dis 204(1):154–163. https://doi.org/10.1093/infdis/jir214

Burdo TH, Lo J, Abbara S, Wei J, DeLelys ME, Preffer F et al (2011b) Soluble CD163, a novel marker of activated macrophages, is elevated and associated with noncalcified coronary plaque in HIV-infected patients. J Infect Dis 204(8):1227–1236. https://doi.org/10.1093/infdis/jir520

Burnham KP, Anderson DR (2002) A practical information-theoretic approach. Model selection and multimodel inference, 2nd edn. Springer, New York, p 2

Byrd DA, Fellows RP, Morgello S, Franklin D, Heaton RK, Deutsch R et al (2011) Neurocognitive impact of substance use in HIV infection. J Acquir Immune Defic Syndr 58(2):154–162. https://doi.org/10.1097/QAI.0b013e318229ba41

Carey CL, Woods SP, Rippeth JD, Gonzalez R, Heaton RK, Grant I (2006) Additive deleterious effects of methamphetamine dependence and immunosuppression on neuropsychological functioning in HIV infection. AIDS Behav 10(2):185–190. https://doi.org/10.1007/s10461-005-9056-4

Centers for Disease Control and Prevention (2019) HIV and Older Americans. National Center for HIV/AIDS, Viral Hepatitis, STD, and TB Prevention – Division of HIV/AIDS Prevention. Retrieved from https://www.cdc.gov/hiv/pdf/group/age/olderamericans/cdc-hiv-older-americans.pdf

Chu K, Tran T, Wei K, Lammering JC, Sondergaard A, Mogadam E et al (2018) Distinguishing brain impact of aging and HIV severity in chronic HIV using multiparametric MR imaging and MR spectroscopy. Open Forum Infect Dis 5(10):ofy243. https://doi.org/10.1093/ofid/ofy243

Clark US, Arce Renteria M, Hegde RR, Morgello S (2018) Early life stress-related elevations in reaction time variability are associated with brain volume reductions in HIV+ adults. Front Behav Neurosci 12:6. https://doi.org/10.3389/fnbeh.2018.00006

Clifford DB, Ances BM (2013) HIV-associated neurocognitive disorder. Lancet Infect Dis 13 (11):976–986. https://doi.org/10.1016/S1473-3099(13)70269-X

Clifford KM, Samboju V, Cobigo Y, Milanini B, Marx GA, Hellmuth JM et al (2017) Progressive brain atrophy despite persistent viral suppression in HIV patients older than 60 years. J Acquir Immune Defic Syndr 76(3):289–297. https://doi.org/10.1097/QAI.0000000000001489

Cohen RA, Harezlak J, Schifitto G, Hana G, Clark U, Gongvatana A et al (2010) Effects of nadir CD4 count and duration of human immunodeficiency virus infection on brain volumes in the highly active antiretroviral therapy era. J Neurovirol 16(1):25–32. https://doi.org/10.3109/13550280903552420

Cohen RA, Seider TR, Navia B (2015) HIV effects on age-associated neurocognitive dysfunction: premature cognitive aging or neurodegenerative disease? Alzheimers Res Ther 7(1):37. https://doi.org/10.1186/s13195-015-0123-4

Cohen RA, Gullett JM, Porges EC, Woods AJ, Lamb DG, Bryant VE et al (2019) Heavy alcohol use and age effects on HIV-associated neurocognitive function. Alcohol Clin Exp Res 43 (1):147–157. https://doi.org/10.1111/acer.13915

Cole JH, Underwood J, Caan MW, De Francesco D, van Zoest RA, Leech R et al (2017) Increased brain-predicted aging in treated HIV disease. Neurology 88(14):1349–1357. https://doi.org/10.1212/WNL.0000000000003790

Collins GS, Reitsma JB, Altman DG, Moons KG (2015) Transparent reporting of a multivariable prediction model for individual prognosis or diagnosis (TRIPOD): the TRIPOD Statement. BMC Med 13(1). https://doi.org/10.1186/s12916-014-0241-z

Cooley SA, Strain JF, Beaumont H, Boerwinkle AH, Doyle J, Morris JC et al (2019) Tau positron emission tomography binding is not elevated in HIV-infected individuals. J Infect Dis 220 (1):68–72. https://doi.org/10.1093/infdis/jiy663

Cote HC, Soudeyns H, Thorne A, Alimenti A, Lamarre V, Maan EJ et al (2012) Leukocyte telomere length in HIV-infected and HIV-exposed uninfected children: shorter telomeres with

uncontrolled HIV viremia. PLoS One 7(7):e39266. https://doi.org/10.1371/journal.pone.0039266
Cysique LA, Maruff P, Darby D, Brew BJ (2006) The assessment of cognitive function in advanced HIV-1 infection and AIDS dementia complex using a new computerised cognitive test battery. Arch Clin Neuropsychol 21(2):185–194. https://doi.org/10.1016/j.acn.2005.07.011
Cysique LA, Murray JM, Dunbar M, Jeyakumar V, Brew BJ (2010) A screening algorithm for HIV-associated neurocognitive disorders. HIV Med 11(10):642–649. https://doi.org/10.1111/j.1468-1293.2010.00834.x
Cysique LA, Moffat K, Moore DM, Lane TA, Davies NW, Carr A et al (2013) HIV, vascular and aging injuries in the brain of clinically stable HIV-infected adults: a (1)H MRS study. PLoS One 8(4):e61738. https://doi.org/10.1371/journal.pone.0061738
Desquilbet L, Jacobson LP, Fried LP, Phair JP, Jamieson BD, Holloway M et al (2007) HIV-1 infection is associated with an earlier occurrence of a phenotype related to frailty. J Gerontol A Biol Sci Med Sci 62(11):1279–1286. https://doi.org/10.1093/gerona/62.11.1279
Devlin KN, Gongvatana A, Clark US, Chasman JD, Westbrook ML, Tashima KT et al (2012) Neurocognitive effects of HIV, hepatitis C, and substance use history. J Int Neuropsychol Soc 18(1):68–78. https://doi.org/10.1017/s1355617711001408
Doyle KL, Morgan EE, Morris S, Smith DM, Little S, Iudicello JE et al (2013) Real-world impact of neurocognitive deficits in acute and early HIV infection. J Neurovirol 19(6):565–573. https://doi.org/10.1007/s13365-013-0218-2
Du Pasquier RA, Jilek S, Kalubi M, Yerly S, Fux CA, Gutmann C et al (2013) Marked increase of the astrocytic marker S100B in the cerebrospinal fluid of HIV-infected patients on LPV/r-monotherapy. AIDS 27(2):203–210. https://doi.org/10.1097/QAD.0b013e32835a9a4a
du Plessis L, Paul RH, Hoare J, Stein DJ, Taylor PA, Meintjes EM, Joska JA (2017) Resting-state functional magnetic resonance imaging in clade C HIV: within-group association with neurocognitive function. J Neurovirol 23(6):875–885. https://doi.org/10.1007/s13365-017-0581-5
Duprez DA, Neuhaus J, Kuller LH, Tracy R, Belloso W, De Wit S et al (2012) Inflammation, coagulation and cardiovascular disease in HIV-infected individuals. PLoS One 7(9):e44454. https://doi.org/10.1371/journal.pone.0044454
Elbirt D, Mahlab-Guri K, Bezalel-Rosenberg S, Gill H, Attali M, Asher I (2015) HIV-associated neurocognitive disorders (HAND). Isr Med Assoc J 17(1):54–59. Retrieved from https://www.ncbi.nlm.nih.gov/pubmed/25739180
Ellis RJ, Badiee J, Vaida F, Letendre S, Heaton RK, Clifford D et al (2011) CD4 nadir is a predictor of HIV neurocognitive impairment in the era of combination antiretroviral therapy. AIDS 25 (14):1747–1751. https://doi.org/10.1097/QAD.0b013e32834a40cd
Erb P, Battegay M, Zimmerli W, Rickenbach M, Egger M (2000) Effect of antiretroviral therapy on viral load, CD4 cell count, and progression to acquired immunodeficiency syndrome in a community human immunodeficiency virus-infected cohort. Swiss HIV Cohort Study. Arch Intern Med 160(8):1134–1140. https://doi.org/10.1001/archinte.160.8.1134
Fazeli PL, Crowe M, Ross LA, Wadley V, Ball K, Vance DE (2014) Cognitive functioning in adults aging with HIV: a cross-sectional analysis of cognitive subtypes and influential factors. J Clin Res HIV AIDS Prev 1(4):155–169. https://doi.org/10.14302/issn.2324-7339.jcrhap-13-191
Ferris MJ, Mactutus CF, Booze RM (2008) Neurotoxic profiles of HIV, psychostimulant drugs of abuse, and their concerted effect on the brain: current status of dopamine system vulnerability in NeuroAIDS. Neurosci Biobehav Rev 32(5):883–909. https://doi.org/10.1016/j.neubiorev.2008.01.004
Fried LP, Tangen CM, Walston J, Newman AB, Hirsch C, Gottdiener J et al (2001) Frailty in older adults: evidence for a phenotype. J Gerontol A Biol Sci Med Sci 56(3):M146–M156. https://doi.org/10.1093/gerona/56.3.m146
Gallant J, Hsue PY, Shreay S, Meyer N (2017) Comorbidities among US patients with prevalent HIV infection-A trend analysis. J Infect Dis 216(12):1525–1533. https://doi.org/10.1093/infdis/jix518

Garvey L, Surendrakumar V, Winston A (2011) Low rates of neurocognitive impairment are observed in neuro-asymptomatic HIV-infected subjects on effective antiretroviral therapy. HIV Clin Trials 12(6):333–338. https://doi.org/10.1310/hct1206-333

Gates TM, Cysique LA (2016) The chronicity of HIV infection should drive the research strategy of NeuroHIV treatment studies: a critical review. CNS Drugs 30(1):53–69. https://doi.org/10.1007/s40263-015-0302-7

Gawron N, Choinski M, Szymanska-Kotwica B, Pluta A, Sobanska M, Egbert AR et al (2019) Effects of age, HIV, and HIV-associated clinical factors on neuropsychological functioning and brain regional volume in HIV+ patients on effective treatment. J Neurovirol 25(1):9–21. https://doi.org/10.1007/s13365-018-0679-4

Gianesin K, Noguera-Julian A, Zanchetta M, Del Bianco P, Petrara MR, Freguja R et al (2016) Premature aging and immune senescence in HIV-infected children. AIDS 30(9):1363–1373. https://doi.org/10.1097/QAD.0000000000001093

Gisslen M, Price RW, Nilsson S (2011) The definition of HIV-associated neurocognitive disorders: are we overestimating the real prevalence? BMC Infect Dis 11:356. https://doi.org/10.1186/1471-2334-11-356

Gisslen M, Price RW, Andreasson U, Norgren N, Nilsson S, Hagberg L et al (2016) plasma concentration of the neurofilament light protein (NFL) is a biomarker of CNS injury in HIV infection: a cross-sectional study. EBioMedicine 3:135–140. https://doi.org/10.1016/j.ebiom.2015.11.036

Gonzalez R, Schuster RM, Vassileva J, Martin EM (2011) Impact of HIV and a history of marijuana dependence on procedural learning among individuals with a history of substance dependence. J Clin Exp Neuropsychol 33(7):735–752. https://doi.org/10.1080/13803395.2011.553584

Grant I, Atkinson JH, Hesselink JR, Kennedy CJ, Richman DD, Spector SA, McCutchan JA (1987) Evidence for early central nervous system involvement in the acquired immunodeficiency syndrome (AIDS) and other human immunodeficiency virus (HIV) infections. Studies with neuropsychologic testing and magnetic resonance imaging. Ann Intern Med 107(6):828–836. https://doi.org/10.7326/0003-4819-107-6-828

Grant RM, Anderson PL, McMahan V, Liu A, Amico KR, Mehrotra M et al (2014) Uptake of pre-exposure prophylaxis, sexual practices, and HIV incidence in men and transgender women who have sex with men: a cohort study. Lancet Infect Dis 14(9):820–829. https://doi.org/10.1016/S1473-3099(14)70847-3

Guaraldi G, Orlando G, Zona S, Menozzi M, Carli F, Garlassi E et al (2011) Premature age-related comorbidities among HIV-infected persons compared with the general population. Clin Infect Dis 53(11):1120–1126. https://doi.org/10.1093/cid/cir627

Guha A, Brier MR, Ortega M, Westerhaus E, Nelson B, Ances BM (2016a) Topographies of cortical and subcortical volume loss in HIV and aging in the cART era. J Acquir Immune Defic Syndr 73(4):374–383. https://doi.org/10.1097/QAI.0000000000001111

Guha A, Wang L, Tanenbaum A, Esmaeili-Firidouni P, Wendelken LA, Busovaca E et al (2016b) Intrinsic network connectivity abnormalities in HIV-infected individuals over age 60. J Neurovirol 22(1):80–87. https://doi.org/10.1007/s13365-015-0370-y

Hachinski V, Iadecola C, Petersen RC, Breteler MM, Nyenhuis DL, Black SE et al (2006) National Institute of neurological disorders and stroke-Canadian stroke network vascular cognitive impairment harmonization standards. Stroke 37(9):2220–2241. https://doi.org/10.1161/01.Str.0000237236.88823.47

Heaps JM, Joska J, Hoare J, Ortega M, Agrawal A, Seedat S et al (2012) Neuroimaging markers of human immunodeficiency virus infection in South Africa. J Neurovirol 18(3):151–156. https://doi.org/10.1007/s13365-012-0090-5

Heaps-Woodruff JM, Joska J, Cabeen R, Baker LM, Salminen LE, Hoare J et al (2018) White matter fiber bundle lengths are shorter in cART naive HIV: an analysis of quantitative diffusion tractography in South Africa. Brain Imaging Behav 12(5):1229–1238. https://doi.org/10.1007/s11682-017-9769-9

Hellmuth J, Colby D, Valcour V, Suttichom D, Spudich S, Ananworanich J et al (2017) Depression and anxiety are common in acute HIV infection and associate with plasma immune activation. AIDS Behav 21(11):3238–3246. https://doi.org/10.1007/s10461-017-1788-4

Horvath S, Levine AJ (2015) HIV-1 infection accelerates age according to the epigenetic clock. J Infect Dis 212(10):1563–1573. https://doi.org/10.1093/infdis/jiv277

Imp BM, Rubin LH, Tien PC, Plankey MW, Golub ET, French AL, Valcour VG (2017) Monocyte activation is associated with worse cognitive performance in HIV-infected women with virologic suppression. J Infect Dis 215(1):114–121. https://doi.org/10.1093/infdis/jiw506

Jacobs DM, Sano M, Albert S, Schofield P, Dooneief G, Stern Y (1997) Cross-cultural neuropsychological assessment: a comparison of randomly selected, demographically matched cohorts of English- and Spanish-speaking older adults. J Clin Exp Neuropsychol 19(3):331–339. https://doi.org/10.1080/01688639708403862

Janssen MA, Bosch M, Koopmans PP, Kessels RP (2015) Validity of the Montreal Cognitive Assessment and the HIV Dementia Scale in the assessment of cognitive impairment in HIV-1 infected patients. J Neurovirol 21(4):383–390. https://doi.org/10.1007/s13365-015-0324-4

Jeon C, Lau C, Kendall CE, Burchell AN, Bayoumi AM, Loutfy M et al (2017) Mortality and health service use following acute myocardial infarction among persons with HIV: a population-based study. AIDS Res Hum Retroviruses 33(12):1214–1219. https://doi.org/10.1089/AID.2017.0128

Joska JA, Fincham DS, Stein DJ, Paul RH, Seedat S (2010) Clinical correlates of HIV-associated neurocognitive disorders in South Africa. AIDS Behav 14(2):371–378. https://doi.org/10.1007/s10461-009-9538-x

Kallianpur KJ, Jahanshad N, Sailasuta N et al (2020) Regional brain volumetric changes despite 2 years of treatment initiated during acute HIV infection. AIDS 34(3):415–426. https://doi.org/10.1097/QAD.0000000000002436

Kim WJ, Ku NS, Lee YJ, Ahn JY, Kim SB, Ahn HW et al (2016) Utility of the Montreal Cognitive Assessment (MoCA) and its subset in HIV-associated neurocognitive disorder (HAND) screening. J Psychosom Res 80:53–57. https://doi.org/10.1016/j.jpsychores.2015.11.006

Kordovski VM, Tierney SM, Woods SP (2019) Conceptualizing and assessing everyday functioning in the context of HIV-associated neurocognitive disorders. Curr Top Behav Neurosci. https://doi.org/10.1007/7854_2018_78

Kore I, Ananworanich J, Valcour V, Fletcher JL, Chalermchai T, Paul R et al (2015) Neuropsychological impairment in acute HIV and the effect of immediate antiretroviral therapy. J Acquir Immune Defic Syndr 70(4):393–399. https://doi.org/10.1097/QAI.0000000000000746

Kramer EL, Sanger JJ (1990) Brain imaging in acquired immunodeficiency syndrome dementia complex. Semin Nucl Med 20(4):353–363. https://doi.org/10.1016/s0001-2998(05)80239-9

Krentz H, Gill MJ (2009) The five-year impact of an evolving global epidemic, changing migration patterns, and policy changes in a regional Canadian HIV population. Health Policy 90 (2-3):296–302. https://doi.org/10.1016/j.healthpol.2008.09.016

Ladak F, Socias E, Nolan S, Dong H, Kerr T, Wood E et al (2019) Substance use patterns and HIV-1 RNA viral load rebound among HIV-positive illicit drug users in a Canadian setting. Antivir Ther 24(1):19–25. https://doi.org/10.3851/IMP3265

Lang S, Mary-Krause M, Simon A, Partisani M, Gilquin J, Cotte L et al (2012) HIV replication and immune status are independent predictors of the risk of myocardial infarction in HIV-infected individuals. Clin Infect Dis 55(4):600–607. https://doi.org/10.1093/cid/cis489

Leeansyah E, Cameron PU, Solomon A, Tennakoon S, Velayudham P, Gouillou M et al (2013) Inhibition of telomerase activity by human immunodeficiency virus (HIV) nucleos(t)ide reverse transcriptase inhibitors: a potential factor contributing to HIV-associated accelerated aging. J Infect Dis 207(7):1157–1165. https://doi.org/10.1093/infdis/jit006

Letendre S (2011) Central nervous system complications in HIV disease: HIV-associated neurocognitive disorder. Top Antivir Med 19(4):137–142. Retrieved from https://www.ncbi.nlm.nih.gov/pubmed/22156215

Levine AJ, Quach A, Moore DJ, Achim CL, Soontornniyomkij V, Masliah E et al (2016) Accelerated epigenetic aging in brain is associated with pre-mortem HIV-associated neurocognitive disorders. J Neurovirol 22(3):366–375. https://doi.org/10.1007/s13365-015-0406-3

Lichtenstein KA, Armon C, Buchacz K, Chmiel JS, Buckner K, Tedaldi EM et al (2010) Low CD4 + T cell count is a risk factor for cardiovascular disease events in the HIV outpatient study. Clin Infect Dis 51(4):435–447. https://doi.org/10.1086/655144

Liner KJ 2nd, Hall CD, Robertson KR (2008) Effects of antiretroviral therapy on cognitive impairment. Curr HIV/AIDS Rep 5(2):64–71. https://doi.org/10.1007/s11904-008-0011-7

Liu JC, Leung JM, Ngan DA, Nashta NF, Guillemi S, Harris M et al (2015) Absolute leukocyte telomere length in HIV-infected and uninfected individuals: evidence of accelerated cell senescence in HIV-associated chronic obstructive pulmonary disease. PLoS One 10(4): e0124426. https://doi.org/10.1371/journal.pone.0124426

Lo J, Abbara S, Shturman L, Soni A, Wei J, Rocha-Filho JA et al (2010) Increased prevalence of subclinical coronary atherosclerosis detected by coronary computed tomography angiography in HIV-infected men. AIDS 24(2):243–253. https://doi.org/10.1097/QAD.0b013e328333ea9e

Loftis JM, Choi D, Hoffman W, Huckans MS (2011) Methamphetamine causes persistent immune dysregulation: a cross-species, translational report. Neurotox Res 20(1):59–68. https://doi.org/10.1007/s12640-010-9223-x

Manly JJ, Byrd DA, Touradji P, Stern Y (2004) Acculturation, reading level, and neuropsychological test performance among African American elders. Appl Neuropsychol 11(1):37–46. https://doi.org/10.1207/s15324826an1101_5

Margolick JB, Ferrucci L (2015) Accelerating aging research: how can we measure the rate of biologic aging? Exp Gerontol 64:78–80. https://doi.org/10.1016/j.exger.2015.02.009

Masia M, Padilla S, Garcia JA, Garcia-Abellan J, Fernandez M, Bernardino I et al (2019) Evolving understanding of cardiovascular, cerebrovascular and peripheral arterial disease in people living with HIV and role of novel biomarkers. A study of the Spanish CoRIS cohort, 2004-2015. PloS One 14(4):e0215507. https://doi.org/10.1371/journal.pone.0215507

McArthur JC, Kumar AJ, Johnson DW, Selnes OA, Becker JT, Herman C et al (1990) Incidental white matter hyperintensities on magnetic resonance imaging in HIV-1 infection. Multicenter AIDS Cohort Study. J Acquir Immune Defic Syndr 3(3):252–259. Retrieved from https://www.ncbi.nlm.nih.gov/pubmed/2406415

McGuire JL, Gill AJ, Douglas SD, Kolson DL, CNS HIV Anti-Retroviral Therapy Effects Research (CHARTER) Group (2015) Central and peripheral markers of neurodegeneration and monocyte activation in HIV-associated neurocognitive disorders. J Neurovirol 21(4):439–448. https://doi.org/10.1007/s13365-015-0333-3

McLaren PJ, Coulonges C, Bartha I, Lenz TL, Deutsch AJ, Bashirova A et al (2015) Polymorphisms of large effect explain the majority of the host genetic contribution to variation of HIV-1 virus load. Proc Natl Acad Sci U S A 112(47):14658–14663. https://doi.org/10.1073/pnas.1514867112

Meade CS, Conn NA, Skalski LM, Safren SA (2011) Neurocognitive impairment and medication adherence in HIV patients with and without cocaine dependence. J Behav Med 34(2):128–138. https://doi.org/10.1007/s10865-010-9293-5

Meade CS, Towe SL, Skalski LM, Robertson KR (2015) Independent effects of HIV infection and cocaine dependence on neurocognitive impairment in a community sample living in the southern United States. Drug Alcohol Depend 149:128–135. https://doi.org/10.1016/j.drugalcdep.2015.01.034

Meyer A-CL, Boscardin WJ, Kwasa JK, Price RW (2013) Is it time to rethink how neuropsychological tests are used to diagnose mild forms of HIV-associated neurocognitive disorders? Impact of false-positive rates on prevalence and power. Neuroepidemiology 41(3-4):208–216. https://doi.org/10.1159/000354629

Milanini B, Samboju V, Cobigo Y, Paul R, Javandel S, Hellmuth J et al (2019) Longitudinal brain atrophy patterns and neuropsychological performance in older adults with HIV-associated neurocognitive disorder compared with early Alzheimer's disease. Neurobiol Aging 82:69–76. https://doi.org/10.1016/j.neurobiolaging.2019.07.006

Miller PJ, Lubke GH, McArtor DB, Bergeman CS (2016) Finding structure in data using multivariate tree boosting. Psychol Methods 21(4):583–602. https://doi.org/10.1037/met0000087

Moons KG, Altman DG, Reitsma JB, Ioannidis JP, Macaskill P, Steyerberg EW et al (2015) Transparent reporting of a multivariable prediction model for individual prognosis or diagnosis (TRIPOD): explanation and elaboration. Ann Intern Med 162(1):W1–W73. https://doi.org/10.7326/M14-0698

Morgello S, Gensler G, Sherman S, Ellis RJ, Gelman BB, Kolson DL et al (2019) Frailty in medically complex individuals with chronic HIV. AIDS 33(10):1603–1611. https://doi.org/10.1097/QAD.0000000000002250

Moulignier A, Savatovsky J, Assoumou L, Lescure FX, Lamirel C, Godin O et al (2018) Silent cerebral small-vessel disease is twice as prevalent in middle-aged individuals with well-controlled, combination antiretroviral therapy-treated human immunodeficiency virus (HIV) than in HIV-uninfected individuals. Clin Infect Dis 66(11):1762–1769. https://doi.org/10.1093/cid/cix1075

Nath A (2010) Human immunodeficiency virus-associated neurocognitive disorder: pathophysiology in relation to drug addiction. Ann N Y Acad Sci 1187:122–128. https://doi.org/10.1111/j.1749-6632.2009.05277.x

Navia BA (1997) Clinical and biologic features of the AIDS dementia complex. Neuroimaging Clin N Am 7(3):581–592. Retrieved from https://www.ncbi.nlm.nih.gov/pubmed/9376969

Navia BA, Cho ES, Petito CK, Price RW (1986) The AIDS dementia complex: II. Neuropathology. Ann Neurol 19(6):525–535. https://doi.org/10.1002/ana.410190603

Nielsen TR, Segers K, Vanderaspoilden V, Bekkhus-Wetterberg P, Minthon L, Pissiota A et al (2018) Performance of middle-aged and elderly European minority and majority populations on a Cross-Cultural Neuropsychological Test Battery (CNTB). Clin Neuropsychol 32 (8):1411–1430. https://doi.org/10.1080/13854046.2018.1430256

Obermeit LC, Beltran J, Casaletto KB, Franklin DR, Letendre S, Ellis R et al (2017) Evaluating the accuracy of self-report for the diagnosis of HIV-associated neurocognitive disorder (HAND): defining "symptomatic" versus "asymptomatic" HAND. J Neurovirol 23(1):67–78. https://doi.org/10.1007/s13365-016-0474-z

Ogishi M, Yotsuyanagi H (2018) Prediction of HIV-associated neurocognitive disorder (HAND) from three genetic features of envelope gp120 glycoprotein. Retrovirology 15(1):12. https://doi.org/10.1186/s12977-018-0401-x

Olsen WL, Longo FM, Mills CM, Norman D (1988) White matter disease in AIDS: findings at MR imaging. Radiology 169(2):445–448. https://doi.org/10.1148/radiology.169.2.3174991

Onen NF, Overton ET, Seyfried W, Stumm ER, Snell M, Mondy K, Tebas P (2010) Aging and HIV infection: a comparison between older HIV-infected persons and the general population. HIV Clin Trials 11(2):100–109. https://doi.org/10.1310/hct1102-100

Ortega M, Heaps JM, Joska J, Vaida F, Seedat S, Stein DJ et al (2013) HIV clades B and C are associated with reduced brain volumetrics. J Neurovirol 19(5):479–487. https://doi.org/10.1007/s13365-013-0202-x

Ortega M, Brier MR, Ances BM (2015) Effects of HIV and combination antiretroviral therapy on cortico-striatal functional connectivity. AIDS 29(6):703–712. https://doi.org/10.1097/QAD.0000000000000611

Paul R (2019) Neurocognitive phenotyping of HIV in the era of antiretroviral therapy. Curr HIV/AIDS Rep 16(3):230–235. https://doi.org/10.1007/s11904-019-00426-9

Paul R, Cohen R, Navia B, Tashima K (2002) Relationships between cognition and structural neuroimaging findings in adults with human immunodeficiency virus type-1. Neurosci Biobehav Rev 26(3):353–359. https://doi.org/10.1016/s0149-7634(02)00006-4

Paul RH, Cooley SA, Garcia-Egan PM, Ances BM (2018) Cognitive performance and frailty in older HIV-positive adults. J Acquir Immune Defic Syndr 79(3):375–380. https://doi.org/10.1097/QAI.0000000000001790

Paul R, Cho K, Mellins C, Malee K, Robbins R, Kerr S et al (2019) Predicting neurodevelopmental outcomes in children with perinatal HIV using a novel machine learning algorithm. bioRxiv:632273

Paul RH, Cho KS, Luckett PM, Strain JF, Belden AC, Bolzenius JD et al (2020) Machine learning analysis reveals novel neuroimaging and clinical signatures of frailty in HIV. J Acquir Immune Defic Syndr. https://doi.org/10.1097/QAI.0000000000002360

Pelchen-Matthews A, Ryom L, Borges AH, Edwards S, Duvivier C, Stephan C et al (2018) Aging and the evolution of comorbidities among HIV-positive individuals in a European cohort. AIDS 32(16):2405–2416. https://doi.org/10.1097/QAD.0000000000001967

Perez-Arce P, Puente AE (1996) Neuropsychological assessment of ethnic-minorities: the case of assessing Hispanics living in North America

Pernas B, Grandal M, Pertega S, Canizares A, Castro-Iglesias A, Mena A et al (2016) Any impact of blips and low-level viraemia episodes among HIV-infected patients with sustained virological suppression on ART? J Antimicrob Chemother 71(4):1051–1055. https://doi.org/10.1093/jac/dkv433

Petersen RC, Aisen PS, Beckett LA, Donohue MC, Gamst AC, Harvey DJ et al (2010) Alzheimer's disease neuroimaging initiative (ADNI): clinical characterization. Neurology 74(3):201–209. https://doi.org/10.1212/WNL.0b013e3181cb3e25

Pfefferbaum A, Zahr NM, Sassoon SA, Kwon D, Pohl KM, Sullivan EV (2018) Accelerated and premature aging characterizing regional cortical volume loss in human immunodeficiency virus infection: contributions from alcohol, substance use, and hepatitis C coinfection. Biol Psychiatry Cogn Neurosci Neuroimaging 3(10):844–859. https://doi.org/10.1016/j.bpsc.2018.06.006

Price RW, Brew BJ (1988) The AIDS dementia complex. J Infect Dis 158(5):1079–1083. https://doi.org/10.1093/infdis/158.5.1079

Price RW, Brew B, Sidtis J, Rosenblum M, Scheck AC, Cleary P (1988) The brain in AIDS: central nervous system HIV-1 infection and AIDS dementia complex. Science 239(4840):586–592. https://doi.org/10.1126/science.3277272

Prosser AT, Tang T, Hall HI (2012) HIV in persons born outside the United States, 2007-2010. JAMA 308(6):601–607. https://doi.org/10.1001/jama.2012.9046

Pumarola-Sune T, Navia BA, Cordon-Cardo C, Cho ES, Price RW (1987) HIV antigen in the brains of patients with the AIDS dementia complex. Ann Neurol 21(5):490–496. https://doi.org/10.1002/ana.410210513

Quiros-Roldan E, Properzi M, Paghera S, Raffetti E, Castelli F, Imberti L (2020) Factors associated with immunosenescence during early adulthood in HIV-infected patients after durable efficient combination antiretroviral therapy. Sci Rep 10(1):10057. https://doi.org/10.1038/s41598-020-67100-8

Robbins GK, Daniels B, Zheng H, Chueh H, Meigs JB, Freedberg KA (2007) Predictors of antiretroviral treatment failure in an urban HIV clinic. J Acquir Immune Defic Syndr 44 (1):30–37. https://doi.org/10.1097/01.qai.0000248351.10383.b7

Ross J, Cunningham CO, Hanna DB (2018) HIV outcomes among migrants from low-income and middle-income countries living in high-income countries: a review of recent evidence. Curr Opin Infect Dis 31(1):25–32. https://doi.org/10.1097/QCO.0000000000000415

Saloner R, Cysique LA (2017) HIV-associated neurocognitive disorders: a global perspective. J Int Neuropsychol Soc 23(9-10):860–869. https://doi.org/10.1017/s1355617717001102

Saloner R, Heaton RK, Campbell LM, Chen A, Franklin D Jr, Ellis RJ et al (2019) Effects of comorbidity burden and age on brain integrity in HIV. AIDS 33(7):1175–1185. https://doi.org/10.1097/QAD.0000000000002192

Samboju V, Philippi CL, Chan P, Cobigo Y, Fletcher JLK, Robb M et al (2018) Structural and functional brain imaging in acute HIV. Neuroimage Clin 20:327–335. https://doi.org/10.1016/j.nicl.2018.07.024

Sanford R, Fernandez Cruz AL, Scott SC, Mayo NE, Fellows LK, Ances BM, Collins DL (2017) Regionally specific brain volumetric and cortical thickness changes in HIV-infected patients in the HAART era. J Acquir Immune Defic Syndr 74(5):563–570. https://doi.org/10.1097/QAI.0000000000001294

Sanford R, Ances BM, Meyerhoff DJ, Price RW, Fuchs D, Zetterberg H et al (2018a) Longitudinal trajectories of brain volume and cortical thickness in treated and untreated primary human immunodeficiency virus infection. Clin Infect Dis 67(11):1697–1704. https://doi.org/10.1093/cid/ciy362

Sanford R, Fellows LK, Ances BM, Collins DL (2018b) Association of brain structure changes and cognitive function with combination antiretroviral therapy in HIV-positive individuals. JAMA Neurol 75(1):72–79. https://doi.org/10.1001/jamaneurol.2017.3036

Schouten J, Wit FW, Stolte IG, Kootstra NA, van der Valk M, Geerlings SE et al (2014) Cross-sectional comparison of the prevalence of age-associated comorbidities and their risk factors between HIV-infected and uninfected individuals: the AGEhIV cohort study. Clin Infect Dis 59 (12):1787–1797. https://doi.org/10.1093/cid/ciu701

Schuster RM, Gonzalez R (2012) Substance abuse, hepatitis C, and aging in HIV: common cofactors that contribute to neurobehavioral disturbances. Neurobehav HIV Med 2012 (4):15–34. https://doi.org/10.2147/NBHIV.S17408

Sheppard DP, Iudicello JE, Morgan EE, Kamat R, Clark LR, Avci G et al (2017) Accelerated and accentuated neurocognitive aging in HIV infection. J Neurovirol 23(3):492–500. https://doi.org/10.1007/s13365-017-0523-2

Shiau S, Strehlau R, Shen J, Violari A, Patel F, Liberty A et al (2018) Biomarkers of aging in HIV-infected children on suppressive antiretroviral therapy. J Acquir Immune Defic Syndr 78 (5):549–556. https://doi.org/10.1097/QAI.0000000000001714

Shirazi TN, Summers AC, Smith BR, Steinbach SR, Kapetanovic S, Nath A, Snow J (2017) Concordance between self-report and performance-based measures of everyday functioning in HIV-associated neurocognitive disorders. AIDS Behav 21(7):2124–2134. https://doi.org/10.1007/s10461-017-1689-6

Soontornniyomkij V, Moore DJ, Gouaux B, Soontornniyomkij B, Tatro ET, Umlauf A et al (2012) Cerebral beta-amyloid deposition predicts HIV-associated neurocognitive disorders in APOE epsilon4 carriers. AIDS 26(18):2327–2335. https://doi.org/10.1097/QAD.0b013e32835a117c

Soontornniyomkij V, Moore DJ, Gouaux B, Soontornniyomkij B, Sinsheimer JS, Levine AJ (2019) Associations of regional amyloid-beta plaque and phospho-tau pathology with biological factors and neuropsychological functioning among HIV-infected adults. J Neurovirol 25(6):741–753. https://doi.org/10.1007/s13365-019-00761-y

Spudich SS (2016) Immune activation in the central nervous system throughout the course of HIV infection. Curr Opin HIV AIDS 11(2):226–233. https://doi.org/10.1097/coh.0000000000000243

Strain JF, Burdo TH, Song SK, Sun P, El-Ghazzawy O, Nelson B et al (2017) Diffusion basis spectral imaging detects ongoing brain inflammation in virologically well-controlled HIV+ patients. J Acquir Immune Defic Syndr 76(4):423–430. https://doi.org/10.1097/QAI.0000000000001513

Su L, Wei D, Yang H, Zeng Y, Hu Y, Yuan D et al (2016) Identification of a novel HIV-1 circulating recombinant form (CRF85_BC) in Sichuan, China. AIDS Res Hum Retroviruses 32 (9):895–899. https://doi.org/10.1089/AID.2016.0053

Tate DF, Delong A, McCaffrey DE, Kertesz K, Paul RH, Conley J et al (2011) Recent clinical history and cognitive dysfunction for attention and executive function among human immunodeficiency virus-infected patients. Arch Clin Neuropsychol 26(7):614–623. https://doi.org/10.1093/arclin/acr065

Tedaldi EM, Minniti NL, Fischer T (2015) HIV-associated neurocognitive disorders: the relationship of HIV infection with physical and social comorbidities. Biomed Res Int 2015:641913. https://doi.org/10.1155/2015/641913

Thames AD, Kuhn TP, Williamson TJ, Jones JD, Mahmood Z, Hammond A (2017) Marijuana effects on changes in brain structure and cognitive function among HIV+ and HIV- adults. Drug Alcohol Depend 170:120–127. https://doi.org/10.1016/j.drugalcdep.2016.11.007

Thomas JB, Brier MR, Snyder AZ, Vaida FF, Ances BM (2013) Pathways to neurodegeneration: effects of HIV and aging on resting-state functional connectivity. Neurology 80(13):1186–1193. https://doi.org/10.1212/WNL.0b013e318288792b

Tu W, Chen PA, Koenig N, Gomez D, Fujiwara E, Gill MJ et al (2019) Machine learning models reveal neurocognitive impairment type and prevalence are associated with distinct variables in HIV/AIDS. J Neurovirol. https://doi.org/10.1007/s13365-019-00791-6

Valcour V, Paul R, Neuhaus J, Shikuma C (2011) The effects of age and HIV on neuropsychological performance. J Int Neuropsychol Soc 17(1):190–195. https://doi.org/10.1017/S1355617710001438

Valcour V, Chalermchai T, Sailasuta N, Marovich M, Lerdlum S, Suttichom D et al (2012) Central nervous system viral invasion and inflammation during acute HIV infection. J Infect Dis 206 (2):275–282. https://doi.org/10.1093/infdis/jis326

Vance DE, Wadley VG, Crowe MG, Raper JL, Ball KK (2011) Cognitive and everyday functioning in older and younger adults with and without HIV. Clin Gerontol 34(5):413–426. https://doi.org/10.1080/07317115.2011.588545

Ver Steeg G, Galstyan A (2014) Discovering structure in high-dimensional data through correlation explanation. Advances in neural information processing systems, pp 577–585. Retrieved from Proceedings of the 27th International Conference on Neural Information

Vinikoor MJ, Napravnik S, Floris-Moore M, Wilson S, Huang DY, Eron JJ (2013) Incidence and clinical features of cerebrovascular disease among HIV-infected adults in the Southeastern United States. AIDS Res Hum Retroviruses 29(7):1068–1074. https://doi.org/10.1089/AID.2012.0334

Wade BS, Valcour VG, Wendelken-Riegelhaupt L, Esmaeili-Firidouni P, Joshi SH, Gutman BA, Thompson PM (2015) Mapping abnormal subcortical brain morphometry in an elderly HIV+ cohort. Neuroimage Clin 9:564–573. https://doi.org/10.1016/j.nicl.2015.10.006

Winston A, Arenas-Pinto A, Stohr W, Fisher M, Orkin CM, Aderogba K et al (2013) Neurocognitive function in HIV infected patients on antiretroviral therapy. PLoS One 8(4): e61949. https://doi.org/10.1371/journal.pone.0061949

Wright PW, Vaida FF, Fernandez RJ, Rutlin J, Price RW, Lee E et al (2015) Cerebral white matter integrity during primary HIV infection. AIDS 29(4):433–442. https://doi.org/10.1097/QAD.0000000000000560

Wright EJ, Grund B, Robertson KR, Cysique L, Brew BJ, Collins GL et al (2018) No neurocognitive advantage for immediate antiretroviral treatment in adults with greater than 500 CD4+ T-cell counts. AIDS 32(8):985–997. https://doi.org/10.1097/QAD.0000000000001778

Yilmaz A, Blennow K, Hagberg L, Nilsson S, Price RW, Schouten J et al (2017) Neurofilament light chain protein as a marker of neuronal injury: review of its use in HIV-1 infection and reference values for HIV-negative controls. Expert Rev Mol Diagn 17(8):761–770. https://doi.org/10.1080/14737159.2017.1341313

Zipursky AR, Gogolishvili D, Rueda S, Brunetta J, Carvalhal A, McCombe JA et al (2013) Evaluation of brief screening tools for neurocognitive impairment in HIV/AIDS: a systematic review of the literature. AIDS 27(15):2385–2401. https://doi.org/10.1097/QAD.0b013e328363bf56

Zoufaly A, an der Heiden M, Kollan C, Bogner JR, Fatkenheuer G, Wasmuth JC et al (2011) Clinical outcome of HIV-infected patients with discordant virological and immunological response to antiretroviral therapy. J Infect Dis 203(3):364–371. https://doi.org/10.1093/jinfdis/jiq055

Reliably Measuring Cognitive Change in the Era of Chronic HIV Infection and Chronic HIV-Associated Neurocognitive Disorders

Lucette A. Cysique, Kaitlin B. Casaletto, and Robert K. Heaton

Contents

Abstract HIV infection has become a chronic disease managed across the life span. In this context, the cognitive health of HIV infection needs to have methods for monitoring over time, in order to better anticipate HIV-associated neurocognitive disorder (HAND) trajectories in relation to biomarkers, and predict prognosis and especially the risk of dementia as People Living with HIV (PLHIV) age. In this chapter, we critically review several statistical frameworks to quantify cognitive change. We then provide a critical review of naturalistic longitudinal studies and

L. A. Cysique (✉)
Faculty of Medicine, Neuroscience Research Australia, School of Medical Sciences, The University of New South Wales, Sydney, Sydney, NSW, Australia
e-mail: lcysique@unsw.edu.au

K. B. Casaletto
Memory and Aging Center, University of California, San Francisco, San Francisco, CA, USA
e-mail: kaitlin.casaletto@ucsf.edu

R. K. Heaton
Department of Psychiatry, UCSD School of Medicine, La Jolla, CA, USA
e-mail: rheaton@ucsd.edu

Curr Topics Behav Neurosci (2021) 50: 271–298
DOI 10.1007/7854_2019_116
Published Online: 27 September 2019

selected randomized clinical trials assessing neurocognitive change as a primary outcome in PLHIV, conducted since the advent of the combined antiretroviral therapy era (censored January 2019). Doing so, we distinguish between PLHIV who were treated early and did not experience AIDS (CDC 1993), versus treated late, after experiencing AIDS and more severe immune compromise. Highlighting strengths and limitations of these studies, we emphasize that issues of reliability pertaining to the use of neuropsychological tests need careful consideration for the robust quantification of cognitive change, including measurement error, practice effect, inter-individual variability, baseline level of functioning, demographic effects, timeframe between testing intervals, normative longitudinal data, and operationalization of clinically meaningful neurocognitive change. In addition, issues pertaining to longitudinal analyses including type, amount and pattern of missing data and/or participant attrition, regression toward the mean, and survivor bias need to be properly addressed. We conclude by proposing future research directions with emphasis on research translation to clinical participants.

Keywords Cognitive change · HIV infection · Longitudinal studies · Neuropsychological tests · Practice effect

1 Introduction: People Living with HIV Infection (PLHIV) Have Different Starting Points for Their Life-Span Cognitive Health

Major improvements in combined antiretroviral treatment (cART) accessibility and successful rates of treatment have caused AIDS-related mortality rates to substantially decline all over the world. At the same time, there has been a steady increase in the overall prevalence and average age of people living with treated chronic HIV infection (UNAIDS 2018). Because of this, People Living with HIV (PLHIV) currently on stable cART can show major differences in their illness profiles and associated differences in duration of infection, ART exposure, and age-related comorbid conditions. Of relevance to life-span cognitive health, we can schematically describe two main groups of well-treated people: (1) PLHIV who have experienced AIDS and severe immune compromise and are now clinically stable and virally suppressed; (2) PLHIV who have never experienced AIDS and severe immune compromise and who were treated early with optimal cART remaining thereafter clinically stable and virally suppressed. Early treatment significantly reduces the likelihood of an AIDS diagnosis and immune suppression below a CD4-T cells count of 200 cp/mL, while controlling viral replication early substantially reduces the odds of developing HIV-associated neurocognitive disorder (HAND). In contrast, late treatment, AIDS diagnosis, a nadir CD4-T-cell count <200 cp/mL, periods of massive HIV replication in the plasma and in the cerebrospinal fluid are known risk factors for HAND (Sacktor 2018; Wright et al. 2015; Smail and Brew 2018). Internationally, both types of patients are well represented

(UNAIDS 2018), and both warrant *equal* focus because independent of the treatment timeline, HAND does *not* occur in all persons with HIV/AIDS (Smail and Brew 2018). This means that a cluster of PLHIV have vulnerabilities for HAND even if treatment was initiated early (i.e., genetic, and other host factors, individual immune response).

NeuroHIV research aimed at describing the neurocognitive trajectories of HAND and neuropsychological (NP) functions in PLHIV, needs to better account for these major treatment, AIDS, and viral replication history differences to accurately determine individual-specific long-term prognoses. Currently, we understand to some extent that the disease/treatment starting point for each PLHIV affects the occurrence of HAND, but we have incomplete knowledge concerning long-term cognitive health prognosis. For example, it is uncertain what the future of PLHIV who have experienced one episode of HAND but who have then at least partially clinically recovered might look like. Similarly, it is uncertain what will happen to chronically infected PLHIV (both with and without a history of HAND) who reach an age where neurodegenerative processes start to become more prevalent (60+). It is only now that there is a substantial number of PLHIV who are reaching this age. Will they be at a higher risk for age-related neurodegenerative processes as is the case for other chronic conditions treated across a large section of the life span (e.g., diabetes type I, sleep apnea, cardiovascular diseases, CVD)? What effects with the latter conditions have as increasingly prevalent comorbid factors in PLHIV? Will the age-related risk for increased immune senescence and chronic immune activation, themselves risk factors for dementia, precipitate neurodegenerative conditions in those who are genetically at risk (e.g., APOE E4 status)? Are the cohorts of survivors from the pre-cART era an adequate prognostic model for those who were treated early with optimal cART? How will the inherent survivor bias of older PLHIV impact our understanding of the future cognitive health of millions with chronic HIV infection? Is there unforeseen cognitive health risk related to long-term cART, including for the most recent versions of ARV drugs and other polypharmacy? Has the increased health attention put on PLHIV led to long-term beneficial effect for cognitive health due to more active monitoring of cardiovascular health and effective treatment of opportunistic infections?

The long-term cognitive health prognosis is also poorly understood in the face of potential systemic or premature, accentuated and/or accelerated brain aging among middle-aged PLHIV. As such, it is not only chronological age that longitudinal NeuroHIV will need to better take into account but also the age at which PLHIV were infected, and biological age (Thurn and Gustafson 2017). Furthermore, with aging, factors that are known to impact on cognitive health in the general population such as CVD and genetic risk for dementia (Iturria-Medina et al. 2016) cannot be overlooked as they may compound persistent HIV-related chronic neuro/immune activation. Commonly seen in the HIV populations is also an elevated prevalence of comorbid psychiatric conditions (Gaynes et al. 2015). These conditions are multiple and will likely impact individual trajectories depending on whether they were optimally treated (depression/anxiety/PTSD, alcohol, substance use) (Cysique et al. 2016; Byrd et al. 2011). Finally, some socio-economic and societal factors that are increasingly recognized to be detrimental to brain health in the general population as well as in PLHIV will likely moderate the future neurocognitive prognosis in each patient (e.g., low education, rural/urban living, poverty, domestic violence, early life trauma, and female sex) (Rubin et al. 2017; Clark et al. 2018).

Keeping this context in mind, we will critically review the existing literature quantifying the longitudinal neurocognitive trajectories and/or the long-term cognitive health prognosis in PLHIV and HAND. This includes research that focused on longitudinal cohorts with no intervention ("naturalistic studies"). Selected randomized clinical trials (RCTs) that have focused on specific samples and/or used specific methods to detect cognitive change will also be described. Prior to reviewing these studies, we will provide an overview of the existing statistical frameworks to quantify cognitive change at the group and at the individual levels. The overview will then be used to appraise the quality of the published studies in regard to their statistical rationale and design. We will emphasize that the conceptual and methodological specificities of using of NP data in a repeated assessment context need to be considered a priori to enhance the validity of longitudinal cognitive health studies in NeuroHIV.

2 Existing Statistical Frameworks to Quantify Cognitive Change at the Group and at the Individual Levels

The need for this section is twofold: (1) Some statistical methods are fundamentally inappropriate for longitudinal analyses of repeated NP data (e.g., methods that do not correct for the inherent time correlation in the data (Bland and Altman 1994)); (2) Among available statistical longitudinal methods, some will perform fundamentally better depending on the study design (e.g., number of retests), as well as types of missing data, and question being addressed (change at the individual level versus group level, measurement of clinically meaningful cognitive change). With the increasing need and use of repeated NP testing in the era of chronic HIV infection, there is a need for best-practice guidance while keeping in mind that this is an active area of research. This section is written to be understood by non-statisticians and non-neuropsychologists, and thus does not replace more specific statistical literature (Hedeker and Gibbons 2006). In particular, conditions where data are not normally distributed and/or outcomes are not continuous were not fully considered due to their significant added complexity and their being beyond the scope of this chapter.

2.1 The Advantage of Longitudinal Studies

Longitudinal studies have many advantages over cross-sectional studies. They are, however, resource intensive and require greater demand on participants. They remain complex to analyze accurately, even when wanting to assess an apparently simple question of whether a participant/patient is *truly* progressing in their disease. Typically, repeated NP measures are not used in isolation but are part of a larger test battery as it is most often used in HIV studies. Depending on the size of the battery, there are preferential approaches (screens versus a comprehensive NP battery) (Kamminga et al. 2017). Each measurement brings additional information on a participant's/patient's performance. Hence, longitudinal designs have greater statistical power than a cross-sectional design. Perhaps the most important advantage of longitudinal designs for the study of NeuroHIV is that it can capture the fluctuating course of HAND (e.g., as may be due to changes in medication effectiveness or toxicity of treatment). Because of this, cross-sectional studies are prone to spurious or biased associations between NP performance and biomarkers. By definition, at a cross-sectional time point each PLHIV will be on their specific cognitive trajectory, so much so that a biomarker may be abnormal before or after the time of the cross-sectional assessment but not during. This could lead to conclusions that a specific biomarker is not associated with the disease when in fact this is not true. In all, cross-sectional studies, at the group level, cannot distinguish between effects due to baseline difference, versus those related to change over time (Hedeker and Gibbons 2006).

2.2 The Critical Importance of the Practice Effect and Its Correction Via Longitudinal Normative Data Strategies

It is often said that in longitudinal studies, one participant becomes their own control. This may be true for non-cognitive-based measurements. However in the context of repeated NP testing, this is significantly complicated by *practice effects (PEs)*, which vary across tests and in relation other factors such as baseline performance, number of retest exposures, and test–retest intervals (Rabbitt et al. 2001; Calamia et al. 2012). PEs represent an improved performance on a NP test with repeated exposure, related to both explicit and implicit learning, as well as greater familiarly with the testing environment. The current best practice strategy to deal with PEs is to reference longitudinal performance in clinical samples to that of an appropriate control sample from which the *expected* PE can be extracted and then corrected for (Cysique et al. 2011a). Appropriate control samples may include healthy controls and clinically stable patient groups. The latter may have the advantage of extending the range of baseline performances, which may affect the expected PE. Normative longitudinal data which depict typical *change* across time enables neuropsychologists to make clinical judgements with a quantitative framework as robust as cross-sectional normative standards (Salthouse and Tucker-Drob 2008). Longitudinal normative standards also have the advantage of providing quantitative and objective methods to disentangle actual cognitive change due to a disease process, or age, from change due to other factors (e.g., PE or normal test–retest variability). Indeed, a neurologically healthy individual given the same test twice, where no true change would be expected, would likely have a difference between test and retest scores. More specifically, in repeat NP assessments, PEs (i.e., previous exposure and familiarity with the test materials) can "artificially" (in the absence of real change in ability) improve test performance at retest (Dikmen et al. 1999). In addition, variation across the testing sessions (such as length of retest interval) and statistical artifacts of repeat assessments (e.g., regression to the mean) can also artificially affect follow-up NP performance (Duff 2012). Thus, without correction using longitudinal normative standards, change or stability in performance may not be accurately detected (Duff 2012). The reader should also keep in mind that quantification and interpretation of PEs is an ongoing, active area of research. Briefly, we summarize below some established knowledge regarding PEs, as well as areas of active inquiry.

1. PEs are commonly seen, yet they are a statistically complex (i.e., non-linear and multi-dimensional) phenomenon: they are often present even when cognitive tests have been designed for repeated assessment (Falleti et al. 2006). PEs vary depending on the cognitive functions assessed and in relation to task novelty and difficulty, which may vary with the cognitive construct being assessed (Heaton et al. 2001). They can also vary with age, such as smaller PEs as people get older (Salthouse 2010), thus representing one of the most robust effects (Calamia et al. 2012). PEs are also not systematic and can be absent due to older age, cognitive impairment, and absence of learning and also due to task simplicity and ceiling effects. There is evidence that the overall baseline cognitive competence (which may be operationally defined as how well the individual performed on the remainder of a test battery relative to the test being examined or normed) is associated with increased or decreased PE (Cysique et al. 2011a). Similarly, there is some partial evidence for increased magnitude of PEs as a function of greater pre-morbid abilities, such as higher IQ and educational level (Heaton et al. 2001). Interestingly, in clinical conditions where patients have been carefully assessed as being clinically stable (e.g., stable HIV disease and treatment, as well as no acute comorbidities of any kind), the magnitude of the PE is the same as that expected in demographically comparable healthy samples providing that baseline performance and overall NP competence are adjusted for (Cysique et al. 2011a). The relation of PE to the test–retest interval is not fully understood, mainly because the majority of studies have included only one follow-up at a fixed interval. This is especially true for investigations associated with standard NP test development in healthy samples. The decision to use only one follow-up was initially justified by the claim that PEs fully subside after the second assessment. This may be the case for some tests (e.g., simple motor processing that includes mostly procedural learning) (Falleti et al. 2006), but it is not the case for many other tests (Cysique et al. 2011a). Indeed, in some instances additional PEs have been shown after up to 7 years on some tasks (Salthouse 2010). Some tests even demonstrate non-linear PEs in relation to test construct and age. There is also evidence that in some tasks, PEs disappear if the test has not been practiced for several years

(Calamia et al. 2012). However, one aspect of PEs that has been consistently found in studies with more than two follow-ups is that the magnitude of the PE is the largest at the first follow-up (Duff 2012; Falleti et al. 2006; Cysique et al. 2011a; Collie et al. 2003).

2. Traditionally, in NP models of cognitive change, PEs have been conceptualized as a source of error in reliably predicting future performance. It is therefore controlled for using the statistical procedures outlined below (Duff 2012; Cysique et al. 2011a). In other words, it is not used as a meaningful cognitive variable in its own right. Several studies have challenged this assumption and have shown that (a) PEs are associated with intact cognitive functioning in elderly adults and is reduced in those with progressive cognitive impairment (i.e., in this case clinical sample with non-stable disease) (Darby et al. 2002; Suchy et al. 2011); (b) in a model predicting cognitive decline across several time points, the addition of a term corresponding to a short-term PE (measured at a 1-week interval in this instance) significantly improves the long-term prediction of cognitive decline in older clinical samples compared to a model where the short-term PE was not included (Duff 2012); (c) the lack of PE is also predictive of cognitive decline several years after in older clinical samples (Darby et al. 2002); and (d) PEs can be associated with greater benefit from the initiation of cognitive-enhancing drugs in schizophrenia and elderly individuals with cognitive impairment (Duff et al. 2010; Watzke et al. 2008). On the other hand, recent work suggests that PEs exist and are comparable at least at the first re-assessment, even among older adults with high vs. low neuropathological burden at autopsy and with mild cognitive impairment (MCI) (Wilson et al. 2018). There are, however, significant limitations to the generalizability of these findings as all studies were based on clinical samples and we know that clinical populations with a progressive condition or even older cognitively healthy people are less likely to show PEs (Calamia et al. 2012).

2.3 *Other Neuropsychological Test Characteristics That Impact the Interpretation of Cognitive Change in Longitudinal Studies*

There are other factors associated with the use of NP tests on repeated occasions that will also impact the detection of "true" change. They include: the degree of test–retest reliability (the degree to which a test produces strongly correlated results over time), measurement error (how well a test is measuring the cognitive construct it is supposed to measure), the impact of demographic effects (age, education, sex, race/ethnicity, pre-morbid IQ can influence PE and true change), test–retest intervals (short and long test–retest interval may impact reliability and PE differently), and test psychometric properties (floor and ceiling effects: screening measures with a floor or ceiling effect are not suited for the measurement of cognitive change) (Kamminga et al. 2017). The cumulative impact of PEs and other specifics of repeated NP testing is so fundamental to the accurate detection of true cognitive change in clinical populations that a normative approach is typically recommended (Cysique et al. 2011a). Put simply, using an individual as his/her own "control" does not work optimally in this particular context. Normative longitudinal standards in cognitively healthy persons and/or clinically stable populations are often required, except perhaps, in the context of RCTs that include both a treatment and matched control group. But even then, more recent research is advocating for the use of normative data in clinical trials (Henry et al. 2017).

2.4 Assessing Cognitive Change Across Groups

Clinical neuroscience and neuropsychological research are increasingly moving away from the repeated measures Analysis of Variance (ANOVA) (multiple time points and continuous data), or standard linear regression (2 time points and continuous data) frameworks, as these methods are fairly rigid: for unbiased estimates, they demand that there is no missing data for each NP measure in a test battery at any of the follow-up time points. A balanced design is always difficult to achieve in longitudinal studies. For example, incomplete data on individual NP measures across a test battery, missing data at an assessment time point, individuals lost to follow-up, or unbalanced design linked to lack of funding are the rule rather than the exception. However, the reason for why some participants do not complete a follow-up assessment or a particular test may carry information that is directly related to the study outcomes so that the context for why data are missing needs to be taken into account in the selection of the appropriate statistical methods (i.e., missing completely at random, MCAR; missing at random, MAR; missing not at random, MNAR) (Hedeker and Gibbons 2006). Altogether, strategies such as simply removing the missing cases, or conducting analyses on only the participants who had complete data (i.e., selected datasets) are not recommended options in the setting of chronic HIV infection and HAND, if we are to correctly estimate disease trajectories. In many instances, it is possible that those with worse prognosis will be the missing cases (i.e., "survival bias"); hence cognitive stability may be falsely concluded. These suboptimal strategies will also produce biased estimates of cognitive change and related associations with biomarkers (e.g., no change detected because only healthy people remained in the longitudinal part of the study, suggesting that there is no incident HAND for people on treatment).

The linear mixed model (LMM) framework (Hedeker and Gibbons 2006) is increasingly favored because it can adjust for different types of missing data and an unbalanced design. In the rare instance where there are no missing data, this model will produce identical results to Repeated Measures ANOVA or the standard linear regression. LMM is an umbrella term for models of varying complexities. The flexibility of this method, however, can also be its drawback. In its simplest form, the subject factor is introduced as a random effect in the model, the group effect is fixed, and in this manner, attrition is adjusted for. More complex models can be determined where the intercept, random, and fixed effects can all be manipulated, but it is highly recommended to work with statisticians who have expertise in these models in such instances (Hedeker and Gibbons 2006). Also, sample size should be considered before using LMM in unbalanced designs, as this type of model is based on large sample theory, and thus inappropriate for small sample (e.g., $N < 50$) (Hedeker and Gibbons 2006). By adjusting for missing data and related unbalanced design, and the use of all available data, LMM has more power, produces more robust and less biased estimates of cognitive change (i.e., smaller error and better data fit), and importantly, subject-specific estimates.

Other group analytic frameworks have been used in NeuroHIV research. The main ones have been the Generalized Estimating Equation (GEE) (Cole et al. 2007), and more recently semi-parametric approaches to form data-driven trajectory clusters (e.g., Group-Based Trajectory Analysis (GBTA) (Brouillette et al. 2016); Mixed Membership Trajectory Model (MMTM) (Molsberry et al. 2015)). GEE is a standard statistical method and its strength is that it is less computationally demanding that LMM. GEE handles missing data with the condition that it is explained by the covariates at hand (Hedeker and Gibbons 2006). This is not always a realistic assumption in longitudinal analyses because the reason for missing data is not always clear. Thus this is a limitation as it renders this model less flexible than LMM (Hedeker and Gibbons 2006). The reader should be aware that there are, however, more flexible versions of the GEE (Hedeker and Gibbons 2006).

Data-driven approaches are more novel and come from other areas of research (e.g., social sciences) and are often applied to neuropsychology by non-neuropsychologists who are unaware of some of the caveats concerning extracting accurate cognitive change (principally the issue of the PE reviewed above). A strength of data-driven approaches such as GBTA and MMTM includes complex modeling of linear and non-linear trajectories. As a whole, given the large inter-individual variability in longitudinal cognitive functioning versus a single prototypical profile of cognitive impairment, application of these methods in NeuroHIV has significantly contributed and helped quantify longitudinal measurement issues in clinical NP more broadly (Jones et al. 2018). These statistical methods can be highly useful in NeuroHIV research because the pathways to dementia in aging PLHIV are unknown; and the long-term prognostic value of one mild episode of HAND is unknown (Molsberry et al. 2015). Data-driven approaches are, by definition, meaningful only when used on *very large* samples so that clustering toward normal aging versus pathological aging patterns can be extracted. There are differences between these models that are beyond the scope of this chapter, and the reader is advised to consult the cited references for further reading. What these methods have in common as a strength (data-driven approaches) is also the basis of their major weakness. They are highly dependent on the quality and type of data used. More precisely, how the original NP data were manipulated/transformed, and which cognitive outcomes were used (global score, domain score, demographically corrected score, PE corrected score or not) have major consequences for what the model is able to demonstrate and, ultimately, its validity. For example, when GBTA is used on screening tools such as the MMSE total raw score (Xie et al. 2011) rather than a demographically corrected total score, the model prediction remains fundamentally linked to the poor psychometric properties and insensitivity to impairment. Perhaps more critical, when cognitive change predictions are made without accounting for PE, fundamental error is embedded on the pattern of change, potentially masking individuals transitioning from normal to mildly impaired cognition (Molsberry et al. 2015). For example, when Brouillette et al. (Brouillette et al. 2016) applied GBTA to the CHARTER longitudinal data without PE corrections on raw scores, the authors extracted a much lower rate of cognitive decline than in the original longitudinal study (Heaton et al. 2015). Unfortunately, results of such methods can sometimes appear as only "descriptive" with major effort needed to improve clinical usefulness. Often data-driven approaches appear to simply restate long-held clinical knowledge (i.e., that inter-individual variability is common); although they provide a quantitative estimate, it is unclear how that estimate can be clinically used at this stage. Furthermore, decisions in the statistical analytical steps to determine "meaningful clusters" have major consequences on the results so that cross-validation would ideally be required when using such methods, hence requiring even larger sample sizes. The predictive power at the individual level of such methods remains therefore to be adequately demonstrated. This is important to note as even if we classified these methods as group-based, it is actually possible to extract individual predictions (Molsberry et al. 2015). However, this should be done with extreme caution. Indeed, it is possible that key characteristics for one particular individual assessed may be missing in the prediction model (e.g., female sex for the only NeuroHIV MMTM study (Molsberry et al. 2015)), which would have a major impact on the validity of that individual prediction. Overall, there is a need for more research to extract what is most clinically useful through these novel approaches. In any case, the caveats concerning how to optimally operationalize NP data in data-driven models should *not* be considered as a secondary issue, but as a primary one.

2.5 Assessing Cognitive Change in Individuals

Several methods have been developed to assist neuropsychologists in measuring cognitive change in individuals (Duff 2012). Of these, standardized regression-based (SRB) change scores have emerged as one of the most reliable and accurate ways to detect "true" change (McSweeny et al. 1993). This approach utilizes regression analysis to predict retest scores such that actual change can be determined by comparing observed retest scores to predicted scores; the change scores are then standardized (as z-scores) by dividing these differences by the error term of the regression model (Duff 2012). The simple SRB change score approach uses only baseline performance to predict a follow-up test score (e.g., McSweeny et al. 1993), while the multiple or multivariate SRB change score approach uses baseline performance in addition to other factors that have shown to impact retest performance (such as demographic information, retest interval, overall baseline cognitive competence, etc., e.g., Dikmen et al. 1999; Heaton et al. 2001; Cysique et al. 2011a). Both approaches are advantageous as they account for variation in baseline scores,

PE, regression to the mean, retest reliability, and "normal" variability in follow-up scores (Duff 2012). A particular advantage of the multiple SRB approach is that by incorporating other relevant information into its prediction models, it can also correct for the effect of included variables on retest scores (e.g., demographics, and particularly baseline overall cognitive competence, operationally defined as average standard score on all other tests in the baseline test battery (Cysique et al. 2011a)) and can therefore be especially sensitive to detecting change in demographically varied or skill diverse samples (provided that the clinically stable normative group is similarly diverse). As such, this approach may be particularly useful in international comparisons in multi-site NeuroHIV studies. SRB "norms for change" have been developed successfully in various groups but have typically been developed for individual test measures and to describe change over a restricted number of retest occasions (e.g., Wechsler 2009; Duff et al. 2005; Sawrie et al. 1996). While it is clearly important to understand typical cognitive change on a particular NP measure, most neuropsychologists in research and clinical settings actually employ numerous measures (or a test battery) to assess multiple cognitive domains. Determining whether an individual has changed on each of many NP measures separately, however, is likely to increase the probability of identifying a significant change by chance (Heaton et al. 2001). To counter this, a method to infer overall cognitive change on a set of tests, and hence reduce the rate of falsely detecting change (improvement or decline), has been proposed (defined as mean SRB change score, and CI around its mean in the normative sample) (Cysique et al. 2011a). As this method summarizes change across the entire battery, it is best suited to detecting change due to diseases which have a diffuse or "spotty" impact on brain functions (e.g., most early dementias and HAND). On the other hand, it may be less useful where localized lesions are likely to produce a focal pattern of NP deficits (e.g., some cases of epilepsy) because tests of abilities that are not affected, when averaged, may "wash out" the relevant signal. As it is unclear how PLHIV may age or progress toward dementia, it will be important to potentially monitor sensitive approaches that evaluate both changes in specific cognitive domains and total battery summary scores.

As noted above, individual-level analyses that have garnered the most clinical applicability to detect cognitive change have a traditional basis in the standard linear regression model, i.e., both strength and weakness (Van der Elst et al. 2013). When based in the simplest test–retest context – that is, when a person is assessed at one follow-up time after a baseline assessment – these methods perform very well (Duff 2012; Collie et al. 2004). Using a statistical benchmark and/or a normative reference corrected for PE and regression toward the mean, in the context of good test–retest reliability (Cysique et al. 2011a), the SRB method is robust in detecting true change (improvement, decline) versus cognitive stability. There is also strong evidence that this method is valid to extract the association of cognitive change to markers of HIV disease (Heaton et al. 2015; Cysique et al. 2010; Gott et al. 2017) and MCI to AD transition as well as dementia progression (Duff 2012). In fact, SRB change scores perform well even under the simplest adjustments (e.g., when change is only adjusted for the baseline performance and nothing else) (Heaton et al. 2001).

Different types of change scores exist (Duff 2012) and reviewing them in detail is beyond the scope of this chapter. However, it is important to understand that all of them will perform poorly when the NP data are not sufficiently reliable. Although there is good evidence that many NP tests are reliable (Calamia et al. 2013), some are less so than others, and individual task reliability in screening tools should ideally be evaluated (Kamminga et al. 2017). When PE correction is not included or poorly estimated, prediction of accurate cognitive change is not possible as PE is a major component of most NP tests (Calamia et al. 2013) and experimental/screening cognitive tests (Falleti et al. 2006). Normally distributed data are a pre-requisite for using SRB change score methods. This condition is often neglected, but it is advisable to use a group-based (LMM) approach for non-parametric data or when continuous data cannot be transformed to approximate the normal distribution (Hedeker and Gibbons 2006). Using confidence intervals around the normative mean (2-tailed, and adequate p-value for a study context), it is still possible to determine who significantly declined or improved versus remained stable. The SRB change score approach compared to other types of reliable change indexes appears to be the most robust under various circumstances, including various levels of baseline performances, and degree and pattern of change (Hinton-Bayre 2016). Relatedly, Temkin et al. (Temkin et al. 1999) observed that when the SRB change score approach was adjusted for relevant covariates, the model demonstrated greater sensitivity to cognitive change (1) used in groups with greater demographic variations, and (2) when baseline scores have a wider range (particularly in the lower spectrum). In the global NeuroHIV context, this finding indicates that optimally adjusted SRB scores will likely be highly applicable in the longitudinal assessment of NP performance where samples are often demographically diverse.

More recently, SRB change score approaches have been challenged (Van der Elst et al. 2013), given that in most longitudinal research, more than one retest will have occurred. Standard linear regression is not adapted to deal with multiple assessments where data is often missing. Our proposed summary-regression-based change score can deal with a few missing data (Cysique et al. 2011a), but it is fair to acknowledge that it cannot optimally deal with major attrition. Van Der Elst et al. (2013) have proposed to expand SRB change score rationale to create linear mixed-effect-based (LMMB) change scores. The method is promising as it can handle various types of missing data and still provide clinically applicable change cut-off predictions. As such, it has been proposed as a way to further develop and integrate LMMB into longitudinal normative data. In addition to the proposed LMMB change scores (Van der Elst et al. 2013), the authors recommend using a Multiple Imputation framework in the case of major attrition and thus further boost the validity of longitudinal normative data development. The proposed LMMB change scores elegantly deals with more complex longitudinal study designs (including common instances of missing data), in addition to yielding individual predictive cognitive change formulae that have clinical applicability. However, further work still needs to be done to extend LMMB approaches for use in multiple cognitive domain testing and large test batteries, as well as screening tools. For example, assessing how PEs should be defined (for cross-sectional classifications across time points) in such a context requires more research. The drawback of such a method is that it is computationally demanding. To address this, Van der Elst et al. (2013) have provided scripts that run the normative data, but this is for a restricted number of tests. Expansion and greater use of this clinically relevant method is warranted.

3 Longitudinal Studies in Early Treated HIV Infection

The long-term prognosis of HIV+ persons who are initiated early on treatment at high CD4 count remains to be studied across the decades on which those patients will be on treatment (Wright et al. 2015, 2018). The START Neurology study, which was designed to address this question among individuals living in Argentina, Australia, Belgium, Brazil, Chile, Germany, Italy, Switzerland, Thailand, the United Kingdom, and the USA, showed that early versus deferred treatment (at CD4 <350/mL) did not lead to different neurocognitive trajectories (Wright et al. 2018). This study lasted 6 years and included an annual NP testing covering speed of cognitive processing, fine motor coordination, verbal memory, and fluency. While 6 years is a fairly long period to assess the effect of early treatment on NP functions, it is still early in the chronic life span of HIV infection. The sample was in their mid-30s and about 20% showed low neurocognitive performance. However, it was difficult to sensitively ascertain who may have had HAND, given that geographically and demographically matched controls were not collected to develop norms for either determination of cross-sectional impairment or longitudinal change. Neurologically vulnerable individuals are always present in primary infection cohorts (Samboju et al. 2018). In the case of START, the majority of participants who probably had normal neurocognitive functioning showed a typical PE of a moderate to large magnitude at the second testing and of a small magnitude afterwards; this effect likely overpowered (in the two arm analysis) the potentially smaller number of participants who may have shown a declining pattern of performance. Another large international trial in resource-limited settings (as part of the ACTG PEARLS study (A5175), including participants in Brazil, India, Malawi, Peru, South Africa, Thailand, and Zimbabwe) was conducted to determine whether three first-line cART regimens may be associated with different NP outcomes over 4 years with an assessment every 24 weeks (Robertson et al. 2012). The authors found no differences in treatment effects but did observe a monotonic improvement in all study arms that they interpreted as "multifactorial, reflecting reduced central nervous system HIV infection, better general health, and practice effects." Therefore, in this study as well, the uncorrected PE likely masks potential abnormal longitudinal performance in some individuals. Overall, future studies should aim to collect data in appropriate control groups to effectively deal with PE and other key factors influencing longitudinal NP performance as reviewed above. It is unfortunately a recurrent issue (Grund et al. 2013), which diminishes the significance and the validity of these very costly trials. Indeed, without appropriate PE correction, it is impossible to determine the magnitude of any treatment effect. Therefore, it is not certain whether these trials truly have negative results, especially in sub-samples who we know are initially neurologically fragile (Samboju et al. 2018). Another issue that we and collaborators have mentioned many times (Cysique et al. 2011b; Gates and Cysique 2016) is the need for a priori testing to verify the presence of neurocognitive impairment at entry of such trials, if the aim is to detect treatment effect as cognitive improvement. This ideally requires classifying HAND using standard diagnostic criteria (Antinori et al. 2007). Ultimately, it means that many of those studies were underpowered to start with, so extreme caution should be applied when interpreting the mostly negative results of these RCTs to the broader population with HAND.

4 Longitudinal Studies in Chronically Treated HIV Infection

Table 1 summarizes 18 longitudinal naturalistic studies (one RCT sub-analysis) that have been conducted on PLHIV on stable cART between 2006 and 2018 (censored January 2019 for the purpose of this chapter), in which cognitive change was investigated as a key outcome. Studies were selected with the following criteria: (1) A majority of the sample (>50%) was clearly reported to be on cART; (2) Studies were original reports that were either the first output for a particular longitudinal cohort and/or the first use of a specific longitudinal methodology to measure neurocognitive change; (3) Only studies including adults (aged >18+) were included; (4) Studies with attrition rate >50% or "small" sample size ($N < 50$) were not included.

Upon review we found the following: No studies were conducted in a resource-limited setting. Half of the studies (9/18) were based on a single follow-up. Other studies included multiple follow-ups with varying intervals (as short as 6 months). Information about baseline HAND was most often provided either in the form of a rate of impairment, group-level *p*-value, or effect size compared to controls in a total of 15/18 studies. There was wide variation in reported baseline impairment rate from 0% (group selected on purpose with no impairment) to 55% impairment rate. The most common rates of impairment ranged between 35 and 45% which is interestingly well within the typically reported prevalence of chronically treated PLHIV HAND when standard NP test methodology is used with some normative corrections (Saloner and Cysique 2017). In fact, just over half of the studies included an HIV− control group (10/18) and in all instances, the control groups were geographically and demographically comparable to their HIV+ counterparts; other studies used locally derived norms. Three studies included a screening battery composed of 2–3 standard NP tests. This means that the majority of studies used a medium to large standard NP battery, inclusive of tests recommended for the assessment of HAND. The majority of studies (15/18) used a global score approach (i.e., the standard NP tests were averaged into a *z*-score or *T*-score of global change score and an

Table 1 Longitudinal studies of cognition on PLHIV on stable cART censored January 2019

Year	First author Cohort	Baseline HAND rate HIV− controls?	Follow-up period	NP tests	Outcome and definition of cognitive decline	Primary outcomes
2006 (Cysique et al. 2006)	Cysique 2002–2004 cART-treated Australian cohort	BL HAND: 38.8% HIV−: yes	Tested at baseline, 6, 15, and 27 months	NART, WAIS-III digit span and similarities, TMT, SDMT, GPT, CVLT, RCFT, COWAT	Global score WSD change score LMM	6 months: 30% HIV+ individuals classified as declined, significantly higher than HIV− controls 15 months: 5–13% HIV+ classified as declined 27 months: 5% HIV+ classified as declined Attrition >10%
2007 (Robertson et al. 2007)	Robertson ACTG ALLRT	BL HAND: 39% HIV−: no	Tested twice across a 48-week period	TMT, WAIS-R digit symbol	Demographically corrected global score Impairment: at least −1D in 2 tests or −2 SD in 1 test No PE correction Sustained impairment: impairment at BL and first follow-up Incident impairment: unimpaired at BL and had at least 1 follow-up impaired	56% had sustained impairment 21% had incident impairment Incident and sustained impairment is not predicted by traditional HIV biomarkers Attrition: unclear and selected sample with data
2007 (Cole et al. 2007)	Cole MACS	BL HAND: NR HIV−: yes	Tested every 6 months over 8 years	TMT, SDMT	Individual test scores GEE model LMM	NS difference between HIV+/− groups Missing cases a priori excluded, therefore no attrition as selected case with data
2010 (Cysique et al. 2010)	Cysique Rural China cART-treated cohort	BL HAND: 34% HIV−: yes	One retest at 12 months	International HNRP battery: color trails, WCST, COWAT, PASAT, WMS-III spatial span, digit span and symbol search, HVLT-R, BVMT-R, GPT, TMT, Stroop	Global score SRB change score Repeated ANOVA	Decline in HIV+ participants (27%) significantly greater than HIV− participants (5%) Attrition < 10%
2014 (Seider et al. 2014)	Seider Florida cART-treated cohort	BL HAND: NR HIV−: yes	One retest at $M = 14.28$ months	BVMT-R, HVLT-R	Individual test scores No PE correction Repeated ANOVA	HIV+ participants more likely to decline on HVLT-R delay score compared to HIV− participants Attrition: NR

(continued)

Table 1 (continued)

Year	First author Cohort	Baseline HAND rate HIV− controls?	Follow-up period	NP tests	Outcome and definition of cognitive decline	Primary outcomes
2015 (Heaton et al. 2015)	Heaton CHARTER	BL HAND: 46% HIV−: no	Tested every 6 months over 16–72 months	CHARTER battery: TMT, WAIS-III digit symbol coding, LNS and symbol search, WCST, COWAT, PASAT, BVMT-R, HVLT-R, GPT	Global score SRB change score LMM	23% HIV+ participants declined across duration of follow-up Attrition > 10%
2015 (Molsberry et al. 2015)	Molsberry MACS	BL HAND: NR HIV−: yes	Multiple tests 2 cohorts 1984–1985 and 1987–1991 2001–2003	MACS battery inclusive of 5 cognitive domains	HAND 2007 classification without PE correction, but data included HIV− Mixed membership trajectory model (MMTM) To be included in the MMTM, participants needed to have at least 1 classification	Three distinct trajectories to cognitive impairment: 'normal aging' (low probability of mild impairment until age 60); 'premature aging' (mild impairment starting at age 45–50); and 'unhealthy' (mild impairment in 20s and 30s) profiles No attrition, but selected sample with data
2015 (Sheppard et al. 2015)	Sheppard HNRP cohort	BL HAND: 0% HIV−: yes	One retest at 12 months	WMS-III logical memory, CVLT-II, WMS-III digit span, TMT, TOL, GPT	HAND diagnosis No PE correction Logistic regression	HAND diagnosis at follow-up significantly higher in HIV+ participants (15.7%) as compared to impairment rate in HIV− participants (3.2%) Attrition: >10%
2015 (Dufouil et al. 2015)	Dufouil ANRS CO3 Aquitaine Study Group	BL HAND in those with follow-up: 54.8% Impaired glycemia associated with MND Diabetes associated with HAD HIV−: no	One retest at 24.1 months ± 3	WAIS-III digit symbol coding; FCSRT; IST; PPT; RCFT; TMT	HAND diagnosis based on Yearly changes in cognitive performances (FU z-score – BL z-score)/delay in years between time points (with adjusted linear regression) and z-score change difference No PE correction	Reduction in HAD rate by 70% and MND rate by 40% at follow-up Increase in NP performance in all but 2 tests Attrition effect assessed and not significant despite 29% attrition rate and greater HAND prevalence in baseline sample
2016 (Brouillette et al. 2016)	Brouillette et al. CHARTER Re-analysis	BL HAND: 46% HIV−: no	Tested every 6 months for 36 months	CHARTER battery	Individual test scores GBTA No PE correction	15.8% declined on at least one NP test, the majority (83.8%) showed decline on only a single test Attrition > 10%

2016 (Marquine et al. 2016)	Marquine VACS	BL HAND: 40% HIV−: no	Multiple tests for up to 6 years	HNRP battery: TMT, WAIS-III digit symbol coding, LNS and symbol search, WCST, COWAT, WMS-III spatial span, BVMT-R, HVLT-R, GPT	Demographically and PE-adjusted global and domain *T*-scores LMM	In time-dependent analyses adjusting for covariates, higher VACS index scores were significantly associated with worse global and domain neurocognitive performance and increased risk for developing NCI in a subgroup of persons who were neurocognitively normal at baseline No attrition, but case selected with data
2016 (Sacktor et al. 2016)	Saktor MACS	BL HAND: 33% HIV−: no	Tested ~every 2 years over 4-year period	TMT, Stroop, SDMT, N-back RT, RAVLT, RCFT, GPT	HAND diagnosis No PE correction Logistic regression	BL to 2 years: 70% of the HIV individuals remained at same HAND stage, 15% deteriorated in HAND stage, 15% improved in HAND stage 2–4 years: 71% of the HIV individuals remained at same HAND stage, 15% deteriorated in HAND stage, 14% improved in HAND stage (e.g., MND to ANI) BL to 4 years: 77% of the HIV individuals remained at same HAND stage, 13% deteriorated in HAND stage, 10% improved in HAND stage No attrition, case selected with data
2017 (Gott et al. 2017)	Gott Australian HIV and Brain Aging Cohort	BL HAND: 55% HIV−: yes	One retest 19.5 months, SD = 7.73 months	TMT, WAIS-III digit symbol coding and LNS, COWAT, WMS-III spatial span, HVLT-R, GPT, DKEFS Stroop	SRB change score Neurocognitive trajectories LMM	14% with clinically significant cognitive decline, none improved significantly Greater continuous cognitive decline in HIV+ versus HIV− 57% of the cohort is undergoing slow evolution of their disease when taking into account historical HAND Attrition: 6%

(continued)

Table 1 (continued)

Year	First author Cohort	Baseline HAND rate HIV− controls?	Follow-up period	NP tests	Outcome and definition of cognitive decline	Primary outcomes
2017 (Rubin et al. 2017)	Rubin WIHS	BL HAND: 23% HIV−: yes	Tested every 2 years for 3 visits between 2009 and 2013	TMT, SDMT, COWAT, WAIS-III LNS, HVLT-R, GPT, Stroop test	Continuous *T*-scores and categorical measures of impairment (*T*-score < 40) LMM PE accounted at group level in LMM with HIV− data	Abnormal NP performance in global cognition persisted over time in virally suppressed HIV+ women vs. HIV− women Motor skills showed decline in HIV+ women
2017 (Vassallo et al. 2017)	Vassallo NICE French treated Cohort	BL HAND: 23% HIV−: no	One retest ~24 months	MMSE, FCSRT, Stroop test, verbal fluency, montreal-toulouse visual agnosia test, PASAT, finger tapping test, GPT, WCST, TMT	HAND 2007 criteria classification change No PE correction	In retested group, 6% improved, 32% deteriorated, 62% were stable Overall 45% had sustained HAND at baseline and follow-up CD4/CD8 decrease more frequently in those with worsening NP performance 36% attrition
2018 (Cole et al. 2018)	Cole COBRA	BL HAND: poorer NP functions in HIV+ compared to HIV− ($p < 0.0001$) HIV−: yes	One retest at 2 years	COBRA battery: TMT, WAIS- III digit symbol coding, LNS and symbol search, WCST, COWAT, PASAT, RAVLT, WAIS-IV visual reproduction, finger tapping test, GPT	Demographically corrected *T*-scores LMM PE accounted at group level in LMM with HIV− data	Stable NP performance between the HIV+ and HIV− group except for attention and memory Improved performance in memory in both groups; for attention, decrease in HIV− Almost no attrition, but selected data
2018 (Elicher et al. 2018)	Elicer MHBB	BL HAND: 17.9% Impairment with other causes: 48.8% HIV−: no	Multiple tests across 10 years	MHBB battery: TMT, WAIS-III symbol search and LNS, WCST, COWAT, WMS-III spatial span, BVMT-R, HVLT-R, GPT, HIV dementia motor scale (HDMS)	Demographically corrected Global *T*-score (GTS) HAND 2007 criteria classification Paired *t*-test Stable group: GTS remained within 1 SD of their BL performance in either direction Improvement group: rise of GTS of 1 SD or greater Decline group was defined as a fall	Stability of HAND of a mild severity in most overall, the GTS tended to improve slightly over time (change = 1.985 points; $d = 0.358$, moderate effect size) Using the criterion of 1 SD, among the 78 participants, $n = 5$ improved, and the rest ($n = 73$) were stable. None decline Declining motor functions associated with cerebrovascular

					of GTS of 1 SD or greater No exact PE correction	disease No attrition, but selected dataset with complete longitudinal data
2018 (Haynes et al. 2018)	Haynes London UK cART-treated cohort	BL HAND: HIV+ poorer NP functions compared to HIV− ($p < 0.0007$, d–0.76) HIV−: yes	One retest 4 years	TMT, WAIS-III digit span, digit symbol coding, matrix reasoning, similarities, vocabulary, symbol search, and block design, WCST, COWAT, WAIS-R logical memory, visual reproduction, RAVLT, word-pair and single word recognition, GPT, Kenna Graded Naming Test	HIV− based z-scores Change scores generated by subtracting the baseline from follow-up z-scores	HIV+ status was associated with greater decreases in executive function and global cognition older HIV+ individuals showing greater global cognitive decline compared to older HIV− individuals

ANOVA analysis of variance, *BL* baseline, *BVMT-R* Brief Visuospatial Memory Test – Revised, *CNS HIV* Anti-Retroviral Therapy Effects Research (CHARTER), Comorbidity in Relation to AIDS (COBRA) Project, *COWAT* Controlled Oral Word Association Test, *CVLT(-II)* California Verbal Learning Test – (Second Edition), *DKEFS* Delis-Kaplan Executive Function System, *FCSRT* Free and Cued Selective Reminding Test, *GBTA* Group-Based Trajectory Analysis, *GEE* Generalized Estimating Equation, *GPT* Grooved Pegboard Test, *HAND* HIV Associated Neurocognitive Disorder, *HNRP* HIV Neurobehavioral Research Program, *HVLT-R* Hopkins Verbal Learning Test – Revised, *IST* Isaac Set Test, *LMM* linear mixed model, *LNS* Letter Number Sequencing, *MHBB* Manhattan HIV Brain Bank, *NART* National Adult Reading Test, *NP* neuropsychological, *NR* not reported, *NS* not significant, *PASAT* Paced Auditory Serial Addition Test, *PPT* Purdue Pegboard Test, *RAVLT* Rey Auditory Verbal Learning Test, *RCFT* Rey Complex Figure Test, *RT* reaction time, *SDMT* Symbol Digit Modalities Test, *SRB* standardized regression-based, *TMT* Trail Making Test, *TOL* Tower of London, Veterans Aging Cohort Study (VACS), *WAIS-III* Wechsler Adult Intelligence Scale – Third Edition, *WCST* Wisconsin Card Sorting Test, *WMS-III* Wechsler Memory Scale – 3rd Edition appropriate HIV− control group) or HAND diagnosis classification as their main outcome. Of those, five studies used a longitudinal normative data approach. Three studies also investigated cognitive change in individual cognitive domains.

The major issue we found upon review was that *studies significantly differed in how they defined and operationalized cognitive change*. A minority of studies (4/18) used a change score approach that corrected for PE and other statistical artifacts affecting cognitive change measurement as reviewed above. Another four studies used an LMM approach with an HIV− control group, controlling at the group level for PE and other factors affecting the interpretation of cognitive change. Among these eight studies, three combined an LMM and a change score approach. All other studies (8/18) used statistical methods and definitions of cognitive change that were suboptimal for PE consideration, regression toward the mean, and variations in test–retest reliability.

In studies with multiple follow-ups, some used a strategy in which cognitive change was operationalized as an absorbing state (wherein once an individual was defined as having declined, this resulting in a permanent classification). Other studies also used this strategy but applied it to the HAND classification (wherein once a change of classification was reached, it was defined as permanent). Four studies operationalized the change definition as a potentially fluctuating state wherein the outcome of change was a trajectory. This included the GBTA and MTMM studies, but those two did not optimally correct for PE. In any case, *apparent* differences in rates of cognitive decline or HAND classification changes should be interpreted carefully. Ultimately, only GEE and LMM with the inclusion of an appropriate HIV− or other comparison group can extract optimal trajectories (Van der Elst et al. 2013), when studying neurocognitive functions in chronic HIV infection over multiple assessments. Data-driven approaches may be useful to derive patterns or clusters of change, but PEs must be corrected for, at the very minimum. Also, in such analyses, which outcome(s) are best for extracting accurate trajectories linked to disease effects remains to be determined.

With this in mind as we inspected studies examining rates of cognitive change, we found that the rate of cognitive decline in individuals with HAND varied between 14 and 30% across a period of 6 months to 3 years. The largest study, CHARTER, using an SRB approach (correcting for PE) and an absorbing rate definition,[1] reported that 23% of HIV+ individuals were classified as decliners who never improved (90% confidence interval 2-tailed) around a normative mean (Heaton et al. 2015). While, Heaton et al. (2015) used the traditional statistical definition of significant cognitive change (90% confidence interval, 2-tailed) via a *p*-value set at $p < 0.05$ 2-tailed. This definition remains the most commonly used in clinical

[1]In the CHARTER study, if an individual declined from baseline and never improved, they were classified as "decliner." If an individual improved from baseline but never declined, they were classified as an "improver." If neither, they were classified as "stable." If both decline and improvement were detected, they were classified as "fluctuant" or were excluded depending on the analysis.

neuropsychology, but there is a lack of evidence as to whether it is optimally sensitive and specific to cognitive change under a variety of conditions.

Two studies that used a group-level strategy on single or multiple retests (one with the LMM and an HIV− control group) did not find that their HIV− and HIV+ participants had different rates of cognitive change, despite evidencing lower starting points (PLHIV being more impaired at baseline). However, studies in the same country (UK, but using a suboptimal definition of cognitive change: simple difference between follow-up score from baseline score) and in the USA in women (using the LMM) detected significantly steeper rates of cognitive decline in their HIV+ group compared to their HIV− controls. Evidence for improvement was less commonly reported and varied between 0 and 15% when reported. But again, caution is needed in interpreting each individual study, as some did not include a control group, and others did not control for PE. Despite the varying rates or evidence for cognitive decline, all studies consistently found that a majority (>60% in most cases) of chronic PLHIV had stable neurocognitive performance. This finding indicates that, at a minimum, many people with HAND at baseline demonstrate mild, but sustained HAND of mild severity in most instances. This was true even in studies using screening battery (Robertson et al. 2007).

Of the two studies that specifically selected long-term medically asymptomatic patients, whose baseline cognitive performance was commensurate with HIV− controls, Sheppard et al. (2015) reported significantly greater incident rate of clinically significant impairment in HIV+ participants. However, Cole et al. (2007) observed stable cognitive performance in HIV+ individuals on an abbreviated battery using two tasks measuring psychomotor speed and mental flexibility.

Among studies that assessed deterioration within the HAND stages (did not correct HIV− adjusted PE and other artifacts affecting longitudinal NP performance), most found lower rates of cognitive decline (10–15%), except one of the French studies (32%) (Vassallo et al. 2017). The Multicenter AIDS Cohort Study (MACS) found that across 4 years of follow-up, 10% of HIV+ individuals progressed to a worse HAND stage (Sacktor et al. 2016), while Sheppard et al. reported that 15% of the cohort progressed to a more severe HAND stage (Sheppard et al. 2015). Dufouil et al. (2015) found that among those with HAD at baseline, almost 70% improved, while 40% of those with MND also improved. However, this latter study in addition to no PE correction and adjustment for regression to the mean had a 29% attrition rate.

Among the studies that examined the effect of aging and HIV on longitudinal cognitive change, outcomes also varied. Some detected an age by HIV interaction, suggesting a disproportionate "cognitive aging" effect in HIV; notably, these studies were specifically designed to assess this question and included a higher number of older participants (60+ years of age) (Haynes et al. 2018). Furthermore, greater cognitive decline linked to HIV and age may be more apparent on some cognitive functions than others. For example, Sieder et al. (2014) observed that on verbal memory measures older HIV+ individuals demonstrated greater decline as compared to the younger HIV+ and HIV− groups.

Finally, attrition and pre-selection of datasets remain a major issue across the cognitive change studies in NeuroHIV. Apart from studies using their entire sample or controlling for the effect of attrition using LMM, all other studies would have been significantly biased in this respect. It is unclear how much impact this may have on the actual rate of cognitive decline detecting in chronic PLHIV, as we can suspect that some of the sickest patients dropped out. On the other hand, busy and healthy patients may also drop out with higher frequencies. Future studies should adequately assess why their participants may be lost to follow-up (or at least report baseline results or dropouts).

5 Conclusions, Clinical Implications, and Future Directions

There are many strengths in the NeuroHIV cognitive change literature, including that the majority of studies used a sufficiently sizeable NP battery measuring at least 5 cognitive domains, as recommended by the 2007 HAND criteria (Antinori et al. 2007). Longer studies with multiple time points are increasingly being conducted; in addition, most studies had either appropriate controls or norms for the detection of HAND at baseline.

Cognitive stability in most treated PLHIV emerged as a consistent finding in the majority of studies despite the methodological caveats mentioned. However, this also means that a subgroup of PLHIV may be vulnerable to decline, and though the exact size of this subgroup remains debatable, it is probably not negligible (currently estimated at 14–30% across 6 months to 3 years). Importantly however, is that when decline is detected, the pattern is not rapidly progressive in the vast majority. Nevertheless, most studies were conducted across relatively short periods of time (<5 years) and included relatively young participants (vast majority being under 60+). Therefore, claiming cognitive stability in the chronic, aging HIV population as a definitive result would likely be premature; for now, it can be interpreted as a relatively tentative result based on the current evidence. Caution is further supported by recent evidence from the longest structural MRI NeuroHIV study (Pfefferbaum et al. 2018) which shows premature/accentuated and accelerating brain atrophy in chronic PLHIV.

Moreover, a substantial subset of PLHIV have mild cognitive impairment while on cART and with viral suppression and are then assessed as cognitively stable, so this means that they have *sustained mild HAND*. If we add these people to the smaller subgroup of PLHIV with declining cognition, this may include close to 50% of chronic PLHIV (Gott et al. 2017). Also, evidence for reliable improvement is in the minority, which estimates range from 0 to 15% cases when provided. In all, the monitoring of chronic cognitive health in PLHIV should remain a health-care priority in HIV research, and especially so as PLHIV age with multiple comorbidities (Aung et al. 2019).

Challenges remaining include conducting longitudinal cognitive change studies in more diverse pools of PLHIV, particularly in resource-limited settings, using cross-culturally valid methods that can also detect cognitive change accurately.

Therefore, international studies with robust methods for detecting cognitive change are still needed to provide globally accurate estimates. This is possible, as international RCTs, and many cross-sectional observational studies have been conducted successfully. In fact, longitudinal NP methods are one of the solutions for handling cross-cultural validity if baseline NP scores are used as a proxy of overall cognitive competence (Cysique et al. 2011a, b).

A standard operationalization of cognitive change is needed with perhaps separate definitions for the clinical (Tierney et al. 2017) and research contexts. However, the definition of the minimum standard should include correction for both the PE and test–retest reliability – these two factors ensure that other factors that affect cognitive change performance will also have correct minimum standards. This is true for both detection of cognitive change (as improved/declined versus stable) and change in HAND diagnostic classification (after PE correction). Group versus individual strategies in the assessment of cognitive change are both perfectly sound, depending on the research question; however, studies should always document their baseline rate of impairment using standard criteria, as it may have a major impact on the level of decline or improvement observed. For research studies, the reporting of all data (complete, missing, attrition, reasons for lost to follow-up) should become a standard, as currently there have been too many studies which have selected only complete datasets, and thus may have produced biased estimates of cognitive change. Ultimately, this proposed strategy resonates with a recent call to action for greater NP methods harmonization by Paul (2019), which is needed to permit innovative, large sample types of analyses.

Last but not least, we need international studies with large and diverse samples to represent the complex mix of comorbidities (e.g., co-infections and baseline chronic immune activation level that are more common in limited-resource setting (Petoumenos et al. 2017)), HIV disease and treatment history, gender, and other demographic effects. This type of data would provide more accurate global estimates of cognitive change as opposed to the current data which is based on a non-representative selected few samples. Furthermore, a global perspective of disease burden, rather than a controlled HIV-specific perspective should take precedent for such research. This is both relevant to the long-term cognitive health as an integral part of the "fourth 90s" defined by good health-related quality of life for PLHIV becomes an outcome as important as early treatment, stable treatment, and viral suppression (Lazarus et al. 2016). In this framework, good health-related quality of life for PLHIV entails attention to two domains: comorbidities and self-perceived quality of life; both aspects are critical to neurological health in aging PLHIV, considering that multiple comorbidities and depressive symptoms are associated with all-type dementia risk (Linn Aung et al. 2019). Aspects that need better representation and examination are age-related CVD effects, chronic immune activation and immune senescence factors, all of which confer added risk for all-type dementia (Paul 2019).

Improved understanding of cognitive trajectories in chronic HIV infection and HAND will improve the testing of treatment effects in RCTs, as well as further contextualize and support the ecological validity of such trials. The inclusion of "prior knowledge data" (e.g., longitudinal normative data) and use of adaptive

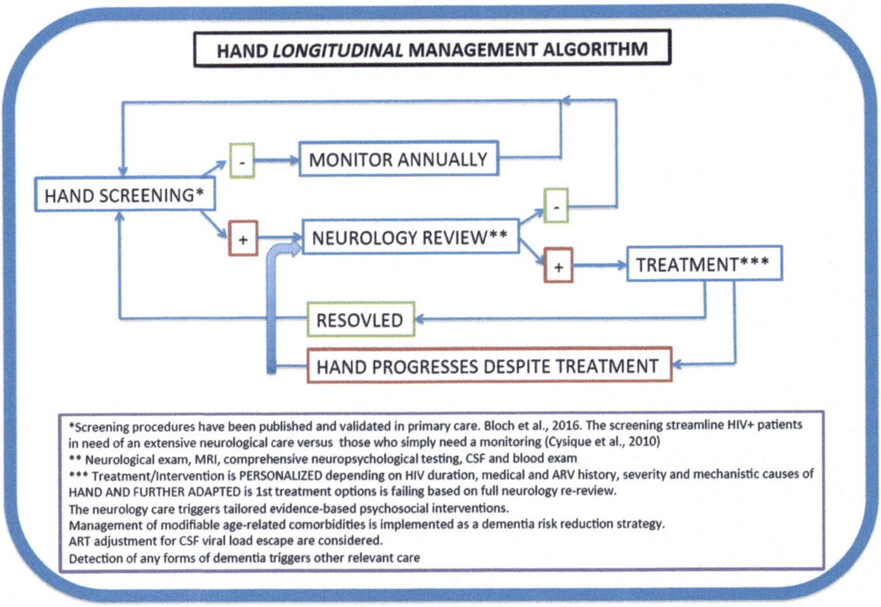

Fig. 1 ∎

randomization (e.g., arm sub-randomized according to key comorbidities such as co-infections, CVD) should now be conceived as priority choices in study design rather than as "luxury extras" to optimally handle the complex nature of cognitive functioning when measured over time. Improved understanding of cognitive trajectories in chronic HIV infection and HAND would also provide stronger evidence for biomarkers, given that NeuroHIV research is increasingly interested in, and could significantly benefit from, factors that identify risk for specific trajectories or clinically relevant "biotypes."

Ultimately, better longitudinal characterization of PLHIV should serve to build a prognostic framework and improve the current diagnostic HAND criteria. Building a prognostic algorithm within the HAND standard nomenclature could help resolve some of the limitations of the Frascati nomenclature by ascertaining whether HAND is "active" or not (Brew 2004), whether a patient is at risk for cognitive decline or not, and whether a patient is at risk of further neurodegeneration as they age. It is also possible that a prognostic nomenclature would help resolve some of the difficulty in ascertaining functional decline (i.e., everyday functioning), which is by definition a time-based phenomenon. Evidence of cognitive decline beyond a longitudinal normative reference may help better discriminate patients with some level of anosognosia (i.e., lack of insight into their cognitive impairment) who tend to minimize their deficits, versus patients who have high cognitive complaints, are often also depressed, and may or may not have ongoing cognitive deficits. Evidence-based methods for detecting cognitive decline at the individual level can also be used in a screening framework as proposed in Fig. 1, to better streamline patients in need of a neurological review versus patients who need some low-level long-term monitoring. This proposed algorithm also would have the benefit of clarifying treatment and intervention pathways in a timely and specific fashion. Altogether, we propose that putting the *assessment of cognitive change* in chronic PLHIV and HAND at the center of neurological and neuropsychological management may bring a better standard of care as both HIV and HAND are fundamentally chronic diseases with possible phases of relapses and remission and lifelong treatment.

References

Antinori A, Arendt G, Becker JT, Brew BJ, Byrd DA, Cherner M et al (2007) Updated research nosology for HIV-associated neurocognitive disorders. Neurology 69(18):1789–1799

Aung HL, Kootar S, Gates TM, Brew BJ, Cysique LA (2019) How all-type dementia risk factors and modifiable risk interventions may be relevant to the first-generation aging with HIV infection? Eur Geriatr Med 10(2):227–238

Bland JM, Altman DG (1994) Statistics notes: correlation, regression, and repeated data. BMJ 308(6933):896

Brew BJ (2004) Evidence for a change in AIDS dementia complex in the era of highly active antiretroviral therapy and the possibility of new forms of AIDS dementia complex. AIDS 18(Suppl 1): S75–S78

Brouillette MJ, Yuen T, Fellows LK, Cysique LA, Heaton RK, Mayo NE (2016) Identifying neurocognitive decline at 36 months among HIV-positive participants in the CHARTER cohort using group-based trajectory analysis. PLoS One 11(5):e0155766

Byrd DA, Fellows RP, Morgello S, Franklin D, Heaton RK, Deutsch R et al (2011) Neurocognitive impact of substance use in HIV infection. J Acquir Immune Defic Syndr 58(2):154–162

Calamia M, Markon K, Tranel D (2012) Scoring higher the second time around: meta-analyses of practice effects in neuropsychological assessment. Clin Neuropsychol 26(4):543–570

Calamia M, Markon K, Tranel D (2013) The robust reliability of neuropsychological measures: meta-analyses of test-retest correlations. Clin Neuropsychol 27(7):1077–1105

Clark US, Arce Renteria M, Hegde RR, Morgello S (2018) Early life stress-related elevations in reaction time variability are associated with brain volume reductions in HIV+ adults. Front Behav Neurosci 12(6):6

Cole MA, Margolick JB, Cox C, Li X, Selnes OA, Martin EM et al (2007) Longitudinally preserved psychomotor performance in long-term asymptomatic HIV-infected individuals. Neurology 69(24):2213–2220

Cole JH, Caan MWA, Underwood J, De Francesco D, van Zoest RA, Wit F et al (2018) No evidence for accelerated aging-related brain pathology in treated human immunodeficiency virus: longitudinal neuroimaging results from the comorbidity in relation to AIDS (COBRA) project. Clin Infect Dis 66(12):1899–1909

Collie A, Maruff P, Darby DG, McStephen M (2003) The effects of practice on the cognitive test performance of neurologically normal individuals assessed at brief test-retest intervals. J Int Neuropsychol Soc 9(3):419–428

Collie A, Maruff P, Makdissi M, McStephen M, Darby DG, McCrory P (2004) Statistical procedures for determining the extent of cognitive change following concussion. Br J Sports Med 38(3):273–278

Cysique LA, Maruff P, Brew BJ (2006) Variable benefit in neuropsychological function in HIV-infected HAART-treated patients. Neurology 66(9):1447–1450

Cysique LA, Letendre SL, Ake C, Jin H, Franklin DR, Gupta S et al (2010) Incidence and nature of cognitive decline over 1 year among HIV-infected former plasma donors in China. AIDS 24: 983–990

Cysique LA, Franklin D Jr, Abramson I, Ellis RJ, Letendre S, Collier A et al (2011a) Normative data and validation of a regression based summary score for assessing meaningful neuropsychological change. J Clin Exp Neuropsychol 7:1–18

Cysique LA, Waters EK, Brew BJ (2011b) Central nervous system antiretroviral efficacy in HIV infection: a qualitative and quantitative review and implications for future research. BMC Neurol 11(148):148

Cysique LA, Dermody N, Carr A, Brew BJ, Teesson M (2016) The role of depression chronicity and recurrence on neurocognitive dysfunctions in HIV-infected adults. J Neurovirol 22(1):56–65

Darby D, Maruff P, Collie A, McStephen M (2002) Mild cognitive impairment can be detected by multiple assessments in a single day. Neurology 59(7):1042–1046

Dikmen SS, Heaton RK, Grant I, Temkin NR (1999) Test-retest reliability and practice effects of expanded Halstead-Reitan neuropsychological test battery. J Int Neuropsychol Soc 5:346–356

Duff K (2012) Evidence-based indicators of neuropsychological change in the individual patient: relevant concepts and methods. Arch Clin Neuropsychol 27(3):248–261

Duff K, Schoenberg MR, Patton D, Paulsen JS, Bayless JD, Mold J et al (2005) Regression-based formulas for predicting change in RBANS subtests with older adults. Arch Clin Neuropsychol 20:281–290

Duff K, Beglinger LJ, Moser DJ, Schultz SK, Paulsen JS (2010) Practice effects and outcome of cognitive training: preliminary evidence from a memory training course. Am J Geriatr Psychiatry 18(1):91

Dufouil C, Richert L, Thiebaut R, Bruyand M, Amieva H, Dauchy FA et al (2015) Diabetes and cognitive decline in a French cohort of patients infected with HIV-1. Neurology 85(12): 1065–1073

Elicher IM, Byrd D, Clark US, Morgello S, Robinson-Papp J (2018) Motor function declines over time in human immunodeficiency virus and is associated with cerebrovascular disease, while HIV-associated neurocognitive disorder remains stable. J Neurovirol 24(4):514–522

Falleti MG, Maruff P, Collie A, Darby DG (2006) Practice effects associated with the repeated assessment of cognitive function using the CogState battery at 10-minute, one week and one month test-retest intervals. J Clin Exp Neuropsychol 28(7):1095–1112

Gates TM, Cysique LA (2016) The chronicity of HIV infection should drive the research strategy of NeuroHIV treatment studies: a critical review. CNS Drugs 30:53–69

Gaynes BN, O'Donnell J, Nelson E, Heine A, Zinski A, Edwards M et al (2015) Psychiatric comorbidity in depressed HIV-infected individuals: common and clinically consequential. Gen Hosp Psychiatry 37(4):277–282

Gott C, Gates T, Dermody N, Brew BJ, Cysique LA (2017) Cognitive change trajectories in virally suppressed HIV-infected individuals indicate high prevalence of disease activity. PLoS One 12(3):e0171887

Grund B, Wright EJ, Brew BJ, Price RW, Roediger MP, Bain MP et al (2013) Improved neurocognitive test performance in both arms of the SMART study: impact of practice effect. J Neurovirol 19(4):383–392

Haynes BI, Pitkanen M, Kulasegaram R, Casey SJ, Schutte M, Towgood K et al (2018) HIV: ageing, cognition and neuroimaging at 4-year follow-up. HIV Med 19(6):376–385

Heaton RK, Temkin NR, Dikmen SS, Avitable N, Taylor MJ, Marcotte TD et al (2001) Detecting change: a comparison of three neuropsychological methods, using normal and clinical samples. Arch Clin Neuropsychol 16(1):75–91

Heaton RK, Franklin DR Jr, Deutsch R, Letendre S, Ellis RJ, Casaletto K et al (2015) Neurocognitive change in the era of HIV combination antiretroviral therapy: the longitudinal CHARTER study. Clin Infect Dis 60(3):473–480

Hedeker D, Gibbons RD (2006) Longitudinal data analysis. Wiley, Hoboken, p 337

Henry D, Tolan P, Gorman-Smith D, Schoeny M (2017) Alternatives to randomized control trial designs for community-based prevention evaluation. Prev Sci 18(6):671–680

Hinton-Bayre AD (2016) Clarifying discrepancies in responsiveness between reliable change indices. Arch Clin Neuropsychol 31(7):754–768

Iturria-Medina Y, Sotero RC, Toussaint PJ, Mateos-Pérez JM, Evans AC, The Alzheimer's Disease Neuroimaging Initiative (2016) Early role of vascular dysregulation on late-onset Alzheimer's disease based on multifactorial data-driven analysis. Nat Commun 7:11934

Jones JD, Kuhn T, Mahmood Z, Singer EJ, Hinkin CH, Thames AD (2018) Longitudinal intra-individual variability in neuropsychological performance relates to white matter changes in HIV. Neuropsychology 32(2):206–212

Kamminga J, Lal L, Wright EJ, Bloch M, Brew BJ, Cysique LA (2017) Monitoring HIV-associated neurocognitive disorder using screenings: a critical review including guidelines for clinical and research use. Curr HIV/AIDS Rep 14(3):83–92

Lazarus JV, Safreed-Harmon K, Barton SE, Costagliola D, Dedes N, Del Amo VJ et al (2016) Beyond viral suppression of HIV – the new quality of life frontier. BMC Med 14(1):94

Linn Aung H, Kootar S, Gates TM, Brew BJ, Cysique LA et al (2019) Eur Geriatr Med 10(2): 227–238

Marquine MJ, Montoya JL, Umlauf A, Fazeli PL, Gouaux B, Heaton RK et al (2016) The Veterans Aging Cohort Study (VACS) index and neurocognitive change: a longitudinal study. Clin Infect Dis 63(5):694–702

McSweeny AJ, Naugle RI, Chelune GJ, Lüders H (1993) T scores for change: an illustration of a regression approach to depicting change in clinical neuropsychology. Clin Neuropsychol 7: 300–312

Molsberry SA, Lecci F, Kingsley L, Junker B, Reynolds S, Goodkin K et al (2015) Mixed membership trajectory models of cognitive impairment in the multicenter AIDS cohort study. AIDS 29(6):713–721

Paul R (2019) Neurocognitive phenotyping of HIV in the era of antiretroviral therapy. Curr HIV/AIDS Rep 16(3):230–235

Petoumenos K, Choi JY, Hoy J, Kiertiburanakul S, Ng OT, Boyd M et al (2017) CD4:CD8 ratio comparison between cohorts of HIV-positive Asians and Caucasians upon commencement of antiretroviral therapy. Antivir Ther 22(8):659–668

Pfefferbaum A, Zahr NM, Sassoon SA, Kwon D, Pohl KM, Sullivan EV (2018) Accelerated and premature aging characterizing regional cortical volume loss in human immunodeficiency virus infection: contributions from alcohol, substance use, and hepatitis C coinfection. Biol Psychiatry Cogn Neurosci Neuroimaging 3(10):844–859

Rabbitt P, Diggle P, Smith D, Holland F, McInnes L (2001) Identifying and separating the effects of practice and of cognitive ageing during a large longitudinal study of elderly community residents. Neuropsychologia 39(5):532–543

Robertson KR, Smurzynski M, Parsons TD, Wu K, Bosch RJ, Wu J et al (2007) The prevalence and incidence of neurocognitive impairment in the HAART era. AIDS 21(14):1915–1921

Robertson K, Jiang H, Kumwenda J, Supparatpinyo K, Evans S, Campbell TB et al (2012) Improved neuropsychological and neurological functioning across three antiretroviral regimens in diverse resource-limited settings: AIDS Clinical Trials Group study a5199, the International Neurological Study. Clin Infect Dis 55(6):868–876

Rubin LH, Maki PM, Springer G, Benning L, Anastos K, Gustafson D et al (2017) Cognitive trajectories over 4 years among HIV-infected women with optimal viral suppression. Neurology 89(15):1594–1603

Sacktor N (2018) Changing clinical phenotypes of HIV-associated neurocognitive disorders. J Neurovirol 24(2):141–145

Sacktor N, Skolasky RL, Seaberg E, Munro C, Becker JT, Martin E et al (2016) Prevalence of HIV-associated neurocognitive disorders in the Multicenter AIDS Cohort Study. Neurology 86(4):334–340

Saloner R, Cysique LA (2017) HIV-associated neurocognitive disorders: a global perspective. J Int Neuropsychol Soc 23(9–10):860–869
Salthouse TA (2010) Influence of age on practice effects in longitudinal neurocognitive change. Neuropsychology 24(5):563–572
Salthouse TA, Tucker-Drob EM (2008) Implications of short-term retest effects for the interpretation of longitudinal change. Neuropsychology 22:800–811
Samboju V, Philippi CL, Chan P, Cobigo Y, Fletcher JLK, Robb M et al (2018) Structural and functional brain imaging in acute HIV. Neuroimage Clin 20:327–335
Sawrie SM, Chelune GJ, Naugle RI, Lüders H (1996) Empirical methods for assessing meaningful neuropsychological change following epilepsy surgery. J Int Neuropsychol Soc 2:556–564
Seider TR, Luo X, Gongvatana A, Devlin KN, de la Monte SM, Chasman JD et al (2014) Verbal memory declines more rapidly with age in HIV infected versus uninfected adults. J Clin Exp Neuropsychol 36(4):356–367
Sheppard DP, Woods SP, Bondi MW, Gilbert PE, Massman PJ, Doyle KL (2015) Does older age confer an increased risk of incident neurocognitive disorders among persons living with HIV disease? Clin Neuropsychol 29(5):656–677
Smail RC, Brew BJ (2018) HIV-associated neurocognitive disorder. Handb Clin Neurol 152:75–97
Suchy Y, Kraybill ML, Franchow E (2011) Practice effect and beyond: reaction to novelty as an independent predictor of cognitive decline among older adults. J Int Neuropsychol Soc 17(1): 101–111
Temkin N, Heaton R, Grant I, Dikmen S (1999) Detecting significant change in neuropsychological test performance: a comparison of four models. J Int Neuropsychol Soc 5(4):357–369
Thurn M, Gustafson DR (2017) Faces of frailty in aging with HIV infection. Curr HIV/AIDS Rep 14(1):31–37
Tierney SM, Sheppard DP, Kordovski VM, Faytell MP, Avci G, Woods SP (2017) A comparison of the sensitivity, stability, and reliability of three diagnostic schemes for HIV-associated neurocognitive disorders. J Neurovirol 23(3):404–421
UNAIDS (2018) Miles to go: closing gaps breaking barriers righting injustices. http://www.unaids.org/sites/default/files/media_asset/miles-to-go_en.pdf
Van der Elst W, Molenberghs G, Van Boxtel MP, Jolles J (2013) Establishing normative data for repeated cognitive assessment: a comparison of different statistical methods. Behav Res Methods 45(4):1073–1086
Vassallo M, Fabre R, Durant J, Lebrun-Frenay C, Joly H, Ticchioni M et al (2017) A decreasing CD4/CD8 ratio over time and lower CSF-penetrating antiretroviral regimens are associated with a higher risk of neurocognitive deterioration, independently of viral replication. J Neurovirol 23(2):216–225
Watzke S, Brieger P, Kuss O, Schoettke H, Wiedl KH (2008) A longitudinal study of learning potential and rehabilitation outcome in schizophrenia. Psychiatr Serv 59(3):248–255
Wechsler D (2009) Advanced clinical solutions for WAIS-IV and WMS-IV. Pearson, San Antonio
Wilson RS, Capuano AW, Yu L, Yang J, Kim N, Leurgans SE et al (2018) Neurodegenerative disease and cognitive retest learning. Neurobiol Aging 66:122–130
Wright EJ, Grund B, Cysique LA, Robertson KR, Brew BJ, Collins G et al (2015) Factors associated with neurocognitive test performance at baseline: a substudy of the INSIGHT Strategic Timing of AntiRetroviral Treatment (START) trial. HIV Med 16(Suppl 1):97–108
Wright EJ, Grund B, Robertson KR, Cysique L, Brew BJ, Collins GL et al (2018) No neurocognitive advantage for immediate antiretroviral treatment in adults with greater than 500 CD4+ T-cell counts. AIDS 32(8):985–997
Xie H, Mayo N, Koski L (2011) Identifying and characterizing trajectories of cognitive change in older persons with mild cognitive impairment. Dement Geriatr Cogn Disord 31(2):165–172

Part III
Comorbidities and Complications

Geriatric Syndromes in People Living with HIV Associated with Ageing and Increasing Comorbidities: Implications for Neurocognitive Complications of HIV Infection

Julian Falutz, Susan Kirkland, and Giovanni Guaraldi

Contents

Abstract Long-term survival of treated people living with HIV (PLWH) currently approaches that of the general population. The average age of PLWH is currently in the mid-50s in resource-rich countries and is predicted that over 40% of PLWH will be older than 60 within a decade. Similar trends have been confirmed in all communities of PLWH with access to antiretroviral therapies. However, the positive impact on survival has been challenged by several developments. Ageing PLWH

J. Falutz (✉)
McGill University Health Centre, Montreal, QC, Canada
e-mail: julian.falutz@mcgill.ca

S. Kirkland
Dalhousie University, Halifax, NS, Canada
e-mail: susan.kirkland@dal.ca

G. Guaraldi
University of Modena and Reggio Emilia, Modena, Italy
e-mail: giovanni.guaraldi@unimore.it

Curr Topics Behav Neurosci (2021) 50: 301–328
DOI 10.1007/7854_2019_119
Published Online: 7 January 2020

have clinical features similar to the general population about 5–10 years older. In addition to the earlier occurrence of common age-related conditions common geriatric syndromes have also impacted this population prematurely. These are often difficult to evaluate and manage conditions usually of multifactorial aetiology. They include polypharmacy, frailty, impaired mobility and falls, sarcopenia, sensory impairment, and increasingly, non-dementing cognitive decline. Cognitive decline is of particular concern to PLWH and their care providers. In the general geriatric population cognitive impairment increases with age and occurs in all populations with a prevalence of over 25% in people over 80. Effective treatments are lacking and therefore minimizing risk factors plays an important role in maintaining healthspan. In the general population geriatric syndromes may increase the risk of cognitive decline. The corollary is that decreasing the risk of their development may limit cognitive impairment. Whether a similar status holds in PLWH is uncertain. This chapter will address the question of whether common geriatric syndromes in PLWH contribute to cognitive impairment. Common risk factors may provide clues to limit or delay cognitive decline.

Keywords Ageing · Antiretroviral therapy · Cognition · Geriatric syndromes · HIV

1 Introduction

1.1 Epidemiology of Ageing in HIV

According to the World Health Organization (WHO), current demographic trends in the ageing population represent an unprecedented societal phenomenon (WHO 2015). The global population of older persons is growing at a rate of 2% per year, much faster than that of the entire population (WHO 2015). Projections estimate that two billion people will be older than 60 years by 2050 (Bellantuono 2018).

Recent demographic changes occurring in people living with HIV (PLWH) are similar to those of the general population. PLWH now live longer due to effective and well-tolerated combinations of antiretroviral therapy (cART). As well, more people acquire HIV at an older age (Chambers et al. 2014; Lazarus and Nielsen 2010). These two distinct groups contribute to the current demographic profile of this population. A deeper understanding of the interaction between HIV, ageing and the changing clinical profile of PLWH is required to inform effective health care provision to PLWH.

1.2 Profile of Ageing People Living with HIV

Ageing PLWH experience heightened risk for multiple noninfectious chronic comorbidities (NICM). The increase in diabetes mellitus, cardiovascular disease, osteoporosis, chronic kidney and liver disease, non-AIDS related malignancies, chronic obstructive pulmonary disease and non-dementing cognitive decline is of multifactorial aetiology. These include a greater prevalence of traditional lifestyle-related risks as well as HIV-specific factors including immune activation and chronic inflammation (Althoff et al. 2019). As well, in ageing PLWH, NICMs often occur concurrently as complex "multi-morbidities" (MM) (Theou and Rockwood 2015).

However, simply diagnosing NICMs and MM does not fully reflect the complexity of ageing as a health condition. In geriatrics it is understood that two people with the exact same comorbid profile can have very different ageing trajectories. This variability has led geriatricians to introduce the concept of geriatric syndromes to better characterize clinical ageing in addition to the traditional evaluation of discrete clinical conditions. These are health conditions of multifactorial aetiology which occur when the accumulated effects of impairments in multiple systems render an older person vulnerable to biologic and environmental challenges (Inouye et al. 2007). For a given geriatric syndrome, multiple risk factors and multiple organ systems are often involved. Standard diagnostic strategies to identify underlying causes can sometimes be ineffective, burdensome, dangerous and costly. Therapeutic management of the clinical manifestations can be helpful even in the absence of a firm diagnosis or clarification of the underlying causes (Guaraldi et al. 2016). In the elderly, geriatric syndromes and measures of physical function are more predictive of self-reported health and mortality risk than diagnoses of specific diseases or MM alone (Erlandson et al. 2014; Koroukian et al. 2016). However, most existing management guidelines remain organ based and do not include formal assessment for geriatric conditions (Guaraldi and Palella 2017; Guaraldi et al. 2019).

In PLWH there is a greater prevalence of typical geriatric syndromes compared to age-matched uninfected persons, including frailty, polypharmacy, falls and dysmobility, impaired cognition and disability (Greene et al. 2015). Several HIV-specific factors contribute to this risk, including chronic inflammation, immune dysregulation, long-term ART toxicity and socio-behavioural risks (Guaraldi and Cossarizza 2017). Clinically it has been shown that treating comorbidities and the early initiation of ART may help to prevent the development of these syndromes (Greene et al. 2015).

1.3 Assessment of Older Patients Living with HIV

Comprehensive health care of the older adult extends beyond the traditional medical management of individual illnesses. It includes evaluation of multiple, often concurrently present issues including physical, cognitive, psychologic, social, financial, environmental and spiritual components. The application of a comprehensive geriatric assessment (CGA) is based on the premise that an interdisciplinary evaluation of at-risk older persons by a team of health professionals may better identify a variety of treatable or manageable health problems. This approach leads to better health outcomes and quality of life (Stuck et al. 1993; Devons 2002). At present, the CGA approach has rarely been used in PLWH (Bitas et al. 2019), but issues regarding multimorbidity, cognitive impairment, frailty and disability are increasingly incorporated into the clinical assessment of older patients with HIV (Guaraldi et al. 2016). Although the amount of information to be evaluated during a CGA may seem overwhelming, various assessment tools used by the interdisciplinary team can reduce this burden (Elsawy and Higgins 2011). A patient-centred approach, in both geriatrics and HIV care, is essential for the success of any treatment plan to ensure it meets the particular needs of individual patients. Closer collaboration between HIV care providers and geriatricians should be considered in the appropriate setting.

2 Geriatric Syndromes

2.1 Polypharmacy

Polypharmacy, commonly defined in the general population as taking five or more different medications on a daily basis (Gnjidic et al. 2012), is an important clinical problem in PLWH (Gleason et al. 2013; Edelman et al. 2013). This is not a new phenomenon in HIV disease as patients have always taken multiple medications in order to maintain their health. In the mid-1990s the first generation of ART regimens generally included three different antiretrovirals (ARVs), each consisting of several tablets, all of which had to be taken three to four times daily, often according to strict dietary restrictions. These ARVs were taken in addition to drugs used for primary or secondary prophylaxis of various opportunistic infections and management of drug-related toxicities.

As in the general population polypharmacy increases the risk of poor adherence to ART (Stone et al. 2004), although adherence is generally better in older compared to younger PLWH. ART has been greatly simplified over the past 10–15 years, and many patients now take single daily regimens containing three distinct ARVs (Geretti and Tsakiroglou 2014). Nevertheless, PLWH still have a greater prevalence of polypharmacy than controls (Greene et al. 2014a; Gimeno-Gracia et al. 2015), given the increased prevalence of NICMs which often require drugs for prevention or therapy (Guaraldi et al. 2011).

Between 15 and 75% of PLWH in their 60s have polypharmacy and 14% of patients older than 65 take four or more non-ARV drugs, most of which are either vitamins and supplements or drugs for cardiovascular or neurologic disorders (Hasse et al. 2011). In a cohort of PLWH with a median age of 64, the average number of drugs taken was 13, of which only four were ARVs (Greene et al. 2014a). In the general population older patients taking these many drugs have an increased risk of falls, frailty and mortality (Gnjidic et al. 2012).

Non-medically prescribed drug use is very common and often under-reported or misrepresented in the general population; this is a particular problem in PLWH. In addition to over-the-counter (OTC) drugs, patients often take recreational drugs and alternative care-related drugs. A large discrepancy exists between what the patient is actually taking and what their provider believes they are taking (Furler et al. 2004).

The consequences of polypharmacy in the elderly population are well known and include altered mental status, falls, increased hospitalization rate and mortality. As well, poor adherence to necessary drugs, increased incidence of adverse drug events (ADEs), drug–drug interactions (DDI) and use of inappropriate medications may also occur. Fifteen per cent of PLWH may take drugs with potential anticholinergic toxicity (Greene et al. 2014a) which may increase risk of falls and altered mental status. PLWH may be particularly susceptible to these complications because of increased rates of renal and hepatic dysfunction leading to altered drug metabolism. Medications have been shown to specifically contribute to increased risk of poor mobility and falls in PLWH (Erlandson et al. 2012; Richert et al. 2014). There are thus several routes by which polypharmacy can increase the risk of confusion and altered mental status in PLWH.

The EACS guidelines provide recommendations on managing ageing PLWH with polypharmacy. Figure 1 stresses the need to consider the impact of polypharmacy with regards to drug–drug interactions (DDI) and Potentially Inappropriate Medications (PIM). The former results from both the interactions between ART with NICM-related treatments as well as to the potential interaction among non-ART medication. DDI can be classified into five categories: class A (no know interaction), B (no action needed), C (monitor therapy), D (consider therapy modification) and X (avoid combination). PIM derives from age-related physiological changes that may impact pharmacokinetics and pharmacodynamics effects of drugs. This in turn can lead to inappropriate drug/dosage use. Assessment of PIM uses the Beers criteria which includes lists of medications and medication classes and drug–disease

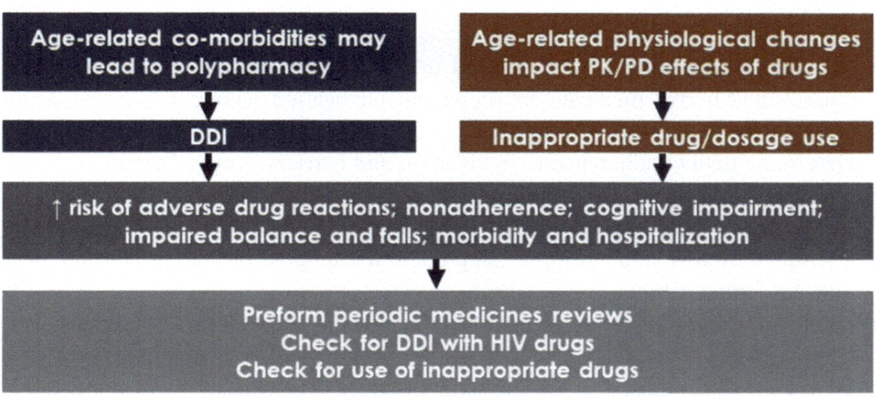

Fig. 1 Management of polypharmacy in OALWH. Freely adapted from EACS guidelines 2017 9.0

interactions that should be avoided in adults age 65 and older (American Geriatrics Society Beers Criteria Update Expert P 2012). DDIs and PIM may also increase risk of adverse drug reaction, non-adherence, cognitive impairment, impaired balance and falls, morbidity and hospitalization.

Although several studies have shown that overall adherence to ART is higher in older PLWH compared to younger controls, risk factors for poor adherence are also increased in older patients. This is a particular problem in PLWH with cognitive impairment, where it has been shown that adherence is decreased primarily in older PLWH who perform poorly on tests of executive function (Ettenhofer et al. 2009).

Sleep disturbances and pain syndromes occur commonly in the elderly, and have also been described in PLWH (Payne et al. 2013; Vosvick et al. 2004). They are frequently treated with drugs associated with increased risk of DDI. The co-administration of opioids and benzodiazepines has been shown to increase mortality in treated PLWH (Weisberg et al. 2015).

The frequency of medication-related problems in older PLWH was recently described in 89 community dwelling PLWH age 60 and older and age and sex-matched HIV-uninfected adults. The median number of medications was 13 (range 9–17) per participant. Compared to PLWH, the HIV-uninfected group was taking a median of 6 (IQR 3–10) medications ($P = 0.03$) with a median of 1 vitamin/herbal medication per participant. Sixty-two participants (70%) had at least one Category D (consider therapy modification) drug–drug interaction with a median of 1 (range 0–15; IQR 0–3) interaction per participant. Ten participants (11%) also had a Category X interaction (avoid combination). Most interactions were between an antiretroviral medication and a non-antiretroviral medication (152 [54%]), although approximately one-third of the interactions (99 [35%]) occurred between two non-antiretroviral agents. Fewer interactions (32 [11%]) occurred between two antiretroviral medications. Of the different drug–drug interaction pairs (101 [60%]), were deemed to be clinically significant by the clinical pharmacist (Greene et al. 2014a).

The following recommendations should be considered in the outpatient setting to help manage polypharmacy in older PLWH. Ideally an experienced pharmacist should be involved.

1. Patients should bring all prescription drugs at every visit, including over-the-counter and herbal medications; an annual medication reconciliation should be done
2. Determine patient adherence to medication and barriers if not adherent
3. Verify appropriate dosage and drug efficacy
4. Review drug–drug interactions
5. Use a "start low and go slow" approach where possible when starting new non-ART medications
6. Establish communication with other healthcare providers (physicians and caregivers)
7. Where possible suggest to use the same pharmacy for current medications
8. Assess for PIM (determine if medication is on Beers List) (Gleason et al. 2013)

In summary, polypharmacy is very common in PLWH and contributes to morbidity, cognitive impairment and mortality and requires increased vigilance on the part of providers involved in the complex care of ageing PLWH.

2.2 *Falls and Impaired Mobility*

One-third of the adults aged 65 and older sustain a fall each year and injuries due to falls are one of the most common causes of emergency room visits and loss of independence among ageing adults (Public Health Agency of Canada 2014; Lee et al. 2013). Falls in the elderly are usually a consequence of multiple interrelated factors including medical comorbidities, physical and sensory impairments, cognitive decline, polypharmacy and frailty (Public Health Agency of Canada 2014; Nobili et al. 2011). However, given that falls are most prevalent in the oldest old (i.e. over the age of 80) we may not fully appreciate the full magnitude of this issue in PLWH until a larger proportion has reached an older threshold. In a study of predominantly male PLWH with a median age of 57, 25.8% reported at least one fall within the previous year, and 12.5% reported a fall that resulted in an injury that required medical attention (Greene et al. 2015). Similarly, a report on falls among 359 PLWH (mean age 52, 85% male, 65% MSM and 21% with a history of IDU) found that 30% had experienced a fall in the previous year, and that among fallers, 61% were repeat fallers (Erlandson et al. 2012). Comparing non-fallers with recurrent fallers, the only demographic characteristics associated with being a recurrent faller were being female and being a smoker but the odds differed by less than 10% when adjusted for other factors. Falls were associated with comorbidities, including CVD, diabetes, neuropathy, arthritis and psychiatric disease; with each additional comorbid condition, the odds of recurrent falls increased by 1.7 (95% CI 1.5–2.1) (Erlandson et al. 2012). In a multivariate logistic model, difficulty with balance, exhaustion, diabetes, being female, unintentional weight loss, opiate use, sedative use, antidepressant use and having ever used didanosine were found to be significantly associated with recurrent falls. Since PLWH are known to have low bone density and increased fracture risk compared to their non-HIV peers, they are also likely to be at greater risk of morbidity when falls occur.

Mobility in the elderly is often assessed using the Short Physical Performance Battery (SPPB) which includes elements of lower extremity function such as balance, chair rise and gait speed (Guralnik et al. 2000). The SPPB predicts mortality in PLWH (Greene et al. 2014b). Physical performance assessments such as the SPPB can capture physical limitations prior to full disability among PLWH.

HIV-Associated Neurocognitive Disorder (HAND), discussed in detail elsewhere in this book series, refers to a spectrum of neurocognitive impairments in PLWH which includes executive dysfunction, memory impairment and motor dysfunction, plus psychomotor slowing, bradykinesia, coordination abnormalities and gait imbalance (Saylor and Sacktor 2016; Wendelken and Valcour 2012; Zamudio-Rodriguez et al. 2018). Risk factors for dysmobility and falls may also predispose to HAND, and are in fact part of the definition of the outcome itself.

Despite a documented association, the temporal nature of the relationship between mobility, falls and cognitive function remains unclear, and few studies have directly assessed the impact of mobility and falls on cognition among PLWH. Indeed, most studies in the general population assess how cognitive function impacts falls, or they consider falls as the outcome. Falls are affected by the presence of cognitive function in the presence of dementia (Taylor et al. 2017), and mobility, falls and cognitive function are inter-related domains (Herman et al. 2010). The role of executive functioning in particular is thought to be critical in compensating for age-associated declines in motor function and the maintenance of falls-free gait in everyday situations that require complex integration of motor and cognitive processes. Among community dwelling older adults, those with poorer executive functioning were more likely to fall during a 5-year follow-up period compared to those with the highest executive functioning, and recurrent falls happened sooner for those with poorer executive functioning (Mirelman et al. 2012).

The relationship between vascular risk factors and neurocognitive performance in PLWH may also be relevant in this context, in that cardiovascular fitness influences mobility. In a cardiovascular disease (CVD) sub-study of the MACS Cohort, increased carotid intima medial thickness was associated with slower psychomotor speed and reduced memory (Becker et al. 2009). In the CHARTER cohort, waist circumference was associated with cognitive impairment (Becker et al. 2009), and in the START trial, higher Framingham-score determined cardiovascular risk was associated with worse neurocognitive test performance (Wright et al. 2015). More directly, studies have demonstrated a relationship between aerobic fitness and cognitive ability among older adults (Netz et al. 2011), and PLWH are known to have reduced aerobic fitness and physical function relative to their non-HIV peers (Oursler et al. 2006). CVD risk factors were associated with slower processing speed based on neuropsychological tests in a group of PLWH, and, compared to those with CVD who were treated with medication, those who were untreated had poorer test performance on processing speed, learning and memory, and executive function (Foley et al. 2010). In a small cross-sectional study of PLWH, those with a higher peak oxygen uptake on a treadmill test, signifying better fitness, were less likely to have mild neurocognitive disorder (OR 0.65, $p = 0.01$) and HIV-associated dementia (OR $= 0.65$, $p = 0.0006$) (Mapstone et al. 2013).

Polypharmacy impacts the risk of mobility impairment and falls in PLWH. One study found that falls risk increased 1.4 times (CI 1.3–1.6) with each prescribed medication, and that beta-blockers, opiates, antidepressants, antipsychotics and sedatives were significantly more commonly used among those who had previously had a fall compared to non-fallers after adjusting for the comorbidity being treated by those medications (Erlandson et al. 2012).

2.3 Frailty

Frailty is a condition of increased vulnerability to biologic and environmental insults, initially described in the elderly. It is generally understood to arise from the degradation of homeostatic mechanisms and is the result of an inability to respond to physiologic stressors. Its presence is associated with increased risk of several negative health outcomes. While the prevalence of frailty is highest in the very old, it can be observed throughout the life-course, and has been increasingly characterized among younger populations, including among individuals with acquired vulnerability states, such as childhood cancer survivors (Ness et al. 2015), diabetics (Cigolle et al. 2011), dialysis patients (Sy and Johansen 2017) and PLWH (Brothers et al. 2014; Desquilbet et al. 2007; Rees et al. 2013). As a geriatric syndrome it partially encompasses the complex overlapping effects of multimorbidity, functional decline and vulnerability to illness (O. A. R. Working Group on HIV and Aging 2012). Notably, mobility impairments and falls are also often incorporated within the concept of frailty.

Two dominant models of operationalizing frailty are commonly used in research settings, although others have been evaluated for their reliability and greater utility in the clinical setting. Both models have demonstrated that frail individuals experience a greater number of hospitalizations, falls, loss of functional independence, institutionalization and death compared to non-frail adults (Cawthon et al. 2007; Clegg et al. 2013).

The Fried frailty phenotype, the most common frailty metric used in the literature, views frailty as a syndrome distinct from ageing and other disease processes. It consists of five specific signs and symptoms: unintentional weight loss, exhaustion, low level of physical activity, slow motor performance and weakness (Fried et al. 2001). The presence of three or more is diagnostic of frailty; one or two of these components denote prefrail status, and their absence indicates a non-frail status (Zamudio-Rodriguez et al. 2018; Fried et al. 2001; Pathai et al. 2014; Thurn and Gustafson 2017).

The cumulative deficit model characterizes frailty as a state of vulnerability capturing an individual's overall general health status, and serves as an integrative marker of biological ageing as opposed to chronological age (Brothers et al. 2014; Rockwood and Mitnitski 2007). In this model, the continued accumulation of nonspecific health deficits across a range of systems – symptoms, disease, functional impairments, disabilities and lab abnormalities – all contribute to frailty (Brothers et al. 2014; Wallace et al. 2017). The benefit of a frailty index is that it can be constructed from any existing health dataset using a standardized procedure (Searle et al. 2008). When a sufficiently large number of readily accessible deficits, usually between 30 and 40, are included, the variables can be selected at random and yield comparable results for the risk of adverse outcomes by presenting the number of deficits accumulated as a proportion of the number of deficits measured in a summary score between 0 and 1. Although best used as a continuous measure, a cut-point of 0.25 has been used to distinguish frail from non-frail (Rockwood et al. 2007; Franconi et al. 2018).

The Veterans Aging Cohort Study Index (VACSI) is specific to PLWH and was initially developed as a mortality index. It is a clinical HIV and general biomarker-based index that includes CD4 count, HIV-1 RNA, haemoglobin, fibrosis-4, estimated glomerular filtration rate and hepatitis C co-infection. Conceptually it has similarities to the cumulative deficit model. However, unlike most frailty scales, age is included in the VACSI. It has been shown to predict hospitalizations, admissions to intensive care and death (Justice et al. 2013). It has also been used as a surrogate marker to identify frail PLWH (Womack et al. 2013; Akgun et al. 2014).

Among ageing PLWH the prevalence of frailty using the phenotype model ranges from 2.9 to 28.6% (Desquilbet et al. 2007; Levett et al. 2016; Zamudio-Rodriguez et al. 2017; Althoff et al. 2014). In the MACS, 12% of HIV+ men aged 50–64 were identified as frail versus 9% of HIV− men. Using similar criteria in the WIHS 17% of HIV+ women and 10% of HIV− women at midlife (mean age 39) were found to be frail (Gustafson et al. 2016). This can be compared with frailty rates of 4.1% for adults aged 50–64 and 17.1% for adults aged 65 and older in SHARE (Santos-Eggimann et al. 2009), and 9.9% in a systematic review of 15 studies (Collard et al. 2012). In studies that have assessed frailty using the frailty index, the mean FI has been found to be higher in PLWH than that seen in the general population, from 0.26 to 0.31 (Guaraldi et al. 2015).

In the pre-cART era, frailty was associated with advanced disease, wasting and low muscle mass, whereas today frailty in PLWH is associated with central adiposity, sarcopenia and the density of muscle fat (Hawkins et al. 2018). The introduction of cART has decreased the prevalence of frailty in PLWH. However, multiple factors associated with frailty persist in the current treatment era (Levett et al. 2016). Some factors include traditional HIV measures such as lower current and nadir CD4 cell count, CD4/CD8 ratio, detectable viral load, history of AIDS and longer time since diagnosis of HIV. Other factors are non-HIV specific and include age, low body mass index (BMI), depressive symptoms, lipodystrophy, hepatitis C coinfection, multiple falls in the previous year and lower cognitive performance (Brothers et al. 2014; Rees et al. 2013; Levett et al. 2016; Escota et al. 2015). Frailty in PLWH has been associated with cardiovascular disease, congestive heart failure, cancer and chronic infection such as cytomegalovirus and is also considered a risk factor for neurocognitive disorders (Zamudio-Rodriguez et al. 2018; Desquilbet et al. 2007). Interestingly, in an era when the majority of HIV+ individuals experience immune recovery, traditional clinical indices of HIV severity such as CD4 cell count, nadir CD4 count and detectable viral load are increasingly found to be poorly associated with frailty and other surrogate markers of vulnerability (Brothers et al. 2014; Guaraldi et al. 2015; Paul et al. 2018). Indeed, a frailty index based on deficit accumulation predicted survival and incident multimorbidity independently of HIV and behavioural risk factors (Guaraldi et al. 2015). This may signify the value of more general health indices in being able to discriminate risk among the immune-reconstituted HIV population of the cART era. Importantly, frailty is amenable to preventive strategies (Thurn and Gustafson 2017). Transitions in frailty have been documented in PLWH, characterizing both changes in frailty states (Gill et al. 2006) and frailty severity using the frailty index (Brothers et al. 2017). In the MHMC cohort, 53% of participants improved (had a lower FI) after 4 years; 18.5% maintained their FI score, and 28.1% had a worse score; 3.0% died (Brothers et al. 2017). Given that frailty can account for interaction and redundancy across multiple systems, it may be increasingly valuable in describing the complex and interacting health problems among PLWH.

As with mobility and falls, it is unclear whether the relationship between frailty and cognitive function is temporal and causal, i.e. frailty impacts neurocognitive function, or the relationship is one of reverse causation, i.e. neurocognitive function affects frailty, or is correlational where both independently occur simultaneously. In the geriatric literature among non-HIV+ individuals, poor executive function and psychomotor speed increase the risk of frailty, whereas memory and language, which are hallmarks for Alzheimer's disease, do not (Yassuda et al. 2012; Macuco et al. 2012; Langlois et al. 2012).

Compared to non-HIV infected individuals, PLWH are nearly twice as likely to exhibit higher rates of cognitive impairment and when matched for age, to exhibit greater severity of frailty (Underwood et al. 2017). In a cross-sectional study of 122 PLWHIV with a mean age 57.5, Paul et al. examined select cognitive domains over and above clinical factors among frail and non-frail individuals using the Fried phenotype (Paul et al. 2018). They found that worse performance in executive function was associated with frailty but other domains were unrelated to frailty. Female sex and depression subscale scores were also significantly associated with frailty, whereas HIV markers of infection such as CD4 count and viral load were not. However, in a stepwise hierarchical model that included age and depressive symptoms there was no longer a significant association between executive function and frailty. Similarly, Zamudio-Rodriguez and colleagues found that after adjustment, pre-frailty was associated with MND, but frailty was not (Zamudio-Rodriguez et al. 2018).

Whereas HAND focuses on neurocognitive deficits, successful cognitive ageing (SCA) focuses on the successful end of the neurocognitive continuum, as measured by the absence of neurocognitive deficits and symptoms on neuropsychological and cognitive functioning tests. A number of studies have examined SCA among PLWH. Moore and colleagues found that 24.9% of PLWH compared to 40.0% of HIV-negative individuals experienced successful cognitive ageing (Moore et al. 2014). Further, a stairstep decline in SCA was noted based on HIV status and age from 47% among young HIV-negative participants, to 18.7% among older PLWH. Similarly, in a study among PLWH followed for at least 5 years, 32% were identified as experiencing successful cognitive ageing, but no association was found with cognitive reserve as has been seen in the general population (Malaspina et al. 2011; Yaffe et al. 2009). Frailty was significantly inversely associated with successful cognitive ageing among PLWH in the MHMC cohort, where successful cognitive ageing was defined as the absence of depressive symptoms, cognitive impairment and functional impairment, and frailty was assessed using a 37-item frailty index (Wallace et al. 2017). For each 0.1 point increase in the frailty index, the odds of successful cognitive ageing were reduced by 36%. This corresponded to a decrease in the odds of successful cognitive ageing by 12% for each additional health deficit experienced. Moreover, the frailty index was the only variable found to be statistically significant in its ability to discriminate SCA (AUC 0.63; $p = 0.02$), including individual disease diagnoses and multimorbidity. Similarly, Oppenheim found that a higher frailty index was associated with worse global neurocognitive functioning after adjustment for age, employment and premorbid intellectual functioning. The individual components of verbal fluency, executive functioning, processing speed and motor skills were also significantly associated with worse frailty scores (Oppenheim et al. 2018).

Apart from HIV-related effects, several other mechanisms have been theorized to explain the connection between frailty and cognitive function. In addition to chronic inflammation, nutrition, cardiovascular risks, mental health, hormones and Alzheimer's disease pathology have all been related to both frailty and cognitive function and may hasten the onset of either for PLWH (Robertson et al. 2013). The presence of both frailty and HAND together may also exacerbate the negative effects of both. It was recently reported that poor health outcomes for PLWH over 2 years of observation were most common among those who were both frail and had neurocognitive impairment, with 74% of participants experiencing at least one of recurrent falls, worsening instrumental activities of daily living (IADL) limitation, or death (Kelly et al. 2018).

In summary, an understanding of the geriatric syndromes of falls, mobility and frailty in relation to neurocognitive functioning can better characterize and explain ageing trajectories of older PLWH. Frailty prevention efforts are essential for successful ageing without experiencing disability or loss of independence. Sarcopenia, depression and vascular risk factors may be targeted for prevention and mitigation. Conceptualizing PLWH in terms of successful cognitive ageing offers targets for intervention and possible preventive strategies. High socioeconomic status, good nutrition, physical activity, social interaction and cognitive remediation have been put forth as positive mediators in conceptual models. Proactive prevention strategies that preserve function should be the priority for the care of older populations living with HIV.

2.4 Sarcopenia

Body composition changes are common complications in PLWH and cause important clinical consequences. Prior to the availability of effective ARVs, significant weight loss, comprising both fat mass and skeletal muscle, occurred often in patients with advanced disease. This condition was known as the AIDS Wasting Syndrome (Grinspoon et al. 2003). Significant loss of skeletal muscle independently predicted mortality (Kotler et al. 1989). Soon after the first generation of effective cART regimens became available in 1996 unexpected body composition changes in PLWH were identified. These were called by various evocative names, including the lipodystrophy syndrome. This term referred to two aetiologically distinct patterns of body fat changes which often developed concurrently in patients soon after starting therapy: peripheral fat wasting, termed peripheral lipoatrophy (LA); and abdominal obesity, termed central lipohypertrophy (LH).

LA refers to the significant, diffuse loss of subcutaneous fat, most easily observed in the face, legs and buttocks. It is due to apoptosis of peripheral adipocytes and was aetiologically related to mitochondrial toxicity caused by first generation thymidine analogue reverse transcriptase inhibitors (azidothymidine and stavudine). The introduction of newer, safer ARVs has essentially resolved this complication as the original offending drugs are now used infrequently (Falutz 2011).

LH refers to the accumulation of ectopic fat, usually occurring as increased visceral adipose tissue, but also in the dorso-cervical fat pad (known commonly, and pejoratively, as "buffalo hump") and intra-mammary fat. LH is of multifactorial origin (Falutz 2011). Current "backbone" ARVs, including protease inhibitors and integrase strand transfer inhibitors, although rarely causing dyslipidemia and glucose homeostasis, complications which occurred with some early ARVs, are however still associated with abdominal obesity (McComsey et al. 2016). LA and LH, both singly and in combination, often led to mood disorders and discontinuation of life-saving ARVs (Shenoy et al. 2014; Crane et al. 2008; Corless et al. 2005).

Older PLWH may also develop physiologic, age-related changes in body composition, including peripheral lipoatrophy, increased central obesity and loss of lean body mass (Kuk et al. 2009; Shaw et al. 2007). These may mimic HIV/cART associated body composition changes.

Sarcopenia, a term introduced over 25 years ago, initially referred only to the loss of skeletal mass. As noted, it also occurred as part of the AWS. Skeletal mass can be determined using anthropometrics, bio-impedance analysis, computerized tomography, magnetic resonance imaging and as recently demonstrated, by ultrasound (Ticinesi et al. 2017) and methods employing tritiated Deuterium (Cawthon et al. 2019). Presently, it is most commonly and reliably quantified by dual-energy X-ray absorptiometry (DXA) (Edwards and Buehring 2015). The clinical assessment of sarcopenia is unreliable. The initial definition of sarcopenia referred only to the loss of skeletal muscle mass, defined as an index of DXA-determined appendicular lean mass (ALM/ht^2) (Baumgartner et al. 1998). Sarcopenia is now defined as the age and gender dependent loss of muscle mass and impaired function, most often determined as decreased hand-grip strength or low gait speed. Primary sarcopenia occurs in the absence of conditions known to cause weight loss and represents physiologic age-related changes. Secondary sarcopenia refers to loss of skeletal mass associated with disuse, disease, poor nutrition or malabsorption (Cruz-Jentoft et al. 2010a). Of concern to some PLWH, ongoing, low-level skeletal mass loss, of multifactorial aetiology, persists despite cART (Wasserman et al. 2014; Abdul Aziz et al. 2018).

In the general elderly population, physiologic sarcopenia occurs in about 30% of persons over 80; 20% over 65 will lose skeletal mass and have related functional impairment (Binkley and Cooper 2015). There is an overlap in risk factors for primary and secondary sarcopenia in the general population with several pathophysiologic conditions also occurring in treated PLWH. These can include genetic factors influencing muscle metabolism, hormonal dysregulation (e.g. hypogonadism and both growth hormone resistance and deficiency), mitochondrial dysfunction, lifestyle (especially tobacco and alcohol use), deconditioning, multimorbidity and polypharmacy (Rolland et al. 2008). Polypharmacy, as noted above, is common in PLWH (Edelman et al. 2013). In the general population, polypharmacy is a risk factor for both sarcopenia (Konig et al. 2017) and cognitive decline (Jyrkka et al. 2011).

The consequences of sarcopenia in the elderly population are clinically nonspecific and thus this condition is often not included in the diagnostic evaluation. In the elderly, sarcopenia is now considered as a geriatric syndrome (Cruz-Jentoft et al. 2010b). It reduces functional status due to decreased endurance and poor mobility leading to increased falls, loss of independence and increased mortality (Visser and Schaap 2011). Sarcopenia is also an independent contributor to frailty (Landi et al. 2015), which, as noted, is itself associated with impaired cognition (Mitnitski et al. 2011). Importantly, sarcopenia may be directly associated with an increased risk of cognitive decline in the elderly (Chang et al. 2016). In ageing PLWH, a recent preliminary analysis, using a functional assessment of sarcopenia, demonstrated a decline of executive function (Montero-Odasso et al. 2019). Frailty, sarcopenia and impaired cognition may be linked via shared pathways including oxidative stress, immunosenescence, sleep disorders, chronic inflammation and insulin resistance. Sarcopenia has also been shown to be a risk factor for overall mortality in PLWH (Scherzer et al. 2011). In summary, the conditions by which sarcopenia may contribute to cognitive decline occur commonly in ageing PLWH.

2.5 Sensory Impairment

Other geriatric syndromes may also increase the risk of cognitive decline in treated PLWH. Sensory impairment, whether due to hearing impairment, decreased visual acuity, olfactory problems or peripheral neuropathy, is common in the elderly and clearly lead to important functional impairments, including cognition. Indeed, the presence of multisensory domain impairments may increase risk of dementia in a stepwise manner in both men and women older than 70 (Brenowitz et al. 2019). The mechanisms contributing to increased cognitive risk are complex and potentially bidirectional. Sensory impairments and cognitive decline may share common pathogenic mechanisms, including ageing-related and vascular risks. Other mechanisms include abnormalities in common anatomic regions (e.g. possible role of the hippocampus in cognition and auditory pathways) and common pathogenic mechanisms (e.g. amyloid-beta is present in Alzheimer's disease and cataracts). It is well known that sensory loss may lead to depression, social isolation and weakness through physical inactivity, all known risk factors for dementia (Whitson et al. 2018).

2.5.1 Visual Impairment

HIV-related Neuro-Retinal Disorder (HIV-NRD) is a novel condition consisting of abnormalities in retinal nerve fibre layers, possibly related to micro-infarcts and microangiopathy associated with chronic inflammation (Ashraf et al. 2015). This may lead to reduced contrast sensitivity, altered colour vision and peripheral visual field loss and has been shown to occur more often in PLWH than controls (Barteselli et al. 2014). Visual quality of life is impaired in PLWH with HIV-NRD (Ashraf et al. 2015) and is therefore a risk factor for cognitive decline.

In PLWH both the prevalence and incidence of macular degeneration is increased compared to controls. Ongoing immune activation and chronic inflammation in PLWH are contributing risks, similar to the general population (Jabs et al. 2015; Jabs et al. 2017, #181).

In the general population, ocular lens density increases with age leading to cataracts, an important cause of blindness and visual impairment worldwide. The human lens has been suggested as a model for ageing, as it may mirror distinct ageing-related processes in other parts of the body. In PLWH, especially untreated persons, the eye is a common site of involvement by opportunistic infections and AIDS-related malignancies. In treated individuals, AIDS manifestations occur much less often, although other complications have been documented. Immune recovery uveitis, chronic steroid use, diabetes and uveitis related to chronic inflammation may increase the risk of cataract formation. A large population-based study in Denmark determined that cataract surgery occurred almost twice as often in PLWH compared to seronegative persons, with the highest risk occurring in those with severe CD4 depletion (<200 cells/mL) (Rasmussen et al. 2011).

2.5.2 Hearing Impairment

In the general population, peripheral hearing impairment (HI or presbycusis) increases with age, is the most common sensory disability in the elderly, occurring in 25–40% of persons older than 65, and has been associated with cognitive impairment in some, but not all studies (Panza et al. 2015). The prospective Health ABC Study found a more than 55% increase in incident dementia in subjects with moderately severe peripheral HI at baseline in a large biracial cohort older than 70 (Deal et al. 2017). Possible mechanisms linking HI with cognitive decline include the resulting social isolation, loneliness, impaired verbal communication and possible effects on cognitive reserve. Vascular risks may predispose to both HI and cognitive impairment. Social isolation has been associated with biologic changes, such as inflammation, which may contribute to both cognitive decline and HI. At present, the link between HI and cognition has not been confirmed to be causal but an important association between the two conditions exists nevertheless.

In PLWH hearing may be impaired due to involvement of the cochlea and eighth cranial nerve. Some antiretroviral drugs cause mitochondrial damage which can affect cochlear function. Recent analyses from the MACS cohort confirm that sensorineural HI is more common among older PLWH than controls, although no association with HIV or treatment-related parameters were found. Of concern, poorer hearing occurred at lower frequencies, which may predispose to increased communication difficulties (Torre et al. 2015). This finding of HI at predominantly lower frequencies was confirmed in another cohort (Luque et al. 2014). Although HI in the elderly impacts quality of life, there was no difference in the reported QoL, as assessed using the SF-36, between male and female PLWH and seronegative controls with HI in a combined analysis from the MACS and WIHS cohorts. The fact that the median age was only in the high 50s may be a possible explanation for the lack of observed association (Duong et al. 2016). No studies have yet evaluated cognition in PLWH with HI, but the risks are clearly present and such analyses are urgently needed. Treatment for HI in the elderly is available and may improve cognitive function (Miller et al. 2015).

2.5.3 Olfaction

The sensations of smell and taste are infrequently assessed clinically in the absence of specific complaints. The Health ABC Study assessed smell in over 1,800 subjects between age 70 and 79 and found abnormal results in 50%. Persons with abnormal results were more likely to have dementia as determined by use of typical medications, admission to hospital with a diagnosis of dementia, or abnormal scores on a modified mini-mental status exam (Brenowitz et al. 2019). Similar findings of impaired olfactory function in older subjects with either confirmed Alzheimer's disease or mild cognitive impairment compared to those with normal cognition were described using comprehensive olfactory evaluations (Peters et al. 2003). A large, prospective study of the relation between sensory impairments and mortality in community-dwelling persons with a mean age of 69 found that olfactory impairment, but not hearing or visual impairment, was independently associated with increased mortality over 13 years (Schubert et al. 2017).

Olfactory function has been infrequently assessed in treated PLWH. Small case series in the pre-cART era suggested that impaired smell did occur, but results may have been influenced by advanced disease and concurrent AIDS-related complications or CNS involvement. In a recent small study of middle-aged female Nigerians and seronegative controls using visual analogue scales, olfactory threshold discrimination and identification were within the normal range in the two groups, but the mean values were significantly lower in the PLWH (Fasunla et al. 2016). Several small studies have suggested that olfactory function is possibly more impaired in PLWH with cognitive impairment (Mueller et al. 2002; Razani et al. 1996).

2.5.4 Peripheral Neuropathy

Sensory neuropathy, of multifactorial aetiology, is common in treated PLWH with an estimated prevalence of 50–60% (Ellis et al. 2010). In a study of middle-aged, predominantly male African-American PLWH, most with advanced disease (50% CD4 < 200, 70% detectable HIV-RNA), and with a significant minority reporting psychologic or anxiety disorder and substance abuse, those with confirmed distal sensory polyneuropathy were more likely to perform worse on timed psychomotor tests, including information processing speed and executive function (Fellows et al. 2012). The authors suggested that neuropsychologic testing should be interpreted with caution in PLWH with sensory neuropathy as this may worsen their objective assessments. Peripheral neuropathy also contributes to chronic pain, impaired functional status, mood disorders and polypharmacy.

2.6 Clinical Implications, Translational Aspects and Future Directions

2.6.1 Clinical Implications

This chapter has presented the perspective that viewing health complications as geriatric syndromes rather than as traditionally determined, discrete conditions will be more helpful when considering the health and well-being of ageing PLWH, particularly in regard to cognitive decline. Assessing clinical problems in the general population in an interdisciplinary and comprehensive manner has proven outcome benefits in regard to patient outcomes, patient and care-giver satisfaction and enduring impact on quality of life.

We propose an adaptation of geriatric principles of assessment and care of the elderly for selected ageing PLWH utilizing a method commonly referred to as the Comprehensive Geriatric Assessment (CGA). The CGA is based on the premise that an interdisciplinary, multidimensional evaluation of at-risk older persons by a team of health professionals may better identify relevant treatable, or manageable health problems, compared to using the traditional "hub and spoke" system of centralized referral to multiple specialists. It also involves a cogent and rational plan for the care of the individual within their particular environment, being mindful of and respecting the diverse influences which may impact an individual. At present, the CGA approach has been infrequently used in PLWH, and its effectiveness is unknown. However, issues regarding the known increase in multimorbidity and its impact on polypharmacy, frailty and disability related to mobility limitations, sensory impairment and sarcopenia are already being incorporated into the clinical assessment of ageing PLWH, and therefore the ground is fertile for the introduction of the geriatric model of care.

2.6.2 Translational Aspects

The mechanisms contributing to increased cognitive risk are complex and potentially bidirectional. Frailty, sarcopenia and impaired cognition may be linked via shared pathways including chronic inflammation, oxidative stress, immune-senescence and insulin resistance. Sensory impairments, frailty and related disorders may share common pathogenic mechanisms with cognitive decline, including ageing-related and vascular risks. Other mechanisms include abnormalities in common anatomic regions (e.g. possible role of the hippocampus in cognition and auditory pathways) and common pathogenic mechanisms (e.g. amyloid-beta is present in Alzheimer's disease and cataracts). It is well known that sensory loss may lead to depression, social isolation and weakness through physical inactivity, all known risk factors for dementia. Sarcopenia and impaired mobility are often associated and may lead to cognitive decline by limiting social interaction and isolation. Complications associated with polypharmacy (drug–drug interactions, adverse drug effects, particularly those increasing the anticholinergic burden, and inappropriate drug prescribing) contribute to cognitive impairment.

Current "backbone" ARVs, including protease inhibitors and possibly integrase strand transfer inhibitors, although rarely causing dyslipidemia and glucose homeostasis, are still associated with abdominal obesity, which may contribute to metabolic complications. The investigation of this association is ongoing.

2.7 Future Directions in Terms of Basic Science/Health Services Delivery/Public Health

Research in ageing PLWH to elucidate the role of vascular risks related to metabolic complications and obesity, as well as the effects of social isolation and polypharmacy on ageing PLWH in predisposing to early cognitive decline, potentially augmented by the ongoing chronic inflammation, will harness emerging techniques ranging from the molecular to novel imaging modalities.

Studying whether minimizing risk factors play a role in maintaining healthspan in ageing PLWH is imperative as is the investigation of the predictors of successful ageing in this population. In this regard, understanding those factors which contribute to maximizing cognitive reserve is also necessary.

Interventions addressing risks of polypharmacy among HIV-infected persons are in their infancy and tools to identify overprescribing are lacking. Nevertheless, considerable experience exists from the early HAART era regarding management of and adherence to complex first-generation ARV drugs. Ideally, the historical lessons learned are adaptable to the management of PLWH receiving treatment for multiple comorbidities, which can lead to potentially complex drug–drug interactions and adverse drug effects, including but not exclusive to use of cART. The availability of easy to use mobile "Apps" has greatly facilitated the ability to assess risk for drug interactions, for both clinicians and patients. The process of medication reconciliation is often enhanced by the involvement of clinical pharmacists working in conjunction with the HIV care team. Such collaborations are crucial and include the non-judgemental awareness on the part of care providers that recreational substance use, particularly alcohol and marijuana, is common among HIV-infected persons and can impact risks associated with polypharmacy.

In developing a public-health response to ageing, it is important not to just consider the approaches that alleviate losses associated with older age, but also those that may reinforce recovery, adaptation and psychosocial growth. In this perspective, WHO has recently built on a conceptual framework to consider the health status in an ageing individual. The term "healthy ageing" has been proposed to promote a positive approach to ageing that relies on reserves and preserved capacities in an individual, rather than accumulation of deficits. In 2015, the World Report on Ageing and Health attempted to combine clinical and public health outcomes for ageing by defining healthy ageing as the process of developing and maintaining the functional ability that enables well-being in older age. This construct derives from the relationship of two entities: "intrinsic capacity", defined as the composite of all cognitive and physical functioning of the individual, and "the environment".

Assessing intrinsic capacity is both a multidisciplinary and a multidimensional process, also intrinsic to the CGA as outlined above, and is designed to evaluate the individual's biology on the basis of five functional domains: locomotion, cognition, psychology, vitality and sensory. The intrinsic capacity construct might be considered an evolution of the frailty concept, taking into special consideration the need for the wide implementation of prevention, in relation to the continuum of the ageing process, and the opportunities offered by novel technologies. The investigation of this novel concept in ageing PLWH will be instructive.

3 Conclusions

In just over 35 years infection with HIV has been transformed from a nearly 100% fatal condition into a chronic disease. This has significantly altered the demographics of PLWH such that most persons with access to cART and who take the drugs reliably will have near-normal long-term survival. However, some of these ageing PLWH will be at increased risk of the premature onset of common bio-psycho-social consequences of ageing. This chapter has presented the perspective that viewing health complications as geriatric syndromes rather than as traditional discrete biomedical conditions will be more helpful when considering the health of ageing PLWH, particularly in regard to cognitive decline, one of the most common problems that all elderly persons face. In the general elderly population, assessing clinical problems in an interdisciplinary and comprehensive manner (using the CGA as detailed above) has proven outcome benefits. Its adaptation for selected ageing PLWH will be important in maintaining a successful healthspan.

In addition to the premature development of ageing-related disorders older PLWH are at increased risk of multimorbidity, as documented in the NA-ACCORD study, and its prevalence has increased over time (Wong et al. 2018). Hypercholesterolemia and chronic kidney disease were among the most common occurring comorbidities. In the general population, a longitudinal study of a healthy cohort with normal baseline cognition showed that the accumulation of new chronic diseases over time, particularly hypercholesterolemia and chronic kidney disease, was associated with deterioration in several cognitive domains, including verbal fluency, itself related to executive dysfunction and problem solving (Fabbri et al. 2016). Elsewhere in this issue, it is noted that executive dysfunction is a common finding in PLWH with HAND, with profound implications for maintaining independence and quality of life. From a management perspective, it is important to recall that many ageing-related comorbidities are lifestyle related and thus may be either preventable or modifiable with appropriate interventions. Some of these may also be appropriate in ageing PLWH.

At present, ageing PLWH remain at increased risk of diverse complications which may affect all aspects of their lives, particularly cognition, and which contribute to a shorter survival and poorer quality of life compared to the general population. It is the responsibility of everyone affected by HIV to continue to be proactive and to assure that evolving challenges will be met with the same determination and resourcefulness, which have been the characteristics of the response to HIV as it continues to impact society in profound ways.

References

Abdul Aziz SA, McStea M, Ahmad Bashah NS, Chong ML, Ponnampalavanar S, Syed Omar SF et al (2018) Assessment of sarcopenia in virally suppressed HIV-infected Asians receiving treatment. AIDS 32(8):1025–1034

Akgun KM, Tate JP, Crothers K, Crystal S, Leaf DA, Womack J et al (2014) An adapted frailty-related phenotype and the VACS index as predictors of hospitalization and mortality in HIV-infected and uninfected individuals. J Acquir Immune Defic Syndr 67(4):397–404

Althoff KN, Jacobson LP, Cranston RD, Detels R, Phair JP, Li X et al (2014) Age, comorbidities, and AIDS predict a frailty phenotype in men who have sex with men. J Gerontol A Biol Sci Med Sci 69(2):189–198

Althoff KN, Gebo KA, Moore RD, Boyd CM, Justice AC, Wong C et al (2019) Contributions of traditional and HIV-related risk factors on non-AIDS-defining cancer, myocardial infarction, and end-stage liver and renal diseases in adults with HIV in the USA and Canada: a collaboration of cohort studies. Lancet HIV 6(2):e93–e104

American Geriatrics Society Beers Criteria Update Expert P (2012) American Geriatrics Society updated beers criteria for potentially inappropriate medication use in older adults. J Am Geriatr Soc 60(4):616–631

Ashraf DC, May KP, Holland GN, Van Natta ML, Wu AW, Thorne JE et al (2015) Relationship between human immunodeficiency virus neuroretinal disorder and vision-specific quality of life among people with AIDS. Ophthalmology 122(12):2560–2567

Barteselli G, Chhablani J, Gomez ML, Doede AL, Dustin L, Kozak I et al (2014) Visual function assessment in simulated real-life situations in HIV-infected subjects. PLoS One 9(5):e97023

Baumgartner RN, Koehler KM, Gallagher D, Romero L, Heymsfield SB, Ross RR et al (1998) Epidemiology of sarcopenia among the elderly in New Mexico. Am J Epidemiol 147 (8):755–763

Becker JT, Kingsley L, Mullen J, Cohen B, Martin E, Miller EN et al (2009) Vascular risk factors, HIV serostatus, and cognitive dysfunction in gay and bisexual men. Neurology 73 (16):1292–1299

Bellantuono I (2018) Find drugs that delay many diseases of old age. Nature 554(7692):293–295

Binkley N, Cooper C (2015) Sarcopenia, the next frontier in fracture prevention: introduction from the guest editors. J Clin Densitom 18(4):459–460

Bitas C, Jones S, Singh HK, Ramirez M, Siegler E, Glesby M (2019) Adherence to recommendations from comprehensive geriatric assessment of older individuals with HIV. J Int Assoc Provid AIDS Care 18. https://doi.org/10.1177/2325958218821656

Brenowitz WD, Kaup AR, Lin FR, Yaffe K (2019) Multiple sensory impairment is associated with increased risk of dementia among black and white older adults. J Gerontol A Biol Sci Med Sci 74:890–896

Brothers TD, Kirkland S, Guaraldi G, Falutz J, Theou O, Johnston BL et al (2014) Frailty in people aging with human immunodeficiency virus (HIV) infection. J Infect Dis 210(8):1170–1179

Brothers TD, Kirkland S, Theou O, Zona S, Malagoli A, Wallace LMK et al (2017) Predictors of transitions in frailty severity and mortality among people aging with HIV. PLoS One 12(10): e0185352

Cawthon PM, Marshall LM, Michael Y, Dam TT, Ensrud KE, Barrett-Connor E et al (2007) Frailty in older men: prevalence, progression, and relationship with mortality. J Am Geriatr Soc 55 (8):1216–1223

Cawthon PM, Blackwell T, Cummings SR, Orwoll ES, Duchowny DK, Kado D et al (2019) The association between muscle mass assessed by D3CR diltution with incident ADL and IADL disability in community dwelling older men. In: 9th international conference on frailty and sarcopenia research, Miami, Florida

Chambers LA, Wilson MG, Rueda S, Gogolishvili D, Shi MQ, Rourke SB et al (2014) Evidence informing the intersection of HIV, aging and health: a scoping review. AIDS Behav 18 (4):661–675

Chang KV, Hsu TH, Wu WT, Huang KC, Han DS (2016) Association between sarcopenia and cognitive impairment: a systematic review and meta-analysis. J Am Med Dir Assoc 17(12):1164 e7–1164e15

Cigolle CT, Lee PG, Langa KM, Lee YY, Tian Z, Blaum CS (2011) Geriatric conditions develop in middle-aged adults with diabetes. J Gen Intern Med 26(3):272–279

Clegg A, Young J, Iliffe S, Rikkert MO, Rockwood K (2013) Frailty in elderly people. Lancet 381 (9868):752–762

Collard RM, Boter H, Schoevers RA, Oude Voshaar RC (2012) Prevalence of frailty in community-dwelling older persons: a systematic review. J Am Geriatr Soc 60(8):1487–1492

Corless IB, Kirksey KM, Kemppainen J, Nicholas PK, McGibbon C, Davis SM et al (2005) Lipodystrophy-associated symptoms and medication adherence in HIV/AIDS. AIDS Patient Care STDs 19(9):577–586

Crane HM, Grunfeld C, Harrington RD, Uldall KK, Ciechanowski PS, Kitahata MM (2008) Lipoatrophy among HIV-infected patients is associated with higher levels of depression than lipohypertrophy. HIV Med 9(9):780–786

Cruz-Jentoft AJ, Baeyens JP, Bauer JM, Boirie Y, Cederholm T, Landi F et al (2010a) Sarcopenia: European consensus on definition and diagnosis: report of the European working group on sarcopenia in older people. Age Ageing 39(4):412–423

Cruz-Jentoft AJ, Landi F, Topinkova E, Michel JP (2010b) Understanding sarcopenia as a geriatric syndrome. Curr Opin Clin Nutr Metab Care 13(1):1–7

Deal JA, Betz J, Yaffe K, Harris T, Purchase-Helzner E, Satterfield S et al (2017) Hearing impairment and incident dementia and cognitive decline in older adults: the health ABC study. J Gerontol A Biol Sci Med Sci 72(5):703–709

Desquilbet L, Jacobson LP, Fried LP, Phair JP, Jamieson BD, Holloway M et al (2007) HIV-1 infection is associated with an earlier occurrence of a phenotype related to frailty. J Gerontol A Biol Sci Med Sci 62(11):1279–1286
Devons CA (2002) Comprehensive geriatric assessment: making the most of the aging years. Curr Opin Clin Nutr Metab Care 5(1):19–24
Duong N, Torre P 3rd, Springer G, Cox C, Plankey MW (2016) Hearing loss and quality of life (QOL) among human immunodeficiency virus (HIV)-infected and uninfected adults. J AIDS Clin Res 7(12):645
Edelman EJ, Gordon KS, Glover J, McNicholl IR, Fiellin DA, Justice AC (2013) The next therapeutic challenge in HIV: polypharmacy. Drugs Aging 30(8):613–628
Edwards MH, Buehring B (2015) Novel approaches to the diagnosis of sarcopenia. J Clin Densitom 18(4):472–477
Ellis RJ, Rosario D, Clifford DB, McArthur JC, Simpson D, Alexander T et al (2010) Continued high prevalence and adverse clinical impact of human immunodeficiency virus-associated sensory neuropathy in the era of combination antiretroviral therapy: the CHARTER study. Arch Neurol 67(5):552–558
Elsawy B, Higgins KE (2011) The geriatric assessment. Am Fam Physician 83(1):48–56
Erlandson KM, Allshouse AA, Jankowski CM, Duong S, MaWhinney S, Kohrt WM et al (2012) Risk factors for falls in HIV-infected persons. J Acquir Immune Defic Syndr 61(4):484–489
Erlandson KM, Schrack JA, Jankowski CM, Brown TT, Campbell TB (2014) Functional impairment, disability, and frailty in adults aging with HIV-infection. Curr HIV/AIDS Rep 11 (3):279–290
Escota GV, Patel P, Brooks JT, Bush T, Conley L, Baker J et al (2015) Short communication: the Veterans Aging Cohort Study Index is an effective tool to assess baseline frailty status in a contemporary cohort of HIV-infected persons. AIDS Res Hum Retrovir 31(3):313–317
Ettenhofer ML, Hinkin CH, Castellon SA, Durvasula R, Ullman J, Lam M et al (2009) Aging, neurocognition, and medication adherence in HIV infection. Am J Geriatr Psychiatry 17 (4):281–290
Fabbri E, An Y, Zoli M, Tanaka T, Simonsick EM, Kitner-Triolo MH et al (2016) Association between accelerated multimorbidity and age-related cognitive decline in older Baltimore longitudinal study of aging participants without dementia. J Am Geriatr Soc 64(5):965–972
Falutz J (2011) HIV infection, body composition changes and related metabolic complications: contributing factors and evolving management strategies. Curr Opin Clin Nutr Metab Care 14 (3):255–260
Fasunla AJ, Daniel A, Nwankwo U, Kuti KM, Nwaorgu OG, Akinyinka OO (2016) Evaluation of olfactory and gustatory function of HIV infected women. AIDS Res Treat 2016:2045383
Fellows RP, Byrd DA, Elliott K, Robinson-Papp J, Mindt MR, Morgello S et al (2012) Distal sensory polyneuropathy is associated with neuropsychological test performance among persons with HIV. J Int Neuropsychol Soc 18(5):898–907
Foley J, Ettenhofer M, Wright MJ, Siddiqi I, Choi M, Thames AD et al (2010) Neurocognitive functioning in HIV-1 infection: effects of cerebrovascular risk factors and age. Clin Neuropsychol 24(2):265–285
Franconi I, Theou O, Wallace L, Malagoli A, Mussini C, Rockwood K et al (2018) Construct validation of a frailty index, an HIV index and a protective index from a clinical HIV database. PLoS One 13(10):e0201394
Fried LP, Tangen CM, Walston J, Newman AB, Hirsch C, Gottdiener J et al (2001) Frailty in older adults: evidence for a phenotype. J Gerontol A Biol Sci Med Sci 56(3):M146–M156
Furler MD, Einarson TR, Walmsley S, Millson M, Bendayan R (2004) Polypharmacy in HIV: impact of data source and gender on reported drug utilization. AIDS Patient Care STDs 18 (10):568–586
Geretti AM, Tsakiroglou M (2014) HIV: new drugs, new guidelines. Curr Opin Infect Dis 27 (6):545–553

Gill TM, Gahbauer EA, Allore HG, Han L (2006) Transitions between frailty states among community-living older persons. Arch Intern Med 166(4):418–423
Gimeno-Gracia M, Crusells-Canales MJ, Javier Armesto-Gomez F, Rabanaque-Hernandez MJ (2015) Prevalence of concomitant medications in older HIV+ patients and comparison with general population. HIV Clin Trials 16(3):117–124
Gleason LJ, Luque AE, Shah K (2013) Polypharmacy in the HIV-infected older adult population. Clin Interv Aging 8:749–763
Gnjidic D, Hilmer SN, Blyth FM, Naganathan V, Waite L, Seibel MJ et al (2012) Polypharmacy cutoff and outcomes: five or more medicines were used to identify community-dwelling older men at risk of different adverse outcomes. J Clin Epidemiol 65(9):989–995
Greene M, Steinman MA, McNicholl IR, Valcour V (2014a) Polypharmacy, drug-drug interactions, and potentially inappropriate medications in older adults with human immunodeficiency virus infection. J Am Geriatr Soc 62(3):447–453
Greene M, Covinsky K, Astemborski J, Piggott DA, Brown T, Leng S et al (2014b) The relationship of physical performance with HIV disease and mortality. AIDS 28(18):2711–2719
Greene M, Covinsky KE, Valcour V, Miao Y, Madamba J, Lampiris H et al (2015) Geriatric syndromes in older HIV-infected adults. J Acquir Immune Defic Syndr 69(2):161–167
Grinspoon S, Mulligan K, Department of Health and Human Services Working Group on the Prevention and Treatment of Wasting and Weight Loss (2003) Weight loss and wasting in patients infected with human immunodeficiency virus. Clin Infect Dis 36(Suppl 2):S69–S78
Guaraldi G, Cossarizza A (2017) Geriatric-HIV medicine: a science in its infancy. Virulence 8 (5):504–507
Guaraldi G, Palella FJ Jr (2017) Clinical implications of aging with HIV infection: perspectives and the future medical care agenda. AIDS 31(Suppl 2):S129–SS35
Guaraldi G, Orlando G, Zona S, Menozzi M, Carli F, Garlassi E et al (2011) Premature age-related comorbidities among HIV-infected persons compared with the general population. Clin Infect Dis 53(11):1120–1126
Guaraldi G, Brothers TD, Zona S, Stentarelli C, Carli F, Malagoli A et al (2015) A frailty index predicts survival and incident multimorbidity independent of markers of HIV disease severity. AIDS 29(13):1633–1641
Guaraldi G, Falutz J, Mussi C, Silva A (eds) (2016) Managing the older adult patient with HIV. Springer, Cham
Guaraldi G, Marcotullio S, Maserati R, Gargiulo M, Milic J, Franconi I et al (2019) The Management of Geriatric and Frail HIV Patients. A 2017 update from the Italian guidelines for the use of antiretroviral agents and the diagnostic clinical management of HIV-1 infected persons. J Frailty Aging 8(1):10–16
Guralnik JM, Ferrucci L, Pieper CF, Leveille SG, Markides KS, Ostir GV et al (2000) Lower extremity function and subsequent disability: consistency across studies, predictive models, and value of gait speed alone compared with the short physical performance battery. J Gerontol A Biol Sci Med Sci 55(4):M221–M231
Gustafson DR, Shi Q, Thurn M, Holman S, Minkoff H, Cohen M et al (2016) Frailty and constellations of factors in aging HIV-infected and uninfected women – the Women's interagency HIV study. J Frailty Aging 5(1):43–48
Hasse B, Ledergerber B, Furrer H, Battegay M, Hirschel B, Cavassini M et al (2011) Morbidity and aging in HIV-infected persons: the Swiss HIV cohort study. Clin Infect Dis 53(11):1130–1139
Hawkins KL, Zhang L, Ng DK, Althoff KN, Palella FJ Jr, Kingsley LA et al (2018) Abdominal obesity, sarcopenia, and osteoporosis are associated with frailty in men living with and without HIV. AIDS 32(10):1257–1266
Herman T, Mirelman A, Giladi N, Schweiger A, Hausdorff JM (2010) Executive control deficits as a prodrome to falls in healthy older adults: a prospective study linking thinking, walking, and falling. J Gerontol A Biol Sci Med Sci 65(10):1086–1092
Inouye SK, Studenski S, Tinetti ME, Kuchel GA (2007) Geriatric syndromes: clinical, research, and policy implications of a core geriatric concept. J Am Geriatr Soc 55(5):780–791

Jabs DA, van Natta ML, Sezgin E, Pak JW, Danis R, Studies of the Ocular Complications of ARG (2015) Prevalence of intermediate-stage age-related macular degeneration in patients with acquired immunodeficiency syndrome. Am J Ophthalmol 159(6):1115–22.e1

Jabs DA, Van Natta ML, Pak JW, Danis RP, Hunt PW (2017) Incidence of intermediate-stage age-related macular degeneration in patients with acquired immunodeficiency syndrome. Am J Ophthalmol 179:151–158. https://doi.org/10.1016/j.ajo.2017.05.004

Justice AC, Modur SP, Tate JP, Althoff KN, Jacobson LP, Gebo KA et al (2013) Predictive accuracy of the Veterans Aging Cohort Study Index for mortality with HIV infection: a North American cross cohort analysis. J Acquir Immune Defic Syndr 62(2):149–163

Jyrkka J, Enlund H, Lavikainen P, Sulkava R, Hartikainen S (2011) Association of polypharmacy with nutritional status, functional ability and cognitive capacity over a three-year period in an elderly population. Pharmacoepidemiol Drug Saf 20(5):514–522

Kelly SG, Wu K, Tassiopoulos K, Erlandson KM, Koletar SL, Palella FJ et al (2018) Frailty is an independent risk factor for mortality, cardiovascular disease, bone disease, and diabetes among aging adults with human immunodeficiency virus. Clin Infect Dis 69(8):1370–1376. https://doi.org/10.1093/cid/ciy110

Konig M, Spira D, Demuth I, Steinhagen-Thiessen E, Norman K (2017) Polypharmacy as a risk factor for clinically relevant sarcopenia: results from the Berlin aging study II. J Gerontol A Biol Sci Med Sci 73(1):117–122

Koroukian SM, Schiltz N, Warner DF, Sun J, Bakaki PM, Smyth KA et al (2016) Combinations of chronic conditions, functional limitations, and geriatric syndromes that predict health outcomes. J Gen Intern Med 31(6):630–637

Kotler DP, Tierney AR, Wang J, Pierson RNJ (1989) Magnitude of body-cell-mass depletion and the timing of death from wasting in AIDS. Am J Clin Nutr 50(3):444–447

Kuk JL, Saunders TJ, Davidson LE, Ross R (2009) Age-related changes in total and regional fat distribution. Ageing Res Rev 8(4):339–348

Landi F, Calvani R, Cesari M, Tosato M, Martone AM, Bernabei R et al (2015) Sarcopenia as the biological substrate of physical frailty. Clin Geriatr Med 31(3):367–374

Langlois F, Vu TT, Kergoat MJ, Chasse K, Dupuis G, Bherer L (2012) The multiple dimensions of frailty: physical capacity, cognition, and quality of life. Int Psychogeriatr 24(9):1429–1436

Lazarus JV, Nielsen KK (2010) HIV and people over 50 years old in Europe. HIV Med 11 (7):479–481

Lee A, Lee KW, Khang P (2013) Preventing falls in the geriatric population. Perm J 17(4):37–39

Levett TJ, Cresswell FV, Malik MA, Fisher M, Wright J (2016) Systematic review of prevalence and predictors of frailty in individuals with human immunodeficiency virus. J Am Geriatr Soc 64(5):1006–1014

Luque AE, Orlando MS, Leong UC, Allen PD, Guido JJ, Yang H et al (2014) Hearing function in patients living with HIV/AIDS. Ear Hear 35(6):e282–e290

Macuco CR, Batistoni SS, Lopes A, Cachioni M, da Silva Falcao DV, Neri AL et al (2012) Mini-mental state examination performance in frail, pre-frail, and non-frail community dwelling older adults in Ermelino Matarazzo, Sao Paulo, Brazil. Int Psychogeriatr 24(11):1725–1731

Malaspina L, Woods SP, Moore DJ, Depp C, Letendre SL, Jeste D et al (2011) Successful cognitive aging in persons living with HIV infection. J Neurovirol 17(1):110–119

Mapstone M, Hilton TN, Yang H, Guido JJ, Luque AE, Hall WJ et al (2013) Poor aerobic fitness May contribute to cognitive decline in HIV-infected older adults. Aging Dis 4(6):311–319

McComsey GA, Moser C, Currier J, Ribaudo HJ, Paczuski P, Dube MP et al (2016) Body composition changes after initiation of raltegravir or protease inhibitors: ACTG A5260s. Clin Infect Dis 62(7):853–862

Miller G, Miller C, Marrone N, Howe C, Fain M, Jacob A (2015) The impact of cochlear implantation on cognition in older adults: a systematic review of clinical evidence. BMC Geriatr 15:16

Mirelman A, Herman T, Brozgol M, Dorfman M, Sprecher E, Schweiger A et al (2012) Executive function and falls in older adults: new findings from a five-year prospective study link fall risk to cognition. PLoS One 7(6):e40297

Mitnitski A, Fallah N, Rockwood MR, Rockwood K (2011) Transitions in cognitive status in relation to frailty in older adults: a comparison of three frailty measures. J Nutr Health Aging 15 (10):863–867

Montero-Odasso M, Sakurai R, Bray NW, Pieruccini-Faria F (2019) Dynapenia is associated with executive dysfunction. Results from the gait and brain study. In: 9th international conference on frailty and sarcopenia research, Miami, Florida

Moore RC, Fazeli PL, Jeste DV, Moore DJ, Grant I, Woods SP et al (2014) Successful cognitive aging and health-related quality of life in younger and older adults infected with HIV. AIDS Behav 18(6):1186–1197

Mueller C, Temmel AF, Quint C, Rieger A, Hummel T (2002) Olfactory function in HIV-positive subjects. Acta Otolaryngol 122(1):67–71

Ness KK, Armstrong GT, Kundu M, Wilson CL, Tchkonia T, Kirkland JL (2015) Frailty in childhood cancer survivors. Cancer 121(10):1540–1547

Netz Y, Dwolatzky T, Zinker Y, Argov E, Agmon R (2011) Aerobic fitness and multidomain cognitive function in advanced age. Int Psychogeriatr 23(1):114–124

Nobili A, Licata G, Salerno F, Pasina L, Tettamanti M, Franchi C et al (2011) Polypharmacy, length of hospital stay, and in-hospital mortality among elderly patients in internal medicine wards. The REPOSI study. Eur J Clin Pharmacol 67(5):507–519

O. A. R. Working Group on HIV and Aging (2012) HIV and aging: state of knowledge and areas of critical need for research. A report to the NIH Office of AIDS Research by the HIV and Aging Working Group. J Acquir Immune Defic Syndr 60(Suppl 1):S1–S18

Oppenheim H, Paolillo EW, Moore RC, Ellis RJ, Letendre SL, Jeste DV et al (2018) Neurocognitive functioning predicts frailty index in HIV. Neurology 91(2):e162–ee70

Oursler KK, Sorkin JD, Smith BA, Katzel LI (2006) Reduced aerobic capacity and physical functioning in older HIV-infected men. AIDS Res Hum Retrovir 22(11):1113–1121

Panza F, Solfrizzi V, Logroscino G (2015) Age-related hearing impairment-a risk factor and frailty marker for dementia and AD. Nat Rev Neurol 11(3):166–175

Pathai S, Bajillan H, Landay AL, High KP (2014) Is HIV a model of accelerated or accentuated aging? J Gerontol A Biol Sci Med Sci 69(7):833–842

Paul RH, Cooley SA, Garcia-Egan PM, Ances BM (2018) Cognitive performance and frailty in older HIV-positive adults. J Acquir Immune Defic Syndr 79(3):375–380

Payne BA, Hateley CL, Ong EL, Premchand N, Schmid ML, Schwab U et al (2013) HIV-associated fatigue in the era of highly active antiretroviral therapy: novel biological mechanisms? HIV Med 14(4):247–251

Peters JM, Hummel T, Kratzsch T, Lotsch J, Skarke C, Frolich L (2003) Olfactory function in mild cognitive impairment and Alzheimer's disease: an investigation using psychophysical and electrophysiological techniques. Am J Psychiatry 160(11):1995–2002

Public Health Agency of Canada (2014) Seniors' falls in Canada: second report. Ottawa

Rasmussen LD, Kessel L, Molander LD, Pedersen C, Gerstoft J, Kronborg G et al (2011) Risk of cataract surgery in HIV-infected individuals: a Danish Nationwide population-based cohort study. Clin Infect Dis 53(11):1156–1163

Razani J, Murphy C, Davidson TM, Grant I, McCutchan A (1996) Odor sensitivity is impaired in HIV-positive cognitively impaired patients. Physiol Behav 59(4–5):877–881

Rees HC, Ianas V, McCracken P, Smith S, Georgescu A, Zangeneh T et al (2013) Measuring frailty in HIV-infected individuals. Identification of frail patients is the first step to amelioration and reversal of frailty. J Vis Exp 77:e50537

Richert L, Brault M, Mercie P, Dauchy FA, Bruyand M, Greib C et al (2014) Decline in locomotor functions over time in HIV-infected patients. AIDS 28(10):1441–1449

Robertson DA, Savva GM, Kenny RA (2013) Frailty and cognitive impairment – a review of the evidence and causal mechanisms. Ageing Res Rev 12(4):840–851

Rockwood K, Mitnitski A (2007) Frailty in relation to the accumulation of deficits. J Gerontol A Biol Sci Med Sci 62(7):722–727

Rockwood K, Andrew M, Mitnitski A (2007) A comparison of two approaches to measuring frailty in elderly people. J Gerontol A Biol Sci Med Sci 62(7):738–743

Rolland Y, Czerwinski S, Abellan Van Kan G, Morley JE, Cesari M, Onder G et al (2008) Sarcopenia: its assessment, etiology, pathogenesis, consequences and future perspectives. J Nutr Health Aging 12(7):433–450

Santos-Eggimann B, Cuenoud P, Spagnoli J, Junod J (2009) Prevalence of frailty in middle-aged and older community-dwelling Europeans living in 10 countries. J Gerontol A Biol Sci Med Sci 64(6):675–681

Saylor D, Sacktor N (2016) Cognitive impairment among older individuals with HIV infection. Curr Geriatrics Rep 5:63–70

Scherzer R, Heymsfield SB, Lee D, Powderly WG, Tien PC, Bacchetti P et al (2011) Decreased limb muscle and increased central adiposity are associated with 5-year all-cause mortality in HIV infection. AIDS 25(11):1405–1414

Schubert CR, Fischer ME, Pinto AA, Klein BEK, Klein R, Tweed TS et al (2017) Sensory impairments and risk of mortality in older adults. J Gerontol A Biol Sci Med Sci 72(5):710–715

Searle SD, Mitnitski A, Gahbauer EA, Gill TM, Rockwood K (2008) A standard procedure for creating a frailty index. BMC Geriatr 8:24

Shaw KA, Srikanth VK, Fryer JL, Blizzard L, Dwyer T, Venn AJ (2007) Dual energy X-ray absorptiometry body composition and aging in a population-based older cohort. Int J Obes 31 (2):279–284

Shenoy A, Ramapuram JT, Unnikrishan B, Achappa B, Madi D, Rao S et al (2014) Effect of lipodystrophy on the quality of life among people living with HIV (PLHIV) on highly active antiretroviral therapy. J Int Assoc Provid AIDS Care 13(5):471–475

Stone VE, Jordan J, Tolson J, Miller R, Pilon T (2004) Perspectives on adherence and simplicity for HIV-infected patients on antiretroviral therapy: self-report of the relative importance of multiple attributes of highly active antiretroviral therapy (HAART) regimens in predicting adherence. J Acquir Immune Defic Syndr 36(3):808–816

Stuck AE, Siu AL, Wieland GD, Adams J, Rubenstein LZ (1993) Comprehensive geriatric assessment: a meta-analysis of controlled trials. Lancet 342(8878):1032–1036

Sy J, Johansen KL (2017) The impact of frailty on outcomes in dialysis. Curr Opin Nephrol Hypertens 26(6):537–542

Taylor ME, Lord SR, Delbaere K, Kurrle SE, Mikolaizak AS, Close JCT (2017) Reaction time and postural sway modify the effect of executive function on risk of falls in older people with mild to moderate cognitive impairment. Am J Geriatr Psychiatry 25(4):397–406

Theou O, Rockwood K (2015) Comparison and clinical applications of the frailty phenotype and frailty index approaches. Interdiscip Top Gerontol Geriatr 41:74–84

Thurn M, Gustafson DR (2017) Faces of frailty in aging with HIV infection. Curr HIV/AIDS Rep 14(1):31–37

Ticinesi A, Meschi T, Narici MV, Lauretani F, Maggio M (2017) Muscle ultrasound and sarcopenia in older individuals: a clinical perspective. J Am Med Dir Assoc 18(4):290–300

Torre P 3rd, Hoffman HJ, Springer G, Cox C, Young MA, Margolick JB et al (2015) Hearing loss among HIV-seropositive and HIV-seronegative men and women. JAMA Otolaryngol Head Neck Surg 141(3):202–210

Underwood J, De Francesco D, Post FA, Vera JH, Williams I, Boffito M et al (2017) Associations between cognitive impairment and patient-reported measures of physical/mental functioning in older people living with HIV. HIV Med 18(5):363–369

Visser M, Schaap LA (2011) Consequences of sarcopenia. Clin Geriatr Med 27(3):387–399

Vosvick M, Gore-Felton C, Ashton E, Koopman C, Fluery T, Israelski D et al (2004) Sleep disturbances among HIV-positive adults: the role of pain, stress, and social support. J Psychosom Res 57(5):459–463

Wallace LM, Ferrara M, Brothers TD, Garlassi S, Kirkland SA, Theou O et al (2017) Lower frailty is associated with successful cognitive aging among older adults with HIV. AIDS Res Hum Retrovir 33(2):157–163

Wasserman P, Segal-Maurer S, Rubin DS (2014) High prevalence of low skeletal muscle mass associated with male gender in midlife and older HIV-infected persons despite CD4 cell reconstitution and viral suppression. J Int Assoc Provid AIDS Care 13(2):145–152

Weisberg DF, Gordon KS, Barry DT, Becker WC, Crystal S, Edelman EJ et al (2015) Long-term prescription of opioids and/or benzodiazepines and mortality among HIV-infected and uninfected patients. J Acquir Immune Defic Syndr 69(2):223–233

Wendelken LA, Valcour V (2012) Impact of HIV and aging on neuropsychological function. J Neurovirol 18(4):256–263

Whitson HE, Cohen HJ, Schmader KE, Morey MC, Kuchel G, Colon-Emeric CS (2018) Physical resilience: not simply the opposite of frailty. J Am Geriatr Soc 66(8):1459–1461

WHO (2015) World report on ageing and health

Womack JA, Goulet JL, Gibert C, Brandt CA, Skanderson M, Gulanski B et al (2013) Physiologic frailty and fragility fracture in HIV-infected male veterans. Clin Infect Dis 56(10):1498–1504

Wong C, Gange SJ, Moore RD, Justice AC, Buchacz K, Abraham AG et al (2018) Multimorbidity among persons living with human immunodeficiency virus in the United States. Clin Infect Dis 66(8):1230–1238

Wright EJ, Grund B, Cysique LA, Robertson KR, Brew BJ, Collins G et al (2015) Factors associated with neurocognitive test performance at baseline: a substudy of the INSIGHT Strategic Timing of AntiRetroviral Treatment (START) trial. HIV Med 16(Suppl 1):97–108

Yaffe K, Fiocco AJ, Lindquist K, Vittinghoff E, Simonsick EM, Newman AB et al (2009) Predictors of maintaining cognitive function in older adults: the health ABC study. Neurology 72(23):2029–2035

Yassuda MS, Lopes A, Cachioni M, Falcao DV, Batistoni SS, Guimaraes VV et al (2012) Frailty criteria and cognitive performance are related: data from the FIBRA study in Ermelino Matarazzo, Sao Paulo, Brazil. J Nutr Health Aging 16(1):55–61

Zamudio-Rodriguez A, Aguilar-Navarro S, Avila-Funes JA (2017) Cognitive impairment among older adults living with HIV/AIDS and frailty. Gac Med Mex 153(5):598–607

Zamudio-Rodriguez A, Belaunzaran-Zamudio PF, Sierra-Madero JG, Cuellar-Rodriguez J, Crabtree-Ramirez BE, Alcala-Zermeno JL et al (2018) Association between frailty and HIV-associated neurodegenerative disorders among older adults living with HIV. AIDS Res Hum Retrovir 34(5):449–455

Conceptualizing and Assessing Everyday Functioning in the Context of HIV-Associated Neurocognitive Disorders

Victoria M. Kordovski, Savanna M. Tierney, and Steven Paul Woods

Contents

Abstract Combination antiretroviral therapy has reduced the rates of severe HIV-associated neurocognitive disorders (HAND), but the prevalence of milder forms of HAND that can affect everyday functioning remains high. As HIV-infected adults approach near-normal life expectancies, they may become increasingly susceptible to declines in everyday functioning secondary to a variety of physical and mental factors, including HAND. Although impairments in everyday functioning are a hallmark of HAND diagnoses and can adversely influence quality of life, there are no gold standard measures of this fundamentally important and complex construct. This chapter provides a brief review of the various self-report, clinician-rated, and performance-based methods by which everyday functioning is measured in the setting of HIV disease, including global activities of daily living and specific domains of medication adherence, financial management, automobile driving, and vocational functioning.

Keywords Activities of daily living · Everyday functioning · Functional living skills · HIV · Performance-based assessment · Self-assessment

Everyday functioning is defined as the constellation of behaviors in which individuals engage on a regular basis, ranging from basic activities of daily living (ADLs) that are necessary for survival (e.g., eating) to more instrumental ADLs that tend to involve more complex, higher-order behaviors (e.g., cooking and medication management).

V. M. Kordovski, S. M. Tierney, and S. P. Woods (✉)
University of Houston, Houston, TX, USA
e-mail: vkordovs@Central.UH.EDU; smtiern2@Central.UH.EDU; spwoods@Central.UH.EDU

Curr Topics Behav Neurosci (2021) 50: 329–346
DOI 10.1007/7854_2018_78
Published Online: 5 January 2019

Reliable and valid assessment of everyday functioning is important in the setting of Human Immunodeficiency Virus Type-I (HIV) disease for several reasons. First, the decreased incidence and prevalence of frank dementia in the era of cART and corresponding increase in milder forms of HAND (e.g., Heaton et al. 2011) have raised questions regarding the relevance of such mild-to-moderate neurocognitive impairment to everyday functioning for HIV-positive persons and healthcare providers alike (Gisslén et al. 2011). Second, as adults living with HIV are living longer in the combination antiretroviral therapy (cART) era (CDC 2016), they are becoming increasingly susceptible to age-related physical and mental conditions (e.g., non-HIV-associated neurocognitive disorders) that can adversely impact everyday functioning. Third, the accurate classification of HIV-associated neurocognitive disorders (HAND) as either syndromic (e.g., Mild Neurocognitive Disorder) or subsyndromic (i.e., Asymptomatic Neurocognitive Impairment) depends heavily on the extent to which observed neurocognitive impairments affect everyday functioning. And yet there is currently no gold standard measurement for this fundamentally important and complex construct.

This chapter provides a brief review of the various self-report, performance-based, and clinician-rated methods by which everyday functioning is commonly measured in the setting of HIV disease, including global activities of daily living and specific domains of medication adherence, financial management, vocational functioning, and automobile driving. To accomplish this goal, we performed a systematic review in accordance with the PRISMA guidelines. We searched article databases (e.g., PubMed and Web of Science) from January 2003 to March 2018 using Boolean global and domain-specific search terms (e.g., "HIV" AND "neurocognitive" AND "everyday functioning"). Key terms were searched only within article titles and abstracts. We also searched well-known neuropsychology journals and reference sections of relevant papers using the same criteria. Article inclusion criteria were: (1) at least one measure of everyday functioning, (2) at least one performance-based measure of neurocognitive functioning, (3) a group of human participants with HIV disease, and (4) published in English. The search and article selection processes were completed by two independent raters, with disagreements resolved by a third rater. With these criteria in mind, the reader will notice that other real-world outcomes relevant to HAND such as mortality (e.g., Sevigny et al. 2007), cognitive symptoms (e.g., van Gorp et al. 1991), psychosocial functioning (e.g., Clark et al. 2010), and quality of life (e.g., Burgoyne and Renwick 2004) are not reviewed here due to space considerations.

1 Global Everyday Functioning

It has been estimated that between 50% and 75% of HIV-infected persons demonstrate "global" impairment everyday functioning (e.g., Blackstone et al. 2013). Everyday functioning declines in HIV are most commonly observed on complex ADLs (e.g., employment, medication adherence, and financial management), whereas declines in basic ADLs (e.g., grooming or dressing) are less common, but of course still occur in 10–20% of cases (e.g., Blackstone et al. 2013; Crystal et al. 2000). Because many variables affect global everyday functioning (e.g., HIV disease severity, lower socioeconomic status, older age, mood and substance use disorders, and larger non-HIV-associated comorbidity burden; Morgan et al. 2012; Malaspina et al. 2011), the isolation of neuropsychological declines as a specific contributing factor is a challenging task (see Obermeit et al. 2017). Most diagnostic systems for HAND consider whether there has been at least a minor decline in two or more aspects of everyday functioning that can be – at least partly – attributable to neurocognitive impairment (Antinori et al. 2007; American Psychiatric Association 2013). Such global everyday functioning declines may be ascertained by self-report, informant report, and/or performance-based measures, which ideally are multimodal and have sound psychometric properties, sociocultural normative standards, and strong evidence of construct validity (e.g., Antinori et al. 2007). As is often the case in vulnerable clinical populations with complex healthcare needs, the lofty ideal is rarely achieved in clinic or research.

Self-Report Questionnaires Across the HIV literature, global everyday functioning is most commonly measured by way of a single self-report measure that assesses a wide range of basic and instrumental ADLs. Participants are typically classified as either ADL dependent or independent on the basis of perceived need for assistance in two or more areas. The most widely utilized measure is the Heaton et al. (2004) modification of the classic Lawton and Brody (1969) ADL scale on which individuals rate both their current and best level of functioning across 16 domains, as well as the timing and attribution of any perceived ADL problems. This measure is comprehensive in its content coverage and shows excellent construct validity (e.g., Heaton et al. 2004). Other similar, although less frequently utilized measures of ADL dependence include the original Lawton and Brody (1969) ADL questionnaire and adaptations (e.g., McDonnell et al. 2014) and the Neurobehavioral Function and Activities of Daily Living Scale (Saykin et al. 1991). While self-report measures of ADLs are reliable (e.g., Woods et al. 2004), inexpensive, and relatively simple to administer and score, they possess several drawbacks. Most of these scales were developed in geriatric populations and thus might not cover domains related to younger and middle-aged adults (e.g., Internet-based tasks; Woods et al. 2017). From a psychometric perspective, they often have highly positively skewed distributions and there are few studies of their factor structure, measurement invariance, etc. Of course, they are also susceptible to biases related to mood and awareness (e.g., Laverick et al. 2017; Woods et al. 2004).

Informant-Rated Measures Informant-rated ADL questionnaires can help to overcome some of the limitations of self-report measures, but are usually difficult to obtain in HIV disease and are not commonly used in research. Instead, clinician-rated approaches such as the Karnofsky Performance Status Scale (KPSS; Karnofsky et al. 1948) are more often used in HIV. Akin to the recently discontinued Global Assessment of Functioning (American Psychiatric Association 2000), the KPSS is a measure of an individual's broad current functioning that ranges from 0 (deceased) to 100 (no disability) that is rated by a knowledgeable health professional. Like the ADL questionnaires reviewed above, the KPSS is typically used as a categorical variable (i.e., dependent vs. independent) based on an *a priori* or sample-driven cut-point. Advantages of informant approaches are that they are quick, inexpensive, and are less susceptible to patient bias. However, the interrater reliability of these scales may vary, as they are susceptible to observer bias and are still based on the clinician's perception of patient self-report rather than direct observation in the patient's everyday environment.

Performance-Based Task Batteries A handful of brief batteries of performance-based everyday functioning tasks have been used in HIV. Rather than relying on self-report or observation-based approaches, these tasks provide a global measure of functional capacity by combining scores across a variety of everyday skills (e.g., financial and medication management and communication-related activities). Among the most commonly cited tests available are the Everyday Functioning Battery (Heaton et al. 2004), University of San Diego Performance Based Skills Assessment (UPSA; Sheppard et al. 2018), Timed Instrumental Activities of Daily Living (Owsley et al. 2002), and the Observed Tasks of Daily Living (Diehl et al. 1995). In contrast to many of the self-report assessments, scores on these tests may be interpreted by domain (see below) in addition to an overall estimate of functional ability. As compared to self-report scales, performance-based measures can demonstrate stronger relationships to manifest functioning and are less subject to self-report biases (Blackstone et al. 2012). However, drawbacks to the use of performance-based tasks include (1) limited commercial availability, (2) lengthy administration time, with some tests requiring 60+ min (e.g., Everyday Functioning Battery), and (3) emphasis on capacity (i.e., what an individual can do) rather than manifest functioning (i.e., what an individual actually does). Finally, assessment of broad ADL functioning may limit the ability to make inferences regarding specific underlying functional domain-specific mechanisms, such as those involved in food preparation as compared to those involved in medication management.

Multimodal Approaches Several studies have used multimodal approaches to measuring global everyday functioning in HIV, which systematically integrate self-report questionnaires of ADLs and cognitive symptoms, informant-rated measures, performance-based tasks, and objective indicators (e.g., unemployment). For example, these approaches create continuous summary scores regarding the number of functional domains that are impaired or dichotomous (i.e., dependent vs. independent) classifications based on the presence of two or more everyday functioning areas being affected. Such multimodal approaches potentially increase sensitivity by allowing one to overcome the weaknesses associated with any singular approach (see Blackstone et al. 2013; Doyle et al. 2013). As with any global composite approach, however, such methods may also be susceptible to Type II error at the level of individual everyday functioning domains.

Medication Management Medication adherence involves the integration of several multistep processes and is critical to the effective long-term management of

HIV disease. Successful medication adherence requires a general understanding of a treatment plan, organizing medications, prospectively remembering to take the prescription, and dosing the correct number of pills under the prescribed conditions (e.g., with or without food). Indeed, effective management of the immunovirological aspects of HIV may involve complex, cumbersome, and habitual cART regimens (e.g., Krentz et al. 2012). HIV-infected adults who are highly adherent (e.g., taking over 90% of their prescribed doses) tend to have better disease outcomes (e.g., Perno et al. 2002) and lower rates of mortality (e.g., Lima et al. 2007). Yet rates of non-adherence to cART can be quite high, ranging from 10 to 60% depending on the subpopulation of HIV-infected persons (Glass et al. 2015), with notable risk factors including younger age, depression, and active substance abuse (e.g., Langebeek et al. 2014). Neurocognitive impairment is also a risk factor for non-adherence to cART (e.g., Hinkin et al. 2002). Notably, medication management has consistently been associated with executive functions (e.g., Thames et al. 2013) and strongly linked to verbal and visual learning (e.g., Patton et al. 2012). Interestingly, fine motor skills, which one may hypothesize are important for pill-dispensing, have shown mostly weak associations with medication management (e.g., Albert et al. 1999).

There is considerable variability in the measurement and operationalization of medication management and adherence in HIV disease. The presumptive gold standard of measuring adherence is direct observation of therapy; however, such methods are very resource intensive and thus infrequently used in research. Instead, medication adherence and management are most often measured via self-report, behavioral sampling, and performance-based capacity tasks. Self-report measures of actual medication adherence such as the Acquired Immune Deficiency Syndrome (AIDS) Clinical Trials Group questionnaire and various visual analogue scales are quick, inexpensive, and easy to administer and score, but they are also subject to under-reporting (for review see Simoni et al. 2006), commonly have highly skewed distributions, do not consistently relate to measures of performance-based medication management or adherence, and only show weak relationships with performance-based measures of neuropsychological functioning (e.g., Patton et al. 2012). In contrast, "performance-based" measures of medication management capacity, such as the Medication Management Test Revised (Heaton et al. 2004), more reliably correlate with neurocognitive impairment (e.g., Albert et al. 1999; Patton et al. 2012). These performance-based simulations typically evaluate medication management by asking participants to dispense pills and answer questions related to a mock medication regimen and associated props (e.g., pill bottles and pillbox). Unlike self-report measures of adherence, performance-based capacity measures are not susceptible to reporting bias, tend to have reasonable psychometric properties, and allow for analyses of component processes. However, they should not be mistaken for measures of manifest functioning, that is, these tasks are measures of capacity that are necessary but not sufficient for actual adherence. In other words, one may have the ability to correctly calculate and dose a medication regimen, but still not take the medication as prescribed (Blackstone et al. 2013). Furthermore, some tasks may be too simple for some subsets of highly educated individuals (i.e., resulting in ceiling effects; Kordovski et al. 2017a) and may be unable to detect subtle impairments.

Therefore, many studies have used a more naturalistic behavioral sampling of measurement of adherence, for example, Medication Event Monitoring System (MEMS) caps. Behavioral methods such as MEMS commonly use an electronic, pressure-activated medication bottle that tracks times and dates of openings of a sentinel medication that can be mapped onto an individual's prescribed regimen to glean detailed information about adherence. MEMS reliably show significant associations between non-adherence and neurocognitive impairment in HIV (e.g., Hinkin et al. 2004). This measurement approach allows researchers to gather valuable adherence data in real-time, which can be collected and stored over months without requiring the patient to return to clinic. Drawbacks of utilizing a MEMS cap include its cost, high rates of unreturned caps, and its incompatibility with liquid medication bottles and pill boxes. Additionally, there are differences in the ways in which adherence performance is characterized. Participants may be grouped dichotomously as "good adherers" or "poor adherers" based on either prespecified or data-driven cut points (e.g., greater than 95% adherent; Barclay et al. 2007), or viewed continuously by calculating the percent of doses taken relative to total doses prescribed (e.g., Ettenhofer et al. 2009). Further, the length of medication adherence monitoring also ranges from days to years, which can affect compliance with the MEMS themselves.

Across the medication management literature, the most robust relationships are observed between adherence and clinical tests of executive functions (e.g., Hinkin et al. 2002) and episodic memory (e.g., Obermeit et al. 2015). Intriguingly, there may be reciprocal relationships between cognition and medication adherence (Ettenhofer et al. 2010) whereby neurocognitive deficits can increase risk of non-adherence, which in turn exacerbates HIV disease and thus adversely impacts neurocognitive functioning. Emerging studies are also taking into consideration the cognitive architecture of medication management (e.g., see Wilson and Park 2008) in trying to identify potentially remediable targets to bolster adherence. For example, intra-individual variability in cognitive performance is a marker of executive dyscontrol that may map onto missed medication doses in daily life in HIV disease (e.g., Thaler et al. 2015). Likewise, prospective memory (i.e., remembering to remember) shows independent and strong associations with self-reported medication management (e.g., Woods et al. 2008) and adherence as measured by MEMS (e.g., Woods et al. 2009). Moving forward, drawing from pharmacokinetic-based methods (e.g., Tu et al. 2017) which allow for direct measurement of drug adherence and effectiveness, or the potential for single-dose monitoring methods (Günthard et al. 2016) may be beneficial to overcome the limitations associated with MEMS caps and self-report.

2 Financial Management

The ability to handle one's own finances involves complex cognitive processes, such as counting and keeping track of how much money one has, completing appropriate calculations to ensure bills are paid, and remembering to make payments at specific times of the month. Not surprisingly, the loss of ability to handle one's own finances is a commonly queried domain of everyday functioning, both in clinical interviews and widely used measures of basic and instrumental ADLs (Lawton and Brody 1969). In HIV, about 15% of cognitively normal and 55% of cognitively impaired adults show impairment in financial management capacity (Heaton et al. 2004). Global cognitive impairment is generally associated with medium-to-large deficits in financial management capacity (Gandhi et al. 2011; cf. Thames et al. 2011) and manifest (i.e., actual) functioning (Benedict et al. 2000). Performance on financial management tasks appears to rely on spatial processing and executive processes, although these relationships may vary across the lifespan (e.g., Thames et al. 2011). Several instruments are available for the performance-based measurement of financial capacity (for review, see Moore et al. 2007), though only four have been examined in HIV. The Advanced Finances test component of Everyday Functioning Battery (Heaton et al. 2004) requires participants to pay fictional bills and manage a fictitious checkbook. Money management subtests are included within other functional tests such as the UPSA (Sheppard et al. 2018) and have been embedded within the Everyday Multitasking Test (Fazeli et al. 2017). These tests generally evidence good psychometric properties, though they have primarily been used for research purposes. To our knowledge, no other commonly used measures of financial capacity, such as Financial Capacity Instrument (Marson et al. 2000) and Independent Living Scales (Loeb 1996), have been examined in HIV. Measuring manifest financial management is a much more difficult undertaking. As noted above, many global self-report measures of ADL include at least one item involving financial management, but few studies in HIV parse out these domain-level items or use specific financial management scales (e.g., Money Mismanagement Measure; Conrad et al. 2006). Moreover, the performance-based financial management capacity measures (e.g., check writing) do not necessarily reflect recent changes to ways in which people handle money in the modern, developed world in which it is a commonplace practice to utilize online banking and automatic deductions from accounts. Very recently, online tasks that more closely mirror realistic modern banking scenarios (e.g., transferring funds between accounts; Woods et al. 2017) have been used in HIV, though the psychometric properties and clinical utility of these approaches have not yet been established. It can be quite difficult to discern when a cognition-related change in functioning occurs in instances where baseline functioning is not clear, for example, in a common clinical scenario where a patient's spouse has historically managed finances for the household. Other arrangements such as when a patient receives government assistance, appoints a trusted fiduciary, or uses automatized web-based banking further complicate the assessment of financial management abilities. Strategies to overcome some practical limitations may include querying for errors in recent transactions, obtaining collateral information from a knowledgeable informant, or assessing the ability of an individual to explain bill payment methods (Benedict et al. 2000).

3 Automobile Driving

It is estimated that approximately 20% of individuals with HIV engage in unsafe driving (Marcotte et al. 2004). The ability to drive an automobile is often a primary factor in maintaining functioning independence across the lifespan (e.g., Marottoli et al. 2000). Driving allows HIV-infected adults to more easily remain independent in other areas of everyday functioning such as engaging in social activities and attending healthcare appointments. For experienced operators, driving is largely an automated behavior (Norman and Shallice 1986). However, many aspects of driving require active, complex engagement of both lower- and higher-order cognitive functions. Non-cognitive factors such as state- (e.g., affective distress) and trait- (e.g., aggressiveness) based personal characteristics (e.g., Beck et al. 2013) as well as medical conditions (e.g., osteoporosis; Ackerman et al. 2010) have also been linked to reduced driving ability.

Drivers must be able to visually scan, perceive, and attend to numerous areas of the environment, make rapid decisions, and plan and follow prescribed sequential motor procedures (Marcotte et al. 1999), making the assessment of driving behavior a complex, multi-faceted task. Global neurocognitive impairment due to HAND also confers a significant risk for diminished driving skills, performance, and safety (e.g., Marcotte et al. 1999, 2004), even when considered alongside other relevant predictors of these outcomes (i.e., demographic characteristics, HIV disease severity, driving history). Consistent with HIV's predilection for disrupting frontostriatal circuitry, the strongest associations between neurocognitive impairment and driving are typically observed in the domain of executive functions, including tests reliant on visual concept formation, novel problem solving, inhibition, complex sequencing, set shifting, and speeded word generation abilities. Not surprisingly, visual attention deficits are also associated with driving accidents among cognitively impaired adults living with HIV (Marcotte et al. 2006). In contrast, verbal functioning and episodic memory demonstrate relatively weaker associations with driving performance in HIV (Marcotte et al. 1999).

Self-report measures of automobile driving and transportation allow one to ascertain a general sense of manifest functioning, but as with all self-report measures they are limited by potential bias and low base rates of the phenomenon of interest (e.g., accidents) and exhibit poor associations with performance-based measures of capacity. Some commonly used self-report measures of global ADL function, such as the Lawton-Brody ADL Scale (Lawton and Brody 1969), include single items that inquire about driving ability or transportation (e.g., Mode of Transportation: ranging from able to "travel independently on public transportation or drives own car" to "does not travel at all"). Others, such as the Mobility Questionnaire, specifically assess driving habits and participant perception of driving quality (Vance et al. 2006). However, such methods do not reliably relate to neuropsychological function, disease characteristics, and manifest function, as many older adults rate their driving as "very good or excellent" which creates ceiling effects (Vance et al. 2014). The number of accidents reported on a cursory self-report measure demonstrated a trend level association with neuropsychological impairment in HIV (Marcotte et al. 2006). Still, brief questionnaires such as these do not permit investigation of the many possible factors that contribute to maintaining independence in automobile driving. Further, it is not unusual for individuals with impaired driving abilities to limit or to altogether stop driving, either of their own volition or by way of encouragement from significant others or medical professionals (e.g., Rebok et al. 1995), which restricts the information gathered by asking a general question about transport or accident history.

Computerized simulators may provide valuable insight into cognitive factors associated with automobile driving in HIV disease. Indeed, computer-based driving simulators can independently predict real-world driving ability (e.g., on-road driving evaluations, reported number of accidents, e.g., Marcotte et al. 1999, 2004, 2006). Measures such as the Routine and Emergency Driving simulation (Rosenthal et al. 1995), the Advanced Routine and Emergency Driving (Marcotte et al. 2004) task, Virtual City (Marcotte et al. 2004), and the Truck Operator Performance System (TOPS, abbreviated from Stein et al. 1992) require navigation of common obstacles faced on the road, such as pedestrians and other cars, multilane highways, or operation of a vehicle within specific parameters (e.g., set speed limit, TOPS). These simulator tasks have been associated with neuropsychological impairment among HIV-positive adults (e.g., Marcotte et al. 1999, 2004). Nevertheless, Marcotte et al. (1999) found that simulator performance was not related to self-reported crash history, which may suggest that these capacity-based tasks either have poor predictive validity or may reflect the low base rate problem of crashes versus other driving inefficiencies and risks. Certainly, subtle neuropsychological impairments in HIV may affect driving performance and safety when challenged with an emergency. As it is not practical, or ethical, to create these situations using a real vehicle with risk of real accidents, the use of a computer driving simulator thus provides a safe, and perhaps sensitive, method of assessing the degree to which neurobehavioral deficits impair driving ability.

One may assume that direct, on-road evaluations would provide the most useful information of manifest automobile driving ability; however, there are limitations to the implementation of this methodology. Real-life driving evaluation of cognitively impaired individuals can be dangerous, resource intensive, and unreliable (Croft and Jones 1987). In an on-road evaluation performed by Marcotte et al. (2004), 36.4% of neuropsychologically impaired HIV-positive adults were deemed as "unsafe" drivers compared to only 5% of HIV-negative participants and 6.9% of the HIV-positive cognitively normal counterparts. The loss of driving privileges can significantly impact quality of life, and this decision should be made following thorough assessment of driving abilities. Unfortunately, laboratory-based evaluations and self-report present significant limitations in accurately classifying impaired drivers. The dearth in the study of automobile driving represents a limitation in the extant literature and an opportunity for future development of examination as well as intervention. Furthermore, as individuals with HIV are living longer, older HIV-positive adults are a particularly important target population to study. The effects of age and HIV may interact and negatively impact various aspects of driving and route-planning (Foley et al. 2013). Moreover, neurocognitive functioning, notably attention and visuospatial abilities, accounts for over 50% of variability in driving performance in older HIV-positive adults (Foley et al. 2013) and is important in predicting driving performance longitudinally (Thames et al. 2013).

4 Employment and Vocational Functioning

Gainful employment affords individual's income for survival, a sense of purpose in life, a rich quality of life, and some protection against declines in mental health. Yet, in North America, estimates of unemployment in HIV-positive populations range from 40 to 80% (Morgan et al. 2012; Rabkin et al. 2004). Unemployment is associated with disease-specific factors such as shorter duration of infection, lower CD4 count, and medical comorbidity (Burns et al. 2006). Neurocognition is also cited as an independent predictor of employment status and abilities (e.g., Weber et al. 2012). Within the HIV literature, the relationship between employment status and neurocognition yields generally medium-sized effects. Specifically, regardless of measurement method, employment outcomes consistently show small-to-medium sized relationships with retrospective learning, working memory, and executive functions, including inhibition, perseverative thinking, visual planning, verbal fluency, and speeded set-shifting (Cattie et al. 2012; van Gorp et al. 1999; Weber et al. 2012; Woods et al. 2011). Additionally, learning and memory appear to show medium-sized effects on return to work after a period of unemployment (van Gorp et al. 2007). Motor functioning has also been associated with employment, though the effect tends to be small to medium, if present at all (Weber et al. 2012; van Gorp et al. 2007; Twamley et al. 2006).

The measurement of employment and work efficiency (when employed) is highly variable across the literature in HIV. The most common method is utilizing a broad dichotomous approach to self-reported employment status. Others, however, have utilized a more granular approach, further differentiating between unemployment and retirement or disability (e.g., Iudicello et al. 2014; Kordovski et al. 2017b), skilled and unskilled labor (Barber et al. 2017), volunteer work as a qualifier or as a level of vocational engagement (Benedict et al. 2000; Chernoff et al. 2010), or number of paid hours worked (Rabkin et al. 2004). Among these studies, these aspects of employment status were differentially related to cognition which suggests that the dichotomous approach to classifying employment status may oversimplify the variability in actual functioning among persons with and without gainful employment. Indeed, one study that utilized a measure of self-reported decreases in job-related activities in actively employed HIV-positive adults reported a large-sized effect of neurocognitive impairment on perceived decline (Heaton et al. 1994). In a more recent study by Kordovski et al. (2017b), neurocognitive functioning did not differ between unemployed (not due to disability) and employed older HIV-positive adults. Furthermore, older HIV-positive adults who remained gainfully employed reported stable functioning at work that was broadly comparable to that of younger or HIV-negative adults despite evidencing worse overall cognitive functioning. Therefore, while employment status itself can be a sufficient indicator of change in everyday functioning, it is crucial to consider not only the current employment status of the individual but also the context in which the individual is operating, including extracurricular activities, relative level of occupational engagement, and the potential match between the areas of cognition affected and the demands of the job itself. Performance-based measures of vocational potential are relatively less common in neuropsychological research (cf. vocational rehabilitation settings), and even less frequent in the HIV literature. To date, there are only a few studies that have evaluated vocational skill in HIV using standardized work samples (MESA SF2, Valpar International Corporation 1986) and vocational computerized assessment software (COMPASS, Valpar International Corporation 1992). Those studies have shown large effects of employment status on work performance on work-related abilities (Twamley et al. 2006), which support the construct validity of these measures. Moreover, they appear to be quite sensitive to effects of neurocognitive impairment (Heaton et al. 2004). However, some drawbacks of such techniques are the cost, time demands (i.e., 30–60 min of administration time), limited availability of services, requirement for computer, and limited literature supporting the validity of the tools in HIV. Additionally, measures to assess employment functioning lose relevancy after an older HIV-positive adult retires. Other activities such as volunteer work may then provide many of the same positive qualities that gainful employment does, including structure, purpose, and opportunity for social networking (Samson et al. 2009).

5 Clinical Implications of Current Research

Declines in everyday functioning are frequently observed among individuals with HIV, show significant relationships to neurocognitive functioning, and are a critical feature of diagnosing HAND. However, the measurement of everyday functioning presents a challenge, as each method has its various advantages and disadvantages and there is currently no gold-standard approach for its measurement in HIV. This chapter described the current state of knowledge of the measurement of several of the most commonly measured functional domains, including global functioning (i.e., ADLs), automobile driving, medication management, vocational functioning, and financial management. In doing so, we considered the strengths and limitations of each assessment with the aim to enhance the reader's understanding of the various tools that are available.

Within this context, we encourage readers to consider the broader role of everyday functioning in clinical neuropsychology. First, we emphasize the importance of considering everyday functioning in the setting of HIV research and clinical practice. While neurocognitive functioning is often of primary interest to a clinician, it is not the final point in the neurocognitive sequelae of HIV disease. Through the eyes of the patient, the relative importance of neurocognitive functioning may be defined by the degree to which it interferes with daily life. Second, we underscore the distinction between functional capacity and manifest, or "real-world," functioning. Functional capacity refers to what an individual has the capacity to do and is often best captured by performance-based tasks. In contrast, manifest functioning refers to what an individual actually does in real life and is often measured by queries about an individual's day-to-day life. The distinction is akin to the difference between traditional clinical measures of impulsivity (e.g., Stroop Color-Word Interference Test) and the scope of impulsive or dysexecutive behaviors that an individual displays at home or in social settings (e.g., risky behavior and poor decision-making). Though there is a relationship between functional capacity and manifest functioning, the degree of the relationship is certainly imperfect and is influenced by individual differences, the method and limitations of measurement, and the severity of the observed impairments. Third, we caution clinicians to avoid conflating domain-level functioning with global functioning. An individual who demonstrates reduced capacity to manage medications may or may not show comparable reductions in their ability to manage finances or adequately perform employment-related duties. Finally, when researching everyday functioning, we urge readers utilize a theoretical model to assist in conceptualizing everyday functioning in clinical populations and in formulating research questions. Operating within framework of a conceptual model can help ensure that the construct is comprehensively measured, and that the influence of important cofactors (e.g., comorbid psychiatric conditions, socioeconomic considerations, and cultural variability) and pathways (e.g., mechanisms of complex behaviors and role of motivational factors) is appropriately considered. Models presently exist for medication management (e.g., Wilson and Park 2008) and general real-world functioning (Blackstone Casaletto et al. 2017, Fig. 1).

Although significant progress in understanding HIV in the cART era has been made in the past 20 years, there are still substantial gains to be made regarding the operationalization and measurement of everyday functioning. One particularly relevant next step for the field may be to consider the component processes involved in complex everyday tasks. The examination of the work within cognitive rehabilitation literature may help researchers parse, understand, and create measures for specific behaviors involved in successful implementation of a complex task. Doing so would also facilitate identification of potential target areas for clinical intervention. Furthermore, researchers may consider other functional domains that have been extensively studied in other clinical populations wherein everyday functioning deficits are common but have been minimally examined in HIV. Clinicians may, for

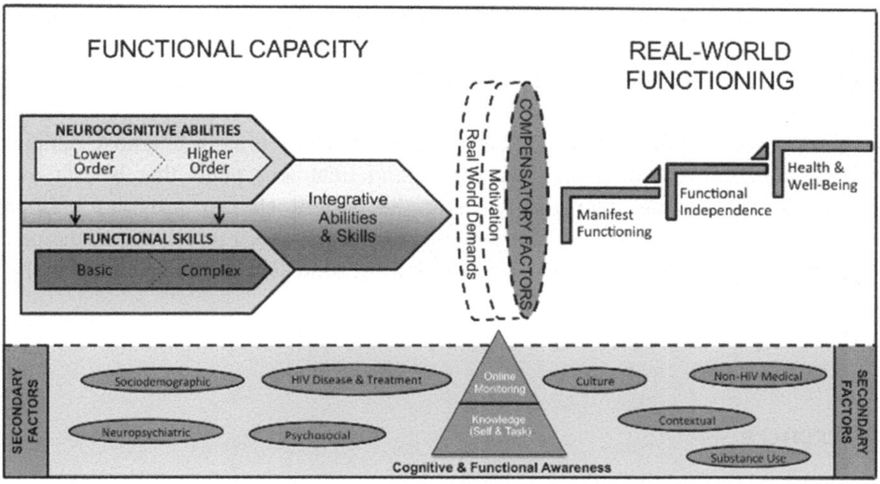

Fig. 1 Integrative conceptual model of everyday functioning depicting the role of HIV-associated neurocognitive deficits in real-world outcomes. Reprinted with permission from Blackstone Casaletto, Weber, Iudicello, & Woods contribution to *Changes in the brain: Impact on daily life.* (p. 214), by N.D. Chiaravalloti, & Y. Goverover, 2016, New York, NY: Springer. Copyright 2017 by Springer

example, consider borrowing well-validated tools that have been developed among populations wherein social functioning is commonly assessed (e.g., schizophrenia). Finally, a more developed definition of mild and major impairment in daily functioning, as described by Antinori et al. (2007), would help improve our current approach to the diagnosis of HAND. To date, we are unaware of any studies that have examined the operationalization and utility of the current diagnostic criteria outlined in the Frascati criteria.

Modernization of tools is also needed as the increased use of technology and the Internet has increasingly changed the ways in which individuals function in their communities. For instance, with the rise of Internet banking and cashless purchasing, assessments that implement handwritten checks and paper currency are becoming increasingly less valid, particularly among younger generations. Some self-report instruments that contain items that reflect a more technologically integrated way of functioning (e.g., Amsterdam Instrumental Activities of Daily Living Questionnaire, Sikkes et al. 2013; Everyday Technology Use Questionnaire, Rosenberg et al. 2009) have been created but have not yet been validated in a HIV population. As described above, some research groups have developed realistic Internet-based functional tasks (e.g., buying airline tickets, shopping, and banking online) though the translation of this work to clinical practice has not yet been established (Goverover et al. 2010; Woods et al. 2017). Furthermore, consideration of the demographic (e.g., aging and prospective memory paradox) and cultural factors (e.g., familial structures) will be important in the development of representative tests and normative data. Relatedly, as the ways in which we interact with real-world environments change, it is essential to consider how patients define "normal" functioning. What aspects of daily activities are most important to our patients? How do these considerations change in different subsets of patients (e.g., older versus younger adults and adults with low socioeconomic status)? Most importantly, how do we, as neuropsychologists, develop measures to reflect these changes in order to accurately diagnose deficits in daily function and develop intervention and treatment plans that benefit our patients as individuals? Our hope is that the current chapter provides a brief overview of the challenges we face moving forward, as well as recommendations for overcoming such obstacles and working toward addressing these questions in the future.

References

Ackerman ML, Crowe M, Vance DE, Wadley VG, Owsley C, Ball KK (2010) The impact of feedback on self-rated driving ability and driving self-regulation among older adults. Gerontologist 51(3):367–378

Albert SM, Weber CM, Todak G, Polanco C, Clouse R, McElhiney M, Rabkin J, Stern Y, Marder K (1999) An observed performance test of medication management ability in HIV: relation to neuropsychological status and medication adherence outcomes. AIDS Behav 3(2):121–128

American Psychiatric Association, DSM-5 Task Force (2013) Diagnostic and statistical manual of mental disorders: DSM-5

American Psychiatric Association, DSM-IV-TR (2000) Diagnostic and statistical manual of mental disorders: DSM-IV-TR. 4th edn. Text revision

Antinori A, Arendt G, Becker JT, Brew BJ, Byrd DA, Cherner M, Clifford DB, Cinque P, Epstein LG, Goodkin K, Gisslen M (2007) Updated research nosology for HIV-associated neurocognitive disorders. Neurology 69(18):1789–1799

Barclay TR, Hinkin CH, Castellon SA, Mason KI, Reinhard MJ, Marion SD, Levine AJ, Durvasula RS (2007) Age-associated predictors of medication adherence in HIV-positive adults: health beliefs, self-efficacy, and neurocognitive status. Health Psychol 26(1):40

Beck KH, Daughters SB, Ali B (2013) Hurried driving: relationship to distress tolerance, driver anger, aggressive and risky driving in college students. Accid Anal Prev 51:51–55

Benedict RH, Mezhir JJ, Walsh K, Hewitt RG (2000) Impact of human immunodeficiency virus type-1-associated cognitive dysfunction on activities of daily living and quality of life. Arch Clin Neuropsychol 15(6):535–544

Blackstone Casaletto K, Weber E, Iudicello JE, Woods SP (2017) Real-world impact of HIV-associated neurocognitive impairment. In: Chiaravalloti ND, Goverover Y (eds) Changes in the brain: impact on daily life. Springer, New York, pp 211–245

Blackstone K, Moore DJ, Heaton RK, Franklin DR, Woods SP, Clifford DB, Collier AC, Marra CM, Gelman BB, McArthur JC, Morgello S (2012) Diagnosing symptomatic HIV-associated neurocognitive disorders: self-report versus performance-based assessment of everyday functioning. J Int Neuropsychol Soc 18(1):79–88

Blackstone K, Iudicello JE, Morgan EE, Weber E, Moore DJ, Franklin DR, Ellis RJ, Grant I, Woods SP, TMARC Group (2013) HIV infection heightens concurrent risk of functional dependence in persons with chronic methamphetamine use. J Addict Med 7(4):255

Burgoyne R, Renwick R (2004) Social support and quality of life over time among adults living with HIV in the HAART era. Soc Sci Med 58(7):1353–1366

Burns SM, Young LR, Maniss S (2006) Predictors of employment and disability among people living with HIV/AIDS. Rehabil Psychol 51(2):127

Cattie JE, Doyle K, Weber E, Grant I, Woods SP, HIV Neurobehavioral Research Program (HNRP) Group (2012) Planning deficits in HIV-associated neurocognitive disorders: component processes, cognitive correlates, and implications for everyday functioning. J Clin Exp Neuropsychol 34(9):906–918

Centers for Disease Control and Prevention (2016) HIV surveillance report, 2015, vol 27

Chernoff RA, Martin DJ, Schrock DA, Huy MP (2010) Neuropsychological functioning as a predictor of employment activity in a longitudinal study of HIV-infected adults contemplating workforce reentry. J Int Neuropsychol Soc 16(1):38–48

Clark US, Cohen RA, Westbrook ML, Devlin KN, Tashima KT (2010) Facial emotion recognition impairments in individuals with HIV. J Int Neuropsychol Soc 16(6):1127–1137

Conrad KJ, Matters MD, Luchins DJ, Hanrahan P, Quasius DL, Lutz G (2006) Development of a money mismanagement measure and cross-validation due to suspected range restriction. J Appl Meas 7(2):206

Croft D, Jones RD (1987) The value of off-road tests in the assessment of driving potential of unlicensed disabled people. Br J Occup Ther 50(10):357–361

Crystal S, Fleishman JA, Hays RD, Shapiro MF, Bozzette SA (2000) Physical and role functioning among persons with HIV: results from a nationally representative survey. Med Care:1210–1223

Diehl M, Willis SL, Schaie KW (1995) Everyday problem solving in older adults: observational assessment and cognitive correlates. Psychol Aging 10(3):478

Doyle KL, Morgan EE, Morris S, Smith DM, Little S, Iudicello JE, Blackstone K, Moore DJ, Grant I, Letendre SL, Woods SP (2013) Real-world impact of neurocognitive deficits in acute and early HIV infection. J Neurovirol 19(6):565–573

Ettenhofer ML, Hinkin CH, Castellon SA, Durvasula R, Ullman J, Lam M, Myers H, Wright MJ, Foley J (2009) Aging, neurocognition, and medication adherence in HIV infection. Am J Geriatr Psychiatry 17(4):281–290

Ettenhofer ML, Foley J, Castellon SA, Hinkin CH (2010) Reciprocal prediction of medication adherence and neurocognition in HIV/AIDS. Neurology 74(15):1217–1222

Fazeli PL, Casaletto KB, Woods SP, Umlauf A, Scott JC, Moore DJ, HNRP Group (2017) Everyday multitasking abilities in older HIV+ adults: neurobehavioral correlates and the mediating role of metacognition. Arch Clin Neuropsychol 32(8):917–928

Foley JM, Gooding AL, Thames AD, Ettenhofer ML, Kim MS, Castellon SA, Marcotte TD, Sadek JR, Heaton RK, Gorp WV, Hinkin CH (2013) Visuospatial and attentional abilities predict driving simulator performance among older HIV-infected adults. Am J Alzheimer's Dis Other Demen 28(2):185–194

Gandhi NS, Skolasky RL, Peters KB et al (2011) A comparison of performance-based measures of function in HIV-associated neurocognitive disorders. J Neurovirol 17(2):159–165

Gisslén M, Price RW, Nilsson S (2011) The definition of HIV-associated neurocognitive disorders: are we overestimating the real prevalence? BMC Infect Dis 11:356

Glass TR, Sterne JA, Schneider MP, De Geest S, Nicca D, Furrer H, Günthard HF, Bernasconi E, Calmy A, Rickenbach M, Battegay M (2015) Self-reported nonadherence to antiretroviral therapy as a predictor of viral failure and mortality. AIDS 29(16):2195–2200

Goverover Y, O'Brien AR, Moore NB, DeLuca J (2010) Actual reality: a new approach to functional assessment in persons with multiple sclerosis. Arch Phys Med Rehabil 91(2):252–260

Günthard HF, Saag MS, Benson CA, Del Rio C, Eron JJ, Gallant JE, Hoy JF, Mugavero MJ, Sax PE, Thompson MA, Gandhi RT (2016) Antiretroviral drugs for treatment and prevention of HIV infection in adults: 2016 recommendations of the International Antiviral Society–USA panel. JAMA 316(2):191–210

Heaton RK, Velin RA, McCutchan JA, Gulevich SJ, Atkinson JH, Wallace MR, Godfrey HP, Kirson DA, Grant I (1994) Neuropsychological impairment in human immunodeficiency virus-infection: implications for employment. Psychosom Med 56(1):8

Heaton RK, Marcotte TD, Mindt MR, Sadek J, Moore DJ, Bentley H, McCutchan JA, Reicks C, Grant I, HNRC Group (2004) The impact of HIV-associated neuropsychological impairment on everyday functioning. J Int Neuropsychol Soc 10(3):317–331

Heaton RK, Franklin DR, Ellis RJ, McCutchan JA, Letendre SL, LeBlanc S, Corkran SH, Duarte NA, Clifford DB, Woods SP, Collier AC (2011) HIV-associated neurocognitive disorders before and during the era of combination antiretroviral therapy: differences in rates, nature, and predictors. J Neurovirol 17(1):3–16

Hinkin CH, Castellon SA, Durvasula RS, Hardy DJ, Lam MN, Mason KI, Thrasher D, Goetz MB, Stefaniak M (2002) Medication adherence among HIV+ adults effects of cognitive dysfunction and regimen complexity. Neurology 59(12):1944–1950

Hinkin CH, Hardy DJ, Mason KI, Castellon SA, Durvasula RS, Lam MN, Stefaniak M (2004) Medication adherence in HIV-infected adults: effect of patient age, cognitive status, and substance abuse. AIDS 18(Suppl 1):S19

Iudicello JE, Morgan EE, Gongvatana A, Letendre SL, Grant I, Woods SP, Translational Methamphetamine AIDS Research Center (TMARC) Group (2014) Detrimental impact of remote methamphetamine dependence on neurocognitive and everyday functioning in older but not younger HIV+ adults: evidence for a legacy effect. J Neurovirol 20(1):85–98

Karnofsky D, Abelman WH, Craver LF, Burchenal JH (1948) The use of nitrogen mustards in the palliative tratment of cancer. Cancer 1:634–456

Kordovski VM, Woods SP, Avci G, Verduzco M, Morgan EE (2017a) Is the newest vital sign a useful measure of health literacy in HIV disease? J Int Assoc Provid AIDS Care 16(6):595–602

Kordovski VM, Woods SP, Verduzco M, Beltran J (2017b) The effects of aging and HIV disease on employment status and functioning. Rehabil Psychol 62(4):591

Krentz HB, Cosman I, Lee K, Ming JM, Gill MJ (2012) Pill burden in HIV infection: 20 years of experience. Antivir Ther 17(5):833–840

Langebeek N, Gisolf EH, Reiss P, Vervoort SC, Hafsteinsdóttir TB, Richter C, Sprangers MA, Nieuwkerk PT (2014) Predictors and correlates of adherence to combination antiretroviral therapy (ART) for chronic HIV infection: a meta-analysis. BMC Med 12(1):142

Laverick R, Haddow L, Daskalopoulou M, Lampe F, Gilson R, Speakman A, Antinori A, Bruun T, Vassilenko A, Collins S, Rodger A (2017) Self-reported decline in everyday function, cognitive symptoms, and cognitive function in people with HIV. J Acquir Immune Defic Syndr 76(3): e74–e83

Lawton MP, Brody EM (1969) Assessment of older people: self-maintaining and instrumental activities of daily living. The Gerontologist 9(3_Part_1):179–186

Lima VD, Geller J, Bangsberg DR, Patterson TL, Daniel M, Kerr T (2007) The effect of adherence on the association between depressive symptoms and mortality among HIV-infected individuals first initiating HAART. AIDS 21(9):1175–1183

Loeb PA (1996) Independent living scales. The Psychological Corporation, San Antonio

Malaspina L, Woods SP, Moore DJ, Depp C, Letendre SL, Jeste D, Grant I, HIV Neurobehavioral Research Programs (HNRP) Group (2011) Successful cognitive aging in persons living with HIV infection. J Neurovirol 17(1):110–119

Marcotte TD, Heaton RK, Wolfson T, Taylor MJ, Alhassoon O, Arfaa K, Grant I (1999) The impact of HIV-related neuropsychological dysfunction on driving behavior. J Int Neuropsychol Soc 5 (07):579–592

Marcotte TD, Wolfson T, Rosenthal TJ, Heaton RK, Gonzalez R, Ellis RJ, Grant I (2004) A multimodal assessment of driving performance in HIV infection. Neurology 63(8):1417–1422

Marcotte TD, Lazzaretto D, Cobb Scott J, Roberts E, Woods SP, Letendre S, The HNRC Group (2006) Visual attention deficits are associated with driving accidents in cognitively-impaired HIV-infected individuals. J Clin Exp Neuropsychol 28(1):13–28

Marottoli RA, de Leon CFM, Glass TA, Williams CS, Cooney LM Jr, Berkman LF (2000) Consequences of driving cessation: decreased out-of-home activity levels. J Gerontol B Psychol Sci Soc Sci 55(6):S334–S340

Marson DC, Sawrie SM, Snyder S, McInturff B, Stalvey T, Boothe A, Aldridge T, Chatterjee A, Harrell LE (2000) Assessing financial capacity in patients with Alzheimer disease: a conceptual model and prototype instrument. Arch Neurol 57(6):877–884

McDonnell J, Haddow L, Daskalopoulou M, Lampe F, Speakman A, Gilson R et al (2014) Minimal cognitive impairment in UK HIV-positive men who have sex with men: effect of case definitions and comparison with the general population and HIV-negative men. J Acquir Immune Defic Syndr 67(2):120–127

Moore DJ, Palmer BW, Patterson TL, Jeste DV (2007) A review of performance-based measures of functional living skills. J Psychiatr Res 41(1–2):97–118

Morgan EE, Iudicello JE, Weber E, Duarte NA, Riggs PK, Delano-Wood L, Ellis R, Grant I, Woods SP, HIV Neurobehavioral Research Program (HNRP) Group (2012) Synergistic effects of HIV infection and older age on daily functioning. J Acquir Immune Defic Syndr 61(3):341

Norman W, Shallice T (1986) Attention to action. In: Davidson RJ, Schwartz GE, Shapiro D (eds) Consciousness and self regulation: advances in research and theory. Plenum, New York, pp 1–18

Obermeit LC, Morgan EE, Casaletto KB, Grant I, Woods SP, HIV Neurobehavioral Research Program (HNRP) Group (2015) Antiretroviral non-adherence is associated with a retrieval profile of deficits in verbal episodic memory. Clin Neuropsychol 29(2):197–213

Obermeit LC, Beltran J, Casaletto KB, Franklin DR, Letendre S, Ellis R, Fennema-Notestine C, Vaida F, Collier AC, Marra CM, Clifford D (2017) Evaluating the accuracy of self-report for the diagnosis of HIV-associated neurocognitive disorder (HAND): defining "symptomatic" versus "asymptomatic" HAND. J Neurovirol 23(1):67–78

Owsley C, Sloane M, McGwin D Jr, Ball K (2002) Timed instrumental activities of daily living tasks: relationship to cognitive function and everyday performance assessments in older adults. Gerontology 48(4):254–265

Patton DE, Woods SP, Franklin D Jr, Cattie JE, Heaton RK, Collier AC et al (2012) Relationship of medication management test-revised (MMT-R) performance to neuropsychological functioning and antiretroviral adherence in adults with HIV. AIDS Behav 16(8):2286–2296

Perno CF, Cozzi-Lepri A, Balotta C, Bertoli A, Violin M, Monno L, Zauli T, Montroni M, Ippolito G, d'Arminio-Monforte A (2002) Low prevalence of primary mutations associated with drug resistance in antiviral-naive patients at therapy initiation. AIDS 16(4):619–624

Rabkin JG, McElhiney M, Ferrando SJ, Van Gorp W, Lin SH (2004) Predictors of employment of men with HIV/AIDS: a longitudinal study. Psychosom Med 66(1):72–78

Rebok GW, Bylsma FW, Keyl PM, Brandt J, Folstein SE (1995) Automobile driving in Huntington's disease. Mov Disord 10(6):778–787

Rosenberg L, Nygård L, Kottorp A (2009) Everyday technology use questionnaire: psychometric evaluation of a new assessment of competence in technology use. Occup Part Health 29(2):52–62

Rosenthal TJ, Parseghian Z, Allen RW, Stein AC (1995) STISIM: the low-cost driving simulator. Version 8:1985–1996

Samson A, Lavigne RM, MacPherson P (2009) Self-fulfillment despite barriers: volunteer work of people living with HIV. AIDS Care 21(11):1425–1431

Saykin AJ, Janssen RS, Sprehn GC, Kaplan JE, Spira TJ, O'Connor B (1991) Longitudinal evaluation of neuropsychological function in homosexual men with HIV infection: 18-month follow-up. J Neuropsychiatry Clin Neurosci 3(2):286

Sevigny JJ, Albert SM, McDermott MP, Schifitto G, McArthur JC, Sacktor N, Conant K, Selnes OA, Stern Y, McClernon DR, Palumbo D (2007) An evaluation of neurocognitive status and markers of immune activation as predictors of time to death in advanced HIV infection. Arch Neurol 64(1):97–102

Sheppard DP, Woods SP, Verduzco M, HIV Neurobehavioral Research Center (HNRC) Group (2018) Construct validity of the UCSD performance-based skills assessment-brief version (UPSA-B) in HIV disease. Appl Neuropsychol Adult 25(6):543–554

Sikkes SA, Knol DL, Pijnenburg YA, De Lange-de Klerk ES, Uitdehaag BM, Scheltens P (2013) Validation of the Amsterdam IADL Questionnaire©, a new tool to measure instrumental activities of daily living in dementia. Neuroepidemiology 41(1):35–41

Simoni JM, Frick PA, Huang B (2006) A longitudinal evaluation of a social support model of medication adherence among HIV-positive men and women on antiretroviral therapy. Health Psychol 25(1):74

Stein AC, Allen RW, Parseghian Z (1992) The use of lowcost driving simulation to detect impaired drivers. Paper presented at the IMAGE VI conference, Scottsdale, AZ

Thaler NS, Sayegh P, Arentoft A, Thames AD, Castellon SA, Hinkin CH (2015) Increased neurocognitive intra-individual variability is associated with declines in medication adherence in HIV-infected adults. Neuropsychology 29(6):919

Thames AD, Kim MS, Becker BW, Foley JM, Hines LJ, Singer EJ, Heaton RK, Castellon SA, Hinkin CH (2011) Medication and finance management among HIV-infected adults: the impact of age and cognition. J Clin Exp Neuropsychol 33(2):200–209

Thames AD, Arentoft A, Rivera-Mindt M, Hinkin CH (2013) Functional disability in medication management and driving among individuals with HIV: a 1-year follow-up study. J Clin Exp Neuropsychol 35(1):49–45

Tu W, Nyandiko WM, Liu H, Slaven JE, Scanlon ML, Ayaya SO, Vreeman RC (2017) Pharmacokinetics-based adherence measures for antiretroviral therapy in HIV-infected Kenyan children. J Int AIDS Soc 20(1):21157

Twamley EW, Narvaez JM, Sadek JR, Jeste DV, Grant I, Heaton RK (2006) Work-related abilities in schizophrenia and HIV infection. J Nerv Ment Dis 194(4):268–274

van Gorp WG, Satz P, Hinkin C, Selnes O, Miller EN, McArthur J, Cohen B, Paz D (1991) Metacognition in HIV-1 seropositive asymptomatic individuals: self-ratings versus objective neuropsychological performance. J Clin Exp Neuropsychol 13(5):812–819

van Gorp WG, Baerwald JP, Ferrando SJ, McElhiney MC, Rabkin JG (1999) The relationship between employment and neuropsychological impairment in HIV infection. J Int Neuropsychol Soc 5(6):534–539

van Gorp WG, Rabkin JG, Ferrando SJ, Mintz J, Ryan E, Borkowski T, Mcelhiney M (2007) Neuropsychiatric predictors of return to work in HIV/AIDS. J Int Neuropsychol Soc 13 (1):80–89

Vance DE, Roenker DL, Cissell GM, Edwards JD, Wadley VG, Ball KK (2006) Predictors of driving exposure and avoidance in a field study of older drivers from the state of Maryland. Accid Anal Prev 38(4):823–831

Vance DE, Fazeli PL, Ball DA, Slater LZ, Ross LA (2014) Cognitive functioning and driving simulator performance in middle-aged and older adults with HIV. J Assoc Nurses AIDS Care 25 (2):e11–e26

Weber E, Blackstone K, Iudicello JE, Morgan EE, Grant I, Moore DJ, Woods SP, Translational Methamphetamine AIDS Research Center (TMARC) Group (2012) Neurocognitive deficits are associated with unemployment in chronic methamphetamine users. Drug Alcohol Depend 125 (1–2):146–153

Wilson EAH, Park D (2008) Prospective memory and health behaviors: context trumps cognition. In: Kliegel M, McDaniel MA, Einstein GO (eds) Prospective memory: cognitive, neuroscience, developmental, and applied perspectives. Erlbaum, Mahwah, NJ, pp 391–410

Woods SP, Rippeth JD, Frol AB, Levy JK, Ryan E, Soukup VM, Hinkin CH, Lazzaretto D, Cherner M, Marcotte TD, Gelman BB (2004) Interrater reliability of clinical ratings and neurocognitive diagnoses in HIV. J Clin Exp Neuropsychol 26(6):759–778

Woods SP, Iudicello JE, Moran LM, Carey CL, Dawson MS, Grant I (2008) HIV-associated prospective memory impairment increases risk of dependence in everyday functioning. Neuropsychology 22(1):110

Woods SP, Dawson MS, Weber E, Gibson S, Grant I, Atkinson JH (2009) Timing is everything: antiretroviral nonadherence is associated with impairment in time-based prospective memory. J Int Neuropsychol Soc 15(1):42–52

Woods SP, Weber E, Weisz BM, Twamley EW, Grant I (2011) Prospective memory deficits are associated with unemployment in persons living with HIV infection. Rehabil Psychol 56(1):77

Woods SP, Iudicello JE, Morgan EE, Verduzco M, Smith TV, Cushman C, HIV Neurobehavioral Research Program (HNRP) Group (2017) Household everyday functioning in the internet age: online shopping and banking skills are affected in HIV-associated neurocognitive disorders. J Int Neuropsychol Soc 23(7):605–615

Neuropsychiatric Disorders, Emotional Disturbances, and Their Associations with HIV-Associated Neurocognitive Disorder

Jose A. Muñoz-Moreno, Lucette A. Cysique, and Sean B. Rourke

Contents

J. A. Muñoz-Moreno (✉)
Infectious Diseases Service, Fundació Lluita contra la SIDA (FLS), Hospital Universitari Germans Trias i Pujol, Badalona, Catalonia, Spain

Department of Psychology and Sciences Education, Universitat Oberta de Catalunya (UOC), Barcelona, Catalonia, Spain
e-mail: jmunoz@flsida.org

L. A. Cysique
MAP Centre for Urban Health Solutions, Li Ka Shing Knowledge Institute, St. Michael's Hospital, Toronto, ON, Canada

Department of Psychology, The University of New South Wales, Sydney, NSW, Australia

S. B. Rourke
MAP Centre for Urban Health Solutions, Li Ka Shing Knowledge Institute, St. Michael's Hospital, Toronto, ON, Canada

Department of Psychiatry, University of Toronto, Toronto, ON, Canada

Curr Topics Behav Neurosci (2021) 50: 347–366
https://doi.org/10.1007/7854_2021_233
Published Online: 4 June 2021

Abstract The current chapter provides a critical and narrative review of recent research on the neuropsychiatric disorders, emotional disturbances, and their associations with neurocognitive functioning in people living with HIV infection. We review a range of neuropsychiatric disorders including depression and anxiety disorders, but also emotional disturbances, which can be partly distinguished from depression and anxiety (apathy, alexithymia, and emotional processing impairment). While reviewing the research into the neuropsychiatric disorders and HIV-associated neurocognitive disorders, we also cover the questions of self-reported cognitive symptoms evaluation and interpretation. The chapter includes research on the role of coping skills, perceived stress and response to stressful life events, and connections to neurocognitive impairment in people living with HIV. Promising non-pharmacological interventions are highlighted. The chapter concludes with the clinical implications on how to best consider neuropsychiatric disorders and cognitive symptoms for the diagnosis of HIV-associated neurocognitive disorders, as well as future research directions.

Keywords Coping · Emotional disorders · HIV-associated neurocognitive disorders · Neurocognitive impairment · Neuropsychiatric symptoms

1 Introduction

The accurate evaluation of neuropsychiatric disorders and cognitive symptoms has major clinical and research implications for the correct diagnosis of HIV-associated neurocognitive disorder (HAND) (Antinori et al. 2007). HAND can happen in people free of neuropsychiatric comorbidities (Blackstone et al. 2012; Rourke et al. 1999a). However, HAND is more common and more severe in people with neuropsychiatric conditions (Heaton et al. 2010). Importantly, neuropsychiatric disorders are common in people living with HIV. Reasons for higher prevalence of neuropsychiatric disorders in HIV population are multiple. Internationally, key populations affected by HIV are often composed of vulnerable individuals who more often belong to sexual minorities, racial minorities, women in patriarchal societies, people who use drugs, and people affected by a high burden of health and socio-economic disparities and inequities (World Health Organization 2019).

The first section of this chapter is dedicated to the epidemiology of neuropsychiatric disorders with a focus on depressive disorder as the most common disorder and studied in people with HIV, followed by anxiety disorders and apathy. The second section of the chapter is dedicated to the screening of self-reported cognitive symptoms and their interpretation. The third section concentrates on the associations between neuropsychiatric disorders and neurocognitive functioning or neurocognitive impairment (NCI). Within the same section, we integrate research on self-reported cognitive symptoms. The fourth section of the chapter considers relevant research into coping skills, perceived stress and response to stressful life events, and neurocognitive functioning. The fifth section provides highlights on the most promising non-pharmacological interventions for ameliorating neuropsychiatric disorders in people living with HIV. The chapter concludes with some clinical implications on how to best consider neuropsychiatric disorders and cognitive complaints in the assessment and differential diagnosis of HAND.

2 Epidemiology

Major Depressive Disorders (MDD) are prevalent in people with HIV (Cysique et al. 2007; Rubin and Maki 2019; Kamat et al. 2015; Mao et al. 2008; Nakasujja 2010; Rabkin 2008), where the rate is at least twice that observed in healthy community samples (Ciesla and Roberts 2001); however, the rates in key populations such as men who have sex with men (MSM), a major risk group for HIV infection, are more comparable (Mao et al. 2008; Atkinson et al. 2008). Lifetime major depression affects one in two persons living with HIV. Current MDD can range from 17 to 37% in people with HIV (Rubin and Maki 2019; Mao et al. 2008; Atkinson et al. 2008; Judd et al. 2005; Relf et al. 2013).

People living with HIV who have symptomatic HIV disease are at an increased likelihood of experiencing a Major Depressive Episode (MDE) when compared to those with asymptomatic HIV infection (Atkinson et al. 2008). Further, HIV disease progression, but not HIV infection itself, increases the intermediate-term risk of MDD (Kelso-Chichetto et al. 2018). Prior psychiatric history more strongly has been shown to predict future vulnerability (Atkinson et al. 2008; Judd et al. 2005). A study on the 10-year risk of experiencing an MDE in HIV-positive women and men demonstrated that women living with HIV at risk were 2.10 more likely (95% CI 1.63–2.70) to experience a moderate MDE and 1.96 more likely (95% CI 1.33–2.9) to experience a severe MDE compared to HV-low risk HIV+ women (Kelso-Chichetto et al. 2018). Recent data from the Women's Interagency HIV study (WIHS) also show that depressive symptoms are severe in up to 80% of women living with HIV (Rubin and Maki 2019). Men who are living with HIV and at risk for depression have a 3.23 increased odds (95% CI 2.22–4.69) of having a moderate MDE compared to low risk men (Kelso-Chichetto et al. 2018). Depressive symptoms are also associated with greater cardiovascular risk, lower antiretroviral therapy use, and unsuppressed viral load (Kelso-Chichetto et al. 2018; Stewart et al. 2020; Crockett et al. 2020).

A critical review and integrative review completed in 2017 represents the best synthesis of the anxiety research in HIV infection. The review shows that anxiety disorders have been the focus of less research relative to depressive disorders in people with HIV (Brandt et al. 2017). This is despite evidence that anxiety disorders are prevalent in this group, with rates reported to be even double that of depression in some studies (33% vs 15%, respectively) (Robertson et al. 2014). While anxiety can co-occur with depression, there are distinct symptoms in each, with underlying neurobiological substrate and treatments. Anxiety disorders are the most common psychiatric disorders. While they can have major life repercussions, they are also the most treatable psychiatric conditions. According to the DSM-5, anxiety disorders include specific phobia, social anxiety disorders, panic disorder, Agoraphobia, and general anxiety disorder (GAD). Of note the DSM-5 does not include post-traumatic stress disorder (PTSD) and obsessive-compulsive disorder (OCD) under the anxiety disorders umbrella. As is the case with depressive disorders, the choice of methods for assessing anxiety impacts the prevalence rates in HIV research. Median rate of anxiety disorders in people with living with HIV across high-, low-, and middle-income countries using clinical interview is 23% (range: 1–47%), while the rate increased to 33% (range 20–44%) when using anxiety scales (Brandt et al. 2017). These prevalence rates are greater than the noted in the general population (Brandt et al. 2017). Anxiety disorders are also more prevalent in people living with HIV from low-middle income countries [median = 28%] as compared with high income countries [median 23%]. This review also highlights that people with HIV who use drugs report higher symptoms of panic and less symptoms of anxiety compared to people with HIV who do not use drugs. Finally, this review highlights that most of the data collected come from studies composed of men who have sex with men (MSM) and that women are not adequately represented.

The relationship of anxiety to HIV disease is complex and bi-directional (Hellmuth et al. 2017). Premorbid anxiety disorders may interact with the immune system and lead to less favorable outcomes (Paul 2019). The importance of these early HIV infection dynamics has not been systematically considered in cross-sectional studies attempting to link anxiety symptoms to HIV disease markers. This may explain the inconsistent findings where, for example, current CD4 cell count is inconsistently associated with anxiety symptoms (Brandt et al. 2017). In contrast, the relationship of increased anxiety associated with lower medication adherence appears to be more consistent across the literature. However, because of the methodological limitations (self-report methodology, cross-sectional analysis) further study is still needed to fully explain these relationships and the potential mechanisms.

Apathy is a multidimensional syndrome that refers to diminished motivation that leads to a reduction in goal related thoughts, actions, and emotional responses. Apathy is typically increased in person with HIV as compared to HIV-negative individuals, but prevalence rates largely vary (between 11% and 50% in individuals with HIV) due most likely to inconsistent methods between studies (Cysique and Brew 2019).

3 Connection to HAND and Relation to Cognitive Symptoms

3.1 Depression

It is the area of mood and depressive symptoms that has generated the most substantial research in the field of neuropsychiatric disorders and neurocognitive functioning in HIV infection. Overall, this research has centered on exploring depressive symptoms rather than "depression" as a disorder, although there are a few exceptions (Rubin and Maki 2019). Several reports have found a link between depression and HAND or presence of NCI in people living with HIV (Bragança and Palha 2011; Gibbie et al. 2006; Shimizu et al. 2011; Vázquez-Justo et al. 2003; Waldrop-Valverde et al. 2005; Weiland-Fiedler et al. 2004), while others have not found this link to be consistent (Cysique et al. 2007, 2016; Ammassari et al. 2004; Au et al. 2008; Castellon et al. 2006; Goggin et al. 1997; Mapou et al. 1993; Richardson et al. 2005; Carter et al. 2003). To address this problem, several authors have focused on specific components or dimensions of depression in an attempt to elucidate the core features while identifying potential confounding factors. For example, Castellon and collaborators found that a core depressive component that was more related to motivation (i.e., mood-motivation disturbance) explained the connections with neurocognitive performance better than two other depressive components of self-reproach and somatic disturbance (Castellon et al. 2006). This effect is most evident for verbal memory, executive functioning, and motor speed. By means of the Beck Depression Inventory (BDI), one of the instruments most widely used to assess depressive symptoms in HIV research, other authors have found an inconstant relationship between depressive symptoms detected by this measure and cognitive performance, suggesting that the tools which detect mostly somatic symptoms may not be accurate for many depressive symptoms (Rabkin et al. 2000).

The clinical features and presentation of depression in individuals with HIV have also been suggested as a possible reason for the discrepancies in the results. For example, depression is known to be more prevalent during acute infection and by not accounting for the stage of the infection, this may lead to inaccurate conclusions (Hellmuth et al. 2017; Gold et al. 2014). Demographic features should also be considered; for example, a study found that potential relationship between depression and NCI may be mediated by age – in this study, depression only was detected in young people with HIV, in contrast to older individuals (Shimizu et al. 2011). Moreover, there is evidence that mental health burden may decompensate in older people with HIV when isolated (Harris et al. 2020). There also continues to be a dearth in the research of depression in HIV as it relates to sex and gender. In the HIV population, in fact, depression and depressive complaints are more commonly reported in women than in men. Also, in the studies on women, there appear to be stronger links between higher depressive symptoms and lower cognitive functioning that have been more reliably observed (Wisniewski et al. 2005). However, the mechanisms to explore these differences still need further exploration and also need to be interpreted through the prism and contexts of cumulative socio-economic, socio-cultural, and health disadvantages that women living with HIV typically face more than men (Rubin and Maki 2019).

Lifetime major depressive disorder is more robustly associated with a higher rate of self-reported cognitive symptoms and complaints than NCI (Cysique et al. 2007, 2016). Research shows that it is important to take into account MDD chronicity, and instability on treatment to better understand its potential association with NCI. In well-treated people living with HIV with access to psychological care, depressive symptoms tend to reduce over time, and they are not associated with neuropsychological performance; that is, depressive disorders may be distinct from HAND in a significant proportion of people living with HIV. These results do highlight the importance of considering the historical characteristics and determinants of depressive symptoms and chronic psychological comorbidities as key features to understand the etiology (causes and consequences) of neurocognitive functioning in people living with HIV. It is important to note, however, the possibility that in some persons depressive symptoms may in fact be a direct expression of or precursor to HAND (Rubin and Maki 2019). More recent studies have begun to show that depressive symptoms can be a risk factor for cognitive decline in people living with HIV – a finding that is comparable to the studies examining risk factors for dementia in non-HIV populations (Byers and Yaffe 2011). More specifically, in several large longitudinal studies depressive symptoms were shown to be a strong predictor of cognitive decline in people with HIV, as was shown in a recent and thorough review of this literature by Rubin and Maki (Rubin and Maki 2019). And in recent extensive studies from Canada and the USA (published after the by Rubin and Maki (2019)), there is evidence to further confirm this finding (Rourke et al. 2020; Paolillo et al. 2020). It is important to note that Rubin and collaborators also showed that men and women are different in terms of rates of MDD, with higher and more severe forms of MDD found more commonly in women (Rubin and Maki 2019). Cumulation of financial stress, socio-economic disadvantages, and gender-violence appear to be major contributors to these increased rates and their associated consequences (Yousuf et al. 2020). Taking note from these findings, we recommend that in the assessment of depressive disorders in NeuroHIV research, it is important to systematically assess gender and cultural effects and historical depressive factors – as these can affect how depression is experienced and how it may differentially impact cognitive functions (Morris et al. 2020).

There is a clear and reliable relationship documented between self-reported cognitive symptoms to depressive symptoms, i.e., with increases in depressive symptoms, there are corresponding increases in cognitive symptoms (Rourke et al. 1999b; Thames et al. 2011; Muñoz-Moreno et al. 2014; Moore et al. 1997; Yoo-Jeong et al. 2018; Woods et al. 2007), as seen on a variety of self-report measures studied. This strong relationship between depressive and cognitive symptoms has significant clinical implications as cognitive symptoms or complaints have been suggested as a screening method for detecting HAND (Simioni et al. 2010; European AIDS Clinical Society 2019; Laverick et al. 2017). As such, when using cognitive symptoms as a rapid measure for a potential sign(s) of HAND, as often necessary for streamlining care in clinical setting (European AIDS Clinical Society 2019), it is important to remember that this approach can lead to erroneous screening results as it is confounded by depressive symptoms and their association with emotional status.

3.2 Anxiety

As early as the late 1980s, the symptoms of both depression and anxiety were considered in the guidelines for the assessment and management of subtle neuropsychological alterations associated with HIV (Butters et al. 1990; Grant et al. 1987; Selnes et al. 1990). Studies assessing the connection between anxiety and NCI in people living with HIV are relatively sparse (Brandt et al. 2017). Some cross-sectional studies have found that anxiety disorders are not associated with NCI in people with HIV, while others have found an association (Brandt et al. 2017; Au et al. 2008; Woods et al. 2007; Laverick et al. 2017; Micali et al. 2011). Similar to depressive symptoms, anxiety levels have been demonstrated to more robustly and incrementally linked to self-reported cognitive symptoms and complaints rather than objective neuropsychological performance (Au et al. 2008; Woods et al. 2007; Laverick et al. 2017). Woods and collaborators found a positive relationship between anxiety and increased frequency of prospective memory complaints (Woods et al. 2007). One small longitudinal study ($N = 30$) has shown that anxiety was associated with cognitive decline covering an 18-month follow-up (Micali et al. 2011). However, larger longitudinal studies are needed to replicate these findings as well as controlling for pre-morbid anxiety levels to better understand potential relationship of anxiety to HAND. As earlier noted, more diverse sample inclusive of women living with HIV are needed in this area of research. Finally, we would recommend that both anxiety and depressive symptoms are assessed concurrently in people with HIV considering the high prevalence of anxiety symptoms in this population, particularly at the early stages of HIV-associated dementia (Antinori et al. 2007, 2013).

3.3 Apathy

Apathy is a common symptom of depression, and while the two can co-occur they are not systematically associated with each other in HIV (Rabkin et al. 2000; Castellon et al. 1998; McIntosh et al. 2015a; Bryant et al. 2015; Shapiro et al. 2014). The presence of apathy may signal underlying NCI and a distinct neuropsychiatric feature of HIV disease (Kamat et al. 2015). This is supported by neuroimaging findings, where different neuroimaging techniques have provided converging evidence suggesting that the severity of apathy symptoms is linked to the propensity of the virus to replicate within frontostriatal brain circuits, the latter being involved in emotional regulation (McIntosh et al. 2015a; Kamat et al. 2014). In this model, it has been shown in a cross-sectional study that there is an association between apathy and NCI, and also with HIV plasma levels, and functional disability (Shapiro et al. 2014) while other studies have not replicated these findings (Cysique and Brew 2019). Interestingly, Shapiro and collaborators also reported an interactive effect of age and apathy on neuropsychological performance (Shapiro et al. 2014). Apathy has also been linked to decline in daily functioning in HIV as well as self-reported cognitive complaints (Kamat et al. 2012). Apathy assessment may help to differentiate people with "pseudo-depressive symptoms" who do not respond to antidepressive treatment and present in fact an apathic form of HAND (Cysique and Brew 2019). Studies on apathy in HIV have been limited in the size and diversity of their study samples and it is recommended that larger longitudinal studies inclusive of more women and diverse people with HIV are needed to determine whether apathy should be considered as a pathognomonic feature of HIV-associated NCI.

3.4 Alexithymia

Alexithymia is defined as an impairment in the processing of affective and cognitive emotions. This trait is considered multidimensional, and is characterized, mainly, by presenting difficulty in the identification of feelings and sensations, as well as in the description of those feelings. Its frequency has been determined in specific non-prevalence studies, ranging from 35 to 54% in people living with HIV (Bogdanova et al. 2010; Mcintosh et al. 2014). Existing research on alexithymia has revealed a dissociation between apathy and alexithymia, with the two conditions originated in overlapping but distinct neural substrates within frontostriatal circuits (McIntosh et al. 2015a). Neuropathological substrates have been tied to alexithymia in people with HIV, specifically in terms of lower levels of amygdala activation (Clark et al. 2017). Alexithymia has been also linked to a greater vulnerability for HIV progression, due to a worse psychoneuroimmunological profile presented in people with HIV, which may also signal neurocognitive worsening (Mcintosh et al. 2014). A study that considered different dimensions for HIV-related alexithymia found that two of its processing components (i.e., difficulty in describing feelings and externally oriented thinking) correlated with performance on measures of executive and visuospatial abilities, but not depression (Bogdanova et al. 2010). This is consistent with dysfunction of the frontostriatal circuits and their cortical projections. The manifestation of alexithymia affects the behavioral level as well, since deficiencies in the identification and communication of thoughts, feelings, and emotions are important for early and long-term psychological adaptation to HIV. Alexithymia may represent an important factor to consider when interpreting self-report of emotional, cognitive, and functional symptoms in people with HIV.

3.5 Emotional Processing

Deficits in emotion processing have been found in people living with HIV, including those on successful antiretroviral treatment with aviremia (González-Baeza et al. 2016). These deficits in emotional processing have been found to be associated with NCI (Lane et al. 2012; Baldonero et al. 2013). Negative emotions and fear appear to be particularly affected. The alteration in fear recognition is thought to be related to disrupted functioning of neural networks in the fronto-basal-amygdala circuits (Baldonero et al. 2013), which can be compromised due to HIV-related injury. A study assessing facial expressions also showed that lower premorbid abilities and coinfection with HCV predicted worse precision in recognizing specific emotions (González-Baeza et al. 2016). Another study found that individuals with a formal diagnosis of HAND were almost 10 times more likely to be impaired in their emotion processing accuracy than individuals without HAND (Grabyan et al. 2017). A study with women living with HIV has shown that the impairment of neural systems associated with the cognitive reappraisal of emotions (i.e., the capability to change the way one thinks about potentially emotion-eliciting events) can be accentuated in advanced clinical stages of HIV infection independently from psychiatric function (Mcintosh et al. 2015b). Further, a study applying neurophysiological assessment methodology yielded event-related potential results which suggested that women with HIV show impairments in attention to emotionally-laden stimuli, and that this impairment might be related to a loss of affective priming (Tartar et al. 2014). Impaired emotion processing has been also associated with decline in everyday functioning capacity in people with HAND (Grabyan et al. 2017), highlighting emotional status as an important factor to consider when evaluating real-world activities in people living with HIV. Clinically, it is recommended that emotional processing be evaluated when there is evidence of cognitive decline in people living with HIV. In addition, further research is needed in this area with links to the field of social cognition and use of more advanced emotional processing testing methods (McDonald 2017).

4 Connection to Perceived Stress and Stressful Life Events

4.1 Perceived Stress

Perceived stress induces activation of neuroimmunological processes, specifically involving the hypothalamic-pituitary-adrenal axis, which is responsible for adrenergic dysregulation and greater release of cortisol (Russell and Lightman 2019). These disrupted neuroimmune responses that are connected to neuroinflammation (Russell and Lightman 2019) and brain inflammation are considered to be one possible mechanism for the development of HAND despite access to modern HIV treatment. Perceived stress has been proven to be independently associated with impaired learning and memory in women with HIV, an association that is more robust than in women without HIV infection (Rubin et al. 2015). Another study demonstrated that highly perceived stress was associated with verbal fluency deficits, but was not associated with depressive symptoms (Rubin et al. 2017a). In women without effective antiretroviral treatment or viral suppression, daily stress was linked with greater cognitive decline, and decline in learning in particular. One interesting study investigated the potential effects of low-dose hydrocortisone on neurocognition in men living with HIV (Rubin et al. 2017b). The authors found that compared with a placebo, low-dose hydrocortisone intensified salivary cortisol, and this was associated with better performance in verbal learning. However, the benefits were only short-term in the 4 h post-administration, suggesting rapid mechanisms of neurocognitive improvement, but no sustained benefits. Measures of perceived stress may be useful as time-covarying measure of depressive symptoms on longitudinal NeuroHIV studies, where these measures may serve as a surrogate of coping.

4.2 Stressful Life Events

Over the past decade, there have been a number of interesting studies examining the links between stress related to past life adversities and NCI in people with HIV. Though this type of stress and daily perceived stress are interrelated constructs, the chronic character of the former may determine different health consequences in terms of inflammation, immune response, or neurologic status. Regarding physical health, there is substantial evidence that stressful life events and passive coping strategies (e.g., denial or avoidance) may have an accentuating effect on HIV disease progression and possible association with NCI progression (Leserman 2003). A number of studies have been conducted to identify the connections between traumatic life events and NCI in HIV infection. In 2003, Pukay-Martin and collaborators demonstrated that, even after adjusting for other emotional variables (i.e., depression and anxiety) and demographic factors (i.e., age and educational level), past stressful events were independently associated with greater odds of NCI (Pukay-Martin et al. 2003). More recently in a study using comprehensive evaluation of neuropsychological measures and neuroimaging brain parameters in South African women with and without HIV infection (Spies et al. 2016; Spies et al. 2017), there were differences observed in information processing speed and executive functioning which depended on the presence or absence of past traumatic events. Differences in neuroimaging outcomes were also observed, specifically in volumetric measures of the right anterior cingulate cortex, bilateral hippocampi, corpus callosum, left and right caudate, and left and right putamen. The authors observed that volumes of those areas were lowest in HIV-positive women with stressful life events compared to all other study groups. When they examined the impact of antiretroviral therapy initiation, the authors also reported neurocognitive improvements in abstract reasoning, speed of information processing, and verbal fluency as a result of therapy initiation, though the effects of HIV and childhood trauma remained evident at 12-month follow-up.

With regard to post-traumatic stress disorder (PTSD), a study by Rubin and colleagues showed an independent impact on learning memory and verbal fluency in people living with HIV (Rubin et al. 2017a). Other investigations have found similar outcomes, establishing a connection between early stressful events and increased presence of current neuropsychiatric symptoms (Clark et al. 2017). Those observations also link lower levels of brain activation in the amygdala to the disruption associated with early adversities (Clark et al. 2017). Such findings in this area underpin the clinical implications for the management of stressful experiences and emphasize the need for interventions to alleviate the effects of past stressful adversities on those living with HIV and HAND. Additional research in stressful life events examining the neuropsychiatric and cognitive changes in people living with HIV is warranted. There is a specific need for longitudinal studies that include more diversity of the sample studies (in sociodemographics) and where childhood and lifetime trauma can be systematically accounted for as there is increasing evidence in particular that these factors represent significant risk factors for several types of dementia (Wang et al. 2019).

5 The Role of Coping

Issues related to coping have a long research tradition in HIV infection but in HIV social science and psychology research rather than in NeuroHIV research. Inadequate coping strategies may lead to an increased risk of acquiring the infection (Woodward et al. 2017). Additionally, becoming seropositive and having to confront new emotional, social, or clinical situations puts pressure on coping-related responses. Two types of approaches to coping have been described, a positive, active approach and a negative, avoidant style (Brien et al. 2018). The first approach refers to strategies that include active efforts to manage distress or modify the stressor or its meaning, through positive refocusing, positive reappraisal, adjusting goals to make them more realistic and achievable. The second approach refers to strategies that fail to directly address the stressor, such as denial, avoidance, or blaming others.

Research in samples of people living with HIV shows that positive coping strategies are associated with fewer symptoms of depression and anxiety (Brien et al. 2018). Importantly, it has been seen that adult men living with HIV who use positive active coping can even experience diminished disease progression over 1 year (Mulder et al. 1995). Coping also refers to personal resources deployed to deal with difficult situations in physical health conditions, with the aim of diminishing levels of distress (Dempster et al. 2015). In this regard, coping skills may influence the connection between perceived stress and brain health (Salama et al. 2013). Indeed, in an investigation that explored whether coping skills and neuropsychological functioning interacted with psychological status in youth with HIV, authors found that positive adaptive coping patterns and preserved executive functioning led independently to better psychosocial functioning (Salama et al. 2013). This is supported by research which has found a link between a greater perception of illness and dysfunctional coping strategies (e.g., passive coping and alcohol use), behaviors that can also influence brain health (Norcini Pala and Steca 2015).

There are only a few studies that have assessed the possible connections between coping and neurocognitive functioning in people living with HIV. Cody and collaborators investigated potential neurocognitive influence on proactive coping behaviors in adults with HIV (Cody et al. 2016). Participants underwent neurocognitive and psychosocial tests to determine whether neurocognitive functioning and other factors that have been associated with coping in other populations influenced proactive coping behaviors. The authors concluded that one psychological dimension, spirituality/religiosity, was a better predictor of proactive coping than neurocognitive functioning. This suggests that intervention oriented around beliefs may assist in managing HAND. In another study involving young people living with HIV, depressive symptoms were predicted by a combination of negative coping skills and poor neuropsychological functioning (Salama et al. 2013). In the same study, cognitive inflexibility and negative coping skills were also linked to symptoms of conduct disorder. However, the capacity to engage in proactive coping may be affected by disruption in the frontal-striatal-thalamic system (Cody et al. 2016), brain circuits that are responsible for abstraction and planning tasks (i.e., executive functioning) and often compromised in adults with HAND (Antinori et al. 2013). Because of this, future research would be needed to assess whether HAND progression is associated with worse coping strategies and how this may be remediated. Other research in people who are at risk of HAND is needed to demonstrate whether fostering improved coping strategies may be associated with less cognitive decline.

6 Non-pharmacological Interventions

There is evidence that achieving better mood and psychological status can lead to better overall mental health and improved neurocognitive functioning (Antinori et al. 2013). In addition, treatment of emotional dysregulations as the consequence of NCI and HIV-related brain injury or co-morbid to HIV infection may halt neurocognitive decline, via behavioral cognitive stimulation and an enriched environment (Vance et al. 2013).

Evidence for effective non-pharmacological behavioral strategies in the treatment of depressive and anxiety disorders encompasses a wide range of possible behavioral techniques which have demonstrated benefits for general health or brain health specifically, whether coming from experience with HIV infection or otherwise (Hahn and Andel 2011; Montoya et al. 2019).

An extensive systematic and meta-analysis of psychosocial interventions for people living with HIV found small pooled effect sizes for symptom-oriented interventions (cognitive and/or behavioral therapy, stress management, interpersonal therapy) ($\hat{g} = 0.19$), supportive interventions (peer/support and psycho-education), ($\hat{g} = 0.21$), and meditation interventions (mindfulness, meditation, or relaxation) ($\hat{g} = 0.20$) (van Luenen et al. 2018). However, there were important moderators; the pooled effect size was larger when mental health was the primary focus of the intervention and when the treatment duration was longer. Also, the pooled effect size was in the medium range when <40% of participants had current AIDS ($\hat{g} = 0.54$), when participants had a positive depression screen ($\hat{g} = 0.46$), and when the intervention was delivered by psychologists ($\hat{g} = 0.42$).

While evidence of peer support for specialized interventions is small, there is, however, robust evidence for showing that peer support is beneficial for long-term health outcomes in HIV and non-HIV population with chronic conditions (Galdas et al. 2015). An investigation developed in the USA found that ageing persons with HIV are isolated from informal networks due to the stigma of HIV/AIDS and ageism (Shippy and Karpiak 2005). Because social support is a multifaceted experience, social support study design recommendations include that social support should be comprehensively examined for emotional support, tangible support, informational support, and companionship support, while assessment of stigma perception remain key in ageing persons with HIV. The link between better mental health outcomes in various chronic conditions is an active area of research. Briefly, there are two dominant theories (Strom and Egede 2012). First, the *buffering hypothesis* asserts that social support is protective (or "buffering") during stressful events. Stressful events will have a greater negative impact on those with lower levels of social support. Second, the *direct effects hypothesis* states that people with high levels of social support are in better health than people with low social support, regardless of the stress. In this model, there is research to suggest that high levels of social support lead to better health, fewer psychological issues, and speedy recoveries from chronic diseases (Strom and Egede 2012). Finally, certain factors influence the social--support-health relationship. Positive influences include size, availability, and type of social support needed as well as peers' sociodemographics. This social support research has gained attention in persons living with HIV and programs have been designed to train peer/buddy navigators. These are interventions worth pursuing particularly to address aging and HAND, particularly as positive psychological factors such as social support and satisfaction with life have been associated with successful aging in older people living with HIV (Moore et al. 2017).

7 Conclusions and Clinical Implications

Neuropsychiatric disorders are central to consider in the differential diagnostic procedure for HAND. At a minimum, screening of current depressive symptoms is required with a thorough history taking (Antinori et al. 2007, 2013; European AIDS Clinical Society 2019; Butters et al. 1990; Cysique et al. 2012; Underwood and Winston 2016). However, we also advise assessing anxiety and stress with available screening tools already translated in multiple languages. Importantly, we strongly recommend that cognitive symptoms should not be interpreted in isolation of mental health status and specifically depressive symptoms. Ideally, a HAND diagnostic procedure will include a detailed history of lifetime neuropsychiatric conditions and treatments, in addition to an assessment of current assessment of anxiety, apathy, perceived daily stress, and/or lifetime traumatic events. When possible, a mix of standard depression, anxiety and stress scales, and psychiatric semi-structured interviews are recommended. If this is not possible at the initial screening stage, due to limitations on economic resources or time availability, for example, a more comprehensive mental health assessment could be made part of a subsequent comprehensive neuropsychological assessment or recommended when referring the patient further to a psychologist or a psychiatrist. When possible, using informant to corroborate the patient's neuropsychiatric history is advised.

A best practice guideline is that mental health symptoms should be treated if detected before undergoing a comprehensive neuropsychological assessment. However, this guideline often tends to be limited to depression. In the light of the current

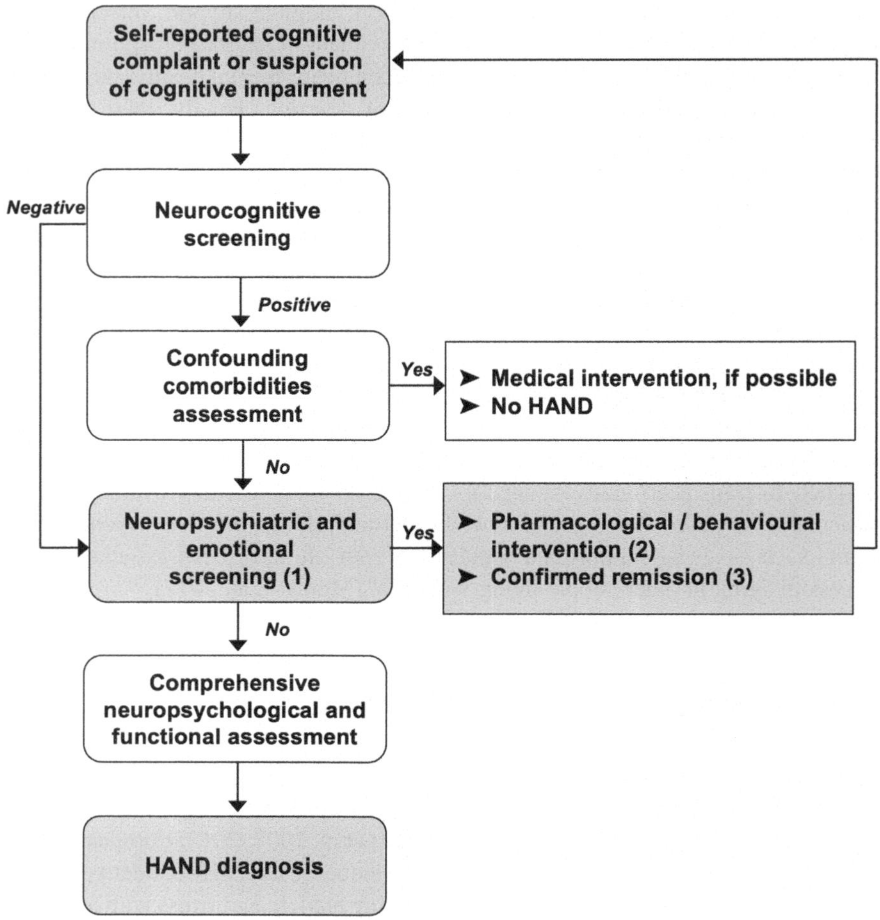

Fig. 1 Conceptual algorithm as proposal for considering neuropsychiatric and emotional disturbances in HAND diagnosis
1. Neuropsychiatric and emotional screening: Amplify what is traditionally covered (i.e., depressive and anxiety symptoms)
2. Pharmacological/behavioral intervention: Select strategies among multiple available according to: (a) empirical evidence, (b) availability, (c) cost-effectiveness
3. Confirmed remission: Confirmation of improvement of the symptoms is needed to reinitiate the algorithm
Observations:
• Most optimal choice is always directly undergoing comprehensive neuropsychological and functional assessment, mainly due to the extensive information provided and amount of scientific evidence
• In the event that emotional symptoms persist after treatment, discuss and weigh choices with patient
• This algorithm may be adapted in accordance with the specific needs of each center/population/region. From here, it is a general conceptual proposal essentially based on clinically relevant steps

knowledge, this recommendation should be extended to all forms of treatable neuropsychiatric conditions. This means that a wider range of therapeutic options should be considered depending on the patient's psychiatric profile and history. Ideally, follow-up of treatment outcomes should systematically be carried out using the same instruments as for the initial psychiatric assessment. If the previously detected mental health symptoms have reduced, the diagnosis process can be continued, with a consequent enhanced accuracy for detecting further HAND. Figure 1 provides an algorithm of the proposed clinical approach.

Clinically, it is best practice to tailor the mental health intervention to the individual, for this reason a variety of management approaches should be considered in order to able to draw from a broad range of potentially effective interventions in the clinical context.

Potential Conflicts of Interest None to be declared.

References

Ammassari A, Antinori A, Aloisi MS et al (2004) Depressive symptoms, neurocognitive impairment, and adherence to highly active antiretroviral therapy among HIV-infected persons. Psychosomatics. https://doi.org/10.1176/appi.psy.45.5.394

Antinori A, Arendt G, Becker JT et al (2007) Updated research nosology for HIV-associated neurocognitive disorders. Neurology 69:1789–1799. https://doi.org/10.1212/01.WNL.0000287431.88658.8b

Antinori A, Arendt G, Grant I et al (2013) Assessment, diagnosis and treatment of human immunodeficiency virus (HIV)-associated neurocognitive disorders (HAND): a consensus report of the mind exchange program. Clin Infect Dis 56:1004–1017. https://doi.org/10.1093/cid/cis975

Atkinson JH, Heaton RK, Patterson TL et al (2008) Two-year prospective study of major depressive disorder in HIV-infected men. J Affect Disord. https://doi.org/10.1016/j.jad.2007.10.017

Au A, Cheng C, Chan I et al (2008) Subjective memory complaints, mood, and memory deficits among HIV/AIDS patients in Hong Kong. J Clin Exp Neuropsychol 30:338–348. https://doi.org/10.1080/13803390701416189

Baldonero E, Ciccarelli N, Fabbiani M et al (2013) Evaluation of emotion processing in HIV-infected patients and correlation with cognitive performance. BMC Psychol. https://doi.org/10.1186/2050-7283-1-3

Blackstone K, Moore DJ, Heaton RK et al (2012) Diagnosing symptomatic HIV-associated neurocognitive disorders: self-report versus performance-based assessment of everyday functioning. J Int Neuropsychol Soc 18:79–88. https://doi.org/10.1017/S135561771100141X

Bogdanova Y, Díaz-Santos M, Cronin-Golomb A (2010) Neurocognitive correlates of alexithymia in asymptomatic individuals with HIV. Neuropsychologia 48:1295–1304. https://doi.org/10.1016/j.neuropsychologia.2009.12.033

Bragança M, Palha A (2011) Depression and neurocognitive performance in portuguese patients infected with HIV. AIDS Behav. https://doi.org/10.1007/s10461-011-9973-3

Brandt C, Zvolensky MJ, Woods SP et al (2017) Anxiety symptoms and disorders among adults living with HIV and AIDS: a critical review and integrative synthesis of the empirical literature. Clin Psychol Rev 51:164–184. https://doi.org/10.1016/j.cpr.2016.11.005

Brien KKO, Dagenais M, Solomon P et al (2018) Use of living strategies among adults aging with HIV in Canada: comparison by age-group using data from the HIV. Health Rehabil Surv 17:1–16. https://doi.org/10.1177/2325958218774041
Bryant AK, Ellis RJ, Umlauf A et al (2015) Antiretroviral therapy reduces neurodegeneration in HIV infection. AIDS 29:323–330. https://doi.org/10.1097/QAD.0000000000000553
Butters N, Grant I, Haxby J et al (1990) Assessment of AIDS-related cognitive changes: recommendations of the NIMH workshop on neuropsychological assessment approaches. J Clin Exp Neuropsychol 12:963–978
Byers AL, Yaffe K (2011) Depression and risk of developing dementia. Nat Rev Neurol 7:323–331. https://doi.org/10.1038/nrneurol.2011.60
Carter SL, Murji S, Rourke SB et al (2003) Cognitive complaints, depression, medical symptoms, and their association with neuropsychological functioning in HIV infection: a structural equation model analysis. Neuropsychology. https://doi.org/10.1037/0894-4105.17.3.410
Castellon SA, Hinkin CH, Wood S, Yarema KT (1998) Apathy, depression, and cognitive performance in HIV-1 infection. J Neuropsychiatry Clin Neurosci 10:320–329. https://doi.org/10.1176/jnp.10.3.320
Castellon SA, Hardy DJ, Hinkin CH et al (2006) Components of depression in HIV-1 infection: their differential relationship to neurocognitive performance. J Clin Exp Neuropsychol 28:420–437. https://doi.org/10.1080/13803390590935444
Ciesla JA, Roberts JE (2001) Meta-analysis of the relationship between HIV infection and risk for depressive disorders. Am J Psychiatry. https://doi.org/10.1176/appi.ajp.158.5.725
Clark US, Sweet LH, Morgello S et al (2017) High early life stress and aberrant amygdala activity: risk factors for elevated neuropsychiatric symptoms in HIV+ adults. Brain Imaging Behav 11:649–665. https://doi.org/10.1007/s11682-016-9542-5
Cody SL, Fazeli PL, Moneyham LD, Vance DE (2016) The influence of neurocognitive functioning on proactive coping behaviors in adults with HIV. J Neurosci Nurs 48:285–294. https://doi.org/10.1097/JNN.0000000000000216
Crockett KB, Entler KJ, Brodie E et al (2020) Linking depressive symptoms to viral non-suppression among women with HIV through adherence self-efficacy and ART adherence. J Acquir Immune Defic Syndr. https://doi.org/10.1097/QAI.0000000000002268
Cysique LA, Brew BJ (2019) Comorbid depression and apathy in HIV-associated neurocognitive disorders in the era of chronic HIV infection. Handb Clin Neurol 165:71–82. https://doi.org/10.1016/B978-0-444-64012-3.00006-X
Cysique LA, Deutsch R, Atkinson JH et al (2007) Incident major depression does not affect neuropsychological functioning in HIV-infected men. J Int Neuropsychol Soc 13:1–11. https://doi.org/10.1017/S1355617707070026
Cysique L, Bain L, Brew B (2012) Management issues in HIV-associated neurocognitive disorders. Neurobehav HIV Med:63. https://doi.org/10.2147/NBHIV.S30466
Cysique LA, Dermody N, Carr A (2016) The role of depression chronicity and recurrence on neurocognitive dysfunctions in HIV-infected adults. J Neurovirol 22:56–65. https://doi.org/10.1007/s13365-015-0368-5
Dempster M, Howell D, McCorry NK (2015) Illness perceptions and coping in physical health conditions: a meta-analysis. J Psychosom Res 79:506–513. https://doi.org/10.1016/j.jpsychores.2015.10.006
European AIDS Clinical Society. European AIDS Clinical Society Guidelines v10.0 November 2019. https://www.eacsociety.org/files/2019_guidelines-10.0_final.pdf. Accessed 1 Sept 2020
Galdas P, Fell J, Bower P et al (2015) The effectiveness of self-management support interventions for men with long-term conditions: a systematic review and meta-analysis. BMJ Open 5: e006620. https://doi.org/10.1136/bmjopen-2014-006620
Gibbie T, Mijch A, Ellen S et al (2006) Depression and neurocognitive performance in individuals with HIV/AIDS: 2-year follow-up. HIV Med 7:112–121. https://doi.org/10.1111/j.1468-1293.2006.00350.x

Goggin KJ, Zisook S, Heaton RK et al (1997) Neuropsychological performance of HIV-1 infected men with major depression. HNRC Group HIV Neurobehavioral Research Center. J Int Neuropsychol Soc 3:457–464

Gold JA, Grill M, Peterson J et al (2014) Longitudinal characterization of depression and mood states beginning in primary HIV infection. AIDS Behav 18(6):1124–1132. https://doi.org/10.1007/s10461-013-0688-5

González-Baeza A, Carvajal F, Bayon C (2016) Facial emotion processing in Aviremic HIV-infected adults. Arch Clin Neuropsychol:1–10. https://doi.org/10.1093/arclin/acw023

Grabyan JM, Morgan EE, Cameron MV et al (2017) Deficient emotion processing is associated with everyday functioning capacity in HIV-associated neurocognitive disorder. Arch Clin Neuropsychol:1–10. https://doi.org/10.1093/arclin/acx058

Grant I, Atkinson JH, Hesselink JR et al (1987) Evidence for early central nervous system involvement in the acquired immunodeficiency syndrome (AIDS) and other human immunodeficiency virus (HIV) infections. Studies with neuropsychologic testing and magnetic resonance imaging. Ann Intern Med 107:828–836. https://doi.org/10.7326/0003-4819-107-6-828

Hahn EA, Andel R (2011) Nonpharmacological therapies for behavioral and cognitive symptoms of mild cognitive impairment. J Aging Health 23:1223–1245. https://doi.org/10.1177/0898264311422745

Harris M, Brouillette MJ, Scott SC et al (2020) Impact of loneliness on brain health and quality of life among adults living with HIV in Canada. J Acquir Immune Defic Syndr 84:336–344. https://doi.org/10.1097/QAI.0000000000002355

Heaton RK, Clifford DB, Franklin DR et al (2010) HIV-associated neurocognitive disorders persist in the era of potent antiretroviral therapy: charter study. Neurology 75:2087–2096. https://doi.org/10.1212/WNL.0b013e318200d727

Hellmuth J, Colby D, Valcour V et al (2017) Depression and anxiety are common in acute HIV infection and associate with plasma immune activation. AIDS Behav. https://doi.org/10.1007/s10461-017-1788-4

Judd F, Komiti A, Chua P et al (2005) Nature of depression in patients with HIV/AIDS. Aust N Z J Psychiatry 39:826–832. https://doi.org/10.1080/j.1440-1614.2005.01659.x

Kamat R, Woods SP, Marcotte TD et al (2012) Implications of apathy for everyday functioning outcomes in persons living with HIV infection. Arch Clin Neuropsychol 27:520–531. https://doi.org/10.1093/arclin/acs055

Kamat R, Brown GG, Bolden K et al (2014) Apathy is associated with white matter abnormalities in anterior, medial brain regions in persons with HIV infection. J Clin Exp Neuropsychol 36:854–866. https://doi.org/10.1080/13803395.2014.950636

Kamat R, Cattie JE, Marcotte TD et al (2015) Incident major depressive episodes increase the severity and risk of apathy in HIV infection. J Affect Disord 175:475–480. https://doi.org/10.1016/j.jad.2015.01.010

Kelso-Chichetto NE, Okafor CN, Cook RL et al (2018) Association between depressive symptom patterns and clinical profiles among persons living with HIV. AIDS Behav. https://doi.org/10.1007/s10461-017-1822-6

Lane TA, Moore DM, Batchelor J et al (2012) Facial emotional processing in HIV infection: relation to neurocognitive and neuropsychiatric status. Neuropsychology 26:713–722. https://doi.org/10.1037/a0029964

Laverick R, Haddow L, Daskalopoulou M et al (2017) Self-reported decline in everyday function, cognitive symptoms, and cognitive function in people with HIV. J Acquir Immune Defic Syndr. https://doi.org/10.1097/QAI.0000000000001468

Leserman J (2003) The effects of stressful life events, coping, and cortisol on HIV infection. CNS Spectr 8:25–30. https://doi.org/10.1017/S1092852900023439

Mao L, Kippax SC, Newman CE et al (2008) Rates of depression among men attending high-HIV-caseload general practices in Australia. Ment Health Fam Med 5:79–83

Mapou RL, Law WA, Martin A et al (1993) Neuropsychological performance, mood, and complaints of cognitive and motor difficulties in individuals infected with the human immunodeficiency virus. J Neuropsychiatry Clin Neurosci:86–93. https://doi.org/10.1176/jnp.5.1.86

McDonald S (2017) What's new in the clinical management of disorders of social cognition? Brain Impair 18:2–10. https://doi.org/10.1017/BrImp.2017.2

Mcintosh RC, Ironson G, Antoni M et al (2014) Alexithymia is linked to neurocognitive, psychological, neuroendocrine, and immune dysfunction in persons living with HIV. Brain Behav Immun 36:165–175. https://doi.org/10.1016/j.bbi.2013.10.024

McIntosh RC, Rosselli M, Uddin LQ, Antoni M (2015a) Neuropathological sequelae of human immunodeficiency virus and apathy: a review of neuropsychological and neuroimaging studies. Neurosci Biobehav Rev 55:147–164. https://doi.org/10.1016/j.neubiorev.2015.04.008

Mcintosh RC, Ph D, Tartar JL et al (2015b) Negative attention bias and processing deficits during the cognitive reappraisal of unpleasant emotions in HIV + women. J Neuropsychiatry Clin Neurosci 27(1):e32–e39. https://doi.org/10.1176/appi.neuropsych.13090222

Micali E, Zirilli A, Abbate C (2011) Minor cognitive deficits and mood states in HIV-1 infected men. Illn Crisis Loss 19:363–377. https://doi.org/10.2190/IL.19.4.e

Montoya J, Henry B, Moore D (2019) Behavioral and physical activity interventions for HAND. Curr Top Behav Neurosci. https://doi.org/10.1007/7854_2018_79

Moore LH, van Gorp WG, Hinkin CH et al (1997) Subjective complaints versus actual cognitive deficits in predominantly symptomatic HIV-1 seropositive individuals. J Neuropsychiatry Clin Neurosci 9:37–44. https://doi.org/10.1176/jnp.9.1.37

Moore DJ, Fazeli PL, Moore RC et al (2017) Positive psychological factors are linked to successful cognitive aging among older persons living with HIV/AIDS. AIDS Behav 22(5):1551–1561. https://doi.org/10.1007/s10461-017-2001-5

Morris EP, Byrd D, Summers AC et al (2020) Depressive symptoms differentially predict Neurocognition in Latinx and non-Hispanic white people living with HIV. J Int Neuropsychol Soc:1–12. https://doi.org/10.1017/S1355617720000855

Mulder CL, Antoni MH, Duivenvoorden HJ et al (1995) Active confrontational coping predicts decreased clinical progression over a one-year period in HIV-infected homosexual men. J Psychosom Res. https://doi.org/10.1016/0022-3999(95)00062-3

Muñoz-Moreno JA, Fuster-Ruiz de Apodaca MJ, Fumaz CR et al (2014) Cognitive complaints in people with human immunodeficiency virus in Spain: prevalence and related variables. Med Clin (Barc) 142:438–444. https://doi.org/10.1016/j.medcli.2013.02.037

Nakasujja N, L Skolasky R, Musisi S, et al (2010) Depression symptoms and cognitive function among individuals with advanced HIV infection initiating HAART in Uganda. BMC Psychiatry doi: https://doi.org/10.1186/1471-244X-10-44

Norcini Pala A, Steca P (2015) Illness perceptions and coping strategies among individuals diagnosed with HIV. J Behav Med. https://doi.org/10.1007/s10865-015-9639-0

Paolillo EW, Pasipanodya EC, Moore RC et al (2020) Cumulative burden of depression and neurocognitive decline among persons with HIV: a longitudinal study. J Acquir Immune Defic Syndr 84:304–312. https://doi.org/10.1097/QAI.0000000000002346

Paul R (2019) Neurocognitive phenotyping of HIV in the era of antiretroviral therapy. Curr HIV/AIDS Rep 16:230–235yo. https://doi.org/10.1007/s11904-019-00426-9

Pukay-Martin ND, Cristiani SA, Saveanu R, Bornstein RA (2003) The relationship between stressful life events and cognitive function in HIV-infected men. J Neuropsychiatry Clin Neurosci. https://doi.org/10.1176/jnp.15.4.436

Rabkin JG (2008) HIV and depression: 2008 review and update. Curr HIV/AIDS Rep 5:163–171. https://doi.org/10.1007/s11904-008-0025-1

Rabkin JG, Ph D, Ferrando SJ et al (2000) Relationships among apathy, depression, and cognitive impairment in HIV/AIDS. J Neuropsychiatry Clin Neurosci 12(4):451–457. https://doi.org/10.1176/jnp.12.4.451

Relf MV, Eisbach S, Okine KN, Ward T (2013) Evidence-based clinical practice guidelines for managing depression in persons living with HIV. J Assoc Nurses AIDS Care. https://doi.org/10.1016/j.jana.2012.05.007

Richardson JL, Nowicki M, Danley K et al (2005) Neuropsychological functioning in a cohort of HIV- and hepatitis C virus-infected women. AIDS 19:1659–1667. https://doi.org/10.1097/01.aids.0000186824.53359.62

Robertson K, Bayon C, Molina J-M et al (2014) Screening for neurocognitive impairment, depression, and anxiety in HIV-infected patients in Western Europe and Canada. AIDS Care 26:1555–1561. https://doi.org/10.1080/09540121.2014.936813

Rourke SB, Halman MH, Bassel C (1999a) Neuropsychiatric correlates of memory-metamemory dissociations in HIV-infection. J Clin Exp Neuropsychol 21:757–768. https://doi.org/10.1076/jcen.21.6.757.852

Rourke SB, Halman MH, Bassel C (1999b) Neurocognitive complaints in HIV-infection and their relationship to depressive symptoms and neuropsychological functioning. J Clin Exp Neuropsychol. https://doi.org/10.1076/jcen.21.6.737.863

Rourke SB, Bekele T, Rachlis A et al (2020) Asymptomatic neurocognitive impairment is a risk for symptomatic decline in a large HIV Canadian cohort study. AIDS 35(1):63–72. https://doi.org/10.1097/QAD.0000000000002709

Rubin LH, Maki PM (2019) HIV, depression, and cognitive impairment in the era of effective antiretroviral therapy. Curr HIV/AIDS Rep 16:82–95. https://doi.org/10.1007/s11904-019-00421-0

Rubin LH, Cook JA, Weber KM et al (2015) The association of perceived stress and verbal memory is greater in HIV-infected versus HIV-uninfected women. J Neurovirol 21:422–432. https://doi.org/10.1007/s13365-015-0331-5

Rubin LH, Cook JA, Springer G et al (2017a) Perceived and posttraumatic stress is associated with decreased learning, memory, and fluency in HIV-infected women. AIDS. https://doi.org/10.1097/qad.0000000000001625

Rubin LH, Phan KL, Keating SM et al (2017b) Low-dose hydrocortisone has acute enhancing effects on verbal learning in HIV-infected men. J Acquir Immune Defic Syndr 75:e65–e70. https://doi.org/10.1097/QAI.0000000000001303

Russell G, Lightman S (2019) The human stress response. Nat Rev Endocrinol 15:525–534. https://doi.org/10.1038/s41574-019-0228-0

Salama C, Morris M, Armistead L et al (2013) Depressive and conduct disorder symptoms in youth living with HIV: the independent and interactive roles of coping and neuropsychological functioning. AIDS Care 25:160–168. https://doi.org/10.1080/09540121.2012.687815

Selnes OA, Miller E, McArthur J et al (1990) HIV-1 infection: no evidence of cognitive decline during the asymptomatic stages. Neurology 40:204–208. https://doi.org/10.1212/wnl.40.2.204

Shapiro ME, Mahoney JR, Peyser D et al (2014) Cognitive reserve protects against apathy in individuals with human immunodeficiency virus. Arch Clin Neuropsychol 29:110–120. https://doi.org/10.1093/arclin/act071

Shimizu SM, Chow DC, Valcour V et al (2011) The impact of depressive symptoms on neuropsychological performance tests in HIV-infected individuals: a study of the Hawaii aging with HIV cohort. World J AIDS 1:139–145. https://doi.org/10.4236/wja.2011.14020

Shippy RA, Karpiak SE (2005) The aging HIV/AIDS population: fragile social networks. Aging Ment Health 9:246–254. https://doi.org/10.1080/13607860412331336850

Simioni S, Cavassini M, Annoni J-M et al (2010) Cognitive dysfunction in HIV patients despite long-standing suppression of viremia. AIDS 24:1243–1250. https://doi.org/10.1097/QAD.0b013e3283354a7b

Spies G, Ahmed-Leitao F, Fennema-Notestine C et al (2016) Effects of HIV and childhood trauma on brain morphometry and neurocognitive function. J Neurovirol 22:149–158. https://doi.org/10.1007/s13365-015-0379-2

Spies G, Fennema-Notestine C, Cherner M, Seedat S (2017) Changes in cognitive function in women with HIV infection and early life stress. AIDS Care 29:14–23. https://doi.org/10.1080/09540121.2016.1204417

Stewart JC, Polanka BM, So-Armah KA et al (2020) Associations of total, cognitive/affective, and somatic depressive symptoms and antidepressant use with cardiovascular disease-relevant biomarkers in HIV: veterans aging cohort study. Psychosom Med 82:461–470. https://doi.org/10.1097/PSY.0000000000000808
Strom JL, Egede LE (2012) The impact of social support on outcomes in adult patients with type 2 diabetes: a systematic review. Curr Diab Rep 12:769–781. https://doi.org/10.1007/s11892-012-0317-0
Tartar JL, Mcintosh RC, Rosselli M et al (2014) Clinical neurophysiology HIV-positive females show blunted neurophysiological responses in an emotion – attention dual task paradigm. Clin Neurophysiol 125:1164–1173. https://doi.org/10.1016/j.clinph.2013.09.048
Thames AD, Becker BW, Marcotte TD et al (2011) Depression, cognition, and self-appraisal of functional abilities in HIV: an examination of subjective appraisal versus objective performance. Clin Neuropsychol 25:224–243. https://doi.org/10.1080/13854046.2010.539577
Underwood J, Winston A (2016) Guidelines for evaluation and management of cognitive disorders in HIV-positive individuals. Curr HIV/AIDS Rep 13:235–240. https://doi.org/10.1007/s11904-016-0324-x
van Luenen S, Garnefski N, Spinhoven P et al (2018) The benefits of psychosocial interventions for mental health in people living with HIV: a systematic review and meta-analysis. AIDS Behav 22:9–42. https://doi.org/10.1007/s10461-017-1757-y
Vance DE, Fazeli PL, Grant JS et al (2013) The role of neuroplasticity and cognitive reserve in aging with HIV: recommendations for cognitive protection and rehabilitation. J Neurosci Nurs 45:306–316. https://doi.org/10.1097/JNN.0b013e31829d8b29
Vázquez-Justo E, Rodríguez Alvarez M, Ferraces Otero MJ (2003) Influence of depressed mood on neuropsychologic performance in HIV-seropositive drug users. Psychiatry Clin Neurosci 57:251–258. https://doi.org/10.1046/j.1440-1819.2003.00113.x
Waldrop-Valverde D, Ownby RL, Kumar M (2005) Influence of depression and HIV serostatus on the neuropsychological performance of injecting drug users. Psychiatry Clin Neurosci 59:372–378. https://doi.org/10.1111/j.1440-1819.2005.01388.x
Wang X-J, Xu W, Li J-Q et al (2019) Early-life risk factors for dementia and cognitive impairment in later life: a systematic review and meta-analysis. J Alzheimers Dis 67:221–229. https://doi.org/10.3233/JAD-180856
Weiland-Fiedler P, Erickson K, Waldeck T et al (2004) Evidence for continuing neuropsychological impairments in depression. J Affect Disord. https://doi.org/10.1016/j.jad.2003.10.009
Wisniewski AB, Apel S, Selnes OA et al (2005) Depressive symptoms, quality of life, and neuropsychological performance in HIV/AIDS: the impact of gender and injection drug use. J Neurovirol 11:138–143. https://doi.org/10.1080/13550280590922748
Woods SP, Carey CL, Moran LM et al (2007) Frequency and predictors of self-reported prospective memory complaints in individuals infected with HIV. Arch Clin Neuropsychol 22:187–195. https://doi.org/10.1016/j.acn.2006.12.006
Woodward EN, Banks RJ, Marks AK, Pantalone DW (2017) Identifying resilience resources for HIV prevention among sexual minority men: a systematic review. AIDS Behav 21:2860–2873. https://doi.org/10.1007/s10461-016-1608-2
World Health Organization (2019) Progress report on HIV, viral hepatitis and sexually transmitted infections, 2019. Accountability for the global health sector strategies, 2016–2021. WHO/CDS/HIV/19.7. Accessed 1 Oct 2020
Yoo-Jeong M, Anderson A, Rahman AF et al (2018) Associations of mood on objective and subjective cognitive complaints in persons living with HIV/AIDS. J HIV AIDS 4. https://doi.org/10.16966/2380-5536.146
Yousuf A, Musa R, Isa MLM, Arifin SRM (2020) Anxiety and depression among women living with HIV: prevalence and correlations. Clin Pract Epidemiol Ment Health 16:59–66. https://doi.org/10.2174/1745017902016010059

Metabolic Syndrome and Cardiovascular Disease Impacts on the Pathophysiology and Phenotype of HIV-Associated Neurocognitive Disorders

Antoine Moulignier and Dominique Costagliola

Contents

Abstract Evidence from epidemiological studies on the general population suggests that midlife cardiovascular disease (CVD) and/or metabolic syndrome (MetS) are associated with an increased risk of cognitive impairment and dementia later in life. In the modern combined antiretroviral therapy (cART) era, as in the general population, CVD and MetS were strongly and independently associated with poorer cognitive performances of sustained immunovirologically controlled persons living with human immunodeficiency viruses (PLHIVs). Those findings suggest that CV/metabolic comorbidities could be implicated in the pathogenesis of HIV-associated neurocognitive disorders (HAND) and might be more important

A. Moulignier (✉)
Department of Neurology, Memory Clinic, Fondation Adolphe de Rothschild, Paris, France
e-mail: amoulignier@for.paris

D. Costagliola (✉)
INSERM, Sorbonne Université, Institut Pierre-Louis d'Epidémiologie et de Santé Publique (IPLESP), Paris, France
e-mail: dominique.costagliola@iplesp.upmc.fr

Curr Topics Behav Neurosci (2021) 50: 367–400
DOI 10.1007/7854_2019_123
Published Online: 28 January 2020

than factors related to HIV infection or its treatment, markers of immunocompetence, or virus replication. The association between CVD/MetS and cognition decline is driven by still not well-understood mechanisms, but risk might well be the consequence of increased brain inflammation and vascular changes, notably cerebral small-vessel disease. In this review, we highlight the correspondences observed between the findings concerning CVD and MetS in the general population and virus-suppressed cART-treated PLHIVs to evaluate the real brain-aging processes. Indeed, incomplete HIV control mainly reflects HIV-induced brain damage described during the first decades of the pandemic. Given the growing support that CVD and MetS are associated with HAND, it is crucial to improve early detection and assure appropriate management of these conditions.

Keywords Adiposity · AIDS · Atrial fibrillation · Body mass index · Cardiovascular diseases · Cardiovascular risk factors · Carotid intima–media thickness · Cerebral small-vessel disease · Cognitive impairment · Coronary artery disease · Diabetes · Dyslipidemia · HIV, human immunodeficiency virus · HIV-associated neurocognitive disorders (HAND) · Hypertension · LDL cholesterol · Metabolic syndrome · Obesity · Stroke · Vascular cognitive impairment (VCI) · White-matter hyperintensities

Abbreviations

AF	Atrial fibrillation
AIDS	Acquired immunodeficiency syndrome
BBB	Blood–brain barrier
BMI	Body mass index
BP	Blood pressure systolic (SBP) and diastolic (DBP)
CAD	Coronary artery disease
cART	Combined antiretroviral therapy
cIMT	Carotid intima–media thickness
CSF	Cerebrospinal fluid
CSVD	Cerebral small-vessel disease
CV	Cardiovascular
CVD	CV disease
HAND	HIV-associated neurocognitive disorders
HCV	Hepatitis C virus
HDL-C	High-density lipoprotein cholesterol
HIV	Human immunodeficiency virus
LDL-C	Low-density lipoprotein cholesterol
MetS	Metabolic syndrome
MRI	Magnetic resonance imaging
plVL	Plasma HIV load
PLHIV	Person living with HIV
VCI	Vascular cognitive impairment
WMH	White-matter hyperintensity

Combined antiretroviral therapy (cART) has made it possible for persons living with human immunodeficiency viruses (PLHIVs) to reach advanced age, and half of patients in high-income countries are now 50 years old or older (Costagliola 2014). Nevertheless, aging PLHIVs' life expectancy persistently lags behind that of the general population, predominantly because of their heightened risk for age-related comorbidities, to which they might be more vulnerable (Cohen et al. 2015; Wang et al. 2015). Among those age-related comorbidities, cardiovascular disease (CVD) and risk factors (i.e., hypertension, diabetes, obesity, dyslipidemia, atherosclerosis, and coronary artery disease (CAD)) are highly prevalent (Calò et al. 2013; Lang et al. 2015; Boccara and Cohen 2016). The pathophysiology underlying CVD and the metabolic syndrome (MetS) in this population is complex and involves traditional risk factors, and infection-relateds and ART-related parameters. Mounting evidence obtained from virus-suppressed cART-treated adult PLHIVs has underscored that CVD and MetS count more than conventional HIV-related factors for persistent HIV-associated neurocognitive disorders (HAND) (Bonnet et al. 2013; Wright et al. 2010; Fabbiani et al. 2013).

Indeed, as in the general population, CVD and MetS were strongly and independently associated with poorer cognitive performance in aviremic PLHIVs, suggesting that CV/metabolic comorbidities could play relevant role(s) in HAND pathogenesis in the modern cART era (Foley et al. 2010; Nakamoto et al. 2012). The hypothesis that HIV-attributable brain injury would occur mainly during untreated/uncontrolled infection, and may reflect a combination of historical effects of HIV itself and immunosuppression, as well as the burden of systemic factors, e.g., CVD/MetS, lifestyle factors, and ongoing neuroinflammation, is more and more accepted (Sanford et al. 2019; van Zoest et al. 2018; Makinson et al. 2019). Indeed, most of today's aging PLHIVs are survivors of the pre-cART era, with long durations of HIV infection, more profound past immunodepression, less effective ART, and less strict plasma HIV-RNA control available at that time. All those factors affected epidemiological studies in older PLHIVs due to the immortal bias. Early cART initiation and full viral suppression may preserve long-term brain health for more recently diagnosed and successfully treated PLHIVs. Hence, the inflammation, encephalitis, and neurodegeneration that had been the forces driving brain pathophysiology before the advent of cART might no longer seem to fit the situation seen in successfully cART-treated PLHIVs. CV/metabolic changes – potentially reversible and primarily targeting the neurovascular unit – are now seen as one of the main drivers of HAND clinical–neuropathological manifestations (Gelman 2015). Thus, incomplete virus control mainly reflects HIV-induced brain damage described during the first decades of the pandemic (Sanford et al. 2019; Fleming et al. 2019). In addition to traditional vascular risk factors, current low CD4+ T-cell count and its nadir, viral detectability, and some previous ART were independently and strongly associated with increased CVD risk (Gutierrez et al. 2017). Unfortunately, although newer ART seemed significantly less metabolically toxic than early ART, integrase-strand–transfer inhibitors have a significant deleterious effect on body weight, body mass index (BMI), and glucose metabolism (Kerchberger et al. 2019; Lagathu et al. 2019; Hill et al. 2019).

Management of PLHIVs' traditional CVD risk factors is suboptimal in primary and secondary preventive settings (Hatleberg et al. 2017). In the AGEhIV cohort, blood pressure (BP) and cholesterol levels were above target levels for 42 and 57% of PLHIVs eligible for secondary prevention (Hatleberg et al. 2017). Because cognitive functions seem to be more severely impacted and widespread by CVD and/or MetS effects when their control is insufficient, it has become extremely important to diagnose them, so as to initiate lifestyle changes and to potentially retard those risk factors' influence on cognitive decline in the modern cART era (Akbaraly et al. 2018; Barberger-Gateau et al. 2007). The 2019 UNAIDS world epidemiological data show that 79% of PLHIVs are aware of their seropositivity, 78% of PLHIVs knowing their HIV status are cART-treated, and 86% of those cART-treated PLHIVs have a plasma HIV load (plVL) below the detection threshold (unaids.org). Moreover, it was recently demonstrated that low plVL of 51–200 and 201–500 copies/mL were strongly associated with virological failure (Fleming et al. 2019). Those findings provide support for the European definition of virological failure as persistent plVL of >50 copies/mL (eacsociety.org). Hence, reporting results concerning virologically uncontrolled cART-treated PLHIVs is not really suitable. According to those worldwide results, including for low-income countries, in this review, we draw parallels between CVD and MetS findings based on numerous published studies conducted on the general population and those in cART-treated PLHIVs with sustained immunovirological control to evaluate the real brain-aging processes.

1 Cardiovascular Diseases

The absolute CVD risk increases with aging, and a growing body of evidence is showing a heightened CVD risk for middle-aged PLHIVs, compared to age-matched HIV-uninfected individuals, even after correcting for traditional CVD risk factors (Boccara et al. 2013). If the relative risk of all CVDs is ~1.5-fold higher for PLHIVs than age-matched, HIV-uninfected individuals, a trend toward its decline in virus-suppressed PLHIVs has been observed over recent calendar periods in the modern cART era (Klein et al. 2015). For example, the more recent myocardial infarction incidence for cART-treated male and female PLHIVs in France is 1.12- and 1.99-fold higher, respectively, than the general population (Baldé et al. 2019). Both HIV-related factors and antiretrovirals might contribute independently to enhancing the CVD risk, together with overrepresentation of CVD risk factors (Lang et al. 2012). In the general population, CVDs – including hypertension, CAD, atrial fibrillation (AF), and chronic heart failure – are associated with cognitive impairment and dementia (Abete et al. 2014). Primary and secondary CVD and risk factor prevention might have been factors contributing to the decline of the dementia incidence over the last three decades (Satizabal et al. 2016). At the same time, emerging evidence suggests that cART-treated PLHIVs have worse cognitive performances that might partly be explained by current or past CVD (Wright et al. 2010;

Becker et al. 2009; Schouten et al. 2016). Notably, PLHIVs with well-controlled infections and preexisting CVD had sixfold higher odds of having neurocognitive impairment, after adjustment for age, sex, race/ethnicity, education, location, prior acquired immunodeficiency syndrome (AIDS), and total cholesterol (Wright et al. 2010).

1.1 Atrial Fibrillation

AF has been recognized as the most prevalent sustained arrhythmia (Sepehri Shamloo et al. 2019). In the general population, AF is associated with cognitive impairment independently of stroke and/or a number of shared risk factors (Sepehri Shamloo et al. 2019; Madhavan et al. 2018). According to the Rotterdam study (Ott et al. 1997), for a person with AF, the odds ratios (OR) were 1.7 (95% confidence interval (CI): 1.2–2.5) for developing cognitive impairment or 2.3 (95% CI: 1.4–3.7) for dementia. The advanced mechanism underlying the AF–cognitive impairment relationship is the high frequency of silent cerebral infarcts, but altered cerebral blood flow, cerebral small-vessel diseases (CSVDs), and especially, cerebral microbleeds, vascular inflammation, and genetic factors have also been underscored (Pastori et al. 2019).

According to epidemiological data, the AF incidence for PLHIVs is increasing, likely due to their advancing age and increasing rates of left ventricular hypertrophy (Hsu et al. 2013). Indeed, the AF frequency was higher for a recent PLHIV cohort than matched, uninfected controls in unadjusted analyses (OR, 1.27; 95% CI, 0.99–1.64). However, that association was no longer significant after adjusting for demographic parameters and CVD risk factors and was mainly driven by CD4+ T-cell nadir <200 μL (Sanders et al. 2018). We did not find any study evaluating the AF impact on HAND. A recent meta-analysis found 2.0–5.13% AF prevalence in PLHIVs, with an incidence rate of 3.6/1,000 person-years (Pastori et al. 2019). Low CD4+ T-cell counts (<200–250 cells/μL) and high plVLs were predictors of AF (Pastori et al. 2019).

1.2 Hypertension

In the general population, hypertension is a major public health problem affecting millions of adults worldwide. Major effects of arterial hypertension on the nervous system are related to CSVD, which represents the damage done to small perforating arteries, a well-known pathway to recurrent strokes, and cognitive loss (Blanco et al. 2017). However, the association of BP with cognitive impairment varies considerably with age and duration of follow-up (O'Callaghan and Kenny 2016). Moreover, plotting that association yields a U-shaped curve (Abete et al. 2014; McNicholas et al. 2018). Indeed, both hypertension and hypotension are associated with poorer cognitive performances (Abete et al. 2014; McNicholas et al. 2018). Higher diastolic blood pressure (DBP) – but not systolic blood pressure (SBP) – was previously associated with cognitive impairment (Abete et al. 2014; Tsivgoulis et al. 2009). In contrast, a recent meta-analysis showed that only midlife higher SBP – but not DBP – was associated with the risk of Alzheimer's disease (Lennon et al. 2019). The mechanism by which hypertension leads independently to cognitive decline remains unclear, and sustained hypertension-induced arterial stiffness has been advanced as a possible cause. Atherosclerosis, hypotension, and excessive hypertensive treatment(s) may also induce cerebral hypoperfusion, ischemia, and hypoxemia, in turn leading to neurodegeneration and cognitive impairment (Abete et al. 2014). General population results from recent randomized clinical trials indicate that intensive BP-lowering attenuates cognitive decline (Yang and Williamson 2019). However, continued follow-up of SPRINT-MIND–trial participants is still crucial to evaluate the full spectrum of the effect of intensive BP control (Ambrosius et al. 2014; The SPRINT MIND, Investigators for the SPRINT Research Group, et al. 2019).

The hypertension–HAND relationship in cART-treated adult PLHIVs appears to be as complex as in the general population and also related to the burden of brain white-matter hyperintensities (WMHs), a marker of CSVD (Su et al. 2016). According to SMART study results, antihypertensive drug use was significantly associated with poorer cognitive performance (Wright et al. 2010). However, in another study, multivariable analyses failed to retain the significant hypertension–poorer cognitive performances association found in univariable analyses (Fabbiani et al. 2013). Also, in cross-sectional (Bonnet et al. 2013) and longitudinal studies (Dufouil et al. 2015) on cART-treated PLHIVs with undetectable HIV, hypertension was not associated with HAND in fully adjusted models. Higher DBP was associated with more extensive cerebral WMHs, one of the cardinal primary CSVD lesions (Ambrosius et al. 2014).The role of hypertension in HAND has not yet been fully elucidated. The apparent implication of vascular disease and its risk factors in HAND might suggest that strategies targeting the vascular system might provide effective mechanisms to prevent cognitive loss in PLHIVs.

1.3 Coronary Artery Disease

In the general population, CAD or myocardial infarction is associated with poorer general cognition and loss of verbal fluency, but the pathogenetic mechanism of that association remains elusive (Burkauskas et al. 2018; Sundbøll 2018). CAD per se may lead to AF and heightened platelet activation, which, in turn, might trigger perivascular inflammation in the brain (Burkauskas et al. 2018; Frazier et al. 2014). Overall, it is estimated that cART-treated PLHIVs are at 1.5-fold higher risk of clinical and subclinical CAD, with men having a lower relative risk (1.12) than women (1.99), compared to HIV-negative individuals (Boccara et al. 2013; Baldé et al. 2019). However, a recent study showed that the myocardial infarction relative risk is no longer elevated among PLHIVs with immune recovery and controlled plVL on cART, be they men or women (Baldé et al. 2019). This decreasing trend of the myocardial infarction risk might be due to earlier treatment initiation with less toxic drugs and the resulting improvement of immunological status. Determinations of coronary artery-calcium accumulation indicate that the mean vascular age of >40% of PLHIVs exceeds by 15 years their chronological age (Guaraldi et al. 2009). Results of the Multicenter AIDS Cohort Study (MACS) and CVD substudy showed that abnormally high coronary artery-calcification levels increased the risk of poorer cognitive performance (Becker et al. 2009). No specific study has evaluated the CAD–HAND association.

1.4 Carotid Intima–Media Thickness (cIMT)

In the general population, despite conflicting findings, cIMT, a subclinical marker of atherosclerosis, has usually been associated with diminished cognitive performance in individuals without vascular and neurological diseases (Frazier et al. 2014; Zhong et al. 2012). Some findings suggested an accelerated-atherosclerosis process in cART-treated middle-aged PLHIVs attributable to multiple factors, including higher prevalence (compared to HIV-uninfected individuals) of conventional risk factors, emerging risk factors (chronic inflammation, immune activation, and HIV-infection-related senescence), and a deleterious cART role (Lang et al. 2015).

Indeed, using cIMT measurements to assess PLHIVs' increased atherosclerosis risk, it was found that ~25% of them had carotid plaques significantly associated with three independent risk factors: older age, hypertension, or higher low-density lipoprotein cholesterol (LDL-C) levels (Jeong et al. 2013). A recent meta-analysis showed that middle-aged and aged PLHIVs and their matched HIV-uninfected individuals had similar cIMTs, suggesting that the HIV effect on carotid structure varies primarily according to age (Hanna et al. 2016). However, those findings associated with older PLHIVs may be explained by a survival effect, that is to say, a switch to favorable health behaviors or other parameters enabling their long-term survival despite HIV infection (Hanna et al. 2016). Nonetheless, that parallel in the PLHIVs is in keeping with recent general population findings showing that cIMT is strongly and linearly age-related and that the relationship is not affected by CVD or CV risk factors (van den Munckhof et al. 2018). cIMT was independently associated with PLHIVs' poorer cognitive performance and an independent risk factor for memory impairment at 2 years of follow-up (Fabbiani et al. 2013; Becker et al. 2009; Ciccarelli et al. 2014).

2 Metabolic Syndrome

MetS is characterized by the combination of central obesity, dysglycemia, dyslipidemia, and arterial hypertension. Among the MetS definitions that have been devised, the National Cholesterol Education Program Adult Treatment Panel III (NCEP/ATP III) is applied the most. According to NCEP/ATP III, individuals are considered to have MetS when they meet at least three of the following criteria: (1) waist circumference, as measure of abdominal obesity, of ≥102 cm for men and ≥88 cm for women; (2) triglycerides ≥1.70 mmol/L or being treated for elevated triglyceride levels; and (3) high-density lipoprotein cholesterol (HDL-C) <1.03 mmol/L for men and <1.29 mmol/L for women or being treated for low HDL-C (Calò et al. 2013); BP ≥130/85 mm Hg or on antihypertensive medication; and (Lang et al. 2015) fasting plasma glucose ≥5.6 mmol/L or taking anti-hyperglycemic agents (Daskalopoulou et al. 2006). Unlike the NCEP/ATP III definition, the World Health Organization (WHO) criteria (Alberti and Zimmet 1998) and the European Group for the Study of Insulin Resistance (EGIR) criteria (Balkau et al. 2002) include insulin resistance or hyperinsulinemia, because clear evidence strongly supports that they play a causal role in most patients' MetSs (Reaven 1988).

Therefore, MetS prevalence strongly depends on the studied population (age, sex, ethnicity, etc.) and the definition applied. If approximately one-third of the American and one-fifth of the European general and adult HIV-infected populations have MetSs (Pal and Ellis 2010; Calza et al. 2017), according to the National Health and Nutrition Examination Survey (NHANES) 2003–2006 cohort, MetS prevalence was found to differ according to sex and to increase with age and for men and women, respectively: 20 and 16% <40 years old, 41 and 37% 40–59 years old, and 52 and 54% ≥60 years (Ervin 2009). Their respective MetS prevalences in the French general population are 23.0% and 16.9% (Grundy 2008). It rises even more dramatically as the BMI increases, and overweight men and women were found, respectively, to be >6 and >5.5 times as likely to meet MetS criteria compared to underweight and normal-weight individuals (Ervin 2009; Grundy 2008).

Most studies on the general population reported that MetS and its components negatively impact cognition (Yates et al. 2012; Muller et al. 2007; Yaffe et al. 2004, 2007). However, findings may vary by sex, with men being more affected in some reports (Cavalieri et al. 2010), women in others (Schuur et al. 2010), and some reporting no differences (Hassenstab et al. 2010), was not only not observed in other studies (Muller et al. 2007), but MetS might even be protective against cognitive decline in older adults (van den Berg et al. 2007). Findings were hampered by the short follow-up and, above all, the failure to include or seek the following confounding factors affecting outcomes in many studies: a significantly higher MetS risk for PLHIVs is associated with demographic and behavioral factors, such as sedentary lifestyle and/or absence of leisure activities; having lower education level and/or socioeconomic status; experiencing financial difficulties; or living without a partner (Costa et al. 2019; Wijndaele et al. 2009; Nobre et al. 2018).

Hence, adjustment for socioeconomic factors, rarely included in statistical models, significantly attenuated the MetS–cognition association, highlighting the importance of socioeconomic stratum in identifying and targeting risk factors for an individual's cognitive decline (Akbaraly et al. 2010).

The relative contributions of MetS's primary components to cognition have varied across studies. Hypertension, diabetes, obesity, hypertriglyceridemia, and impaired glucose tolerance have been individually associated with cognitive impairment, but the relationship between each metabolic risk factor and cognition is complex (Noble et al. 2012; Dye et al. 2017; Czuriga-Kovács et al. 2016; Morley 2014). It is unknown if MetS relationships with cognition simply reflect the effects of one or two dominant MetS components rather than a synergistic effect of several of them (Crichton et al. 2012). Hence, the MetS–cognitive impairment association might be mainly driven by the inclusion of individuals with diabetes (Creavin et al. 2012). The underlying mechanisms by which MetS is thought to affect cognition are not completely understood (Borshchev et al. 2019). As recently analyzed in the general population, the clustering of MetS factors may induce structural and functional alterations in the brain vasculature, resulting in the development of CSVD with potential effects on cerebral blood flow and cognition (Mellendijk et al. 2015; Alberti et al. 2009).

One study on PLHIVs found that MetS had an independent significant effect on global neurocognitive deficits, not found among the HIV-uninfected controls (Yu et al. 2019). In that study, MetS was more strongly associated with learning, fine motor skills, and executive function (Yu et al. 2019). However, it was hampered by well-known and powerful confounding effects of several variables on vascular disease markers and cognition (Yu et al. 2019). Sex, ethnicity, educational level (all $p = 0.02$), any lifetime substance abuse disorder, and current and lifetime major depression (both $p < 0.001$) all differed significantly between PLHIVs and the control group (Yu et al. 2019).

2.1 *Diabetes*

Among MetS criteria, diabetes has most constantly been associated with the development of cognitive dysfunction in non-HIV elderly. Although the mechanisms driving that diabetes-related cognitive dysfunction in the general population are not completely elucidated, the main hypothesis is that insulin resistance can somehow lead to CSVD (Yin et al. 2014). However, recent studies have shown a bidirectional association between severe mood disorders and diabetes mellitus, both factors strongly associated with cognitive impairment (Khaledi et al. 2019; Atique-Ur-Rehman and Neill 2019). In the majority of the studies, the potential confounding effect of concomitant depression has not been well-controlled in diabetic patients.

For PLHIVs, ART is the most incriminated risk factor for the development of diabetes mellitus through diverse mechanisms, depending on the drugs, leading to insulin resistance and increased inflammatory status (Noubissi et al. 2018).

However, the high frequency of type 2 diabetes mainly seen with first-generation ART has been mostly resolved with newer molecules (Lagathu et al. 2019). However, insulin resistance is still prevalent, possibly as a result of previous fat alterations, ongoing weight gain observed worldwide, and/or truncal adiposity associated with aging (Lagathu et al. 2019). Diabetes and insulin resistance have been associated with PLHIVs' poorer cognitive abilities, and that relationship was stronger for older adults (Dufouil et al. 2015; Yu et al. 2019; McCutchan et al. 2012; Wright et al. 2015; Yang et al. 2018; Khuder et al. 2019). Increased insulin resistance was associated with more severe cognitive impairment in the Hawaii Aging with HIV Cohort studies (Valcour et al. 2005, 2006). However, in those studies, 50% of PLHIVs had detectable plVL – a situation that hampers extrapolation to current virus-controlled PLHIVs (Khuder et al. 2019; Valcour et al. 2005, 2006). Diabetes in cART-treated PLHIVs, with plVL <500 copies/mL, seems to negatively impact all cognitive domains, including memory, executive functions, attention, psychomotor speed, language, and manual dexterity (Dufouil et al. 2015; Yu et al. 2019). Notably, PLHIVs with well-controlled plasma virus replication (84% with HIV RNA <50 copies/mL) exhibited the same findings, with the exception of the language domain (Fabbiani et al. 2013). Over 2 years of follow-up and based on only a limited number of tests, diabetic PLHIVs had slightly greater deterioration of executive and memory functions (Dufouil et al. 2015). As for the general population, the mechanism through which diabetes can engender cognitive decay remains to be fully elucidated in PLHIVs: damage to the cerebral arteries or impaired brain-glucose metabolism. Another possible explanation is altered permeability of the blood–brain barrier (BBB) by HIV proteins, thereby potentially increasing brain exposure to higher glucose levels (McCutchan et al. 2012). Nakamoto et al. (2012) observed abnormal microstructural caudate and hippocampus changes associated with irregular glucose metabolism in aging PLHIVs. According to a recent study, diabetic PLHIVs have significantly greater WMH volumes, a CSVD feature, than nondiabetic PLHIVs and HIV-uninfected individuals (Wu et al. 2018). WMHs are associated with worse cognitive performances by cART-treated PLHIVs (Su et al. 2016).

2.2 *Obesity*

A growing body of evidence shows that, for the general population, being overweight or obese in midlife, as assessed with BMI or central adiposity (waist circumference), and a rising BMI curve throughout life, have been associated with brain atrophy, white-matter changes, BBB integrity, and risk of all-cause late-onset dementia and Alzheimer's disease (Kiliaan et al. 2014; Emmerzaal et al. 2015). Hence, an approximately twofold higher risk of Alzheimer's disease is associated with midlife obesity defined as BMI >30 kg/m^2 (Kivipelto et al. 2005). High late-life BMI is associated with a lower risk of cognitive decline and dementia, and the cognition–central obesity relationship for women is complex and nonlinear

(McCutchan et al. 2012; Kerwin et al. 2011; Cova et al. 2016). Executive function and speed-of-information processing are the cognitive domains most commonly associated with high BMI and central adiposity (Okafor et al. 2017).

An increasing obesity rate for cART-treated PLHIVs in the USA was reported, and recent results based mostly on men (70%) showed that two-thirds of PLHIVs were either overweight (~37%) or obese (~28%) (Becofsky et al. 2016). In France, the obesity frequency is much lower, 5% for men and 15% for women (Pourcher et al. 2015). The relationship between elevated BMI obesity and neurocognitive impairment in HIV-infected persons is complex and sometimes contradictory (McCutchan et al. 2012; Gustafson et al. 2013), probably because the odds of having diabetes and hypertension are three and two times higher, respectively, for obese PLHIVs than normal-weight PLHIVs (Becofsky et al. 2016). However, one study's authors considered abdominal obesity per se to be associated with HAND in PLHIVs because the relationship between overweight or obesity and neurocognitive impairment was not attenuated when adjusting for these cardiometabolic factors (Sattler et al. 2015). Obesity – but not overweight – was associated with slowed cognitive processing compared to normal-weight PLHIVs (Okafor et al. 2017). However, that report did not specify the percentage of cART-treated aviremic PLHIVs, whose sample included 40% of alcohol- and illicit drug-addicted PLHIVs and 34% with current hepatitis C virus (HCV) infection.

A study on the MACS cohort, which included immunovirologically well-controlled cART-treated PLHIVs, identified no association between regional adipose tissue or anthropometric measurements and neuropsychological evaluation parameters (Lake et al. 2015). In contrast, a recent study also on the MACS cohort found, in its cross-sectional part, that the higher the adiposity, the lower the motor-speed-based cognitive functions (Rubin et al. 2019). Therein, BMI and central obesity were solely associated with motor-function decline evaluated with only one test, whereas it is highly recommended that each cognitive domain be evaluated with at least two different tests (Carey et al. 2004). Analysis of CHARTER study data found a protective effect of higher BMI on a global assessment of neurocognitive impairment, whereas larger waist circumference was predictive of neurocognitive impairment after correcting for a diagnosis of AIDS, diabetes, or elevated serum triglycerides (McCutchan et al. 2012; Sattler et al. 2015). However, McCutchan et al. (2012) reported detectable plasma and cerebrospinal fluid (CSF) HIV loads in 35% and 17% of cART-treated PLHIVs, respectively. Sattler et al. (2015) found that 20% of PLHIVs had detectable plVLs, but obese and nonobese groups may not have been well-matched, thereby introducing bias, because sex, plVLs, and current CD4+ T-cell counts differed significantly between groups. Moreover, the third nonobese group's BMI quartile was 31 kg/m^2 as opposed to the first obese group's quartile of 22 kg/m^2 (Sattler et al. 2015).

Another study that evaluated specific neurocognitive domains of HIV-infected women found no significant association between midlife obesity, compared to normal weight, for processing speed (Gustafson et al. 2013). However, obesity was associated therein with better performance on the Trail-Making Test B, but worse performance on the Stroop Interference task, two assessments exploring executive functions. A study on HIV-infected men in the MACS cohort found longitudinal associations between BMI obesity and less decline in motor function with increasing age versus those with a normal ($p = 0.04$) or overweight ($p = 0.05$) BMI (Rubin et al. 2019). A similar trend ($p = 0.07$) was observed between central obesity and motor function (Rubin et al. 2019). A critical review of that MACS report also raised its hypothetical nature because of the absence of a temporal association between HIV-disease markers and adiposity (Cysique et al. 2019). All those findings may also be explained by differences in study populations and the presence of other cofounding comorbidities, including current HCV coinfection, for instance (Okafor et al. 2017). Systemic and central nervous system immune-inflammatory factors were thought to mediate the relationship between increased body fat or obesity and cognitive impairment (Sattler et al. 2015), but that hypothesis was not confirmed by other studies (Okafor et al. 2017). Therefore, further studies are warranted to determine the potential impact of BMI on PLHIVs' cognition.

2.3 *Dyslipidemia*

The dyslipidemia–cognition relationship in the general population is complex and varies considerably, depending mainly on the age at which cholesterol is measured. Longitudinal and cross-sectional studies have revealed associations between high LDL-C or total cholesterol levels and cognitive impairment (Yaffe et al. 2002; Ma et al. 2017). In contrast, other large studies on older adults found an association between higher triglycerides (Yin et al. 2012) or LDL-C and better cognitive performance (Lv et al. 2016).

Dyslipidemia is common in PLHIVs. Hypercholesterolemia was associated with poorer cognitive performance in the INSIGHT SMART study (Wright et al. 2010). Longitudinal study results showed an association between dyslipidemia and cognitive impairment in cART-treated PLHIVs with well-controlled HIV infection (Ciccarelli et al. 2014; Sacktor et al. 2016; Mukerji et al. 2016). Indeed, within the MACS cohort, hypercholesterolemia almost tripled the risk for worsening HAND stage during the 4 years of follow-up (Sacktor et al. 2016). For PLHIVs with elevated cholesterol levels, higher HDL-C and statin use were associated with a slower rate of cognitive decline (Mukerji et al. 2016). Hyperlipidemia was weakly associated with HAND in older (>55 years) but not younger PLHIVs in a cross-sectional study on PLHIVs, whose cART failed to achieve virus suppression (Fogel et al. 2015). However, blood HDL- and LDL-C levels were not significantly associated with HAND in the CHARTER cohort (McCutchan et al. 2012). Notably, ancillary CHARTER studies often included high percentages of PLHIVs with unsuppressed plasma and CSF HIV RNA. However, in that study, one-third of the patients had detectable plasma and 17% CSF HIV loads. Pertinently, those studies may not be representative of PLHIVs in Western countries, the vast majority of whom receive suppressive cART. Moreover, 86% of PLHIVs on cART achieved viral suppression in the 2019 worldwide UNAIDS data (aidsinfo.unaids.org), confirming that their data are not applicable to a majority of countries today.

3 The Concept of Vascular Cognitive Impairment (VCI)

VCI refers to a syndrome with evidence of clinical stroke or subclinical vascular brain pathologies that contribute to any degree of cognitive impairment affecting at least one cognitive domain, and ranging from subjective cognitive decline to dementia, the most severe form of VCI (van der Flier et al. 2018; Gorelick et al. 2011). Diagnosing VCI involves two steps: establishing the presence of a cognitive impairment by a comprehensive neuropsychological assessment and confirming that it results directly from brain vascular injury (Gorelick et al. 2011). However, VCI diagnosis is hampered by the lack of standardization of diagnostic criteria (van der Flier et al. 2018; Wiederkehr et al. 2008a, b; Sachdev et al. 2014). Hence, the prevalence of early-onset vascular dementia (in patients <65 years old) ranges from 3.1 to 44% in several studies (Cosentino et al. 2004), with those differences mainly reflecting inconsistent diagnostic criteria, diverse sampling methods, and patient or country demographic disparities (Khan et al. 2016).

To date, VCI-diagnostic criteria have been proposed by at least six organizations, each with disparate levels of proof and agreement: the National Institute of Neurological Disorders and Stroke (NINDS) and the Association Internationale pour la Recherche et l'Enseignement en Neurosciences (AIREN), the State of California Alzheimer's Disease Diagnostic and Treatment Centers (ADDTC), Diagnostic and Statistical Manual of Mental Disorder-5 (DSM-5), the American Heart Association (AHA) and the American Stroke Association (ASA) scientific statement on vascular contributions to cognitive impairment and dementia, the Vascular Impairment of Cognition Classification Consensus Study (VICCCS-2), and the Vascular Behavioral and Cognitive Disorders (VASCOG) (van der Flier et al. 2018; Gorelick et al. 2011; Sachdev et al. 2014; Khan et al. 2016). The patient, an informant, or clinician must note "subjective cognitive complaints" as a recent requirement specific to the VASCOG and DSM-5 criteria.

The neuroimaging criteria, available for VASCOG, NINDS–AIREN, and ADDTC criteria, differ markedly. Reviews of clinical–pathological studies comparing various diagnostic criteria for vascular dementia showed that their sensitivity (average 50–56%) and specificity (range 64–98%, average 87%) varied widely, with interrater reliability differing broadly, and thus their incidence, prevalence, and frequency rates as well (Wiederkehr et al. 2008a, b; Jellinger 2013). Devised after those reviews, the VASCOG and DSM-5 criteria appear to be more sensitive, while the NINDS–AIREN criteria are more specific and recommended for research purposes (van der Flier et al. 2018; Sachdev et al. 2014). Aiming for simplification, the VICCS-2 criteria have proposed referring only to mild or major VCI (Skrobot et al. 2018), with mild VCI defined as impairment of at least one cognitive domain and mild-to-no impairment of activities of daily living, and major VCI referring to significant deficits in at least one cognitive domain and severe disruption of those activities. However, several teams invalidated the impairment in at least one cognitive domain criterion for VCI because it resulted in a high false-positive rate, even when using the fifth percentile cutoff (i.e., 1.65 standard deviation [SD]) to dichotomize each domain score and proposed retaining at least two impaired cognitive domains (Barbay et al. 2018a; Godefroy et al. 2013). The choice of the threshold also significantly influences the true prevalence of cognitive impairment, with a strong benefit for the fifth percentile cutoff (Barbay et al. 2018a, 2018b). The frequently used 1.5-SD cutoff provides similar estimated prevalences at the cost of slightly more false positives (Barbay et al. 2018b). However, the 1-SD cutoff, used to define HAND, induces an unreasonably high false-positive VCI rate (>20–30%) (Barbay et al. 2018b; Antinori et al. 2007; Underwood et al. 2019).

3.1 Stroke

Cerebrovascular diseases are a leading cause of death and disability worldwide. Vascular risk-factor impact on the brain can result in clinically silent disease or overt stroke, all of which lead to cognitive impairment and frailty. During HIV infection, ischemic strokes are much more common than cerebral hemorrhage, the latter affecting mainly PLHIV intravenous-drug abusers (Rasmussen et al. 2011). Currently, the TOAST (Trial of ORG 10172 in acute stroke treatment) classification scheme is the most widely accepted ischemic stroke-subtyping system internationally (Adams and Biller 2015). This simple system describes five major stroke subtypes: large artery atherosclerosis, cardioembolism, CSVD, stroke of other determined cause, and stroke of undetermined cause. The authors of several studies reported an association between HIV infection and the risk of stroke (Rasmussen et al. 2011; Benjamin et al. 2012; Chow et al. 2014a, b; Sico et al. 2015; Ovbiagele and Nath 2011; Moulignier et al. 2015). According to a cART-era meta-analysis, the rate of stroke during a 4-year median period was 1.78% for PLHIVs, whereas ischemic stroke incidence for non-HIV individuals <64 years old was 0.08% per year and 0.24% at 4 years (D'Ascenzo et al. 2015). However, the results of those studies were rather heterogeneous in terms of methodology, ethnicity, HIV–HCV coinfection, and drug abuse.

Ethnic differences in stroke risk are widely recognized and are an important parameter of incidence variability, even greater for younger populations (Chong and Sacco 2005). Hence, in a large clinical cohort study conducted in the Southeastern USA with a majority of African–American patients, the ischemic stroke incidence was 1.5 times the rate of a population-based cohort in North Carolina (Vinikoor et al. 2013). In Europe, PLHIVs have a slightly lower stroke incidence than in the USA, consistent with lower rates seen in the general population of Europe compared to the USA (Vinikoor et al. 2013; Corral et al. 2009). More recent studies revealed a decline in the stroke risk among PLHIVs in recent years after the introduction of cART (Alvaro-Meca et al. 2017; Lin et al. 2019). There may be many reasons for this decline: adverse events associated with older antiretroviral regimens or more effective cART in the recent era, higher CD4+ T-cell nadir, lower frequency of opportunistic infections, and overestimated misdiagnoses due to limited diagnostic tests during the earlier period of the infection. Hence, as observed for myocardial infarction (Baldé et al. 2019), ischemic stroke incidence in PLHIVs with high CD4+ T-cell counts or low HIV RNA is similar to that of HIV-negative individuals (Marcus et al. 2014).

Hypertensive CSVD and large artery atherosclerosis are the main risk factors for these infarcts in PLHIVs (Du and Xu 2019). A higher risk of stroke of undetermined etiology in cART-treated PLHIVs compared to age-matched uninfected individuals found in one study (Chow et al. 2017) was hampered by the failure to comply with the thorough investigations recommended in this setting (Vizzardi et al. 2013; Hart et al. 2017), with 63% immunovirologically uncontrolled PLHIVs within the 6 months preceding the event, 47% of African–American individuals and potential factors contributing to stroke (e.g., 38% cocaine and 17% methamphetamine current users). Indeed, according to the Baltimore–Washington Cooperative Young Stroke Study, any drug use was found for 22% of strokes occurring in young individuals, and African–Americans were significantly more likely to have recent drug use than Caucasians (Chong and Sacco 2005; Sloan et al. 1998).

The results of numerous studies on the general population have highlighted that the prevalence of cognitive impairment after stroke is high, especially after stroke recurrence, and demographic factors, like age, education, and occupation, as well as hypertension, diet, and physical activity, all play significant roles (Lo Coco et al. 2016). Most recent findings support the notion that cognitive impairment in elderly individuals with stroke is likely the result of a synergistic interplay between neurodegenerative changes that progressively occur in many cerebral regions and vascular injuries. We did not find any study evaluating the stroke impact on HAND, but it should not be different from what is observed in the general population.

3.2 *Cerebral Small-Vessel Disease*

CSVD is defined by a range of neuroimaging, pathological, and associated clinical abnormalities, which may be associated with subtle physical and cognitive function deficits that often go unnoticed, thought to originate from a disease process modifying perforating cerebral arterioles, capillaries, and venules (Wardlaw et al. 2013a; Cuadrado-Godia et al. 2018; Li et al. 2018). CSVD is one of the major mechanisms underlying cognitive decline, frailty, and altered gait, and it is the second cause of dementia in seniors (Wardlaw et al. 2013a; Pantoni 2010; Charidimou et al. 2016; Greenberg 2006). It has been implicated in 25% of strokes and as more than doubling the risk of stroke recurrence (Pantoni 2010). CSVD is significantly associated with neurodegenerative diseases (e.g., Alzheimer's disease) and might even be the primary mechanism of cognitive decline attributed to vascular risk factors (Wardlaw et al. 2013a). Although the best known magnetic resonance imaging

(MRI) characteristics of CSVD are WMHs of presumed vascular origin, silent brain infarcts and cerebral microbleeds, recent small subcortical infarcts, and prominent perivascular spaces were more recently identified as being attributable to CSVD (Wardlaw et al. 2013b). To better characterize and differentiate CSVD-surrogate WMHs from WMHs of another origin, the STandards for ReportIng Vascular changes on Euroimaging (STRIVE) criteria have attempted to standardize CSVD neuroimaging (Wardlaw et al. 2013b). For the most part, the pathogenetic mechanism underlying CSVD remains unknown, and data are still insufficient to establish whether inflammation is causal of, or secondary to, CSVD (Low et al. 2019).

Age and hypertension are the most frequently cited major risk factors, while CVD and MetS are the more discussed risk factors (Wardlaw et al. 2013a; Pantoni 2010). The role of dyslipidemia has not yet been clearly elucidated, with hyperlipidemia associated with either a higher (Lin et al. 2017) or lower WMH burden (Jimenez-Conde et al. 2010). A proof-of-principle study found no difference between WMH volumes in diabetic and matched nondiabetic individuals (Utrecht Vascular Cognitive Impairment Study Group et al. 2018). Recent evidence strongly suggests that CSVD development and progression may result from altered immune homeostasis in the central nervous system, leading to endothelial failure, astrogliosis, and microglial activation and, thus, to neurovascular unit dysfunction (Kaiser et al. 2014; Shoamanesh et al. 2015).

The authors of numerous publications have described a vascular WMH–cognitive deterioration association in the general population (Pantoni et al. 2015; Chui and Ramirez 2017; Cannistraro et al. 2019). The results of longitudinal studies suggested that individuals with less white-matter integrity of vascular origin at baseline are more likely to progress to cognitive impairment or Alzheimer's disease than are those with more white-matter integrity (Zhuang et al. 2012; Wardlaw et al. 2015). However, while extensive WMHs (Fazekas grade 3) are associated with VCI, the relationship between mild-to-moderate WMH burden (Fazekas grade 1–2) and cognition, especially when it is the sole CSVD parameter, is much less clear (Jellinger 2013; Dey et al. 2019). Moreover, as mentioned above, in the general population, no uniform diagnostic criteria exist for CSVD-associated cognitive impairment, and predicting which individuals with CSVD will progress to cognitive decay has proven difficult (Cannistraro et al. 2019). MRI and serum biomarkers are currently being investigated but are not yet available for routine use in clinical practice. Cognitive reserve significantly affects the strength of the association between WMHs and cognitive change over time, with less education associated with a stronger influence of WMHs on processing speed, while higher education attenuated this relationship (Jokinen et al. 2009, 2016). Finally, differentiating CSVD cognitive decline from mood disorder is also a clinical challenge. Hence, as in the general population, acknowledging WMHs as the link to cognitive decay in PLHIVs requires controlling numerous confounding factors.

CSVD prevalence is doubled in cART-treated immunovirologically suppressed middle-aged PLHIVs compared to age-matched HIV-uninfected individuals (Moulignier et al. 2018; Trentalange et al. 2018). Advancing age and hypertension – the two major risk factors for WMHs in the general population – were also identified in the majority of studies on PLHIVs (Su et al. 2016; Moulignier et al. 2018; van Zoest et al. 2018), sometimes along with the effects of AIDS or prior profound immunodeficiency (Moulignier et al. 2018; van Zoest et al. 2018), neuroinflammation (van Zoest et al. 2018), or diabetes (Wu et al. 2018). The lack of a detrimental effect of exposure to any ART-drug class on CSVD risk was reported in two studies analyzing the determinants of WMHs of presumed vascular origin in cART-treated PLHIVs with sustained and suppressed plVLs and immune recovery, after accounting for traditional and HIV-specific CV risk factors (Su et al. 2016; Januel et al. 2019). On the other hand, a significant association between protease inhibitors and CSVD was found in two autopsy studies (Soontornniyomkij et al. 2014; Morgello et al. 2014). However, those studies were differently hampered by the lack of a matched control group (Soontornniyomkij et al. 2014) and the time since HIV diagnosis/on cART (Morgello et al. 2014), insufficient data on CD4+ T-cell counts and/or plVLs (Soontornniyomkij et al. 2014), the high percentage of immunovirologically uncontrolled PLHIVs (Soontornniyomkij et al. 2014; Morgello et al. 2014), and potential contributors to CSVD (e.g., current cocaine use and HCV coinfection) (Soontornniyomkij et al. 2014; Morgello et al. 2014).

MRI-visualized white-matter injury in PLHIVs is associated with different underlying pathological features, and WMHs of presumed vascular origin, strictly defined by STRIVE criteria, are only one of them (Su et al. 2016; Moulignier et al. 2018). CVSD-surrogate WMHs are usually distributed bilaterally in the white matter, including the pons and brainstem, and are also seen in deep gray matter. They appear hyperintense compared to the normal brain on T2-weighted or fluid-attenuated inversion recovery (FLAIR) MRI and can be patchy or confluent depending on their severity stage (Fazekas et al. 1987; Shi and Wardlaw 2016). Imaging studies of HIV encephalopathy show diffuse bilateral and symmetric periventricular WMHs, preferentially affecting the more central white matter, without mass effect or enhancement, far different from CSVD (Sarbu et al. 2016). White-matter changes noted during the pre- and early-ART era were associated with the AIDS–dementia complex, defined as cognitive, motor, and behavioral neurological impairments not attributable to opportunistic infections (Navia et al. 1986), and were histologically characterized by myelin pallor, gliosis, and leukoencephalopathy (Davies et al. 1997; Everall et al. 2005). Should those changes remain predominant in the post-cART era, they might possibly reflect the remaining legacy of HIV infection preceding the cART era, ART toxicities, or medical comorbidities (Jensen et al. 2019). Those WMHs have been correlated with lower fractional anisotropy values, consistent with disrupted organization of white-matter tracts (Kochunov et al. 2007). Diffusion-tensor imaging and fractional anisotropy-characterized WMHs in cART-treated PLHIVs with fully controlled HIV infections were independently associated with hypertension and also higher concentrations of monocyte-activation biomarkers in CSF, suggesting that WMHs likely reflect the influence of ongoing neuroinflammation independent of HIV infection (van Zoest et al. 2018). Those findings have been reproduced in transgenic rodent models, and increased mean diffusivity in diffusion-tensor imaging and lower fractional anisotropy values within the corpus callosum were associated with demyelination and increased space between white-matter tracts and pathological hallmarks of HIV encephalopathy (Lentz et al. 2014). Advanced etiologies include demyelination, neuroinflammation/oxidative stress, synaptodendritic injury, ART toxicities, and microvascular alterations (Alakkas et al. 2019).

The neuropathological features of CVSD-surrogate WMHs include axonal loss, enlargement of perivascular spaces, thicker vessel walls, gliosis, myelin loss, increased expression of several hypoxia markers, afferent arteriolar tortuosity, decreased vessel densities, and microglial activation (Gouw et al. 2011). They are caused by arteriosclerosis, lipohyalinosis, and fibrinoid necrosis of small vessels, thereby describing endothelial proliferation, tunica media degeneration, and overall small-vessel wall-thickening (Paradise and Sachdev 2019; Schreiber et al. 2019). Only a few studies have been able to clearly differentiate CVSD-surrogate WMHs from other potential causes of WHMs in HIV-infected persons (Su et al. 2016; Moulignier et al. 2018; Watson et al. 2017). Hence, according to Watson et al. (2017), WMHs typical or atypical of CSVD and mixed distribution types were, respectively, observed in 41%, 32%, and 27% of PLHIVs.

Clinical MRI studies have often focused on separate aspects of CSVD, such as WMHs, and found only weak associations with clinical symptoms (Wardlaw et al. 2013a, 2015; Shi and Wardlaw 2016; Gouw et al. 2011; Paradise and Sachdev 2019). The results of several studies documented the cognitive impact of WMHs on middle-aged and aged cART-treated PLHIVs with long-term virus suppression (Sanford et al. 2019; Su et al. 2016; van Zoest et al. 2018; Watson et al. 2017; Schouten et al. 2016). However, in most people, MRI brain-damage markers of CSVD do not occur alone. Hence, numerous studies on the general population have emphasized that not only WMHs but also lacunes and microbleeds contribute to clinical symptoms, like cognitive decline (Gouw et al. 2011). According to Wardlaw et al. (Wardlaw et al. 2013b, 2015; Shi and Wardlaw 2016), the effect(s) of their combined presence are a better predictor of the cumulative CSVD impact on the brain than each feature alone. The WMHs of presumed vascular origin are the main CSVD feature in PLHIVs, and microbleeds and/or silent brain infarcts are less frequent (Moulignier et al. 2018). Therefore, CSVD should be considered a "whole-brain disease" and MRI markers assessed together with frequently coexisting large-vessel infarcts, to improve understanding of the mechanisms involved in VCI (Gouw et al. 2011).

3.3 *Clinical Implications for HAND Criteria*

The discrepancies among findings, some showing a deleterious and others a beneficial (or no impact) of CVD and/or MetS on cognitive function, may be partly explained by the majority of studies on cART-treated PLHIVs having been conducted using HAND criteria. According to those criteria, estimates of the burden of mild forms of cognitive impairment in PLHIV-cohort studies ranged from 5% to >80% (Vivithanaporn et al. 2010; Simioni et al. 2010). The main stumbling block of those Frascati criteria (Antinori et al. 2007) is that they label >30% of a normative reference population as cognitively impaired (Underwood et al. 2018), which yields a substantial overestimation of the true proportion of affected participants in a study population, due to their lower-than-expected specificity (On Behalf of the POPPY Study Group et al. 2016; McDonnell et al. 2014; Meyer et al. 2013). Hence, evidence is mounting in support of not using the HAND criteria in the modern cART era in resource-rich settings, and updated standards, with a threshold of impairment referring to validated normative datasets, are sorely needed (Schouten et al. 2016; Underwood et al. 2018; McDonnell et al. 2014; Meyer et al. 2013; Gisslén et al. 2011; Torti et al. 2011; Nightingale et al. 2014; Saloner and Cysique 2017). Indeed, HIV infection is the only disease for which specific cognitive norms have been proposed (<1 SD). For other diseases with potential cognitive impact also affecting young and middle-aged patients, e.g., multiple sclerosis or inflammatory systemic diseases, standardized neurologically defined norms are currently used (Palta et al. 2016).

Cerebrovascular lesions are frequent in elderly PLHIVs. However, their impact on cognition is not that clear, and while VCI-prevalence rates seem to be high in clinical studies, CSVD is rarely found to correspond neuropathologically to clinical cognitive impairment in postmortem studies. Indeed, CSVD was not described in the National NeuroAIDS Tissue Consortium (NNTC) paper, based on 589 brain samples collected from 1999 to 2008, and only minimal histological changes were correlated to HAND (Everall et al. 2009).

As in the general population, the pathophysiology behind characteristic MRI CSVD findings is still subject to controversy (McAleese et al. 2016). Whether or not CSVD is the leading cause of VCI in the general population and frequent in PLHIVs, works on the mechanisms of inflammation-induced neuronal insults cannot be ignored on the basis of those findings (Caruana et al. 2017). Recent evidence suggests that brain exposure to HIV can directly or indirectly modulate the amyloid and tau pathways (Canet et al. 2018). Neuropathology studies on aged PLHIVs often demonstrated the presence of insoluble protein aggregates that are found in aged brains of the general population, such as β-amyloid, hyperphosphorylated tau, or α-synuclein, in addition to vascular pathologies (Canet et al. 2018). Studies evaluating WMHs and cognition are mainly cross-sectional, so they can only show association rather than directly examine causality.

The numerous VCI criteria have not been validated in the setting of HIV infection. Moreover, recent findings showing a WMH–HAND association tend to confuse CVSD-surrogate WMHs of presumed vascular origin and WMHs of another origin (Lescure et al. 2013; Langford et al. 2002; Gongvatana et al. 2011; Nir et al. 2014). Indeed, WMHs are heterogeneous, and structural white-matter abnormalities observed on brain MR images are not always linked with CSVD (Desai and Mullins 2014; Kanekar and Devgun 2014; Moritani et al. 2014).

With increasing availability of brain MRI, numerous WMHs are being visualized, and distinguishing WMHs due to CSVD from those of multiple sclerosis, other inflammatory brain diseases or metabolic leukodystrophies can be challenging, in the general population and PLHIVs (Alber et al. 2019). While brain MRI is valuable to determine the extent of WMHs, its use at elucidating what might be the HIV-related neuropathology substrate in virally suppressed cART-treated PMHIVs is less clear. Some studies on PLHIVs clearly excluded WMHs of presumed vascular origin, retaining only diffuse WMHs, and showed a significant association among the latter, HIV escape in the CSF and cognitive impairment (Kugathasan et al. 2017).

Studies exploring radiological–neuropathological relationships have shown that white-matter loss and WMHs are closely correlated with cytokines and synaptodendritic injury markers, like neuronal microtubule-associated protein-2 (MAP-2), particularly when HIV encephalitis is present (Ellis et al. 2007). A recent meta-analysis of diffusion-tensor imaging, widely used to assess HIV effects on white-matter microarchitecture, has revealed high heterogeneity between studies and relatively small differences (O'Connor et al. 2017). Resting-state functional MRI showed that HAND in virus-suppressed PLHIVs are associated with significantly decreased brain-connectivity networks in the absence of vascular injury (Chaganti et al. 2017). Hence, findings in cART-treated aviremic PLHIVs underlined that widespread WMHs are more suggestive of white-matter demyelination, as described with HIV encephalitis/encephalopathy, than WMH of presumed vascular origin (van Zoest et al. 2018; Underwood et al. 2017).

In the healthy general population, isolated WMHs are also highly prevalent, and it remains unclear whether WMHs are always pathological or clinically insignificant and part of physiological aging (Das et al. 2019). The subcortical or periventricular WMH-distribution pattern is also important for the diagnosis of their vascular origin and rarely described in the HIV literature (Wardlaw et al. 2013b; Alber et al. 2019; Das et al. 2019). Punctate WMHs probably result from a variety of causes and have relatively low risk for further progression (Das et al. 2019). On the other hand, confluent WMHs are likely to progress more aggressively (Das et al. 2019). Only the STRIVE criteria are recognized as the gold standard for CSVD-associated WMHs (Wardlaw et al. 2013b). Finally, mild cognitive impairments according to HAND and VCI criteria differ inherently and do not overlap. Indeed, mild forms of HAND are defined by at least two cognitive domains with $\geq$1.0 SD of cognitive impairment, mild VCI by only at least one cognitive domain but with $\geq$1.65 SD of cognitive impairment in the absence of functional impairment for both criteria (Skrobot et al. 2018; Antinori et al. 2007). To date, no argument supports preferentially choosing VICCS-2 criteria for evaluating HIV-associated cognitive decay rather than the five others.

4 Conclusion

With growing evidence that CVD and MetS probably contribute to cognitive impairment in cART-treated PLHIVs with well-sustained virological control, it is essential to improve early detection and encourage appropriate management of these conditions. However, their precise roles remain to be elucidated because several studies included PLHIVs not virologically suppressed and with major confounding cognitive impairment risk factors, e.g., ethnically diverse cohorts, heterogeneous ages at which reading was acquired and years of education, current HCV infection, or illicit drug abuse (Saloner and Cysique 2017; Marquine et al. 2018; Haddow et al. 2015). That is why we think that only immunovirologically controlled PLHIVs with plVL <50 copies/mL can serve to optimally evaluate the true brain consequences of chronic HIV infection. Moreover, cognitive decline should not be evaluated with HAND criteria but rather with defined neuropsychological norms, as widely acknowledged by neurocognitive specialists who use them in memory clinics (Azam et al. 2016). To date, none of the numerous VCI criteria has been validated for PLHIVs or may be preferentially chosen.

A better definition of WMHs must also be used to identify different WMH subtypes, possibly with different etiologies, outcomes, and clinical significance, which emphasizes the need to comply with the STRIVE neuroimaging standards used in neurovascular trials for CVSD-surrogate WMHs (Wardlaw et al. 2013b). Future studies should also take into account all neuroimaging CSVD markers and include advanced multimodal structural and functional MR sequences with high spatial resolution, to better assess the VCI risk in PLHIVs and its potential participation in cognitive impairment. As Brew (2016) questioned in his editorial in AIDS on Su et al.'s (2016) findings: If HAND is synonymous with VCI, why does evidence, albeit somewhat spotty, support the efficacy of neuro-cART? And why would maraviroc intensification of cART attenuate HAND in sustained immunovirologically suppressed PLHIVs?

Elucidation of cerebrovascular lesion pathophysiology, clarification of characteristic brain-MRI findings, and understanding of the impact of combined parameters are strongly needed to improve the diagnostic accuracy of HIV-associated cognitive impairment. The hypothesis of a unifying pathological mechanism in aviremic aging PLHIVs, based on VCI, albeit attractive, seems too restrictive. Indeed, in the general population, pure VCI forms or neurodegenerative cognitive impairment is very rare, and a mixture of both is the dominant pathology. CSVD- and HIV-driven neurodegenerative processes may interact, either independently of each other or through additive or synergistic effects on PLHIVs' cognitive decline. Does CSVD merely reduce the cognitive threshold needed for overt clinical cognitive impairment, or do both factors potentiate HAND-specific pathophysiological pathways? A more definite answer as to the relative contributions of these factors at a population level, however, will only come from further large, well-designed, longitudinal studies, supported by neuropathological analyses (Brew 2016).

References

Abete P, Della-Morte D, Gargiulo G, Basile C, Langellotto A, Galizia G et al (2014) Cognitive impairment and cardiovascular diseases in the elderly. A heart–brain continuum hypothesis. Ageing Res Rev 18:41–52

Adams HP, Biller J (2015) Classification of subtypes of ischemic stroke: history of the trial of org 10172 in acute stroke treatment classification. Stroke 46(5):114–117
Akbaraly TN, Kivimaki M, Shipley MJ, Tabak AG, Jokela M, Virtanen M et al (2010) Metabolic syndrome over 10 years and cognitive functioning in late midlife: the Whitehall II study. Diabetes Care 33(1):84–89
Akbaraly T, Sexton C, Zsoldos E, Mahmood A, Filippini N, Kerleau C et al (2018) Association of long-term diet quality with hippocampal volume: longitudinal cohort study. Am J Med 131 (11):1372–1381
Alakkas A, Ellis RJ, Watson CW-M, Umlauf A, Heaton RK, Letendre S et al (2019) White matter damage, neuroinflammation, and neuronal integrity in HAND. J Neurovirol 25(1):32–41
Alber J, Alladi S, Bae H-J, Barton DA, Beckett LA, Bell JM et al (2019) White matter hyperintensities in vascular contributions to cognitive impairment and dementia (VCID): knowledge gaps and opportunities. Alzheimers Dement 5:107–117
Alberti KG, Zimmet PZ (1998) Definition, diagnosis and classification of diabetes mellitus and its complications. Part 1: diagnosis and classification of diabetes mellitus provisional report of a WHO consultation. Diabet Med J 15(7):539–553
Alberti KGMM, Eckel RH, Grundy SM, Zimmet PZ, Cleeman JI, Donato KA et al (2009) Harmonizing the metabolic syndrome: a joint interim statement of the international diabetes federation task force on epidemiology and prevention; national heart, lung, and blood institute; American Heart Association; World Heart Federation; International Atherosclerosis Society; and International Association for the Study of Obesity. Circulation 120(16):1640–1645
Alvaro-Meca A, Berenguer J, Díaz A, Micheloud D, Aldámiz-Echevarría T, Fanciulli C et al (2017) Stroke in HIV-infected individuals with and without HCV coinfection in Spain in the combination antiretroviral therapy era. Plos One 12(6):e0179493
Ambrosius WT, Sink KM, Foy CG, Berlowitz DR, Cheung AK, Cushman WC et al (2014) The design and rationale of a multicenter clinical trial comparing two strategies for control of systolic blood pressure: The Systolic Blood Pressure Intervention Trial (SPRINT). Clin Trials J 11 (5):532–546
Antinori A, Arendt G, Becker JT, Brew BJ, Byrd DA, Cherner M et al (2007) Updated research nosology for HIV-associated neurocognitive disorders. Neurology 69(18):1789–1799
Atique-Ur-Rehman H, Neill JC (2019) Cognitive dysfunction in major depression: from assessment to novel therapies. Pharmacol Ther 202:53–71
Azam B, Whitfield TJ, Radford D, Dontham SG, Stevens T, Dannhauser T et al (2016) Trends in referred patient profiles in a memory clinic over 20 years. Dementia 15(4):789–797
Baldé A, Lang S, Wagner A, Ferrières J, Montaye M, Tattevin P et al (2019) Trends in the risk of myocardial infarction among HIV-1-infected individuals relative to the general population in France: impact of gender and immune status. Plos One 14(1):e0210253
Balkau B, Charles M-A, Drivsholm T, Borch-Johnsen K, Wareham N, Yudkin JS et al (2002) Frequency of the WHO metabolic syndrome in European cohorts, and an alternative definition of an insulin resistance syndrome. Diabetes Metab 28(5):364–376
Barbay M, Taillia H, Nédélec-Ciceri C, Bompaire F, Bonnin C, Varvat J et al (2018a) Prevalence of poststroke neurocognitive disorders using National Institute of Neurological disorders and Stroke-Canadian Stroke Network, VASCOG Criteria (Vascular Behavioral and Cognitive Disorders), and optimized criteria of cognitive deficit. Stroke 49(5):1141–1147
Barbay M, Diouf M, Roussel M, Godefroy O, Group G Study (2018b) Systematic review and meta-analysis of prevalence in post-stroke neurocognitive disorders in hospital-based studies. Dement Geriatr Cogn Disord 46(5–6):322–334
Barberger-Gateau P, Raffaitin C, Letenneur L, Berr C, Tzourio C, Dartigues JF et al (2007) Dietary patterns and risk of dementia: the three-city cohort study. Neurology 69(20):1921–1930
Becker JT, Kingsley L, Mullen J, Cohen B, Martin E, Miller EN et al (2009) Vascular risk factors, HIV serostatus, and cognitive dysfunction in gay and bisexual men. Neurology 73 (16):1292–1299

Becofsky KM, Wing EJ, Wing RR, Richards KE, Gillani FS (2016) Obesity prevalence and related risk of comorbidities among HIV+ patients attending a New England ambulatory centre. Obes Sci Pract 2(2):123–127

Benjamin LA, Bryer A, Emsley HC, Khoo S, Solomon T, Connor MD (2012) HIV infection and stroke: current perspectives and future directions. Lancet Neurol 11(10):878–890

Blanco PJ, Müller LO, Spence JD (2017) Blood pressure gradients in cerebral arteries: a clue to pathogenesis of cerebral small vessel disease. Stroke Vasc Neurol 2(3):108–117

Boccara F, Cohen A (2016) HIV and heart disease: what cardiologists should know. Rev Esp Cardiol (Engl Ed) 69(12):1126–1130

Boccara F, Lang S, Meuleman C, Ederhy S, Mary-Krause M, Costagliola D et al (2013) HIV and coronary heart disease. J Am Coll Cardiol 61(5):511–523

Bonnet F, Amieva H, Marquant F, Bernard C, Bruyand M, Dauchy F-A et al (2013) Cognitive disorders in HIV-infected patients: are they HIV-related? AIDS 27(3):391–400

Borshchev YY, Uspensky YP, Galadgudza MM (2019) Pathogenetic pathways of cognitive dysfunction and dementia in metabolic syndrome. Life Sci 237:116932

Brew BJ (2016) Has HIV-associated neurocognitive disorders now transformed into vascular cognitive impairment? AIDS 30(15):2379–2380

Burkauskas J, Lang P, Bunevičius A, Neverauskas J, Bučiūtė-Jankauskienė M, Mickuvienė N (2018) Cognitive function in patients with coronary artery disease: a literature review. J Int Med Res 46(10):4019–4031

Calò LA, Caielli P, Maiolino G, Rossi G (2013) Arterial hypertension and cardiovascular risk in HIV-infected patients. J Cardiovasc Med Hagerstown 14(8):553–558

Calza L, Colangeli V, Magistrelli E, Rossi N, Rosselli Del Turco E, Bussini L et al (2017) Prevalence of metabolic syndrome in HIV-infected patients naive to antiretroviral therapy or receiving a first-line treatment. HIV Clin Trials 18(3):110–117

Canet G, Dias C, Gabelle A, Simonin Y, Gosselet F, Marchi N et al (2018) HIV neuroinfection and Alzheimer's disease: similarities and potential links? Front Cell Neurosci 12:307

Cannistraro RJ, Badi M, Eidelman BH, Dickson DW, Middlebrooks EH, Meschia JF (2019) CNS small vessel disease: a clinical review. Neurology 92(24):1146–1156

Carey CL, Woods SP, Gonzalez R, Conover E, Marcotte TD, Grant I et al (2004) Predictive validity of global deficit scores in detecting neuropsychological impairment in HIV infection. J Clin Exp Neuropsychol 26(3):307–319

Caruana G, Vidili G, Serra PA, Bagella P, Spanu A, Fiore V et al (2017) The burden of HIV-associated neurocognitive disorder (HAND) in post-HAART era: a multidisciplinary review of the literature. Eur Rev Med Pharmacol Sci 21(9):2290–2301

Cavalieri M, Ropele S, Petrovic K, Pluta-Fuerst A, Homayoon N, Enzinger C et al (2010) Metabolic syndrome, brain magnetic resonance imaging, and cognition. Diabetes Care 33(12):2489–2495

Chaganti JR, Heinecke A, Gates TM, Moffat KJ, Brew BJ (2017) Functional connectivity in virally suppressed patients with HIV-associated neurocognitive disorder: a resting-state analysis. Am J Neuroradiol 38(8):1623–1629

Charidimou A, Pantoni L, Love S (2016) The concept of sporadic cerebral small vessel disease: a road map on key definitions and current concepts. Int J Stroke 11(1):6–18

Chong JY, Sacco RL (2005) Epidemiology of stroke in young adults: race/ethnic differences. J Thromb Thrombolysis 20(2):77–83

Chow FC, Bacchetti P, Kim AS, Price RW, Hsue PY (2014a) Effect of CD4+ cell count and viral suppression on risk of ischemic stroke in HIV infection. AIDS 28(17):2573–2577

Chow FC, He W, Bacchetti P, Regan S, Feske SK, Meigs JB et al (2014b) Elevated rates of intracerebral hemorrhage in individuals from a US clinical care HIV cohort. Neurology 83 (19):1705–1711

Chow FC, Price RW, Hsue PY, Kim AS (2017) Greater risk of stroke of undetermined etiology in a contemporary HIV-infected cohort compared with uninfected individuals. J Stroke Cerebrovasc Dis 26(5):1154–1160

Chui HC, Gomez LR (2017) Vascular contributions to cognitive impairment in late life. Neurol Clin 35(2):295–323
Ciccarelli N, Grima P, Fabbiani M, Baldonero E, Borghetti A, Milanini B et al (2014) Baseline CD4+ T-cell count and cardiovascular risk factors predict the evolution of cognitive performance during 2 years follow-up in HIV-infected patients. Antivir Ther 20(4):433–440
Cohen RA, Seider TR, Navia B (2015) HIV effects on age-associated neurocognitive dysfunction: premature cognitive aging or neurodegenerative disease? Alzheimers Res Ther 7(1):37
Corral I, Quereda C, Moreno A, Pérez-Elías M-J, Dronda F, Casado J-L et al (2009) Cerebrovascular ischemic events in HIV-1-infected patients receiving highly active antiretroviral therapy: incidence and risk factors. Cerebrovasc Dis 27(6):559–563
Cosentino SA, Jefferson AL, Carey M, Price CC, Davis-Garrett K, Swenson R et al (2004) The clinical diagnosis of vascular dementia: a comparison among four classification systems and a proposal for a new paradigm. Clin Neuropsychol 18(1):6–21
Costa CRB, Melo ES, Antonini M, Jesus GJ, Pontes PS, Gir E et al (2019) Association between sociodemographic and behavioral factors with metabolic syndrome in people living with HIV. Rev Gaucha Enferm 40:e20180379
Costagliola D (2014) Demographics of HIV and aging. Curr Opin HIV AIDS 9(4):294–301
Cova I, Clerici F, Maggiore L, Pomati S, Cucumo V, Ghiretti R et al (2016) Body mass index predicts progression of mild cognitive impairment to dementia. Dement Geriatr Cogn Disord 41 (3-4):172–180
Creavin ST, Gallacher J, Bayer A, Fish M, Ebrahim S, Ben-Shlomo Y (2012) Metabolic syndrome, diabetes, poor cognition, and dementia in the Caerphilly prospective study. J Alzheimers Dis 28 (4):931–939
Crichton GE, Elias MF, Buckley JD, Murphy KJ, Bryan J, Frisardi V (2012) Metabolic syndrome, cognitive performance, and dementia. J Alzheimers Dis 30(Suppl 2):S77–S87
Cuadrado-Godia E, Dwivedi P, Sharma S, Ois Santiago A, Roquer Gonzalez J, Balcells M et al (2018) Cerebral small vessel disease: a review focusing on pathophysiology, biomarkers, and machine learning strategies. J Stroke 20(3):302–320
Cysique LA, Messinis L, Albert SM (2019) Could excess body weight be good for cognitive health in chronic HIV infection? Neurology 93(3):95–96
Czuriga-Kovács KR, Czuriga D, Csiba L (2016) Influence of hypertension, alone and in combination with other vascular risk factors on cognition. CNS Neurol Disord Drug Targets 15 (6):690–698
D'Ascenzo F, Quadri G, Cerrato E, Calcagno A, Omedè P, Grosso Marra W et al (2015) A meta-analysis investigating incidence and features of stroke in HIV-infected patients in the highly active antiretroviral therapy era. J Cardiovasc Med 16(12):839–843
Das AS, Regenhardt RW, Vernooij MW, Blacker D, Charidimou A, Viswanathan A (2019) Asymptomatic cerebral small vessel disease: insights from population-based studies. J Stroke 21(2):121–138
Daskalopoulou SS, Athyros VG, Kolovou GD, Anagnostopoulou KK, Mikhailidis DP (2006) Definitions of metabolic syndrome: where are we now? Curr Vasc Pharmacol 4(3):185–197
Davies J, Everall IP, Weich S, McLaughlin J, Scaravilli F, Lantos PL (1997) HIV-associated brain pathology in the United Kingdom: an epidemiological study. AIDS 11(9):1145–1150
Desai NK, Mullins ME (2014) An imaging approach to diffuse white matter changes. Radiol Clin North Am 52(2):263–278
Dey AK, Stamenova V, Bacopulos A, Jeyakumar N, Turner GR, Black SE et al (2019) Cognitive heterogeneity among community-dwelling older adults with cerebral small vessel disease. Neurobiol Aging 77:183–193
Du J, Xu Q (2019) Neuroimaging studies on cognitive impairment due to cerebral small vessel disease. Stroke Vasc Neurol 4(2):99–101
Dufouil C, Richert L, Thiébaut R, Bruyand M, Amieva H, Dauchy F-A et al (2015) Diabetes and cognitive decline in a French cohort of patients infected with HIV-1. Neurology 85 (12):1065–1073

Dye L, Boyle NB, Champ C, Lawton C (2017) The relationship between obesity and cognitive health and decline. Proc Nutr Soc 76(04):443–454
Ellis R, Langford D, Masliah E (2007) HIV and antiretroviral therapy in the brain: neuronal injury and repair. Nat Rev Neurosci 8(1):33–44
Emmerzaal TL, Kiliaan AJ, Gustafson DR (2015) 2003–2013: a decade of body mass index, Alzheimer's disease, and dementia. J Alzheimers Dis 43(3):739–755
Ervin RB (2009) Prevalence of metabolic syndrome among adults 20 years of age and over, by sex, age, race and ethnicity, and body mass index: United States, 2003–2006. Natl Health Stat Rep 13:1–7
Everall IP, Hansen LA, Masliah E (2005) The shifting patterns of HIV encephalitis neuropathology. Neurotox Res 8(1-2):51–61
Everall I, Vaida F, Khanlou N, Lazzaretto D, Achim C, Letendre S et al (2009) Cliniconeuropathologic correlates of human immunodeficiency virus in the era of antiretroviral therapy. J Neurovirol 15(5-6):360–370
Fabbiani M, Ciccarelli N, Tana M, Farina S, Baldonero E, di Cristo V et al (2013) Cardiovascular risk factors and carotid intima-media thickness are associated with lower cognitive performance in HIV-infected patients: cardiovascular risk and cognition in HIV. HIV Med 14(3):136–144
Fazekas F, Chawluk JB, Alavi A, Hurtig HI, Zimmerman RA (1987) MR signal abnormalities at 1.5 T in Alzheimer's dementia and normal aging. Am J Roentgenol 149(2):351–356
Fleming J, Mathews WC, Rutstein RM, Aberg J, Somboonwit C, Cheever LW et al (2019) Low level viremia and virologic failure in persons with HIV infection treated with antiretroviral therapy. AIDS 33(13):2005–2012
Fogel GB, Lamers SL, Levine AJ, Valdes-Sueiras M, McGrath MS, Shapshak P et al (2015) Factors related to HIV-associated neurocognitive impairment differ with age. J Neurovirol 21(1):56–65
Foley J, Ettenhofer M, Wright MJ, Siddiqi I, Choi M, Thames AD et al (2010) Neurocognitive functioning in HIV-1 infection: effects of cerebrovascular risk factors and age. Clin Neuropsychol 24(2):265–285
Frazier DT, Seider T, Bettcher BM, Mack WJ, Jastrzab L, Chao L et al (2014) The role of carotid intima-media thickness in predicting longitudinal cognitive function in an older adult cohort. Cerebrovasc Dis 38(6):441–447
Gelman BB (2015) Neuropathology of HAND with suppressive antiretroviral therapy: encephalitis and neurodegeneration reconsidered. Curr HIV/AIDS Rep 12(2):272–279
Gisslén M, Price RW, Nilsson S (2011) The definition of HIV-associated neurocognitive disorders: are we overestimating the real prevalence? BMC Infect Dis 11:356
Godefroy O, Leclercq C, Bugnicourt J-M, Roussel M, Moroni C, Quaglino V et al (2013) Neuropsychological assessment and cerebral vascular disease: the new standards. Rev Neurol (Paris) 169(10):779–785
Gongvatana A, Cohen RA, Correia S, Devlin KN, Miles J, Kang H et al (2011) Clinical contributors to cerebral white matter integrity in HIV-infected individuals. J Neurovirol 17(5):477–486
Gorelick PB, Scuteri A, Black SE, Decarli C, Greenberg SM, Iadecola C et al (2011) Vascular contributions to cognitive impairment and dementia: a statement for healthcare professionals from the American Heart Association/American Stroke Association. Stroke 42(9):2672–2713
Gouw AA, Seewann A, van der Flier WM, Barkhof F, Rozemuller AM, Scheltens P et al (2011) Heterogeneity of small vessel disease: a systematic review of MRI and histopathology correlations. J Neurol Neurosurg Psychiatry 82(2):126–135
Greenberg SM (2006) Small vessels, big problems. N Engl J Med 354(14):1451–1453
Grundy SM (2008) Metabolic syndrome pandemic. Arterioscler Thromb Vasc Biol 28(4):629–636
Guaraldi G, Zona S, Alexopoulos N, Orlando G, Carli F, Ligabue G et al (2009) Coronary aging in HIV-infected patients. Clin Infect Dis 49(11):1756–1762
Gustafson DR, Mielke MM, Tien PC, Valcour V, Cohen M, Anastos K et al (2013) Anthropometric measures and cognition in middle-aged HIV-infected and uninfected women. The Women's Interagency HIV Study. J Neurovirol 19(6):574–585

Gutierrez J, Albuquerque ALA, Falzon L (2017) HIV infection as vascular risk: a systematic review of the literature and meta-analysis. PLoS One 12(5):e0176686

Haddow LJ, Pierce KJ, Daskalopoulou M, Lampe F, Rodger AJ (2015) Neurocognitive change observed in the charter HIV cohort could be due to chance, and may be a cause as well as a consequence of detectable viremia. Clin Infect Dis 60(9):1441–1442

Hanna DB, Guo M, Bůžková P, Miller TL, Post WS, Stein JH et al (2016) HIV infection and carotid artery intima-media thickness: pooled analyses across 5 cohorts of the NHLBI HIV-CVD collaborative. Clin Infect Dis 63(2):249–256

Hart RG, Catanese L, Perera KS, Ntaios G, Connolly SJ (2017) Embolic stroke of undetermined source: a systematic review and clinical update. Stroke 48(4):867–872

Hassenstab JJ, Sweat V, Bruehl H, Convit A (2010) Metabolic syndrome is associated with learning and recall impairment in middle age. Dement Geriatr Cogn Disord 29(4):356–362

Hatleberg CI, Lundgren JD, Ryom L (2017) Are we successfully managing cardiovascular disease in people living with HIV? Curr Opin HIV AIDS 12(6):594–603

Hill A, Waters L, Pozniak A (2019) Are new antiretroviral treatments increasing the risks of clinical obesity? J Virus Erad 5(1):41–43

Hsu JC, Li Y, Marcus GM, Hsue PY, Scherzer R, Grunfeld C et al (2013) Atrial fibrillation and atrial flutter in human immunodeficiency virus-infected persons. J Am Coll Cardiol 61 (22):2288–2295

Januel E, Godin O, Moulignier A, Lescure F-X, Savatovsky J, Lamirel C et al (2019) Brief report: impact of ART classes on the increasing risk of cerebral small-vessel disease in middle-aged, well-controlled, cART-treated, HIV-infected individuals. J Acquir Immune Defic Syndr 81 (5):547–551

Jellinger KA (2013) Pathology and pathogenesis of vascular cognitive impairment – a critical update. Front Aging Neurosci 5:17

Jensen BK, Roth LM, Grinspan JB, Jordan-Sciutto KL (2019) White matter loss and oligodendrocyte dysfunction in HIV: a consequence of the infection, the antiretroviral therapy or both? Brain Res 1724:146397

Jeong SJ, Kim HW, Ku NS, Han SH, Kim CO, Choi JY et al (2013) Clinical factors associated with carotid plaque and intima-medial thickness in HIV-infected patients. Yonsei Med J 54 (4):990–998

Jimenez-Conde J, Biffi A, Rahman R, Kanakis A, Butler C, Sonni S et al (2010) Hyperlipidemia and reduced white matter hyperintensity volume in patients with ischemic stroke. Stroke 41 (3):437–442

Jokinen H, Kalska H, Ylikoski R, Madureira S, Verdelho A, van der Flier WM et al (2009) Longitudinal cognitive decline in subcortical ischemic vascular disease – The LADIS Study. Cerebrovasc Dis 27(4):384–391

Jokinen H, Melkas S, Madureira S, Verdelho A, Ferro JM, Fazekas F et al (2016) Cognitive reserve moderates long-term cognitive and functional outcome in cerebral small vessel disease. J Neurol Neurosurg Psychiatry 87(12):1296–1302

Kaiser D, Weise G, Möller K, Scheibe J, Pösel C, Baasch S et al (2014) Spontaneous white matter damage, cognitive decline and neuroinflammation in middle-aged hypertensive rats: an animal model of early-stage cerebral small vessel disease. Acta Neuropathol Commun 2:169

Kanekar S, Devgun P (2014) A pattern approach to focal white matter hyperintensities on magnetic resonance imaging. Radiol Clin North Am 52(2):241–261

Kerchberger AM, Sheth AN, Angert CD, Mehta CC, Summers NA, Ofotokun I et al (2019) Weight gain associated with integrase stand transfer inhibitor use in women. Clin Infect Dis Off Publ Infect Dis Soc Am. https://doi.org/10.1093/cid/ciz853

Kerwin DR, Gaussoin SA, Chlebowski RT, Kuller LH, Vitolins M, Coker LH et al (2011) Interaction between body mass index and central adiposity and risk of incident cognitive impairment and dementia: results from the Women's Health Initiative Memory Study. J Am Geriatr Soc 59(1):107–112

Khaledi M, Haghighatdoost F, Feizi A, Aminorroaya A (2019) The prevalence of comorbid depression in patients with type 2 diabetes: an updated systematic review and meta-analysis on huge number of observational studies. Acta Diabetol 56(6):631–650

Khan A, Kalaria RN, Corbett A, Ballard C (2016) Update on vascular dementia. J Geriatr Psychiatry Neurol 29(5):281–301

Khuder SS, Chen S, Letendre S, Marcotte T, Grant I, Franklin D et al (2019) Impaired insulin sensitivity is associated with worsening cognition in HIV-infected patients. Neurology 92(12): e1344–e1353

Kiliaan AJ, Arnoldussen IAC, Gustafson DR (2014) Adipokines: a link between obesity and dementia? Lancet Neurol 13(9):913–923

Kivipelto M, Ngandu T, Fratiglioni L, Viitanen M, Kareholt I, Winblad B et al (2005) Obesity and vascular risk factors at midlife and the risk of dementia and Alzheimer disease. Arch Neurol 62 (10):1556–1560

Klein DB, Leyden WA, Xu L, Chao CR, Horberg MA, Towner WJ et al (2015) Declining relative risk for myocardial infarction among HIV-positive compared with HIV-negative individuals with access to care. Clin Infect Dis 60(8):1278–1280

Kochunov P, Thompson PM, Lancaster JL, Bartzokis G, Smith S, Coyle T et al (2007) Relationship between white matter fractional anisotropy and other indices of cerebral health in normal aging: tract-based spatial statistics study of aging. Neuroimage 35(2):478–487

Kugathasan R, Collier DA, Haddow LJ, El Bouzidi K, Edwards SG, Cartledge JD et al (2017) Diffuse white matter signal abnormalities on magnetic resonance imaging are associated with human immunodeficiency virus type 1 viral escape in the central nervous system among patients with neurological symptoms. Clin Infect Dis 64(8):1059–1065

Lagathu C, Béréziat V, Gorwood J, Fellahi S, Bastard J-P, Vigouroux C et al (2019) Metabolic complications affecting adipose tissue, lipid and glucose metabolism associated with HIV antiretroviral treatment. Expert Opin Drug Saf 18(9):829–840

Lake JE, Vo QT, Jacobson LP, Sacktor N, Miller EN, Post WS et al (2015) Adiponectin and interleukin-6, but not adipose tissue, are associated with worse neurocognitive function in HIV-infected men. Antivir Ther 20(2):235–244

Lang S, Mary-Krause M, Simon A, Partisani M, Gilquin J, Cotte L et al (2012) HIV replication and immune status are independent predictors of the risk of myocardial infarction in HIV-infected individuals. Clin Infect Dis 55(4):600–607

Lang S, Boccara F, Mary-Krause M, Cohen A (2015) Epidemiology of coronary heart disease in HIV-infected versus uninfected individuals in developed countries. Arch Cardiovasc Dis 108 (3):206–215

Langford TD, Letendre SL, Marcotte TD, Ellis RJ, McCutchan JA, Grant I et al (2002) Severe, demyelinating leukoencephalopathy in AIDS patients on antiretroviral therapy. AIDS 16 (7):1019–1029

Lennon MJ, Makkar SR, Crawford JD, Sachdev PS (2019) Midlife hypertension and Alzheimer's disease: a systematic review and meta-analysis. J Alzheimers Dis 71(1):307–316

Lentz MR, Peterson KL, Ibrahim WG, Lee DE, Sarlls J, Lizak MJ et al (2014) Diffusion tensor and volumetric magnetic resonance measures as biomarkers of brain damage in a small animal model of HIV. PLoS One 9(8):e105752

Lescure F-X, Moulignier A, Savatovsky J, Amiel C, Carcelain G, Molina J-M et al (2013) CD8 encephalitis in HIV-infected patients receiving cART: a treatable entity. Clin Infect Dis 57 (1):101–108

Li Q, Yang Y, Reis C, Tao T, Li W, Li X et al (2018) Cerebral small vessel disease. Cell Transplant 27(12):1711–1722

Lin Q, Huang W-Q, Ma Q-L, Lu C-X, Tong S-J, Ye J-H et al (2017) Incidence and risk factors of leukoaraiosis from 4683 hospitalized patients: a cross-sectional study. Medicine (Baltimore) 96 (39):e7682

Lin H-L, Muo C-H, Lin C-Y, Chen H-J, Chen P-C (2019) Incidence of stroke in patients with HIV infection: a population-based study in Taiwan. Plos One 14(5):e0217147

Lo Coco D, Lopez G, Corrao S (2016) Cognitive impairment and stroke in elderly patients. Vasc Health Risk Manag 12:105–116
Low A, Mak E, Rowe JB, Markus HS, O'Brien JT (2019) Inflammation and cerebral small vessel disease: a systematic review. Ageing Res Rev 53:100916
Lv Y-B, Yin ZX, Chei C-L, Brasher MS, Zhang J, Kraus VB et al (2016) Serum cholesterol levels within the high normal range are associated with better cognitive performance among Chinese elderly. J Nutr Health Aging 20(3):280–287
Ma C, Yin Z, Zhu P, Luo J, Shi X, Gao X (2017) Blood cholesterol in late-life and cognitive decline: a longitudinal study of the Chinese elderly. Mol Neurodegener 12(1):24
Madhavan M, Graff-Radford J, Piccini JP, Gersh BJ (2018) Cognitive dysfunction in atrial fibrillation. Nat Rev Cardiol 15(12):744–756
Makinson A, Dubois J, Eymard-Duvernay S, Leclercq P, Zaegel-Faucher O, Bernard L et al (2019) Increased prevalence of neurocognitive impairment in aging people living with human immunodeficiency virus: the ANRS EP58 HAND 55–70 study. Clin Infect Dis. https://doi.org/10.1093/cid/ciz670
Marcus JL, Leyden WA, Chao CR, Chow FC, Horberg MA, Hurley LB et al (2014) HIV infection and incidence of ischemic stroke. AIDS 28(13):1911–1919
Marquine MJ, Heaton A, Johnson N, Rivera-Mindt M, Cherner M, Bloss C et al (2018) Differences in neurocognitive impairment among HIV-infected Latinos in the United States. J Int Neuropsychol Soc 24(02):163–175
McAleese KE, Alafuzoff I, Charidimou A, de Reuck J, Grinberg LT, Hainsworth AH et al (2016) Post-mortem assessment in vascular dementia: advances and aspirations. BMC Med 14(1):129
McCutchan JA, Marquie-Beck JA, Fitzsimons CA, Letendre SL, Ellis RJ, Heaton RK et al (2012) Role of obesity, metabolic variables, and diabetes in HIV-associated neurocognitive disorder. Neurology 78(7):485–492
McDonnell J, Haddow L, Daskalopoulou M, Lampe F, Speakman A, Gilson R et al (2014) Minimal cognitive impairment in UK HIV-positive men who have sex with men: effect of case definitions and comparison with the general population and HIV-negative men. J Acquir Immune Defic Syndr 67(2):120–127
McNicholas T, Tobin K, Carey D, O'Callaghan S, Kenny RA (2018) Is baseline orthostatic hypotension associated with a decline in global cognitive performance at 4-year follow-up? Data from TILDA (The Irish Longitudinal Study on Ageing). J Am Heart Assoc 7(19):e008976
Mellendijk L, Wiesmann M, Kiliaan A (2015) Impact of nutrition on cerebral circulation and cognition in the metabolic syndrome. Nutrients 7(11):9416–9439
Meyer A-CL, Boscardin WJ, Kwasa JK, Price RW (2013) Is it time to rethink how neuropsychological tests are used to diagnose mild forms of HIV-associated neurocognitive disorders? Impact of false-positive rates on prevalence and power. Neuroepidemiology 41(3-4):208–216
Morgello S, Murray J, van der Elst S, Byrd D (2014) HCV, but not HIV, is a risk factor for cerebral small vessel disease. Neurol Neuroimmunol Neuroinflamm 1(3):e27
Moritani T, Capizzano A, Kirby P, Policeni B (2014) Viral infections and white matter lesions. Radiol Clin North Am 52(2):355–382
Morley JE (2014) Cognition and nutrition. Curr Opin Clin Nutr Metab Care 17(1):1–4
Moulignier A, Lamirel C, Mazighi M (2015) Stroke and HIV infection. Sang Thromb Vaiss 2:84–95
Moulignier A, Savatovsky J, Assoumou L, Lescure F-X, Lamirel C, Godin O et al (2018) Silent cerebral small-vessel disease is twice as prevalent in middle-aged individuals with well-controlled, combination antiretroviral therapy-treated human immunodeficiency virus (HIV) than in HIV-uninfected individuals. Clin Infect Dis 66(11):1762–1769
Mukerji SS, Locascio JJ, Misra V, Lorenz DR, Holman A, Dutta A et al (2016) Lipid profiles and apoe4 allele impact midlife cognitive decline in HIV-Infected men on antiretroviral therapy. Clin Infect Dis 63(8):1130–1139
Muller M, Tang M-X, Schupf N, Manly JJ, Mayeux R, Luchsinger JA (2007) Metabolic syndrome and dementia risk in a multiethnic elderly cohort. Dement Geriatr Cogn Disord 24(3):185–192

Nakamoto BK, Jahanshad N, McMurtray A, Kallianpur KJ, Chow DC, Valcour VG et al (2012) Cerebrovascular risk factors and brain microstructural abnormalities on diffusion tensor images in HIV-infected individuals. J Neurovirol 18(4):303–312

Navia BA, Cho ES, Petito CK, Price RW (1986) The AIDS dementia complex: II. Neuropathology. Ann Neurol 19(6):525–535

Nightingale S, Winston A, Letendre S, Michael BD, McArthur JC, Khoo S et al (2014) Controversies in HIV-associated neurocognitive disorders. Lancet Neurol 13(11):1139–1151

Nir TM, Jahanshad N, Busovaca E, Wendelken L, Nicolas K, Thompson PM et al (2014) Mapping white matter integrity in elderly people with HIV. Hum Brain Mapp 35(3):975–992

Noble JM, Manly JJ, Schupf N, Tang M-X, Luchsinger JA (2012) Type 2 diabetes and ethnic disparities in cognitive impairment. Ethn Dis 22(1):38–44

Nobre N, Pereira M, Roine RP, Sutinen J, Sintonen H (2018) HIV-related self-stigma and health-related quality of life of people living with HIV in Finland. J Assoc Nurses AIDS Care 29 (2):254–265

Noubissi EC, Katte J-C, Sobngwi E (2018) Diabetes and HIV. Curr Diab Rep 18(11):125

O'Callaghan S, Kenny RA (2016) Neurocardiovascular instability and cognition. Yale J Biol Med 89(11):59–71

O'Connor EE, Jaillard A, Renard F, Zeffiro TA (2017) Reliability of white matter microstructural changes in HIV infection: meta-analysis and confirmation. Am J Neuroradiol 38(8):1510–1519

Okafor CN, Kelso NE, Bryant V, Burrell LE, Míguez MJ, Gongvatana A et al (2017) Body mass index, inflammatory biomarkers and neurocognitive impairment in HIV-infected persons. Psychol Health Med 22(3):289–302

On Behalf of the POPPY Study Group, de Francesco D, Underwood J, Post FA, Vera JH, Williams I et al (2016) Defining cognitive impairment in people-living-with-HIV: the POPPY study. BMC Infect Dis 16(1):617

Ott A, Breteler M, de Bruyne M, van Harskamp F, Grobbee D, Hofman A (1997) Atrial fibrillation and dementia in a population-based study. The Rotterdam Study. Stroke 28(2):316–321

Ovbiagele B, Nath A (2011) Increasing incidence of ischemic stroke in patients with HIV infection. Neurology 76(5):444–450

Pal S, Ellis V (2010) The chronic effects of whey proteins on blood pressure, vascular function, and inflammatory markers in overweight individuals. Obesity 18(7):1354–1359

Palta P, Snitz B, Carlson MC (2016) Neuropsychologic assessment. Handb Clin Neurol 138:107–119

Pantoni L (2010) Cerebral small vessel disease: from pathogenesis and clinical characteristics to therapeutic challenges. Lancet Neurol 9(7):689–701

Pantoni L, Fierini F, Poggesi A, LADIS Study Group (2015) Impact of cerebral white matter changes on functionality in older adults: an overview of the LADIS Study results and future directions: LADIS study. Geriatr Gerontol Int 15:10–16

Paradise MB, Sachdev PS (2019) Vascular cognitive disorder. Semin Neurol 39(2):241–250

Pastori D, Mezzaroma I, Pignatelli P, Violi F, Lip GYH (2019) Atrial fibrillation and human immunodeficiency virus type-1 infection: a systematic review. Implications for anticoagulant and antiarrhythmic therapy. Br J Clin Pharmacol 85(3):508–515

Pourcher G, Costagliola D, Martinez V (2015) Obesity in HIV-infected patients in France: prevalence and surgical treatment options. J Visc Surg 152(1):33–37

Rasmussen LD, Engsig FN, Christensen H, Gerstoft J, Kronborg G, Pedersen C et al (2011) Risk of cerebrovascular events in persons with and without HIV: a Danish nationwide population-based cohort study. AIDS 25(13):1637–1646

Reaven GM (1988) Banting lecture 1988. Role of insulin resistance in human disease. Diabetes 37 (12):1595–1607

Rubin LH, Gustafson D, Hawkins KL, Zhang L, Jacobson LP, Becker JT et al (2019) Midlife adiposity predicts cognitive decline in the prospective Multicenter AIDS Cohort Study. Neurology 93(3):e261–e271

Sachdev P, Kalaria R, O'Brien J, Skoog I, Alladi S, Black SE et al (2014) Diagnostic criteria for vascular cognitive disorders: a VASCOG statement. Alzheimer Dis Assoc Disord 28 (3):206–218
Sacktor N, Skolasky RL, Seaberg E, Munro C, Becker JT, Martin E et al (2016) Prevalence of HIV-associated neurocognitive disorders in the Multicenter AIDS Cohort Study. Neurology 86 (4):334–340
Saloner R, Cysique LA (2017) HIV-associated neurocognitive disorders: a global perspective. J Int Neuropsychol Soc 23(9-10):860–869
Sanders JM, Steverson AB, Pawlowski AE, Schneider D, Achenbach CJ, Lloyd-Jones DM et al (2018) Atrial arrhythmia prevalence and characteristics for human immunodeficiency virus-infected persons and matched uninfected controls. Plos One 13(3):e0194754
Sanford R, Strain J, Dadar M, Maranzano J, Bonnet A, Mayo NE et al (2019) HIV infection and cerebral small vessel disease are independently associated with brain atrophy and cognitive impairment. AIDS 33(7):1197–1205
Sarbu N, Shih RY, Jones RV, Horkayne-Szakaly I, Oleaga L, Smirniotopoulos JG (2016) White matter diseases with radiologic-pathologic correlation. Radiogr Rev 36(5):1426–1447
Satizabal CL, Beiser AS, Chouraki V, Chêne G, Dufouil C, Seshadri S (2016) Incidence of dementia over three decades in the Framingham Heart Study. N Engl J Med 374(6):523–532
Sattler FR, He J, Letendre S, Wilson C, Sanders C, Heaton R et al (2015) Abdominal obesity contributes to neurocognitive impairment in HIV-infected patients with increased inflammation and immune activation. J Acquir Immune Defic Syndr 68(3):281–288
Schouten J, Su T, Wit FW, Kootstra NA, Caan MWA, Geurtsen GJ et al (2016) Determinants of reduced cognitive performance in HIV-1-infected middle-aged men on combination antiretroviral therapy. AIDS 30(7):1027–1038
Schreiber S, Wilisch-Neumann A, Schreiber F, Assmann A, Scheumann V, Perosa V et al (2019) Invited review: the spectrum of age-related small vessel diseases: potential overlap and interactions of amyloid and nonamyloid vasculopathies. Neuropathol Appl Neurobiol. https://doi.org/10.1111/nan.12576
Schuur M, Henneman P, van Swieten JC, Zillikens MC, de Koning I, Janssens ACJW et al (2010) Insulin-resistance and metabolic syndrome are related to executive function in women in a large family-based study. Eur J Epidemiol 25(8):561–568
Sepehri Shamloo A, Dagres N, Müssigbrodt A, Stauber A, Kircher S, Richter S et al (2019) Atrial fibrillation and cognitive impairment: new insights and future directions. Heart Lung Circ. https://doi.org/10.1016/j.hlc.2019.05.185
Shi Y, Wardlaw JM (2016) Update on cerebral small vessel disease: a dynamic whole-brain disease. Stroke Vasc Neurol 1(3):83–92
Shoamanesh A, Preis SR, Beiser AS, Vasan RS, Benjamin EJ, Kase CS et al (2015) Inflammatory biomarkers, cerebral microbleeds, and small vessel disease Framingham Heart Study. Neurology 84(8):825–832
Sico JJ, Chang C-CH, So-Armah K, Justice AC, Hylek E, Skanderson M et al (2015) HIV status and the risk of ischemic stroke among men. Neurology 84(19):1933–1940
Simioni S, Cavassini M, Annoni J-M, Rimbault Abraham A, Bourquin I, Schiffer V et al (2010) Cognitive dysfunction in HIV patients despite long-standing suppression of viremia. AIDS 24 (9):1243–1250
Skrobot OA, Black SE, Chen C, DeCarli C, Erkinjuntti T, Ford GA et al (2018) Progress toward standardized diagnosis of vascular cognitive impairment: guidelines from the Vascular Impairment of Cognition Classification Consensus Study. Alzheimers Dement J Alzheimers Assoc 14 (3):280–292
Sloan MA, Kittner SJ, Feeser BR, Gardner J, Epstein A, Wozniak MA et al (1998) Illicit drug-associated ischemic stroke in the Baltimore-Washington Young Stroke Study. Neurology 50 (6):1688–1693

Soontornniyomkij V, Umlauf A, Chung SA, Cochran ML, Soontornniyomkij B, Gouaux B et al (2014) HIV protease inhibitor exposure predicts cerebral small vessel disease. AIDS 28 (9):1297–1306

Su T, Wit FWNM, Caan MWA, Schouten J, Prins M, Geurtsen GJ et al (2016) White matter hyperintensities in relation to cognition in HIV-infected men with sustained suppressed viral load on combination antiretroviral therapy. AIDS 30(15):2329–2339

Sundbøll J (2018) Depression, stroke, and dementia in patients with myocardial infarction. Dan Med J 65(4):B5423

The SPRINT MIND, Investigators for the SPRINT Research Group, Williamson JD, Pajewski NM, Auchus AP, Bryan RN, Chelune G et al (2019) Effect of intensive vs standard blood pressure control on probable dementia: a randomized clinical trial. JAMA 321(6):553–561

Torti C, Focà E, Cesana BM, Lescure FX (2011) Asymptomatic neurocognitive disorders in patients infected by HIV: fact or fiction? BMC Med 9:138

Trentalange A, Prochet A, Imperiale D, Cusato J, Tettoni M, Nunnari G et al (2018) Cerebral white matter hyperintensities in HIV-positive patients. Brain Imaging Behav. https://doi.org/10.1007/s11682-018-9966-1

Tsivgoulis G, Alexandrov AV, Wadley VG, Unverzagt FW, Go RCP, Moy CS et al (2009) Association of higher diastolic blood pressure levels with cognitive impairment. Neurology 73(8):589–592

Underwood J, Cole JH, Caan M, de Francesco D, Leech R, van Zoest RA et al (2017) Gray and white matter abnormalities in treated human immunodeficiency virus disease and their relationship to cognitive function. Clin Infect Dis 65(3):422–432

Underwood J, de Francesco D, Leech R, Sabin CA, Winston A, Pharmacokinetic and Clinical Observations in PeoPle Over fiftY (POPPY) Study (2018) Medicalising normality? Using a simulated dataset to assess the performance of different diagnostic criteria of HIV-associated cognitive impairment. PLoS One 13(4):e0194760

Underwood J, de Francesco D, Cole JH, Caan MW, van Zoest RA, Schmand BA, et al. Validation of a novel multivariate method of defining HIV-associated cognitive impairment. In: Open forum infectious diseases. Oxford University Press, Oxford; 2019. 6(6):ofz198.

Utrecht Vascular Cognitive Impairment Study Group, de Bresser J, Kuijf HJ, Zaanen K, Viergever MA, Hendrikse J et al (2018) White matter hyperintensity shape and location feature analysis on brain MRI; proof of principle study in patients with diabetes. Sci Rep 8(1):1893

Valcour VG, Shikuma CM, Shiramizu BT, Williams AE, Watters MR, Poff PW et al (2005) Diabetes, insulin resistance, and dementia among HIV-1-infected patients. J Acquir Immune Defic Syndr 38(1):31–36

Valcour VG, Sacktor NC, Paul RH, Watters MR, Selnes OA, Shiramizu BT et al (2006) Insulin resistance is associated with cognition among HIV-1-infected patients: the Hawaii Aging with HIV cohort. J Acquir Immune Defic Syndr 43(4):405–410

van den Berg E, Biessels GJ, de Craen AJ, Gussekloo J, Westendorp RGJ (2007) The metabolic syndrome is associated with decelerated cognitive decline in the oldest old. Neurology 69 (10):979–985

van den Munckhof ICL, Jones H, Hopman MTE, de Graaf J, Nyakayiru J, van Dijk B et al (2018) Relation between age and carotid artery intima-medial thickness: a systematic review. Clin Cardiol 41(5):698–704

van der Flier WM, Skoog I, Schneider JA, Pantoni L, Mok V, Chen CLH et al (2018) Vascular cognitive impairment. Nat Rev Dis Primers 4:18003

van Zoest RA, Underwood J, de Francesco D, Sabin CA, Cole JH, Wit FW et al (2018) Structural brain abnormalities in successfully treated HIV infection: associations with disease and cerebrospinal fluid biomarkers. J Infect Dis 217(1):69–81

Vinikoor MJ, Napravnik S, Floris-Moore M, Wilson S, Huang DY, Eron JJ (2013) Incidence and clinical features of cerebrovascular disease among HIV-infected adults in the Southeastern United States. AIDS Res Hum Retroviruses 29(7):1068–1074

Vivithanaporn P, Heo G, Gamble J, Krentz HB, Hoke A, Gill MJ et al (2010) Neurologic disease burden in treated HIV/AIDS predicts survival: a population-based study. Neurology 75 (13):1150–1158

Vizzardi E, Gelsomino S, D'Aloia A, Lorusso R (2013) Aortic atheromas and stroke: review of literature. J Invest Med 61(6):956–966

Wang T, Yi R, Green LA, Chelvanambi S, Seimetz M, Clauss M (2015) Increased cardiovascular disease risk in the HIV-positive population on ART: potential role of HIV-Nef and Tat. Cardiovasc Pathol 24(5):279–282

Wardlaw JM, Smith C, Dichgans M (2013a) Mechanisms of sporadic cerebral small vessel disease: insights from neuroimaging. Lancet Neurol 12(5):483–497

Wardlaw JM, Smith EE, Biessels GJ, Cordonnier C, Fazekas F, Frayne R et al (2013b) Neuroimaging standards for research into small vessel disease and its contribution to ageing and neurodegeneration. Lancet Neurol 12(8):822–838

Wardlaw JM, Valdés Hernández MC, Muñoz-Maniega S (2015) What are white matter hyperintensities made of? Relevance to vascular cognitive impairment. J Am Heart Assoc 4 (6):001140

Watson C, Busovaca E, Foley JM, Allen IE, Schwarz CG, Jahanshad N et al (2017) White matter hyperintensities correlate to cognition and fiber tract integrity in older adults with HIV. J Neurovirol 23(3):422–429

Wiederkehr S, Simard M, Fortin C, van Reekum R (2008a) Comparability of the clinical diagnostic criteria for vascular dementia: a critical review. Part I. J Neuropsychiatry Clin Neurosci 20 (2):150–161

Wiederkehr S, Simard M, Fortin C, van Reekum R (2008b) Validity of the clinical diagnostic criteria for vascular dementia: a critical review. Part II. J Neuropsychiatry Clin Neurosci 20 (2):162–177

Wijndaele K, Duvigneaud N, Matton L, Duquet W, Delecluse C, Thomis M et al (2009) Sedentary behaviour, physical activity and a continuous metabolic syndrome risk score in adults. Eur J Clin Nutr 63(3):421–429

Wright EJ, Grund B, Robertson K, Brew BJ, Roediger M, Bain MP et al (2010) Cardiovascular risk factors associated with lower baseline cognitive performance in HIV-positive persons. Neurology 75(10):864–873

Wright EJ, Grund B, Cysique LA, Robertson KR, Brew BJ, Collins G et al (2015) Factors associated with neurocognitive test performance at baseline: a substudy of the INSIGHT Strategic Timing of AntiRetroviral Treatment (START) trial. HIV Med 16(Suppl 1):97–108

Wu M, Fatukasi O, Yang S, Alger J, Barker PB, Hetherington H et al (2018) HIV disease and diabetes interact to affect brain white matter hyperintensities and cognition. AIDS 32 (13):1803–1810

Yaffe K, Barrett-Connor E, Lin F, Grady D (2002) Serum lipoprotein levels, statin use, and cognitive function in older women. Arch Neurol 59(3):378–384

Yaffe K, Kanaya A, Lindquist K, Simonsick EM, Harris T, Shorr RI et al (2004) The metabolic syndrome, inflammation, and risk of cognitive decline. JAMA 292(18):2237–2242

Yaffe K, Haan M, Blackwell T, Cherkasova E, Whitmer RA, West N (2007) Metabolic syndrome and cognitive decline in elderly Latinos: findings from the Sacramento Area Latino Study of Aging study. J Am Geriatr Soc 55(5):758–762

Yang M, Williamson J (2019) Blood pressure and statin effects on cognition: a review. Curr Hypertens Rep 21(9):70

Yang J, Jacobson LP, Becker JT, Levine A, Martin EM, Munro CA et al (2018) Impact of glycemic status on longitudinal cognitive performance in men with and without HIV infection. AIDS 32 (13):1849–1860

Yates KF, Sweat V, Yau PL, Turchiano MM, Convit A (2012) Impact of metabolic syndrome on cognition and brain: a selected review of the literature. Arterioscler Thromb Vasc Biol 32 (9):2060–2067

Yin Z-X, Shi X-M, Kraus VB, Fitzgerald SM, Qian H-Z, Xu J-W et al (2012) High normal plasma triglycerides are associated with preserved cognitive function in Chinese oldest-old. Age Ageing 41(5):600–606

Yin Z-G, Cui M, Zhou S-M, Yu M-M, Li R, Zhou H-D (2014) Association between metabolic syndrome and white matter lesions in middle-aged and elderly patients. Eur J Neurol 21 (7):1032–1039

Yu B, Pasipanodya E, Montoya JL, Moore RC, Gianella S, McCutchan A et al (2019) Metabolic syndrome and neurocognitive deficits in HIV infection. J Acquir Immune Defic Syndr 81 (1):95–101

Zhong W, Cruickshanks KJ, Schubert CR, Acher CW, Carlsson CM, Klein BEK et al (2012) Carotid atherosclerosis and 10-year changes in cognitive function. Atherosclerosis 224 (2):506–510

Zhuang L, Sachdev PS, Trollor JN, Kochan NA, Reppermund S, Brodaty H et al (2012) Microstructural white matter changes in cognitively normal individuals at risk of amnestic MCI. Neurology 79(8):748–754

HIV-Associated Neurocognitive Disorder (HAND): Relative Risk Factors

Sindhura Kompella, Thabit Al-Khateeb, Ossama Abid Riaz, Sylvester Olubolu Orimaye, Patrick Olumuyiwa Sodeke, Adeola Olubukola Awujoola, Joseph Ikekwere, and Karl Goodkin

Contents

S. Kompella
Aventura Hospital, Aventura, FL, USA

T. Al-Khateeb and O. A. Riaz
Department of Psychiatry and Behavioral Sciences, James H. Quillen College of Medicine, East Tennessee State University, Johnson City, TN, USA

S. O. Orimaye
Department of Health Services Management and Administration, College of Public Health, East Tennessee State University, Johnson City, TN, USA

P. O. Sodeke and A. O. Awujoola
Department of Biostatistics and Epidemiology, College of Public Health, East Tennessee State University, Johnson City, TN, USA

J. Ikekwere
Pediatrics Psychiatry, Michigan Medicine, Ann Arbor, MI, USA

K. Goodkin (✉)
Department of Psychiatry, University of Nebraska Medical Center in Omaha, Omaha, NE, USA
e-mail: karl.goodkin@unmc.edu

Curr Topics Behav Neurosci (2021) 50: 401–426
https://doi.org/10.1007/7854_2020_131
Published Online: 28 July 2020

Abstract This chapter will address the issue of risk for HIV-associated neurocognitive disorder (HAND), focusing on HIV-associated dementia (HAD), among persons living with HIV in relationship to the risk for other dementias. Advances in effective antiretroviral therapy (ART) have led to an increase in the prevalence of older persons surviving with HIV – in addition to older persons who become infected by HIV later in life. Hence, HIV is no longer a disease of younger persons, and additional attention has been brought to bear against the plight of older persons living with HIV – not only as it pertains to treatment but also to prevention. The additional risk caused by aging among older persons living with HIV is complex to asses, and HIV infection is a research area that requires a robust approach to multiple other factors causing neurocognitive impairment with older age. The long-term and potentially neurotoxic exposure to ART and the deleterious consequences of chronic infection with HIV and its associated neuro-inflammation have been described for health. This aids in the understanding of dementia risk factors in this patient population, but the comorbidities (HIV- and non-HIV-associated) occurring among older persons living with HIV must also be addressed to properly assess the overall impact on dementia risk in this group. This need also warrants our examination of the risk factors for other dementias (and comorbid dementias) in persons living with HIV versus the general population through the assessment and quantification of modifiable and non-modifiable risk factors identified as major contributors toward dementia.

Keywords Aging · Dementia · HIV-associated neurocognitive disorder (HAND) · Major neurocognitive disorder · Neurocognitive impairment · Risk factors

Current estimates show that more than 50% of persons living with HIV are 50 years of age or older in the USA – including approximately 7% who are 65 years of age or older. Studies forecast a stable, annual increase of 2% in the proportion of persons aged 50 and older living with HIV going forward. Persons aged 50 and older are projected to comprise 17% of the population of persons with newly acquired HIV. While the increased longevity of persons living with HIV is a tremendous success in terms of biomedical advances, it comes with added clinical implications that are specific to living with HIV as a chronic illness and aging (Pfefferbaum et al. 2014). The added risk for dementia in aging persons with HIV is an area that requires a robust "call to arms" as more and more people enter into the older age range. The potentially neurotoxic effects of long-term exposure to effective ART and the consequences of chronic infection with HIV are not well-known in the highest end of the older age range. Greater life expectancy in persons living with HIV has led to a greater risk of developing age-related medical comorbidities involving cardiovascular, metabolic, respiratory, infectious, and neurodegenerative sequelae. Neurodegenerative disorders of persons aging with HIV discussed here will include Alzheimer's disease (AD), dementia with Lewy bodies (DLB), vascular cognitive impairment and dementia (VCID), frontotemporal dementia (FTD), and Parkinson's disease (PD) with dementia.

1 Neurodegeneration and Types of Dementia

Neurodegeneration involves progressive accumulation of toxic proteins, neuronal loss due to damage of selective synaptic circuitries, astrogliosis, myelin degradation, microgliosis, and vascular alterations as a result of neuro-inflammation. In terms of dementia type, AD exhibits abnormal accumulations of amyloid and Tau protein in the brain beyond age-related changes (Reas 2017). Further research has shown that age-related factors play a role with amyloid and Tau protein accumulation in the neocortex and hippocampus. An accumulation of α-synuclein, in the nigrostriatal pathway – a component of the sequence of pathways known as the cortico-basal ganglia-thalamo-cortical loop, is associated with DLB and PD with dementia. In FTD, aggregates of Tau or TDP-43 accumulations are found. Aggregates of TDP-43 have also been associated with age-related hippocampal sclerosis (Cykowski et al. 2017). Hence, this alludes to a possible multidimensional neuropathophysiology manifested by the overlap of protein aggregates across types of dementia, suggesting a possible underlying linkage across the dementia spectrum.

In VCID, there is ischemic disease that can lead to the accumulation of amyloid. While the proteinopathies underlying the protein aggregates characteristic of these forms of dementia are not fully understood, genetic and environmental factors have been found to play a role in these imbalanced protein accumulations (Mackiewicz et al. 2019). The role of other dementias as it pertains to HAND, and HAD particularly, is not fully understood. We will examine common mechanisms involved in favor of and against the risk for HAND by examining the risk factors associated with other causes of neurocognitive disorder – with an emphasis on modifiable (accounting for up to 50% of all-dementia risk) vs. non-modifiable risk factors.

2 Non-modifiable Risk Factors

2.1 The Aging Process

The aging process represents the strongest predictor for the development of dementia, regardless of type. Studies show a significant increase in the percentage of persons living with dementia after 65 years of age – with an exponential increase for every decade thereafter. The greatest percentage of persons with dementia is in the age range over 85, regardless of dementia type. The increased survival time of persons living with HIV has raised concerns for current treatment options, since these options have not proven sufficient to protect against the occurrence of HAND. US data shows that most persons living with HIV are in the fifth decade of their lives. Age is an independent risk factor for HIV-associated neurocognitive impairment and disorder and is associated with the severity of disorder observed. Several studies have found that aging increases the likelihood of neurocognitive impairment for persons living with HIV. According to a study in sub-Saharan Africa, there is significant evidence that each additional 10 years in age results in a twofold increase in the risk for HIV-associated dementia (HAD) (Wong et al. 2007). The authors attributed this to both the chronicity of neuro-inflammation and to a lack of antiretroviral (ARV) adherence.

Another study showed an association of the apolipoprotein E4 allele and increased amyloid plaques in brains infected by HIV. Age alone independently increased the likelihood of amyloid plaque deposition in the brain in that study (Fjell et al. 2014). Not only does the risk of developing HAND increase with aging (Goodkin et al. 2001, 2017) but prominently does the potential risk for other neurodegenerative dementias, such as AD as well as PD with dementia. Similarly, another case control study compared performance on neuropsychological testing by age in groups of persons living with and without HIV. According to this study, persons living with HIV had poorer performance in both information processing speed and executive functioning compared to persons living without HIV. Age plays an important role in the overall level of neurocognitive impairment manifested in persons living with HIV – perhaps, due to an increase in the duration of HIV infection and to associated neuro-inflammation combined with increased permeability of the blood-brain barrier (BBB) usually seen with the normal aging process. Studies also suggest a premature, accelerated systemic aging process in persons living with HIV, though there is more evidence for the CNS than the systemic aging effect. To better delineate aging effects, studies have focused on biological over chronological aging. The biological aging process is associated with disability, frailty, comorbidities, and polypharmacy that manifest at an earlier age in the persons living with HIV than in the general population (Kaul 2009).

2.2 Genetics

Most age-related dementias map to a class of proteinopathy, i.e., a type of protein that deposits in the brain and accumulated abnormally, often within specific neuropathological structures that dictate the associated symptoms observed. The key classes of proteinopathy across dementias are amyloidopathies, tauopathies, synucleinopathies, TDP43-opathies, prionopathies, and polyglutaminopathies. On the spectrum of genetic versus environmental influences, some dementias are predominantly genetically driven, such as Huntington's disease, with mild environmental influences still playing a role. In contrast, there are dementias on the other end of the spectrum where environmental influences play a predominant role, such as chronic, traumatic encephalopathy (CTE). Diseases like AD, FTD, VCID, PD with dementia, and DLB fall more toward the middle of this spectrum, reflecting a balanced impact of genetic and environmental influences. The environmental/genetics continuum becomes more complicated when the overlap of major dementia types is also considered. For instance, AD is an amyloidopathy and tauopathy, while DLB is characterized by deposits of α-synuclein aggregates in the brain also accompanied by both amyloid plaques and Tau (Irwin and Hurtig 2018). Moreover, TDP-43 inclusions seen in FTD are also found in other dementia types, including AD. Hence, classifying dementia into strict pathological categories is not feasible, arguing for a "mixed-type" characterization of dementia overall (Paulson and Igo 2011).

With respect to dementia genetics in persons living with HIV, the data is as yet inconclusive, due to a critical lack of reproducibility among various studies. Genetic linkage studies of HAND have been limited by the heterogeneity and ambiguity associated with measuring HAND outcomes and the associated variability among HAND presentations. Some studies have linked the apoE4 allele associated with AD to HAND, but other studies have not (Becker et al. 2015). The inconsistent results observed across studies are due to age being an independent risk factor and to the variable role played by apoe4 status in persons living with HIV. In addition, due to the multifactorial nature of HAND neuropathogenesis, it is not likely that a single genetic variant is responsible for the diverse manifestations of HAND. Variations in the genes encoding the chemokine CCL3L1 and the HIV co-receptor, CCR5, influence cell-mediated immunity in both persons living with and without HIV. CCL3L1 and CCR5 genotypes also modify the clinical course of HIV (Dolan et al. 2007). CCL3L1 is a copy number variant in the MIP-1α gene; an increased CCL3L1 copy number hinders HIV binding and infection of cells by increasing MIP-1α, which then out-competes HIV for CCR5 binding. However, a specific study of CCL3L1 copy number has not shown that it is higher in persons living without HAND than in persons living with HAND (Brown et al. 2012). Also of note, the CCR5 receptor is a crucial co-receptor for HIV entry into monocytes and microglia, and mutations in CCR5 have been documented to affected HIV disease progression and neurocognitive impairment in the era prior to effective ART. In fact, the CCR5 Δ32 allele [arising from a deletion of 32 base pairs within the CCR5 gene] slowed disease progression and decreased neurocognitive impairment. It was also reported to be associated with a lower prevalence of HAD. However, this finding appears to be specific to the era prior to effective ART, as subsequent studies did not uphold these effects (Carroll and Brew 2017). Hence, it might be concluded that more research is required to identify common pathways in HAND pathogenesis that may shed light on genetic linkages (Olivier et al. 2018), particularly in the era of effective ART.

2.3 *Ethnicity and Gender*

Establishing relationships with ethnicity across dementias is a challenging task, as modifiable dementia risk factors, such as educational attainment, social opportunities, life style choices, life event stressor burden, social support dynamics, and coping strategy patterns, all play a role in the manifestation of dementia risk. Controlling for these factors is important yet difficult to achieve, in order for an ethnicity factor to be well established. In addition, international studies have lacked cross-cultural harmonization in establishing uniform diagnostic criteria for dementia, standardized assessment protocols, and/or consensus guidelines to establish contributory ethnicity factors globally (Azad et al. 2007). In a recent study from Switzerland of neurocognitive impairment in persons living with HIV, the Neurocognitive Assessment in the Metabolic and Aging Cohort (NAMACO) study showed a generalized factor of "non-Caucasian ethnicity" that was noted to be contributory (Métral et al. 2019) as did Goodkin et al. (2017). To date, HAND does not appear to manifest any specific pattern of prevalence by ethnicity. In addition to inadequately controlled studies across geographical and cultural landscapes, another issue is that the lack of ethnic associations documented to date may be due to HAND itself being a relatively novel disease process. Overall, the results highlight the need for cross-cultural and global health study initiatives to specifically power future studies to address ethnicity as a HAND risk factor in persons living with HIV.

In terms of sex assigned at birth and the role it plays in dementia pathology, the results vary across a wide spectrum by dementia type. Women are at higher risk for AD, while men carry a higher risk for VCID. Regarding ethnicity, the data show that persons of African descent have higher rates of dementia than their Caucasian counterparts, who, in turn, have higher dementia rates than persons of Asian descent. No gender differences have been clearly delineated in terms of HAND risk, overall.

Regarding persons living with HIV, one early retrospective review of 6,548 AIDS cases conducted by the AIDS in Europe Study Group found that the risk of then "AIDS dementia complex" (ADC) in Europe from 1979 to 1989 was twice as high in women than men (Chiesi et al. 1996). However, another study did not demonstrate that sex differences were a significant dementia risk factor among persons living with HIV (Robertson et al. 2004). Yet another study reported that females were more likely to have persistent neurocognitive impairment that was less reversible compared to males (23.7% vs. 8.6%) (Tozzi et al. 2007). While sex differences in cognitive function are well-documented, few studies of persons living with HIV have had adequate sample sizes to confirm sex differences in neurocognitive impairment with sufficient statistical power. A systematic review of recent studies of sex differences in neurocognitive impairment showed that studies with sufficient power to address sex differences in neurocognitive impairment have reported evidence of a greater frequency in women living with HIV (Rubin et al. 2019). This is especially true for the domains of memory, information processing speed, and motor function. As was the case with ethnicity, closer attention to confirm a potential differential HAND prevalence among women living with HIV is warranted in future studies sufficiently powered to achieve this objective, particularly in terms of controlling for differential hormonal regulatory effects on cognition at the point of assessment as well as for a differentially higher rate of major depressive disorder and post-traumatic stress disorder.

3 Modifiable Risk Factors

3.1 *Cardiovascular Health*

Cardiovascular risk factors are relevant to all types of dementia and play a role in persons living with HIV from mid-life onward. Hypertension shows a positive association with late-life dementia overall, and specifically with AD (Lennon et al. 2019) and VCID. Antihypertensive treatment options have been shown to prevent cognitive decline and decrease the risk for dementia. Diabetes in mid-life and CVA have shown powerful associations with increased risk of dementia overall, and VCID – as might be expected. As was the case for hypertension, medications for diabetes reduced the risk for dementia overall (Wium-Andersen et al. 2019), indicating the importance of glucose control. In terms of life style choices, smoking is a risk factor for dementia overall and for AD. Physical exercise and a diet high in fruits and vegetables have shown protective effects against dementia – as negative risk factors overall. Specific nutritional diets, such as the Mediterranean diet, the DASH diet (Dietary Approaches to Stop Hypertension), and MIND diet (Mediterranean-DASH Intervention for Neurodegenerative Delay), have shown promising results in terms of reducing dementia risk, overall – generalizing to HAND. Specifically, persons living with HIV have shown improvements in psychomotor speed when on the DASH diet versus a normal diet (Baumgart et al. 2015; Aung et al. 2019).

The mechanisms that link prior cardiovascular disease (CVD) to HAND are multifactorial. Ongoing pro-inflammatory cytokine release in the face of completely suppressed plasma viral load and associated ongoing release of inflammatory mediators, such as C-reactive protein, are associated with neurocognitive impairment in persons living with HIV. Prior CVD also independently increases the likelihood of VCID and is frequently associated with neurodegenerative dementias. According to Wright et al. (2010), patients living with HIV who were adherent to their ARV medications and had high CD4 cell counts were nevertheless found to have an elevated frequency of neurocognitive impairment in the setting of CVD. This double blind, controlled study found that 14% of the participants with prior CVD had neurocognitive impairment. Lower neurocognitive performance was associated with prior CVD, hypertension, and hypercholesterolemia. The source of neurocognitive impairment that leads to HAND is not completely established; however, the chronicity of ongoing neuro-inflammation occurring in the face of controlled viral replication as reflected by the peripheral blood likely ultimately contributes significantly.

It should be noted that there has been inconsistency with respect to the implications of CVD for HAND. Becker et al. (2012) used MRI in the Multicenter AIDS Cohort Study (MACS) to acquire high-resolution neuroanatomic data in 160 men aged 50 years and over, including 84 persons living with HIV and 76 persons living without as control participants. Both age and HIV status had a significant effect on both gray matter (GM) and white matter (WM) volume. The age-related GM atrophy was primarily seen in the superior temporal and inferior frontal regions; the

HIV-related GM loss included the posterior and inferior temporal lobes, the parietal lobes, and the cerebellum. Contrary to expectation, the CVD disease variables were not linked to brain volume in statistically adjusted models. Further, similar findings were obtained in a subsequent MACS study in which total volume of GM, WM, and CSF were significantly associated with intraindividual variability in cognitive testing reflected by the degree of dispersion across the five performance variables for each individual, in which higher values refer to greater across-task cognitive variability, and smaller values reflect a flatter, more consistent profile of cognitive abilities. Increased intraindividual cognitive variability was associated with abnormal brain structure; however, again, there was no effect of CVD variables (Hines et al. 2016). The contribution of CVD and cardiovascular risk factors to the cognition of persons living with HIV continues to warrant further investigation.

3.2 Depressive and Substance Use Disorders

The onset of major depressive disorder (MDD) later in life has been shown to be a significant risk factor for the development of dementia, with the risk for AD at 90% and for all-type dementia in the 85–100% range. Other psychopathology and deleterious psychosocial factors also play key roles as risk factors for dementia. In terms of substance use disorder being a risk factor for dementia overall, alcohol use disorder has been the most studied substance. Alcohol use disorder has been correlated with a higher risk of dementia, overall. Studies pertaining to use disorders of other substances are less frequent, and more research is needed to explain the correlation of the specific substance use disorder of interest with the development of dementia (Aung et al. 2019). Psychostimulant substance use – methamphetamine (Theodore et al. 2006) and cocaine (Bagetta et al. 2004) in particular – have been shown to increase the likelihood of neurocognitive impairment in persons living with HIV. Methamphetamine and cocaine use both increase dopaminergic transmission in the CNS acutely but decreases it chronically. The specific interactions involved between the opioids and HAND risk appear to be more complex (El-Hage et al. 2015). The use of stimulants and opioids has been associated with enhanced macrophage activation and with pro-inflammatory cytokine release. In turn, these factors have been associated with neuronal apoptosis that increases HAND risk. Moreover, cocaine and opioids use can also disrupt the integrity of the BBB, since these substances alter tight junctions and membrane permeability. Substance use in persons living with HIV can further disrupt the BBB beyond the effect caused by HIV itself, promoting viral entry into the brain and neuro-inflammation in response as well as trafficking of HIV-infected cells. Specifically, cocaine has been associated with increased TNF-alpha release and decreased integrity of the BBB. In addition, persons living with HIV who use alcohol and other psychoactive substances are more likely to be less adherent to their ARVs, which can promote the development of clinically demonstrable neurocognitive impairment and other HIV-associated morbidities. Surprisingly, alcohol use has received relatively limited attention as a possible co-factor in HAND pathogenesis and progression, though data to date suggest that alcohol might be highly deleterious (Pfefferbaum et al. 2018). Since psychoactive substance use is strongly associated with neurocognitive impairment in persons living without HIV, it is important to regularly screen persons living with HIV who use alcohol and other psychoactive substances as a particularly high-risk group for HAND.

Prevalence estimates for MDD vary from 20 to 37% among persons living with HIV. This range of estimates is greater than three times the rate of MDD in the general population, which varies from 5 to 12% (Simoni et al. 2011). Thus, it is important to screen patients living with HIV for MDD regularly, since MDD includes cognitive dysfunction as one of diagnostic criteria and can result in delayed medication initiation and low subsequent ARV adherence – ultimately resulting in more rapid clinical HIV disease progression. MDD in persons living with HIV is also more frequently accompanied by comorbid alcohol and substance use disorder. Depressive disorders as a group remain underdiagnosed and generally undertreated, especially among persons living with HIV (Do et al. 2014). This finding is especially concerning in the light of data showing that the treatment of depressive disorders may be associated with better ARV adherence (Uthman et al. 2014). Coleman et al. (2012) have reported on a 124-patient clinical chart review that evaluated the effectiveness of a collaborative, measurement-based approach to depression care, including psychopharmacologic and ancillary psychological therapies in patients living with HIV. Pre- vs. posttreatment analyses revealed significant reductions in depression and plasma HIV RNA as well as significant increases in the CD4 cell count. The authors concluded that a collaborative, measurement-based approach to depression care appears to be an effective method of improving MDD as well as virologic and immune outcomes. However, these broader results of intervention on MDD have not always been observed. Pence et al. (2012) conducted a study with multiple different antidepressants (ten drugs) and concluded that viral load outcomes were not improved but clinically significant MDD improvement and increased depression-free days was achieved. However, it must be pointed out that this study had high baseline adherence, used a highly heterogeneous intervention (multiple antidepressants), and included participants with concomitant substance use and anxiety disorders. The foregoing comorbid psychopathology with MDD might have interfered with study outcomes. This study also had a large amount of missing data, did not examine CD4 cell count or clinical outcomes, and only measured adherence by unannounced telephone-based pill counts. Further, the effect of treating MDD was not analyzed in women, since their participants were mostly men.

Kennard et al. (2014) reported a feasibility pilot study of a cognitive behavioral therapy (CBT) intervention and did show an effect of CBT on improved adherence. A follow-up study by the same group (Brown et al. 2016) of a 24-week manualized, measurement-guided CBT psychotherapy intervention, and a medication management algorithm tailored for youth living with HIV found that these programs were more effective in achieving and sustaining remission from depression compared with treatment as usual at HIV care clinic sites. In this latter relatively short study, CD4 cell count and plasma viral load were not differentially improved. In summary, the wide range of responses to treatment of depressive disorders across studies to date is due to differences in the samples accrued, in assessment tools, and in the specific techniques for depression intervention. The complex interplay between psychological and biomedical factors known to exist with MDD, specifically, in persons living with HIV can make both its diagnosis and treatment challenging.

The pathophysiology of MDD involves an imbalance in the release of pro-inflammatory and counter-regulatory cytokines. These mechanisms overlap with the pathophysiology of HAND. The neuro-inflammatory pathways, more generally, stimulated in MDD are also over-stimulated by HIV infection itself. The concomitant inflammation occurring with MDD in persons living with HIV might be expected to at least be additive and possibly synergistic in increasing the severity of clinically demonstrable neurocognitive impairment, though this has not ostensibly been well demonstrated. The role of pro-inflammatory mediators in the pathophysiology of MDD regarding CNS cytokines released (particularly IL-1-beta, IL-6, and TNF-alpha) upregulates the underlying depressive symptomatology of MDD. These pro-inflammatory cytokines interact with the HPA axis and other stress-responsive neuroendocrine systems, bringing about dysfunctional and excessive levels of cortisol. Excessive cortisol, in turn, is associated with a more decreased Th1/Th2 cytokine balance and ultimately with neurotoxicity that might interact with that of HIV itself.

Mitochondrial bioenergetic dysfunction has also been implicated in the pathogenesis of MDD – particularly, alterations in brain acylcarnitines that subserve the stabilization of a neuroprotective milieu. Cassol et al. (2015) used an untargeted approach to characterize the plasma metabolome of three independent cohorts and identified a set of six metabolites altered in depressed persons living with and without HIV. These metabolites (phenyl acetate, 4-hydroxyphenylacetate, propionylcarnitine, isobutyrylcarnitine, isovalerylcarnitine, and 2-methylbutyrylcarnitine) correlated with depressive symptom severity. They also mapped to pathways not only involved in mitochondrial function but also in monoamine metabolism and in inflammation. Relationships among interferon (IFN) responses, kynurenine, the K:T ratio, and metabolites associated with mitochondrial function were identified, suggesting that augmented IFN responses and increased tryptophan catabolism may devolve from the pathophysiology of MDD. Together, these results suggest that subtle alterations in the metabolism of monoamines (tryptophan and its anabolism to serotonin; phenylalanine and its anabolism to dopamine; and trace amines) and mitochondrial energetics might contribute to mechanisms of MDD pathogenesis. Further, these same processes might be influenced by inflammation during HIV infection. Moreover, ARV usage causes mitochondrial toxicity that could enhance the impact of this mechanism over the duration of HIV care.

Examining the primary level of HIV load, the CNS HIV Antiretroviral Therapy Effects Research (CHARTER) cohort was noteworthy; it is a six-center US-based prospective cohort study with biannual follow-up of 674 participants. As part of that study, Hammond et al. (2016) fit linear mixed models ($N = 233$) and discrete-time survival models ($N = 154$; 832 observations) to evaluate trajectories of Beck Depression Inventory (BDI) II scores and the incidence of new-onset moderate-to-severe depressive symptoms (a BDI-II score $\geq$ 17) suggestive of the presence of MDD among participants on effective ART who were free of depression at study entry and had received a minimum of three CSF examinations over 2,496 person-months of follow-up. Detectable CSF HIV RNA (at a lower limit of sensitivity of 50 copies/ml) at any visit was associated with a 4.7-fold increase in new-onset depression at subsequent visits – adjusted for plasma HIV RNA and ARV treatment adherence [hazard ratio = 4.76, ($p = 0.006$)]. If CSF HIV RNA was detectable at a prior study visit in fully adjusted models [including age, sex, race, educational level, plasma HIV RNA, duration, and adherence to effective ART] and lifetime depression diagnosis made using Diagnostic Statistical Manual of Mental Disorders (DSM-IV) criteria, then BDI-II scores were 2.53 points higher ($p = 0.02$) over 6 months. Persistent CSF (but not plasma) HIV RNA was associated with this increased risk for new-onset depression, suggesting that the overlap of neuro-inflammatory changes with HIV and MDD may be reflected in CNS-specific viral burden changes.

3.3 Educational Level and Socioeconomic Status

Higher educational attainment is a negative risk factor for dementia overall, putatively through diversification and enhancement of neuronal networks and optimization of mental stimulation – in line with the predictions of the theory of cognitive reserve. Higher educational attainment is associated with higher socioeconomic status (SES), which putatively leads to increased awareness of healthy nutrition and diet, access to higher quality of medical care, and a lower incidence of cardiovascular risk factors and CVD. Being socially engaged (associated with higher SES) also allows people to be further stimulated. The effect of education al attainment may be quite similar to that of people engaged in cognitively stimulating activities, such as reading, writing, and problem solving. These activities lower overall dementia risk (Aung et al. 2019). In terms of cognitive training RCTs, the Advanced Cognitive Training for Independent and Vital Elderly (ACTIVE) trial showed that the intervention group improved their cognitive skills of reasoning and information processing speed but unfortunately did not improve memory function (Kivipelto et al. 2018).

As opposed to the relatively sparse research supporting sex differences, lower educational level is a well-documented risk factor for HAND. According to one study, there are persistent neurocognitive deficits in persons living with HIV with less than 10 years of education compared to those with at least 10 years of education. The results were significant, and this can be explained by an increase in cognitive reserve with higher educational level, which allows for a later presentation of neurocognitive deficits in patients with higher educational attainment. Persons living with HIV having a lower educational level are more likely to present with cognitive deficits earlier in systemic HIV clinical disease progression. As systemic disease progression is permissive for CNS HIV disease progression, they are more likely to develop HAND. Likewise, the pathologic changes seen with HAND are masked for a longer time in persons living with HIV having attained a higher educational level. As expected, such persons present with the signs and symptoms of HAND later in the course of HIV infection, when they present (Tozzi et al. 2007; Satz et al. 1993; Stern et al. 1992; Ryan et al. 2005).

Despite effective ART, studies have reported that HIV-associated neuropathogenesis and HAND persist among persons living with HIV, particularly those of older age. Commonalities between HAND (including HAD), AD, and PD with dementia have been described. The aging process represents a common underlying factor. Neuropsychological, physiological, and immunological alterations are shared between HAND and the aging process that are manifested by neuropathologic manifestations and indices of HIV disease progression such as highs level of neuro-inflammation, impaired protein degradation pathways, and oxidative and nitrosative stress. Another similarity between normal aging and HAND involves functional deficits of the proteasome-ubiquitin complex and autophagy. In addition, αβ amyloid plaques specific to AD, ubiquitinylated deposits found in FTD, and α-synuclein aggregation in the substantia nigra characteristic of DLB have each been observed in persons living with HIV (Kaul 2009).

4 Risk Factors and Diagnosis for HAND and for All-Type Dementia

The known risk factors for HAND include older age; low educational level; CVD; diabetes; hypertension; HCV co-infection; methamphetamine, cocaine and alcohol use, traumatic brain injury; low CD4 cell nadir; MDD; and specific nutrient and vitamin deficiencies (Moore et al. 2011). ARVs that cause neurocognitive impairment as a toxicity have been associated with HAND as well – specifically efavirenz and dolutegravir. In addition to a medical risk factor assessment, there are several neuropsychological tests that are used to screen for the neurocognitive impairment occurring in HAND. These tests include the HIV Dementia Scale (HDS) (Power et al. 1995), the International HIV Dementia Scale (IHDS) (Sacktor et al. 2005), the NEU screen, and the EXIT interview. The mini-mental state examination (MMSE) and the Montreal Cognitive Assessment (MoCA) have not been proven to be particularly valuable for HAND screening. A standardized, neuropsychological test battery representing diverse cognitive domains continues to be the gold standard for the diagnosis of HAND. Quantified neuropsychological test scores are required to identify the neurocognitive impairment criterion for HAND. Unfortunately, no single screening test has proven to be sufficiently sensitive and specific to make a HAND diagnosis with high level of accuracy (Goodkin et al. 2014). Currently, the best prospect appears to be to investigate the combination of a HAND screening test with a standardized neuropsychological test that could be easily administered and scored by clinicians.

In elderly populations, AD is the most frequent cause of dementia, while HIV is the major cause of dementia among persons less than 65 years of age. Of note, the prevalence of mild cognitive impairment (MCI) prior to AD in the general population of older persons is approximately 15%. While the prevalence of MND in HAND overall is similar, when asymptomatic neurocognitive impairment (ANI) is accounted for as another HAND condition preceding HAD, the total pre-HAD neurocognitive condition prevalence rises to approximately 48% – well beyond that of MCI preceding AD. As persons living with HIV are more commonly over 65 years of age today (with 50% now at 50 years of age or older in the USA), it has become better appreciated that there might be a significant pathophysiological overlap between AD and HAND. This is based upon an increased association of amyloid plaques with both aging in the general population (Fjell et al. 2014) and with HIV infection (Soontornniyomkij et al. 2012). Increased β-amyloid deposition, extracellular amyloid plaques, and decreased CSF β-amyloid levels seen in AD have been reported in persons with HAD in association with apolipoprotein ε4 alleles, although an interaction of the two has not been consistently observed (Becker et al. 2015).

Likewise, microvascular impairment has been associated with HAND as well as VCID, and both are related to diabetes and hypertension as known risk factors. The direct impact of ischemic events related to minor and major vascular compromise has been documented in HV for some time. Dyslipidemia occurs with increased duration of effective, although the mechanism(s) have not been resolved. Of specific note, the protease inhibitors (PIs) were associated with an increased incidence of myocardial infarction at a younger age in persons living with HIV, and the nucleoside reverse transcriptase inhibitors were likewise implicated in coronary artery disease inducing effects, albeit to a lesser extent than the PIs. Similar effects have been described in the cerebrovasculature, with increases noted in abnormal cerebrovascular vasomotor reactivity by transcranial Doppler studies; in carotid intima-media thickness; and in the frequency of internal carotid artery plaques. Moreover, these effects were shown to be associated with increased clinical events (transient ischemic attacks and CVAs). A recent review of VCID and HAND took this focus on the vascular factor one step further. Citing the recently updated VCID nomenclature, it demonstrated that the neuropsychological and neuroimaging characteristics of VCID and HAND overlap significantly. Further, this review (Cysique and Brew 2019) linked underlying mechanisms of VCID and HAND – promulgating a potentially groundbreaking hypothesis that the neurovascular unit itself may constitute the primary target for injury to the brain today.

PD is also a consideration in persons living with HIV due to the significance of the dopaminergic deficits common to both diseases. In fact, PD motor symptoms have been noted to occur with HAND from the early years of the epidemic. However, there has been less documentation of this pattern of motor symptoms in the era of effective ART, although there have been reports that Parkinsonism persists among persons living with HIV today. Yet, the association of neurocognitive impairment with PD motor symptoms has not been focused upon in the HIV literature, rendering comparisons with risk for PD with dementia difficult to ascertain. In summary, it is not yet possible to determine how the prevalence of HAND varies independently from that of AD, VCIM, and PD with dementia, in particular. While other neurocognitive disorders having specific, respective, identified, pathogenic risk factors are more readily separated from HAND, comparative studies of dementia risk for those disorders in persons living with HIV are not yet available. Thus, further research needs to be undertaken before comparative risk factors for dementias in the general population can be effectively and comprehensively compared to the population of persons living with HIV.

5 Animal Models for Dementia

Since HIV-1 infection is species-specific and HAND frequently affects higher-order cognitive functions, it is more difficult to simulate the illness in a rodent model. Initially, simian immunodeficiency virus (SIV) models were used to approximate HIV disease progression, inflammation, and pathogenesis. Vaccine experiments using a SIV challenge utilize SIV-based immunogens, invalidating the direct testing of HIV-specific antigens for protection prior to their testing in the clinic. Therefore, the field turned to the use of genetic chimeras termed “SHIV” that utilize the backbone of one virus (SIV here) and have one or more genes swapped with HIV to acquire the HIV-associated function needed. Such SHIV models are typically nonpathogenic unless passed in vivo between nonhuman primates to increase fitness and viral persistence (Kumar et al. 2016). The use of state-of-the-art nonhuman primate (NHP) models of SIV and SHIV infection together with effective ART in combination with measures of viral reservoir size and function, along with novel therapeutic “anti-reservoir” approaches, represents a formidable tool for basic and translational studies to develop a cure for HIV infection. Recently, rodent models have incorporated designs that are more suitable for the recapitulation of human HIV disease and are more cost-effective for evaluating disease progression and treatment response (Saylor et al. 2016; Akkina et al. 2016; Kraft-Terry et al. 2009; Donahoe and Vlahov 1998). Unfortunately, studies on CNS HIV infection and the viral reservoir in brain have been hampered by a relative dearth of small animal models. Yet, SCID mice reconstituted with a human immune system are susceptible to HIV infection and have been proven to be useful tools.

The recapitulation of human CNS HIV infection has been hampered by a lack of human glial cells generally in currently available rodent models. In fact, perivascular macrophages, microglia, and astrocytes are all needed to mimic the human brain reservoir of HIV and its pathogenesis. Recently, a new immunodeficient strain supplemented with the human IL-34 transgene to support microglial development in humanized mice has been evaluated. Human CD34+ hematopoietic stem cell-transplanted mice develop human microglia-like cells in the presence of the tissue-specific ligand-IL-34. To identify relationships between HIV-1 infection of microglia and neuropathology, mice at 6 months of age were infected with HIV-1ADA, and brain tissues were subjected for histopathologic (glial and neuronal) and transcriptomic (mouse and human) profiling by next-generation sequencing. CD34-NOG-hIL-34 mice showed sustained plasma viremia with CD4 cell count decrements and productive human microglial infection. Reductions in neuronal architecture and synaptic integrity were observed in the brain subregions by reduced expression of synaptophysin, MAP 2, and neurofilament-H (Mathews et al. 2019). Astrogliosis and microgliosis were evident. Molecular profiling of these infected brains revealed a significant increase in human genes, such as IFIT1-5, ISG15, MX2, and OAS1 – pertaining to interferon signaling. The other pathways which were upregulated included toll-like receptors and pattern-recognition receptors, indicating activation of the innate immune response and increased inflammation. The analyses of mouse genes indicated that ERK, integrin, MAPK, and apoptosis signaling were differentially regulated in association with neurodegeneration. The human microglial mouse closely reflects the pathophysiology of human HIV-1 infection with astrogliosis, microgliosis, productive viral infection of microglia, synaptic alterations, and inflammatory responses. This model will prove useful for studies of neural-glial cross talk and studies designed to locate and eliminate the viral reservoir.

Excision of HIV-1 proviral DNA in animal models was reported in 2017 (Yin et al. 2017). Researchers have since demonstrated the feasibility and efficiency of disrupting HIV-1 provirus using an all-in-one adeno-associated virus (AAV) combined with multiplex sgRNAs and SaCas9 in three different animal models. Recently, the birth of gene-edited babies in 2018 has aroused widespread criticism across the scientific disciplines. This was the first time that the CCR5 gene-edited human embryos by clustered regularly interspaced short palindromic repeat (CRISPR)/CRISPR-associated nuclease 9 (Cas9) were implanted into women with the intent of documenting resistance to HIV infection in the infants. The CRISPR-Cas9 system has been engineered as an effective gene-editing technology with the potential to treat HIV infection. It can be used to target cellular co-factors or the HIV-1 genome itself to reduce HIV-1 infection and clear the provirus, as well as to induce transcriptional activation of latent HIV from viral reservoirs for the purpose of eradication (Xiao et al. 2019).

Mutations in the amyloid precursor protein (APP), presenilin 1 (PSEN1), and presenilin 2 (PSEN2) are the main causes of autosomal dominant early-onset AD, while the APOEε4 allele is a major risk factor for late-onset AD (Dawson et al. 2018). Genes associated with APP, PSEN1, and PSEN 2 have been targeted in transgenic mice models of AD focusing on a single-gene mutation, but the short rodent life cycle has been limiting due to the longevity required to contract AD. Pigs have been used as an alternative to mice due to improved neuroanatomical correlates and greater similarities between human and porcine APP, including identical secretase cleavage sites, and similar production of αβ40 and αβ42. However, minipigs cloned expressing APP695 or PSEN1 using SCNT technology have failed to show pathologic changes up to 3 years after production. This time period is likely premature to discover any relevant outcome, as dementia symptoms occur much later in the life cycle. However, interest has been garnered through in vitro analysis of radial glial cells from mutant APP minipigs that have shown early deficits in alterations of ribosomal expression and cell cycle genes and increased astrogenesis and hyperphosphorylation of Tau. The use of NHPs, as utilized for HAD, unfortunately has not been similarly successfully undertaken in AD using NHPs (Heuer et al. 2012), who – while developing senile plaques and cognitive deficits – have not shown the widespread neuronal loss typifying AD.

PD has been induced in NHPs, ovine, porcine, and feline models through intravenous delivery of the dopamine toxin, MPTP, a lipophilic toxin that easily crosses the blood-brain barrier. While successful in inducing PD symptomatology and pathology in animal models, the disease effects have been transient in most mammals – apart from the Gottingen minipig and NHPs – with the NHP model being the only one to exhibit Lewy bodies with dopaminergic depletion in the substantia nigra. Due to the fact that no single-gene mutation has been linked to PD, single-gene knock-out models have been evaluated. They have shown variability in PD expression. However, single-gene edit PD models have failed in larger animal models. Yet, in the case of HIV, a polygenetic porcine model with a triple PD gene mutation targeting parkin/DJ-1/PINK1 gene loci using CRISPR/Cas9 technology in pronuclear embryos has been discovered. If this triple targeted genetically engineered manipulation proves to significantly parallel human pathology, rodent models would no longer be needed. This would be a groundbreaking step forward for meaningfully studying therapeutic modalities aimed at human polygenetic diseases.

6 Biological Measures to Detect All-Type Dementia Risk

Before the era of effective ART, the CD4 cell count, and plasma viral load were routinely used to screen clinically for neurocognitive impairment and HAND. However, in the era of effective ART, both CD4 cell count and plasma viral load measures have shown no clear-cut relationships with the presence of HIV-associated neurocognitive impairment; this may be due to the persistence of neuroinflammation regardless of suppression of plasm viral load to non-detectable levels and to higher, sustained CD4 cell counts on effective ART (Saylor et al. 2016). Some measures that are used to detect neurocognitive impairment today include CSF measures (e.g., TNF-α, IL-1, IL-6, MCP-1, CXCL10, sCD14, sCD163, neopterin, β2 microglobulin, and NFL), which variably may also reflect the presence of neuroinflammation, the HIV reservoir, and neurodegeneration. On neuroimaging with magnetic resonance spectroscopy (MRS), there may be changes in glutamate and/or glutamine levels in the frontal lobe and/or basal ganglia – as well as decreased NAA peaks and increased choline and myo-inositol peaks – depending on the region of interest selected. Although this technique is clinically available with structural MRI, this metabolic profile has not yet been conclusively proven to detect HAND and differentiate it from all other diseases with similar neurocognitive manifestations and metabolic profiles, particularly progressive multifocal leukoencephalopathy (PML), for example. Of note, another more recent neuroimaging technique applicable to neuro-inflammation may prove to be of critical utility to HAND diagnosis and treatment response assessment in the future – i.e., TSPO PET scanning.

At the time of the review of the criteria for HAND during the Frascati Conference in 2005, a thorough review of measures indicative of HAND was conducted; however, the conclusion at the time was that no positive proof could be offered for any specific measure for inclusion in the HAND diagnostic criteria. Currently, three classes of HAND measures are being pursued to establish objective laboratory-based definitions of HIV-associated neurologic injury – CSF, blood, and neuroimaging measures. While there have been great efforts since the Frascati diagnostic consensus conference in the identification of HAND, it currently remains the case that no reliable and validated measure or combination of measures for HAND have yet been confirmed to be diagnostic (Peterson et al. 2014; Rahimian and He 2017). Nevertheless, the need for reliable measures of HAND is expected to increase further, as persons living with HIV continue to age and their vulnerability to neurodegenerative diseases beyond HAND grows – particularly for AD. It might be concluded that there remains a major need for a reliable and valid measure (or, perhaps more likely, a combination of measures) to be established in order to improve the detection (as well as the prevention) of HAND.

Measures that are more clearly delineated for AD include neuroimaging showing global brain atrophy with early disproportionate, symmetrical involvement of medial temporal lobe structures, including the hippocampi. Progressive atrophy of the parietal and occipital lobes is also indicative of AD and particularly helpful in distinguishing AD from FTD, with visualization of posterior atrophy improving the distinction of AD from other causes of dementia more generally. In addition, the temporal lobe atrophy pattern helps to differentiate AD from the aging process, DLB, and VCID with high rates of sensitivity and specificity. It should be noted that the medial temporal lobe atrophy does not differentiate AD from FTD – though temporal lobe atrophy is helpful in differentiating subgroups of FTD. The CSF is also an important diagnostic source in AD with decreased levels of β-amyloid and increased levels of Tau and phosphorylated Tau 181 being diagnostic for AD. Future measures under consideration in AD include novel MRI techniques such as diffusion tensor imaging (DTI) and resting-state functional MRI (rsfMRI) probing white matter tracts. The rsfMRI can measure resting brain state conditions observed through changes in blood flow in the brain creating a blood oxygen level-dependent (BOLD) signal. Arterial spin labeling (ASL) is another MRI technique that shows cerebral blood flow patterns with a nonionizing and completely noninvasive MRI technique using magnetically labeled arterial blood water protons as an endogenous tracer. Quantified Tau PET imaging has been added quantified amyloid PET imaging as a focus in AD. The combination of PET imaging (a hybrid neuroimaging technique) may prove valuable as a future diagnostic, prognostic, and treatment response resource in AD. In terms of CSF measures in AD, a recent focus has been on the import of using the concentration ratio of Aβ42 to Aβ40 (Aβ42/40 Ratio) over the concentration of Aβ42 alone when identifying patients with AD. Neuroinflammatory markers have also contributed with S100B elevations being of interest in mild/moderate AD; however, S100B is of interest in a number of diseases, including HAND. Glycoprotein YKL-40 – activated by microglia – might also prove to play a role as a relevant measure in AD and FTD in the future. CSF neurofilament light chain (NFL) elevation is a measure of axonal degeneration and is robustly elevated in the blood of many neurological and neurodegenerative conditions, including both HAND and AD.

6.1 A Focus on FTD

FTD comprises a group of neurodegenerative disorders characterize by frontal and temporal lobe atrophy. It was one of the lesser known dementias until advances were made revealing its genetic and pathological foundations. FTD is traditionally difficult to diagnose due to the heterogeneity in its signs and symptoms. There is currently no biochemical test or marker of disease activity for FTD, and the clinical diagnosis rests on the opinion of an experienced neurologist. There are three groups based on the functional neuroanatomy of the frontal and temporal lobes: (1) behavioral variant frontotemporal dementia (bvFTD), (2) semantic dementia (SD), and (3) progressive, nonfluent aphasia (PNFA) (Cardarelli et al. 2010). BvFTD is characterized by changes in social behavior and conduct, with loss of social awareness and poor impulse control. SD FTD is characterized by the loss of semantic understanding, resulting in impaired word comprehension, although speech remains fluent and grammatically intact PNFA FTD is characterized by progressive difficulties in speech production, which overlaps with AD.

FTD has two well-defined types of clinical variants – the behavioral variant (bvFTD), wherein personality and social deterioration are prominent, and the primary progressive aphasia (PPA) variant, wherein language decline is prominent. The latter clinical form, PPA, has clinical subtypes including semantic dementia (SD), progressive nonfluent aphasia (PNFA), logogenic aphasia (LPA), and progressive apraxia of speech. Structural neuroimaging plays an important role in differentiating the clinical types of FTD. For example, bvFTD shows medial frontal, anterior insula cortical, and orbitofrontal atrophy, and SD shows bilateral but focused left-sided atrophy of the anterior temporal lobes and anterior fusiform gyri. The MRI quantification of atrophy rates of the whole brain and lobar volumes is useful for objective measures of FTD disease progression, overall. Structural neuroimaging is critical in differentiating FTD from other diseases as well. The clinical forms of FTD show region-specific changes overlapping with Pick's disease, progressive supranuclear palsy, and cortico-basal degeneration – all of which can present clinically as FTD-like syndromes.

There are also histopathologic subtypes of FTD. The main histopathologic subtypes of FTD found at autopsy are (1) FTD-TDP, (2) FTD-Tau, and (3) FTD-FUS (Bahia et al. 2013). FTD-TDP is the most common histopathologic class of FTD and is associated with the TDP-43 protein. FTD-Tau is considered slightly less common, and FTD-FUS (fused in sarcoma protein) is rare. TDP-43 is the main ubiquitinated peptide in Tau-negative FTD. It is typically a nuclear protein, and its aggregation and cytoplasmic translocation may represent steps in FTD pathogenesis as a "TDP-43 proteinopathy." Increased levels of TDP-43 have been found in the CSF of persons with FTD, but the significance of this finding is, as yet, unclear (Feneberg et al. 2018).

FTD-TDP, specifically, is characterized by TDP-43 inclusions that exhibit considerable pathologic, clinical, and genetic heterogeneity. Patients present with a broad spectrum of clinical signs and symptoms, ranging from executive dysfunction (with or without motor neuron disease) to language dysfunction, including semantic and nonfluent/agrammatic variants of PPA. In addition, 20–40% of FTD-TDP patients have a family history of neurodegenerative disease. This is in part due to a high proportion associated with disease-causing mutations. Recent harmonization of the FTD-TDP subtype has demarcated four histopathologic types A to D. Type A is bvFTD or nonfluent/agrammatic PPA, together with a disease-causing mutation in the GRN gene (granulin precursor). Type B is associated with bvFTD including patients with motor neuron disease, together with other disease-causing mutations in C9orf72 (chromosome 9, open reading frame 72). Type C is associated with bvFTD or semantic variant PPA, is typically sporadic FTD, and has no association with disease-causing genetic mutations. Type D is associated with mutation in the valosin-containing protein that causes familial inclusion body myositis, Paget's disease of bone, FTD, and/or ALS – called "multisystem proteinopathy." Another histopathologic pattern reveals an apparently distinct form of FTD-TDP pathology with a wide neuroanatomic distribution and a very rapid clinical course from onset to death within 3 years (Lee et al. 2017), which may become designated as "type E."

Functional neuroimaging using FDG-PET shows frontal and temporal lobe hypometabolism in FTD and is helpful in differentiating FTD from AD – particularly when amyloid pathology is present. It is also helpful in identifying patients with behavioral changes having normal structural neuroimaging results. Furthermore, as with AD, diffusion tensor imaging (DTI) and resting-state functional (rsf) MRI have shown considerable promise to detect presymptomatic and disease-specific network breakdown for the different types of FTD. In addition, CSF biomarkers are also helpful in distinguishing FTD from AD – particularly with CSF total-Tau (T-tau) levels being lower in FTD than those seen in AD and with phosphorylated Tau (P-Tau) level elevations seen in AD versus reductions P-Tau level seen in FTD.

To our knowledge, no specific associations have been reported between FTD and HAND. However, AD, DLB (see below), PD with dementia, and FTD share the process of pathogenic protein accumulation that also occurs in HAND. Hence, there may be a common underlying pathogenic mechanism related to defective protein clearance involved in each disease. One critical protein clearance pathway is autophagy, which has been reported across dementia types and may contribute to an association between the foregoing dementias.

6.2 A Focus on DLB

As another dementia of import, DLB shows preserved volume of the medial temporal lobes – relative to what is seen in AD. In addition, hippocampal sparing distinguishes DLB from AD. However, the utilization of MRI techniques as proof positive for the differentiation of DLB from AD has not yet been established. In 2005, imaging of the dopamine transporter (DAT) was added to the diagnostic criteria for DLB – as a supporting piece of evidence (van der Zande et al. 2016). Functional PET/SPECT measures of dopaminergic loss in the basal ganglia are suggestive of DLB, and 123 I-FP CIT SPECT imaging may aid in differentiating DLB from AD and from some forms of FTD with accuracy. However, both DLB and atypical PD show similar nigrostriatal abnormalities, and their differentiation is less accurate. In fact, amyloid PET imaging correlates with amyloid neuropathology in PD patients (Akhtar et al. 2016). Emerging evidence suggests that amyloid imaging predicts dementia in PD. At autopsy, an elevated level of amyloid is commonly associated with other pathologies – not only AD but also VCID. To date, we could find no evidence that DLB or PD with dementia overlaps HAND. Hence, the exploration of this potentially significant dementia comorbidity remains an area for future research.

7 Clinical Implications, Translational Aspects, and Future Directions

More people successfully aging with HIV has led to a call for added attention to examining not only the associated risk for HAND, which is increased, but also the risk for comorbidities with other types of dementia. While effective ART has permitted persons living with HIV to survive longer, it has also added to the complexity and uncertainty of establishing risk relationships for HAND (and HAD specifically) with other types of dementias – as well as with normal age-associated cognitive impairment, which itself shares some relationships with HAND. "Dementia," overall, has become a burgeoning field where research and education are still sorely needed. More focused methodologies are particularly lacking to address different aspects of the relative risk for HAND (and HAD) versus other types of dementia. This is an even greater concern in examining relative risk factors for the lower levels of HAND (ANI and MND) and MCI – as the predecessors of HAD and AD, respectively. Pattern profiling of neurocognitive domains in dementia has shown significant variability when comparing one dementia to another. Currently, specific neurocognitive screening tools have not been developed that are sufficient in reliably differentiating HAND (including HAD) from other types of dementia to be clinically useful in and of themselves. Moreover, broader dementia pattern profiles vary from one dementia to another with a variability beyond cognition to other phenotypic variables, functional outcomes, and, as expected, markers of disease onset and progression. Hence, more specific neurobehavioral as well as neurocognitive tools are needed for dementia pattern profile studies – both in general and among persons living with HIV. The potential for the prevention of HAND among persons living with HIV is an area yet more distant from the implications of current research results.

The FINGER trial (Finnish Geriatric Intervention study to Prevent Cognitive Impairment and Disability) focused on life style changes and showed that positive life style changes have beneficial effects on the metabolic, vascular, and cognitive abilities of patients (Aung et al. 2019). Improved medication adherence, more available supportive monitoring, dementia-focused mental health care, substance use reductions, stigmatization reductions, and higher overall quality of care for persons with dementia could accelerate a reduction in the functional impact of a dementia diagnosis – not just for a person with dementia but also for her/his caregivers. Most importantly, physicians themselves need to be more cognizant of the improvements in treatment responses documented among patients across dementias due to a higher quality of patient care, overall. In line with these findings, those caring for older persons living with HIV need to emphasize the critical importance of quality of care overall as well – just as is true for any person living with dementia. Stigma against persons living with HIV should not dictate the quality of care received by these patients any more than stigma should impact the care of any persons living with dementia, which nevertheless remains a common issue. Proper provisions regarding avoiding misdiagnosis of dementia remains a concern, especially for an older person living with HIV. Complicating matters yet further are the depression and isolation more commonly experienced by older persons in the population as a whole – in both the social and medical spheres. While organizations such as the Alzheimer's Association have made tremendous strides in pursuing personalized, wellness-oriented care for persons living with dementia in general, more specific educational programs are required to be directed at older persons living with HIV, as this group continues to progress further into the aging demographic – along with the need for additional insights into factors influencing the stages of HAND as they progress with aging.

8 Summary

Despite many advances that have been made since the advent of the era of effective ART in controlling systemic HIV infection, the control of CNS HIV infection has not advanced in parallel. In fact the proportion of persons living with HIV manifesting with HAND has persisted at a high rate. Treatment gains in HAND have remained limited. Linkages within the pathophysiology of HAND in terms of its neuro-inflammatory, virologic, and neurodegenerative aspects are important to understand and further investigate in the future to effect a change in treatment efficacy. Recent advances in rodent and NHP animal models can help contribute to the extended knowledge base required to achieve this treatment goal. A greater depth of understanding of the risk factors among dementias will elucidate how HAND relates to the other dementias in both risk and disease progression. Currently, the concerns are greatest with respect to the comorbidity with AD, where absolute numbers of persons having both diseases will increase, although there is not yet a true interaction effect demonstrated with HAND. Another high level of concern involves the comorbidity of HAND with VCID, as a high potential for a true interaction with HAND exists but has not yet been adequately investigated. There are also concerns related to a common problem with protein clearance and other degenerative types of dementia, possibly mediated by deficits in autophagy. Improved identification of reliable, valid, and specific pathophysiologic and/or diagnostic measures for HAND along with a greater focus on research in this area may hold the key to achieving consistent early detection and diagnoses of dementia by specific type in the future. In summary, this area of study is critical for the proper determination of disease-specific versus shared pathogenic factors among the types of dementia and, therefore, for defining the risk for comorbidities of dementia types among persons living with HIV.

References

Akhtar RS, Xie SX, Brennan L, Pontecorvo MJ, Hurtig HI, Trojanowski JQ, Weintraub D, Siderowf AD (2016) Amyloid-beta positron emission tomography imaging of Alzheimer's pathology in Parkinson's disease dementia. Mov Disord Clin Pract 3:367–375

Akkina R, Allam A, Balazs A et al (2016) Improvements and limitations of humanized mouse models for HIV research: NIH/NIAID "meet the experts" 2015 workshop summary. AIDS Res Hum Retrovir 32(2):109–119. https://doi.org/10.1089/aid.2015.0258

Aung HL et al (2019) How all-type dementia risk factors and modifiable risk interventions may be relevant to the first-generation aging with HIV infection? Eur Geriatr Med 10(2):227–238. https://doi.org/10.1007/s41999-019-00164-6

Azad NA et al (2007) Gender differences in dementia risk factors. Gend Med 4(2):120–129. https://doi.org/10.1016/s1550-8579(07)80026-x

Bagetta G, Piccirilli S, Del Duca C et al (2004 Feb) Inducible nitric oxide synthase is involved in the mechanisms of cocaine enhanced neuronal apoptosis induced by HIV-1 gp120 in the neocortex of rat. Neurosci Lett 356(3):183–186. https://doi.org/10.1016/j.neulet.2003.11.065

Bahia VS, Takada LT, Deramecourt V (2013) Neuropathology of frontotemporal lobar degeneration: a review. Dement Neuropsychol 7(1):19–26. https://doi.org/10.1590/S1980-57642013DN70100004

Baumgart M et al (2015) Summary of the evidence on modifiable risk factors for cognitive decline and dementia: a population-based perspective. Alzheimers Dement 11(6):718–726. https://doi.org/10.1016/j.jalz.2015.05.016

Becker JT, Maruca V, Kingsley LA et al (2012) Factors affecting brain structure in men with HIV disease in the post-HAART era. Neuroradiology 54(2):113–121

Becker JT, Martinson JJ, Penugonda S et al (2015) J Neurovirol 21:24. https://doi.org/10.1007/s13365-014-0290-2

Brown A, Sacktor N, Marder K et al (2012) CCL3L1 gene copy number in individuals with and without HIV-associated neurocognitive disorder. Curr Biomark Find 2012(2):1–6. https://doi.org/10.2147/CBF.S27685

Brown LK, Kennard BD, Emslie GJ et al (2016) Effective treatment of depressive disorders in medical clinics for adolescents and young adults living with HIV: a controlled trial. J Acquir Immune Defic Syndr 71(1):38–46

Cardarelli R, Kertesz A, Knebl J (2010) Frontotemporal dementia: a review for primary care physicians. Am Fam Physician 82(11):1372–1377. https://www.aafp.org/afp/2010/1201/p1372.html

Carroll A, Brew B (2017) HIV-associated neurocognitive disorders: recent advances in pathogenesis, biomarkers, and treatment. F1000Res 6:312. https://doi.org/10.12688/f1000research.10651.1

Cassol E, Misra V, Morgello S, Kirk GD, Mehta SH, Gabuzda D (2015) Altered monoamine and acylcarnitine metabolites in HIV-positive and HIV-negative subjects with depression. J Acquir Immune Defic Syndr 69(1):18–28. https://doi.org/10.1097/QAI.0000000000000551

Chiesi A, Vella S, Dally LG et al (1996) Epidemiology of AIDS dementia complex in Europe. J Acquir Immune Defic Syndr Hum Retrovirol 11(1):39–44. https://doi.org/10.1097/00042560-199601010-00005

Coleman SM, Blashill AJ, Gandhi RT, Safren SA, Freudenreich O (2012) Impact of integrated and measurement-based depression care: clinical experience in an HIV clinic. Psychosomatics 53 (1):51–57

Cykowski MD, Powell SZ, Schulz PE et al (2017) Hippocampal sclerosis in older patients: practical examples and guidance with a focus on cerebral age-related TDP-43 with sclerosis. Arch Pathol Lab Med 141(8):1113–1126

Cysique LA, Brew BJ (2019) Vascular cognitive impairment and HIV-associated neurocognitive disorder: a new paradigm. J Neurovirol 25(5):710–721

Dawson TM et al (2018) Animal models of neurodegenerative diseases. Nat Neurosci 21 (10):1370–1379. https://doi.org/10.1038/s41593-018-0236-8

Do AN, Rosenberg ES, Sullivan PS, Beer L, Strine TW, Schulden JD, Fagan JL, Freedman MS, Skarbinski J (2014 Mar 24) Excess burden of depression among HIV-infected persons receiving medical care in the United States: data from the medical monitoring project and the behavioral risk factor surveillance system. PLoS One 9(3):e92842. https://doi.org/10.1371/journal.pone.0092842

Dolan M, Kulkarni H, Camargo J et al (2007) CCL3L1 and CCR5 influence cell-mediated immunity and affect HIV-AIDS pathogenesis via viral entry-independent mechanisms. Nat Immunol 8:1324–1336

Donahoe R, Vlahov D (1998) Opiates as potential cofactors in progression of HIV-1 infections to AIDS. J Neuroimmunol 83(1–2):77–87. https://doi.org/10.1016/s0165-5728(97)00224-5

El-Hage N, Rodriguez M, Dever SM, Masvekar RR, Gewirtz DA, Shacka JJ (2015) HIV-1 and morphine regulation of autophagy in microglia: limited interactions in the context of HIV-1 infection and opioid abuse. J Virol 89(2):1024–1035. https://doi.org/10.1128/JVI.02022-14

Feneberg E, Gray E, Ansorge O, Talbot K, Turner MR (2018) Towards a TDP-43-based biomarker for ALS and FTLD. Mol Neurobiol 55(10):7789–7801. https://doi.org/10.1007/s12035-018-0947-6

Fjell AM, McEvoy L, Holland D, Dale AM, Walhovd KB, Alzheimer's Disease Neuroimaging Initiative (2014) What is normal in normal aging? Effects of aging, amyloid and Alzheimer's disease on the cerebral cortex and the hippocampus. Prog Neurobiol 117:20–40. https://doi.org/10.1016/j.pneurobio.2014.02.004

Goodkin K, Wilkie FL, Concha M, Hinkin CH, Symes S, Baldewicz TT, Asthana D, Fujimura RK, Lee D, van Zuilen MH, Khamis I, Shapshak P, Eisdorfer C (2001) Aging and neuro-AIDS conditions and the changing spectrum of HIV-1-associated morbidity and mortality. J Clin Epidemiol 54(Suppl 1):S35–S43. https://doi.org/10.1016/S0895-4356(01)00445-0

Goodkin K, Hardy DJ, Singh D, Lopez E (2014) Diagnostic utility of the international HIV dementia scale for HIV-associated neurocognitive impairment and disorder in South Africa. J Neuropsychiatry Clin Neurosci 26(4):352–358. https://doi.org/10.1176/appi.neuropsych.13080178

Goodkin K, Miller E, Cox C, Reynolds S, Becker JT, Martin E, Selnes O, Ostrow D, Sacktor N (2017) Effect of aging on neurocognitive function by stage of HIV infection: evidence from the multi-center AIDS cohort study. Lancet HIV 4(9):e411–e422

Hammond ER, Crum RM, Treisman GJ et al (2016) Persistent CSF but not plasma HIV RNA is associated with increased risk of new-onset moderate-to-severe depressive symptoms; a prospective cohort study. J Neurovirol 22(4):479–487. https://doi.org/10.1007/s13365-015-0416-1

Heuer E, Rosen RF, Cintron A, Walker LC (2012) Nonhuman primate models of Alzheimer-like cerebral proteopathy. Curr Pharm Des 18(8):1159–1169

Hines LJ, Miller EN, Hinkin CH et al (2016) Brain Imaging Behav 10:640. https://doi.org/10.1007/s11682-015-9441-1

Irwin DJ, Hurtig HI (2018) The contribution of tau, amyloid-beta and alpha-synuclein pathology to dementia in lewy body disorders. J Alzheimers Dis Parkinsonism 8(4):444

Kaul M (2009) HIV-1 associated dementia: update on pathological mechanisms and therapeutic approaches. Curr Opin Neurol 22(3):315–320. https://doi.org/10.1097/wco.0b013e328329cf3c

Kennard B, Brown L, Hawkins L et al (2014) Development and implementation of health and wellness CBT for individuals with depression and HIV. Cogn Behav Pract 21(2):237–246. https://doi.org/10.1016/j.cbpra.2013.07.003

Kivipelto M, Mangialasche F, Ngandu T (2018) Lifestyle interventions to prevent cognitive impairment, dementia and Alzheimer disease. Nat Rev Neurol 14(11):653–666. https://doi.org/10.1038/s41582-018-0070-3

Kraft-Terry SD, Buch SJ, Fox HS, Gendelman HE (2009) A coat of many colors: neuroimmune crosstalk in human immunodeficiency virus infection. Neuron 64(1):133–145. https://doi.org/10.1016/j.neuron.2009.09.042

Kumar N, Chahroudi A, Silvestri G (2016) Animal models to achieve an HIV cure. Curr Opin HIV AIDS 11(4):432–441. https://doi.org/10.1097/COH.0000000000000290

Lee EB, Porta S, Michael Baer G et al (2017) Expansion of the classification of FTLD-TDP: distinct pathology associated with rapidly progressive frontotemporal degeneration. Acta Neuropathol 134(1):65–78

Lennon MJ, Makkar SR, Crawford JD, Sachdev PS (2019) Midlife hypertension and Alzheimer's disease: a systematic review and meta-analysis. J Alzheimers Dis 71(1):307–316. https://doi.org/10.3233/jad-190474

Mackiewicz MM et al (2019) Pathogenesis of age-related HIV neurodegeneration. J Neurovirol 25 (5):622–633. https://doi.org/10.1007/s13365-019-00728-z

Mathews S, Branch Woods A, Katano I, Makarov E, Thomas MB, Gendelman HE et al (2019) Human Interleukin-34 facilitates microglia-like cell differentiation and persistent HIV-1 infection in humanized mice. Mol Neurodegener 14(1):12. https://doi.org/10.1186/s13024-019-0311-y

Métral M, Darling K, Locatelli I et al (2019) The neurocognitive assessment in the metabolic and aging cohort (NAMACO) study: baseline participant profile. HIV Med 21(1):30–42. https://doi.org/10.1111/hiv.12795

Moore DJ, Arce M, Moseley S et al (2011) Family history of dementia predicts worse europsychological functioning among HIV-infected persons. J Neuropsychiatry Clin Neurosci 23(3):316–323. https://doi.org/10.1176/jnp.23.3.jnp316

Olivier I et al (2018) Risk factors and pathogenesis of HIV-associated neurocognitive disorder: the role of host genetics. Int J Mol Sci 19(11):3594. https://doi.org/10.3390/ijms19113594

Paulson H, Igo I (2011) Genetics of dementia. Semin Neurol 31(05):449–460. https://doi.org/10.1055/s-0031-1299784

Pence BW, Gaynes BN, Williams Q et al (2012) Assessing the effect of measurement-based care depression treatment on HIV medication adherence and health outcomes: rationale and design of the SLAM DUNC study. Contemp Clin Trials 33(4):828–838. https://doi.org/10.1016/j.cct.2012.04.002

Peterson J, Gisslen M, Zetterberg H, Fuchs D, Shacklett BL, Hagberg L, Yiannoutsos CT, Spudich SS, Price RW (2014) Cerebrospinal fluid (CSF) neuronal biomarkers across the spectrum of

HIV infection: hierarchy of injury and detection. PLoS One 9(12):e116081. https://doi.org/10.1371/journal.pone.0116081

Pfefferbaum A et al (2014) Accelerated aging of selective brain structures in human immunodeficiency virus infection: a controlled, longitudinal magnetic resonance imaging study. Neurobiol Aging 35(7):1755–1768. https://doi.org/10.1016/j.neurobiolaging.2014.01.008

Pfefferbaum A, Zahr NM, Sassoon SA, Kwon D, Pohl KM, Sullivan EV (2018) Accelerated and premature aging characterizing regional cortical volume loss in human immunodeficiency virus infection: contributions from alcohol, substance use, and hepatitis C coinfection. Biol Psychiatry Cogn Neurosci Neuroimaging 3(10):844–859

Power C, Selnes OA, Grim JA, Mcarthur JC (1995) HIV dementia scale. PsycTESTS Dataset. https://doi.org/10.1037/t55375-000

Rahimian P, He JJ (2017) HIV/neuroAIDS biomarkers. Prog Neurobiol 157:117–132. https://doi.org/10.1016/j.pneurobio.2016.04.003

Reas ET (2017) Amyloid and tau pathology in normal cognitive aging. J Neurosci 37(32):7561–7563

Robertson KR, Kapoor C, Robertson WT, Fiscus S, Ford S, Hall CD (2004) No gender differences in the progression of nervous system disease in HIV infection. J Acquir Immune Defic Syndr 36(3):817–822. https://doi.org/10.1097/00126334-200407010-00008

Rubin LH, Neigh GN, Sundermann EE et al (2019) Curr Psychiatry Rep 21:94. https://doi.org/10.1007/s11920-019-1089-x

Ryan EL, Baird R, Mindt MR et al (2005) Neuropsychological impairment in racial/ethnic minorities with HIV infection and low literacy levels: effects of education and reading level in participant characterization. J Int Neuropsychol Soc 11(7):889–898. https://doi.org/10.1017/s1355617705051040

Sacktor NC, Wong M, Nakasujja N et al (2005) The international HIV dementia scale: a new rapid screening test for HIV dementia. AIDS 19(13):1367–1374. https://doi.org/10.1097/01.aids.0000180790.77379.3a

Satz P, Morgenstern H, Miller EN et al (1993) Low education as a possible risk factor for cognitive abnormalities in HIV-1. J Acquir Immune Defic Syndr 6(5):503–511. https://doi.org/10.1097/00126334-199305000-00011

Saylor D, Dickens AM, Sacktor N et al (2016) HIV-associated neurocognitive disorder–pathogenesis and prospects for treatment [published correction appears in Nat Rev Neurol. 2016 May;12(5):309]. Nat Rev Neurol 12(4):234–248. https://doi.org/10.1038/nrneurol.2016.27

Simoni JM, Safren SA, Manhart LE, Lyda K, Grossman CI, Rao D, Mimiaga MJ, Wong FY, Catz SL, Blank MB, DiClemente R, Wilson IB (2011) Challenges in addresing depression in HIV research: assessment, cultural context, and methods. AIDS Behav 15(2):376–388

Soontornniyomkij V, Moore DJ, Gouaux B et al (2012) Cerebral β-amyloid deposition predicts HIV-associated neurocognitive disorders in APOE ε4 carriers. AIDS 26(18):2327–2335. https://doi.org/10.1097/qad.0b013e32835a117c

Stern Y, Andrews H, Pittman J et al (1992) Diagnosis of dementia in a heterogeneous population: development of a neuropsychological paradigm-based diagnosis of dementia and quantified correction for the effects of education. Arch Neurol 49(5):453–460. https://doi.org/10.1001/archneur.1992.00530290035009

Theodore S, Cass WA, Nath A et al (2006) Inhibition of tumor necrosis factor-alpha signaling prevents human immunodeficiency virus-1 protein Tat and methamphetamine interaction. Neurobiol Dis 23(3):663–668. https://doi.org/10.1016/j.nbd.2006.05.005

Tozzi V, Balestra P, Vlassi C, Salvatori MF (2007) Author's response to "Persistence of neuropsychologic deficits despite long-term highly active antiretroviral therapy in patients with HIV-related neurocognitive impairment". J Acquir Immune Defic Syndr 46(5):657–658. https://doi.org/10.1097/qai.0b013e318157b0f0

Uthman OA, Magidson JF, Safren SA, Nachega JB (2014) Depression and adherence to antiretroviral therapy in low-, middle- and high-income countries: a systematic review and meta-analysis. Curr HIV/AIDS Rep 11(3):291–307. https://doi.org/10.1007/s11904-014-0220-1

van der Zande JJ, Booij J, Scheltens P, Raijmakers PG, Lemstra AW (2016) [(123)]FP-CIT SPECT scans initially rated as normal became abnormal over time in patients with probable dementia with Lewy bodies. Eur J Nucl Med Mol Imaging 43(6):1060–1066. https://doi.org/10.1007/s00259-016-3312-x

Wium-Andersen IK, Osler M, Jørgensen MB et al (2019) Antidiabetic medication and risk of dementia in patients with type 2 diabetes: a nested case–control study. Eur J Endocrinol 181 (5):499–507. https://doi.org/10.1530/eje-19-0259

Wong MH, Robertson K, Nakasujja N et al (2007) Frequency of and risk factors for HIV dementia in an HIV clinic in sub-Saharan Africa. Neurology 68(5):350–355. https://doi.org/10.1212/01.wnl.0000252811.48891.6d

Wright EJ, Grund B, Robertson K et al (2010) Cardiovascular risk factors associated with lower baseline cognitive performance in HIV-positive persons. Neurology 75(10):864–873. https://doi.org/10.1212/WNL.0b013e3181f11bd8

Xiao Q, Guo D, Chen S (2019) Application of CRISPR/Cas9-based gene editing in HIV-1/AIDS therapy. Front Cell Infect Microbiol 9:69. https://doi.org/10.3389/fcimb.2019.00069

Yin C, Zhang T, Qu X et al (2017) In vivo excision of HIV-1 provirus by saCas9 and multiplex single-guide RNAs in animal models. Mol Ther 25(5):1168–1186. https://doi.org/10.1016/j.ymthe.2017.03.012

Part IV
Screening, Interventions and Clinical Management

Screening for HIV-Associated Neurocognitive Disorders: Sensitivity and Specificity

Reuben N. Robbins, Travis M. Scott, Hetta Gouse, Thomas D. Marcotte, and Sean B. Rourke

Contents

R. N. Robbins (✉)
HIV Center for Clinical and Behavioral Studies, New York State Psychiatric Institute and Columbia University Medical Center, New York, NY, USA
e-mail: rnr2110@cumc.columbia.edu

T. M. Scott
Department of Psychology, Fordham University, The Bronx, NY, USA

VA Palo Alto Health Care System, Sierra Pacific MIRECC, Palo Alto, CA, USA
e-mail: tshivleyscott@fordham.edu

H. Gouse
Department of Psychiatry and Mental Health, University of Cape Town, Cape Town, South Africa
e-mail: hetta.gouse@uct.ac.za

T. D. Marcotte
HIV Neurobehavioral Research Program, Center for Medicinal Cannabis Research, University of California, San Diego, San Diego, CA, USA
e-mail: tmarcotte@ucsd.edu

S. B. Rourke
Centre for Urban Health Solutions, St Michael's Hospital, Toronto University, Toronto, ON, Canada
e-mail: sean.rourke@utoronto.ca

Curr Topics Behav Neurosci (2021) 50: 429–478
DOI 10.1007/7854_2019_117
Published Online: 17 July 2020

Abstract HIV-associated neurocognitive disorder (HAND) remains prevalent among people living with HIV (PLWH), especially the mild forms, even those with well-controlled HIV. Recommendations from the literature suggest routine and regular screening for HAND to detect it early and manage it effectively and adjust treatments, if warranted, when present. However, screening for HAND is not routinely done, as there are no current guidelines on when to screen and which test or tests to use. Furthermore, many of the available screening tools for HAND often cannot accurately detect the mild forms of HAND and require highly trained healthcare professionals to administer and score the tests, a requirement that is not feasible for those low- and middle-income countries with the highest HIV incidence and prevalence rates. The purpose of this chapter was to review recent research on screening tests to detect HAND and report on the strengths, limitations, and psychometric properties of those tests to detect HAND.

Keywords HAND · HIV · Neurocognitive impairment · Psychometrics · Screening · Test

1 Screening for HAND: Needs Versus Practice

The neurocognitive impairment (NCI) related to HIV-associated neurocognitive disorder (HAND) is one of the most common sequelae and comorbid conditions of HIV infection and has significant medical, functional, and public health consequences (Hinkin et al. 2002, 2004; Heaton et al. 2004; Marcotte et al. 2004; Van Gorp et al. 2007; Gorman et al. 2009; Ettenhofer et al. 2009, 2010; Vivithanaporn et al. 2010; Schouten et al. 2011; Umaki et al. 2013). NCI can be caused by HIV infection of the central nervous system (CNS; resulting in neuropathology of the basal ganglia and white matter) or by other comorbid or preexisting conditions (e.g., chronic substance abuse, head trauma, CNS opportunistic infections) and/or a combination of both (Grant 2008; Heaton et al. 2010, 2011; Grant and Sacktor 2012). NCI in HIV typically causes impairments in mental processing speed, learning, memory, attention and concentration, higher-order executive functions, and motor speed (Grant 2008; Heaton et al. 2011; Grant and Sacktor 2012). Collectively known as HAND, HIV-related NCI ranges in severity from its mild forms, asymptomatic neurocognitive impairment (ANI) and mild neurocognitive disorder (MND), to its most severe form: HIV-associated dementia (HAD) (Antinori et al. 2007).

Diagnosing HAND requires neuropsychological testing across the domains of language, attention/working memory, abstraction/executive, memory [learning and recall], speed of information processing, and sensory-perceptual, motor skills (Antinori et al. 2007). Specifically, an ANI diagnosis requires acquired NCI in at least two domains at least one standard deviation below the mean for age- and education-adjusted norms on the neuropsychological tests. Furthermore, the NCI does not interfere with everyday functioning. A diagnosis of MND requires the same acquired NCI in at least two domains at least one standard deviation below the normative mean as ANI but causes mild interference in daily functioning by self-report and/or observation by knowledgeable others. Diagnosis of HAD requires marked acquired NCI in at least two domains at least two standard deviations below the normative mean with the NCI causing marked interference with everyday functioning. For ANI, MND, and HAD, the NCI does not meet criteria for delirium (or dementia for ANI and MND) and cannot be accounted for by a preexisting condition.

The mild forms (ANI and MND) are by far the most common and even occur in people living with HIV (PLWH) on antiretroviral therapy (ART) with well-controlled viremia; prevalence ranges from 22 to 70% (Becker et al. 2004; Simioni et al. 2010; Heaton et al. 2010; Grant and Sacktor 2012; Bonnet et al. 2013; NNTC Public Reports 2014) depending on the study, participant characteristics, and method of assessing neurocognitive and functional status. One study, with over 1,500 PLWH, estimated rates of HAND at about 50% – excluding those with other comorbidities that could better explain the NCI present (Heaton et al. 2010). Prevalence of HAD is much lower, from 2 to 9% (Becker et al. 2004; Simioni et al. 2010; Heaton et al. 2010, 2011; Grant and Sacktor 2012; Bonnet et al. 2013). Compared to younger PLWH, those 50 years of age and over have been shown to have higher rates of NCI and are at greater risk for developing it (Becker et al. 2004; Valcour et al. 2004; High et al. 2006; Ettenhofer et al. 2009; Mateen and Mills 2012). Prevalence of NCI among PLWH in the presence of multiple confounders ranges from 7% to as high as 33%, depending on the sample (Heaton et al. 2010; NNTC Public Reports 2014). Less is known about the prevalence of HAND in children, adolescents, and adults who acquired HIV at birth (Hoare et al. 2016). Having even mild (or asymptomatic) HAND has been associated with increased risk for developing more severe HAND and mortality (Vivithanaporn et al. 2010). Research has also established a strong relationship between NCI and worse ART adherence (Hinkin et al. 2002, 2004; Ettenhofer et al. 2009, 2010), thus jeopardizing positive health outcomes. NCI in HIV is also associated with work difficulties, impaired activities of daily living, (e.g., planning, driving, finance management), worse overall quality of life, and need for more social services (Heaton et al. 2004; Marcotte et al. 2004; Van Gorp et al. 2007; Gorman et al. 2009; Umaki et al. 2013). Finally, neurocognition deficits are associated with poor decision-making and greater HIV transmission risk behaviors (e.g., unprotected sex; Wardle et al. 2010; Thames et al. 2012; Iudicello et al. 2013).

Routine screening for HAND – especially those newly diagnosed with HIV – has been recommended as good clinical practice (Valcour et al. 2011; Cysique et al. 2012; The Mind Exchange Working Group et al. 2012; Haddow et al. 2013; Morley et al. 2013; Zipursky et al. 2013; Barber et al. 2014; Kim et al. 2014). As a first step in the clinical decision-making process, screening can enable providers to determine who is most likely to have HAND, detect early signs of HAND, allocate limited resources more effectively, track and monitor neurocognitive function, and educate patients about the impact of HAND and ways to minimize it – all of which can improve health outcomes (Cysique et al. 2012; The Mind Exchange Working Group et al. 2012). Once adjuvant behavioral therapy and/or pharmacotherapy become available, screening for HAND will assist providers in appropriate treatment referrals. However, it rarely occurs (Valcour et al. 2011; Haddow et al. 2013; Morley et al. 2013; Kim et al. 2014).

Globally, there are 36.7 million people living with HIV (Joint United Nations Programme on HIV/AIDS 2016), and millions have and are at risk for developing HAND. Hence, screening for HAND is critical. Even as greater numbers of PLWH gain access to effective ART and live longer, healthier lives, HAND persists (Becker et al. 2004; Simioni et al. 2010; Heaton et al. 2010; Grant and Sacktor 2012; Bonnet et al. 2013). Furthermore, as PLWH grow older, they are at a higher risk than the general population of developing dementia. Screening for and detecting HAND early can play a critical role in the lives of PLWH through understanding its impact on health and health outcomes and developing strategies to optimally manage it.

There are numerous reasons why screening does not routinely occur in clinical practice. Most screening tests developed specifically for HAND were either designed to detect only the most severe form (HAD) or lack accuracy to detect the more common milder forms to be clinically useful (i.e., maximizing true positives and minimizing false positives; Power et al. 1995; Smith et al. 2003; Sacktor et al. 2005; Bottiggi et al. 2007; Valcour et al. 2011; Muñoz-Moreno et al. 2013; Zipursky et al. 2013). Many of the currently available, conventional, paper-and-pencil screening tests and short batteries for HAND require specialized skills and knowledge to properly administer, score, and interpret them (Valcour et al. 2011; Zipursky et al. 2013). Some screeners are cumbersome, requiring additional proprietary test forms and expensive equipment (Valcour et al. 2011; Zipursky et al. 2013). Furthermore, conventional, paper-and-pencil screening instruments are prone to human error in administration and scoring, nor are they well-suited for easy integration with electronic medical records (test data must be manually entered into electronic systems – a task also prone to error). Two studies have demonstrated that computer-based neurocognitive testing may be useful to detect mild HAND; however, it requires expensive laptop or desktop computers plus proprietary software, and more research is needed to determine its accuracy in detecting HAND, whether it is feasible to use in the clinical setting and the facilitators and barriers to its incorporation into clinical practice (Becker et al. 2011; Overton et al. 2011).

2 Screening Tests for HAND

Numerous screening tests have been developed for and/or evaluated to detect HAND. Two recent review articles by Kamminga et al. (2013) and Zipursky et al. (2013) summarized results across a total of 36 different studies that examined 40 different screening tests, subtests, or short batteries of individual neuropsychological tests as screening instruments for HAND. We have presented the tables with permission from Kamminga and Zipursky (see Tables in the Appendix to this chapter). Most of the studies ($N = 20$) were conducted in the United States, with three from South Africa, three from Australia, one from an AIDS Clinical Trials Group study from multiple countries, and the rest from Europe, Asia, and sub-Saharan Africa. The most widely reviewed screening tests were the HIV Dementia Scale (HDS) (Power et al. 1995) and the International HIV Dementia Scale (IHDS) (Sacktor et al. 2005). Results greatly varied between studies with some evidence to suggest some individual screening tests, such as the HDS and IHDS, may be adequate to detect only HAD in some settings (Haddow et al. 2013), and combinations of individual tests may be useful in the detection of more mild forms of HAND (Kamminga et al. 2013, 2017; Monteiro de Almeida et al. 2017). For example, one study found that different pairs of six different neuropsychological measures (i.e., WAIS-III Digit Symbol, HVLT-R Total Recall, PASAT-50, and Grooved Pegboard, and Trail Making Test Part B) were more accurate than the HDS in classifying NCI in HIV+ individuals (Carey et al. 2004). One additional scoping review of screening tests for HAND in sub-Saharan Africa is not reviewed or presented here due to it being an open-access and open peer review study with only two reviews, one of which requires revisions (the study can be viewed at https://aasopenresearch.org/articles/1-28/v1).

Since the Kamminga et al. (2013) and Zipursky et al. (2013) articles, 23 additional studies from the past 5 years presented data on 18 different screening tests for HAND and/or the NCI associated with it, several of which were novel computer or mobile device-based screeners (see Table 1). One study published before the Kamminga et al. and Zipursky et al. reviews but not mentioned in those reviews is also included here (Kvalsund et al. 2009). Nine of the studies were conducted solely in the United States. One study was conducted in both the United States and South Africa. Other countries represented across the studies are Australia, Brazil, Malaysia, the Netherlands, Nigeria, South Korea, Uganda, the United Kingdom, and Zambia.

All but one of the studies were of adults living with HIV with a mean age across adult studies of about 46 years. Three studies had participants with mean ages of 50 years or older. One study had perinatally HIV-infected children and adolescents between 9 and 12 years of age (Phillips et al. 2019).

The Montreal Cognitive Assessment (MoCA; Nasreddine et al. 2005) was the most widely examined screening tool (included in 11 of the 23 studies), and it was used in several countries (e.g., South Africa, Zambia, the Netherlands, and the United States). The IHDS was examined in six of the studies and the HDS in three of the studies. One study included combinations of tests from a comprehensive neuropsychological test battery (Monteiro de Almeida et al. 2017).

Nineteen of the studies compared screening test results to a comprehensive neuropsychological test battery. About half of the studies ($n = 13$) did not include an assessment of functional ability and hence did not provide HAND diagnostics. Seven of the studies included HIV-uninfected comparison groups. Sample sizes ranged from a minimum of $N = 39$ to a maximum of $N = 342$. The mean sample size was 126 ($SD = 83$). Base rates for NCI and HAND (based on a comprehensive neuropsychological test battery [for NCI] and a functional assessment [for HAND])

Table 1 HAND screening tests

Authors (year)	Country	Screening test(s)	Cutoff score(s)	Comparison test	HAND/ NCI criteria	Functional assessment	Sample size HIV+	HIV–	Test administrator	Minutes to complete	Sensitivity	Specificity	Base rate
Anderson et al. (2016)	United States	NCAD	<75.44	Neuropsychological test battery (8 tests)	Antinori et al. (2007) criteria	NR	39	–	NR	20	67%	83%	69%
		HDS	>9								85%	NR	
Bloch et al. (2016)	Australia	CogState	GDS > 0.5	Neuropsychological test battery	Blackstone et al. (2019) criteria	IADLs	254	72	Research staff	20	76%	71%	31%
Brouillette et al. (2015)	United States	MoCA	≤ 2.82 logits	Neuropsychological test battery (9 tests)	Antinori et al. (2007) criteria	NR	200	–	NR		74%	68%	74%
Chartier et al. (2015)	United States	MoCA	<26	NR	NR	NR	44	–	NR		NR	NR	NR
		IHDS	<10								NR	NR	NR
Fazeli et al. (2019)	United States	MoCA	≤26	Neuropsychological test battery (15 tests)	Antinori et al. (2007) criteria	IADLs, PAOFI, Karnofsky	100	–		10–15	84%	56%	57%
Hasbun et al. (2012)	United States	MoCA	<26	Neuropsychological test battery (6 tests)	Gisslén et al. (2011)	NR	100	–	Nurse or physician	10–15	85%	40%	75%
Janssen et al. (2015)	Netherlands	MoCA	<26	Neuropsychological test battery (17 tests)	Antinori et al. (2007) and CIND criteria	NR	102	–	Neuropsychologists	10–15	56% (Frascati criteria) 55% (CIND criteria)	63% (Frascati criteria) 58% (CIND criteria)	41%
		HDS	<11							NR	26% (Frascati criteria) 36% (CIND criteria)	96% (Frascati criteria) 95% (CIND criteria)	

Joska et al. (2016)	South Africa	IHDS	≤10	Neuropsychological test battery (9 tests)	American Academy of Neurology criteria	ART adherence, ADLs, symptom questions	156	–	Trained research assistant	4	41%	86%	41%
	United States	MoCA	≤26							13	89%	23%	
		CAT-rapid	≤10							7	64%	52%	
		Simioni questions	≥1							3	78%	32%	
		MMSE	≤24							12	24%	98%	
		IHDS/CAT-rapid combo	≤16							NR	43%	82%	
Kamminga et al. (2017)	Australia	CogState	GDS >0.5	Neuropsychological test battery (11 tests)	Blackstone et al. (2019) criteria	ADLs	53	22	NR	15–20	76% (GDS) 72% (domain rating)	71% (GDS) 57% (domain rating)	51–57%
Kim et al. (2016)	South Korea	MoCA	≤26	Neuropsychological test battery (10 tests)	Antinori et al. (2007) criteria	Karnofsky and custom Korean questions	194	–	NR	10–15			26%
Ku et al. (2014)	South Korea	Four questions	NR	Neuropsychological test battery (10 tests)	Antinori et al. (2007) criteria	Karnofsky and custom Korean questions	184	–	NR	NR	NR	NR	26%
		IHDS	≤10							NR	73%	61%	
		MoCA	≤25							NR	53%	72%	
		Grooved Pegboard	NR							NR	90%	72%	
		Wisconsin Card Sorting Test	NR							NR	61%	84%	
Kvalsund et al. (2009)	Zambia	MMSE	NR	NR	Maj et al. (1994) and Sebit (1995) criteria	NR	48	15	Non-physician healthcare workers	NR	NR	NR	NR
		HDS	NR							NR	NR	NR	
		Color trails 1	NR							NR	NR	NR	

(continued)

Table 1 (continued)

Authors (year)	Country	Screening test(s)	Cutoff score(s)	Comparison test	HAND/NCI criteria	Functional assessment	Sample size HIV+	HIV−	Test administrator	Minutes to complete	Sensitivity	Specificity	Base rate
Milanini et al. (2014)	United States	MoCA	≤25	Neuropsychological test battery	Antinori et al. (2007) criteria	Clinical Dementia Rating scale	67	–	NR	NR	72%	67%	40%
Monteiro de Almeida et al. (2017)	Brazil	IHDS	≤10	Neuropsychological test battery (17 tests)	Antinori et al. (2007) criteria	ADLs	60	48	Neurologist	10	36%	75%	60%
		IHDS	≤11.5								72%	58%	
		NP test combination (Trail Making Test, Part A, WAIS-III Digit Symbol Coding, HVLT-R Total Recall)	GDS > 0.5								91%	96%	
		NP test combination (WAIS-III Digit Symbol Coding, BVMT-R Total, Grooved Pegboard non-dominant hand)	GDS > 0.5								94%	91%	
Moore et al. (2017)	United states	UCSD UPSA-B	≤79	Neuropsychological test battery (15 tests)	Antinori et al. (2007) criteria	IADLs	103	91	Research staff and neuropsychologist	10–15	70%	73%	39%
Mukherjee et al. (2018)	Malaysia	MoCA	≤26	NR	Antinori et al. (2007) criteria	IADLs	342	243		10–15	NR	NR	NR

Overton et al. (2013)	United States	MoCA	≤25	Neuropsychological test battery (8 tests)	Antinori et al. (2007) criteria	NR	200	–	NR		63%	71%	
		AD-8	≥2								61%	51%	
		MoCA/AD-8 combo	NR								40%	82%	
Parry et al. (2017)	United Kingdom	NCI-3Q	NR	NR	Simioni-style questions	NR	95	–	NR	NR	NR	NR	NR
Phillips et al. (2019)	South Africa	y-IHDS	≤10	Neuropsychological test battery (15 tests)	Hoare et al. (2016) criteria	Academics	203	–	Trained research assistants	3–5	94%	24%	46%
Robbins et al. (2014)	United States	NeuroScreen	≤0.9	Neuropsychological test battery (13 tests)	Antinori et al. (2007)	NR	44	–	Trained research assistants	20–25	94%	64%	75%
Robbins et al. (2018)	South Africa	NeuroScreen (all tests)	NR	Neuropsychological test battery (15 tests)	Blackstone et al. (2019)	NR	102	–	Lay counselors	20–25	81%	81%	27%
		NeuroScreen (four tests: visual discrimination 1 and 2, trail making 1, and number span total)	NR	Neuropsychological test battery (15 tests)	Blackstone et al. (2019)	NR	102		Lay counselors	12	93%	71%	27%
Royal et al. (2012)	Nigeria	IHDS	≤ 9	Neuropsychological test battery (13 tests)	NR	NR	60	56	NR	20	100%	79%	NR
Yechoor et al. (2017)	Uganda	CogState Brief	≥1 SD below the mean on 1+ tasks	Neuropsychological test battery (5 tests)	Blackstone et al. (2019)	NR	181	–	NR	25	57%	77%	38%

NR none or not reported, *MoCA* Montreal Cognitive Assessment, *HDS* HIV Dementia Scale, *IHDS* International HIV Dementia Scale, *MMSE* Mini-Mental Status Exam, *CAT-rapid* cognitive assessment tool – rapid, *NCAD* novel computerized cognitive assessment device, *NCI-3Q* neurocognitive impairment 3 questionnaire, *y-IHDS* youth International HIV Dementia Scale, *ADLs* activities of daily living, *IADLs* instrumental activities of daily living, *Karnofsky* Karnofsky scale of performance status, *PAOFI* patient's assessment of own functioning inventory ranged widely from 26 to 75%, depending on how NCI was classified (e.g., global domain scores vs. clinical ratings).

Most of the screening tests examined were reported to take between 10 and 20 min to complete. Ten studies did not report on who administered the screening tests. Of those that did, screening test administrators ranged from lay counselors (e.g., Robbins et al. 2018), to non-physician healthcare workers (e.g., Kvalsund et al. 2009), to trained research assistants, to neuropsychologists (e.g., Janssen et al. 2015), and to physicians (e.g., Hasbun et al. 2013).

Joska et al. (2016), comparing adults living with HIV in the United States and South Africa, found that assessing for HAD the cognitive assessment tool – Rapid (CAT-Rapid) demonstrated good sensitivity and weak specificity (94% and 52%; cutoff score $\leq$ 10), the IHDS showed fair sensitivity and good specificity (68% and 86%; cutoff score $\leq$ 10), and the MoCA showed excellent sensitivity but poor specificity (100% and 22%; cutoff score $\leq$ 26. The Mini-Mental Status Exam (Folstein et al. 1983) and Simioni symptom questionnaire (Simioni et al. 2010) did not demonstrate sufficient psychometric properties to detect HAND in the sample. None of the five tools were sufficient to adequately detect less severe forms of HAND.

Several international, single-country studies explored the ability of adapted versions of the IHDS and MoCA in Brazil, Nigeria, South Africa, and Korea to detect HAND (Royal et al. 2012; Ku et al. 2014; Monteiro de Almeida et al. 2017). Two of the studies concluded that the IHDS had enough sensitivity to be effective in detecting ANI and MND (Ku et al. 2014) and more generally HAND (Royal et al. 2012) in these populations, but it had poor specificity. In Brazil, Monteiro de Almeida et al. (2017) reported that a combination of several gold standard neuropsychological tests (see Table 1 for list of test combinations) was more accurate in detecting HAND than the IHDS. Ku et al. (2014) and Mukherjee et al. (2018) also evaluated the utility of the MoCA in detecting HAND in Korea and Malaysia, respectively. Both studies reported that the MoCA was effective at detecting HAND, with the Mukherjee et al. study concluding so only after accounting for demographic factors.

The computerized CogState Brief (Cysique et al. 2006; Maruff et al. 2009) was examined in Uganda and Australia (Bloch et al. 2016; Yechoor et al. 2017). A Ugandan study evaluated the CogState Brief in detecting HAND (Yechoor et al. 2017). In Uganda, the CogState had adequate specificity to detect HAND but poor sensitivity. Yechoor et al. concluded that additional research is required to identify tools with high sensitivity in detecting HAND in resource-limited settings. In Australia, Bloch et al. (2016) evaluated an updated CogState screener that specifically targets the NCI observed in HAND. The updated CogState had 76% sensitivity and 71% specificity to detect any HAND defined by a gold standard neuropsychological evaluation. CogState's classification accuracy was increased to 100% sensitivity and 98% specificity when only MND and HAD were considered.

Two studies examined a mobile device-based screening tool, NeuroScreen (Robbins et al. 2014, 2018). The first study examined a large format smartphone version of NeuroScreen to detect NCI among adults living with HIV in the United

States (Robbins et al. 2014) and found that sensitivity ranged from 89 to 94% and specificity ranged from 63 to 64%, depending on the method of NCI classification used. The second study examined the same screening tool using a tablet-based device administered by lay counselors to detect NCI among adults living with HIV in South Africa (Robbins et al. 2018). Sensitivity ranged from 82 to 93% and specificity from 75 to 81% depending on the combination of NeuroScreen subtests used.

A recent American study investigated whether a functional capacity measure, the UCSD Performance-Based Skills Assessment – Brief (UPSA-B), was able to detect functional dependence and neurocognitive impairment in adults in San Diego, California (Moore et al. 2017). This study found that the UPSA-B had sensitivity, specificity, and accuracy rates in 70%, but scores on this measure were unrelated to self-reported functional dependence.

Only one study examined a HAND screening tool for youth living with HIV. Phillips et al. (2019) examined an adapted for youth IHDS (y-IHDS) among perinatally HIV-infected children and adolescents between 9 and 12 years of age in South Africa. The y-IHDS had good sensitivity but poor specificity (94% and 24%, ≤ 10) for assessing HAND in children and adolescents. The authors concluded that while the y-IHDS has both clinical practice and research value in low-resource settings, further research to optimize the tool would be beneficial.

Finally, a recent meta-analysis of the MoCA's accuracy to detect HAND (not included in Table 1) that includes several of the studies cited herein examined sensitivity and specificity across multiple cutoff scores (Rosca et al. 2019). The authors recommend using a modified cutoff score of ≤ 23 (versus the standard ≤ 25) to define impairment. Using this cutoff best balances true and false positives (see Sect. 2.2) and yields sensitivity of 44% and specificity of 79%.

2.1 *Need for Functional Assessment*

Screening tests for HAND focus on neurocognitive impairment. However, to fully diagnose HAND, a functional assessment is required. Only one of the reviewed screening tools for HAND included a performance-based functional assessment (see UPSA-B above), and none of the others included a functional assessment within the screening tool or procedure. Of the 11 studies with a functional assessment, the functional assessment was most often used in conjunction with the comprehensive neuropsychological test battery to define the base rate of HAND. In fact, there is a dearth of research on functional assessment tools for HAND across countries and contexts, and few tools to assess functional ability are available for those regions and countries most affected by HIV. Levels of everyday functioning vary greatly within and across countries depending on socioeconomic and cultural factors, for example, differences in gender roles across societies. These factors need to be accounted for when doing functional assessments in local and international settings. Furthermore, most of the commonly used assessment tools for everyday functioning, like Lawton and Brody's Instrumental Activities of Daily Living (Lawton and Brody 1969), are self-report and often contain items that are not appropriate for certain contexts, like ability to manage finances via check writing or managing bank accounts. Many individuals in the most affected communities of sub-Saharan Africa do not have bank accounts, let alone use checks. Hence, this activity would need to be appropriately adapted to the context in which the measure is being used. Though self-report assessments do not provide an objective assessment of functional abilities, they can be useful to screen individuals, as many are fairly quick.

2.2 Screening Test Psychometrics

There are several important factors that need to be considered when choosing a screening test: (1) sensitivity and specificity; (2) positive and negative predictive values, and (3) base rates of the disease the screening test is trying to identify. Sensitivity refers to a test's ability to correctly classify a patient as having the disease or disorder, whereas specificity refers to a test's ability to correctly identify a patient as being disease or disorder free (Rosenfeld et al. 2000). A test with high sensitivity and low specificity will accurately detect those who have the disease (true positives) but also produce more false positives (those without the disease but who screen positive for it). Similarly, if the sensitivity is lower than the specificity, the test will be better at detecting those who do not have the disease (true negatives) than those who do. Positive predictive value (PPV) is the percentage of patients who test positive and who do in fact have the disorder. Higher PPV indicates that the screening test is more able to detect those patients who have the disorder. PPV is related to the prevalence of the disorder in the population and will increase as prevalence increases. Negative predictive value (NPV) is the percentage of patients with a negative test who do not have the disease. Higher NPV indicates that the screening tests are more accurately classifying those people who do not have the disorder as disorder free.

Base rates of the disease or disorder further complicate the issue of a test's predictive ability. A screening test with 80% sensitivity and 70% specificity used in a population where the known base rate of HAND is 50% would result in a PPV of 73% and a NPV of 78%, which would result in a 73% probability that a person who screens positive actually has HAND. However, if the base rate of HAND is actually 30% in the population but the screening test was calibrated on a base rate of 50%, then the psychometric properties change dramatically such that PPV would be 53%, NPV would be 89%, and the probability that a positive screen is truly positive would be 53%.

Choosing and interpreting a screening test requires a thorough understanding of these psychometric properties. Few, if any, screening tests are perfect. Hence, clinicians and clinics must consider the ethics and costs of a screening test's limitations. What are the consequences of missing a patient with the disorder (lower sensitivity/higher false-negative rate) and providing additional resources for someone who screens positive but does not truly have NCI (lower specificity/high false-positive rate)?

3 Global Perspective

The burden of HIV is in low- and middle-income countries (LMICs), particularly those in sub-Saharan Africa where, in some countries, the national prevalence of HIV is ~12% (Joint United Nations Programme on HIV/AIDS 2016). To date, most of the research on HAND screening has been focused on adult populations with HIV and populations in the United States. Given that the burden of the disease is disproportionally represented in sub-Saharan Africa (~27 million PLWH; Joint United Nations Programme on HIV/AIDS 2016), more studies are needed to evaluate properly adapted or locally developed screening tools in the most affected countries. Furthermore, there are approximately 2.1 million children under 15 who are living with HIV (Joint United Nations Programme on HIV/AIDS 2016), yet little attention has been paid to screening tools for them.

LMICs, such as South Africa, are facing massive resource shortages of healthcare professionals to test for and manage HIV, let alone screen and assess for HAND. For example, in the United States and Australia, there are 25.95 and 35.88 physicians per 10,000 population, whereas in South Africa there are only 9.01 physicians per 10,000 population, and in Zimbabwe there are only 0.76 physicians per 10,000 population (World Health Organization 2019). To address these shortages, many LMICs have been practicing task shifting where certain aspects of the HIV care continuum have been shifted to lay health professionals, such as community healthcare workers (Callaghan et al. 2010). Task shifting for HAND screening has been considered in South Africa. Unfortunately, the research has shown that certain HAND screening tools (i.e., IHDS) while developed to be used in any culture/country, when used by lay professionals, may highly over- or underestimate rates of HAD due to administration and interpretation errors (Robbins et al. 2011; Breuer et al. 2012).

Given the global distribution of HIV, considerations for the use of any screening test for HAND must take into account the psychometric validity of the test in the new population, the test's cultural and demographic appropriateness, and who will be administrating the test. For example, the MoCA (Nasreddine et al. 2005) has test items (e.g., cube drawing) that are not well designed for some populations in South Africa and could erroneously indicate NCI (see Robbins et al. 2013).

Another challenge in LMICs is the issue of normative performance on cognitive tests. Many countries with the highest burden of HIV do not have formally validated and normed cognitive tests, let alone screening tests, to detect the NCI associated with HAND. This creates problems in understanding how individual performance compares to the general population. For example, using the MoCA's North

American norms in a South African population suggested very high rates of dementia among normal adults (see Robbins et al. 2013).

4 Future Directions

Computer- and tablet-based screening tests may provide a platform that greatly increases the feasibility and accuracy of NCI screening. The Robbins et al. (2018) study demonstrated that a lay counselor administered tablet-based screening test had robust sensitivity and specificity to detect NCI among South African PLWH. This creates an opportunity to make screening more feasible and widely available in resource-limited settings. With their ease of use, low unit cost, reliability, and accuracy, tablet-based cognitive tests may also make the collection of normative performance data more feasible and less expensive across larger segments of societies, as it would not require highly trained neuropsychologists or psychometrists to administer a lengthy paper-and-pencil battery. Furthermore, because data from tablets could be easily linked to electronic databases, new opportunities to use big data science approaches to examine prevalence and predictors of NCI among and across populations will arise.

Regardless of using a tablet- or paper-based screening test, a first step that is critical to developing and implementing screening and referral programs is research to evaluate any potential program's acceptability and feasibility from patient, provider, and clinic system perspectives. Understanding what the implementation challenges will be to making screening in clinical settings routine, such as time and space constraints, training requirements, and insurance reimbursement, is an essential step to developing a scalable and sustainable program. Finally, as screening programs are developed and implemented, they will also have to demonstrate positive impact on clinical care and patient outcomes.

5 Summary

HAND remains prevalent among PLWH. There are numerous tests available to screen for HAND, though many have a variety of limitations that make them less appealing for routine use. Common limitations among most available tests include who can administer it, how well it can detect the range of HAND severity (most lack accuracy to detect mild forms), and lack of norms for specific populations. Short neuropsychological test combination screeners often have better performance criterion validity than individual screening tests, but they too have limitations in who can administer, score, and interpret them, as well as costs for testing materials and equipment (Kamminga et al. 2013). Neuropsychological tests must be interpreted by a qualified neuropsychologist. In most LMIC settings, where the burden of HIV is the greatest, qualified staff are rarely available and not easily accessible.

Computerized screeners, such as CogState (Cysique et al. 2006) and NeuroScreen (Robbins et al. 2014, 2018), offer new possibilities to screen for HAND. While the requirements of staff administrators for these tools are greatly reduced (and may be appropriate for community health workers and lay professionals) and the hardware for using them (i.e., tablets and computers) are becoming more affordable and ubiquitous, there are challenges to using these types of tools in many settings. Software and technical support costs may be prohibitive. Moreover apps for mobile devices require ongoing programming maintenance as mobile devices constantly receive operating system updates and changes in form factors which require developers to provide timely updates. Moreover, computers and mobile devices are also dependent on infrastructure, such as electricity to charge devices and Wi-Fi for software updates. These may be limiting factors in some settings. Paper-and-pencil tests do not suffer from these limitations.

Despite these limitations, newer screening tools may help to make screening more widely available to larger segments of society. Routine screening for NCI among PLWH – especially those newly diagnosed – constitutes good clinical practice. It can enable providers to detect early signs of NCI, determine if and when to adjust ART regimens, track and monitor neurocognitive function, and educate patients about the impact of NCI and ways to minimize it – all of which can improve health outcomes. The impact of NCI on ART adherence can be minimized through behavioral planning, and detecting NCI in highly infectious PLWH (i.e., those with detectable viral load) may help in tailoring transmission prevention strategies. Furthermore, if/when pharmacotherapies or emerging behavioral interventions become widely available for HIV-related NCI, screening will be essential to link patients to appropriate services.

Appendix

Table 2 Neuropsychological tests used in each study by cognitive domain

Study	Premorbid ability	Motor/fine motor	Attention/ working memory	Psychomotor speed/ reaction time	Learning and memory	Executive function	Language	Visuo-construction
Chalermchai et al. (2013)		GP FT		TMTA CT1 Coding, SS	BVMT AVLT	CT2 First names fluency	SemFl	Block design
Sakamoto et al. (2013)		GP	LNS PASAT	TMTA Coding SS	SMLT FMLT	WCST LetFl TMTB	SemFl	
Moore et al. (2012)		GP	PASAT Digits	TMTA Coding SS Stroop Color	HVLT BVMT	Stroop CWT WCST LetFl ActFl TMTB	SemFl	
Becker et al. (2011)				X	[a]X	X	X	X
Joska et al. (2011)		FT GP		Coding CT1 TMTA	BVMT HVLT	CT2 CWT MAT MCT WCST		
Levine et al. (2011)		GP	LNS PASAT	Coding SS	BVMT HVLT	CWT TMTB LetFl WSCT		
Simioni et al. (2010)	NART		Digits SpWM	RT TMTA VIP		TMTB		

Skinner et al. (2009)		GP		[b]CalCAP SDMT TMTA		TMTB		
Morgan et al. (2008)		GP	LNS PASAT	Coding SS TMTA	BVMT HVLT	TMTB LetFl WCST		
Singh et al. (2008)			Digits	TMTA		TMTB		RCFT copy
Bottiggi et al. (2007)	NART	FO GP TG	Ruff 2 and 7 Seq RT	SDMT TMTA	FMLT RAVLT	CWT TMTB LetFl		
Wojna et al. (2007)		GP		SDMT, TMTA VCRT	RAVLT	CWT TMTB		
Cysique et al. (2006)		GP	Digit span	SDMT TMTA	CVLT RCFT 3 min	Similarities TMTB LetFl	BNT SemFl	RCFT copy
Ellis et al. (2005)	Vocabulary	TG GP	PASAT	SS	HVLT			
Richardson et al. (2005)		GP	Digit span	[b]CalCAP TMTA	RAVLT	CWT TMTB		RCFT copy
Sacktor et al. (2005) (American)		GP		[b]CalCAP SDMT	RAVLT	Odd man out LetFl		RCFT copy
Sacktor et al. (2005) (Ugandan)		GP TG	Digit span	CT1 SDMT	VLT	CT2		
Carey et al. (2004)		GP	LNS PASAT	Coding SS TMTA	BVMT HVLT	CWT HCT TMTB LetFl WCST	SemFl	

(continued)

Table 2 (continued)

Study	Premorbid ability	Motor/fine motor	Attention/ working memory	Psychomotor speed/ reaction time	Learning and memory	Executive function	Language	Visuo-construction
Gonzalez et al. (2003)		GP	Digit span PASAT VisSp	CT1 SDMT TMTA	RAVLT	CT2 CWT HCT TMTB LetFl		Block design
Smith et al. (2003)		GP	LNS	bCalCAP CT1	CVLT Faces	CT2 CWT		

NART national adult reading test, *GP* grooved pegboard, *FT* finger tapping, *FO* finger oscillation, *TG* timed gait, *LNS* letter number sequencing; *PASAT* paced auditory serial addition test, *SpWM* spatial working memory, CANTAB, *SeqRT* sequential reaction time, CalCAP, *VisSp* visual span, *TMTA* trial making test A, *CT1* color trails test 1, *SS* symbol search, *SDMT* symbol digit modalities test, *RT* reaction time, CANTAB, *VIP* visual information processing, CANTAB, *VCRT* visual choice reaction time, *HVLT* hopkins verbal learning test, *BVMT* brief visuo-spatial memory test, *SMLT* story memory learning test, *FMLT* figure memory learning test, *RAVLT* rey auditory verbal learning test, *CVLT* california verbal learning test, *RCFT 3 min* rey complex figure test, 3 min delay, *VLT* World Health Organisation/University of California verbal learning test, *MAT* mental alternation test, *MCT* mental control test, *CT2* color trails 2, *CWT* stroop color word interference test, *WCST* Wisconsin card sort test, *LetFl* letter fluency, *TMTB* trail making test B, *HCT* halstead category test, *SemFl* semantic fluency, *BNT* boston naming test, *RCFT copy* rey complex figure test copy

"X" test not specified

[a]Verbal and visuo-spatial learning and memory

[b]Individual tests not specified, assumed that the whole battery was administered

Table 3 Screen versus gold standard NP impairment rates and standard criterion validity indexes across studies (%)

Study	Screen	NP IR	Screen IR	Cut-off	Sub-sample	Sensitivity	Specificity	Accuracy
Morgan et al. (2008)	HDS	43[a]	-	*T*<40	HAD only	93	73	-
Moore et al. (2012)	4 NP tests	19[b]	-	4 tests *T* < 40, *or* 2 tests *T* < 40 + 1 test *T* < 35, *or* 2 tests *T* < 35, *or* 1 test *T* < 40 + 1 test *T*< 30, *or* 1 test *T* < 25	-	87	87	-
Moore et al. (2012)	3 NP tests	19[b]	-	3 tests *T* < 40 *or* 1 test *T* < 40 + 1 test *T* < 35, *or* 1 test *T* < 30	-	87	76	-
Cysique et al. (2006)	CogState	62[c]	62[c]	-	-	81	70	-
Carey et al. (2004)	NP tests (HVLTR & ndGP)	29[c]	34[c]	*T*<40 on 1 test *or* T<35 on 2 tests	-	78	85	83
Morgan et al. (2008)	HDS	43[c]	-	*T*<40	MND only	77	73	-
Carey et al. (2004)	NP tests (HVLTR & Cod)	29[c]	-	*T*<40 on 1 test *or* T<35 on 2 tests	-	75	92	87
Moore et al. (2012)	2 NP tests	19[c]	-	2 tests *T* < 40 *or* 1 test *T* ≤ 35	-	73	83	-
Sacktor et al. (2005)	IHDS(American)	38[c]	-	≤10.5	-	71	79	-
Becker et al. (2011)	CAMCI	31[c]	30[c]	-	-[g]	72	98	-
Singh et al. (2008)	IHDS	80[c,f]	-	≤10.5	-	94	25	-
Bottiggi et al. (2007)	HDS	52[c]	-	≤10	-	93	38	-
Simioni et al. (2010)	HDS	74[c]	-	≤14	No self reported cognitive complaints	88	67	-
Singh et al. (2008)	IHDS	80[c,f]	-	≤10	-	88	50	-
Sacktor et al. (2005)	IHDS (Ugandan)	31[c]	-	≤9.5	-	88	48	-

(continued)

Table 3 (continued)

Study	Screen	NP IR	Screen IR	Cut-off	Sub-sample	Sensitivity	Specificity	Accuracy
Wojna et al. (2007)	HDS	68[b]	48[b,e]	≤13	-	87	46	-
Chalermchai et al. (2013)	IHDS	51[c]	-	≤10	-	86	-	-
Simioni et al. (2010)	HDS	74[c]	-	≤14	Self reported cognitive complaints	83	63	-
Joska et al. (2011)	IHDS	81[b]	-	≤11	-	81	54	-
Sacktor et al. (2005)	IHDS (American)	38[b]	-	≤10	-	80	57	-
Sacktor et al. (2005)	IHDS (Ugandan)	31[c]	-	≤10	-	80	55	-
Skinner et al. (2009)	IHDS	40[b]	-	≤10	-	77	65	70
Sakamoto et al. (2013)	HDS	51[c]	12[b]	≤10	moderate-severe impairment	77	-	-
Levine et al. (2011)	NP tests	52[b]	71[c]	*T* score (varied)	-	75	61	
Sakamoto et al. (2013)	HDS	51[c]	56[c]	$T < 40$	-	69	56	63
Gonzalez et al. (2003)	CalCAP	57[c]	49[c]	Average Deficit score >0.5	-	68	77	72
Levine et al. (2011)	HDS	52[b]	62[c]	≤10	-	67	50	-
Sakamoto et al. (2003)	HDS	51[c]	-	≤14	-	66	61	-
Sakamoto et al. (2003)	HDS	51[c]	-	≤10	virologically suppressed only	66	55	61
Sakamoto et al. (2003)	HDS	51[c]	19[b]	≤10	mild-moderate impairment	65	-	-
Wojna et al. (2007)	HDS	68[b]	48[b,e]	≤12	-	63	84	-
Sakamoto et al. (2003)	HDS	51[c]	20[b]	≤10	mild impairment	63	-	-

Skinner et al. (2009)	HDS	40[b]	-	<11	-	62	80	-
Ellis et al. (2005)	3 NP tests	56[c,d]	-	Averaged z-scores ≤ 0.5	-	58	82	68
Bottiggi et al. (2007)	HDS	52[c]	-	≤10	Moderate & Severe impairment	57	84	-
Richardson et al. (2005)	HDS	50[c]	40[c]	≤10	-	55	75	63
Simioni et al. (2010)	HDS	74[c]	-	≤10	-	54	96	-
Chalermchai et al. (2013)	IHDS	51[c]	-	≤10	-	53	89	-
Morgan et al. (2008)	HDS	43[b]	-	$T<40$	ANI only	50	73	-
Skinner et al. (2009)	HDS	40[b]	-	≤10	-	46	80	67
Skinner et al. (2009)	MMSE	40[b]	-	≤27	-	46	55	-
Joska et al. (2011)	IHDS	81[b]	-	≤10	-	45	80	56
Ellis et al. (2005)	3 NP tests	56[c,d]	32[c]	≤1 SD 1 test	-	44	84	-
Smith et al. (2003)	HDS	49[b]	-	≤10	-	39	85	-
Morgan et al. (2008)	HDS	43[b]	-	≤10	HAD only	36	94	-
Bottiggi et al. (2007)	HDS	52[c]	-	≤10	Severe impairment only	36	94	-
Chalermchai et al. (2013)	IHDS	51[c]	-	≤10	-	34	87	-
Ellis et al. (2005)	3 NP tests	56[c,d]	15[c]	≤1 SD on 1 test / ≤ 2 SD on 2 tests	-	24	98	-
Sakamoto et al. (2013)	HDS	51[c]	17[c]	≤10	-	24	92	57
Morgan et al. (2008)	HDS	43[b]	-	≤10	MND only	23	94	-

(continued)

Table 3 (continued)

Study	Screen	NP IR	Screen IR	Cut-off	Sub-sample	Sensitivity	Specificity	Accuracy
Morgan et al. (2008)	HDS	43[b]	-	≤10	-	17	94	-
Carey et al. (2004)	HDS	29[b]	4[b]	≤11	-	9	98	-
Levine et al. (2011)	MMSE	52[b]	14[c]	≤25	-	8	88	-
Ellis et al. (2005)	3 NP test	56[c,d]	-	Clinical rating ≥5	-	2	100	42
Morgan et al. (2008)	HDS	43[b]	-	≤10	ANI only	0	94	-
N = 55	**Median**	**51**	**33**			**66**	**80**	**65**
	Range	**19-81**	**4-71**			**0-93**	**25-100**	**42-87**

See reference list for numbered studies. Rounded to nearest whole value. Ordered with balance of highest sensitivity and specificity. Studies with ≥70% sensitivity and specificity are bold. Studies ≤ 70% sensitivity and specificity are in gray. Values are based on standard cut-off scores unless otherwise stated. Values represent percentages unless otherwise stated and are to the nearest decimal reported. *NP* neuropsychology, *IR* impairment rate, *PPP* positive predictive power, *NPP* negative predictive power, *Acc* classification accuracy, *AUC* area under the curve, – not reported. Sensitivity refers to the proportion of individuals identified as impaired on both the gold standard and the screen. Specificity refers to the proportion of individuals identified as unimpaired on both the gold standard and the screen. PPP relates to specificity and is defined as the probability that an individual has HAND given their positive screen result. NPP relates to sensitivity and is the probability that an individual does not have HAND given their negative screen result.

[a]Proportion

[b]Manually calculated

[c]Values reported

[d]Based on test scores only

[e]Based on a cut-off score of 12 or less

[f]"Any neurocognitive impairment"

[g]Sample excluded ANI category

Table 4 Screen versus gold standard NP impairment rates and standard criterion validity indexes for studies where sensitivity and specificity are 70% or higher (%)

Study	Screen	NP IR	Screen IR	Cut-off	Sample	Sensitivity	Specificity	Accuracy
Morgan et al. (2008)	HDS	43[a]	–	$T < 40$	*HAD only*	93	73	–
Moore et al. (2012)	4 NP tests	19[b]	–	4 tests $T < 40$, *or* 2 tests $T < 40$ + 1 test $T < 35$, *or* 2 tests $T < 35$, *or* 1 test $T < 40$ + 1 test $T < 30$, *or* 1 test $T < 25$	Entire sample	87	87	–
Moore et al. (2012)	3 NP tests	19[b]	–	3 tests $T < 40$ *or* 1 test $T < 40$ + 1 test $T < 35$, *or* 1 test $T < 30$	Entire sample	87	76	–
Cysique et al. 2012	CogState	62[a]	62[a]	–	Entire sample	81	70	–
Carey et al. (2004)	NP tests (HVLT-R and ndGP)	29[a]	34[a]	$T<40$ on 1 test *or* $T<35$ on 2 tests	Entire sample	78	85	83
Morgan et al. (2008)	HDS	43[a]	–	$T<40$	*MND only*	77	73	–
Carey et al. (2004)	NP tests (HVLT-R and Cod)	29[a]	–	$T<40$ on 1 test *or* $T<35$ on 2 tests	Entire sample	75	92	87
Moore et al. (2012)	2 NP tests	19[b]	–	2 tests $T < 40$ *or* 1 test $T \leq 35$	Entire sample	73	83	–

(continued)

Table 4 (continued)

Study	Screen	NP IR	Screen IR	Cut-off	Sample	Sensitivity	Specificity	Accuracy
Sacktor et al. (2005) (American)	IHDS	38[a]	–	$\leq$10.5	Entire sample	71	79	–
Becker et al. (2011)	CAMCI	31[a]	30[b]	–	Entire sample[c]	72	98	–

Consideration of the sample in question is needed for the proper selection of the screen. See reference list for numbered studies. Rounded to nearest whole value. Ordered with balance of highest sensitivity and specificity. *NP* neuropsychology, *IR* impairment rate, *Accuracy* overall correct classification accuracy, – not reported. Sub-sample refers to the subset of the overall sample upon which criterion validity indexes were calculated. Sensitivity refers to the proportion of individuals identified as impaired on both the gold standard and the screen. Specificity refers to the proportion of individuals identified as unimpaired on both the gold standard and the screen. *HAD* HIV-associated dementia, *MND* mild neurocognitive disorder, *HVLR-R* Hopkins verbal learning test, revised, *ndGP* nondominant hand grooved pegboard, *Cod* coding

[a]Manually calculated

[b]Values reported

[c]Sample excluded ANI category

Table 5 Characteristics of studies on brief screening tools for HIV-associated neurocognitive disorders

Author, date; location	Design	Recruitment method	Participants (mean age; mean years education; percent male)	Tools examined
Becker et al. (2011); USA	Cross-sectional	Convenience sampling	Age: 51 (HIV+); 51 (HIV−) Education: 15 (HIV+); 15 (HIV−) Male: 96% (HIV+); 78% (HIV−)	The computer assessment of mild cognitive impairment (CAMCI)
Bottiggi et al. (2007); Kentucky, USA	Cross-sectional	Convenience sampling	Age: 39 Education: 14 Male: 87%	HIV dementia scale
Carey et al. (2004); San Diego, USA	Cross-sectional	Convenience sampling	Age: 41 Education: 14 Male: 84%	Hopkins verbal learning test revised and grooved pegboard test nondominant hand (PND) pair; Hopkins verbal learning test revised and WAIS-III digit symbol (DS) subtest pair; HIV dementia scale
Cysique et al. (2006); Sydney, Australia	Cross-sectional	Random sampling	Age: 47 (Advanced HIV); 48 (ADC); 49 (HIV−) Education: 14 (Advanced HIV); 14 (ADC); 15 (HIV−) Male: 98% (Advanced HIV); 100% (ADC); 100% (HIV−)	CogState
Cysique et al. (2010); Sydney, Australia	Cross-sectional	Convenience sampling	Age: 49 (HIV+); 47 (HIV−) Education: 14 (HIV+); 15 (HIV−) Male: 100% (HIV+); 100% (HIV−)	Screening algorithm
Ellis et al. (2005); 15 sites, USA	Cross-sectional	Convenience sampling	Age: 44 Education: 14 Male: 86%	The NeuroScreen – brief neurocognitive screen (BNCS) and brief peripheral neuropathy screen (BPNS)

(continued)

Table 5 (continued)

Author, date; location	Design	Recruitment method	Participants (mean age; mean years education; percent male)	Tools examined
Fogel (1991); USA	Cross-sectional	NR	Age: NR Education: NR Male: 92%	Brief Cognitive Screen (BCS) – memory subtest, verbal fluency items, and conflicting stimulus test of the high sensitivity cognitive screen (HSCS)
Ganasen et al. (2008); Western Cape, South Africa	Cross-sectional	Convenience sampling	Age: 34 Education: 10 Male: 26%	HIV dementia scale
Garvey et al. (2009); London, UK	Cross-sectional	Convenience sampling	Age: 48 Education: NR Male: 84%	Prospective and retrospective memory questionnaire (PRMQ)
Gonzalez et al. (2003); USA	Cross-sectional	Not reported	Age: 40 Education: 14 Male: 100%	California computerized assessment package (CalCAP), mini-version
Jones et al. (1993); Baltimore, USA	Cross-sectional	Convenience sampling	Age: NR Education: 13 Male: 79%	Mental alternation test
Joska et al. (2011); Cape Town, South Africa	Cross-sectional	Convenience sampling	Age: 30 (HIV+); 25 (HIV−) Education: 10 (HIV+); 11 (HIV−) Male: 21% (HIV+); 38% (HIV−)	International HIV dementia scale
Knippels et al. (2002); Netherlands and Flanders, Belgium	Cross-sectional	Convenience sampling	Age: 39 Education: NR Male: 100%	Medical outcomes study HIV (MOS-HIV), four-item version in Dutch
Kwasa et al. (2012); Kisumu, Kenya	Cross-sectional	Convenience sampling	Age: 39 Education: NR Male: 57%	HIV dementia diagnostic tool

Lyon et al. (2009); Washington, DC; USA	Cross-sectional	Convenience sampling	Age: 17 (Dementia); 17 (No Dementia) Education: NR Male: 83% (Dementia); 45% (No Dementia)	HIV dementia scale; mini-mental state examination
Maruff et al. (2009); Melbourne, Australia	Cross-sectional	Convenience sampling	Age: 46 (ADC); 47 (controls, AIDS but no ADC or MND) Education: 13 (ADC); 12 (controls, AIDS but no ADC or MND) Male: 89% (ADC); 91% (controls, AIDS but no ADC or MND)	CogState (brief battery)
Minor et al. (2010); USA	Cross-sectional	Convenience sampling	Age: 38 Education: 12 Male: 52%	Coin rotation test
Morgan et al. (2008); San Diego, USA	Cross-sectional	Random sampling	Age: 40 (HIV+); 37 (HIV−) Education: 13 (HIV+); 14 (HIV−) Male: 83% (HIV+); 68% (HIV−)	HIV dementia scale
Muniyandi et al. (2012); Thanjavur, India	Cross-sectional	Convenience sampling	Age: NR Education: NR Male: 61%	Mini-mental state examination (MMSE); Bender gestalt test (BGT); Wechsler memory scale; International HIV dementia scale (IHDS)
Overton et al. (2011); St. Louis, USA	Cross-sectional	Convenience sampling	Age: 40 (median) Education: NR Male: 72%	CogState
Parsons et al. (2007); USA	Cross-sectional	Convenience SAMPLING	Age: 24 Education: 13 Male: 66%	Motor battery (timed gait, grooved pegboard, finger-tapping)

(continued)

Table 5 (continued)

Author, date; location	Design	Recruitment method	Participants (mean age; mean years education; percent male)	Tools examined
Power et al. (1995); Baltimore, USA	Cross-sectional	Convenience sampling	Age: 42 (HIV−); 37 (Asymptomatic HIV+); 36 (Non-Demented AIDS); 39 (Mildly Demented AIDS); 39 (Severely Demented AIDS) Education: 15 (HIV−); 14 (Asymptomatic HIV+); 14 (Non-Demented AIDS); 14 (Mildly Demented AIDS); 11 (Severely Demented AIDS) Male: 89%	HIV dementia scale; mini-mental state examination; grooved pegboard
Revicki et al. (1998); Baltimore-Washington, USA	Longitudinal	Convenience sampling	Age: 37 Education: NR Male: 66%	4-item cognitive function scale (CF4) from HIV health survey and complete 6-item cognitive function scale (CF6) from medical outcomes study (MOS)
Richardson et al. (2005); Boston Area, USA	Cross-sectional	Convenience sampling	Age: 41 Education: 12 Male: 65%	HIV dementia scale
Robertson et al. (2006); ACTG sites all over the world NYC, Chapel Hill and 42 sites around the world	Cross-sectional	Convenience sampling	Age: 38 (AIDS); 39 (SX); 35 (ASX); 35 (HIV−) Education: 14 (AIDS); 15 (SX); 15 (ASX); 16 (HIV−) Male: 87% (AIDS); 96% (SX); 91% (ASX); 56% (HIV−)	Timed gait test
Sacktor et al. (2005); Baltimore, USA and Uganda	Cross-sectional	Convenience sampling	Age: US HIV+: 43 (no impairment); 44 (subclinical impairment); 47 (mild dementia); 44 (moderate dementia); 49 (severe dementia); Uganda: 37 (HIV+); 31 (HIV−) Education: US HIV+:	International HIV dementia scale

			14 (no impairment); 13 (subclinical impairment); 13 (mild dementia); 12 (moderate dementia); 13 (severe dementia); Uganda: 9 (HIV+); 10 (HIV−) Male: NR	
Simioni et al. (2010); Geneva, Switzerland	Cross-sectional	Convenience sampling	Age: NR Education: NR Male: 72%	HIV dementia scale
Singh et al. (2008); Durban, South Africa	Cross-sectional	Convenience sampling	Age: 34 (median) Education: NR Male: 40%	International HIV dementia scale
Smith et al. (2003); USA	Cross-sectional	Not reported	Age: 41 (NP normal); 41 (NP abnormal) Education: 14 (NP normal); 13 (NP abnormal) Male: NR	HIV dementia scale
Von Giesen et al. (2005); Dusseldorf, Germany	Cross-sectional	Convenience sampling	Age: 45 (mildly demented); 41 (not demented) Education: NR Male: 100% (mildly demented); 100% (not demented)	HIV dementia scale
Wojna et al. (2007); Puerto Rico, USA	Cross-sectional	Convenience sampling	Age: 36 (HIV+); 34 (HIV−) Education: 12 (HIV+); 13 (HIV−) Male: 0%	HIV dementia scale, Spanish

AAN 2007 HAND Criteria: Asymptomatic Neurocognitive Disorder (ANI): prevalence of 30–35% and characterized by at least mild neurocognitive impairment in two domains but with no evidence of difficulty with day-to-day functioning; Mild Neurocognitive Disorder (MND): prevalence of 20–25% and characterized by at least mild neurocognitive impairment in two domains and with mild interference with day-to-day functioning; and HIV-Associated Dementia (HAD): prevalence of 2–3% with generally moderate to severe impairments in neurocognitive functioning and marked difficulties with everyday functioning [3]

ADC AIDS dementia complex, *MND* mild neurocognitive disorder, *ASX* asymptomatic; *SX* symptomatic, *NR* not reported

Table 6 Study outcome in literature on screening tools for HIV-associated neurocognitive disorders

Study (author, date; location)	Sample size (total; by group)	Impairment evaluated (type(s); classification system)	Tool characteristics (person can administer; time to administer; materials needed)	Reference test	Reference test details (size of battery; objective assessment; domains assessed; language of administration)	Sensitivity; specificity	Main findings
HIV Dementia Scale (HDS)							
Bottiggi et al. (2007); Kentucky, USA	46	Types: MCMD, HAD, neurocognitive deficits or impairment (memory, attention, psychomotor, and construction) Classification: 1991	Person: NR Time: NR Materials: NR	NP battery	Size: large Objective: NR Domains: intelligence, attention/concentration, memory, language, executive functioning, visuo-spatial, speed of processing, motor Language: English	Cut-off<=10: Sensitivity: 0.36 Specificity: 0.94	HDS is not efficient in predicting the presence of subtle and mild HIV Dementia
Carey et al. (2004); San Diego, USA	190	Type: NP impaired and unimpaired Classification: DSM-IV 1994, AAN 1991, Grant and Atkinson 1995	Person: NR Time: 5 min Materials: NR	NP battery	Size: large Objective: NR Domains: intelligence, attention/concentration, memory, language, executive functioning. visuo-spatial, speed of processing, motor Language: English	Cut-off <11: Sensitivity: 0.09 Specificity: 0.98	HDS is less accurate than paired NP test combinations – Hopkins Verbal Learning Test Revised (HVLTR; Total Recall) and the Grooved Pegboard Test nondominant hand (PND) pair, and the HVLTR and WAIS-III Digit Symbol (DS) subtest pair – in classifying HIV+ participants as NP impaired or not

Ganasen et al. (2008); Western Cape, South Africa	474	Type: NP impaired and unimpaired Classification: MMSE used as the gold standard	Person: NR Time: NR Materials: NR	Mini-mental state examination (MMSE)	Size: small Objective: NR Domains: attention/concentration, memory, motor Language: Xhosa, Afrikaans	Cut-off ≤10 Sensitivity: 0.80 Specificity: 0.80	HDS and MMSE were significantly correlated and showed significant agreement. Nonetheless, the HDS identified more participants that demonstrated cognitive impairment than the MMSE. HDS cut-off of ≤10 yielded a sensitivity of 80%, specificity of 80%, and discriminated between the presence and absence of cognitive impairment 90% of the time
Lyon et al. (2009); Washington, DC, USA	71	Type: HAD, HIV encephalopathy in adolescents Classification: 1991	Person: NR Time: 10 min Materials: NR	ANI 1991 criteria	Size: NR Objective: NR Domains: intelligence, attention/memory, language Language: NR	Cut-off of ≤10: Sensitivity: 0.83 Specificity: 0.79 Cut-off of ≤9: Sensitivity: 0.88 Specificity: 0.83	No statistically significant differences in sensitivity and specificity between the HDS and MMSE. Using standard cut-offs, HDS had 83% sensitivity and 79% specificity, although MMSE had 50% sensitivity and 92% specificity. The optimal cut-off score for the HDS, producing the highest sensitivity and specificity, was ≤9, providing 88% sensitivity and 83% specificity (87% correct classification)

(continued)

Table 6 (continued)

Study (author, date; location)	Sample size (total; by group)	Impairment evaluated (type(s); classification system)	Tool characteristics (person can administer; time to administer; materials needed)	Reference test	Reference test details (size of battery; objective assessment; domains assessed; language of administration)	Sensitivity; specificity	Main findings
Morgan et al. (2008); San Diego, California	317; 135 (HIV+), 182 (HIV−)	Type: ANI, MCMD, HAD Classification: 1991	Person: NR Time: 5–10 min Materials: NR	Modified AAN 1991 criteria and Grant and Atkinson 1995 criteria	Size: large Objective: yes Domains: NR Language: NR	Demographically adjusted T-score <40: Sensitivity: 0.71 Specificity: 0.74 Raw cut-off score ≤ 10: Sensitivity: 0.17 Specificity: 0.94	In comparison to the traditional HDS cut-off score (raw score total ≤ 10), use of the demographically adjusted normative standards significantly improved the sensitivity (from 17 to 71%) and overall classification accuracy (increasing the odds ratio from 3 to approximately 6). The application of demographically adjusted normative standards on the HDS improves the clinical applicability and accuracy
Power et al. (1995); Baltimore, USA	130	Type: HAD Classification: 1991	Person: NR Time: NR Materials: NR	Memorial sloan kettering dementia evaluation; mini-mental state exam (MMSE); grooved pegboard (PB)	Size: small Objective: NR Domains: attention/ concentration, memory, executive functioning, motor Language: English	Cut-off ≤ 10: Sensitivity: 0.80 Specificity: 0.91	HDS demonstrated greater efficiency in identifying HIV dementia compared to Grooved Pegboard and the Mini-Mental State Examination

Richardson et al. (2005); Boston Area, USA	40	Types: neurocognitive deficit or impairment (impairment in attention and concentration, psychomotor functioning, behavioral inhibition, constructional praxis) Classification: performance at least 2 standard deviation units below established norms on one or more independent NP measures in two or more domains of functioning	Person: NR Time: 10 min Materials: NR	NP battery	Size: medium Objective: yes Domains: memory, executive functioning, motor Language: English	Cut-off <=10: Sensitivity: 0.55 Specificity: 0.75	HDS prediction resulted in modest sensitivity and moderate specificity. In ROC curve analysis, area under the curve was only modestly better than chance (0.58). Optimal cut-off for the HDS is ≤10
Simioni et al. (2010), Geneva, Switzerland	100 (50 with cognitive complaints, 50 without cognitive complaints)	Types: ANI, MND, HAD Classification: 2007	Person: NR Time: NR Materials: NR	NP battery	Size: medium Objective: no Domains: intelligence, attention/concentration, memory, language, executive functioning, speed of processing, motor Language: NR	Cut-off 10: Sensitivity: 0.54 Specificity: 0.96 Cut-off 14: Complaining: Sensitivity: 0.83 Specificity: 0.63 Non-Complaining: Sensitivity: 0.88 Specificity: 0.67	Prevalence of HAND is high even in long-standing aviremic HIV-positive patients. HAND without functional repercussion in daily life is most frequent. Cut-off of 14 points or less seemed to provide a useful tool to screen for HANDs
Smith et al. (2003); USA	90	Type: neurocognitive deficit or impairment (subtle HIV-related cognitive dysfunction) Classification: cognitive "abnormality" defined as performance that deviated at least 2 SD units below established norms on at least two independent NP measures	Person: NR Time: NR (noted "brief") Materials: NR	NP battery	Size: medium Objective: no Domains: intelligence, memory, executive functioning, visuospatial, speed of processing, motor Language: English	Cut-off<=10: Sensitivity: 0.39 Specificity: 0.85	HDS lacks sufficient sensitivity to screen for NP abnormality beyond frank dementia. Intact performance (i.e., performance above established cut-off levels) contributes to a significant number of false-negative errors, suggesting need for NP battery for subtle neurocognitive deficits

(continued)

Table 6 (continued)

Study (author, date; location)	Sample size (total; by group)	Impairment evaluated (type(s); classification system)	Tool characteristics (person can administer; time to administer; materials needed)	Reference test	Reference test details (size of battery; objective assessment; domains assessed; language of administration)	Sensitivity; specificity	Main findings
Von Giesen et al. (2005); Dusseldorf, Germany	266; 55 (mildly demented), 211 (not demented)	Types: mild dementia, no dementia Classification: mild dementia (HDS score $\leq$ 10), no dementia (HDS score >10)	Person: NR Time: NR (noted "brief") Materials: NR	NP battery	Size: small Objective: NR Domains: motor Language: German	Sensitivity: NR Specificity: NR	Patients with mild dementia showed significant slowing of most rapid alternating movement (MRAM) and significantly prolonged contraction time compared to non-demented patients. Motor performance correlated significantly with time-dependent HDS sub-scores for psychomotor speed and construction
Wojna et al. (2007); Puerto Rico, USA	96; 60 (HIV+), 36 (HIV−)	Types: asymptomatic cognitive impairment and symptomatic impairment (MCMD, HAD) Classification: modified 1991	Person: NR Time: NR (noted "rapid") Materials: NR	NP battery	Size: medium Objective: yes Domains: intelligence, memory, executive functioning, speed of processing, motor Language: Spanish	Cut-off $\leq$12: Sensitivity: 0.63 Specificity: 0.84 Cut-off $\leq$13: Sensitivity: 0.87 Specificity: 0.46	HDS-Spanish total score and sub-scores for psychomotor speed and memory recall showed significant difference between HIV-negative women and HIV-positive women with dementia and between HIV-positive women with normal cognition and with dementia. Optimal cut-off point was $\leq$13

International HIV Dementia Scale							
Joska et al. (2011); Cape Town, South Africa	190; 96 (HIV+); 94 (HIV−)	Types: ANI, MND, HAD Classification: 2007	Person: NR Time: NR Materials: NR	NP battery	Size: medium Objective: no Domains: attention/ concentration, memory, executive function, visuo-spatial, motor Language: isiXhosa, Afrikaans	Cut-off of 10: Sensitivity=0.45 Specificity=0.79	In ART naïve sample, HIV+ individuals displayed greater impairment compared to HIV− on IHDS and range of NP tests. With ROC analysis, the area under curve was 0.64. This data suggests that the IHDS may have limitations as a tool to screen for HAD in South Africa
Muniyandi et al. (2012); Thanjavur, India	33	Type: ANI, MND, HAD Classification: 2007	Person: NR Time: NR Materials: NR	NP Battery	Size: NR Objective: yes Domains: NR Language: NR	NR	Tests which assess cognitive and motor speed may be more helpful than clinical psychiatric interview to spot the AIDS patients who have cognitive impairment. The International HIV Dementia Scale was the most sensitive instrument
Sacktor et al. (2005); Uganda/ Baltimore, USA	247; 66 (HIV+ USA), 81 (HIV+ Uganda), 100 (HIV− control, Uganda)	Type: HAD Classification: 1991	Person: Non-neurologist Time: 2–3 min Materials: watch with a second hand	Memorial sloan kettering dementia staging; NP Battery	Size: medium Objective: yes Domains: attention/ concentration, memory, language, executive functioning, speed of processing, motor Language: English, Luganda	Cut-off <=10: USA: Sensitivity: 0.80 Specificity: 0.57 Uganda: Sensitivity: 0.80 Specificity: 0.55	IHDS may be a useful screening test to identify individuals at risk for HIV dementia in both industrialized and developing world. Full NP testing should be performed to confirm diagnosis of HIV dementia
Singh et al. (2008); Durban, South Africa	20	Type: neurocognitive deficit or impairment (moderate and severe) Classification: Moderate – beyond the norms on at least 2 tests; Severe – three or more tests abnormal	Person: non-specialist Time: 2–3 min Materials: none	NP battery	Size: small Objective: NR Domains: attention, memory, executive functioning, motor Language: English, isiZulu	Cut-off of 10: Specificity=0.88 Sensitivity=0.50	Low specificity may limit clinical utility of IHDS. Research needed to verify the high burden on neurocognitive impairment among people with low CD4. Larger study needed to validate IHDS in South Africa

(continued)

Table 6 (continued)

Study (author, date; location)	Sample size (total; by group)	Impairment evaluated (type(s); classification system)	Tool characteristics (person can administer; time to administer; materials needed)	Reference test	Reference test details (size of battery; objective assessment; domains assessed; language of administration)	Sensitivity; specificity	Main findings
CogState							
Cysique et al. (2006); Sydney, Australia	81	Types: ADC; neurocognitive deficits or impairments (psychomotor speed, working memory, attention, learning) Classification System: ≤ -2 SD in 2 of 14 neuropsychological measures (Cysique 2004)	Person: NR Time: 10–15 min Materials: desktop computer	NP battery	Size: large Objective: yes Domains: intelligence, attention/concentration, memory, language, executive functioning, visuo-spatial, motor Language: English	Sensitivity: 0.81 Specificity: 0.70	Study supports utility of brief computerized battery in detection of HIV-associated neurocognitive impairment. Good agreement between standard neuropsychological tests and the CogState indices in identifying neurocognitive impairment
Maruff et al. (2009); Melbourne, Australia	293; 20 (ADC), 20 (ADC controls); 253 healthy adults	Type: ADC Definition: Price and Brew, 1988	Person: NR Time: 8–10 min Materials: personal computer	NP battery	Size: medium Objective: NR Domains: memory, visuo-spatial processing, motor Language: English	Sensitivity: NR Specificity: NR	Brief CogState battery has adequate construct validity and is sensitive to subtle cognitive impairment in ADC. Recommends that assessment of attention, processing speed, memory, and working memory based only on CogState can support solely on broad conclusions

Overton et al. (2011); St. Louis, USA	46	Type: neurocognitive deficit or impairment (mild to moderate impairment) Classification: Carey et al. (2004)	Person: NR Time: 12–15 min Materials: Computer and computer program	NP battery	Size: medium Objective: NR Domains: attention/ concentration, memory, language, executive functioning, motor Language: English	GDS≥0.5 Sensitivity: NR Specificity: NR	Measures of both simple detection tests and identification task correlated with GDS and had highest level of correlation with tests in CHARTER battery. Other tests correlated poorly with NP testing. Composite score of 5 tests (significant in ROC analyses) correctly classified 90% of individuals according to impairment
Mini-Mental State Examination (MMSE)							
Lyon et al. (2009); Washington, DC, USA	71	Type: HIV-encephalopathy in adolescents Classification: 1991	Person: NR Time: 10 min Materials: NR	ANI 1991 criteria	Size: NR Objective: NR Domains: NR Language: NR	Cut-off <24 Sensitivity: 0.50 Specificity: 0.92	MMSE identified 3 of 6 cases of encephalopathy. HDS appeared to be more clinically useful
Muniyandi et al. (2012); Thanjavur, India	33	Type: ANI, MND, HAD Classification: 2007	Person: NR Time: NR Materials: NR	NP battery	Size: NR Objective: yes Domains: NR Language: NR	NR	Tests which assess cognitive and motor speed may be more helpful than clinical psychiatric interview to spot the AIDS patients who have cognitive impairment. The International HIV Dementia Scale was the most sensitive instrument
Power et al. (1995); Baltimore, USA	130	Type: HAD Classification: 1991	Person: NR Time: NR Materials: NR	Memorial sloan kettering dementia evaluation; HIV dementia scale (HDS); grooved pegboard (PB)	Size: Small Objective: NR Domains: attention/ concentration, memory, executive functioning, motor Language: English	Cut-off≤28: Sensitivity: 0.50 Specificity: 0.88	MMSE was less efficient at identifying HIV Dementia than the HDS and the Grooved Pegboard

(continued)

Table 6 (continued)

Study (author, date; location)	Sample size (total; by group)	Impairment evaluated (type(s); classification system)	Tool characteristics (person can administer; time to administer; materials needed)	Reference test	Reference test details (size of battery; objective assessment; domains assessed; language of administration)	Sensitivity; specificity	Main findings
Brief Neurocognitive Screen (Component NeuroScreen)							
Ellis et al. (2005); 15 sites, USA	301	Type: MCMD, HAD, neurocognitive deficits or impairment (speed of information processing, mental flexibility, working memory) Classification: 1991	Person: non-neurologist Time: 12–15 min Materials: NR	NP battery	Size: medium Objective: yes Domains: attention/ concentration, memory, language, speed of processing, motor Language: English	Sensitivity: 0.65 Specificity: 0.72	Designed to estimate the frequency of HIV-associated neurocognitive disorders. In ROC, when compared to NP battery the area under the curve was 0.74. Yields (NeuroScreen as a whole) a substantial number of false positives and negatives so more useful for tracking prevalence in large cohorts rather than individual patients
Bender Gestalt Test (BGT)							
Muniyandi et al. (2012); Thanjavur, India	33	Type: ANI, MND, HAD Classification: 2007	Person: NR Time: NR Materials: NR	NP battery	Size: NR Objective: yes Domains: NR Language: NR	NR	Tests which assess cognitive and motor speed may be more helpful than clinical psychiatric interview to spot the AIDS patients who have cognitive impairment. The International HIV Dementia Scale was the most sensitive instrument

California Computerized Assessment Package (CalCAP) Mini-Battery							
Gonzalez et al. (2003); California, USA	82	Type: neurocognitive deficits (normal, mild, mild-moderate, moderate, moderate-severe, severe) or impairments (attention and speed of information processing, abstraction, learning) Heaton 1995	Person: NR Time: 10 min Materials: computer	NP battery	Size: large Objective: No Domains: attention/concentration, memory, executive functioning, visuo-spatial, speed of processing, motor Language: English	Cut-off D-score of ≥ 0.5: Sensitivity: 0.68 Specificity: 0.77	Traditional NP batteries and computerized reaction time tests do not measure the same thing. They are not interchangeable in examining HIV-related NP impairment
Coin Rotation Test							
Minor et al. (2010); Louisiana, USA	204	Type: neurocognitive deficits or impairment (psychomotor performance) Classification: NR	Person: Not specified Time: 1 min Materials: NR (infer coin and timer)	Psychomotor speed subscale of the modified HIV dementia scale (MHDS-PS)	Size: small Objective: NR Domains: memory, executive functioning, motor Language: English	Cut-off of 20 Rotations: Sensitivity: 0.72 Specificity: 0.61	Good convergent validity between Coin Rotation Test and Modified HDS. Gender did not significantly affect CRT performance but did affect modified HDS performance. CRT performance was less affected by education than MHDS performance
The Computer Assessment of Mild Cognitive Impairment							
Becker et al. (2011); USA	59; 29 (HIV+), 30 (HIV−)	Type: neurocognitive deficits or impairment (normal, borderline, impaired) Classification: Global Impairment Rating (Woods 2004)	Person: health professionals Time: 20 min Materials: tablet with touch screen	NP battery	Size: large Objective: NR Domains: memory, language, executive functioning, visuo-spatial, speed of processing, motor Language: English	Sensitivity: 0.72 Specificity: 0.97	Detected mild impairment and median stability over 12 and 24 weeks of follow-up were 0.32 and 0.46 (did not differ as a function of serostatus). Discriminate functional analysis (6 CAMCI scores) correctly classified 90% of subjects

(continued)

Table 6 (continued)

Study (author, date; location)	Sample size (total; by group)	Impairment evaluated (type(s); classification system)	Tool characteristics (person can administer; time to administer; materials needed)	Reference test	Reference test details (size of battery; objective assessment; domains assessed; language of administration)	Sensitivity; specificity	Main findings
Four-item scale from Health Survey; Six-item scale from MOS							
Revicki et al. (1998); Baltimore, USA	162 (baseline) 131 follow-up	Type: Mild, severe impairment Classification: TMT manual (Reitan 1992)	Person: NR Time: Less than 2 min Materials: NR	Trail Making Test (TMT)	Size: Small Objective: No Domains: Executive Functioning, Speed of Processing, Motor Language: English	Sensitivity: NR Specificity: NR	Logistic regression analysis showed both four-item version (CF4) included in the HIV Health Survey and the complete six-item scale (CF6) from the Medical Outcomes Study (MOS) predicted mild cognitive impairment based on TMT scores ($p = 0.046 - 0.008$) and severe cognitive impairment based on TMT scores ($p = 0.0012 - 0.0003$). Baseline significant differences in mean CF6 and CF4 scores for mildly impaired compared to less than mildly impaired and severely impaired and less than severely impaired
Grooved Pegboard							
Power et al. (1995); Baltimore, USA	130	Type: HAD Classification: 1991	Person: NR Time: NR Materials: NR	Memorial sloan kettering dementia evaluation; HIV dementia scale (HDS); Mini-mental state exam (MMSE);	Size: small Objective: NR Domains: attention/ concentration, memory, executive functioning, motor Language: English	Cut-off $\geq$ 90 sec: Sensitivity: 0.91 Specificity: 0.82	Was efficient in detecting HAD (86%) compared with the HDS (84%) and MMSE (72%)

Brief Cognitive Screen (BCS) – memory subtest, verbal fluency items, and conflicting stimulus test of the High Sensitivity Cognitive Screen (HSCS)							
Fogel (1991); 3 locations, USA	156	Type: neurocognitive deficits or impairment (memory, verbal fluency, conflicting stimulus) Classification: abnormality on the brief dementia screen	Person: NR Time: NR Materials: NR	Standardized dementia screen (registration and delayed memory for three simple words, months in reverse, five serial sevens, and orientation to month and year)	Size: Small Objective: No Domains: memory, language, orientation Language: English	Sensitivity: NR Specificity: NR	Patients with abnormal scores on the Brief Cognitive Screen showed greater symptoms and functional impairment
HIV Dementia Diagnostic Test							
Kwasa et al. (2012); Kisumu, Kenya	26; 14 (ANI, MND), 6 (HAD)	Type: ANI, MDN, HAD Classification: 2007	Person: non-physician health-care worker Time: NR Materials: NR	NP battery	Size: medium Objective: yes Domains: attention/concentration, memory, language, executive functioning, speed of processing, motor Language: English, Dhulou	Cut-off ≤ 22: Sensitivity: 0.63 Specificity: 0.67	Moderate sensitivity and specificity for HAD. Reliability was poor, suggesting that substantial training and formal evaluations of training adequacy will be critical
Hopkins Verbal Test Revised and Grooved Pegboard Test Nondominant Hand							
Carey et al. (2004); San Diego, USA	190	Type: neurocognitive deficits or impairment (NP impaired or unimpaired) Classification: DSM-IV 1994, AAN 1991, Grant and Atkinson 1995	Person: trained psychometrist Time: 5 min Materials: NR	NP battery	Size: large Objective: NR Domains: intelligence, attention/concentration, memory, language, executive functioning, visuo-spatial, speed of processing, motor Language: English	Sensitivity: 0.78 Specificity: 0.85	The combination of Hopkins verbal test revised and grooved pegboard test nondominant hand was more accurate than the HIV dementia scale (HDS) in classifying HIV + participants as NP impaired or unimpaired
Hopkins Verbal Learning Test/WAIS-III Digit Symbol							
Carey et al. (2004); San Diego, USA	190	Type: neurocognitive deficits or impairment (NP impaired and unimpaired) Classification: DSM-IV 1994, AAN 1991, Grant and Atkinson 1995	Person: trained psychometrist Time: 5 min Materials: NR	NP battery	Size: large Objective: NR Domains: intelligence, attention/concentration, memory, language, executive functioning, visuo-spatial, speed of processing, motor Language: English	Sensitivity: 0.75 Specificity: 0.92	The combination of Hopkins verbal test revised and WAIS-III digit symbol (DS) subtest was more accurate than the HIV dementia scale (HDS) in classifying HIV + participants as NP impaired or unimpaired

(continued)

Table 6 (continued)

Study (author, date; location)	Sample size (total; by group)	Impairment evaluated (type(s); classification system)	Tool characteristics (person can administer; time to administer; materials needed)	Reference test	Reference test details (size of battery; objective assessment; domains assessed; language of administration)	Sensitivity; specificity	Main findings
Medical Outcome Study HIV (MOS-HIV) Health Survey							
Knippels et al. (2002); Netherlands and Flanders, Belgium	82	Type: neurocognitive deficits or impairment (neuropsychological impairment) Classification: NR	Person: completed at home Time: NR Materials: questionnaire	NP battery	Size: medium Objective: No Domains: attention/ concentration, memory, language, executive functioning, visuo-spatial, motor Language: Dutch	Sensitivity: NR Specificity: NR	Showed significant associations with NP test performance overall and, specifically, with the domains of abstraction, language, visuo-spatial abilities (controlling for CD4 cell count and CDC disease stage). Trend toward significance in memory domain. Seems particularly sensitive to changes in NP test performance in early HIV-infection
Mental Alternation Test							
Jones et al. (1993); Baltimore, USA	62	Type: neurocognitive deficits or impairment (orientation, memory, concentration, language, praxis, psycho or speed, sequencing ability) Classification: Abnormal performance on MMSE and Trail making (Crum 1993, Bornstein 2005)	Person: NR Time: 60 sec Materials: NR	Mini-mental state examination; trail making part A and B	Size: small Objective: NR Domains: intelligence, attention/concentration, memory, language, executive functioning, speed of processing Language: English	With MMSE: Sensitivity: 0.95 Specificity: 0.79 With Trail making: Sensitivity: 0.78 Specificity: 0.93	Scores correlated significantly with MMSE and trail making test part B scores when controlled for confounders. ROC curve showed cut-off of 15 yielded best results for detection of abnormal performance on MMSE and trail making test Part B

Motor Battery (Timed Gait, Grooved Pegboard, Finger-tapping)							
Parsons et al. (2007); USA	361	Type: neurocognitive deficits or impairment (attention, executive, figural memory, verbal memory, language) Classification: NR	Person: NR Time: NR (noted "brief") Materials: NR	NP battery	Size: Large Objective: NR Domains: attention/concentration, memory, language, executive functioning, visuo-spatial, speed of processing, motor Language: English	Cut-off of −0.42: Sensitivity: 0.79 Specificity: 0.76	Significant correlation with comprehensive battery (52% variance). Increased variance to 73% when Digit Symbol and Trail making added. Motor battery may have broader utility to diagnose and monitor HIV-related neurocognitive disorder in international settings
Prospective and Retrospective Memory Questionnaire (PRMQ)							
Garvey et al. (2009); London, UK	45	Types: neurocognitive deficits or impairment (asymptomatic neurocognitive impairment aNCI, memory) Classification: aNCI-performance score more than 1 SD below the normative mean in at least 2 domains of CogState	Person: NR Time: 10 min Materials: questionnaire, writing utensil	CogState	Size: small Objective: NR Domains: attention/concentration, memory, executive functioning, motor, learning Language: English	Sensitivity: NR Specificity: NR	No statistically significant associations between PRMQ and CogState; questionnaire should not be used as a screening tool. Association between PRMQ and set-shifting task of CogState; questionnaire able to capture part of the executive function deterioration in HIV-associated NCI
Screening Algorithm							
Cysique et al. (2010); Sydney, Australia	127; 97 (HIV+), 30 (HIV−)	Types: neurocognitive deficits or impairment (HAND) Classification: 2007	Person: clinical individual Time: 3 min Materials: patient's clinical characteristics	NP battery	Size: large Objective: yes Domains: intelligence, attention/concentration, memory, language, speed of processing, motor, reasoning Language: English	Sensitivity: 0.78 Specificity: 0.70	Good overall prediction accuracy and specificity. Proved useful to identify HIV-infected patients with advanced disease at high risk of HAND who require more formal assessment. Recommended for HIV-infected Caucasian men with advanced disease

(continued)

Table 6 (continued)

Study (author, date; location)	Sample size (total; by group)	Impairment evaluated (type(s); classification system)	Tool characteristics (person can administer; time to administer; materials needed)	Reference test	Reference test details (size of battery; objective assessment; domains assessed; language of administration)	Sensitivity; specificity	Main findings
Timed Gait Test							
Robertson et al. (2006); ACTG sites all over the world	1549; 1122 (AIDS), 113 (symptomatic), 165 (asymptomatic), 87 (HIV−)	Type: ADC Classification: neurologic history (Robertson 1997)	Person: NR Time: NR Materials: stopwatch, recording sheet	ADC Staging; NP Battery	Size: NR Objective: yes Domains: memory, language, executive functioning, visuo-spatial processing, motor Language: English	Sensitivity: 0.83 Specificity: 0.59	Good sensitivity and moderate specificity for detection of ADC. Abnormal Timed Gait scores were also significantly correlated with abnormal scores on neurological and neuropsychological evaluations. Cutting scores on the Timed Gait procedure resulted in reasonably good classification rates of ADC staging, especially for use as a screening tool
Wechsler Memory Scale							
Muniyandi et al. (2012); Thanjavur, India	33	Type: ANI, MND, HAD Classification: 2007	Person: NR Time: NR Materials: NR	NP battery	Size: NR Objective: Yes Domains: NR Language: NR	NR	Tests that assess cognitive and motor speed may be more helpful than clinical psychiatric interview to spot the AIDS patients who have cognitive impairment. The International HIV Dementia Scale was the most sensitive instrument

MCMD minor cognitive-motor disorder, *HAD* HIV-associated dementia, *ANI* asymptomatic neurocognitive disorder, *MND* mild neurocognitive disorder, *ADC* AIDS dementia complex; *NP* neuropsychological; *aNCI* asymptomatic neurocognitive impairment; *NR* not reported

Criteria for battery size: "short" battery/criterion was defined by testing that took <30 min to administer; "medium" battery/criterion was defined by testing that took 30–90 min to administer; and "long and comprehensive" battery/criterion was defined by testing which took >90 min to administer

References

Anderson AM, Lennox JL, Nguyen ML, Waldrop-Valverde D, Tyor WR, Loring DW (2016) Preliminary study of a novel cognitive assessment device for the evaluation of HIV-associated neurocognitive impairment. J Neurovirol 22(6):816–822

Antinori A, Arendt G, Becker JT et al (2007) Updated research nosology for HIV-associated neurocognitive disorders. Neurology 69:1789–1799. https://doi.org/10.1212/01.WNL.0000287431.88658.8b

Barber TJ, Bradshaw D, Hughes D et al (2014) Screening for HIV-related neurocognitive impairment in clinical practice: challenges and opportunities. AIDS Care 26:160–168

Becker JT, Lopez OL, Dew MA, Aizenstein HJ (2004) Prevalence of cognitive disorders differs as a function of age in HIV virus infection. AIDS 18:11–18

Becker JT, Dew MA, Aizenstein HJ et al (2011) Concurrent validity of a computer-based cognitive screening tool for use in adults with HIV disease. AIDS Patient Care STDs 25:351–357

Blackstone K, Moore DJ, Heaton RK, Franklin DR, Woods SP, Clifford DB et al (2019) Diagnosing symptomatic HIV-associated neurocognitive disorders: self-report versus performance-based assessment of everyday functioning. J Int Neuropsychol Soc 18:79–88. https://doi.org/10.1017/S135561771100141X

Bloch M, Kamminga J, Jayewardene A et al (2016) A screening strategy for HIV-associated neurocognitive disorders that accurately identifies patients requiring neurological review. Clin Infect Dis 63:687–693. https://doi.org/10.1093/cid/ciw399

Bonnet F, Amieva H, Marquant F et al (2013) Cognitive disorders in HIV-infected patients: are they HIV-related? AIDS 27:391–400

Bottiggi KA, Chang JJ, Schmitt FA, Avison MJ, Mootoor Y, Nath A et al (2007) The HIV dementia scale: predictive power in mild dementia and HAART. J Neurol Sci 260:11–15

Breuer E, Stoloff K, Myer L et al (2012) Reliability of the lay adherence counsellor administered substance abuse and mental illness symptoms screener (SAMISS) and the international HIV dementia scale (IHDS) in a primary care HIV clinic in Cape Town, South Africa. AIDS Behav 16:1464–1471

Brouillette MJ, Mayo N, Fellows LK, Lebedeva E, Higgins J, Overton ET et al (2015) A better screening tool for HIV-associated neurocognitive disorders: is it what clinicians need? AIDS 29 (8):895–902. https://doi.org/10.1097/QAD.0000000000000152

Callaghan M, Ford N, Schneider H (2010) A systematic review of task-shifting for HIV treatment and care in Africa. Hum Resour Health 8:8

Carey CL, Woods SP, Rippeth JD et al (2004) Initial validation of a screening battery for the detection of HIV-associated cognitive impairment. Clin Neuropsychol 18:234–248

Chalermchai T, Sithinamsuwam P, Clifford D et al (2013) Trail making test a improve performance characteristics of the international HIV dementia scale to identify symptomatic HAND. J Neurovirol 19:137–143

Chartier M, Crouch PC, Tullis V, Catella S, Frawley E, Filanosky C et al (2015) The Montreal Cognitive Assessment: a pilot study of a brief screening tool for mild and moderate cognitive impairment in HIV-positive veterans. J Int Assoc Provid AIDS Care 14(3):197–201. https://doi.org/10.1177/2325957414557270

Cysique LA, Maruff P, Darby D, Brew BJ (2006) The assessment of cognitive function in advanced HIV-1 infection and AIDS dementia complex using a new computerised cognitive test battery. Arch Clin Neuropsychol 21:185–194

Cysique LA, Murray JM, Dunbar M, Jeyakumar V, Brew BJ (2010) A screening algorithm for HIV-associated neurocognitive disorders. HIV Med 11:642–649

Cysique LA, Bain MP, Lane TA, Brew BJ (2012) Management issues in HIV-associated neurocognitive disorders. Neurobehav HIV Med 4:63–73

Ellis RJ, Evans SR, Clifford DB, Moo LR, McArthur JC, Collier AC et al (2005) Clinical validation of the NeuroScreen. J Neurovirol 11:503–511

Ettenhofer ML, Hinkin CH, Castellon SA et al (2009) Aging, neurocognition, and medication adherence in HIV infection. Am J Geriatr Psychiatry 17:281–290. https://doi.org/10.1097/JGP.0b013e31819431bd

Ettenhofer ML, Foley J, Castellon SA, Hinkin CH (2010) Reciprocal prediction of medication adherence and neurocognition in HIV/AIDS. Neurology 74:1217–1222. https://doi.org/10.1212/WNL.0b013e3181d8c1ca

Fazeli PL, Woods AJ, Pope CN, Vance DE, Ball KK (2019) Effect of transcranial direct current stimulation combined with cognitive training on cognitive functioning in older adults with HIV: a pilot study. Appl Neuropsychol Adult 26(1):36–47. https://doi.org/10.1080/23279095.2017.1357037

Fogel BS (1991) The high sensitivity cognitive screen. Int Psychogeriatr 3:273–288

Folstein MF, Robins LN, Helzer JE (1983) The mini-mental state examination. Arch Gen Psychiatry 40:812

Ganasen KA, Fincham D, Smit J, Seedat S, Stein D (2008) Utility of the HIV dementia scale (HDS) in identifying HIV dementia in a South African sample. J Neurol Sci 269:62–64

Garvey LJ, Yerrakalva D, Winston A (2009) Correlations between computerized battery testing and a memory questionnaire for identification of neurocognitive impairment in HIV type 1-infected subjects on stable antiretroviral therapy. AIDS Res Hum Retrovir 25:765–769

Gisslén M, Price RW, Nilsson S (2011) The definition of HIV-associated neurocognitive disorders: are we overestimating the real prevalence? BMC Infect Dis 11. https://doi.org/10.1186/1471-2334-11-356

Gonzalez R, Heaton RK, Moore DJ, Letendre S, Ellis RJ, Wolfson T et al (2003) Computerized reaction time battery versus a traditional neuropsychological battery: detecting HIV-related impairments. J Int Neuropsychol Soc 9:64–71

Gorman AA, Foley JM, Ettenhofer ML et al (2009) Functional consequences of HIV-associated neuropsychological impairment. Neuropsychol Rev 19:186–203

Grant I (2008) Neurocognitive disturbances in HIV. Int Rev Psychiatry 20:33–47

Grant I, Sacktor N (2012) HIV-associated neurocognitive disorders, 3rd edn. Oxford University Press, New York

Haddow LJ, Cpsychoil AA, Cartledge JD et al (2013) Routine detection and management of neurocognitive impairment in HIV-positive patients in a UK centre. Int J STD AIDS 24:217–219

Hasbun R, Eraso J, Ramireddy S, Wainwright D, Salazar L, Grimes R et al (2012) Screening for neurocognitive impairment in HIV individuals: the utility of the Montreal Cognitive Assessment test. J AIDS Clin Res 3(10). https://doi.org/10.4172/2155-6113.1000186

Hasbun R, Eraso J, Ramireddy S et al (2013) Screening for neurocognitive impairment in hiv individuals: the utility of the Montreal cognitive assessment test. J AIDS Clin Res 3:186. https://doi.org/10.4172/2155-6113.1000186

Heaton RK, Marcotte TD, Rivera Mindt M et al (2004) The impact of HIV-associated neuropsychological impairment on everyday functioning. J Int Neuropsychol Soc 10:317–331

Heaton RK, Clifford DB, Franklin DR et al (2010) HIV-associated neurocognitive disorders persist in the era of potent antiretroviral therapy: CHARTER study. Neurology 75:2087–2096. https://doi.org/10.1212/WNL.0b013e318200d727

Heaton RK, Franklin DR, Ellis RJ et al (2011) HIV-associated neurocognitive disorders before and during the era of combination antiretroviral therapy: differences in rates, nature, and predictors. J Neurovirol 17:3–16

High KP, Valcour V, Paul R (2006) HIV infection and dementia in older adults. Clin Infect Dis 42:1449–1454

Hinkin CH, Castellon SA, Durvasula RS et al (2002) Medication adherence among HIV+ adults: effects of cognitive dysfunction and regimen complexity. Neurology 59:1944–1950

Hinkin CH, Hardy DJ, Mason KI et al (2004) Medication adherence in HIV-infected adults: effect of patient age, cognitive status, and substance abuse. AIDS 18(Suppl 1):S19–S25

Hoare J, Phillips N, Joska JA et al (2016) Applying the HIV-associated neurocognitive disorder diagnostic criteria to HIV-infected youth. Neurology 87:86–93

Iudicello JE, Woods SP, Cattie JE et al (2013) Risky decision-making in HIV-associated neurocognitive disorders (HAND). Clin Neuropsychol 27:256–275. https://doi.org/10.1080/13854046.2012.740077

Janssen MAM, Bosch M, Koopmans PP, Kessels RPC (2015) Validity of the Montreal cognitive assessment and the HIV dementia scale in the assessment of cognitive impairment in HIV-1 infected patients. J Neurovirol 21:383–390. https://doi.org/10.1007/s13365-015-0324-4

Joint United Nations Programme on HIV/AIDS (2016) Global report: UNAIDS report on the global AIDS epidemic 2013

Jones BN, Teng EL, Folstein MF, Harrison KS (1993) A new bedside test of cognition for patients with HIV infection. Ann Intern Med 119:1001–1004

Joska JA, Westgarth-Taylor J, Hoare J, Thomas KGF, Paul R, Myer L et al (2011) Validity of the international HIV dementia scale in South Africa. AIDS Patient Care STDs 25:95–101

Joska JA, Witten J, Thomas KG et al (2016) A comparison of five brief screening tools for HIV-associated neurocognitive disorders in the USA and South Africa. AIDS Behav 20:1621–1631

Kamminga J, Cysique LA, Lu G et al (2013) Validity of cognitive screens for HIV-associated neurocognitive disorder: a systematic review and an informed screen selection guide. Curr HIV/AIDS Rep 10:342–355. https://doi.org/10.1007/s11904-013-0176-6

Kamminga J, Bloch M, Vincent T et al (2017) Determining optimal impairment rating methodology for a new HIV-associated neurocognitive disorder screening procedure. J Clin Exp Neuropsychol 39:753–767. https://doi.org/10.1080/13803395.2016.1263282

Kim S, Ades M, Pinho V et al (2014) Patterns of HIV and mental health service integration in New York state. AIDS Care 26:1027–1031

Kim WJ, Ku NS, Lee Y-J, Ahn JY, Kim SB, Ahn H-W et al (2016) Utility of the Montreal Cognitive Assessment (MoCA) and its subset in HIV-associated neurocognitive disorder (HAND) screening. J Psychosom Res 80:53–57. https://doi.org/10.1016/j.jpsychores.2015.11.006

Knippels HMA, Goodkin K, Weiss JJ, Wilkie FL, Antoni MH (2002) The importance of cognitive self-report in early HIV-1 infection: validation of a cognitive functional status subscale. AIDS 16:259–267

Ku NS, Lee Y, Ahn JY et al (2014) HIV-associated neurocognitive disorder in HIV-infected Koreans: the Korean NeuroAIDS project. HIV Med 15:470–477. https://doi.org/10.1111/hiv.12137

Kvalsund MP, Haworth A, Murman DL et al (2009) Short report: closing gaps in antiretroviral therapy access: human immunodeficiency virus-associated dementia screening instruments for non-physician healthcare workers. Am J Trop Med Hyg 80:1054–1059

Kwasa J, Cettomai D, Lwanya E, Osiemo D, Oyaro P, Birbeck GL et al (2012) Lessons learned developing a diagnostic tool for HIV-associated dementia feasible to implement in resource-limited settings: pilot testing in Kenya. PLoS One 7

Lawton MP, Brody EM (1969) Assessment of older people: self-maintaining and instrumental activities of daily living. Gerontologist 9:179–186

Levine AJ, Palomo M, Hinkin CH et al (2011) A comparison of screening batteries in the detection of neurocognitive impairment in HIVinfected Spanish speakers. Neurobehav HIV Med 2011:79–86

Lyon ME, McCarter R, D'Angelo LJ (2009) Detecting HIV associated neurocognitive disorders in adolescents: what is the best screening tool? J Adolesc Health 44:133–135

Maj M, Janssen R, Starace F, Zaudig M, Satz P, Sughondhabirom B et al (1994) WHO neuropsychiatric AIDS study, cross-sectional phase I. Study design and psychiatric findings. Arch Gen Psychiatry 51(1):39–49. https://doi.org/10.1001/archpsyc.1994.03950010039006

Marcotte TD, Wolfson T, Rosenthal TJ et al (2004) A multimodal assessment of driving performance in HIV infection. Neurology 63:1417–1422

Maruff P, Thomas E, Cysique L, Brew B, Collie A, Snyder P et al (2009) Validity of the CogState brief battery: relationship to standardized tests and sensitivity to cognitive impairment in mild traumatic brain injury, schizophrenia, and AIDS dementia complex. Arch Clin Neuropsychol 24:165–178

Mateen FJ, Mills EJ (2012) Aging and HIV-related cognitive loss. JAMA 308:349–350

Milanini B, Wendelken LA, Esmaeili-Firidouni P, Chartier M, Crouch P-C, Valcour V (2014) The Montreal Cognitive Assessment to screen for cognitive impairment in HIV patients older than 60 years. J Acquir Immune Defic Syndr 67(1):67–70. https://doi.org/10.1097/QAI.0000000000000220

Minor KS, Jones GN, Stewart DW, Hill BD, Kulesza M (2010) Comparing two measures of psychomotor performance in patients with HIV: the coin rotation test and the modified HIV dementia screen. J Acquir Immune Defic Syndr 55:225–227

Monteiro de Almeida S, Kamat R, Cherner M et al (2017) Improving detection of HIV-associated cognitive impairment: comparison of the international HIV dementia scale and a brief screening battery. J Acquir Immune Defic Syndr 74:332

Moore DJ, Roediger MJ, Eberly LE et al (2012) Identification of an abbreviated test battery for detection of HIV-associated neurocognitive impairment in an early-managed HIV-infected cohort. PLoS One 7:e47310

Moore RC, Paolillo EW, Heaton A et al (2017) Clinical utility of the UCSD performance-based skills assessment – brief (UPSA-B) in adults living with HIV: associations with neuropsychological impairment and patient-reported everyday functioning difficulties. PLoS One 12:1–14. https://doi.org/10.1371/journal.pone.0183614

Morgan EE, Woods SP, Scott JC, Childers M, Beck JM, Ellis RJ et al (2008) Predictive validity of demographically adjusted normative standards for the HIV dementia scale. J Clin Exp Neuropsychol 30:83–90

Morley D, McNamara P, Kennelly S et al (2013) Limitations to the identification of HIV-associated neurocognitive disorders in clinical practice. HIV Med 14:497–502

Mukherjee T, Sakthivel R, Fong HY et al (2018) Utility of using the Montreal cognitive assessment (MoCA) as a screening tool for HIV-associated neurocognitive disorders (HAND) in multi-ethnic Malaysia. AIDS Behav 22:3226–3233. https://doi.org/10.1007/s10461-018-2073-x

Muniyandi K, Venkatesan J, Arutselvi T, Jayaseelan V (2012) Study to assess the prevalence, nature and extent of cognitive impairment in people living with AIDS. Indian J Psychiatry 54:149–153

Muñoz-Moreno JA, Prats A, Pérez-Álvarez N et al (2013) A brief and feasible paper-based method to screen for neurocognitive impairment in HIV-infected patients: the NEU screen. J Acquir Immune Defic Syndr 63:585–592

Nasreddine ZS, Phillips NA, Bédirian V et al (2005) The Montreal cognitive assessment, MoCA: a brief screening tool for mild cognitive impairment. J Am Geriatr Soc 53:695–699

NNTC Public Reports (2014) National NeuroAIDS Tissue Consortium

Overton E, Kauwe J, Paul R, Tashima K, Tate D, Patel P et al (2011) Performances on the CogState and standard neuropsychological batteries among HIV patients without dementia. AIDS Behav 15:1902–1909

Overton ET, Azad TD, Parker N, Demarco Shaw D, Frain J, Spitz T et al (2013) The Alzheimer's disease-8 and Montreal Cognitive Assessment as screening tools for neurocognitive impairment in HIV-infected persons. J Neurovirol 19(1):109–116. https://doi.org/10.1007/s13365-012-0147-5

Parsons TD, Rogers S, Hall C, Robertson K (2007) Motor based assessment of neurocognitive functioning in resource-limited international settings. J Clin Exp Neuropsychol 29:59–66

Parry S, Zetler S, Kentridge A, Petrak J, Barber T (2017) Simple screening for neurocognitive impairment in routine HIV outpatient care: is it deliverable? AIDS Care 29(10):1275–1279. https://doi.org/10.1080/09540121.2017.1300632

Phillips NJ, Thomas KGF, Sacktor N et al (2019) Screening for HIV-associated neurocognitive disorders in perinatally infected adolescents. AIDS 33(5):815–824. https://doi.org/10.1097/qad.0000000000002144

Power C, Selnes OA, Grim JA, McArthur JC (1995) HIV dementia scale: a rapid screening test. J Acquir Immune Defic Syndr Hum Retrovirol 8:273–278

Revicki DA, Chan K, Gevirtz F (1998) Discriminant validity of the medical outcomes study cognitive function scale in HIV disease patients. Qual Life Res 7:551–559

Richardson MA, Morgan EE, Vielhauer MJ, Cuevas CA, Buondonno LM, Keane TM (2005) Utility of the HIV dementia scale in assessing risk for significant HIV-related cognitive-motor deficits in a high-risk urban adult sample. AIDS Care 17:1013–1021

Robertson K, Wong M, Musisi S, Ronald A, Katabira E, Robertson KR et al (2006) Timed gait test: normative data for the assessment of the AIDS dementia complex. J Clin Exp Neuropsychol 28:1053–1064

Robbins RN, Remien RH, Mellins CA et al (2011) Screening for HIV-associated dementia in South Africa: potentials and pitfalls of task-shifting. AIDS Patient Care STDs 25:587–593. https://doi.org/10.1089/apc.2011.0154

Robbins RN, Joska JA, Thomas KGF et al (2013) Exploring the utility of the Montreal cognitive assessment to detect HIV-associated neurocognitive disorder: the challenge and need for culturally valid screening tests in South Africa. Clin Neuropsychol 27:437–454

Robbins RN, Brown H, Ehlers A et al (2014) A smartphone app to screen for HIV-related neurocognitive impairment. J Mob Technol Med 3:23–36. https://doi.org/10.7309/jmtm.3.1.5

Robbins RN, Gouse H, Brown HG et al (2018) A mobile app to screen for neurocognitive impairment: preliminary validation of neuroscreen among HIV-infected South African adults. J Med Internet Res 6:e5. https://doi.org/10.2196/mhealth.9148

Rosca EC, Albarqouni L, Simu M (2019) Montreal cognitive assessment (MoCA) for HIV-associated neurocognitive disorders. Neuropsychol Rev 29(3):313–327. https://doi.org/10.1007/s11065-019-09412-9

Rosenfeld B, Sands SA, Van Gorp WG (2000) Have we forgotten the base rate problem? Methodological issues in the detection of distortion. Arch Clin Neuropsychol 15:349–359

Royal WI, Cherner M, Carr J et al (2012) Clinical features and virological correlates of neurocognitive impairment among HIV-infected individuals in Nigeria. J Neurovirol 18:191

Sakamoto M, Marcotte TD, Umlauf A et al (2013) Concurrent classification accuracy of the HIV dementia scale for HIV-associated neurocognitive disorders in the CHARTER cohort. J Acquir Immune Defic Syndr Hum Retrovirol 62:36–42

Sacktor NC, Wong M, Nakasujja N, Skolasky RL, Selnes OA, Musisi S et al (2005) The international HIV dementia scale: a new rapid screening test for HIV dementia. AIDS 19:1367–1374

Schouten J, Cinque P, Gisslen M et al (2011) HIV-1 infection and cognitive impairment in the cART era: a review. AIDS 25:561–575

Sebit MB (1995) Neuropsychiatric HIV-1 infection study: in Kenya and Zaire cross-sectional phase I and II. Cent Afr J Med 41(10):315–322. http://www.ncbi.nlm.nih.gov/pubmed/8556777

Singh D, Sunpath H, John S, Eastham J, Gouden R (2008) The utility of a rapid screening tool for depression and HIV dementia amongst patients with low CD4 counts - a preliminary report. Afr J Psychiatry 11:282–286

Skinner S, Adewale AJ, DeBlock L et al (2009) Neurocognitive screening tools in HIV/AIDS: comparative performance among patients exposed to antiretroviral therapy. HIV Med 10:246–252

Simioni S, Cavassini M, Annoni JM, Rimbault Abraham A, Bourquin I, Schiffer V et al (2010) Cognitive dysfunction in HIV patients despite long-standing suppression of viremia. AIDS 24:1243–1250

Smith CA, van Gorp WG, Ryan ER, Ferrando SJ, Rabkin J (2003) Screening subtle HIV-related cognitive dysfunction: the clinical utility of the HIV dementia scale. J Acquir Immune Defic Syndr 33:116–118

Thames AD, Streiff V, Patel SM et al (2012) The role of HIV infection, cognition, and depression in risky decision-making. J Neuropsychiatry Clin Neurosci 24:340–348. https://doi.org/10.1176/appi.neuropsych.11110340

The Mind Exchange Working Group, Antinori A, Arendt G et al (2012) Assessment, diagnosis, and treatment of HIV-associated neurocognitive disorder: a consensus report of the mind exchange program. Clin Infect Dis 56:1004–1017
Umaki TM, Gangcuangco LMA, Chow DC et al (2013) Poorer neuropsychological performance increases risk for social services among HIV-infected individuals. Hawaii J Med Public Health 72:422
Valcour V, Shikuma C, Shiramizu B et al (2004) Higher frequency of dementia in older HIV-1 individuals: the Hawaii Aging with HIV-1 Cohort. Neurology 63:822–827
Valcour V, Paul R, Chiao S et al (2011) Screening for cognitive impairment in human immunodeficiency virus. Clin Infect Dis 53:836–842
Von Giesen HJ, Haslinger BA, Rohe S, Koller H, Arendt G (2005) HIV dementia scale and psychomotor slowing - the best methods in screening for neuro-AIDS. J Neuropsychiatry Clin Neurosci 17:185–191
Van Gorp WG, Rabkin JG, Ferrando SJ et al (2007) Neuropsychiatric predictors of return to work in HIV/AIDS. J Int Neuropsychol Soc 13:80–89
Vivithanaporn P, Heo G, Gamble J et al (2010) Neurologic disease burden in treated HIV/AIDS predicts survival: a population-based study. Neurology 75:1150–1158
Wardle MC, Gonzalez R, Bechara A, Martin-Thormeyer EM (2010) Iowa Gambling Task performance and emotional distress interact to predict risky sexual behavior in individuals with dual substance and HIV diagnoses. J Clin Exp Neuropsychol 32:1110–1121. https://doi.org/10.1080/13803391003757833
Wojna V, Skolasky RL, McArthur JC, Maldonado E, Hechavarria R, Mayo R et al (2007) Spanish validation of the HIV dementia scale in women. AIDS Patient Care STDs 21:930–941
World Health Organization (2019) Global health observatory (GHO) data. WHO, Geneva
Yechoor N, Towe SL, Robertson KR et al (2017) Utility of a brief computerized battery to assess HIV-associated neurocognitive impairment in a resource-limited setting. J Neurovirol 22:808–815. https://doi.org/10.1007/s13365-016-0456-1
Zipursky AR, Gogolishvili D, Rueda S et al (2013) Evaluation of brief screening tools for neurocognitive impairment in HIV/AIDS: a systematic review of the literature. AIDS 27:2385

Behavioral and Physical Activity Interventions for HAND

Jessica L. Montoya, Brook Henry, and David J. Moore

Contents

Abstract Approximately 30–50% of persons living with HIV manifest some degree of neurocognitive impairment. Even mild-to-moderate forms of HIV-associated neurocognitive disorders (HAND) can result in difficulties with everyday functioning, such as suboptimal medication adherence and impaired driving. Despite the pervasive presence and consequences of HAND, there is a significant unmet need to develop

J. L. Montoya and B. Henry
Department of Psychiatry, University of California San Diego, La Jolla, CA, USA

D. J. Moore (✉)
Department of Psychiatry, University of California San Diego, La Jolla, CA, USA

HIV Neurobehavioral Research Program, San Diego, CA, USA
e-mail: djmoore@ucsd.edu

Curr Topics Behav Neurosci (2021) 50: 479–502
DOI 10.1007/7854_2018_79
Published Online: 30 January 2019

effective behavioral strategies to reduce the incidence and consequences of HAND. Although there is an absence of evidence-based behavioral interventions specific to HAND, the literature reviewed in this chapter suggest the following modifiable lifestyle factors as intervention targets: physical activity, diet, sleep, and antiretroviral medication adherence. Adoption and maintenance of these healthy lifestyle factors may reduce inflammation and oxidative stress, which, in turn, may reduce the incidence and/or severity of HAND.

Keywords Antiretroviral therapy adherence · Cognition · Diet · Exercise · Nutrition · Physical activity · Sleep

1 HIV-Associated Neurocognitive Disorders

Approximately 30–50% of persons living with Human Immunodeficiency Virus (HIV) Type-1 manifest some degree of neurocognitive impairment, typically characterized by deficits in executive function, attention, and memory referred to as HIV-associated neurocognitive disorders (HAND) (Heaton et al. 2010). HAND per Frascati diagnostic criteria includes three categories: asymptomatic neurocognitive impairment (ANI), HIV-associated mild neurocognitive disorder (MND), which includes self-reported deficits in everyday functioning, and the most severe of the disorders, HIV-associated dementia (HAD) (Antinori et al. 2007). While HAD prevalence has declined with the use of effective antiretroviral therapy, ANI and MND, marked by impaired performance in at least two neurocognitive domains (greater than one standard deviation below the mean of normative scores), remain a prevalent and significant public health concern. Even mild-to-moderate forms of HAND translate to higher risk of impaired driving, difficulties with instrumental activities of daily living, and poorer antiretroviral therapy adherence (Thames et al. 2011; Vance et al. 2011; Marcotte et al. 2004). Despite the pervasive presence and consequences of HAND, there is a significant unmet need to develop effective behavioral strategies to reduce the incidence and consequences of HAND (Woods et al. 2009b). Among existing behavioral interventions, promotion of physical activity, a healthy diet, improved sleep, and antiretroviral therapy adherence may benefit neurocognitive function.

2 Physical Activity as a Behavioral Intervention for HAND

2.1 Physical Activity Levels Among People Living with HIV

Physical activity, traditionally defined as the movement of skeletal muscles that requires energy expenditure, has been recommended as a safe therapeutic strategy to sustain and maintain the health of persons living with HIV (Botros et al. 2012). Physical activity interventions among people living with HIV are reported to improve body composition, muscle strength, aerobic fitness, and quality of life; however, there are notable barriers to physical activity among people living with HIV, including disease symptoms, antiretroviral therapy side effects, depression, and pain (Henry and Moore 2016; Vancampfort et al. 2018b). Surprisingly, relatively few studies have examined physical activity among people living with HIV. A 2012 review indicated that the diversity and inconsistency of methods used to assess physical activity (typically a variety of self-report questionnaires) precluded any pooled estimate of overall physical activity among people living with HIV, although the range of estimates suggest that 19–73% of people living with HIV are "sedentary," as defined by various criteria (Schuelter-Trevisol et al. 2012). Webel and colleagues reported that an Ohio cohort of persons living with HIV tended to exercise regularly, but the mean level of physical activity was lower in women than men, and below recommended levels (i.e., 150 min per week of moderate-to-vigorous physical activity) (Webel et al. 2015). A sample of 50 African-American females living with HIV from the southern United States exhibited low levels of physical activity, with only one person exceeding 150 min per week of moderate-to-vigorous physical activity (Rehm and Konkle-Parker 2016). In contrast, two-thirds of a cohort of people living with HIV from Vietnam were rated as physically active on the International Physical Activity Questionnaire, although this group did include rural participants with potentially more endogenous physical activity than urban residents (Dang et al. 2018). A recent meta-analysis reviewed physical activity data from 24 studies of people living with HIV conducted across the world (Vancampfort et al. 2018a). To optimize consistency, the authors included physical activity results obtained only through the International Physical Activity Questionnaire or objective measures, such as an accelerometer. The findings from this seminal study showed that approximately 50% of people living with HIV failed to meet suggested physical activity guidelines (i.e., exhibiting less than 150 min per week of moderate-to-vigorous physical activity). Furthermore, average daily steps, when collected, was about 5,800 per day, which is far below the 10,000 steps recommended for adults and close to a sedentary range (typically defined as below 5,000 steps per day). In summary, it is clear that physical activity habits among people living with HIV may vary significantly depending upon geographic and demographic factors, but studies to date suggest that a large proportion of people living with HIV are not sufficiently physically active. As a result, physical activity interventions may be beneficial for numerous persons living with HIV.

2.2 *Relationship Between Physical Activity and Neurocognition*

Moderate- and vigorous-intensity physical activity has beneficial effects on neurocognitive function (Engeroff et al. 2018). Among healthy individuals and children, physical activity is shown to improve neurocognitive functioning (Davis et al. 2007; Best et al. 2015; Mura et al. 2015; Guiney and Machado 2013; Gill et al. 2015).

Aerobic exercise is associated with improvement in multiple neurocognitive domains, including attention, processing speed, executive function, and memory (Smith et al. 2010). Physical activity may also benefit global neurocognition in patients with mild neurocognitive impairment (Song et al. 2018) or history of stroke (Vanderbeken and Kerckhofs 2017), and may slow neurocognitive decline in Alzheimer's disease (Farina et al. 2014). In the context of HIV, higher engagement in physical activity is associated with better executive function (Ortega et al. 2015) and reduced likelihood of neurocognitive impairment in cross-sectional studies (Dufour et al. 2013; Fazeli et al. 2015). Longitudinally, persons living with HIV who consistently engage in physical activity begin with, and maintain, significantly better neurocognitive function compared to persons living with HIV who do not engage in physical activity or do so inconsistently (Dufour et al. 2018). In a longitudinal Multicenter AIDS Cohort Study, high engagement in physical activity was associated with lower odds of impairment in the domains of learning, memory, and motor function, and these effects were found to be more pronounced among the sample of men living with HIV compared to the pooled sample of men living with HIV and HIV-uninfected men (Monroe et al. 2017). In a scoping review of the effect of physical activity on neurocognitive function among people living with HIV (Quigley et al. 2018), noninterventional studies reported a positive association between physical activity engagement in neurocognitive function, whereas there was a dearth of positive outcomes of aerobic interventions on neurocognition. Null results of interventional studies potentially may be a result of methodological factors (e.g., low prescribed doses of physical activity, recruitment of relatively young people living with HIV who are at lower risk for neurocognitive impairment compared to older people living with HIV, and reliance on self-reported versus objective measures of neurocognitive function). Thus, future randomized controlled trials should prescribe doses of physical activity consistent with current recommendations (150 min per week of moderate-to-vigorous physical activity), recruit participants most likely to benefit from physical activity interventions, and utilize comprehensive neurocognitive assessments.

2.3 *Underlying Biologic Mechanisms Linking Physical Activity and Neurocognition*

Chronic inflammation is established early in HIV infection and is postulated to contribute to neurocognitive impairment across the age span of persons living with HIV (Sattler et al. 2015; Hong and Banks 2015; Tavazzi et al. 2014; Kapetanovic et al. 2010, 2014; Ancuta et al. 2008; Gannon et al. 2011). Physical activity is known to exert anti-inflammatory effects, which may be a consequence of reduced visceral fat and decreased release of adipokines (d'Ettorre et al. 2014). In addition to reduced inflammation, some of the underlying biologic mechanisms linking physical activity to neurocognition include promotion of cerebral angiogenesis, improved cerebral and peripheral vascular reactivity (i.e., increased maximal oxygen consumption), upregulation of neurotrophins, increased neurogenesis, and decreased hippocampus apoptosis (Stimpson et al. 2018). Imaging studies indicate that engagement in physical activity can increase brain volume, and biomarker studies show that physical activity leads to upregulation of brain-derived neurotrophic factor (BDNF) (Firth et al. 2017). A randomized controlled trial showed that among individuals with schizophrenia, those who participated in an aerobic exercise intervention (compared to treatment as usual) showed improvements in physical fitness and neurocognitive function (Kimhy et al. 2015). Furthermore, enhancement of physical fitness and increases in BDNF accounted for much of the variance in neurocognitive improvement, supporting the hypothesis that physical activity-induced upregulation of BDNF may contribute to improved neurocognitive outcomes.

2.4 *Interventions to Increase Physical Activity, Thereby Benefiting Neurocognition*

Low-to-moderate intensity walking interventions have demonstrated effectiveness in improving neurocognitive performance in older adults (Rosenberg et al. 2012) and neurocognitively impaired populations (Kemoun et al. 2010). Compared to moderate physical activity, high intensity interval training, which involves short periods of exercise performed at high intensity (greater than 80–85% peak oxygen uptake), may have greater benefits for improving cardiometabolic function (e.g., improved insulin and glucose regulation and reduced inflammation markers, such as interleukin-6 and C-reactive protein) (Tjonna et al. 2009; Munk et al. 2011; Ramos et al. 2015). Interventions employing a combination of aerobic and resistance exercise also show evidence of improving body composition and lipid profile in people living with HIV (Hand et al. 2009; O'Brien et al. 2010). However, the effect of a combined aerobic and resistance exercise regimens on neurocognition has not been investigated as an intervention strategy for HAND.

Limitations of many physical activity interventions are issues of feasibility and scalability. Many intervention studies are conducted at gym facilities, which may not be accessible in nonintervention contexts to many persons living with HIV who may face limited financial resources (Montoya et al. 2015). To date, few studies have specifically examined the effect of a physical activity intervention on neurocognition among people living with HIV. One recent 16-week protocol required participants to attend three aerobic exercise sessions at a gym facility each week, but did not observe any effects of the physical activity intervention on neurocognition relative to a control group (McDermott et al. 2017). However, the sample size was extremely small; only 11 people participated in total (6 control and 5 intervention participants completed the study out of 57 persons living with HIV screened), individuals did not attend 40% of the planned physical activity training sessions, and the neurocognitive assessment was limited, including only the Montreal Cognitive Assessment and

Trail-making A and B tests. These issues highlight some of the challenges involved in conducting physical activity interventions to improve neurocognitive outcomes. Although physical activity that is gym-based may not be feasible for many persons living with HIV, increasing walking/step count may be an appropriate treatment target. Walking may reduce sedentary behavior and improve metabolic function (Healy et al. 2008; Manson et al. 1999). For example, an increase in mean daily step counts by approximately 2,000 improves lipid profile (Sugiura et al. 2002).

In the general population, mobile health interventions have been developed to promote engagement in physical activity, such as walking (Fjeldsoe et al. 2009). An ongoing randomized controlled trial is investigating whether a novel and personalized text messaging intervention (iSTEP) can significantly increase moderate physical activity in people living with HIV (R21MH100968) (Henry and Moore 2016). Second, the iSTEP intervention aims to evaluate the effect of physical activity engagement on neurocognitive performance. Preliminary data supports high text message response rates and positive participant feedback (Henry and Moore 2016).

3 Diet as a Behavioral Intervention for HAND

3.1 Diet and Neurocognitive Function in the General Population

Epidemiological research indicates that a healthy diet, such as the Mediterranean diet and consumption of omega-3 fatty acids (e.g., docosahexaenoic acid, DHA), may help prevent neurocognitive decline (Barak and Aizenberg 2010). The Mediterranean diet is characterized by high intake of vegetables, legumes, fruits, nuts, cereals, and olive oil; moderate intake of fish and alcohol; low-to-moderate intake of dairy products; and low intake of saturated lipids and meat (Loughrey et al. 2017). A recent systematic review indicated a positive, concurrent association between the Mediterranean diet and global neurocognition; however, the association between the Mediterranean diet and specific neurocognitive domains is less established (Knight et al. 2017). Methodological differences in relation to neuropsychological assessment are a likely factor contributing to the lack of consensus among studies on the relationship between the Mediterranean diet and neurocognitive function (Knight et al. 2017). Similar to the Mediterranean diet, diets rich in polyphenols – found in fruits, vegetables, tea, wine, juices, plants, and some herbs – may promote better performance in neurocognitive abilities in a dose-dependent manner (Nurk et al. 2009) and slower rates of neurocognitive decline (Devore et al. 2012) in older persons. Results from the Framingham Heart Study (Schaefer et al. 2006) indicate individuals with higher DHA levels had a relative risk of 0.61 of developing Alzheimer's disease, compared to individuals with lower

DHA levels, after adjustment for relevant covariates (i.e., sex, apolipoprotein E allele, plasma homocysteine concentration, and education level). Higher DHA levels translated to a mean DHA intake of 180 mg/day and a mean fish intake of 3.0 servings per week.

3.2 Diet and Nutrition Concerns in the Context of HIV

The introduction of effective antiretroviral therapy resulted in improvements in nutritional status and weight gain among people living with HIV (Leyes et al. 2008). However, body composition changes (i.e., lipodystrophy) persist even with effective antiretroviral therapy, particularly among people living with HIV on protease inhibitors (Carr et al. 1998). Body composition changes may include intra-abdominal fat accumulation (Miller et al. 1998), which is of relevance to pathogenesis of HAND given that the HIV CHARTER cohort study has found a link between abdominal obesity and neurocognitive function (Sattler et al. 2015). Multiple studies indicate that cardiometabolic dysfunction (e.g., abdominal obesity, hyperglycemia, dyslipidemia, and hypertension) confers risk for HAND and correlates with imaging markers of neurochemical abnormalities and neuroinflammation among persons living with HIV (Cysique et al. 2013; McCutchan et al. 2012; Saylor et al. 2016; Valcour et al. 2005).

The nutritional status of persons living with HIV can influence body composition changes, but the efficacy of diet interventions to treat metabolic conditions, thereby improving neurocognitive outcomes is limited (Leyes et al. 2008). Despite limitations in the literature to be able to tease apart the various contributors of metabolic changes observed among people living with HIV, cross-sectional studies indicate that diets high in cholesterol and saturated and trans fats play a role in metabolic disturbances (Shah et al. 2005; Hadigan et al. 2001). Although diet and nutrient deficiencies have been extensively documented for persons living with HIV in low-income countries, with a focus on treatment with supplements such as vitamins (Duran et al. 2008), less work has been performed to characterize the diet quality for persons living with HIV in developed countries such as the United States. A couple of reports indicate that youth living with HIV demonstrate a lower Healthy Eating Index score, higher fat intake, and lower micronutrient consumption (including vitamins A, E, calcium, and potassium) compared to an age-matched HIV-seronegative comparison group (Kruzich et al. 2004; Ziegler et al. 2014). Pregnant women living with HIV in their third trimester also exhibited low Healthy Eating Index scores, although individuals born outside the United States had better scores, perhaps indicative of more unhealthy diet practices adopted by people living in the United States (Miller et al. 2017). One cohort of adults living with HIV (53% male) also consumed higher levels of saturated fat and lower amounts of polyunsaturated fat and fiber than recommended by the National Cholesterol Education Program (Capili and Anastasi 2008). While more data needs to be obtained, the existing evidence suggests that interventions that address diet content as well as physical activity may be helpful for people living with HIV in the United States.

3.3 Diet Interventions Specific to People Living with HIV

Current international guidelines recommend dietary intervention as a first-line treatment for HIV-related dyslipidemia (Dube et al. 2003), but the results of clinical trials have been inconsistent. Persons living with HIV who receive macronutrient (protein/carbohydrate) or micronutrient (vitamin) supplements in randomized controlled trials often do not show a significant reduction in morbidity, mortality, or disease progression (e.g., CD4 count and viral load) (Grobler et al. 2013; Visser et al. 2017). Interventions that have promoted low-fat diets are reported to have beneficial but limited effects, such as a reduction in triglycerides but not cholesterol (Stradling et al. 2012). Some nutrition studies have focused on polyunsaturated fatty acids (PUFA) and omega-3 PUFA (n-3 PUFA), which may benefit brain health and function and are emphasized as part of the Mediterranean diet (Poulose et al. 2014). PUFAs may contribute to optimal brain function by reducing oxidative stress and inflammation, maintaining neuronal membrane integrity, and attenuating protein aggregation implicated in neurodegenerative diseases and age-related neurocognitive decline (Poulose et al. 2014). Along these lines, a large 12-month study (PREDIMED) [7 years and 8,000 participants (Zazpe et al. 2008)] indicated that adoption of the Mediterranean diet provided significantly more benefits compared to control or a low-fat diet condition, including a reduction in cardiovascular disease events, decreased markers of oxidative stress, and improved neurocognition (Estruch et al. 2013; Schroder et al. 2014; Fito et al. 2014; Mitjavila et al. 2013; Martinez-Lapiscina et al. 2013). Despite potential neurocognitive benefits, the Mediterranean diet has not been evaluated as a strategy for preventing or treating HAND, although greater Mediterranean diet adherence is associated with improved metabolic function (lower insulin resistance and higher high-density lipoprotein cholesterol) in people living with HIV (Tsiodras et al. 2009). Future studies may examine the effect of the Mediterranean diet on cardiovascular risk and neurocognitive deficits associated with HIV.

4 Sleep as a Behavioral Intervention for HAND

4.1 Sleep Disturbance in the Context of HIV

Sleep disturbance and fatigue are common symptoms reported by persons living with HIV, with up to 75% of people living with HIV experiencing sleep disturbance according to the Pittsburgh Sleep Quality Index (Rubinstein and Selwyn 1998) and up to 88% experience fatigue (Jong et al. 2010). A meta-analysis of sleep disturbances among people living with HIV calculated that the overall prevalence of self-reported sleep disturbance was 58% (Wu et al. 2015). Furthermore, more than half of a sample population of people living with HIV reported symptoms related to sleep disturbance and fatigue, such as lack of energy (65%), drowsiness (57%), and difficulty sleeping (56%) (Lee et al. 2009). In a Taiwanese study, people living with HIV had a 3.74-fold higher risk of sleep disturbances compared to a general population control group (Chen et al. 2017). Among cohorts of people living with HIV, sleep disturbance and/or fatigue severity are associated with depression and anxiety symptoms (Millikin et al. 2003; Jong et al. 2010); antiretroviral therapy medication types and family and social support (Ren et al. 2018); high levels of HIV-RNA (greater than 1,000 copies/mL) (Womack et al. 2017); and substantial night-to-night variability in bedtime and risetime (Taibi et al. 2013).

4.2 *Effect of Sleep Disturbance on Neurocognitive Function*

Studies on the association between self-reported sleep disturbance and neurocognition have been mixed but suggest a link between poor sleep quality and worse neurocognitive outcomes (Yaffe et al. 2014). A meta-analysis of the impact of short-term sleep deprivation on neurocognition indicated that sleep deprivation had the largest effect on simple attention tasks (e.g., Psychomotor Vigilance Test and other simple reaction time tests), with tasks of greater complexity being affected to a lesser degree after sleep deprivation (Lim and Dinges 2010). On the other end of the spectrum, longer total sleep time (e.g., greater than or equal to 10 h of sleep/night) is also reported to significantly influence neurocognition, such as lower scores on the Mini-Mental Status Examination (Faubel et al. 2009) and lower performance on a recall test (Xu et al. 2011). Other measures of sleep disturbance (i.e., lower sleep efficiency, higher levels of wake after sleep onset, and a higher number of long wake episodes) have been related to decline in executive function as measured by the Trails B Test over an average study period of 3.4 years in a community-dwelling cohort of older men (mean age 76.0) (Blackwell et al. 2014). In a cohort of people living with HIV, 63% reported poor sleep quality based on a cutoff point of 5 on the Pittsburgh Sleep Quality Index (Byun et al. 2016). In this same sample, poorer subjective sleep quality, shorter or longer total sleep time measured by actigraphy (i.e., fewer than 7 h or greater than 8 h vs 7–8 h), and greater morning fatigue were associated with self-reported problems with neurocognitive function (e.g., difficulties with reasoning, concentration and thinking, confusion, memory, attention, and psychomotor function), even after controlling for covariates including age, gender, education, and sleep medication use. A study of sleep and neurocognition in a cohort of people living with HIV (75% were neurocognitively impaired based on Frascati criteria) that involved a more comprehensive neurocognitive assessment found that better performance on tasks of attention, frontal/executive function, and psychomotor/motor speed were associated with better polysomnogram sleep parameters, including reduced wake-after sleep, greater sleep efficiency, greater sleep latency, and greater total sleep time (Gamaldo et al. 2013). Thus, these results demonstrate that confounding factors, such as sleep disturbances, might influence the presentation of HAND.

Sleep disturbance may contribute to neurocognitive decline via impaired metabolism and decreased cerebral perfusion. Sleep deprivation may particularly impair metabolism of the prefrontal cortex, a brain region implicated in executive function (Durmer and Dinges 2005). Insufficient sleep duration has been shown to result in decreased cerebral blood flow in the frontal lobes and in worse performance on the Continuous Performance Test and driving performance (i.e., break reaction time in a harsh-braking test) (Miyata et al. 2010). Evidence from epidemiological and experimental studies indicate that sleep disturbance may impair amyloid beta clearance and increase tau phosphorylation, as well as impair synaptic plasticity via disruption of pathways involving gamma-aminobutyric acid (GABA) and cyclin adenosine monophosphate (cAMP) (Yaffe et al. 2014). In a study specific to persons living with HIV, higher inflammation levels (measured by C-reactive protein and interleukin-6) were observed among those with poor sleep characteristics (i.e., later sleep onset and lower total sleep time) and low engagement in moderate-vigorous physical activity (Wirth et al. 2015). These results indicate that disturbances in sleep and low levels of physical activity are associated with inflammation, which is implicated in the pathogenesis of HAND. Thus, improving sleep indices and maintaining regular participation in physical activity may help reduce the risk of inflammation, which, in turn, may reduce the incidence of HAND.

4.3 Behavioral Strategies to Improve Sleep Quality

Cognitive behavioral therapy for insomnia is the first-line treatment for insomnia based on guidelines of the American College of Physicians for management of chronic insomnia (Qaseem et al. 2016). Cognitive behavioral therapy for insomnia is a multimodal cognitive behavioral therapy that can be delivered in individual or group therapy, telephone- or web-based modules, or self-help books, and includes the behavioral strategies of sleep restriction and stimulus control. Although cognitive behavioral therapy for insomnia is associated with robust, long-term improvements in sleep parameters, cognitive behavioral therapy for insomnia has small-to-moderate effects on subjective measures of neurocognitive functioning, and there is insufficient data to determine the effect of cognitive behavioral therapy for insomnia on objective measures of neurocognition in the general population (Herbert et al. 2018).

SystemCHANGE-HIV – a 10-week intervention grounded in a socioecological model that covers different topics of HIV management, including sleep hygiene and behavioral modification strategies – was tested to determine its effectiveness at improving sleep outcomes, including sleep duration, sleep fragmentation index, sleep efficiency, and self-reported sleep quality (Webel et al. 2013). SystemCHANGE-HIV had high levels of engagement (e.g., participants attended 71% of all intervention sessions on average). Although nonsignificant, the pilot study provided preliminary data indicating that SystemCHANGE-HIV leads to improvement in sleep efficiency and sleep fragmentation. A major limitation of the SystemCHANGE-HIV study in regard to its applicability for the treatment of HAND was the absence of neurocognitive assessments. Given the prevalence and association between sleep disturbances and neurocognitive impairment in the context of HIV, a future research direction is to determine whether strategies for improving sleep (e.g., cognitive behavioral therapy for insomnia, SystemCHANGE-HIV) may lead to improvement in subjective and objective measures of neurocognitive functioning among people living with HIV.

5 Promotion of Antiretroviral Therapy Adherence as a Behavioral Intervention for HAND

5.1 Antiretroviral Therapy Adherence and Neurocognitive Performance

Early detection and initiation of antiretroviral therapy is a strategy for preventing significant immune compromise and protecting against neurocognitive decline (D'Antoni et al. 2018). In addition to initiating antiretroviral therapy during acute HIV infection, antiretroviral therapy adherence is crucial for the prevention of HAND (Martin et al. 1999; Suarez et al. 2001). Adherence is generally defined as the extent to which patients take medications as prescribed and is typically reported as the percentage of prescribed doses taken over a specified period, which may include consideration of specific dose timing (Osterberg and Blaschke 2005). Achieving consistently higher levels of antiretroviral therapy adherence in order to derive therapeutic benefit (e.g., sustained viral suppression and immune reconstitution) is a challenge for many people living with HIV (Kirtane et al. 2016). Generally, adherence rates are lower for chronic compared to acute conditions, and even under the controlled settings of clinical trials, average adherence rates range between 43 and 67% (Osterberg and Blaschke 2005).

Higher levels of antiretroviral therapy adherence are predictive of improvements in neurocognitive performance in the domains of information processing speed, attention, executive function, and motor function (Ettenhofer et al. 2010). Among patients who initiate antiretroviral therapy or change to a more effective regimen, improvements in neurocognitive function have been observed after 6 months for patients who achieved successful viral suppression (Parsons et al. 2006). Although antiretroviral therapy may improve neurocognitive function, a major complication is that people living with HIV with neurocognitive impairment, including deficits in prospective memory, are particularly at risk for antiretroviral therapy non-adherence (Hinkin et al. 2002; Woods et al. 2009a). Thus, behavioral strategies to promote antiretroviral therapy adherence among people living with HIV with neurocognitive deficits are particularly pertinent for the treatment of HAND.

5.2 *Strategies to Promote Antiretroviral Therapy Adherence*

Strategies to promote optimal adherence among people living with HIV include simpler dosing demands (e.g., fewer pills and once-a-day single tablet regiments) (Mohd Salleh et al. 2018). Given the ubiquity of phones, text message interventions have also been developed to prompt people living with HIV to adhere to their antiretroviral therapy regimen (Horvath et al. 2012). Such interventions are particularly effective in promoting antiretroviral therapy adherence in low-resource settings and may be helpful for promoting better antiretroviral therapy dose timing in patients with severe mental illness, such as bipolar disorder (Moore et al. 2015).

Adherence to antiretroviral therapy is influenced by psychosocial stressors, such as the experience of negative life events and depression, as well as individual characteristics like behavioral impulsivity (Salmoirago-Blotcher et al. 2017). Mindfulness training, which may allay distress and lessen impulsivity, is currently being investigated to determine its potential to help persons living with HIV adhere to their antiretroviral therapy regimen (Salmoirago-Blotcher et al. 2017). Given that depression has consistently shown a robust association with poor antiretroviral therapy adherence, several interventions grounded in cognitive behavioral therapy have been developed and evaluated for the treatment of depression and/or to promote medication adherence (Balfour et al. 2006; Safren et al. 2009; Simoni et al. 2007; Olatunji et al. 2006). A small pilot study investigated the effect of one-session behavioral activation treatment for depression designed specifically for people living with HIV (Tull et al. 2018). Although nonsignificant, persons living with HIV who received the one-session of behavioral activation treatment showed improvements in medication adherence of medium effect (Tull et al. 2018). Additional findings indicate that low levels of physical activity were predictive of poor antiretroviral therapy adherence and higher viral load, but importantly, this relationship was mediated by depression symptoms (Blashill et al. 2013). These results highlight the need to develop cost-effective interventions to simultaneously address symptoms of depression and antiretroviral therapy adherence among people living with HIV in order to prevent or treat HAND, and suggest that physical activity may have multifaceted benefits for people living with HIV.

6 Clinical Implications

In this section, intervention approaches to increase engagement in health behaviors (i.e., physical activity, healthy eating, sleep, and antiretroviral therapy adherence) are reviewed. Such approaches have the potential to improve neurocognitive outcomes among people living with HIV by increasing health behaviors that may protect against the development of non-communicable conditions (Lim et al. 2012) that adversely impact neurocognitive outcomes.

Health care providers have a prominent role in integrating health behavior promotion into routine HIV care (Webel et al. 2017). Health care providers can leverage their relationships with patients living with HIV to emphasize a holistic concept of well-being that includes engagement in health behaviors (Webel et al. 2017). Despite evidence supporting the effectiveness of health behavior promotion in primary health care, implementation has been slow (Brotons et al. 2012). One institution-level approach to target health behaviors more effectively and consistently in routine care is to integrate primary care and behavioral health in a single clinic; however, integration requires substantial reengineering of practice (Cifuentes et al. 2015). Another viable solution for targeting health behaviors in routine clinical care has been to train health care providers in Motivational Interviewing, a client-centered, evidence-based behavior change consultation style (Rollnick et al. 2008). Adoption of a Motivational Interviewing approach may be particularly helpful for clinicians providing HIV care, with some evidence indicating the effectiveness of Motivational Interviewing for improving antiretroviral therapy adherence (Golin et al. 2006; Beach et al. 2015).

Beyond the patient–provider relationship, engagement in a health behavior is influenced by a complex system of determinants, including intrapersonal (e.g., cognitive and emotional factors), interpersonal/community, institutional (e.g., access to specialty medical care), environmental (e.g., neighborhood characteristics and employment conditions), cultural, and policy/legislation factors (Dahlgren and Whitehead 2006). For persons living with HIV, readiness to engage in a health behavior is a dynamic and fluctuating construct, which may also be influenced by the episodic nature of HIV and multimorbidity (Simonik et al. 2016). Identification of contextual factors that have a strong relationship with the presence (or absence) of a health behavior is necessary to identify potential modifiable targets of behavioral health interventions (Michie et al. 2011; Wight et al. 2016). Although the field of behavior change research in persons living with HIV acknowledges that interventions need to target more than just factors at the intrapersonal level (Albarracin et al. 2010), there is only a small literature on multi-level models of intervention (Kaufman et al. 2014). Thus, behavior change efforts may be guided by multiple existing theories that, in combination, target various levels of influence (Kaufman et al. 2014).

Ideally, development of behavioral health interventions involves an interdisciplinary group representing relevant expertise (e.g., clinical healthcare, psychology, epidemiology, and policy) and key stakeholders (e.g., patients, caregivers, healthcare professionals, policymakers, and funders) to ensure interventions are evidence-based and acceptable to (1) those for whom the intervention is developed and (2) those involved in the adoption and implementation of the intervention (Eldredge et al. 2016; Witteman et al. 2017; Araújo-Soares et al. 2018). To optimize the effectiveness of a behavioral health intervention, researchers and/or interventionists must consider the appropriateness of various behavior change techniques, mode(s) of intervention delivery, provider(s), location, timing, and dosing of an intervention (Araújo-Soares et al. 2018).

Given the complexity of behavior change interventions, it is important to identify (1) effective behavior change techniques and (2) the processes through which behavior change occurs (i.e., the mechanisms of action) (Connell et al. 2018). For researchers and interventionists, various protocols and taxonomies have been developed to aid in behavior change intervention development. For example, Intervention Mapping describes an iterative process for developing theory- and evidence-based health promotion programs (Kok et al. 2004). In addition, an extensive taxonomy of consensually agreed, distinct behavior change techniques has been developed by a large international network of behavior change experts, which is intended to be used in combination with Intervention Mapping (Michie et al. 2013). Additional research investigating the effectiveness of specific behavior change techniques for promoting the adoption and maintenance of healthy lifestyle factors is needed. Such intervention efforts may reduce inflammation and oxidative stress, which, in turn, may reduce the incidence and/or severity of HAND.

7 Conclusion

Currently, there is no gold-standard treatment for the prevention or treatment of HAND. Although there is an absence of evidence-based neurocognitive interventions for people living with HIV, the literature reviewed in this chapter suggest potential targets for intervention. Modifiable lifestyle factors, such as physical activity, diet, sleep, and antiretroviral therapy adherence, may benefit neurocognitive function among people living with HIV. These healthy lifestyle factors reduce inflammation and oxidative stress, which, in turn, may reduce the incidence of HAND.

References

Albarracin D, Rothman AJ, di Clemente R, del Rio C (2010) Wanted: a theoretical roadmap to research and practice across individual, interpersonal, and structural levels of analysis. AIDS Behav 14(Suppl 2):185–188. https://doi.org/10.1007/s10461-010-9805-x

Ancuta P, Kamat A, Kunstman KJ, Kim EY, Autissier P, Wurcel A et al (2008) Microbial translocation is associated with increased monocyte activation and dementia in AIDS patients. PLoS One 3(6):e2516. https://doi.org/10.1371/journal.pone.0002516

Antinori A, Arendt G, Becker JT, Brew BJ, Byrd DA, Cherner M et al (2007) Updated research nosology for HIV-associated neurocognitive disorders. Neurology 69(18):1789–1799. https://doi.org/10.1212/01.WNL.0000287431.88658.8b

Araújo-Soares V, Hankonen N, Presseau J, Rodrigues A, Sniehotta F (2018) Developing behavior change interventions for self-management in chronic illness. Eur Psychol. https://doi.org/10.1027/1016-9040/a000330

Balfour L, Kowal J, Silverman A, Tasca GA, Angel JB, Macpherson PA et al (2006) A randomized controlled psycho-education intervention trial: improving psychological readiness for successful HIV medication adherence and reducing depression before initiating HAART. AIDS Care 18 (7):830–838. https://doi.org/10.1080/09540120500466820

Barak Y, Aizenberg D (2010) Is dementia preventable? Focus on Alzheimer's disease. Expert Rev Neurother 10(11):1689–1698. https://doi.org/10.1586/ern.10.159

Beach MC, Roter DL, Saha S, Korthuis PT, Eggly S, Cohn J et al (2015) Impact of a brief patient and provider intervention to improve the quality of communication about medication adherence among HIV patients. Patient Educ Couns 98(9):1078–1083. https://doi.org/10.1016/j.pec.2015.05.011

Best JR, Chiu BK, Liang Hsu C, Nagamatsu LS, Liu-Ambrose T (2015) Long-term effects of resistance exercise training on cognition and brain volume in older women: results from a randomized controlled trial. J Int Neuropsychol Soc 21(10):745–756. https://doi.org/10.1017/S1355617715000673

Blackwell T, Yaffe K, Laffan A, Ancoli-Israel S, Redline S, Ensrud KE et al (2014) Associations of objectively and subjectively measured sleep quality with subsequent cognitive decline in older community-dwelling men: the MrOS sleep study. Sleep 37(4):655–663. https://doi.org/10.5665/sleep.3562

Blashill AJ, Mayer KH, Crane H, Magidson JF, Grasso C, Mathews WC et al (2013) Physical activity and health outcomes among HIV-infected men who have sex with men: a longitudinal mediational analysis. Ann Behav Med 46(2):149–156. https://doi.org/10.1007/s12160-013-9489-3

Botros D, Somarriba G, Neri D, Miller TL (2012) Interventions to address chronic disease and HIV: strategies to promote exercise and nutrition among HIV-infected individuals. Curr HIV/AIDS Rep 9(4):351–363. https://doi.org/10.1007/s11904-012-0135-7

Brotons C, Bulc M, Sammut MR, Sheehan M, Manuel da Silva Martins C, Bjorkelund C et al (2012) Attitudes toward preventive services and lifestyle: the views of primary care patients in Europe. The EUROPREVIEW patient study. Fam Pract 29(Suppl 1):i168–i176. https://doi.org/10.1093/fampra/cmr102

Byun E, Gay CL, Lee KA (2016) Sleep, fatigue, and problems with cognitive function in adults living with HIV. J Assoc Nurses AIDS Care 27(1):5–16. https://doi.org/10.1016/j.jana.2015.10.002

Capili B, Anastasi JK (2008) Body mass index and nutritional intake in patients with HIV and chronic diarrhea: a secondary analysis. J Am Acad Nurse Pract 20(9):463–470. https://doi.org/10.1111/j.1745-7599.2008.00349.x

Carr A, Samaras K, Burton S, Law M, Freund J, Chisholm DJ et al (1998) A syndrome of peripheral lipodystrophy, hyperlipidaemia and insulin resistance in patients receiving HIV protease inhibitors. AIDS 12(7):F51–F58

Chen YC, Lin CY, Strong C, Li CY, Wang JS, Ko WC et al (2017) Sleep disturbances at the time of a new diagnosis: a comparative study of human immunodeficiency virus patients, cancer patients, and general population controls. Sleep Med 36:38–43. https://doi.org/10.1016/j.sleep.2017.04.005

Cifuentes M, Davis M, Fernald D, Gunn R, Dickinson P, Cohen DJ (2015) Electronic health record challenges, workarounds, and solutions observed in practices integrating behavioral health and primary care. J Am Board Fam Med 28(Suppl 1):S63–S72. https://doi.org/10.3122/jabfm.2015.S1.150133

Connell LE, Carey RN, de Bruin M, Rothman AJ, Johnston M, Kelly MP et al (2018) Links between behavior change techniques and mechanisms of action: an expert consensus study. Ann Behav Med. https://doi.org/10.1093/abm/kay082

Cysique LA, Moffat K, Moore DM, Lane TA, Davies NW, Carr A et al (2013) HIV, vascular and aging injuries in the brain of clinically stable HIV-infected adults: a (1)H MRS study. PLoS One 8(4):e61738. https://doi.org/10.1371/journal.pone.0061738

D'Antoni ML, Byron MM, Chan P, Sailasuta N, Sacdalan C, Sithinamsuwan P et al (2018) Normalization of soluble CD163 after institution of antiretroviral therapy during acute HIV infection tracks with fewer neurological abnormalities. J Infect Dis 218(9):1453–1463. https://doi.org/10.1093/infdis/jiy337

d'Ettorre G, Ceccarelli G, Giustini N, Mastroianni CM, Silvestri G, Vullo V (2014) Taming HIV-related inflammation with physical activity: a matter of timing. AIDS Res Hum Retrovir 30(10):936–944. https://doi.org/10.1089/AID.2014.0069

Dahlgren G, Whitehead M (2006) European strategies for tackling social inequities in health: levelling up Part 2. World Health Organization, Copenhagen

Dang AK, Nguyen LH, Nguyen AQ, Tran BX, Tran TT, Latkin CA et al (2018) Physical activity among HIV-positive patients receiving antiretroviral therapy in Hanoi and Nam Dinh, Vietnam: a cross-sectional study. BMJ Open 8(5):e020688. https://doi.org/10.1136/bmjopen-2017-020688

Davis CL, Tomporowski PD, Boyle CA, Waller JL, Miller PH, Naglieri JA et al (2007) Effects of aerobic exercise on overweight children's cognitive functioning: a randomized controlled trial. Res Q Exerc Sport 78(5):510–519. https://doi.org/10.1080/02701367.2007.10599450

Devore EE, Kang JH, Breteler MM, Grodstein F (2012) Dietary intakes of berries and flavonoids in relation to cognitive decline. Ann Neurol 72(1):135–143. https://doi.org/10.1002/ana.23594

Dube MP, Stein JH, Aberg JA, Fichtenbaum CJ, Gerber JG, Tashima KT et al (2003) Guidelines for the evaluation and management of dyslipidemia in human immunodeficiency virus (HIV)-infected adults receiving antiretroviral therapy: recommendations of the HIV Medical Association of the Infectious Disease Society of America and the Adult AIDS Clinical Trials Group. Clin Infect Dis 37(5):613–627. https://doi.org/10.1086/378131

Dufour CA, Marquine MJ, Fazeli PL, Henry BL, Ellis RJ, Grant I et al (2013) Physical exercise is associated with less neurocognitive impairment among HIV-infected adults. J Neurovirol 19 (5):410–417. https://doi.org/10.1007/s13365-013-0184-8

Dufour CA, Marquine MJ, Fazeli PL, Umlauf A, Henry BL, Zlatar Z et al (2018) A longitudinal analysis of the impact of physical activity on neurocognitive functioning among HIV-infected adults. AIDS Behav 22(5):1562–1572. https://doi.org/10.1007/s10461-016-1643-z

Duran AC, Almeida LB, Segurado AA, Jaime PC (2008) Diet quality of persons living with HIV/AIDS on highly active antiretroviral therapy. J Hum Nutr Diet 21(4):346–350. https://doi.org/10.1111/j.1365-277X.2008.00886.x

Durmer JS, Dinges DF (2005) Neurocognitive consequences of sleep deprivation. Semin Neurol 25 (1):117–129. https://doi.org/10.1055/s-2005-867080

Eldredge LKB, Markham CM, Ruiter RAC, Kok G, Fernandez ME, Parcel GS (2016) Planning health promotion programs: an intervention mapping approach. Wiley, Hoboken

Engeroff T, Ingmann T, Banzer W (2018) Physical activity throughout the adult life span and domain-specific cognitive function in old age: a systematic review of cross-sectional and longitudinal data. Sports Med 48(6):1405–1436. https://doi.org/10.1007/s40279-018-0920-6

Estruch R, Ros E, Salas-Salvado J, Covas MI, Corella D, Aros F et al (2013) Primary prevention of cardiovascular disease with a Mediterranean diet. N Engl J Med 368(14):1279–1290. https://doi.org/10.1056/NEJMoa1200303

Ettenhofer ML, Foley J, Castellon SA, Hinkin CH (2010) Reciprocal prediction of medication adherence and neurocognition in HIV/AIDS. Neurology 74(15):1217–1222. https://doi.org/10.1212/WNL.0b013e3181d8c1ca

Farina N, Rusted J, Tabet N (2014) The effect of exercise interventions on cognitive outcome in Alzheimer's disease: a systematic review. Int Psychogeriatr 26(1):9–18. https://doi.org/10.1017/S1041610213001385

Faubel R, Lopez-Garcia E, Guallar-Castillon P, Graciani A, Banegas JR, Rodriguez-Artalejo F (2009) Usual sleep duration and cognitive function in older adults in Spain. J Sleep Res 18 (4):427–435. https://doi.org/10.1111/j.1365-2869.2009.00759.x

Fazeli PL, Marquine MJ, Dufour C, Henry BL, Montoya J, Gouaux B et al (2015) Physical activity is associated with better neurocognitive and everyday functioning among older adults with HIV disease. AIDS Behav 19(8):1470–1477. https://doi.org/10.1007/s10461-015-1024-z

Firth J, Cotter J, Carney R, Yung AR (2017) The pro-cognitive mechanisms of physical exercise in people with schizophrenia. Br J Pharmacol 174(19):3161–3172. https://doi.org/10.1111/bph.13772

Fito M, Estruch R, Salas-Salvado J, Martinez-Gonzalez MA, Aros F, Vila J et al (2014) Effect of the Mediterranean diet on heart failure biomarkers: a randomized sample from the PREDIMED trial. Eur J Heart Fail 16(5):543–550. https://doi.org/10.1002/ejhf.61

Fjeldsoe BS, Marshall AL, Miller YD (2009) Behavior change interventions delivered by mobile telephone short-message service. Am J Prev Med 36(2):165–173. https://doi.org/10.1016/j.amepre.2008.09.040

Gamaldo CE, Gamaldo A, Creighton J, Salas RE, Selnes OA, David PM et al (2013) Evaluating sleep and cognition in HIV. J Acquir Immune Defic Syndr 63(5):609–616. https://doi.org/10.1097/QAI.0b013e31829d63ab

Gannon P, Khan MZ, Kolson DL (2011) Current understanding of HIV-associated neurocognitive disorders pathogenesis. Curr Opin Neurol 24(3):275–283. https://doi.org/10.1097/WCO.0b013e32834695fb

Gill SJ, Friedenreich CM, Sajobi TT, Longman RS, Drogos LL, Davenport MH et al (2015) Association between lifetime physical activity and cognitive functioning in middle-aged and older community dwelling adults: results from the brain in motion study. J Int Neuropsychol Soc 21(10):816–830. https://doi.org/10.1017/S1355617715000880

Golin CE, Earp J, Tien HC, Stewart P, Porter C, Howie L (2006) A 2-arm, randomized, controlled trial of a motivational interviewing-based intervention to improve adherence to antiretroviral therapy (ART) among patients failing or initiating ART. J Acquir Immune Defic Syndr 42 (1):42–51. https://doi.org/10.1097/01.qai.0000219771.97303.0a

Grobler L, Siegfried N, Visser ME, Mahlungulu SS, Volmink J (2013) Nutritional interventions for reducing morbidity and mortality in people with HIV. Cochrane Database Syst Rev 2: CD004536. https://doi.org/10.1002/14651858.CD004536.pub3

Guiney H, Machado L (2013) Benefits of regular aerobic exercise for executive functioning in healthy populations. Psychon Bull Rev 20(1):73–86. https://doi.org/10.3758/s13423-012-0345-4

Hadigan C, Jeste S, Anderson EJ, Tsay R, Cyr H, Grinspoon S (2001) Modifiable dietary habits and their relation to metabolic abnormalities in men and women with human immunodeficiency virus infection and fat redistribution. Clin Infect Dis 33(5):710–717. https://doi.org/10.1086/322680

Hand GA, Lyerly GW, Jaggers JR, Dudgeon WD (2009) Impact of aerobic and resistance exercise on the health of HIV-infected persons. Am J Lifestyle Med 3(6):489–499. https://doi.org/10.1177/1559827609342198

Healy GN, Dunstan DW, Salmon J, Cerin E, Shaw JE, Zimmet PZ et al (2008) Breaks in sedentary time: beneficial associations with metabolic risk. Diabetes Care 31(4):661–666

Heaton RK, Clifford DB, Franklin DR Jr, Woods SP, Ake C, Vaida F et al (2010) HIV-associated neurocognitive disorders persist in the era of potent antiretroviral therapy: CHARTER Study. Neurology 75(23):2087–2096. https://doi.org/10.1212/WNL.0b013e318200d727

Henry BL, Moore DJ (2016) Preliminary findings describing participant experience with iSTEP, an mHealth intervention to increase physical activity and improve neurocognitive function in people living with HIV. J Assoc Nurses AIDS Care 27(4):495–511. https://doi.org/10.1016/j.jana.2016.01.001

Herbert V, Kyle SD, Pratt D (2018) Does cognitive behavioural therapy for insomnia improve cognitive performance? A systematic review and narrative synthesis. Sleep Med Rev 39:37–51. https://doi.org/10.1016/j.smrv.2017.07.001

Hinkin CH, Castellon SA, Durvasula RS, Hardy DJ, Lam MN, Mason KI et al (2002) Medication adherence among HIV+ adults: effects of cognitive dysfunction and regimen complexity. Neurology 59(12):1944–1950

Hong S, Banks WA (2015) Role of the immune system in HIV-associated neuroinflammation and neurocognitive implications. Brain Behav Immun 45:1–12. https://doi.org/10.1016/j.bbi.2014.10.008

Horvath T, Azman H, Kennedy GE, Rutherford GW (2012) Mobile phone text messaging for promoting adherence to antiretroviral therapy in patients with HIV infection. Cochrane Database Syst Rev 3:Cd009756. https://doi.org/10.1002/14651858.Cd009756

Jong E, Oudhoff LA, Epskamp C, Wagener MN, van Duijn M, Fischer S et al (2010) Predictors and treatment strategies of HIV-related fatigue in the combined antiretroviral therapy era. AIDS 24(10):1387–1405. https://doi.org/10.1097/QAD.0b013e328339d004

Kapetanovic S, Leister E, Nichols S, Miller T, Tassiopoulos K, Hazra R et al (2010) Relationships between markers of vascular dysfunction and neurodevelopmental outcomes in perinatally HIV-infected youth. AIDS 24(10):1481–1491. https://doi.org/10.1097/QAD.0b013e32833a241b

Kapetanovic S, Griner R, Zeldow B, Nichols S, Leister E, Gelbard HA et al (2014) Biomarkers and neurodevelopment in perinatally HIV-infected or exposed youth: a structural equation model analysis. AIDS 28(3):355–364. https://doi.org/10.1097/QAD.0000000000000072

Kaufman MR, Cornish F, Zimmerman RS, Johnson BT (2014) Health behavior change models for HIV prevention and AIDS care: practical recommendations for a multi-level approach. J Acquir Immune Defic Syndr 66(Suppl 3):S250–S258. https://doi.org/10.1097/QAI.0000000000000236

Kemoun G, Thibaud M, Roumagne N, Carette P, Albinet C, Toussaint L et al (2010) Effects of a physical training programme on cognitive function and walking efficiency in elderly persons with dementia. Dement Geriatr Cogn Disord 29(2):109–114. https://doi.org/10.1159/000272435

Kimhy D, Vakhrusheva J, Bartels MN, Armstrong HF, Ballon JS, Khan S et al (2015) The impact of aerobic exercise on brain-derived neurotrophic factor and neurocognition in individuals with schizophrenia: a single-blind, randomized clinical trial. Schizophr Bull 41(4):859–868. https://doi.org/10.1093/schbul/sbv022

Kirtane AR, Langer R, Traverso G (2016) Past, present, and future drug delivery systems for antiretrovirals. J Pharm Sci 105(12):3471–3482. https://doi.org/10.1016/j.xphs.2016.09.015

Knight A, Bryan J, Murphy K (2017) The Mediterranean diet and age-related cognitive functioning: a systematic review of study findings and neuropsychological assessment methodology. Nutr Neurosci 20(8):449–468. https://doi.org/10.1080/1028415X.2016.1183341

Kok G, Schaalma H, Ruiter RA, van Empelen P, Brug J (2004) Intervention mapping: protocol for applying health psychology theory to prevention programmes. J Health Psychol 9(1):85–98. https://doi.org/10.1177/1359105304038379

Kruzich LA, Marquis GS, Wilson CM, Stephensen CB (2004) HIV-infected US youth are at high risk of obesity and poor diet quality: a challenge for improving short- and long-term health outcomes. J Am Diet Assoc 104(10):1554–1560. https://doi.org/10.1016/j.jada.2004.07.031

Lee KA, Gay C, Portillo CJ, Coggins T, Davis H, Pullinger CR et al (2009) Symptom experience in HIV-infected adults: a function of demographic and clinical characteristics. J Pain Symptom Manag 38(6):882–893. https://doi.org/10.1016/j.jpainsymman.2009.05.013

Leyes P, Martinez E, Forga Mde T (2008) Use of diet, nutritional supplements and exercise in HIV-infected patients receiving combination antiretroviral therapies: a systematic review. Antivir Ther 13(2):149–159

Lim J, Dinges DF (2010) A meta-analysis of the impact of short-term sleep deprivation on cognitive variables. Psychol Bull 136(3):375–389. https://doi.org/10.1037/a0018883

Lim SS, Vos T, Flaxman AD, Danaei G, Shibuya K, Adair-Rohani H et al (2012) A comparative risk assessment of burden of disease and injury attributable to 67 risk factors and risk factor clusters in 21 regions, 1990–2010: a systematic analysis for the Global Burden of Disease Study 2010. Lancet 380(9859):2224–2260. https://doi.org/10.1016/s0140-6736(12)61766-8

Loughrey DG, Lavecchia S, Brennan S, Lawlor BA, Kelly ME (2017) The impact of the Mediterranean diet on the cognitive functioning of healthy older adults: a systematic review and meta-analysis. Adv Nutr 8(4):571–586. https://doi.org/10.3945/an.117.015495

Manson JE, Hu FB, Rich-Edwards JW, Colditz GA, Stampfer MJ, Willett WC et al (1999) A prospective study of walking as compared with vigorous exercise in the prevention of coronary heart disease in women. N Engl J Med 341(9):650–658

Marcotte TD, Wolfson T, Rosenthal TJ, Heaton RK, Gonzalez R, Ellis RJ et al (2004) A multimodal assessment of driving performance in HIV infection. Neurology 63(8):1417–1422

Martin EM, Pitrak DL, Novak RM, Pursell KJ, Mullane KM (1999) Reaction times are faster in HIV-seropositive patients on antiretroviral therapy: a preliminary report. J Clin Exp Neuropsychol 21(5):730–735. https://doi.org/10.1076/jcen.21.5.730.867

Martinez-Lapiscina EH, Clavero P, Toledo E, Estruch R, Salas-Salvado J, San Julian B et al (2013) Mediterranean diet improves cognition: the PREDIMED-NAVARRA randomised trial. J Neurol Neurosurg Psychiatry 84(12):1318–1325. https://doi.org/10.1136/jnnp-2012-304792

McCutchan JA, Marquie-Beck JA, Fitzsimons CA, Letendre SL, Ellis RJ, Heaton RK et al (2012) Role of obesity, metabolic variables, and diabetes in HIV-associated neurocognitive disorder. Neurology 78(7):485–492. https://doi.org/10.1212/WNL.0b013e3182478d64

McDermott A, Zaporojan L, McNamara P, Doherty CP, Redmond J, Forde C et al (2017) The effects of a 16-week aerobic exercise programme on cognitive function in people living with HIV. AIDS Care 29(6):667–674. https://doi.org/10.1080/09540121.2016.1263723

Michie S, van Stralen MM, West R (2011) The behaviour change wheel: a new method for characterising and designing behaviour change interventions. Implement Sci 6:42. https://doi.org/10.1186/1748-5908-6-42

Michie S, Richardson M, Johnston M, Abraham C, Francis J, Hardeman W et al (2013) The behavior change technique taxonomy (v1) of 93 hierarchically clustered techniques: building an international consensus for the reporting of behavior change interventions. Ann Behav Med 46 (1):81–95. https://doi.org/10.1007/s12160-013-9486-6

Miller KD, Jones E, Yanovski JA, Shankar R, Feuerstein I, Falloon J (1998) Visceral abdominal-fat accumulation associated with use of indinavir. Lancet 351(9106):871–875. https://doi.org/10.1016/S0140-6736(97)11518-5

Miller TL, Jacobson DL, Somarriba G, Neri D, Kurtz-Vraney J, Graham P et al (2017) A multicenter study of diet quality on birth weight and gestational age in infants of HIV-infected women. Matern Child Nutr 13(4):e12378. https://doi.org/10.1111/mcn.12378

Millikin CP, Rourke SB, Halman MH, Power C (2003) Fatigue in HIV/AIDS is associated with depression and subjective neurocognitive complaints but not neuropsychological functioning. J Clin Exp Neuropsychol 25(2):201–215. https://doi.org/10.1076/jcen.25.2.201.13644

Mitjavila MT, Fandos M, Salas-Salvado J, Covas MI, Borrego S, Estruch R et al (2013) The Mediterranean diet improves the systemic lipid and DNA oxidative damage in metabolic syndrome individuals. A randomized, controlled, trial. Clin Nutr 32(2):172–178. https://doi.org/10.1016/j.clnu.2012.08.002

Miyata S, Noda A, Ozaki N, Hara Y, Minoshima M, Iwamoto K et al (2010) Insufficient sleep impairs driving performance and cognitive function. Neurosci Lett 469(2):229–233. https://doi.org/10.1016/j.neulet.2009.12.001

Mohd Salleh NA, Richardson L, Kerr T, Shoveller J, Montaner J, Kamarulzaman A et al (2018) A longitudinal analysis of daily pill burden and likelihood of optimal adherence to antiretroviral therapy among people living with HIV who use drugs. J Addict Med 12(4):308–314. https://doi.org/10.1097/ADM.0000000000000403

Monroe AK, Zhang L, Jacobson LP, Plankey MW, Brown TT, Miller EN et al (2017) The association between physical activity and cognition in men with and without HIV infection. HIV Med 18(8):555–563. https://doi.org/10.1111/hiv.12490

Montoya JL, Wing D, Knight A, Moore DJ, Henry BL (2015) Development of an mHealth intervention (iSTEP) to promote physical activity among people living with HIV. J Int Assoc Provid AIDS Care 14(6):471–475. https://doi.org/10.1177/2325957415601505

Moore DJ, Poquette A, Casaletto KB, Gouaux B, Montoya JL, Posada C et al (2015) Individualized texting for adherence building (iTAB): improving antiretroviral dose timing among HIV-infected persons with co-occurring bipolar disorder. AIDS Behav 19(3):459–471. https://doi.org/10.1007/s10461-014-0971-0
Munk PS, Breland UM, Aukrust P, Ueland T, Kvaloy JT, Larsen AI (2011) High intensity interval training reduces systemic inflammation in post-PCI patients. Eur J Cardiovasc Prev Rehabil 18 (6):850–857. https://doi.org/10.1177/1741826710397600
Mura G, Vellante M, Nardi AE, Machado S, Carta MG (2015) Effects of school-based physical activity interventions on cognition and academic achievement: a systematic review. CNS Neurol Disord Drug Targets 14(9):1194–1208
Nurk E, Refsum H, Drevon CA, Tell GS, Nygaard HA, Engedal K et al (2009) Intake of flavonoid-rich wine, tea, and chocolate by elderly men and women is associated with better cognitive test performance. J Nutr 139(1):120–127. https://doi.org/10.3945/jn.108.095182
O'Brien K, Nixon S, Tynan AM, Glazier R (2010) Aerobic exercise interventions for adults living with HIV/AIDS. Cochrane Database Syst Rev 8:CD001796. https://doi.org/10.1002/14651858.CD001796.pub3
Olatunji BO, Mimiaga MJ, O'Cleirigh C, Safren SA (2006) Review of treatment studies of depression in HIV. Top HIV Med 14(3):112–124
Ortega M, Baker LM, Vaida F, Paul R, Basco B, Ances BM (2015) Physical activity affects brain integrity in HIV+ individuals. J Int Neuropsychol Soc 21(10):880–889. https://doi.org/10.1017/S1355617715000879
Osterberg L, Blaschke T (2005) Adherence to medication. N Engl J Med 353(5):487–497. https://doi.org/10.1056/NEJMra050100
Parsons TD, Braaten AJ, Hall CD, Robertson KR (2006) Better quality of life with neuropsychological improvement on HAART. Health Qual Life Outcomes 4:11. https://doi.org/10.1186/1477-7525-4-11
Poulose SM, Miller MG, Shukitt-Hale B (2014) Role of walnuts in maintaining brain health with age. J Nutr 144(4 Suppl):561S–566S. https://doi.org/10.3945/jn.113.184838
Qaseem A, Kansagara D, Forciea MA, Cooke M, Denberg TD (2016) Management of chronic insomnia disorder in adults: a clinical practice guideline from the American College of Physicians. Ann Intern Med 165(2):125–133. https://doi.org/10.7326/M15-2175
Quigley A, O'Brien K, Parker R, MacKay-Lyons M (2018) Exercise and cognitive function in people living with HIV: a scoping review. Disabil Rehabil:1–12. https://doi.org/10.1080/09638288.2018.1432079
Ramos JS, Dalleck LC, Tjonna AE, Beetham KS, Coombes JS (2015) The impact of high-intensity interval training versus moderate-intensity continuous training on vascular function: a systematic review and meta-analysis. Sports Med 45(5):679–692. https://doi.org/10.1007/s40279-015-0321-z
Rehm KE, Konkle-Parker D (2016) Physical activity levels and perceived benefits and barriers to physical activity in HIV-infected women living in the deep south of the United States. AIDS Care 28(9):1205–1210. https://doi.org/10.1080/09540121.2016.1164802
Ren J, Zhao M, Liu B, Wu Q, Hao Y, Jiao M et al (2018) Factors associated with sleep quality in HIV. J Assoc Nurses AIDS Care 29(6):924–931. https://doi.org/10.1016/j.jana.2018.04.006
Rollnick S, Miller WR, Butler CC (2008) Motivational interviewing in health care. Guilford Press, New York
Rosenberg DE, Kerr J, Sallis JF, Norman GJ, Calfas K, Patrick K (2012) Promoting walking among older adults living in retirement communities. J Aging Phys Act 20(3):379–394. https://doi.org/10.1123/japa.20.3.379
Rubinstein ML, Selwyn PA (1998) High prevalence of insomnia in an outpatient population with HIV infection. J Acquir Immune Defic Syndr Hum Retrovirol 19(3):260–265
Safren SA, O'Cleirigh C, Tan JY, Raminani SR, Reilly LC, Otto MW et al (2009) A randomized controlled trial of cognitive behavioral therapy for adherence and depression (CBT-AD) in HIV-infected individuals. Health Psychol 28(1):1–10. https://doi.org/10.1037/a0012715

Salmoirago-Blotcher E, Rich C, Rosen RK, Dunsiger S, Rana A, Carey MP (2017) Phone-delivered mindfulness training to promote medication adherence and reduce sexual risk behavior among persons living with HIV: design and methods. Contemp Clin Trials 53:162–170. https://doi.org/10.1016/j.cct.2016.12.016

Sattler FR, He J, Letendre S, Wilson C, Sanders C, Heaton R et al (2015) Abdominal obesity contributes to neurocognitive impairment in HIV-infected patients with increased inflammation and immune activation. J Acquir Immune Defic Syndr 68(3):281–288. https://doi.org/10.1097/QAI.0000000000000458

Saylor D, Dickens AM, Sacktor N, Haughey N, Slusher B, Pletnikov M et al (2016) HIV-associated neurocognitive disorder – pathogenesis and prospects for treatment. Nat Rev Neurol 12(5):309. https://doi.org/10.1038/nrneurol.2016.53

Schaefer EJ, Bongard V, Beiser AS, Lamon-Fava S, Robins SJ, Au R et al (2006) Plasma phosphatidylcholine docosahexaenoic acid content and risk of dementia and Alzheimer disease: the Framingham Heart Study. Arch Neurol 63(11):1545–1550. https://doi.org/10.1001/archneur.63.11.1545

Schroder H, Salas-Salvado J, Martinez-Gonzalez MA, Fito M, Corella D, Estruch R et al (2014) Baseline adherence to the Mediterranean diet and major cardiovascular events: Prevencion con Dieta Mediterranea trial. JAMA Intern Med 174(10):1690–1692. https://doi.org/10.1001/jamainternmed.2014.3463

Schuelter-Trevisol F, Wolff FH, Alencastro PR, Grigoletti S, Ikeda ML, Brandao AB et al (2012) Physical activity: do patients infected with HIV practice? How much? A systematic review. Curr HIV Res 10(6):487–497

Shah M, Tierney K, Adams-Huet B, Boonyavarakul A, Jacob K, Quittner C et al (2005) The role of diet, exercise and smoking in dyslipidaemia in HIV-infected patients with lipodystrophy. HIV Med 6(4):291–298. https://doi.org/10.1111/j.1468-1293.2005.00309.x

Simoni JM, Pantalone DW, Plummer MD, Huang B (2007) A randomized controlled trial of a peer support intervention targeting antiretroviral medication adherence and depressive symptomatology in HIV-positive men and women. Health Psychol 26(4):488–495. https://doi.org/10.1037/0278-6133.26.4.488

Simonik A, Vader K, Ellis D, Kesbian D, Leung P, Jachyra P et al (2016) Are you ready? Exploring readiness to engage in exercise among people living with HIV and multimorbidity in Toronto, Canada: a qualitative study. BMJ Open 6(3):e010029. https://doi.org/10.1136/bmjopen-2015-010029

Smith PJ, Blumenthal JA, Hoffman BM, Cooper H, Strauman TA, Welsh-Bohmer K et al (2010) Aerobic exercise and neurocognitive performance: a meta-analytic review of randomized controlled trials. Psychosom Med 72(3):239–252. https://doi.org/10.1097/PSY.0b013e3181d14633

Song D, Yu DSF, Li PWC, Lei Y (2018) The effectiveness of physical exercise on cognitive and psychological outcomes in individuals with mild cognitive impairment: a systematic review and meta-analysis. Int J Nurs Stud 79:155–164. https://doi.org/10.1016/j.ijnurstu.2018.01.002

Stimpson NJ, Davison G, Javadi AH (2018) Joggin' the noggin: towards a physiological understanding of exercise-induced cognitive benefits. Neurosci Biobehav Rev 88:177–186. https://doi.org/10.1016/j.neubiorev.2018.03.018

Stradling C, Chen YF, Russell T, Connock M, Thomas GN, Taheri S (2012) The effects of dietary intervention on HIV dyslipidaemia: a systematic review and meta-analysis. PLoS One 7(6): e38121. https://doi.org/10.1371/journal.pone.0038121

Suarez S, Baril L, Stankoff B, Khellaf M, Dubois B, Lubetzki C et al (2001) Outcome of patients with HIV-1-related cognitive impairment on highly active antiretroviral therapy. AIDS 15 (2):195–200

Sugiura H, Sugiura H, Kajima K, Mirbod SM, Iwata H, Matsuoka T (2002) Effects of long-term moderate exercise and increase in number of daily steps on serum lipids in women: randomised controlled trial [ISRCTN21921919]. BMC Womens Health 2(1):3

Taibi DM, Price C, Voss J (2013) A pilot study of sleep quality and rest-activity patterns in persons living with HIV. J Assoc Nurses AIDS Care 24(5):411–421. https://doi.org/10.1016/j.jana.2012.08.001

Tavazzi E, Morrison D, Sullivan P, Morgello S, Fischer T (2014) Brain inflammation is a common feature of HIV-infected patients without HIV encephalitis or productive brain infection. Curr HIV Res 12(2):97–110

Thames AD, Kim MS, Becker BW, Foley JM, Hines LJ, Singer EJ et al (2011) Medication and finance management among HIV-infected adults: the impact of age and cognition. J Clin Exp Neuropsychol 33(2):200–209. https://doi.org/10.1080/13803395.2010.499357

Tjonna AE, Stolen TO, Bye A, Volden M, Slordahl SA, Odegard R et al (2009) Aerobic interval training reduces cardiovascular risk factors more than a multitreatment approach in overweight adolescents. Clin Sci (Lond) 116(4):317–326. https://doi.org/10.1042/CS20080249

Tsiodras S, Poulia KA, Yannakoulia M, Chimienti SN, Wadhwa S, Karchmer AW et al (2009) Adherence to Mediterranean diet is favorably associated with metabolic parameters in HIV-positive patients with the highly active antiretroviral therapy-induced metabolic syndrome and lipodystrophy. Metabolism 58(6):854–859. https://doi.org/10.1016/j.metabol.2009.02.012

Tull MT, Berghoff CR, Bardeen JR, Schoenleber M, Konkle-Parker DJ (2018) An initial open trial of a brief behavioral activation treatment for depression and medication adherence in HIV-infected patients. Behav Modif 42(2):196–209. https://doi.org/10.1177/0145445517723901

Valcour VG, Shikuma CM, Shiramizu BT, Williams AE, Watters MR, Poff PW et al (2005) Diabetes, insulin resistance, and dementia among HIV-1-infected patients. J Acquir Immune Defic Syndr 38(1):31–36

Vancampfort D, Mugisha J, De Hert M, Probst M, Firth J, Gorczynski P et al (2018a) Global physical activity levels among people living with HIV: a systematic review and meta-analysis. Disabil Rehabil 40(4):388–397. https://doi.org/10.1080/09638288.2016.1260645

Vancampfort D, Stubbs B, Mugisha J, Firth J, Schuch FB, Koyanagi A (2018b) Correlates of sedentary behavior in 2,375 people with depression from 6 low- and middle-income countries. J Affect Disord 234:97–104. https://doi.org/10.1016/j.jad.2018.02.088

Vance DE, Wadley VG, Crowe MG, Raper JL, Ball KK (2011) Cognitive and everyday functioning in older and younger adults with and without HIV. Clin Gerontol 34(5):413–426. https://doi.org/10.1080/07317115.2011.588545

Vanderbeken I, Kerckhofs E (2017) A systematic review of the effect of physical exercise on cognition in stroke and traumatic brain injury patients. NeuroRehabilitation 40(1):33–48. https://doi.org/10.3233/NRE-161388

Visser ME, Durao S, Sinclair D, Irlam JH, Siegfried N (2017) Micronutrient supplementation in adults with HIV infection. Cochrane Database Syst Rev 5:CD003650. https://doi.org/10.1002/14651858.CD003650.pub4

Webel AR, Moore SM, Hanson JE, Patel SR, Schmotzer B, Salata RA (2013) Improving sleep hygiene behavior in adults living with HIV/AIDS: a randomized control pilot study of the SystemCHANGE(TM)-HIV intervention. Appl Nurs Res 26(2):85–91. https://doi.org/10.1016/j.apnr.2012.10.002

Webel AR, Barkley J, Longenecker CT, Mittelsteadt A, Gripshover B, Salata RA (2015) A cross-sectional description of age and gender differences in exercise patterns in adults living with HIV. J Assoc Nurses AIDS Care 26(2):176–186. https://doi.org/10.1016/j.jana.2014.06.004

Webel AR, Perazzo JD, Dawson-Rose C, Smith C, Nicholas PK, Rivero-Mendez M et al (2017) A multinational qualitative investigation of the perspectives and drivers of exercise and dietary behaviors in people living with HIV. Appl Nurs Res 37:13–18. https://doi.org/10.1016/j.apnr.2017.07.002

Wight D, Wimbush E, Jepson R, Doi L (2016) Six steps in quality intervention development (6SQuID). J Epidemiol Community Health 70(5):520–525. https://doi.org/10.1136/jech-2015-205952

Wirth MD, Jaggers JR, Dudgeon WD, Hebert JR, Youngstedt SD, Blair SN et al (2015) Association of markers of inflammation with sleep and physical activity among people living with HIV or AIDS. AIDS Behav 19(6):1098–1107. https://doi.org/10.1007/s10461-014-0949-y
Witteman HO, Presseau J, Nicholas Angl E, Jokhio I, Schwalm JD, Grimshaw JM et al (2017) Negotiating tensions between theory and design in the development of mailings for people recovering from acute coronary syndrome. JMIR Hum Factors 4(1):e6. https://doi.org/10.2196/humanfactors.6502
Womack JA, Murphy TE, Bathulapalli H, Akgun KM, Gibert C, Kunisaki KM et al (2017) Trajectories of sleep disturbance severity in HIV-infected and uninfected veterans. J Assoc Nurses AIDS Care 28(3):431–437. https://doi.org/10.1016/j.jana.2017.02.006
Woods SP, Dawson MS, Weber E, Gibson S, Grant I, Atkinson JH et al (2009a) Timing is everything: antiretroviral nonadherence is associated with impairment in time-based prospective memory. J Int Neuropsychol Soc 15(1):42–52. https://doi.org/10.1017/S1355617708090012
Woods SP, Moore DJ, Weber E, Grant I (2009b) Cognitive neuropsychology of HIV-associated neurocognitive disorders. Neuropsychol Rev 19(2):152–168. https://doi.org/10.1007/s11065-009-9102-5
Wu J, Wu H, Lu C, Guo L, Li P (2015) Self-reported sleep disturbances in HIV-infected people: a meta-analysis of prevalence and moderators. Sleep Med 16(8):901–907. https://doi.org/10.1016/j.sleep.2015.03.027
Xu L, Jiang CQ, Lam TH, Liu B, Jin YL, Zhu T et al (2011) Short or long sleep duration is associated with memory impairment in older Chinese: the Guangzhou Biobank Cohort Study. Sleep 34(5):575–580
Yaffe K, Falvey CM, Hoang T (2014) Connections between sleep and cognition in older adults. Lancet Neurol 13(10):1017–1028. https://doi.org/10.1016/S1474-4422(14)70172-3
Zazpe I, Sanchez-Tainta A, Estruch R, Lamuela-Raventos RM, Schroder H, Salas-Salvado J et al (2008) A large randomized individual and group intervention conducted by registered dietitians increased adherence to Mediterranean-type diets: the PREDIMED study. J Am Diet Assoc 108(7):1134–1144. Discussion 1145. https://doi.org/10.1016/j.jada.2008.04.011
Ziegler TR, McComsey GA, Frediani JK, Millson EC, Tangpricha V, Eckard AR (2014) Habitual nutrient intake in HIV-infected youth and associations with HIV-related factors. AIDS Res Hum Retrovir 30(9):888–895. https://doi.org/10.1089/AID.2013.0282

Targeting HIV-Related Neurocognitive Impairments with Cognitive Training Strategies: Insights from the Cognitive Aging Literature

David E. Vance, Pariya L. Fazeli, John Cheatwood, Chance Nicholson, Shannon Morrison, and Linda D. Moneyham

Contents

Abstract Approximately 50% of older adults with HIV meet the Frascati diagnostic criteria of HIV-associated neurocognitive disorders (HAND) which can interfere with everyday function such as medication adherence, employment, and driving ability, thus reducing quality of life. As the number of older adults with HIV continues to grow, many will become vulnerable to cognitive frailty, especially as they experience multimorbidities, polypharmacy, and geriatric syndromes. Healthcare professionals need strategies to prevent, remediate, and compensate for cognitive losses observed in memory, language, executive functioning, and speed of processing. Sadly, there are no standard protocols or accepted treatment/intervention guidelines to address HAND at this time. Fortunately, evidence from the cognitive aging literature indicates that cognitive training can protect and improve cognition in normal older adults and may even reduce the incidence of dementia/MCI. This article provides the scientific context in which computerized cognitive training approaches have been successfully used in older adults and provides examples of how these approaches have been translated to adults with HIV. Evidence from ongoing clinical

D. E. Vance (✉), P. L. Fazeli, J. Cheatwood, C. Nicholson, S. Morrison, and L. D. Moneyham
School of Nursing, University of Alabama at Birmingham, Birmingham, AL, USA
e-mail: devance@uab.edu

Curr Topics Behav Neurosci (2021) 50: 503–516
DOI 10.1007/7854_2018_80
Published Online: 2 February 2019

trials are also presented that suggest that reversing a diagnosis of HAND may be possible. Recommendations for clinical practice and research are provided.

Keywords Cognitive aging · Cognitive efficiency · Cognitive training · HIV/AIDS · NeuroHIV · Neuroplasticity

1 Introduction

Non-pathological cognitive aging is a normal process in which cognitive decline is observed over time in the domains of speed of processing (SOP), memory, language, and other fluid cognitive abilities (Ball and Vance 2007). By 2020 nearly 70% of adults with HIV in the United States will be 50 and older (United States Senate Special Committee on Aging 2013). Thus, as people age with this disease, the prevalence and severity of HIV-associated neurocognitive disorders (HAND; Antinori et al. 2007) will likely increase (Goodkin et al. 2017; Kinai et al. 2017). In fact, in a sample of 182 younger and older adults with HIV, Valcour et al. (2004) observed that HIV-associated dementia (HAD) was three times more prevalent in these older adults. Given the concern of greater cognitive decline as people age with HIV, devising ways to prevent or ameliorate decline in cognitive function and protect and maintain cognitive reserve are needed. Fortunately, the evidence surrounding cognitive training across diverse populations is encouraging (Lampit et al. 2014).

The basic premise supporting cognitive training programs is that mental stimulation provided by cognitive exercises promotes neuroplastic changes in the brain which, in turn, increase cognitive reserve and support existing cognitive function (Cody and Vance 2016; Vance et al. 2017a). With advancements in technology-based graphics and programming, computerized cognitive training programs have become more popular, easier to use, commercially available, and relatively inexpensive to implement. In the cognitive aging literature, cognitive training has been found to improve cognitive functioning as well as everyday functioning (Lampit et al. 2014; Rebok et al. 2014). Given its established efficacy in older adults, cognitive training may be most suited for other clinical populations with a cognitively vulnerable phenotype, including chronic HIV infection (Cody et al. 2015).

The purpose of this chapter is to provide the scientific context of cognitive training as a valid strategy to address the cognitive deficits associated with HIV, with a focus on computerized cognitive training programs. In this discussion, the term cognitive training is used to refer to a systematic, formalized approach in which people engage in specialized neurocognitive exercise to maintain, improve, or remediate cognitive functioning, often targeting a specific cognitive domain. Thus, this review will focus on restorative cognitive training interventions as opposed to compensatory interventions (Weber et al. 2012). In the following, a brief review of the major cognitive training studies in relation to aging and their applicability to other conditions is presented. This is followed by examples of the few existing cognitive training studies in adults with HIV, with a focus on SOP training. Finally, implications for clinical practice and research are posited.

2 Cognitive Training Across Aging Populations

Numerous cognitive enrichment programs have been developed for older adults. Some approaches implement social activities and social interactions (i.e., solving problems as a team) or action video games with the rationale that increased cognitive and social engagement may improve cognitive function and perhaps even delay dementia (Wang et al. 2016). As these approaches often do not target any specific cognitive domain, they are more appropriately classified as social or environmental enrichment stimulation strategies rather than cognitive training. To avoid confusion, cognitive training is defined as specifically designed targeted neurocognitive exercises with the clear expectation that with continued use and practice, cognitive function will improve, most likely within a specific domain (e.g., verbal memory, executive functioning). Although much of the work on cognitive training programs began with pencil-and-paper or videotape training techniques (Wadley et al. 2006), over the past two decades, computerized versions have been developed. Improvements in computer technology have advanced the standardization of these cognitive training techniques, thus removing interventionist error. Furthermore, many of these computerized approaches keep track of performance errors and adherence and can administer training stimuli and record responses within milliseconds, which is particularly important for cognitive training programs that require rapid information processing.

Using this standard definition of what constitutes computerized cognitive training, Lampit et al. (2014) conducted a systematic review and meta-analysis of effect modifiers on 52 studies in older adults (60 years and older). The investigators pooled the findings of studies specific to their impact on various cognitive domains to derive an overall effect size across studies as to their efficacy in improving cognition in SOP ($g = 0.31$), visuospatial skills ($g = 0.30$), nonverbal memory ($g = 0.24$), working memory ($g = 0.22$), and verbal memory ($g = 0.08$); improvements in all of these domains were significant. Significant effect sizes were not observed for attention and executive functioning; however, many of the individual studies within the pooled analyses did find significant effects in these areas. Also certain training parameters were noted to enhance the effect size, such as training session length, dose (<20 h; >20 h), and frequency of training (1 session/week vs >3 sessions/week).

Included in the aforementioned cognitive training meta-analysis were the results of the Advanced Cognitive Training for Independent and Vital Elderly (ACTIVE) Study, which is the largest randomized clinical trial on cognitive training programs to date (Rebok et al. 2014). In the ACTIVE Study, 2,802 community-dwelling older adults (65 years and older) were randomized to either a no-contact control group or one of the three training groups: (1) memory, (2) reasoning, or (3) SOP. Those in the cognitive training groups received 10 h of cognitive exercises designed to improve the cognitive domain they were assigned. In general, compared to the no-contact control group, those in the cognitive training groups improved on the domain in which they were trained, with the reasoning and SOP groups experiencing more robust cognitive gains than the memory group. In fact, the cognitive gains observed were robust over several years, especially for those in the SOP training group. Although treatment gains within the targeted domain did not transfer to improvements in other cognitive domains, training gains in the SOP group were found to result in less depressive symptomatology, less driving cessation with age, and better self-rated health, locus of control, and health-related quality of life (Vance et al. 2017b). The findings of the ACTIVE Study and those of other studies highlight that cognitive gain may also improve everyday functioning such as instrumental activities of daily living (e.g., driving) (Rebok et al. 2014; Ross et al. 2016). Recently, Edwards et al. (2017) examined the ACTIVE Study data and found that over a 10-year period, those who completed the SOP training had a 29% reduction in the risk of developing dementia, compared to the control group; regarding dosage, such protection from dementia was increased for those who engaged in more SOP training cognitive exercises. Unfortunately, those in the memory training or reasoning training did not experience such therapeutic benefits which suggest that there are unique neurological characteristics of the SOP training that supports cognitive reserve.

Given the efficacy and utility of cognitive training programs in non-pathological community-dwelling older adults, cognitive training has been launched into other clinical populations (e.g., multiple sclerosis) that experience cognitive problems. For example, in the SOAR (Speed of Processing in Middle-Aged and Older Breast Cancer Survivors) Study, 60 middle-aged and older breast cancer survivors were randomly assigned to either a 10-h SOP training group or a no-contact control group and were assessed 6 weeks and 6 months after baseline. This study found that those who received the SOP training improved on measures of SOP and executive functioning at 6 weeks and that the cognitive gains in SOP were maintained at 6 months (Meneses et al. 2018). Further evidence from a systematic review suggests that such cognitive training programs may benefit clinical populations who express a vulnerable cognitive phenotype (Vance et al. 2017c). Since the evidence suggests that such cognitive training programs have therapeutic value in other cognitively vulnerable populations, cognitive training strategies may possess similar efficacy for those aging with HIV.

3 Cognitive Training in HIV

In a systematic review of cognitive training studies in adults with HIV, Vance et al. 2018b, 2019) identified ten studies designed to improve cognitive functioning either globally or in certain domains. Despite most of these studies often being limited by small sample sizes and lacking adequate control groups, when treatment adherence was good, cognitive training generally favored cognitive improvement in the domain

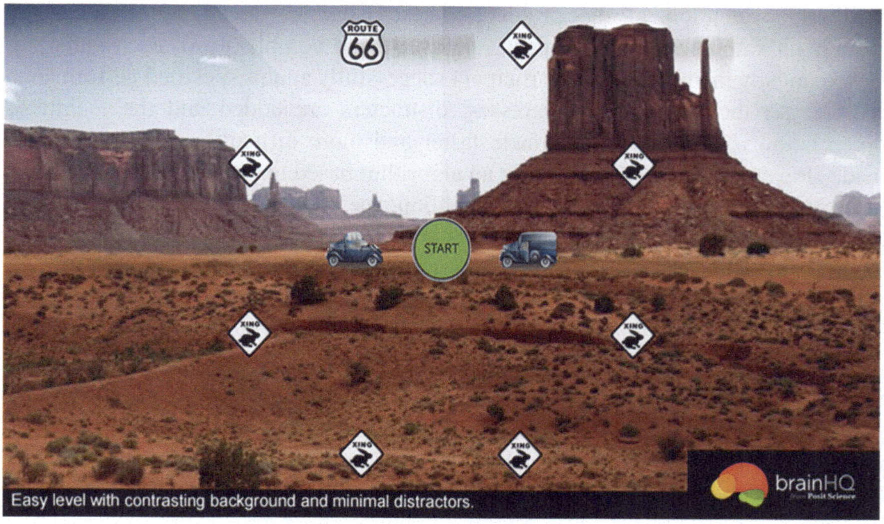

Fig. 1 Answer screenshot of the RoadTour™ training module used for speed of processing training. Permission is provided by Posit Science, Inc., to use this screenshot for publication as it is also provided publicly on their website

targeted for training, usually memory and SOP. Paralleling studies in the cognitive aging literature such as the ACTIVE Study, SOP training in the HIV cognitive training literature has also emerged as a promising research vector.

3.1 Speed of Processing Training

SOP training has been used for nearly 20 decades. In SOP training protocols, participants are presented with individual trials designed to improve the speed and accuracy in which they identify and locate visual information. There are a number of such SOP training modules available, but the traditional SOP training exercise is based upon the Useful Field of View (UFOV®) Test (Edwards et al. 2018) and is incorporated into the module called RoadTour™ from Posit Science, Inc. (brainhq.com). Typically, on each trial, a target vehicle (a car or truck) is quickly (17–500 ms) presented in the center of the screen. At the same time, a visual target (the Route 66 sign) is presented in one of eight locations in the periphery. After this presentation, the answer screen is presented (Fig. 1). Two vehicles are then presented in the center of the screen, one of which was the previously presented target. Using a mouse or a touch screen monitor, a participant must select the most recently presented target vehicle and then select the location where the peripheral target (Route 66 sign) last appeared (in this case the upper left center). Gaming components (e.g., winning points) added within this program make the task enjoyable, thereby facilitating retention to the protocol. These exercises are customized to the participant's ability level; the speed, difficulty, and complexity of each game are systematically increased as the participant successfully attains specified performance criteria. As the participant progresses, distracters are added and the length of presentation is decreased (i.e., made faster and more difficult) or increased (i.e., made slower and less difficult) using an algorithm based upon whether the previous responses are incorrect or correct. In addition, the vehicle pairs morph to become more similar over time; this alteration further increases task difficulty and thus forces the brain to adapt over time (i.e., neuroplasticity). Other training modules from Posit Science, Inc. employ similar gaming features to improve SOP (e.g., Bird Safari).

Using this SOP training protocol adapted from the ACTIVE Study, Vance et al. (2012) randomized 46 middle-aged and older adults with HIV ($M_{age} = 51.55$ years) to either a no-contact control group ($n = 24$) or a laboratory-based SOP training group ($n = 22$) described above. Over 5–6 weeks, participants completed approximately 10 h of assigned training ($M_{hours} = 9.32$; range = 2–10). Compared to the no-contact control group, the SOP training group improved significantly on a measure of visual SOP (UFOV®) with a medium effect size (Cohen's $d = 0.61$). Improvement was also observed on an everyday functioning measure called the Timed Instrumental Activities of Daily Living (TIADL) Test. Twelve (19.30%) participants were lost to attrition; however, no group difference in attrition was observed. Based on self-ratings of one's cognitive functioning at posttest, SOP training participants indicated that the SOP training improved their memory, attention, SOP, and overall mental functioning. In a follow-up within-group study testing a home-based version of SOP training in 20 middle-aged and older adults with HIV ($M_{age} = 50.22$ years), Cody and colleagues also found significant improvements in SOP (UFOV®; Cohen's $d = 0.44$) and self-rated cognitive functioning.

Building on the ACTIVE Study and the SOP training studies in HIV, a study called ThinkFast (1R01MH106366-01A1) is currently ongoing (Vance et al. 2017b). In this longitudinal three-group randomized controlled clinical trial, researchers are randomizing 264 middle-aged (40+ years) and older adults with HAND or borderline HAND to one of the three conditions: (1) 10 h of Internet (sham) training, (2) 10 h of SOP training, or (3) 20 h of SOP training. This study is investigating the effect of a dosage response of SOP training over time. With the positive findings from the ACTIVE Study on the effects of SOP training in everyday functioning, driving, and quality of life indices, ThinkFast is examining the effects of SOP training on cognition, everyday functioning (TIADL Test, medication adherence), driving (driving simulator performance, crash records), and psychological outcomes (locus of control, depression) over time.

In a case comparison of the ThinkFast study, Hossain et al. (2017) examined the first three cases with the same level of HAND (global deficit score = 5; Blackstone et al. 2012) that completed each of the training conditions. Two trends were observed. First, using the UFOV® Test, those who received 10 versus 20 h of SOP training improved by 203 and 596 ms more, respectively, than the control participant. Second, the participant that received 20 h of SOP training no longer met the criteria for HAND (global deficit score = 4). This preliminary finding is encouraging because it demonstrates changes in HAND due to a cognitive training intervention. Although ThinkFast targets the domain of SOP and thus was not designed to change the overall HAND diagnosis, it is interesting to consider the role that SOP training may play in doing so. Indeed, optimal SOP underlies performance and functioning across several other cognitive and everyday domains, and follow-up studies have shown that training gains may translate to executive functioning (Wolinsky et al. 2013).

3.2 *Changing the HAND Diagnosis*

The Training on Purpose Study (TOPS; 1R21NR016632-01) was specifically designed to resolve the HAND diagnosis (Vance et al. 2018a). In this ongoing two-group pre-post clinical trial, researchers are randomizing 146 middle-aged (40+) adults with HAND to either (1) a no-contact control group or (2) the Individualized-Targeted Cognitive Training group. Those in the training group receive targeted cognitive training in the domains that contributed to their HAND diagnosis; however, since there are several domains that may be compromised and contributing to this diagnosis, it is not feasible to administer 10–20 h of cognitive training in each of these domains due to participant burden, time, and expense. In fact, as mentioned earlier from the cognitive training meta-analysis, Lampit et al. (2014) observed that the therapeutic benefit of cognitive training may be reduced after 10–20 h for the domain being trained, perhaps due to participant fatigue. To address these concerns, the TOPS assigns 20 h of training in only two cognitive domains (10 h each) that contribute to the HAND diagnosis using the Individualized-Targeted Training Framework (Vance et al. 2018a).

This framework specifies that if SOP or attention or both are compromised, the priority is to conduct training in these cognitive domains as improvements in them may also support improvements in other cognitive domains. This notion is based on Wickens' Model of Information Processing and the Diminished SOP Theory. According to Wickens' Model of Information Processing, if one does not efficiently attend to the stimuli to be processed, neural representations of those stimuli fail to be usable in other domains (Wickens et al. 2013); thus, improving the ability to attend improves the likelihood that cognitive processes in other domains will be efficiently processed. Similarly, according to the Diminished SOP Theory (Salthouse 1996), if neural representations of stimuli are not quickly processed, these signals are degraded in other cognitive domains. In fact, several studies demonstrate that SOP mediates the effects of HIV and aging on executive functioning, memory, and other cognitive domains (Fellows et al. 2014; Fristoe et al. 1997). Otherwise, if SOP and attention are not impaired, this Individualized-Targeted Cognitive Training Framework selects the cognitive domain(s) that is closest to normal (<1 below demographically adjusted mean). By choosing the least compromised cognitive domain, this implies that there is enough cognitive reserve such that someone may benefit from the training. As the study is ongoing, study results are not yet available; however, a preliminary case study was conducted that is described in the following.

A case comparison study was conducted of six adults with HAND randomized to either the cognitive training group or the no-contact control group (Vance et al. 2018b). Results revealed no change in the HAND diagnosis from baseline to posttest. Regardless, cognitive training in the targeted domains showed improvement in SOP, attention, and spatial learning/memory. Obviously, a larger sample size is needed to derive more conclusive data.

3.3 Variations of Speed of Processing Training

As cognitive training protocols have become more commonly used, they have also been combined with other cognitive interventions such as physical exercise and transcranial direct current stimulation (tDCS) in order to boost their cognitive effects. In a two-group pre-post randomized control study, Fazeli et al. (2018; K99AG048762) randomized 33 adults with HIV on antiretroviral medication to receive either (1) 10 h of SOP training with concurrent active tDCS or (2) 10 h of SOP training with concurrent sham tDCS. For both groups, the anodal electrode was placed on F10 (right inferior frontal cortex), and the cathodal electrode was placed on the contralateral upper arm. SOP training was administered for 1 h per study visit. Those in the active tDCS received 20 mA of stimulation for the first 20 min of SOP training. Those in the sham tDCS received 20 mA of stimulation for the first 30 s of SOP training. Unfortunately, although both groups benefitted from the SOP training as indicated on the UFOV® Test, as expected, group differences were not observed indicating there was no therapeutic cognitive gain by adding this electrical stimulation to this part of the cortex. Albeit, other placements of the electrodes (e.g., left dorsolateral prefrontal cortex) may be more fortunate in augmenting the therapeutic effects of SOP training.

4 Clinical Implications

The clinical implications for treating such cognitive deficits in adults aging with HIV with cognitive training programs are not well established (Vance et al. 2013). Based on the cognitive aging and NeuroHIV literature, there is a precedent on the therapeutic efficacy in using such cognitive training programs to augment cognitive function in normal older adults and adults aging with HIV; however, clearly more rigorous clinical trials with large sample sizes are needed. Yet, prescribing such cognitive training has not reached the level of FDA approval; however, at a practical level, it is a strategy that patients may try on their own as many of these programs are commercially available and relatively inexpensive.

In informing patients of such cognitive training programs, several caveats are recommended. First, not all cognitive training programs have the same therapeutic effects. Caution is needed in determining which programs are scientifically validated.

The companies that produce these cognitive training programs occasionally provide a list of scientific publications. Sometimes these articles are misleading and are not specifically written about the cognitive training product being promoted on their website. Other times, when scientific articles are provided on the cognitive training product, as a lay person, it may be difficult to evaluate the scientific merit of these articles.

Second, even if the cognitive training program is therapeutically effective, studies suggest that there may be U-shaped dose-response effects (Lampit et al. 2014); in other words, after engaging in so many hours of cognitive training exercises in a given time period (e.g., per week), additional training may not yield additional cognitive benefit. Yet, work reported in the cognitive aging literature (e.g., the ACTIVE Study) does indeed show that booster sessions may yield additional cognitive benefits. Thus, it is important to determine the optimal dose to reap the most cognitive benefit while avoiding participant burnout and fatigue. Likewise, such fatigue may reduce adherence/compliance to this cognitive intervention; in turn, this results in reduced dosage and therapeutic benefit.

Third, given the costs of time, effort, and money, patients must be selective as to which cognitive domain to target. While this article focuses on and favors SOP training, it is only one among many cognitive training programs available which target different cognitive domains. Although SOP training clearly has many other touted therapeutic benefits (i.e., safer driving, protection from depression, possible protection from dementia), there are many other cognitive domains that may be targeted for intervention. Patients may select training in the cognitive domain that they feel they are having the most difficulty. Often, trouble with memory is a common complaint; however, memory is more noticeable than deficits in other areas such as verbal fluency, SOP, or executive functioning. In fact, studies suggest that many adults with HIV are not accurate in rating their cognitive functioning and either overestimate or underestimate their cognitive abilities (Vance et al. 2008). This cognitive phenomenon reflects deficits in their metacognitive abilities, the ability to think about thinking; in fact, declines in metacognitive abilities may require intervention as well. Thus, it is necessary to have a neurocognitive evaluation to determine what cognitive deficits are present in order to target them for treatment; a similar approach is being utilized in the TOPS discussed earlier.

5 Research Implications

As the cognitive training science continues to develop, three areas are of particular interest. First, cognitive training may not only improve function in a particular cognitive domain; the use of cognitive training programs over time may protect and maintain cognitive reserve. In that regard, incorporating such cognitive training programs across multiple cognitive domains into one's daily life may be neuroprotective as adults age with HIV. Examining whether such cognitive training over time supports cognitive reserve would require following large groups of older adults with HIV. Unfortunately, a randomized controlled longitudinal study of this nature would require significant resources and logistical support to examine this adequately.

Second, as cognitive reserve is depleted, it is uncertain how much cognitive reserve one must retain in order for someone to benefit from the cognitive training. For example, Rojas et al. (2013) examined the therapeutic efficacy in which cognitive training, teaching cognitive strategies, and using external memory aids can improve cognitive and everyday functioning in normal older adults with mild cognitive impairment (MCI), a preclinical stage of dementia. Yet, those with more severe MCI may not have enough cognitive resources to benefit from the cognitive training, although, in this study by Rojas and colleagues, cognitive benefits were derived from the intervention. The same question can be applied in adults with HAND. In the ThinkFast study, as adults with various degrees of HAND, from borderline HAND to HAD, receive different dosages of SOP training, researchers will be able to examine this question in more detail (Vance et al. 2017a, b, c).

Finally, cognitive training interventions may produce greater therapeutic effects when combined with other cognitive interventions as attempted with tDCS and SOP training (Fazeli et al. 2018). For example, compared to a control group, Ngandu et al. (2015) combined cognitive training with diet, exercise, and vascular risk monitoring to find that this multimodal approach resulted in improved cognitive functioning in older adults at risk of cognitive decline. Clearly, cognitive training may be combined with neurocognitive medications (i.e., psychostimulants), healthy lifestyle behaviors (i.e., physical activity), and other types of cognitive stimulation (i.e., theater training) to bolster the cognitive training effects versus cognitive training alone.

6 Conclusion

As functional and structural declines have been observed in the brain of adults with HIV as they age (Goodkin et al. 2017), it is important to note that cognitive training studies have been shown to change and increase certain brain morphology in normal older adults. Clearly, cognitive training is a logical step in addressing some of these HIV neurocognitive complications, especially as this population ages. Fortunately, much of the foundation of this work has already been laid by research found in the cognitive aging literature (Edwards et al. 2018; Lampit et al. 2014).

Acknowledgments This study was funded by an NIH/NINR R21 award (1R21NR016632-01; ClinicalTrials.gov (NCT03122288); Vance, Principal Investigator) titled "Individualized-Targeted Cognitive Training in Older Adults with HAND," by an NIH/NIMH R01 award (1R01MH106366-01A1; ClinicalTrials.gov (NCT02758093); Vance, Principal Investigator) titled "An RCT of Speed of Processing Training in Middle-aged and Older Adults with HIV"), and by a NIH/NIA P30 award (Edward R. Roybal Center for Translational Research in Aging and Mobility; P30 AG022838).

Disclosures David E. Vance was a consultant for Posit Science, Inc.

All the other authors report no real or perceived vested interest that relates to this article that could be construed as a conflict of interest. David Vance was a paid consultant of Posit Science, Inc. in 2014.

References

Antinori A, Arendt G, Becker JT et al (2007) Updated research nosology for HIV-associated neurocognitive disorder. Neurology 69(18):1879–1799. https://doi.org/10.1212/01.WNL.0000287431.88658.8b

Ball KK, Vance DE (2007) Everyday life applications and rehabilitation of processing speed deficits: aging as a model for clinical populations. In: DeLuca J, Kalmar J (eds) Information processing speed in clinical populations. Taylor & Francis, New York, pp 243–263

Blackstone K, Moore DJ, Franklin DR et al (2012) Defining neurocognitive impairment in HIV: deficit scores versus clinical ratings. Clin Neuropsychol 26(6):894–908. https://doi.org/10.1080/13854046.2012.694479

Cody SL, Vance DE (2016) The neurobiology of HIV and its impact on cognitive reserve: a review of cognitive interventions for an aging population. Neurobiol Dis 92(part B):144–156. https://doi.org/10.1016/j.nbd.2016.01.011

Cody SL, Fazeli PL, Vance DE (2015) Feasibility of a home-based speed of processing training program in middle-aged and older adults with HIV. J Neurosci Nurs 47(4):247–254. https://doi.org/10.1097/jnn.0000000000000147

Edwards JD, Xu H, Clark DO et al (2017) Speed of processing training results in lower risk of dementia. Alzheimers Dement 3:603–611. https://doi.org/10.1016/j.trci.2017.09.002

Edwards JD, Fausto BA, Tetlow AM et al (2018) Systematic review and meta-analyses of useful field of view cognitive training. Neurosci Biobehav Rev 84:72–91. https://doi.org/10.1016/j.neubiorev.2017.11.004

Fazeli PL, Woods AJ, Pope CN et al (2018) The effect of transcranial direct current stimulation combined with cognitive training on cognitive functioning in older adults with HIV: a pilot study. Appl Neuropsychol Adult., epub ahead of print. https://doi.org/10.1080/23279095.2017.1357037

Fellows RP, Byrd DA, Morgello S (2014) Effects of information processing speed on learning, memory, and executive functioning in people living with HIV/AIDS. J Clin Exp Neuropsychol 36(8):806–817. https://doi.org/10.1080/13803395.2014.943696

Fristoe NM, Salthouse TA, Woodard JL (1997) Examination of age-related deficits on the Wisconsin Card Sorting Test. Neuropsychology 11(3):428–436

Goodkin K, Miller EN, Cox C et al (2017) Effect of ageing on neurocognitive function by stage of HIV infection: evidence from the Multicenter AIDS Cohort Study. Lancet HIV 4:e411–e422. https://doi.org/10.1016/S2352-3018(17)30098-X

Hossain S, Fazeli PL, Tende F et al (2017) The potential of computerized cognitive training on HIV-associated neurocognitive disorder: a case comparison study. J Assoc Nurses AIDS Care 28(6):971–976. https://doi.org/10.1016/j.jana.2017.06.011

Kinai E, Komatsu K, Sakamoto M et al (2017) Association of age and time of disease with HIV-associated neurocognitive disorders: a Japanese nationwide multicenter study. J Neurovirol 23(6):864–874. https://doi.org/10.1007/s13365-017-0580-6

Lampit A, Hallock H, Valenzuela M (2014) Computerized cognitive training in cognitively healthy older adults: a systematic review and meta-analysis of effect modifiers. PLoS Med 11(11): e1001756. https://doi.org/10.1371/journal.pmed.1001756

Meneses K, Benz R, Ball JR et al (2018) Speed of processing training in middle-aged and older breast cancer surivіors (SOAR): results of a randomized controlled pilot. Breast Cancer Res Treat 168(1):259–267. https://doi.org/10.1007/s10549-017-4564-2

Ngandu T, Lehtisalo J, Solomon A et al (2015) A 2 year multidomain intervention of diet, exercise, cognitive training, and vascular risk monitoring versus control to prevent cognitive decline in at-risk elderly people (FINGER): a randomize control trial. Lancet 385(9984):2255–2263. https://doi.org/10.1016/S0140-6736(15)60461-5

Rebok GW, Ball K, Guey LT et al (2014) Ten-year effects of advanced cognitive training for independent and vital elderly cognitive training trial on cognition and everyday functioning in older adults. J Am Geriatr Soc 62:16–24. https://doi.org/10.1111/jps.12607

Rojas GJ, Villar V, Iturry M et al (2013) Efficacy of a cognitive intervention program in patients with mild cognitive impairment. Int Psychogeriatr 25(5):825–831. https://doi.org/10.1017/S1041610213000045

Ross LA, Edwards JD, O'Connor ML et al (2016) The transfer of cognitive speed of processing training to older adults' driving mobility across 5 years. J Gerontol B Psychol Sci Soc Sci 71 (1):87–97. https://doi.org/10.1093/geronb/gbv022

Salthouse TA (1996) Constraints on theories of cognitive aging. Psychon Bull Rev 3:287–299. https://doi.org/10.3758/BF03210753

United States Senate Special Committee on Aging (2013) Hearing: Older Americans: the changing face of HIV/AIDS in America. In: One Hundred Thirteenth Congress, First Session, Washington

Valcour V, Shikuma C, Shiramizu B et al (2004) Age, apolipoprotein E4, and the risk of HIV dementia: the Hawaii Aging with HIV Cohort. J Neuroimmunol 157(1–2):197–202. https://doi.org/10.1016/j.neuroim.2004.08.029

Vance DE, Farr K, Struzick TC (2008) Assessing the clinical value of cognitive appraisal in adults aging with HIV. J Gerontol Nurs 34(1):36–41. https://doi.org/10.3928/00989134-20080101-11

Vance DE, Fazeli PL, Ross LA et al (2012) Speed of processing training with middle-age and older adults with HIV: a pilot study. J Assoc Nurses AIDS Care 23(6):500–510. https://doi.org/10.1016/j.jana.2012.01.005

Vance DE, Fazeli PL, Moneyham L et al (2013) Assessing and treating forgetfulness and cognitive problems in adults with HIV. J Assoc Nurses AIDS Care 24(1S):S40–S60. https://doi.org/10.1016/j.jana.2012.03.006

Vance DE, Cody SL, Moneyham L (2017a) Remediating HIV-associated neurocognitive disorders via cognitive training: a perspective on neurocognitive aging. Interdiscip Top Gerontol Geriatr 42:173–186. https://doi.org/10.1159/000448562

Vance D, Fazeli P, Shacka J et al (2017b) Testing a computerized cognitive training protocol in adults aging with HIV-associated neurocognitive disorder: an RCT in the Southern United States. J Med Internet Res 6(4):e68. https://doi.org/10.2196/resprot.6625

Vance DE, Frank JS, Bail J et al (2017c) Interventions for cognitive deficits in breast cancer survivors treated with chemotherapy. Cancer Nurs 40(1):e11–e27. https://doi.org/10.1097/NCC.0000000000000349

Vance DE, Fazeli PL, Azuero A et al (2018a) Can computerized cognitive training reverse the diagnosis of HIV-associated neurocognitive disorder? A research protocol. Res Nurs Health 41 (1):11–18. https://doi.org/10.1002/nur.21841

Vance DE, Jensen M, Tende F et al (2018b) Individualized-targeted computerized cognitive training to treat HIV-associated neurocognitive disorder: a case comparison study. J Assoc Nurses AIDS Care 29(4):604–611. https://doi.org/10.1016/j.jana.2018.04.005

Vance DE, Fazeli PL, Cheatwood J et al (2019) Computerized cognitive training for the neurocognitive complications of HIV infection: a systematic review. J Assoc Nurses AIDS Care 30(1):51–72

Wadley VG, Benz RL, Ball KK et al (2006) Development and evaluation of home-based speed of processing training for older adults. Arch Phys Med Rehabil 87:757–763. https://doi.org/10.1016/j.apmr.2006.02.027

Wang P, Liu HH, Zhu XT et al (2016) Action video game training for healthy adults: a meta-analytic study. Front Psychol 7:907. https://doi.org/10.3389/fpsyg.2016.00907

Weber E, Woods SP, Kellogg E et al (2012) Self-generation enhances verbal recall in individuals with HIV. J Int Neuropsychol Soc 18(1):128–133. https://doi.org/10.1017/S135561771100124X

Wickens CD, Hollands JG, Bandbury S et al (2013) A model of human information processing. In: Engineering psychology and human performance. Pearson, Upper Saddle River

Wolinsky FD, Vander Weg MW, Howren MB et al (2013) A randomized controlled trial of cognitive training using a visual speed of processing intervention in middle aged and older adults. PLoS One 8(5):e61624. https://doi.org/10.1371/journal.pone.0061624

Clinical Treatment Options and Randomized Clinical Trials for Neurocognitive Complications of HIV Infection: Combination Antiretroviral Therapy, Central Nervous System Penetration Effectiveness, and Adjuvants

Shih-Ping Lin, Andrea Calcagno, Scott L. Letendre, and Qing Ma

Contents

Abstract The etiology and pathogenesis of human immunodeficiency virus type-I (HIV)-associated neurocognitive disorders (HAND) remain undetermined and are likely the produce of multiple mechanisms. This can mainly include neuronal injury from HIV, inflammatory processes, and mental health issues. As a result, a variety of

S.-P. Lin
Department of Pharmacy Practice, University at Buffalo, Buffalo, NY, USA

Taichung Veterans General Hospital, Taichung, Taiwan

A. Calcagno
Unit of Infectious Diseases, Department of Medical Sciences, University of Torino, Torino, Italy

S. L. Letendre
Department of Medicine and Psychiatry, HIV Neurobehavioral Research Center, University of California San Diego, San Diego, CA, USA

Q. Ma (✉)
Department of Pharmacy Practice, University at Buffalo, Buffalo, NY, USA
e-mail: qingma@buffalo.edu

Curr Topics Behav Neurosci (2021) 50: 517–546
https://doi.org/10.1007/7854_2020_186
Published Online: 19 February 2021

treatment options have been tested including NeuroHIV-targeted regimens based on the central nervous system (CNS) penetration effectiveness (CPE) of antiretroviral therapy (ART) and adjuvant therapies for HAND. NeuroHIV-targeted ART regimens have produced consistent and statistically significant HIV suppression in the CNS, but this is not the case for cognitive and functional domains. Most adjuvant therapies such as minocycline, memantine, and selegiline have negligible benefit in the improvement of cognitive function of people living with HIV (PLWH) with mild to moderate neurocognitive impairment. Newer experimental treatments have been proposed to target cognitive and functional symptoms of HAND as well as potential underlying pathogenesis. This review aims to provide an analytical overview of the clinical treatment options and clinical trials for HAND by focusing on NeuroHIV-targeted ART regimen development, CPE, and adjuvant therapies.

Keywords Clinical trials · CPE · HIV associated neurocognitive impairment · Treatment options

1 NeuroHIV-Targeted Combination Antiretroviral Therapy

Although combination antiretroviral therapy (ART) has markedly reduced morbidity and mortality among people living with HIV (PLWH) (Cysique et al. 2004; Deutsch et al. 2001), the prevalence of HIV-associated neurocognitive disorders (HAND) remains high, with current estimates ranging from 35% to greater than 50% depending on disease stability and viral suppression status (Brouillette et al. 2015; Calcagno et al. 2018). Several etiological factors may be responsible for such a high prevalence of HAND.

Besides well-studied contributing factors such as lower nadir CD4+ T-cell count, substance use disorders, cardiovascular diseases (CVD), mood disorders, education, and hepatitis C virus (HCV) coinfection, the importance of underlying mechanisms has been increasingly recognized including patient-related (e.g., age and inflammation), virus-related (e.g., persistent viral replication in the central nervous system (CNS) and microglia/perivascular macrophage infection), and treatment-related (e.g., suboptimal ART distribution in the CNS and neurotoxicity) (Ellis and Letendre 2016; Lanman et al. 2019; Letendre 2011). Since people with HAND have a lower quality of life, are more frequently unemployed, and have poor medication adherence and impairment in other activities of daily living in addition to a higher risk of death (Ellis et al. 1997; Sevigny et al. 2007), development of ART with higher CNS penetration effectiveness (CPE or referred more generally as NeuroHIV-targeted ART in this chapter) and optimization of ART for neurocognitive outcomes remain critically needed for the long-term HIV management.

Despite the high prevalence of HAND and the importance of healthy cognition to the quality of life in PLWH, consensus treatment guidelines for NeuroHIV-targeted ART have yet to be formulated, as they require a high level of clinical evidence. In particular, controversy remains about whether antiretroviral penetration into the CNS is clinically important for treating HAND. One of the major hurdles for the development and optimization of NeuroHIV-targeted ART is the accessibility of antiretrovirals across the blood–brain barrier (BBB). The BBB can limit the distribution of ART in the CNS, creating a distinct pharmacologic compartment (Letendre 2011). Differences between antiretrovirals in crossing the BBB and brain concentrations may partially explain inter-individual variances in susceptibility to HAND among treated individuals (Decloedt et al. 2015).

In addition, the CNS can serve as a virologic compartment (Marban et al. 2016), providing potential targets for interventions, in particular, most adjuvant therapies for HAND (Fig. 1). Viral proteins can cross the BBB via transcytosis or paracellularly followed by infections of astrocytes and microglia. HIV-infected T cells and monocytes can also migrate into the brain to induce chronic inflammation, in which inflammatory cytokines further activate astrocytes and microglia. The astrocytes/microglia activation leads to an increased permeability of BBB and release of glutamine and other neurotoxic cytokines, eventually causing neuronal injuries and contributing to HAND (Letendre 2011; Bougea et al. 2019; Hong and Banks 2015). Despite successful peripheral suppression, HIV remains detectable in the cerebrospinal fluid (CSF) in ~5 to 15% individuals receiving ART (Perez Valero et al. 2014), which has been referred as "CSF viral escape" (Eden et al. 2010), suggesting that ART with a better penetration might be associated with a better control of viral replication in the CNS.

One of the major practical attempts to target NeuroHIV is the development and validation of the CPE score system for ART from the CHARTER (Ellis and Letendre 2016; Letendre 2011). The CPE was first established in 2008 (CPE 2008), and antiretroviral drugs were assigned into three ranks from 0 (low), 0.5 (intermediate), to 1 (high) mainly based on their penetration profiles (Letendre et al. 2008). In 2010, CPE was revised to 1, 2, 3, and 4 with larger numbers reflecting better penetration and CNS effectiveness (CPE 2010) (Letendre et al. 2010). The CPE ranks antiretroviral agents according to their physicochemical properties (e.g., molecular weight, protein binding, and octanol-water partition coefficient), CSF concentrations, and efficacy based on CSF virologic suppression (Table 1) (Letendre 2011; Letendre et al. 2008). The score of a combination ART regimen is calculated by summing the values of individual agents. The objective of this review is to provide an analytical overview of relevant data and accumulating evidence on the clinical trials of NeuroHIV-targeted ART regimens with a particular reference to CPE and adjuvant therapies tested as potential options to prevent and treat HAND.

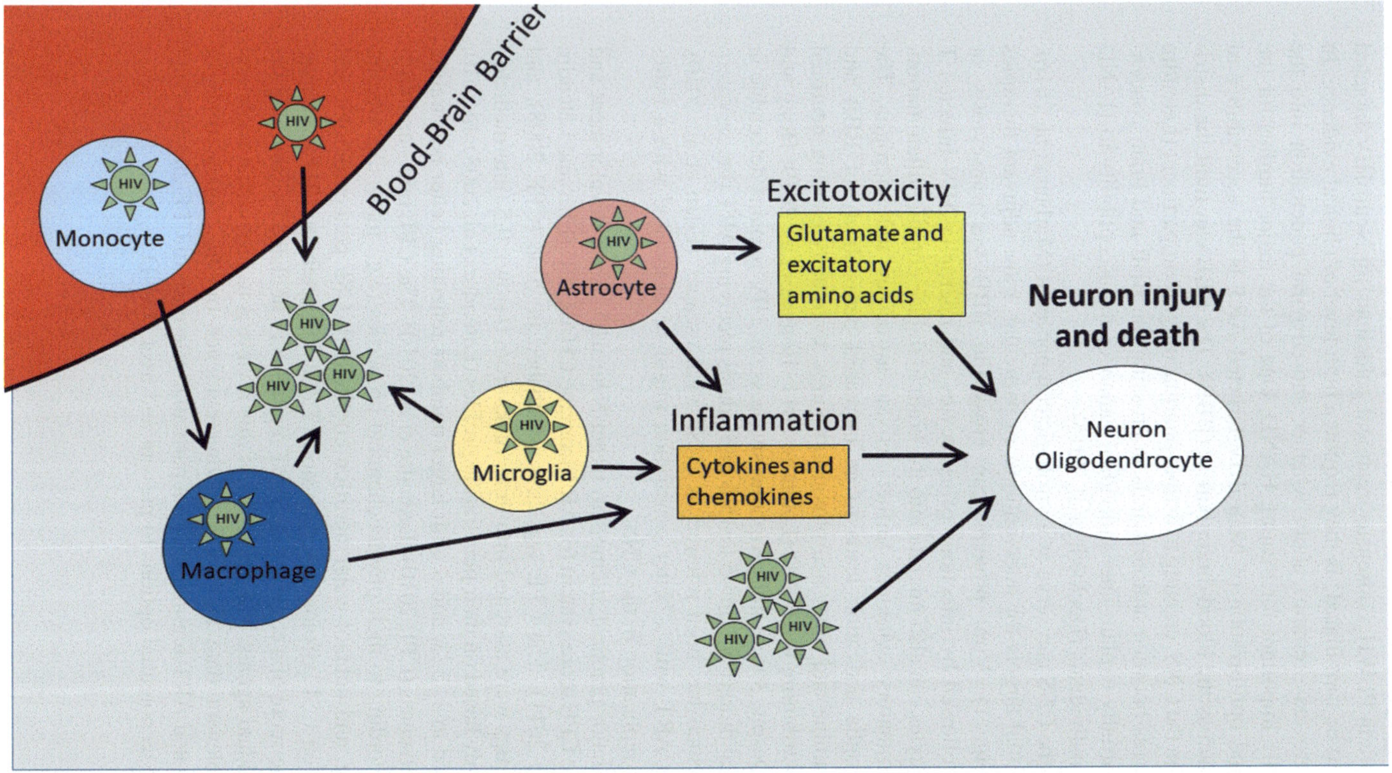

Fig. 1 The effects of HIV in the central nervous system. *HIV entry*: HIV enters the CNS through the blood–brain barrier (BBB) directly or within infected monocytes. This process involves HIV-induced monocyte and endothelial activation, astrocyte dysfunction, and structural impairment of the BBB. *Viral replication*: HIV infects and activates macrophages and microglia followed by viral replication. *Inflammation*: HIV-induced activation of microglia, macrophages, and astrocytes leads to the release of proinflammatory cytokines and chemokines, which cause further influx of immune cells and mediate neuronal injury. *Excitotoxicity*: HIV induces release of glutamate and other excitatory amino acids from neurons and astrocytes, which together with HIV and HIV-induced chemokines, overstimulate N-methyl-D-aspartate (NMDA) receptors, leading to excitotoxicity. *Neuronal injury*. Insults from HIV, excitotoxicity, and inflammation lead to axonal injury and neuronal apoptosis. HIV also affects neural progenitor cells, impeding repair and growth

Table 1 Ranking of commonly prescribed antiretroviral drugs as for CNS penetration effectiveness (CPE) scores and in vitro neurotoxicity

	Abbreviation	Approval year	CPE	In vitro neurotoxicity
NRTI				
Zidovudine	AZT/ZDV	1987	4	+/–
Didanosine	ddI	1991	2	+
Stavudine	d4T	1994	2	+/–
Lamivudine	3TC	1995	2	+/–
Abacavir	ABC	1998	3	++
Tenofovir disoproxil fumarate	TDF	2001	1	+/–
Emtricitabine	FTC	2003	3	+/–
Tenofovir alafenamide fumarate	TAF	2015	1	
NNRTI				
Nevirapine	NVP	1996	4	+
Delavirdine	DLV	1997	3	
Efavirenz	EFV	1998	3	++
Etravirine	ETR	2008	2	+
Rilpivirine	RPV	2011	–	+
Doravirine	DOR	2018	–	
Protease inhibitors				
Saquinavir mesylate	SQV	1995	1	+
Indinavir	IDV	1996	3	+
Nelfinavir mesylate	NFV	1997	1	
Lopinavir	LPV	2000	3	+
Atazanavir sulfate	ATV	2003	2	+
Fosamprenavir calcium	FOS	2003	2	+
Tipranavir	TPV	2005	1	
Darunavir	DRV	2006	3	–
Fusion inhibitors				
Enfuvirtide	T-20	2003	1	
CCR5 co-receptor antagonists				
Maraviroc	MVC	2007	3	–
INSTIs				
Raltegravir	RAL	2007	3	+/–
Dolutegravir	DTG	2013	–	
Elvitegravir	EVG	2014	–	+/–
Bictegravir	BIC	2018	–	
Post-attachment inhibitors				
Ibalizumab	IBA	2018	–	
Pharmacokinetic enhancers				
Ritonavir	RTV	1996	1	+/–
Cobicistat	COBI	2014	–	

NRTIs nucleoside reverse transcriptase inhibitors, NNRTIs non-nucleoside reverse transcriptase inhibitors, INSTIs integrase strand transfer inhibitors, CPE 2010

2 Association Between CPE and HIV Suppression in the CSF

The 12 clinical studies with detailed data on ART regimens and viral load in the plasma and CSF were selected to determine if higher CPE was associated with a better HIV control in the CSF (Table 2). The majority of the clinical studies ($n = 9$, 75%) concluded with positive findings. For instance, among 401 participants from the CHARTER cohort followed up for 34 months, ART regimens with lower CPE were significantly associated with detectable HIV RNA levels in the CSF over time (Livelli et al. 2019). An early study in 142 patients, who underwent lumbar punctures due to neurological complications, demonstrated that low CPE (<2, CPE 2008) was linked with detectable HIV RNA in the CSF despite aviremia (Rawson et al. 2012). These clinical studies included prospective, longitudinal ones with a relatively small sample size and large retrospective and cross-sectional analyses that consistently demonstrated an association between CPE and HIV suppression in the CSF.

Among the three studies showing no association, two were cross-sectional with a relatively small sample size. Eden et al. reported similar mean CPE between 7 CSF viremia and 62 CSF aviremia (7.3 vs. 7.4, CPE 2010), all of whom had undetectable concentrations of HIV RNA in plasma, and concluded that CPE is not a predictor for CSF viremia (Eden et al. 2010). The other study retrospectively analyzed 155 patients from the Frankfurt HIV Cohort, among whom 131 received ART with high CPE (mean 7.3, CPE 2010) versus 24 on boosted dual protease inhibitors (bdPI) with low CPE (mean 4.2, CPE 2010). Although the proportion of undetectable CSF HIV virus was lower in the bdPI group, the median CSF viral load was significantly higher (600 vs. 50 copies/mL, $p = 0.027$), suggesting viral replication in the CNS over time due to the low CPE of dbPI regimens; however, no significant correlation was noted between CPE and quantitative HIV-1 RNA in the CSF (Donath et al. 2016). The only well-designed randomized clinical trial (RCT) of NeuroHIV-targeted ART in 49 patients with HAND was prematurely interrupted due to slow accrual and imbalance among study arms. After 16 weeks follow-up, the use of a NeuroHIV-targeted ART regimens ($n = 26$, mean CPE 2.5, CPE 2008) resulted similar plasma and CSF HIV viral suppression in comparison to that with the non-targeted ones ($n = 23$, mean CPE 1.0, CPE 2008) (Ellis et al. 2014). Importantly, significant limitations presented in these studies, such as relatively small differences in CPE between groups, the cross-sectional nature, complex NeuroHIV-targeted regimens, and the small sample size in the randomized clinical trial, which made interpretations of study findings challenging.

In summary, the majority of the observational studies demonstrated that ART regimens with high CPE were associated with better HIV suppression in the CSF. Therefore, optimization of ART regimens based on CPE remains a practical approach for HIV management in the CNS.

Table 2 Association between CPE and HIV suppression in the CNS

Year	*n*	Design	Patient selection	Age (years)	CD4 cell count (c/mL)	HIV RNA (log c/mL)		CPE		CPE and CSF HIV association	Ref
						Plasma	CSF	Baseline	Cut-off		
2002	50	Longitudinal	LP due to neurologic signs or research purpose	37	59	5.25	3.16	Nil	Nil	ART with higher CSF penetration correlated with a more profound CSF HIV-1 viral load reduction	De Luca et al. (2002)
2002	75 (cross-sectional), 29 (longitudinal)	Cross-sectional, longitudinal	LP due to neurologic signs (37% ART naïve)	39	131	5	3.5	Number of CSF penetrating drug: 2	Nil	A significant difference in CSF HIV-1-RNA reduction was observed according to the use of three or more drugs penetrating the blood–brain barrier	Antinori et al. (2002)
2008	467	Cross-sectional	Receiving ART and having HIV VL measured in both plasma and CSF. LP for research purpose	44	406	1.7	1.7 (CSF detectable group = 2.5)	1.5[a]	1.5[a]	Lower CPE ranks was associated with detectable CSF VL	Letendre et al. (2008)
2009	79	Longitudinal	Initial ART or changing to a new ART. LP for research purpose	39	111	4.89	3.33	2.0[a]	2[a]	CPE score ≥ 2 was associated with 4.10-folds CSF HIV suppression	Marra et al. (2009)

(continued)

Table 2 (continued)

Year	*n*	Design	Patient selection	Age (years)	CD4 cell count (c/mL)	HIV RNA (log c/mL)		CPE		CPE and CSF HIV association	Ref
						Plasma	CSF	Baseline	Cut-off		
2010	69	Cross-sectional	Neurological asymptomatic, ART >6 months with plasma HIV VL < 50 c/mL. LP for research purpose	CSF viremia: 46, control: 45	CSF viremia: 620, control: 525	<50 c/mL	CSF viremia: 121 c/mL, control: <50 c/mL	CSF viremia: 7.3, control: 7.4	Nil	CPE rank was not a predictor of detectable CSF virus	Eden et al. (2010)
2012	142	Retrospective, cross-sectional	LP due to clinical CNS events	45	395	48% detectable	54% detectable	1.5[a]	2[a]	Even with plasma HIV RNA <50 copies/mL, CPE <2 was significantly associated with detectable CSF HIV RNA	Rawson et al. (2012)
2013	83 (CSF undetectable), 4 (CSF detectable)	Longitudinal	ART for at least 6 months and plasma HIV <50 copies/mL. LP for research purpose	44	520 (CSF undetectable); 369 (CSF detectable)	<50 c/mL	Undetectable vs. detectable	CSF undetectable: 2.3[a], 8; CSF detectable:1.0[a],6	2[a], 7	The CPE score was significantly lower in patients with detectable CSF HIV RNA	Cusini et al. (2013)
2014	49	RCT	Initial or change ART. LP for research purpose	CNS-T = 44.9, non-CNS-T = 43.6	CNS-T = 214, non-CNS-T = 306	CNS-T = 4.2, non-CNS-T = 3.5	CNS-T = 3.1, non-CNS-T = 3.1	CNS-T = 2.5[a], non-CNS-T = 1[a]	Nil	CSF viral suppression rate was similar for the 2 arms	Ellis et al. (2014)

2016	155	Cross-sectional, retrospective	Boosted dual protease inhibitor (bdPI) regimen, or any 2NRTI-containing ART. LP for clinical purpose	46.85	174.5	bdPI 115, ART 173	bdPI 600, ART 50	bdPI 4.29, ART 7.53	Nil	No significant correlation between quantitative HIV-1 RNA in CSF and CPE score	Donath et al. (2016)
2017	220	Retrospective, cross-sectional ($n = 220$), longitudinal ($n = 55$)	HIV in plasma and CSF < 50 copies/mL. LP for research purpose	44	503	65.2% plasma RNA > 1 c/mL (low-level viremia)	CSF VL > 1: $n = 93$, CSF VL < 1: $n = 127$	CSF VL > 1:6.8, CSF VL < 1:7.2	7	Lower CPE values were associated with CSF HIV-1 RNA loads of ≥1 copy/mL	Anderson et al. (2017)
2018	71	Retrospective	Under stable ART, plasma HIV VL < 1000 copies/mL, neurological symptom. LP for clinical reason	38	361	71.8% undetectable	4250 c/mL in CSF/plasma HIV discordance	91.5%, ≥6	6	CPE values <6 were more likely to develop CSF/plasma HIV discordance	Dravid et al. (2018)
2019	401	Longitudinal	Under stable ART, HIV VL measurable in plasma and CSF. LP for clinical purpose	44	446	60%, <50 c/mL	87%, <50 c/mL	7.5	Nil	Detectable HIV RNA concentrations in CSF were associated with decreased CPE value	Livelli et al. (2019)

[a]CPE 2008 vision (rank from 0, 0.5, to 1)

CPE central nervous system penetration effectiveness, *LP* lumbar puncture, *ART* antiretroviral therapy, *CSF* cerebrospinal fluid, *CNS* central nervous system, *VL* viral load, *RCT* randomized clinical trial

3 Association Between CPE and Neurocognitive Improvement

Controversies have mostly centered on the associations between CPE and neurocognitive improvement. Although mounting evidence indicated that higher CSF HIV virus load predicted the progression of neurocognitive impairment, it remains largely unclear if the use of ART regimens with high CPE to achieve better HIV control in the CNS would lead to subsequent improved neurocognitive performance (Ellis et al. 2002). Among 18 studies assessing the association between CPE and neuropsychological performance (Table 3), 8 (44%) concluded higher CPE associated with neurocognitive improvement, 3 (17%) showed an inverse relationship, and the rest 7 (39%) showed no interaction. Except for two RCTs, 16 studies (89%) were observational without control, including 8 prospective cohort and 8 cross-sectional analyses. In the AIDS Clinical Trials Group (ACTG) A5175 study, 860 treatment naïve participants were randomized to three groups receiving lamivudine-zidovudine-efavirenz (CPE $= 9$, CPE 2010, $n = 289$), atazanavir-emtricitabine-didanosine (CPE $= 7$, CPE 2010, $n = 293$), and emtricitabine-tenofovir-disoproxil fumarate-efavirenz (CPE $= 7$, CPE 2010, $n = 278$). At week 192 follow-up, neurocognitive performance significantly improved, but no differences were observed between treatment regimens (Robertson et al. 2012). The other RCT with only 49 participants did not show a significant improvement of neurocognitive performance at week 16 in CNS-targeted ART group with high CPE (Ellis et al. 2014).

The largest prospective study was conducted in ACTG Longitudinal Linked Randomized Trials (ALLRT) cohort involving a total of 2636 aviremic participants, who received neurocognitive tests every 48 weeks. After a median 4.7 years of follow-up, the ALLRT study demonstrated that higher CPE was associated with better neurocognitive functioning (Smurzynski et al. 2011). Another French prospective study included 96 participants with a median viral load of 10,760 c/mL at the baseline, 71 (74%) without HAND and 25 (26%) with HAND at baseline, with a median 22 months follow-up. The neurocognitive tests revealed that 6 (6%) improved, 31 (32%) worsen, and 59 (61%) stable. Lower CPE at baseline (6.9 vs. 8.1, CPE 2010) and at the end of follow-up (7.2 vs. 7.8, CPE 2010) were independently associated with clinical neurocognition deterioration (Vassallo et al. 2014). In contrast to the findings from most prospective studies (5 positive, 1 no interaction), Marra et al. reported that high CPE (≥ 2, CPE 2008) was associated with poor neurocognitive performance, despite better HIV viral suppression in the CSF among 79 participants with advanced HIV disease beginning or changing a new ART, who received neurocognitive tests at baseline, week 24, and week 52. One possible and appealing explanation for such a discrepancy was that ART with high CPE exhibited more neurotoxicity in advanced HIV disease (Marra et al. 2009). In fact, neurotoxicity of ART, as summarized in Table 1, has become an increasing concern especially for HAND and elderly PLWH. A theoretical model has been proposed that predicts a cognitive deterioration after an initial improvement due to

Table 3 Association of CPE and neurocognitive performance

Year	*n*	Design	Patient selection	Follow-up	Age (years)	CD4 cell count (c/mL)	Plasma HIV RNA (log c/mL)	CPE		Neurocognitive performance (NP) test	CPE and neurocognition association	Ref
								Baseline	Cut-off			
2009	37	Prospective cohort	Mild to moderate NP impairment, untreated or planned initiation of a new ART	Baseline, w12, w24, w36, w48	40	196	4.90	1.4[a]	2[a]	GDS (based on 6 NP measures)	CPE (≥2) independent factor for NP improvement	Cysique et al. (2009)
2009	185	Prospective cohort	With or suspected HAND, CD4 + <200, initial or change ART	Baseline, 20 months, 39 months	39	293	4.14	1.65[a]	Nil	Composite NP test z scores (6 NP measures)	Higher CPE scores correlated with greater improvements in NP test	Tozzi et al. (2009)
2009	79	Prospective cohort	CD4+ <200, plasma and HIV >2000 copies/mL, or HIV >50,000 c/mL, initial or change ART	Baseline, w24, w52	39	111	4.89	2.0[a]	2[a]	Composite z score for the short battery (NPZ4) and the longer battery (NPZ8)	Higher CPE associated with poorer neurocognitive performance	Marra et al. (2009)
2011	101	Cross-sectional	Stable ART >3 months, plasma HIV <50 c/mL, no neurological symptoms	NP test once	53	525	<50 c/mL	1.5[a], 7.0	Nil	Computerized cognitive test (CogState)	CPE not associated with NP impairment	Garvey et al. (2011)
2011	2,636	Prospective cohort	ART ≥6 weeks, plasma HIV <50 c/mL	Baseline, every 48 weeks	40	244	<50 c/mL	2.0[a]	Nil	NPZ3	Higher CPE associated with better neurocognitive functioning	Smurzynski et al. (2011)

(continued)

Table 3 (continued)

Year	*n*	Design	Patient selection	Follow-up	Age (years)	CD4 cell count (c/mL)	Plasma HIV RNA (log c/mL)	CPE		Neurocognitive performance (NP) test	CPE and neurocognition association	Ref
								Baseline	Cut-off			
2012	860	RCT	CD4+ <300, treatment naïve	Baseline, every 24 weeks till w192	34	173	5	7.0 vs. 9.0	Nil	Grooved pegboard, timed gait, semantic verbal fluency, and finger tapping	No differences in neurological and neuropsychological functioning between regimens with different CPE	Robertson et al. (2012)
2013	111	Prospective cohort	Age 18–35 years, treatment naïve	Baseline, 1 year later	18–35	<350	Nil	>7 ($n = 38$), ≤7 ($n = 31$)	7	GDS	No significant difference in cognitive outcomes between higher and lower CPE regimens	Cross et al. (2013)
2013	101	Cross-sectional cohort	ART, plasma HIV <50 c/mL	NP test once	47	620	<50 c/mL	1.5[a], 6	1.5[a], 6	18 NP measures	CPE ≥ 6 showed a decreased risk of cognitive impairment	Ciccarelli et al. (2013)
2013	54	Cross-sectional cohort	ART for 4–7 years with a stable CPE	NP test once	42	460	Nil	1[a] ($n = 14$), 1.5–2[a] ($n = 18$), 2.5[a] ($n = 25$)	1[a], 1.5–2[a], 2.5[a]	Short neuropsychological battery (4 NP measures)	High CPE scores associated with poorer NP performance	Kahouadji et al. (2013)
2014	49	RCT	Initial ART or change ART regimen	Baseline, week16	CNS-T 45, non-CNS-T 44	CNS-T 214, non-CNS-T 306	CNS-T 4.2, non-CNS-T 3.5	CNS-T 2.5[a], non-CNS-T 1[a]	Nil	GDS	No evidence of neurocognitive benefit for a CNS-targeted strategy	Ellis et al. (2014)
2014	96	Prospective cohort	>18 years, but no limit set for CD4+ count or HIV VL	Baseline, 2 years later	48	551	10,760 c/mL	7.8 (without HAND 7.6, with HAND 8.1)	Baseline: 6.9 vs. 8.1; at f/u 7.2 vs. 7.8	8 NP measures	Clinical deterioration associated with lower CPE at baseline and at the end of follow-up	Vassallo et al. (2014)
2014	229	Cross-sectional	Stable ART >12 months, plasma HIV <50 c/mL > 6 months	NP test once	45	325	<50 c/mL	6.93	7	Global NPZ-4	CPE <7 was associated with a trend to worse neurocognitive performance	Casado et al. (2014)

2014	660	Retrospective, cross-sectional	ART-treated	NP test once	49	586	84%, <40 c/mL	6.6	Nil	14 NP measures	Higher CPE values associated with poor NP performance	Libertone et al. (2014)
2015	64	Cross-sectional	Stable ART for >3 months	NP test once	38	227	1.3	Low CPE ($n = 29$); high CPE ($n = 35$)	7	NPZ-4	No significant differences between different CPE groups	Baker et al. (2015)
2016	417	Prospective, cross-sectional	ART for >90 days	NP test once	47	215	382 c/mL	7	Nil	GDS (6 NP measures)	Higher CPE values correlated with lower prevalence of neurocognitive impairment	Carvalhal et al. (2016)
2017	220	Retrospective, cross-sectional ($n = 220$), longitudinal cohort ($n = 55$)	ART, plasma and CSF HIV <50 c/mL	NP test once	44	503	65.2% plasma RNA > 1 c/mL (low level viremia)	7.1, (CSF VL > 1: 6.8, CSF VL < 1: 7.2)	7	Global neurocognitive performance	Worse neurocognitive performance not associated with CPE	Anderson et al. (2017)
2017	94	Prospective cohort	>18 years but no limit set for CD4 count or HIV VL	Baseline, 2 years later	46	552	69%, <200 c/mL	7.72 (HAND: 7.62, non-HAND: 8)	Nil	8 NP measures	Lower CPE at baseline independent risk factors for cognitive deterioration	Vassallo et al. (2017)
2019	909	Cross-sectional and retrospective	ART, plasma HIV <50 c/mL	NP test once	53	638	<50 c/mL	6.66	7	9 NP measures	No association between neurocognitive impairment and CPE	Santos et al. (2019)

[a]CPE 2008 vision (rank from 0, 0.5, to 1)
CPE central nervous system penetration effectiveness, *NP* neurocognitive performance, *ART* antiretroviral therapy, *GDS* global deficit score, *VL* viral load

the inhibition of viral replication, most likely dictated by long-term ART neurotoxicity (Underwood et al. 2015).

In addition to the conflicting results from these prospective studies, the findings from several large cross-sectional analyses are inconclusive. For instance, one Swiss cohort study involving 909 aviremic participants concluded that CPE was not associated with neurocognitive improvement (Santos et al. 2019). In another study from Italy, 660 participants demonstrated that higher CPE was associated with poor neurocognitive performance (Libertone et al. 2014). More recently, HIV-CAUSAL Collaboration (1998–2013) has concluded that initiation of ART with high CPE increases the risk of HIV dementia, but not of other NeuroHIV conditions after evaluating a total of 61,938 individuals followed for a median of 37 months (Caniglia et al. 2014). Unfortunately, this conclusion suffered significantly from several pitfalls, such as obsolete ART regimens evaluated, focusing on the most severe form of HAND (i.e., HIV-associated dementia, HAD), and a small gap in the CPE scores among different groups. These findings nevertheless provided some useful information, particularly from both the analytical and pharmacological perspectives.

In summary, the association between CPE and neurocognitive performance remains questionable, and many factors might contribute to such an inconclusiveness. First, most studies were observational, and none of them evaluated integrase strand transfer inhibitor (INSTI)-based ART regimens. Second, the inclusion criteria varied significantly (e.g., with and without HAND at baseline, treatment naïve and experienced, and plasma/CSF viremic and aviremic), which makes interpretation and comparison challenging. In addition, the neurocognitive test used were inconsistent across these studies, and there was no consistent CPE cut-off score used across the studies either. Despite all these limitations, the findings in general supported the concept that high CPE was associated with better HIV suppression in the CNS and possible benefit of neurocognitive improvement, but with the caution of potential neurotoxicity from ART due to high CNS concentrations. Thus, large RCT and prospective studies, especially those focusing on INSTI-based regimens, are warranted to further evaluate clinical utility of CPE.

4 Adjuvant Therapies

Because of the persistence of HAND in many individuals despite the use of ART, numerous therapeutic strategies and adjuvant therapies have been investigated. With the discovery of the inferred mechanisms through which HIV might cause HAND (Fig. 1), predominantly neuroprotective strategies have been evaluated (Table 4). A total of 22 trials on various adjuvant therapies are identified and assessed, mostly double-blind RCTs with small sample sizes. Early investigations of potent antioxidants including OPC-14117 and CPI-1189 failed to show a significant improvement in cognitively impaired patients with advanced HIV disease (The Dana Consortium on the Therapy of HIV Dementia and Related Cognitive Disorders 1997; Clifford

Table 4 Adjuvant therapies for HAND

Year	Design	Patient selection	Adjuvant	Dosage	Study arm	n	Follow-up neurocognitive performance change outcome	Main findings	Ref
1997	RCT	Cognitive impairment and under ART for 6 weeks	OPC-14117	OPC-14117:120 mg daily for the initial 6 weeks of the study and 240 mg daily for the remaining 6 weeks	OPC-14117 ($n = 15$), placebo ($n = 15$)	30	Baseline to w12	No benefit	The Dana Consortium on the Therapy of HIV Dementia and Related Cognitive Disorders (1997)
1998	RCT, double-blind	Cognitive impairment and under stable ART for 6 weeks	Deprenyl, thioctic acid	Deprenyl 2.5 mg three times a week oral; thioctic 600 mg twice daily oral	Placebo ($n = 9$), deprenyl ($n = 9$), thioctic acid ($n = 9$), both ($n = 9$)	36	Baseline to w10	1. Subjects receiving deprenyl performed significantly verbal memory improvement2. Thioctic acid: no benefit	Dana Consortium on the Therapy of HIV Dementia and Related Cognitive Disorders (1998)
1998	RCT, double-blind	Cognitive deficit (either no ART for 4 weeks or stable ART use for 12 weeks before study entry)	Peptide T	Peptide T, 2 mg, intranasally 3 times a day for 6 months	Peptide ($n = 66$), placebo ($n = 77$)	143	Baseline to month 6	Peptide T treatment was associated with overall cognitive improvement in patients with baseline global deficit scores (GDS) of at least 0.5	Heseltine et al. (1998)
1999	RCT, double-blind	Cognitive impairment and under ART for 6 weeks	Lexipafant	Lexipafant 250 mg, orally, twice daily	Lexipafant ($n = 16$), placebo ($n = 14$)	30	Baseline to w10	There were trends toward improvement in the Rey auditory verbal learning test and timed gait test	Schifitto et al. (1999)

(continued)

Table 4 (continued)

Year	Design	Patient selection	Adjuvant	Dosage	Study arm	n	Follow-up neurocognitive performance change outcome	Main findings	Ref
2000	RCT, double-blind	Cognitive impairment and under stable ART for 6 weeks	Selegiline transdermal system (STS)	STS 3.1 mg per 24 h	STS ($n = 9$), placebo ($n = 5$)	14	Baseline to w10	The selegiline group performed better on the Rey auditory verbal learning test (RAVLT) delayed recall and the grooved pegboard dominant hand test	Sacktor et al. (2000)
2002	RCT, double-blind	HIV-associated cognitive–motor disorder and under stable ART for 8 weeks	CPI-1189	CPI-1189 50 mg/day, CPI-1189 100 mg/day	CPI-1189 50 mg/day ($n = 21$), CPI-1189 100 mg/day ($n = 22$), placebo ($n = 21$)	64	Baseline to w6 and w10	No benefit	Clifford et al. (2002)
2006	RCT, double-blind	With and without cognitive impairment	Valproic acid (VPA)	VPA 250 mg twice daily	VPA ($n = 11$), placebo ($n = 11$)	22	Baseline to w10	With the exception of the mean reaction time and trial 5 of the Rey auditory verbal memory, all neuropsychological measures favored the impaired subjects in the VPA group	Schifitto et al. (2006)

2006	Single-arm, open-label	Cognitive impairment and under ART for 12 weeks	Lithium	Oral lithium was initiated at 300 mg daily and was titrated to maintain 12-h trough concentrations between 0.4 and 0.8 mEq/L	Lithium ($n = 8$)	8	Baseline to w12	Six of the eight individuals improved sufficiently to reduce their GDS from the impaired to the normal range	Letendre et al. (2006)
2007	RCT, double-blind	Nil	Selegiline transdermal system (STS)	STS 3 mg/24 h, 6 mg/24 h	Nil	86	Baseline to w24	No benefit	Evans et al. (2007)
2007	RCT, double-blind	Cognitive impairment and under ART for 6 weeks	Memantine	Memantine 10 mg per day for 1 week, escalated by 10 mg in weekly increments to 40 mg per day by week 4, or up to the maximum tolerated dose	Memantine ($n = 70$), placebo ($n = 70$)	140	Baseline to w16	No benefit	Schifitto et al. (2007b)
2007	RCT, placebo-control	Cognitive impairment and under stable ART	Selegiline transdermal system (STS)	STS 3 mg/24 h patch daily, STS 6 mg/24 h patch daily	STS 3 mg/24 ($n = 42$), STS 6 mg/24 ($n = 43$), placebo ($n = 43$)	128	Baseline to w24	No benefit	Schifitto et al. (2007a)
2009	RCT, placebo-control	With cognitive impairment	Selegiline transdermal system (STS)	STS 3 mg/24 h patch daily, STS 6 mg/24 h patch daily	STS 3 mg/24 ($n = 19$), STS 6 mg/24 ($n = 18$), placebo ($n = 25$)	62	Baseline to w12, w24	No benefit	Schifitto et al. (2009b)

(continued)

Table 4 (continued)

Year	Design	Patient selection	Adjuvant	Dosage	Study arm	n	Follow-up neurocognitive performance change outcome	Main findings	Ref
2009	Single-arm, open label	Cognitive impairment and under ART for 8 weeks	Lithium carbonate	Lithium carbonate 300 mg PO bid	Lithium carbonate ($n = 15$)	15	Baseline to w10	No benefit	Schifitto et al. (2009a)
2010	RCT, double-blind	With or without ART, ADC (AIDS dementia complex) stage$\geq$1	Memantine	Memantine up to 40 mg/day	Memantine ($n = 51$), placebo ($n = 48$)	99	Baseline to w20 (double-blind phase), w48 (open-label phase)	During the initial 12-week, memantine arm had a statistically significant higher NP test improvement compared to placebo. No statistically significant NP changes were detected during the 48-week extension	Zhao et al. (2010)
2011	RCT, double-blind	Cognitive impairment and under ART for 16 weeks	Minocycline	Minocycline 100 mg orally every 12 h	Minocycline ($n = 52$), placebo ($n = 55$)	107	Baseline to w24	No benefit	Sacktor et al. (2011)
2013	RCT, double-blind	With undetectable plasma and CSF VL and with HAND	Rivastigmine	Rivastigmine 1.5 mg/day and was progressively increased every 2 weeks (3, 4.5, 6, 9, and 12 mg/day)	Rivastigmine ($n = 9$), placebo ($n = 8$)	17	Baseline to w20	Processing speed improved on rivastigmine	Simioni et al. (2013)

2013	RCT, double-blind	Naïve to ART, with CD4 + T cell 250–350/μL, AIDS dementia scale stage 0.5–1, international HIV dementia scale score < 10	Minocycline	Minocycline 100 mg orally every 12 h	Minocycline ($n = 36$), placebo ($n = 37$)	73	Baseline to w24	No benefit	Nakasujja et al. (2013)
2016	RCT, double-blind	Diagnosis of HAND, under stable cART with plasma and CSF HIV VL $\leq$ 50 copies/mL	Maraviroc	Maraviroc 150 mg/300 mg/600 mg twice daily according to background therapy	Maraviroc ($n = 9$), control ($n = 8$)	17	Baseline to month 6, 12	Improved global neurocognitive functioning in maraviroc arm	Gates et al. (2016)
2017	RCT	Age 20–75 years, with cognitive impairment, under stable cART at least 6 months, and undetectable plasma HIV viral load	Rivastigmine, lithium	Rivastigmine (started at 4.6 mg daily and increased to 9.5 mg daily at week 4), lithium (400 mg twice daily, titrated progressively to ensure plasma drug concentrations of between 0.4 and 0.8 mEq/L)	Rivastigmine ($n = 10$), lithium ($n = 11$), control ($n = 8$)	29	Baseline, w12, w48	Better cognitive outcomes were observed in all groups, although there were no significant differences between the arms. The rivastigmine group showed the highest positive trend	Munoz-Moreno et al. (2017)
2018	RCT, double-blind	Age 18–65 years, with cognitive impairment, under	Paroxetine, fluconazole	Paroxetine 20 mg orally every evening per day,	Placebo ($n = 11$, paroxetine ($n = 11$),	45	Baseline to w24	1. HIV+ individuals receiving paroxetine showed improved summary	Sacktor et al. (2018)

(continued)

Table 4 (continued)

Year	Design	Patient selection	Adjuvant	Dosage	Study arm	*n*	Follow-up neurocognitive performance change outcome	Main findings	Ref
		stable cART at least 3 months		fluconazole 100 mg orally every 12 h per day	fluconazole ($n = 11$), paroxetine + fluconazole ($n = 12$)			scores2. Fluconazole: No benefit	
2018	Single-arm, open-label, clinical trial	Age 18–70 years, under stable ART>1 year, with plasma HIV VL < 50 copies/mL, below-normal cognitive performance (<−0.5 SD)	Cenicriviroc (CVC)	Cenicriviroc (CVC) dosage adjusted by each participant's ART regimen	CVC ($n = 17$)	17	Baseline to w24	NP test improvements over 24 weeks	D'Antoni et al. (2018)
2019	RCT	Age > 50 years, stable HIV (CD4 + >350, ART ≥6 months), mild to moderate cognitive impairment	Ketogenic diet (KD)	Low carbohydrate (<50 g/day) and high-fat diet for 12 weeks	KD ($n = 7$), control ($n = 7$)	14	Baseline, w12, w18	At week 12, the KD group performed significantly better on trails A and trails B assessments	Morrison et al. (2020)

ART antiretroviral therapy, *cART* combined ART, *GDS* global deficit score, *RCT* randomized clinical trial, *VL* viral load, *NP* neuropsychological et al. 2002). Likewise, selegiline, a monoamine oxidase B inhibitor with anti-oxidant properties, did not show significant benefit in an initial trial despite some verbal memory improvement at week 10 (Dana Consortium on the Therapy of HIV Dementia and Related Cognitive Disorders 1998). The other potential neuroprotective agent, peptide T, a short peptide derived from HIV envelop protein (gp 120), only showed negligible cognitive improvement in patients with mild impairment at the baseline (Heseltine et al. 1998). The results from clinical trials testing experimental approaches against neuronal apoptosis were mostly disappointing as well. Lexipafant, a platelet-activating factor (PAF) antagonist, showed no significant impact on neurocognition except some trends toward improvement of timed gait and learning in the Rey Auditory Verbal Learning Test (RAVLT) (Schifitto et al. 1999). Valproic acid, which inhibited neuronal apoptosis induced by PAF and glycogen synthesis kinase (GSK)-3β, only showed a weak trend toward improvement in neurocognitive performance (Schifitto et al. 2006). Two single-arm open-label trials on lithium, a mood stabilizer widely used to treat bipolar disorder and a neuroprotectant, providing neurons protection from apoptosis, also yielded inconsistent results (Letendre et al. 2006; Schifitto et al. 2009a).

Because of the initial promising results on selegiline, five additional RCTs were conducted for the transdermal formulation of this agent (Evans et al. 2007; Schifitto et al. 2007a, 2009b), but unfortunately, all showed no significant benefit, except one demonstrating some better performance in RAVLT and Grooved Pegboard Test (dominant hand) (Sacktor et al. 2000). In addition, because of excitotoxicity in the HAND pathogenesis, memantine, a neuroprotectant and first-generation N-methyl-D-aspartate (NMDA) receptor antagonist that is commonly used for moderate to severe Alzheimer's disease, was once considered a promising candidate for HAND treatment. However, a placebo-controlled study of memantine in HAND indicated no association with overall neurocognitive improvement, but only transit and moderate changes (Schifitto et al. 2007b; Zhao et al. 2010).

In the last 10 years, other new adjuvant therapies have been tested, also based on neuroprotection and anti-inflammation strategies. Minocycline, which has both anti-inflammatory and neuroprotective effects, however, showed no benefit in two RCTs (Sacktor et al. 2011; Nakasujja et al. 2013). Rivastigmine, a cholinesterase inhibitor commonly used for Alzheimer's disease, had no overall effects on neurocognition except some improvement in processing speed (Simioni et al. 2013), similar to that observed with lithium (Munoz-Moreno et al. 2017). Paroxetine, a selective serotonin reuptake inhibitor typically used for the treatment of depression, was compared to fluconazole, an anti-fungal agent with potent anti-inflammatory effect, in HAND. Unlike fluconazole that showed no benefit, paroxetine demonstrated some improvement in the neurocognitive test, although the clinical relevance for such a modest improvement remained questionable (Sacktor et al. 2018). Ketogenic diet (KD), which was associated with an improvement in brain metabolism due to its potent anti-inflammatory and antioxidant effects, was tested in HAND, which demonstrated significant better executive function and processing speed; however, the cognitive gains were not sustained after the usual diets resumed (Morrison et al. 2019). Despite those mostly negative, more promising and better-tolerated neuroprotective therapies are being developed. For instance, intranasal insulin therapy, targeting insulin signaling defect-related metabolic dysregulation in the brain, demonstrated a potential benefit by reversing hippocampal dendritic injury and cognitive impairment in a mouse model, and a clinical trial (NCT03277222) is currently ongoing as a treatment for HANDs (Kim et al. 2019). Tesamorelin, a growth hormone-releasing hormone (GHRH), was approved as an injectable medication to treat abdominal fat accumulation in HIV. A randomized clinical trial conducted in PLWH with mild cognitive impairment and healthy elderly showed favorable effect on cognition improvement (Baker et al. 2012). Further phase 2 clinical trial of tesamorelin for cognition in the elderly PLWH is currently ongoing (NCT02572323).

Among age-related comorbidities, cardiovascular disease (CVD) and metabolic syndrome were strongly and independently associated with poor cognitive performance in PLWH (Foley et al. 2010; Wright et al. 2010). Even among the well-controlled with a long-term viral suppression, current CVD risk, past CVD, and age were independent risk factors for neuronal injury and inflammation, suggesting that vascular changes in the CNS lead to cognitive impairment (Cysique et al. 2013). Thus, potential interventions targeting CVD have been tested in different model systems without consistent findings. In HIV-1 transgenic rats, an experimental model for HAND, chronic low-dose aspirin, reduced neuroinflammatory markers and oxidative stress (Blanchard et al. 2015). In vitro studies found that statin treatment decreased CD14+/CD16+ inflammatory monocyte subpopulation, which played a central role in the pathogenesis of HAND (Yadav et al. 2016). However, the analysis of 658 HIV+ patients in CHARTER showed that statin use was not associated with better neurocognitive performance (Letendre et al. 2007). A longitudinal analysis of a cohort nested from the Multicenter AIDS Cohort Study indicated that higher total cholesterol and low-density lipoprotein cholesterol were associated with faster rate of cognition decline. In addition, among patients with elevated cholesterol, statin use was associated with slower rate of cognition decline (Mukerji et al. 2016). In ACTG ALLRT cohort of 3949 participants, neither statin (a median 133-week use) nor angiotensin-converting enzyme inhibitors (ACEI)/angiotensin receptor blockers (ARB) (a median 180-week use) showed a significant effect on neurocognitive function (Erlandson et al. 2017). Future studies are needed for neurocognitive effects of these agents for CVD treatment among PLWH.

ART intensification with maraviroc has been recently suggested as an option for the clinical management of HAND in the context of viral suppression (Carroll and Brew 2017). The rationales behind this suggestion include: (1) CCR5 receptor interactions with CNS reservoir cells are linked with CNS HIV and HAND; (2) Maraviroc, a CCR5 receptor antagonist, has adequate CNS penetration, with low rates of resistance, inhibitory effects on CNS viral replication in monocyte/macrophage cells (Kelly et al. 2013), and anti-inflammatory properties. A recent prospective, open-label pilot randomized controlled trial in participants with viral suppression and stable ART for 12 months found that maraviroc-intensified ART improved global neurocognitive performance without significant side effects (Gates et al. 2016; Robertson et al. 2016). The other study compared standard and maraviroc/raltegravir-intensified ART regimens in 62 acute HIV infection that showed improved CNS-related outcomes but no difference between the two regimens at week 24, suggesting large randomized controlled studies would be necessary to confirm this option as a treatment for HAND in the future (Valcour et al. 2015). ACTG A5324 (https://clinicaltrials.gov/ct2/show/NCT02519777) is an ongoing double-blinded RCT to compare maraviroc (MVC) and dolutegravir (DTG) intensification to the standard ART regimens in aviremic participants with neurocognitive impairment. Cenicriviroc, a dual CCR2 and CCR5 antagonist, demonstrated potent anti-inflammatory effects by decreasing monocyte activity maker (sCD14) (Thompson et al. 2016). A single-arm, open-label clinical trial of cenicriviroc for 24 weeks showed a sizable improvement in the neurocognitive test (D'Antoni et al. 2018).

In summary, although multiple adjuvant therapies have been developed to target various mechanisms of action and studied in small-scale trials, none have shown clear positive effects on HAND. Yes, with a better understanding of the pathogenesis and therapeutic targets, newer and effective interventions would become available in the near future.

5 Reservoirs Eradication

Although invasion of CNS is an early event that occurs during primary HIV infection (Thompson et al. 2011; Valcour et al. 2012), the brain displays chronic neuroinflammation and persistent viral RNA and DNA despite of effective ART. The CNS can serve a reservoir of ongoing HIV replication (Churchill et al. 2006), which limits the opportunity for HIV cure or eradication. While T-cell populations are the main source of CNS HIV in early HIV infection, perivascular macrophage and microglia are considered the primary cells that harbor HIV replication in chronic phase (Joseph et al. 2015). In addition to brain parenchyma, choroid plexus, CSF, and meninges are considered distinct reservoir sites in brain (Petito et al. 1999). Clearance of both latent and productive HIV from the brain would determine successful viral eradication. Numerous approaches have been developed to reduce these HIV reservoirs (e.g., early initiation of ART during acute infection (NCT00796146), gene and cell base therapy for HIV cure (Wang and Cannon 2016), nanotechnology (Cao and Woodrow 2019), and broadly neutralizing HIV antibody (bNAb)) (Lu et al. 2016). Most of these approaches remain experimental, and their effects on the HIV reservoirs in the CNS remain largely unknown. For instance, although the newly developed bNAbs represent a promising treatment entity for viral eradication, due to the large molecular weights, their CSF concentrations are 100-fold to 1000-fold lower than those in the plasma (Prabhakaran et al. 2020). Thus, additional studies are warranted to evaluate the effects of bNAbs on viral eradication in the CNS and their implications in HAND management.

6 Conclusion

HAND remains one of the ongoing challenges for the care of PLWH in the modern ART era, in addition to aging, multimorbidity, polypharmacy, and drug–drug interactions. This review tends to provide an overview of completed studies, understanding of the association between CPE, HIV viral suppression in the CNS, and neurocognitive performance, and status of adjuvant therapy for HAND management.

While the majority of the studies demonstrated that higher CPE was associated with better HIV control in the CSF, the relationship between CPE and neurocognitive performance is largely unclear. The possible reasons for this lack of clarity include neurotoxicity from ART and polypharmacy, lack of standard neurocognitive assessment, variations in study populations, and lack of newer ART regimens. Unfortunately, most of the tested adjuvant therapies showed no significant benefit for HAND, suggesting a better understanding of pathogenesis and therapeutic targets remains warranted.

6.1 Clinical Implications

Since lower nadir CD4+ T-cell count and late presentation are strong predictors for HAND, early ART initiation should be recommended with a NeuroHIV-targeted regimen to reduce CNS reservoir and prevent HIV-associated neurocognitive impairment. In addition, psychosocial interventions may be beneficial because illicit drug use and psychiatric illness are significant risk factors. More evidence has suggested a critical role of age and CVD in the pathogenesis of HAND; thus, conventional approaches such as blood pressure control and statin use should be considered especially for those with a high risk, although the benefit for neurocognitive improvement remains to be determined. Since no specific biomarkers for HAND diagnosis currently are available, nor standard neurocognitive tests for follow-up, practicing physicians particularly HIV specialists should be aware of the high prevalence of HAND and the mild asymptomatic subtypes, and routine neurocognitive and mental health screenings should be recommended. From a practical perspective, in patients presenting with significant neurological symptoms of HAND, ART regimens with high CPE should be recommended to better control the HIV in the CNS.

6.2 Further Perspectives

In the combination ART era, the neurocognitive performance evaluation has been challenged to identify subtle deficits. Therefore, development of biomarkers in combination with neuroimaging and neurocognitive testing should be developed.

PLWH are living longer under stable viral suppression, and they are more likely experiencing multimorbidity and polypharmacy. In the context of HAND, this means that a comprehensive management plan for multimorbidity especially CVD, diabetes, and metabolic syndrome should also be recommended. Although largely unknown, polypharmacy, typically resulting from multimorbidity, may likely contribute to the pathogenesis of HAND, in particular, for the medications with high neurotoxicity. The interactions between aging, polypharmacy, neurotoxicity, and HAND are currently under evaluation. Finally, a growing number of bNAbs are now in development. The CSF penetration profiles of these bNAbs and their effects on viral reservoirs, especially in the CNS, remain largely unknown and warrant further evaluation. The eradication of CNS reservoirs not only is a challenge but also might be the key to the future free of HAND.

Acknowledgments Drs. Scott Letendre and Qing Ma are currently supported in part by NIH grant R01AG063659.

References

Anderson AM, Munoz-Moreno JA, McClernon DR et al (2017) Prevalence and correlates of persistent HIV-1 RNA in cerebrospinal fluid during antiretroviral therapy. J Infect Dis 215:105–113

Antinori A, Giancola ML, Grisetti S et al (2002) Factors influencing virological response to antiretroviral drugs in cerebrospinal fluid of advanced HIV-1-infected patients. AIDS 16:1867–1876

Baker LD, Barsness SM, Borson S et al (2012) Effects of growth hormone-releasing hormone on cognitive function in adults with mild cognitive impairment and healthy older adults: results of a controlled trial. Arch Neurol 69:1420–1429

Baker LM, Paul RH, Heaps-Woodruff JM et al (2015) The effect of central nervous system penetration effectiveness of highly active antiretroviral therapy on neuropsychological performance and neuroimaging in HIV infected individuals. J Neuroimmune Pharmacol 10:487–492

Blanchard HC, Taha AY, Rapoport SI, Yuan ZX (2015) Low-dose aspirin (acetylsalicylate) prevents increases in brain PGE2, 15-epi-lipoxin A4 and 8-isoprostane concentrations in 9 month-old HIV-1 transgenic rats, a model for HIV-1 associated neurocognitive disorders. Prostaglandins Leukot Essent Fatty Acids 96:25–30

Bougea A, Spantideas N, Galanis P, Gkekas G, Thomaides T (2019) Optimal treatment of HIV-associated neurocognitive disorders: myths and reality. A critical review. Ther Adv Infect Dis 6:2049936119838228

Brouillette MJ, Mayo N, Fellows LK et al (2015) A better screening tool for HIV-associated neurocognitive disorders: is it what clinicians need? AIDS 29:895–902

Calcagno A, Barco A, Trunfio M, Bonora S (2018) CNS-targeted antiretroviral strategies: when are they needed and what to choose. Curr HIV/AIDS Rep 15:84–91

Caniglia EC, Cain LE, Justice A et al (2014) Antiretroviral penetration into the CNS and incidence of AIDS-defining neurologic conditions. Neurology 83:134–141

Cao S, Woodrow KA (2019) Nanotechnology approaches to eradicating HIV reservoirs. Eur J Pharm Biopharm 138:48–63

Carroll A, Brew B (2017) HIV-associated neurocognitive disorders: recent advances in pathogenesis, biomarkers, and treatment. F1000Res 6:312

Carvalhal A, Gill MJ, Letendre SL et al (2016) Central nervous system penetration effectiveness of antiretroviral drugs and neuropsychological impairment in the Ontario HIV Treatment Network cohort study. J Neurovirol 22:349–357

Casado JL, Marin A, Moreno A et al (2014) Central nervous system antiretroviral penetration and cognitive functioning in largely pretreated HIV-infected patients. J Neurovirol 20:54–61

Churchill MJ, Gorry PR, Cowley D et al (2006) Use of laser capture microdissection to detect integrated HIV-1 DNA in macrophages and astrocytes from autopsy brain tissues. J Neurovirol 12:146–152

Ciccarelli N, Fabbiani M, Colafigli M et al (2013) Revised central nervous system neuropenetration-effectiveness score is associated with cognitive disorders in HIV-infected patients with controlled plasma viraemia. Antivir Ther 18:153–160

Clifford DB, McArthur JC, Schifitto G et al (2002) A randomized clinical trial of CPI-1189 for HIV-associated cognitive-motor impairment. Neurology 59:1568–1573

Cross HM, Combrinck MI, Joska JA (2013) HIV-associated neurocognitive disorders: antiretroviral regimen, central nervous system penetration effectiveness, and cognitive outcomes. S Afr Med J 103:758–762

Cusini A, Vernazza PL, Yerly S et al (2013) Higher CNS penetration-effectiveness of long-term combination antiretroviral therapy is associated with better HIV-1 viral suppression in cerebrospinal fluid. J Acquir Immune Defic Syndr 62:28–35

Cysique LA, Maruff P, Brew BJ (2004) Prevalence and pattern of neuropsychological impairment in human immunodeficiency virus-infected/acquired immunodeficiency syndrome (HIV/AIDS) patients across pre- and post-highly active antiretroviral therapy eras: a combined study of two cohorts. J Neurovirol 10:350–357

Cysique LA, Vaida F, Letendre S et al (2009) Dynamics of cognitive change in impaired HIV-positive patients initiating antiretroviral therapy. Neurology 73:342–348

Cysique LA, Moffat K, Moore DM et al (2013) HIV, vascular and aging injuries in the brain of clinically stable HIV-infected adults: a (1)H MRS study. PLoS One 8:e61738

D'Antoni ML, Paul RH, Mitchell BI et al (2018) Improved cognitive performance and reduced monocyte activation in virally suppressed chronic HIV after dual CCR2 and CCR5 antagonism. J Acquir Immune Defic Syndr 79:108–116

Dana Consortium on the Therapy of HIV Dementia and Related Cognitive Disorders (1998) A randomized, double-blind, placebo-controlled trial of deprenyl and thioctic acid in human immunodeficiency virus-associated cognitive impairment. Neurology 50:645–651

De Luca A, Ciancio BC, Larussa D et al (2002) Correlates of independent HIV-1 replication in the CNS and of its control by antiretrovirals. Neurology 59:342–347

Decloedt EH, Rosenkranz B, Maartens G, Joska J (2015) Central nervous system penetration of antiretroviral drugs: pharmacokinetic, pharmacodynamic and pharmacogenomic considerations. Clin Pharmacokinet 54:581–598

Deutsch R, Ellis RJ, McCutchan JA et al (2001) AIDS-associated mild neurocognitive impairment is delayed in the era of highly active antiretroviral therapy. AIDS 15:1898–1899

Donath M, Wolf T, Sturmer M et al (2016) HIV-1 replication in central nervous system increases over time on only protease inhibitor therapy. Med Microbiol Immunol 205:575–583

Dravid AN, Natrajan K, Kulkarni MM et al (2018) Discordant CSF/plasma HIV-1 RNA in individuals on virologically suppressive antiretroviral therapy in Western India. Medicine (Baltimore) 97:e9969

Eden A, Fuchs D, Hagberg L et al (2010) HIV-1 viral escape in cerebrospinal fluid of subjects on suppressive antiretroviral treatment. J Infect Dis 202:1819–1825

Ellis R, Letendre SL (2016) Update and new directions in therapeutics for neurological complications of HIV infections. Neurotherapeutics 13:471–476

Ellis RJ, Hsia K, Spector SA et al (1997) Cerebrospinal fluid human immunodeficiency virus type 1 RNA levels are elevated in neurocognitively impaired individuals with acquired immunodeficiency syndrome. HIV Neurobehavioral Research Center Group. Ann Neurol 42:679–688

Ellis RJ, Moore DJ, Childers ME et al (2002) Progression to neuropsychological impairment in human immunodeficiency virus infection predicted by elevated cerebrospinal fluid levels of human immunodeficiency virus RNA. Arch Neurol 59:923–928

Ellis RJ, Letendre S, Vaida F et al (2014) Randomized trial of central nervous system-targeted antiretrovirals for HIV-associated neurocognitive disorder. Clin Infect Dis 58:1015–1022

Erlandson KM, Kitch D, Wester CW et al (2017) The impact of statin and angiotensin-converting enzyme inhibitor/angiotensin receptor blocker therapy on cognitive function in adults with human immunodeficiency virus infection. Clin Infect Dis 65:2042–2049

Evans SR, Yeh TM, Sacktor N et al (2007) Selegiline transdermal system (STS) for HIV-associated cognitive impairment: open-label report of ACTG 5090. HIV Clin Trials 8:437–446

Foley J, Ettenhofer M, Wright MJ et al (2010) Neurocognitive functioning in HIV-1 infection: effects of cerebrovascular risk factors and age. Clin Neuropsychol 24:265–285

Garvey L, Surendrakumar V, Winston A (2011) Low rates of neurocognitive impairment are observed in neuro-asymptomatic HIV-infected subjects on effective antiretroviral therapy. HIV Clin Trials 12:333–338

Gates TM, Cysique LA, Siefried KJ, Chaganti J, Moffat KJ, Brew BJ (2016) Maraviroc-intensified combined antiretroviral therapy improves cognition in virally suppressed HIV-associated neurocognitive disorder. AIDS 30:591–600

Heseltine PN, Goodkin K, Atkinson JH et al (1998) Randomized double-blind placebo-controlled trial of peptide T for HIV-associated cognitive impairment. Arch Neurol 55:41–51

Hong S, Banks WA (2015) Role of the immune system in HIV-associated neuroinflammation and neurocognitive implications. Brain Behav Immun 45:1–12

Joseph SB, Arrildt KT, Sturdevant CB, Swanstrom R (2015) HIV-1 target cells in the CNS. J Neurovirol 21:276–289

Kahouadji Y, Dumurgier J, Sellier P et al (2013) Cognitive function after several years of antiretroviral therapy with stable central nervous system penetration score. HIV Med 14:311–315

Kelly KM, Beck SE, Metcalf Pate KA et al (2013) Neuroprotective maraviroc monotherapy in simian immunodeficiency virus-infected macaques: reduced replicating and latent SIV in the brain. AIDS 27:F21–F28

Kim BH, Kelschenbach J, Borjabad A et al (2019) Intranasal insulin therapy reverses hippocampal dendritic injury and cognitive impairment in a model of HIV-associated neurocognitive disorders in EcoHIV-infected mice. AIDS 33:973–984

Lanman T, Letendre S, Ma Q, Bang A, Ellis R (2019) CNS neurotoxicity of antiretrovirals. J Neuroimmune Pharmacol

Letendre S (2011) Central nervous system complications in HIV disease: HIV-associated neurocognitive disorder. Top Antivir Med 19:137–142

Letendre SL, Woods SP, Ellis RJ et al (2006) Lithium improves HIV-associated neurocognitive impairment. AIDS 20:1885–1888

Letendre SL, Marquie-Beck J, Ellis RJ et al (2007) The role of cohort studies in drug development: clinical evidence of antiviral activity of serotonin reuptake inhibitors and HMG-CoA reductase inhibitors in the central nervous system. J Neuroimmune Pharmacol 2:120–127

Letendre S, Marquie-Beck J, Capparelli E et al (2008) Validation of the CNS penetration-effectiveness rank for quantifying antiretroviral penetration into the central nervous system. Arch Neurol 65:65–70

Letendre SL, Ellis RJ, Ances BM, McCutchan JA (2010) Neurologic complications of HIV disease and their treatment. Top HIV Med 18:45–55

Libertone R, Lorenzini P, Balestra P et al (2014) Central nervous system penetration-effectiveness rank does not reliably predict neurocognitive impairment in HIV-infected individuals. J Int AIDS Soc 17:19655

Livelli A, Vaida F, Ellis RJ et al (2019) Correlates of HIV RNA concentrations in cerebrospinal fluid during antiretroviral therapy: a longitudinal cohort study. Lancet HIV 6:e456–ee62

Lu CL, Murakowski DK, Bournazos S et al (2016) Enhanced clearance of HIV-1-infected cells by broadly neutralizing antibodies against HIV-1 in vivo. Science 352:1001–1004
Marban C, Forouzanfar F, Ait-Ammar A et al (2016) Targeting the brain reservoirs: toward an HIV cure. Front Immunol 7:397
Marra CM, Zhao Y, Clifford DB et al (2009) Impact of combination antiretroviral therapy on cerebrospinal fluid HIV RNA and neurocognitive performance. AIDS 23:1359–1366
Morrison SA, Fazeli PL, Gower B et al (2019) Cognitive effects of a Ketogenic diet on neurocognitive impairment in adults aging with HIV: a pilot study. J Assoc Nurses AIDS Care
Morrison SA, Fazeli PL, Gower B et al (2020) Cognitive effects of a Ketogenic diet on neurocognitive impairment in adults aging with HIV: a pilot study. J Assoc Nurses AIDS Care 31:312–324
Mukerji SS, Locascio JJ, Misra V et al (2016) Lipid profiles and APOE4 allele impact midlife cognitive decline in HIV-infected men on antiretroviral therapy. Clin Infect Dis 63:1130–1139
Munoz-Moreno JA, Prats A, Molto J et al (2017) Transdermal rivastigmine for HIV-associated cognitive impairment: a randomized pilot study. PLoS One 12:e0182547
Nakasujja N, Miyahara S, Evans S et al (2013) Randomized trial of minocycline in the treatment of HIV-associated cognitive impairment. Neurology 80:196–202
Perez Valero I, Gonzalez-Baeza A, Montes Ramirez ML (2014) Central nervous system penetration and effectiveness of darunavir/ritonavir monotherapy. AIDS Rev 16:101–108
Petito CK, Chen H, Mastri AR, Torres-Munoz J, Roberts B, Wood C (1999) HIV infection of choroid plexus in AIDS and asymptomatic HIV-infected patients suggests that the choroid plexus may be a reservoir of productive infection. J Neurovirol 5:670–677
Prabhakaran M. Narpala S, Gama L, et al. (2020) Infiltration of bNAb VRC01 into the cerebrospinal fluid in humans in the RV397 study. CROI 2020, Boston USA. Conference abstract 453
Rawson T, Muir D, Mackie NE, Garvey LJ, Everitt A, Winston A (2012) Factors associated with cerebrospinal fluid HIV RNA in HIV infected subjects undergoing lumbar puncture examination in a clinical setting. J Infect 65:239–245
Robertson K, Jiang H, Kumwenda J et al (2012) Improved neuropsychological and neurological functioning across three antiretroviral regimens in diverse resource-limited settings: AIDS Clinical Trials Group study a5199, the international neurological study. Clin Infect Dis 55:868–876
Robertson KR, Miyahara S, Lee A et al (2016) Neurocognition with maraviroc compared with tenofovir in HIV. AIDS 30:2315–2321
Sacktor N, Schifitto G, McDermott MP, Marder K, McArthur JC, Kieburtz K (2000) Transdermal selegiline in HIV-associated cognitive impairment: pilot, placebo-controlled study. Neurology 54:233–235
Sacktor N, Miyahara S, Deng L et al (2011) Minocycline treatment for HIV-associated cognitive impairment: results from a randomized trial. Neurology 77:1135–1142
Sacktor N, Skolasky RL, Moxley R et al (2018) Paroxetine and fluconazole therapy for HIV-associated neurocognitive impairment: results from a double-blind, placebo-controlled trial. J Neurovirol 24:16–27
Santos GMA, Locatelli I, Metral M et al (2019) Cross-sectional and cumulative longitudinal central nervous system penetration effectiveness scores are not associated with neurocognitive impairment in a well treated aging human immunodeficiency virus-positive population in Switzerland. Open Forum Infect Dis 6:ofz277
Schifitto G, Sacktor N, Marder K et al (1999) Randomized trial of the platelet-activating factor antagonist lexipafant in HIV-associated cognitive impairment. Neurological AIDS Research Consortium. Neurology 53:391–396
Schifitto G, Peterson DR, Zhong J et al (2006) Valproic acid adjunctive therapy for HIV-associated cognitive impairment: a first report. Neurology 66:919–921
Schifitto G, Zhang J, Evans SR et al (2007a) A multicenter trial of selegiline transdermal system for HIV-associated cognitive impairment. Neurology 69:1314–1321

Schifitto G, Navia BA, Yiannoutsos CT et al (2007b) Memantine and HIV-associated cognitive impairment: a neuropsychological and proton magnetic resonance spectroscopy study. AIDS 21:1877–1886

Schifitto G, Zhong J, Gill D et al (2009a) Lithium therapy for human immunodeficiency virus type 1-associated neurocognitive impairment. J Neurovirol 15:176–186

Schifitto G, Yiannoutsos CT, Ernst T et al (2009b) Selegiline and oxidative stress in HIV-associated cognitive impairment. Neurology 73:1975–1981

Sevigny JJ, Albert SM, McDermott MP et al (2007) An evaluation of neurocognitive status and markers of immune activation as predictors of time to death in advanced HIV infection. Arch Neurol 64:97–102

Simioni S, Cavassini M, Annoni JM et al (2013) Rivastigmine for HIV-associated neurocognitive disorders: a randomized crossover pilot study. Neurology 80:553–560

Smurzynski M, Wu K, Letendre S et al (2011) Effects of central nervous system antiretroviral penetration on cognitive functioning in the ALLRT cohort. AIDS 25:357–365

The Dana Consortium on the Therapy of HIV Dementia and Related Cognitive Disorders (1997) Safety and tolerability of the antioxidant OPC-14117 in HIV-associated cognitive impairment. Neurology 49:142–146

Thompson KA, Cherry CL, Bell JE, McLean CA (2011) Brain cell reservoirs of latent virus in presymptomatic HIV-infected individuals. Am J Pathol 179:1623–1629

Thompson M, Saag M, DeJesus E et al (2016) A 48-week randomized phase 2b study evaluating cenicriviroc versus efavirenz in treatment-naive HIV-infected adults with C-C chemokine receptor type 5-tropic virus. AIDS 30:869–878

Tozzi V, Balestra P, Salvatori MF et al (2009) Changes in cognition during antiretroviral therapy: comparison of 2 different ranking systems to measure antiretroviral drug efficacy on HIV-associated neurocognitive disorders. J Acquir Immune Defic Syndr 52:56–63

Underwood J, Robertson KR, Winston A (2015) Could antiretroviral neurotoxicity play a role in the pathogenesis of cognitive impairment in treated HIV disease? AIDS 29:253–261

Valcour V, Chalermchai T, Sailasuta N et al (2012) Central nervous system viral invasion and inflammation during acute HIV infection. J Infect Dis 206:275–282

Valcour VG, Spudich SS, Sailasuta N et al (2015) Neurological response to cART vs. cART plus Integrase inhibitor and CCR5 antagonist initiated during acute HIV. PLoS One 10:e0142600

Vassallo M, Durant J, Biscay V et al (2014) Can high central nervous system penetrating antiretroviral regimens protect against the onset of HIV-associated neurocognitive disorders? AIDS 28:493–501

Vassallo M, Fabre R, Durant J et al (2017) A decreasing CD4/CD8 ratio over time and lower CSF-penetrating antiretroviral regimens are associated with a higher risk of neurocognitive deterioration, independently of viral replication. J Neurovirol 23:216–225

Wang CX, Cannon PM (2016) The clinical applications of genome editing in HIV. Blood 127:2546–2552

Wright EJ, Grund B, Robertson K et al (2010) Cardiovascular risk factors associated with lower baseline cognitive performance in HIV-positive persons. Neurology 75:864–873

Yadav A, Betts MR, Collman RG (2016) Statin modulation of monocyte phenotype and function: implications for HIV-1-associated neurocognitive disorders. J Neurovirol 22:584–596

Zhao Y, Navia BA, Marra CM et al (2010) Memantine for AIDS dementia complex: open-label report of ACTG 301. HIV Clin Trials 11:59–67

The manufacturer's authorised representative in the EU is Springer
Nature Customer Service Centre GmbH, Europaplatz 3, 69115 Heidelberg,
Germany. If you have any concerns regarding our products, please
contact ProductSafety@springernature.com

Printed and bound by CPI Group (UK) Ltd, Croydon, CR0 4YY
30/09/2026
02225506-0017